COLLOIDS IN
DRUG DELIVERY

SURFACTANT SCIENCE SERIES

FOUNDING EDITOR

MARTIN J. SCHICK
1918–1998

SERIES EDITOR

ARTHUR T. HUBBARD
Santa Barbara Science Project
Santa Barbara, California

COLLOIDS IN DRUG DELIVERY

Edited by

Monzer Fanun

Al-Quds University
East Jerusalem, Palestine

CRC Press
Taylor & Francis Group
Boca Raton   London   New York

CRC Press is an imprint of the
Taylor & Francis Group, an **informa** business

CRC Press
Taylor & Francis Group
6000 Broken Sound Parkway NW, Suite 300
Boca Raton, FL 33487-2742

First issued in paperback 2016

© 2010 by Taylor and Francis Group, LLC
CRC Press is an imprint of Taylor & Francis Group, an Informa business

No claim to original U.S. Government works

ISBN 13: 978-1-138-19915-6 (pbk)
ISBN 13:978-1-4398-1825-1 (hbk)

Library of Congress Cataloging-in-Publication Data

Colloids in drug delivery / editor, Monzer Fanun.
    p. ; cm. -- (Surfactant science series ; v. 148)
    Includes bibliographical references and index.
    ISBN 978-1-4398-1825-1 (hardcover : alk. paper)
    1. Colloids. 2. Drug delivery systems. I. Fanun, Monzer. II. Title. III. Series: Surfactant science series ; v. 148.
    [DNLM: 1. Colloids--pharmacokinetics. 2. Drug Delivery Systems. QV 785 C7137 2010]

QD549.C635 2010
615'.19--dc22   2009045584

Visit the Taylor & Francis Web site at
http://www.taylorandfrancis.com

and the CRC Press Web site at
http://www.crcpress.com

Contents

Preface

Colloids have come a long way from the time when Thomas Graham coined the term *colloid* to describe "pseudosolutions." These are a dispersion of one phase in another and their size ranges from 1 nm to 1 µm, providing flexibility in the development of drug delivery. Today colloidal systems are more specifically characterized by their submicron size and unique surface properties. Colloidal drug delivery has been a boon for the difficult to formulate active drug ingredient. It has been used by manufacturers to improve therapeutic performance by targeting drugs to the site of action and frequently to obtain patentability or to increase the patent life of the product. As pharmaceutical carriers, they can be classified as self-assembled lipid systems (emulsions, liposomes, solid lipid nanoparticles, etc.), polymer systems (nanoparticles, micelles, dendrimers, conjugates etc.), drug nanoparticle systems, and procolloidal systems (self-emulsifying oral delivery systems and liquid crystalline systems). Colloidal systems have evolved from use in the enhancement of solubility and protection of labile substances to reductions in the toxicity of the drugs. The efforts have been in the alteration of the surface properties and particle size distribution of drugs incorporated into colloidal carriers to achieve a diverse array of therapeutic delivery objectives such as controlled and targeted drug delivery, poly(ethylene glycol) (PEG)ylation to escape the reticuloendothelial system, active targeting after attaching with biologically reactive material, and antibodies and stimuli responsive systems. Anticancer drugs, such as doxorubicin, have been formulated in various colloidal systems like microemulsions; solid liposomes; PEGylated liposomes; polymeric micelles; solid lipid nanoparticles; polymeric nanoparticles; dendrimer complexes attached to liposomes; tumor-specific antibody tagged liposomes; doxorubicin–PEG–folate conjugates; and thermosensitive liposomes to enable enhanced drug solubility and loading, prolong the residence time in the body, provide controlled and stimuli responsive release, and tumor-specific targeting of the drug. A varied range of pharmaceutical surfactant-based colloidal carriers have been explored. The constituent surfactant and polymers allow for the controlled and targeted drug delivery, enhanced effective solubility of the drug and facilitation of cellular drug uptake, and minimization of drug degradation and toxicities. More recently, various types of polymer-based nanocarriers have been designed, synthesized, and utilized for the formation of self-assembled micellar structures for life science applications. These colloidal systems have been used by the manufacturers for improving therapeutic performance by targeting the drug to the site of action and frequently to obtain patentability and to increase the patent life of the product. Hence, the thorough understanding and consequent implementation of the physicochemical properties and interfacial behavior of surfactants can provide a major impetus for drug delivery.

The introductory chapter by Misra et al. is an attempt to discuss the basics of surfactant and polymer surface activity and self-assembly, the various types of structures formed by such compounds, and their use in drug delivery and biotechnology. It is generally devoted to micelles, liquid crystalline phases, liposomes, microemulsions, emulsions, gels, and solid particles. Because biodegradation of surfactant and block copolymer systems has a particularly important effect on their use in drug delivery, an overview of it is also given in this chapter. Colloid properties have opened new frontiers in the delivery of drugs; chemistry; and nano-, micro-, and biotechnological products leading to an increased surge of interest. This is further facilitated by flexibility in tailoring the internal structure and surface to achieve new delivery modes for existing drugs and to improve their therapeutic efficacy. The characterization of any prepared dispersion is an important prerequisite to confirm the reproducibility and control of dispersion properties. The characteristics of the system provide information on the main features of the dispersion to understand the underlying mechanisms of drug availability at its absorption site. There are numerous techniques available for the characterization of colloids. In addition to the basic techniques of macroscopic and microscopic analyses, there are numerous novel characterization techniques, such as mechanical spectroscopy, neutron scattering, photon correlation spectroscopy, small-angle x-ray scattering, dynamic light scattering, and atomic force microscopy, that have provided an impetus for interface studies of colloidal systems. The aim of Chapter 2 by Saraf is to develop controlled and targeted delivery systems by utilizing these physicochemical, electrical, magnetic, hydrophilic, and lipophilic properties of colloids to predict the interaction between the biomolecules and the particle surface. These properties of colloidal system are an obligatory step in the new processes in colloidal drug delivery.

Chapters 3 to 5 provide recent insights on the use of hard colloidal drug delivery systems formed by surfactants, polymers, proteins, and lipids. Nanocapsules are a particular type of nanoparticle composed of an oily core surrounded by a polymeric wall and stabilized by surfactants at the particle–water interface. These nanocarries based on biodegradable polymers were developed for drug delivery in the 1980s, and since then a huge volume of research has appeared. The aim of Chapter 3 by Guterres et al. is to discuss the main scientific issues and challenges to the development of products based on this technology. Poly(alkylcyanoacrylate) (PACA) nanoparticles emerged as a new class of particulate drug delivery systems in the 1980s. Since then PACA nanoparticles have gained extensive interest and are considered as promising polymer-based colloidal drug delivery systems for controlled and targeted drug delivery. In Chapter 4 Graf et al. review the current status of drug delivery using PACA nanoparticles with an emphasis on protein delivery, and they look at the specific challenges of using these particles for vaccine delivery. The different templates and methods to prepare PACA nanoparticles and their influence on the preparation method and properties of the nanoparticles were explored. Special consideration was given to microemulsions as polymerization templates. An outlook on the future of PACA nanoparticles used as drug delivery systems was provided. Particulate drug delivery systems, such as polymer micro- or nanoparticles, are advantageous vehicles for topical drug delivery to increase bioavailability and provide controlled release to the skin and systemic circulation. Polymer particles are widely used for oral and parenteral administration; they are also being used for topical administration and in cosmetics formulation. Such particulate carriers show a number of advantages for topical application because of their ability to incorporate a large variety of active ingredients, to target specific sites, and to behave as drug reservoirs in the skin for sustained release purposes. The use and formulation of three-dimensional gel systems to be used as gelating agents, matrices in patches, and wound dressings for dermal and transdermal drugs delivery are presented in Chapter 5 by Musial and Kokol. Correlations between various known gel systems and the results of current scientific investigations in dermal and transdermal delivery applications are discussed. The applications of these gel systems in the development of smart textile materials is presented.

Soft colloidal systems include micelles, multiple emulsions, nanoemulsions, microemulsions, liquid crystals, niosomes, and liposomes that are covered in the third part of the book. Chapter 6 by Mohanty et al. considers the current status and possible future directions in the emerging area of multifunctional micellar systems as drug delivery systems. Interest is specifically focused on the potential application of polymeric micelles in the major areas in drug delivery, drug solubilization, controlled drug release, drug targeting, and diagnostic purposes. Multiple emulsions are complex emulsion systems where both oil-in-water and water-in-oil emulsion types exist simultaneously. Water-in-oil-in-water based formulations have been widely used in the pharmaceutical industry as vaccine adjuvant, sustained release, and parental drug delivery systems. Many particulate drug delivery systems (nanoparticles, microspheres, liposomes) have been prepared by using multiple emulsions during one of the developmental steps. The mechanism of its stability, different formulation methods used to enhance monodispersibility, and evaluation of the finished products ensuring quality are presented at the beginning of the chapter in greater detail. Multiple emulsions are versatile drug carriers that have a wide range of applications in controlled drug delivery, particulate-based drug delivery systems, targeted drug delivery, taste masking, bioavailability enhancement, enzyme immobilization, overdosage treatment and detoxification, red blood cell substitutes, lymphatic delivery, and shear-induced drug release formulations for topical application. Some novel applications include encapsulation of a drug by the evaporation of the intermediate phase, leaving behind capsules formed of polymers (polymerosomes), solid particles (colloidosomes), and vesicles.

The chapters by Khurana et al. (Chapter 7) and Cortesi and Esposito (Chapter 8) review the use of multiple emulsions as promising delivery systems. These three chapters also discuss the future direction of research to address the persistent stability problem of water-in-oil-in-water multiple emulsions using hydrodynamic dual stabilization and other innovative approaches. Nanoemulsions are transparent or translucent systems that have a typical dispersed-phase droplet size range of 20–200 nm. These systems are attractive as pharmaceutical formulations and are used drug carrier systems for oral, topical, and parenteral administration. These dosage forms in particular have been recently suggested as carriers for peroral peptide–protein drugs. It was hypothesized that formulating a nanoemulsion of the drug would help to increase the bioavailability of the drug because of the high solubilization capacity as well as the potential for enhanced absorption. The structure and characteristics of nanoemulsions and the use of these systems in the pharmaceutical field are discussed in Chapter 9 by Tirnaksiz et al. Microemulsions were discovered in 1943 by Schulman, and they have been used as new pharmaceutical dosage forms over the last few decades. Their potential as pharmaceutical dosage forms was disclosed in 1980 for the delivery of an antibiotic to the skin. From the early 1980s, microemulsions have been intensively studied because of their great potential in various application domains such as food and pharmaceutical applications. One of the major challenges for pharmaceutical formulators is to discover new methods for delivery of poorly soluble drugs that will be efficient and economically acceptable for drug manufacturers. An important strategy for the enhancement of the delivery of poorly soluble drugs is the concept of microemulsions. Because of their unique properties (thermodynamic stability, heterogeneous microstructure with a dynamic character, and significant capacity for solubilization of hydrophilic and lipophilic drugs), these colloidal systems are currently of interest as templates for the development of nanoparticulate carriers of drugs such as self-microemulsifying drug delivery systems (SMEDDSs).

Recent developments and future directions in the investigations of SMEDDSs are delineated by Djekic and Primorac in Chapter 10. In Chapter 11, I review the solubilization of diclofenac, a nonsteroidal antiinflammatory drug in mixed nonionic surfactant-based microemulsions. Procolloidal systems or self-emulsifying drug delivery systems are reasonably successful in improving the oral bioavailability of poorly water-soluble and lipophilic drugs. Traditional preparation of self-emulsifying drug delivery systems involves the dissolution of drugs in oils and their blending with suitable solubilizing agents. Improvements or alternatives of conventional liquid self-emulsifying drug delivery systems are superior in reducing production cost, simplifying industrial manufacture, and improving stability as well as patient compliance. Most importantly, self-emulsifying drug delivery systems are very flexible to develop various solid dosage forms for oral and parenteral administration. Chapter 12 by Maurya et al. reviews the use of self-emulsifying drug delivery systems.

During the past decade, there has been great interest in liquid crystalline systems as delivery systems in pharmaceuticals that is attributable to the extensive similarity of these colloid systems to those in living organisms. Liquid crystalline systems are characterized by intermediate states of matter or mesophases and combine the properties of both the liquid and solid states; that is, they exhibit in part a structure typical of fluids and in part the structured, crystalline state of solids. Chapter 13 by Patel and Patel introduces different liquid crystalline systems, their characterization methods, and pharmaceutical applications in the field of drug delivery. The formation of dispersed submicron-sized particles with embedded inverted-type mesophases, such as cubosomes, hexosomes, and micellar cubosomes, are receiving much attention in pharmaceutical applications because of their unique physicochemical characteristics like high interfacial area, high drug solubilization capacities, and low viscosity. These nanostructured aqueous dispersions represent an interesting colloidal family of naturally occurring surfactant-like lipids that have excellent potential to solubilize bioactive molecules with different physicochemical properties (hydrophilic, amphiphilic, and hydrophobic molecules). In particular, there is an enormous interest in testing the possibility of forming efficient drug delivery systems based on such dispersions for enhancing the solubilization of poorly water-soluble drugs and improving their bioavailability and efficacy without causing toxicity. Chapter 14 by Yaghmur and Rappolt summarizes recent studies conducted on the possibility of utilizing the colloidal aqueous dispersions with confined inner nanostructures (cubosomes, hexosomes, and micellar cubosomes) as drug nanocarriers. In addition, the authors demonstrate that for the optimal utilization of these dispersions we need to characterize the impact of drug loading on internal nanostructures, to check their stability under different conditions (such as the presence of salt, pH variation), and to fully understand the interaction of these dispersed particles with biological interfaces to ensure the efficient transportation of the solubilized drugs.

Niosomes, a self-assembly of nonionic amphiphiles, have been widely investigated in drug delivery because of their unique properties. During the past two decades, a variety of biocompatible nonionic surfactants have been investigated for formulating niosomal drug delivery systems, and a number of techniques have been developed for niosomal delivery both for conventional drugs and biotechnology-based pharmaceuticals. The products of the latter are typically peptide and protein drugs as well as genetic drugs, and they are usually defined as macromolecular drugs in terms of their molecular weights. They usually encounter the primary challenge for clinical application: delivery to the desired sites in their intact forms. Niosomes represent a promising platform for *in vivo* delivery of macromolecular drugs. Chapter 15 by Huang et al. focuses on the recent development of niosomal delivery of macromolecular drugs. Chapter 16 by Bordi et al. focuses on a new class of colloids built up by the aggregation of cationic liposomes stuck together by oppositely charged linear polyions. These self-assembling supermolecular complexes represent a new class of colloids, whose intriguing properties are not yet completely investigated and are far from being thoroughly understood. These structures should have wide potential biomedical applications. Biocompatibility, stability and, above all, the ability to deliver a broad range of bioactive molecules make these colloidal aggregates a versatile drug delivery system with the possibility of efficient targeting to different organs. Recent results are reviewed concerning some of the hydrodynamic and structural properties of these aggregates in different environmental conditions, with special attention to the authors' own works. This chapter provides a comprehensive overview of electrostatic stabilized polyion-induced liposome aggregates, encouraging the proposed strategy into clinical reality. Recent advances in liposome technologies for conventional and nonconventional drug delivery have provided pharmacologists with a tool to increase the therapeutic index of several drugs by improving the ratio of the therapeutic effect to the drug's side effects. Thanks to liposome versatility, effective formulations that are able to deliver hydrophobic drugs, to prevent drug degradation, to alter biodistribution of their associated drugs, to modulate drug release, and, most importantly, to selectively target the carrier to specific areas have been obtained. Moreover, by gradually increasing the complexity of the formulation, sophisticated membrane models have been developed and new landscapes on characterization possibilities of membrane-associated biomacromolecules have been discovered.

Chapter 17 by Luciani et al. covers the correlation between the different liposomal preparation strategies and the *in vitro* activity of biomacromolecules that takes advantage of a lipidic milieu. Although it can be demonstrated thanks to a series of recent studies, a systematic approach in rationalizing parameters during liposome preparation usually lacks preparation methods that affect physicochemical features and consequently biological responses and biomacromolecule functionality instead. The examined topics are supplemented with examples of the latest developments in the field of liposome applications in pharmaceutics and biotechnology. The successful use of approved liposomal drugs in the treatment of intracellular infections and cancer is now a clinical reality mainly because colloidal nanocarrier systems are able to improve the cellular uptake of drugs and, as a result, are able to increase efficacy and reduce side effects of antibiotics and anticancer drugs.

Chapter 18 by Santos-Magalhães et al. presents and discusses the potential applications of colloidal nanocarriers systems such as liposomes, nanocapsules, and microspheres to improve the antimycobacterial and antitumor activities of usnic acid and to reduce its hepatotoxicity. The successful *in vitro* and *in vivo* results and the issues associated with further exploitation of these colloidal nanocarrier systems are described. *In vitro* and *in vivo* results taken together clearly corroborate that the nano- and microencapsulation of usnic acid in colloidal carrier systems improves its cellular uptake and biological activity and ensures reduced hepatotoxicity. In particular, the development of long-circulating and site-specific nanocarriers containing usnic acid may play a decisive role in tuberculosis therapy within the foreseeable future because this approach can provide a valuable pharmaceutical dosage form for this drug.

Macromolecular self-assembly has been exploited recently to engineer materials for the encapsulation and controlled delivery of therapeutics. Such delivery systems are discussed and exemplified in regard to more traditional drug delivery systems such as micelles, liquid crystalline phases, liposomes, and polymer gels as well as more novel structures such as carbon nanotubes, polyelectrolyte multilayer capsules, and liquid crystalline particles. Dendrimers have a highly branched, nanoscale architecture with very low polydispersity and high functionality, comprising a central core, internal braches, and a number of reactive surface groups. Because of their unique highly adaptable structures, dendrimers have been extensively investigated for drug delivery and have demonstrated great potential for improving therapeutic efficacy. To date, dendrimers have been tailored to deliver a variety of drugs for the treatment of various diseases such as cancers.

Chapter 19 by Yang not only traces the evolution of the field of dendrimers in drug delivery but also reflects on the journey through the subject from inception to contemporary progress as well as product development. In addition, disease-specific treatments based on dendrimer-based drug delivery technology are reviewed. Future directions for dendrimer drug delivery are discussed. There has been considerable interest of late in using protein- or polymer-based microspheres as drug carriers. Several methods of microsphere preparation, like single and double emulsion, phase separation, and coacervation, as well as spray drying, congealing, and solvent evaporation techniques are used currently. The inclusion of drugs in microparticulate carriers clearly holds significant promise for improvement in the therapy of several disease categories. The microspheres are characterized with respect to physical, chemical, and biological parameters. There are numerous applications of microspheres for drug delivery that include dermatology, vaccine adjuvants, ocular delivery, brain targeting, gene therapeutics, cancer targeting, magnetic targeting, and many more. This is confidence that microparticulate technology will take its place, along with other drug delivery technologies, in enhancing the effectiveness, convenience, and general utility of new and existing drugs.

The use of microspheres as colloidal drug delivery systems is reviewed by Samad et al. in Chapter 20. Inhalation drug delivery is gaining increasing popularity in the treatment of lung diseases because of its advantages over oral administration such as less side effects and quicker onset of action. However, the therapeutic results of inhaled medications are dependent upon effective deposition at the target site in the respiratory tract. Determining the regional and local deposition characteristics within the respiratory tract is a critical first step for making accurate predictions of the dose received and the resulting topical and systemic health effects. Furthermore, the diversifying areas of pharmaceutical research and the growing interaction among them make inhalation drug delivery a multidisciplinary effort, necessitating inputs from engineering and computer techniques and from medicine and physiology. Aerosol drug delivery of pharmaceutical colloidal preparations is a novel mode of drug delivery that has shown promise in the treatment of various local and systemic disorders. Noninvasive administration of drugs directly to the lungs by various aerosol delivery techniques results in rapid absorption across bronchopulmonary mucosal membranes. Particle size and size distribution are important factors in efficient aerosol delivery of medicaments into the lungs. A wide variety of medicinal agents are delivered to the lungs as aerosols for the treatment of diverse diseases; however, currently, most of them are for the management of asthma and chronic obstructive pulmonary diseases. Aerosolized drug delivery into the deep lungs is expanding with the increased number of different diseases. Local and systemic delivery of drugs for various diseases is now focused on using aerosol formulations, which have tremendous potential. The future of aerosol delivery of nanoparticles and large molecules for systemic conditions with improved patient compliance is promising.

Chapter 21 by Islam focuses on different techniques in aerosol delivery of pharmaceutical colloids to the lungs for a wide range of local and systemic disorders. He discusses the contribution of colloid science to the development of aerosol drug delivery. Chapter 22 by Xi et al. reviews the latest advances in modeling and simulations of reparatory aerosol dynamics with application to pharmaceutical drug delivery. They also discuss aspects of respiratory physiology that influence the transport and deposition of inhaled medications as well as the aerosol properties that might be used to improve drug delivery efficiency. Specifically, the geometric effect of the oral airway, the effect of breathing maneuvers on targeted nasal drug delivery, and the potential of hygroscopic aerosols in pulmonary drug delivery are examined. The presented results are intended to provide guidance in making appropriate dose–response predictions for either the targeted delivery of inhaled therapeutics or the risk assessment of airborne contaminates.

In the last chapters of the book, subjects such as colloidal nanocarriers for imaging applications, colloidal carriers for the treatment of dental and periodontal diseases, and the classification and application of colloidal drug delivery systems in tumor targeting are discussed. Chapter 23 by Kalaji et al. reviews the biocompatible carriers used for drug delivery in dental tissue engineering with a specific emphasis on colloidal carriers. The different materials used in dental tissue engineering are summarized followed by the development of the methods of microsphere preparation. The main applications of scaffolds or capsules loaded with growth factors in dental tissue engineering are reviewed.

One of the limitations inherent in current cancer chemotherapy is the lack of selectivity of anticancer drugs. The systemic administration of a chemotherapeutic agent results in its distribution throughout the body, leading to serious side effects arising from the cytotoxic actions of most anticancer drugs on normal cells. Chapter 24 by Karasulu et al. focuses on the current status of colloidal drug delivery systems in cancer therapy and discusses the growing and emerging potentialities of these systems, without neglecting possible limitations. Biomedical imaging has revolutionized the field of preventive medicine and cellular biology by enabling early detection of diseases and key cellular events at the molecular level. Imaging contrast agents comprise moieties with diverse natures, ranging from metals such as gadolinium to radionucleotides. Because of such diversity in the physical and chemical properties, efficient and site-specific delivery of imaging contrast agents is a very challenging task. With the advent of nanotechnology, the difficulties in the delivery of imaging agents can be efficiently overcome. The specificity and sensitivity of the contrast agents can be significantly enhanced by encapsulating them in nanocarriers. By modulating various properties of nanocarriers such as the particle size, surface charge, and surface properties, targeting of imaging agents to specific tumors, organs, tissues, and cells can be successfully achieved. These nanocarriers can be used to enhance the diagnosis by different imaging techniques like positron emission tomography, computed tomography, magnetic resonance imaging, single photon emission computed tomography, magnetic particle imaging, or ultrasound. The last chapter by Patravale and Joshi focuses on *in vivo* applications of various nanocarriers such as liposomes; nanoemulsions; polymeric, lipid, and magnetic nanoparticles; dendrimers; quantum dots; and ferrofluids in the delivery of various imaging agents.

This book represents the shared understanding of young and well-known researchers in the field of colloidal drug delivery. We discuss the colloidal pharmaceutical applications that include controlled release drug delivery, the mechanisms of drug release and solubilization in a particular colloidal system, drug targeting, manufacturing particulate products, taste masking, bioavailability enhancement, detoxification, enzyme immobilization, extending therapeutic efficacy, steps for the preparation of particular colloidal systems, and advantages of the discussed colloidal system. This book also includes present discussions of the key issues pertinent to nasal, ocular, vaginal, oral, buccal, gastrointestinal, and colon drug delivery. It covers the depth and breadth of the field, from physical chemistry and assessment of drug permeability to available enhancement technologies to regulatory approval. Some chapters of the book focus on the anatomical and physiological aspects of the skin barrier and the drug absorption and adsorption mechanism of drug adsorption, kinetics of drug delivery, influence of skin and the drug properties, skin enzymes, drug delivery enhancement, penetration enhancement, and permeation enhancement. Other chapters provide detailed reviews of the methods used for the characterization of colloidal drug delivery systems that include microscopy (i.e., transmission electron microscopy, polarized light microscopy), x-ray scattering, rheology, conductivity, differential scanning calorimetry, nuclear magnetic resonance, dynamic light scattering, and others. This book examines topics necessary to the critical evaluation of a drug candidate's potential for delivery. It describes the preparation, classification, interfacial activity, surface modifications, and influence on particle characteristics, drug delivery, and drug targeting. Finally, modeling of the nasal, oral, larynx, and lung respiratory physiologies is also presented in some chapters. Each chapter explains why the system is used for the intended application, how it is made, and how it behaves.

An important feature of this book is that the authors of each chapter have been given the freedom to present, as they see fit, the spectrum of the relevant science from pure to applied in their particular topic. Any author has personal views on, and approaches to, a specific topic, that are molded by his or her own experience. I hope that this book will familiarize the reader with the technological features of colloidal drug delivery and will provide experienced researchers, scientists, and engineers in academic and industry communities with the latest developments in this field.

I thank all of those who contributed as chapter authors, despite their busy schedules. In total, 65 individuals from 15 countries contributed to the work. All of them are recognized and respected experts in the areas that were the subject of their chapters. None of them is associated with any possible errors or omissions; for that I take full responsibility. Special thanks are due to the reviewers for their valuable comments because peer review is required to preserve the highest standard of publication. My appreciation goes to Barbara Glunn of Taylor & Francis for her true awareness in this project.

Monzer Fanun

Editor

Monzer Fanun is a professor in surface and colloid science and the head of the Colloids and Surfaces Research Laboratory at Al-Quds University, East Jerusalem, Palestine. He is the editor of *Microemulsions: Properties and Applications* and author or co-author of more than 50 professional papers. Professor Fanun is a member of the European Colloid and Interface Society and a Fellow of the Palestinian Academy for Science and Technology. Dr. Fanun received his PhD in applied chemistry in 2003 from the Casali Institute of Applied Chemistry, which is a part of the Institute of Chemistry at the Hebrew University of Jerusalem, Israel.

Contributors

Sarbari Acharya
Institute of Life Sciences
Bhubaneswar, Orissa, India

M. S. Akhter
Department of Pharmaceutical Technology
Jadavpur University
Kolkata, India

Seyda Akkus
Department of Pharmaceutical Technology
Gazi University
Etiler Ankara, Turkey

Mohammed Intakhab Alam
Department of Pharmaceutics
Hamdard University
New Delhi, India

Paula J. Anderson
Department of Internal Medicine
University of Arkansas for Medical Sciences
Little Rock, Arkansas

Piero Baglioni
Department of Chemistry
University of Florence and CSGI
Sesto Fiorentino, Firenze, Italy

Debora Berti
Department of Chemistry
University of Florence and CSGI
Sesto Fiorentino, Firenze, Italy

Federico Bordi
Department of Physics
University of Rome "La Sapienza" and
   INFM-CRS SOFT
Rome, Italy

Cesare Cametti
Department of Physics
University of Rome "La Sapienza" and
   INFM-CRS SOFT
Rome, Italy

Nevin Celebi
Department of Pharmaceutical Technology
Gazi University
Etiler Ankara, Turkey

Letícia M. Colomé
College of Pharmacy
Federal University of Rio Grande do Sul
Porto Alegre, RS, Brazil

Rita Cortesi
Department of Pharmaceutical Sciences
University of Ferrara
Ferrara, Italy

Alekha K. Dash
Department of Pharmacy Sciences
Creighton University
Omaha, Nebraska

Ljiljana Djekic
Department of Pharmaceutical Technology
   and Cosmetology
University of Belgrade
Belgrade, Serbia

Elisabetta Esposito
Department of Pharmaceutical Sciences
University of Ferrara
Ferrara, Italy

Monzer Fanun
Colloids and Surfaces Research Laboratory
Al-Quds University
East Jerusalem, Palestine

M. S. Ferraz
Laboratory of Immunopathology Keizo-Asami
Federal University of Pernambuco
Recife, Pernambuco, Brazil

Hatem Fessi
Lyon University
LAGEP/CNRS
Villeurbanne, France

Kiruba Florence
Department of Pharmacy
The Maharaja Sayajirao University of Baroda
Vadodara, Gujarat, India

Anja Graf
School of Pharmacy
University of Otago
Dunedin, New Zealand

Sílvia S. Guterres
College of Pharmacy
Federal University of Rio Grande do Sul
Porto Alegre, Rio Grande do Sul, Brazil

Yongzhuo Huang
Department of Pharmaceutical Sciences
University of Michigan
Ann Arbor, Michigan

Nazrul Islam
Pharmacy Section
Queensland University of Technology
Brisbane, Queensland, Australia

Medha Joshi
Department of Pharmaceutical Technology
Freie Universitte Berlin
Berlin, Germany

Nader Kalaji
Lyon University
LAGEP/CNRS
Villeurbanne, France

M. Abul Kalam
Department of Pharmaceutics
Hamdard University
New Delhi, India

Burcak Karaca
Department of Medical Oncology
University of Ege
Bornova, Izmir, Turkey

Ercüment Karasulu
Department of Biopharmaceutics and
    Pharmacokinetics
University of Ege
Bornova, Izmir, Turkey

H. Yesim Karasulu
Department of Pharmaceutical Technology
University of Ege
Bornova, Izmir, Turkey

Jatin Khurana
Department of Pharmacy Sciences
Creighton University
Omaha, Nebraska

Vanja Kokol
Department of Textile Materials and Design
University of Maribor
Maribor, Slovenia

Karen Krauel-Göllner
Institute of Food, Nutrition and Human Health
Massey University
Wellington, New Zealand

Manisha Lalan
Department of Pharmacy
The Maharaja Sayajirao University of Baroda
Vadodara, Gujarat, India

Wenquan Liang
College of Pharmaceutical Sciences
Zhejiang University
Hangzhou, China

M. C. B. Lira
Laboratory of Immunopathology Keizo-Asami
Federal University of Pernambuco
Recife, Pernambuco, Brazil

P. Worth Longest
Departments of Mechanical Engineering and
    Pharmaceutics
Virginia Commonwealth University
Richmond, Virginia

Paola Luciani
Department of Chemistry
University of Florence and CSGI
Sesto Fiorentino, Firenze, Italy

D. P. Maurya
Department of Pharmaceutics
Hamdard University
New Delhi, India

Ambikanandan Misra
Department of Pharmacy
The Maharaja Sayajirao University of Baroda
Vadodara, Gujarat, India

Chandana Mohanty
Institute of Life Sciences
Bhubaneswar, Orissa, India

Witold Musial
Department of Pharmaceutical Technology
Wroclaw Medical University
Wroclaw, Poland

Rakesh Patel
Department of Pharmaceutics and
    Pharmaceutical Technology
Ganpat University
Mehsana, Gujarat, India

Tanmay N. Patel
Department of Pharmaceutics and
    Pharmaceutical Technology
Ganpat University
Mehsana, Gujarat, India

Vandana Patravale
Department of Pharmaceutical Sciences and
    Technology
Institute of Chemical Technology
Matunga, Mumbai, India

E. C. Pereira
Department of Geography
Federal University of Pernambuco
Recife, Pernambuco, Brazil

Adriana R. Pohlmann
Institute of Chemistry
Federal University of Rio Grande do Sul
Porto Alegre, Rio Grande do Sul, Brazil

Fernanda S. Poletto
Institute of Chemistry
Federal University of Rio Grande do Sul
Porto Alegre, Rio Grande do Sul, Brazil

Marija Primorac
Department of Pharmaceutical Technology
    and Cosmetology
University of Belgrade
Belgrade, Serbia

Thomas Rades
School of Pharmacy
University of Otago
Dunedin, New Zealand

Renata P. Raffin
College of Pharmacy
Federal University of Rio Grande do Sul
Porto Alegre, Rio Grande do Sul, Brazil

Michael Rappolt
Institute of Biophysics and Nanosystems
    Research
Austrian Academy of Sciences
Graz, Austria

Sanjeeb K. Sahoo
Institute of Life Sciences
Bhubaneswar, Orissa, India

Abdus Samad
Pharmaceutical Medicine
Ranbaxy Research Unit
Majeedia Jamia Hamdard
New Delhi, India

N. P. S. Santos
Academic Center of Vitória de Santo Antão
Federal University of Pernambuco
Vitória de Santo Antão, Pernambuco, Brazil

N. S. Santos-Magalhães
Laboratory of Immunopathology Keizo-Asami
Federal University of Pernambuco
Recife, Pernambuco, Brazil

Swarnlata Saraf
Institute of Pharmacy
Pt. Ravishankar Shukla University
Raipur, Chhattisgarh, India

Simona Sennato
Department of Physics
University of Rome "La Sapienza" and
    INFM-CRS SOFT
Rome, Italy

Tapan Shah
Department of Pharmacy
The Maharaja Sayajirao University of Baroda
Vadodara, Gujarat, India

Nida Sheibat-Othman
Lyon University
LAGEP/CNRS
Villeurbanne, France

N. H. Silva
Department of Biochemistry
Federal University of Pernambuco
Recife, Pernambuco, Brazil

Andreza R. Simioni
Department of Chemistry
University of São Paulo
Ribeirão Preto-SP, Brazil

Somnath Singh
Department of Pharmacy Sciences
Creighton University
Omaha, Nebraska

Yasmin Sultana
Department of Pharmaceutics
Hamdard University
New Delhi, India

Mohammad Tariq
Department of Pharmaceutics
Hamdard University
New Delhi, India

Figen Tirnaksiz
Department of Pharmaceutical
    Technology
Gazi University
Etiler Ankara, Turkey

Jinxiang Xi
Department of Systems
    Engineering
University of Arkansas
Little Rock, Arkansas

Anan Yaghmur
Department of Pharmaceutics and
    Analytical Chemistry
University of Copenhagen
Copenhagen, Denmark

Hu Yang
Department of Biomedical
    Engineering
Virginia Commonwealth University
Richmond, Virginia

Faquan Yu
Department of Pharmaceutical
    Sciences
University of Michigan
Ann Arbor, Michigan

1 Surfactants and Block Copolymers in Drug Delivery

Ambikanandan Misra, Kiruba Florence,
Manisha Lalan, and Tapan Shah

CONTENTS

1.1 INTRODUCTION

Surfactants and polymers find widespread and versatile use in drug delivery. The current scientific interest in drug delivery research is on the numerous applications of surfactants and polymers. Their physicochemical properties and behavior have been under systematic scrutiny to maximize their applications. Almost every type of colloidal drug delivery system utilizes surfactants or polymers. The properties of these substances have been exploited and manipulated to obtain the desired rate and extent of drug release and absorption, including drug targeting to the site of action. Drug discovery programs today concentrate more on the potency and high activity of drugs, which often leads to the birth of lipophilic or water-insoluble drug candidates. These drugs have larger volumes of distribution, that is, distribution into nonspecific tissues that leads to untoward side effects. Hence, these high-potency candidates require site-specific targeting. Similarly, the increased applications of biological candidates such as antibodies and proteins into therapeutics have warranted delivery systems that are capable of maintaining their activity and preserving their integrity. This has increased the importance of delivery systems that can tackle the physicochemical and biopharmaceutical needs of therapeutics and has resulted in the advent of colloidal drug delivery. Colloid science refers to the utilization of submicron-sized materials. Although colloidal drug delivery originated several decades ago, significant development has come only in the last 25 years. Colloidal drug delivery has a major influence on the therapeutic efficacy of a drug. These improved delivery techniques that minimize toxicity and improve efficacy offer great potential benefits to patients and open new markets for pharmaceutical and drug delivery companies (Table 1.1).

Colloidal drug delivery provides the formulation scientist with an alternative formulation approach that could enhance solubility; ensure improved dissolution; and provide options for controlling or sustaining drug release, which targets specific sites, and for tailoring the surface properties to modify the pharmacokinetics and dynamics. These multifarious benefits have placed colloidal drug delivery in the limelight and forefront of drug research. The emergence of truly biologically interactive systems represents the latest step forward in colloidal delivery systems. The various systems have evolved from conception to the use of safer excipients, approaching more site-specific drug delivery with a greater assurance of better therapeutic response and a reduction in the associated untoward responses.

Several types of colloidal formulation approaches have been designed, explored, and applied to clinical use. Major thrust areas include micelles, liposomes, nanocarriers, liquid crystalline phases, aerosols, microemulsions, and procolloidal systems. The foundation of all advances in colloidal drug delivery science has been based on the use of surfactants and polymers. In almost every type of such drug carriers, there is a direct or indirect application of these amazing molecules. They are magnificent in the diverse array of applications that are permitted. A brief introduction to surfactants and polymers (Boyd, 2005, 2008; Malmsten, 2002; Kaparissides et al., 2006) is now presented.

1.2 SURFACTANTS

Surfactants or surface-active agents are amphiphilic molecules with a polar, ionic, or zwitterionic hydrophilic part and a nonpolar hydrophobic part that usually comprises a hydrocarbon or fluorocarbon chain. They may be ionic or nonionic in nature. The strong dipole interactions between the hydrophilic part and water render them water soluble, and the balance between the dual properties of hydrophilicity and hydrophobicity endows them with a unique characteristic of surface-active properties in solutions. They tend to accumulate at various interfaces, reducing the contact of the hydrophobic part with aqueous milieu, and tend to lower the free energy of the phase boundary. The amount of surfactant adsorption at the interface is dependent on their structure and the nature of the two phases forming the interface. These molecules also undergo self-assembly and form varied structures, for example, micelles, microemulsions, and liquid crystalline phases. The impetus for the formation of such a structure is the reduced interaction of the hydrophobic part with water. An understanding of the basics of applied surfactant science is necessary to arrive at the right composition and control of the system involved. Surfactants have been classified according to their polar head group as anionic, cationic, nonionic, and zwitterionic. Figure 1.1 gives a schematic illustration of the different types of surfactant structures.

a. *Anionic surfactants:* These are molecules with a negatively charged head group. They are utilized in the majority of industrial applications. Major classes of anionic surfactants include carboxylates ($C_nH_{2n+1}COO^-X$), sulfates ($C_nH_{2n+1}OSO_3^-X$), sulfonates ($C_nH_{2n+1}SO_3^-X$), and

TABLE 1.1
Overview of Colloidal Drug Delivery Systems

Delivery System	Components	Applicable Route of Administration	Advantages	Disadvantages	Size
Micelles	Surfactants, block copolymers	Intravenous	1. Enhancement of solubilization (tropicamide, vancomycin) 2. Controlled and targeted delivery (ADR PEO-polyaspartate) 3. Ligand mediated targeting (folate conjugated PEG-*b*-PCL micelles containing paclitaxel) 4. EPR effect due to small and uniform size 5. Low surfactant concentration 6. Long-term stability	1. Sensitive to dilution	20–100 nm
Liquid crystals	Surfactants, block copolymers	Oral, topical, intramuscular	1. Increase in the mean resident time of delivery system intended for oral cavity (lidocaine) 2. Enhanced release and absorption nimesulide liquid crystal based topical preparation 3. Lyotropic, thermotropic depot formation 4. Sustained release from depot preparation (irinotecan, octreotide)	1. Difficult to prepare 2. Viscous formulations	<100 nm
Emulsion	Oils, surfactants, aqueous phase	Intravenous, oral, topical	1. Taste masking (choloroquine phosphate) 2. Enhancement of oral bioavailability (cefpodoxime proxetil) 3. Wide applications in topical preparations	1. Toxicity due to surfactant concentration 2. Restricted parenteral use due to possible emboli formation	>0.5 µm

	Composition	Route	Advantages	Disadvantages	Size
Microemulsion	Oil, surfactant, cosurfactant, cosolute, aqueous phase	Intravenous, oral, topical, intranasal, transdermal, vaginal	1. Solubilization of poorly water-soluble drugs (diazepam, dexamethasone palmitate) 2. Solubilization of hydrolytically susceptible compounds (lomustine) 3. Reduction of irritation, pain, or toxicity of intravenously administered drugs (diazepam, propofol) 4. Sustained release dosage forms (barbiturates) 5. Site-specific drug delivery to various organs (cytotoxic drugs) 6. Long-term stability (block copolmer microemulsion) 7. Ease of preparation 8. Clarity and stability 9. Low viscosity	1. Toxicity due to high surfactant concentration 2. Limited parenteral use	<100 nm
SEDDS	Oil, surfactant, cosurfactant	Oral, Intravenous	1. Increased rate of absorption (phenytoin) 2. Solubilization of poorly water-soluble drugs (halofantrine, diclofenac, simavastatin) 3. Improvement in the stability of the entrapped drug (BCNU) 4. Faster dissolution and drug release (coenzyme Q_{10})	1. Toxicity due to high surfactant concentration	SEDDS 100–300 nm, self-nanoemulsifying drug delivery system <50 nm
Liposomes	Amphiphilic phospholipids, cholesterol	Intravenous, subcutaneous, intraperitoneal, pulmonary, oral, transdermal, ocular, nasal, vaginal	1. Reduction in toxicity to nontargeted tissue accumulation (nephrotoxicity reduction in liposomal methotrexate and cardiotoxicity in doxorubicin liposomal formulation)	1. Lack of long-term stability	100 nm–0.5 μm

continued

TABLE 1.1 (continued)
Overview of Colloidal Drug Delivery Systems

Delivery System	Components	Applicable Route of Administration	Advantages	Disadvantages	Size
			2. Improved bioavailability (acyclovir) 3. Tumor and brain targeting 4. Gene delivery 5. Excellent biocompatibility 6. Encapsulation of water-soluble drugs 7. Solubilization of lipophilic drugs	2. Difficulty in scale up	
Nanoparticles	Natural, synthetic polymer materials	Intravenous	1. Sustained release (tamoxifen nanoparticles) 2. Targeted and sustained drug delivery for tumor and brain 3. Reduced drug distribution to nontargeted sites	1. Toxicity due to polymeric materials 2. Poor biocompatibility 3. Difficulty in scale up	100 nm–1 μm
Aerosols	Propellant, surfactants, cosolvents	Inhalation, topical	1. Minimization of systemic side effects		1–10 μm

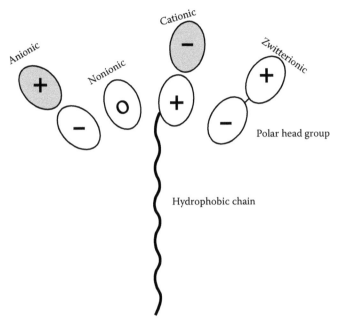

FIGURE 1.1   Different types of surfactants.

phosphates $[C_nH_{2n+1}OPO(OH)O^-X]$. Carboxylates were the earliest soaps, whereas sulfates (e.g., sodium lauryl sulfate) form the most important class of synthetic surfactants. Sulfated fixed oils are used as emulsifiers and solubilizers. Alkyl aryl sulfonates are the most common sulfonates, and the subgroup sulfosuccinates (e.g., AEROSOL®-OT) have been explored in pharmaceutical applications.

b. *Cationic surfactants:* These include molecules that have a positively charged head group. This category mostly includes molecules with an amine-containing head group. They are known to be more toxic compared to other groups of surfactants. Cationic surfactants have been used as bactericidal agents, for example, benzalkonium chloride (Figure 1.2) and cetrimide.

c. *Nonionic surfactants:* These are uncharged molecules that have the most widespread use. This category includes several chemical classes such as alcohol ethoxylates, alkyl phenol ethoxylates, fatty acid ethoxylates, sorbitan ester ethoxylates, ethylene oxide–propylene oxide copolymers (referred to as polymeric surfactants), polyglucosides, and glycerol esters. Sorbitan esters (Figure 1.3) and their ethoxylated derivatives (Spans and Tweens) are widely used in cosmetics and pharmaceuticals. Nonionic surfactants are often sensitive to temperature, but they are not sensitive to salts compared to other groups.

d. *Zwitterionic surfactants:* These contain both positively and negatively charged head groups. They are less common, and their charge is dependent on the pH of the medium. The most common amphoterics are *N*-alkyl betaines. They are used in dermatological products because of their comparatively low potential for irritation.

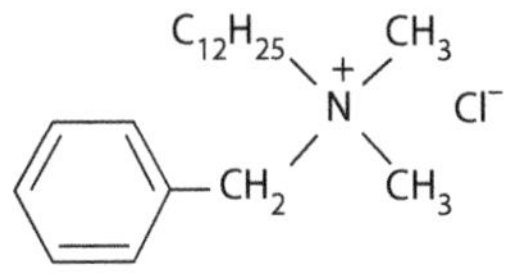

FIGURE 1.2   Benzalkonium chloride.

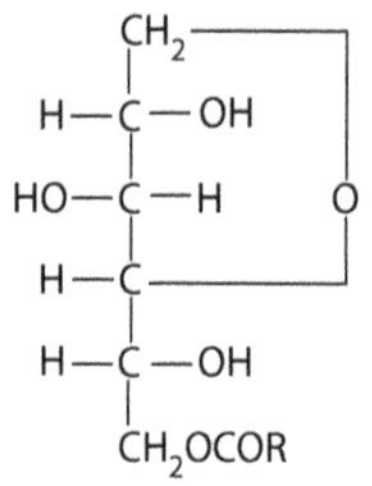

$$
\begin{array}{c}
\mathrm{CH_2} \!-\!\!\!-\!\!\!-\!\!\!-\!\!\!\!\\
| \quad\quad\quad | \\
\mathrm{H} \!-\! \mathrm{C} \!-\! \mathrm{OH} \quad | \\
| \quad\quad\quad\quad | \\
\mathrm{HO} \!-\! \mathrm{C} \!-\! \mathrm{H} \quad\quad \mathrm{O} \\
| \\
\mathrm{H} \!-\! \mathrm{C} \!-\!\!\!-\!\!\!-\!\!\!\\
| \\
\mathrm{H} \!-\! \mathrm{C} \!-\! \mathrm{OH} \\
| \\
\mathrm{CH_2OCOR}
\end{array}
$$

FIGURE 1.3   Sorbitan monoester.

Speciality surfactants: These include fluorocarbon and silicone surfactants, which possess excellent wetting properties, but they are expensive (Malmsten, 2002; Tadros, 2005).

Biosurfactants: A number of biological origin compounds possess surface-active properties; examples include bile salts, phospholipids, lipoproteins–lipopeptides, glycolipids, and polysaccharide–lipid complexes. They exhibit diverse structures, so it is therefore reasonable to expect diverse properties and physiological functions for different groups of biosurfactants. Bile salts are natural emulsifiers that are present in the body. Lipopeptides can act as antibiotics, antiviral and antitumor agents, immunomodulators for specific toxins, and enzyme inhibitors. Glycolipids have been implicated in growth arrest, apoptosis, and the differentiation of malignant melanoma cells. Phospholipids form aggregates, which have been explored in drug delivery applications (Kosaric et al., 1987; Rodrigues et al., 2006).

1.3   SURFACTANT-BASED MICELLAR SYSTEMS

Apart from being adsorbed at the interface to reduce the free energy, amphiphiles also undergo self-association to form micellar structures. This involves the formation of small aggregates in the bulk of solution, where hydrophobic moieties constitute the core and are shielded by a shell of ionic or nonionic polar head groups. The generalized structure can be assumed to be composed of a liquid core formed from the association of n hydrocarbon chains with fully ionized head groups projecting into water. The stern outer layer also contains associated counterions. In polyoxyethylated nonionic micelles, the core is surrounded by a layer composed of poly-(oxyethylene) chains to which solvent molecules may be hydrogen bonded. This region of the micelle is known as the palisade layer. Micelles are dynamic structures with a liquid core, so they cannot be assumed to have a definitive rigid shape. When used for scientific consideration, they are usually regarded as having sphericity. Micelles are formed only when the concentration of the surfactant in the solution increases above the critical micelle concentration (CMC; Attwood and Florence, 1983).

Micellar shape may be affected by factors such as the concentration, temperature, and presence of an added electrolyte. Thus, micelles may undergo a transition when any of the factors are changed. Micelles also exhibit size polydispersity. Polydispersity is expressed as the ratio of the weight average (n_w) to the number average (n_n). A ratio of unity represents a completely monosized system. The size distribution for small surfactants (mostly ionic) is usually close to unity, but the values are higher for polyoxyethylated nonionic surfactants (Mukherjee, 1972). CMC and micellar size are dependent on several parameters. The nature of the hydrophobic group of the surfactant influences both size and CMC. It has been observed for ionic and nonionic micelles that the CMC decreases with increasing length of the hydrophobic chain. The increasing chain length confers higher hydrophobicity and increases micellar size (Arnarson and Elworthy, 1981). The influence of the hydrophilic group is strikingly different in ionic and nonionic surfactants. With identical hydrophobic moieties, nonionic surfactants have a lower CMC that is due to the lack of electrical work required for micelle formation (Anacker et al., 1971). The greater the number of ionized groups present, the higher the CMC. In ionic surfactants, a change in the higher polarizable and valenced counterion leads to a decrease in the CMC. In nonionic surfactants, the addition of an electrolyte reduces the CMC, but the effect is not as significant. The increase in temperature causes an increase in the CMC of ionic surfactants because of the dehydration of monomers followed by the disruption of structured water around the hydrophobic group, which opposes micellization. An opposite temperature effect is observed with nonionic surfactants, ultimately causing phase separation (Balmbra, 1962).

1.3.1 THERMODYNAMICS OF MICELLE FORMATION

The two main approaches are the phase separation model and the mass action model. Both approaches are based on the classical system of thermodynamics.

    a. *Phase separation model:* In this approach, micelles form a separate phase at the CMC. In ionic molecules, it also includes the counterions. The major drawback with this approach is that it predicts the activity of the monomers to remain constant above the CMC, which is not consistent with the experimental results. Moreover, micelles cannot be considered as a phase because they are not uniform throughout (Pethica, 1960).

    b. *Mass action model:* This describes a dynamic equilibrium between the micelle and the unassociated surfactant molecule, which are in association–dissociation equilibria. Although the theory was initially described for ionic surfactants, it was extended to nonionic surfactants. It is a more realistic model and describes the variation in monomer concentration with the total concentration above the CMC (Murray and Hartley, 1935; Corkill, 1964).

1.3.2 KINETICS OF MICELLE FORMATION

Modern analytical techniques have aided in the understanding of micelle kinetics. There are two relaxation times for micelles. The slower one accounts for the micellization and dissolution equilibrium. This is the equilibrium between the dissociation of the complete micelle into n monomers and its complete reformation. The second faster relaxation time is attributed to the exchange of monomers between micellar species and bulk solution (Muller, 1972).

1.3.3 MICELLAR SOLUBILIZATION

The increased solubility of a less soluble organic substance in a surfactant solution has been used in applications for many years. This increased solubility of a solubilizate in surfactant solution is attributable to some form of attachment of the solubilizate to the exterior of the micelle or solution in it. There is a marked difference in the behavior of polar and nonpolar solutes (Lawrence, 1937). The solubilizate may be present at different sites in the micelles, which is dependent on its chemical nature. Usually, the nonpolar solubilizates are accommodated in the hydrocarbon core and the semipolar and polar solubilizates are present in the palisade layer. The location in the palisade layer can be either deeply buried or a short penetration (Reigelman et al., 1958). Figure 1.4 depicts the simplified structure of a micelle with the different locations of the solubilizate. Similar to surfactant molecules, solubilizates also have some freedom of motion, depending on the solubilization site. Several factors affect micellar solubilization. The surfactant structure influences solubilization; for solubilizates located in the micellar core, an increase in alkyl chain length enhances solubilization capacity. In general, the solubilization capacity of surfactants with the same hydrocarbon chain varies in the order anionics < cationics < nonionics. This is attributed to an increase of the head group area, leading to the formation of looser micelles and more accommodation of the solubilizate

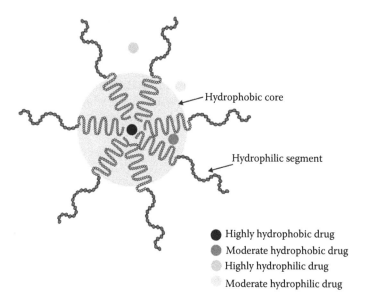

Hydrophobic core

Hydrophilic segment

● Highly hydrophobic drug
● Moderate hydrophobic drug
Highly hydrophilic drug
Moderate hydrophilic drug

FIGURE 1.4 Simplified structure of a micelle with different possible locations of solubilizate.

(Schott, 1967). The polarity, polarizability, chain length and branching, and molecular size and shape of the solubilizate affect its solubilization. The increase in the chain length of the solubilizate decreases its solubility in the surfactant, whereas unsaturation and cyclic compounds exhibit higher solubility (Klevens, 1950).

To summarize, the solubilization capacity and site are dependent on the solubility parameter of the solute as well as the surfactant. In general, solubilization increases with temperature because of changes in its aqueous solubility and changes in the micelles (Kolthoff and Stricks, 1948). The addition of an electrolyte may increase the solubility of the solubilizate present in the core but may decrease it when solubilizate is present in the palisade layer.

1.4 BLOCK COPOLYMERS

Block copolymers are high molecular weight compounds made up of connecting blocks (sequences) of two or more types of monomer that retain their intrinsic properties. Their molecular arrangement may be linear, radial, or both. A simple diblock copolymer AB consists of two types (A and B) of sequences linked end to end. Similarly, the molecular structure can be explained for triblock copolymers ABA or BAB and (AB)*n*. Incorporation of a third sequence C gives an ABC triblock copolymer. Radial arrangements of the units may form star-shaped structures; if the *n* homopolymer units junction at a common point, they can form a heteroarm block copolymer (Riess, 2003). The different block copolymer structures are illustrated in Figure 1.5a and b. Analogous to low molecular weight surfactants, hydrophilic–hydrophobic block copolymers can be classified into three groups that are based on their hydrophilic units.

a. *Nonionic copolymers:* These are mainly based on poly(ethylene oxide) (PEO), such as PEO–poly(propylene oxide) (PPO) and PEO–polystyrene (PS) di- and triblock copolymers.
b. *Anionic copolymers:* Examples of these polymers are those with poly(acrylic acid) (PAA) or poly(methyl methacrylate) (PMMA) blocks.
c. *Cationic copolymers:* These have cationic or cationizable units, for example, poly-2-vinylpyridine, and poly[amino(meth)acrylates].

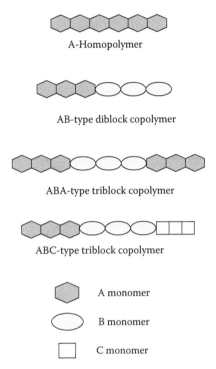

A-Homopolymer

AB-type diblock copolymer

ABA-type triblock copolymer

ABC-type triblock copolymer

A monomer

B monomer

C monomer

FIGURE 1.5a   Different types of block copolymers.

The synthesis of linear block copolymers, whether diblock, triblock, or segmented, follows two general reaction schemes. The first reaction involves generation of an active α site, an active ω site, or both on one of the polymeric sequences (A). This initiates the reaction with polymeric sequence B. The polymerization can proceed through a free radical, anionic, or cationic mechanism. The second scheme is the condensation or popularly known coupling, which involves a chemical reaction between the functional end groups of the two units. A synthesis scheme is adopted based on the polymerization mechanism involved, the structure of the copolymer, the molecular weight range and monodispersity of each block desired, and the required purity of the end product (Riess et al., 1985; Quirk et al., 1989). The concepts developed for the synthesis of linear polymers have been extended to the preparation of block copolymers with complex architectures. The progress made in living polymerization techniques and the use of multifunctional initiators have propelled advancements in block copolymer synthesis.

Scientific interest in these spectacular polymers is mainly based on the characteristic solution and associative properties that are due to their molecular structure. Their segmental incompatibility renders them surface active and leads to the formation of self-assembly structures. When dissolved in a liquid that is a thermodynamically good solvent for one but not for the other, copolymer chains may associate to form micellar aggregates. Micellar assembly is also possible in hydrophilic–hydrophilic copolymers under varying thermal or pH conditions.

1.5   BLOCK COPOLYMER-BASED MICELLAR SYSTEMS

Micellar self-association takes place in dilute solutions of block copolymers in specific solvents at a fixed temperature above the CMC or critical association concentration for polymeric micelles. Below the CMC, the polymers are present as unimers; as the concentration rises above the CMC, multimolecular micelles in dynamic equilibrium with unimers can be observed. In block

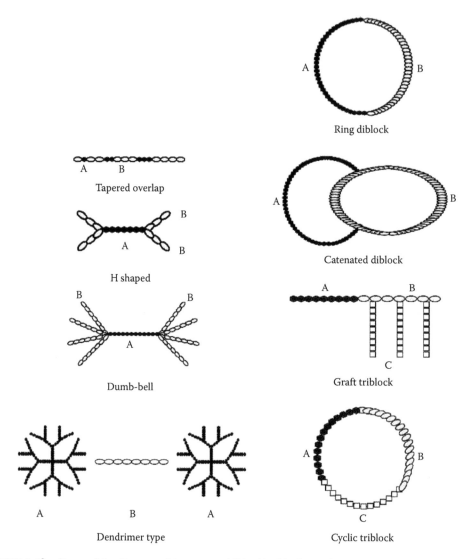

FIGURE 1.5b Some of the diverse architectures exhibited by block copolymers.

copolymers, micellization is greatly influenced by temperature; hence, a critical micelle temperature (CMT) is usually specified for a fixed concentration.

Block copolymer micellar systems are usually prepared with two procedures. The first procedure involves dissolution of the block copolymer in a common solvent followed by a change in
conditions such as the temperature, composition of the solvent system, or a selective precipitant for
one of the blocks in order to initiate micelle formation. At the end, usually a dialysis process is
used to replace the common solvent with the selective solvent (Munk, 1996). In the second procedure, a solid sample of the block copolymer is dissolved in a selective solvent and the micellar
solution undergoes annealing with time or by varying the temperature followed by ultrasonic irradiation (Hurtrez, 1992).

The micelles thus exhibit different morphologies. Two extremes of micellar structures can be
distinguished for diblock copolymers, depending on the relative length of the blocks. If the soluble
block is larger than the insoluble one, the micelles that are formed consist of a small core and a very
large corona and are thus called "star micelles." In contrast, micelles possessing a large insoluble
segment with a short soluble corona are referred to as "crew-cut micelles" (Mofitt et al., 1996).

Apart from spherical micelles, block copolymers form other morphological micellar structures. For instance, Eisenberg and Yu (1998) reported the formation of cylindrical structures by the self-assembly of PS-*b*-PEO diblock copolymers in aqueous solution. Flowerlike micelles have been reported when the ABA block copolymer is dissolved in a selective solvent for the B block. Reverse micelle formation is an enthalpy-driven process in which the hydrophilic blocks form the micellar core in an oil continuous system. Amphiphilic block copolymers with opposite charges have been used to prepare polyion complexes. These are formed by the coupling of polyelectrolytes with oppositely charged ionomers.

Block copolymer micelles are similar to low molecular weight surfactant micelles because both are characterized the following: the micelle has a typical molecular weight, the aggregation number (Z), the simplest morphology is spherical, the dynamic equilibrium between unimers and micelles, the radius of gyration (R_g), the hydrodynamic radius of the micelle (R_h), the micellar core radius (R_c), and the thickness of the corona (L). Similar to nonionic surfactants, micellization is promoted by increased length of the hydrophobic block and decreased length of the hydrophilic one. Usually micellization is promoted by increased temperature, which is due to more efficient packing. Salts do not affect micellization at low to moderate concentrations, but lyotropic salt effects are observed at higher salt concentrations. The localization of the solubilized molecule depends on the physicochemical properties of the solubilizate. The solubilization capacity in such systems is dependent on the molecular architecture of the copolymer, with a lower solubility observed with higher branched copolymers. A major difference in comparison to low molecular weight surfactants is the polydispersity in block copolymers, which affects the micellization process and the composition of micelles. Only at higher concentrations does it resemble the overall composition of the system. The dynamic equilibrium between polymer molecules and the micelle also exhibits a higher residence time of the polymer in the micelles compared to surfactants because of the slower dissociation kinetics in polymeric micelles. This property has significant biological implications for drug delivery (Malmsten, 2002).

1.5.1 DYNAMICS OF MICELLAR SYSTEMS

This encompasses the chain dynamics in the core and corona, the kinetics of micellization that are based on the micellar equilibrium between the unimer and micelles, and the concept of hybridization. Similar to low molecular weight surfactants, block copolymer micelles also exhibit two relaxation times corresponding to unimer ↔ micelle equilibration and a slower process corresponding to association ↔ dissociation equilibration. There are observable differences in di- and triblock copolymers that are attributable to restrained escape of the unimer from a triblock copolymer micelle (Tuzar and Kratochvil, 1993; Kositza et al., 1999). Scientific overviews confirmed the compactness and rigidity of the micellar core, and more recent analytical techniques depict a sharp core–corona interface. However, in compatible blocks, partial mixing and the presence of an interphase are suggested (Zana, 2000). Nuclear magnetic resonance (NMR) studies have provided detailed information on chain mobility. Hybridization is the formation of mixed micelles by the exchange of unimers between two micelle populations, and the driving force for this is the increase in entropy of the system (Munk, 1996).

1.5.2 DIFFERENT THEORIES OF BLOCK COPOLYMER MICELLES

A number of theories have been put forward to predict the structural parameters of micelles as a function of copolymer characteristics. In these theories and approaches, the total free energy (G) of the micelle is expressed as

$$G(\text{micelle}) = G(\text{core}) + G(\text{shell}) + G(\text{interface}). \qquad (1.1)$$

a. *Thermodynamic basis:* The micellization of block copolymers in an organic medium is an enthalpy-driven process. Micellar systems in aqueous media are entropy-driven processes, in which micellization is due to hydrophobic interactions and changes in the water microstructure near the polymer chains. These observations are similar to those in low molecular weight surfactants (Price, 1982; Quintana et al., 1996).

b. *Scaling theories:* Scaling theories have developed a simple model to correlate the core radius, shell thickness, and aggregation number to the block copolymer characteristic. The theory explains two types of micelles: crew-cut micelles in which the core radius is much higher than the shell thickness and hairy micelles in which the opposite is true. The radius of the core (R_c) for crew-cut micelles was proposed as

$$R_c \sim \gamma^{1/3} N_B^{2/3} a \quad \text{and} \quad Z \sim \gamma N_B, \tag{1.2}$$

where γ is the A/B interfacial tension, a is the segment length, and N is the number of repeating units. In the case of hairy micelles, it was suggested that the star polymer radius is

$$R \sim N_A^{3/5} f^{1/5}, \tag{1.3}$$

where f is the number of arms. This theory is more applicable to micellar systems with long polymer chains; they fail to include finite chain effects and polymer–solvent interaction (Daoud and Cotton, 1982; Linse, 2000).

c. *Mean field theory:* The self-consistent mean field theory was first developed by Noolandi and Hong (1982), who predicted the size of micelles at equilibrium and the variation in aggregation number as a function of the degree of polymerization. Their theory takes into consideration the molecular characteristics of the polymer, its concentration in solution, and an estimation of the core–corona interfacial tension. The results are consistent with experimental observations. Further development of this theory allowed its use for other aspects such as the temperature dependence of the hydrodynamic radius and aggregation number, the transition between different morphologies, and the influence of polydispersity.

Israelachvili (1992) used geometrical considerations to develop a very accessible approach that predicts the micellization phenomenon and the resultant morphologies. The approach describes it on the basis of the critical packing parameter (CPP), which is based on the geometry of two blocks. This in turn is governed by hydrophobic interactions and repulsive electrostatic or steric interactions that determine the final structure. The CPP values vary from small values (less than unity) for spherical micelles to approximate unity for bicontinuous bilayers to greater than unity for inverted structures.

1.5.3 MICELLAR SOLUBILIZATION

The most useful property of micelles in drug delivery is the improved solubilization of hydrophobic substances. The micellar core serves as a suitable microenvironment for water-insoluble compounds. The solubilization characteristics are defined in terms of the micelle–water partition coefficient, and the solubilization capacity is expressed as the volume or mass fraction of the solubilizate in the micellar core. PEO–PPO block copolymers have been widely used and investigated for changes in size and shape that are due to solubilization, a change in the CMC, the solubilization capacity, the partition coefficient, and so forth. It was observed that solubilization in the micellar core was correlated to the Flory–Huggins parameter (χ), which is dependent on the solubility parameters for the core forming block and the solubilizate as given in Equation 1.4:

$$\chi = \frac{(\delta_s - \delta_{core})v_s}{kT},$$ (1.4)

where δ represents the solubility parameter for the solubilizate and core, v_s is the molar volume of the solute, k is the Boltzmann constant, and T is the absolute temperature (Nagarajan, 1996).

The localization of the solubilizate depends on its hydrophobicity; the more hydrophobic solutes are located in the core. The molecular volume of the solubilizate governs the amount solubilized. The solubilization capacity is dependent on the molecular architecture, molecular weight, and composition of the block copolymer. Lower solubilities can be expected in branched polymers. A higher aggregation number of the micelle promotes solubilization (Malmsten, 2002).

Scientists have tried to develop a thermodynamic consideration of solubilization. This concept assumes that the micellar core containing the solubilizate is a pseudophase in equilibrium with the solubilizate and block copolymer molecules present in the solution. It allowed for the prediction of solubilization capacity, change in aggregation number, CMC, and so forth, as well as an explanation of the selective solubilization of a given component in the presence of a mixture of solubilizates (Tuzar and Kratochvil, 1993).

1.5.4 STABILIZATION OF MICELLAR SYSTEMS

Enhancement in the stabilization of block copolymer micellar systems can be achieved through functionalization of block copolymers, in addition to improvement in the temporal control for the system. Stabilization can be achieved mainly by crosslinking the micellar core or crosslinking the micellar corona. In the crosslinking of the micellar core, functional groups are present in the hydrophobic block. Similarly, functional groups on the hydrophilic groups enable crosslinking of the micellar corona. Crosslinked micelles exhibit stability at concentrations even below the CMC; as a result, they are stable at physiological dilution and display higher circulation times. Crosslinking is usually carried out by modifying the hydrophobic chains with a polymerizable group or by polymerizing a low molecular weight monomer in the micellar core. Shell crosslinked micelles not only improve micellar stability but also offer the possibility of controlling drug release. Here the micellar corona is fixed in solution after micellization by covalent bond formation, ionic bond formation, or both. Wooley et al. (1996) pioneered the concept of chemically crosslinking the corona of amphiphilic diblock copolymer micelles using a broad range of polymers, including styrene, isoprene, butadiene, ε-caprolactone, and methyl methacrylate. Monomers that have been used for the preparation of the hydrophilic, water-soluble corona include 4-vinylpyridine, methacrylic acid (MAA), and 2-di-methylaminoethyl methacrylate; they possess great potential for the development of advanced drug delivery systems (Rosler et al., 2001).

The polymerization of block copolymer end groups inside the micelle also serves to improve the stability of micelles. The described methodologies are based on free radical crosslinking. However, stabilization can be achieved by chemical crosslinking, crosslinking of the block copolymer in the bulk, through hydrogen bonding, and by using some thermodynamic parameters such as the glass-transition temperature of the polymer (Lecommandoux et al., 2005).

1.6 CHARACTERIZATION OF MICELLAR SYSTEMS

Characterization of micellar systems is important in order to understand the kinetics and dynamics of the system. The main factor to be determined is the CMC or CMT, but other parameters like are being studied such as the aggregation number, micellar shape, hydration state, counterion binding, microviscosity of the micellar core, and micellar dynamics. A number of techniques have been applied to study these systems. These include surface tension measurements, light scattering studies, NMR, fluorescence spectroscopy, calorimetry, osmotometry, conductivity, and solubilization studies.

The physical properties change drastically at the CMC, so these simple measurements are often employed for the determination of the CMC. The most prominent and widely used is the surface tension measurement. Amphiphiles are known to accumulate at surfaces and interfaces, leading to lowered surface tension. At concentrations above the CMC, the additional amphiphile molecules form micelles and the number of free unimers and the surface tension remain constant. This change is recorded to determine the CMC. This test is very sensitive to impurities and thus can be an indirect test for surfactant purity. Calorimetry, osmotometry, and conductivity studies are similarly employed. Dilute solution capillary viscometry measurements can be used to obtain information about the radius of hydration as well as the intrinsic viscosity (Malmsten, 2002).

1.6.1 Light Scattering Studies

Light scattering methods for surfactant solutions have been extensively used for small molecular weight surfactants. However, they have less applicability for block copolymer micelles because of weak signals as the CMC values are low compared to small surfactant molecules (Hadjichristidis et al., 2003). This means that equilibrium conditions are reached only after a long time period. X-rays, light, and neutrons can be used for scattering studies. Static light scattering (Burchard, 1983) is a powerful technique to estimate the average molar masses of self-assembled structures (and their CMCs). The scattering intensity is collected at different angles for a range of concentrations and summarized in a Zimm plot. In the presence of unimers and micelles, static light scattering leads to an apparent weight-average molecular weight (M_{wapp}) given by

$$M_{\text{wapp}} = M_{\text{w(u)}}x + M_{\text{w(m)}}(1 - x), \tag{1.5}$$

where u and m designate unimer and micelle, respectively, and x is the weight fraction of unimers. If scattering from the core and the corona of the micellar system is not very different, R_g can also be calculated. The second virial coefficient B, which is a measure of intermolecular interactions, is also interpreted from the inverse of the scattering intensities. Dynamic (or quasielastic) light scattering (DLS), also called photon correlation spectroscopy (Berne and Pecora, 2000), can be used to estimate the hydrodynamic radius of a block copolymer micellar system from the determination of its diffusion coefficient. In addition, the sensitivity and versatility of DLS can be used to monitor micelle equilibrium with varying temperature, pH, or other parameters. Small-angle x-ray scattering is used in micellar solutions to obtain overall and internal sizes. These measurements are extracted from differences in the electron density of the solvent and solute. Finally, small-angle neutron scattering provides information on the micellar shape, weight average, and cross section.

1.6.2 Nuclear Magnetic Resonance

This is based on the changes occurring in the microenvironment of the nucleus and transport properties as a consequence of micellization. NMR studies are carried out to measure the self-diffusion coefficient, which gives valuable information on free molecules and micelles. In ionic surfactants, the degree of counterion binding can be predicted by studying the diffusion characteristics of the surfactant and counterions. The measurement of the chemical shift of surfactant molecules describes the extent of water penetration in the micellar core. An advantage in these studies is that optical clarity or dilution of the samples is not necessary, and no labeling procedures are required (Malmsten, 2002).

1.6.3 Fluorescence Studies

Fluorescence techniques with either free probes like pyrene solubilization or covalently fixed fluoroprobes are widely used and the preferred method for determination of the CMC and CMT (Kalyanasundaram and Thomas, 1977). Pyrene is the preferred fluorescent probe because of its strong fluorescence in nonpolar domains and its weak radiation in polar media. The shift of the excitation peak can be used to probe the transfer of pyrene molecules into an increasingly nonpolar micellar environment. The ratio of intensities of the excitation maxima at 339 and 333 nm can be plotted as a function of concentration; the crossover value represents the CMC. The technique of covalently fixed probes may yield different information compared to free probes, but it is useful for studying micellar kinetics.

1.6.4 Ultraviolet Absorption Spectroscopy

This method is based on the tautomerism of 1-phenyl-1,3-butadione between the keto and enol forms that possess different absorption maxima, which appear in nonpolar solvents such as cyclohexane and in polar solvents. These are used for the determination of the CMC. When a surfactant is added to a water solution of 1-phenyl-1,3-butadione, the amount of enol form increases rapidly. As the surfactant concentration rises above the CMC, the enolic form is taken up into the core of the micelles, which provide a less polar environment than the external aqueous phase (Dominguez et al., 1997).

Apart from these scattering and spectroscopic methods, other techniques such as transmission electron microscopy and atomic force microscopy provide images that confirm the size, shape, or internal structure of the micelles. Size exclusion chromatography can be used to determine the radius of hydration and is useful for studying micellar dynamics (Forster and Plantenberg, 2002).

Micellar systems are important in the field of drug delivery for numerous reasons. For example, hydrophobic drugs can be loaded in micelles and exceed their aqueous solubility. Block copolymer micelles with a PEO corona can form hydrogen bonds and effectively shield the entrapped drug from protein adsorption and cellular adhesion, thus preventing them from hydrolysis and enzymatic degradation. This stealth property prevents recognition by the reticuloendothelial system (RES) and enhanced blood circulation times. Their chemical composition, molecular weight, and block length ratios can be manipulated to control the size and morphology of the micelles. The limiting factors for the use of micelles as a solubilizer are their limited loading capacity, their acute and chronic toxicity concerns, the concomitant solubilization of other formulation ingredients, and a more precise direction over its temporal and distribution control. A number of block copolymers exhibiting biocompatibility and biodegradability [e.g., poly(D,L-lactide-co-glycolide) (PLGA)] have caught the attention of formulation scientists and are being explored for the development of drug delivery systems.

1.7 LIQUID CRYSTALLINE PHASES

Surfactants and block copolymers form a range of different structures on self-association, including rods, lamellae, and bicontinuous interconnected structures. These liquid crystalline phases exhibit local disorder ("liquidlike" behavior) and are dynamic at the molecular level; however, a long-range order exists, which provides unique rheological, mass-transport, and optical properties. The formation of such structures depends on the packing characteristics of the surfactant–block copolymer. The critical packing parameter (CPP) is defined as:

$$CPP = \frac{v}{a \times l} \qquad (1.6)$$

where v is the volume of hydrophobic tails, l is the length of hydrophobic chains of the surfactant, and a is the polar head group area. As the surfactant aggregate curves toward the oil, there is a lamellar $\rightarrow$ hexagonal $\rightarrow$ micellar association progression.

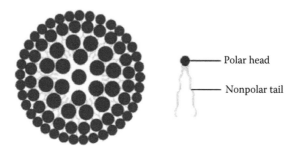

FIGURE 1.6   Normal micelles.

The construction of phase diagrams is a useful tool in studying liquid crystalline phases. When the surfactant is mixed with a solvent (notably water) only, the system contains two components and is described by a binary phase diagram, indicating the different phases formed as a function of composition and temperature. However, oil is also present in most of the cases; hence, such systems are described by a ternary phase diagram (Alexandris, 2000).

The packing properties of amphiphile molecules and the types of self-association structures that are formed are dependent on the structure of the molecule.

Ionic surfactants differ from nonionic ones in several respects. The presence of the charged head group results in head group repulsion, which precludes self-assembly (indicated, e.g., by a higher CMC). However, electrostatic interaction is diminished by the addition of salt; salt has more of an affect on the structures formed in ionic surfactant systems than those formed in nonionic ones.

The more common liquid crystal structures are hexagonal, reversed hexagonal, and lamellar phases; but cubic phases also frequently occur in surfactant systems. Figures 1.6 through 1.10 show the different liquid crystals in dispersion. Such cubic phases can consist of either micelles or reversed micelles that are closely packed in a cubic symmetry or of a bicontinuous structure. These cubic phases present some unique opportunities for drug delivery. They can solubilize large amounts of hydrophilic and hydrophobic drugs. The drug release rate can also be tailored in such systems by controlling the microstructure, which can be put to use for immobilizing protein drugs. They are also ideal for *in situ* depot preparations. Common methods include x-ray diffraction, NMR, and studies with polarizing microscopes (Malmsten, 2002).

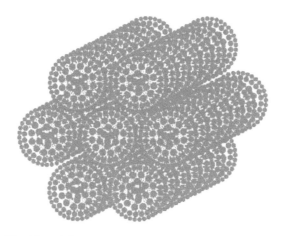

FIGURE 1.7   Hexagonal liquid crystal.

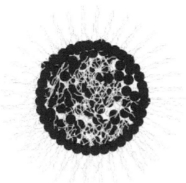

FIGURE 1.8   Reversed micelles.

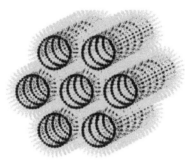

FIGURE 1.9   Reversed hexagonal liquid crystal.

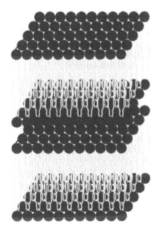

FIGURE 1.10   Lamellar.

1.8   EMULSIONS, MICROEMULSIONS, AND SELF-EMULSIFYING DRUG DELIVERY SYSTEMS

An emulsion is a thermodynamically unstable system consisting of two immiscible liquids: oil and water. One is dispersed as minute globules in the other continuous medium. The system is stabilized by the presence of surfactants.

Emulsion formation is the complex process consisting of the generation of a new oil–water interface followed by the stabilization of the interface. The dispersion of one immiscible liquid into the other requires high energy input, and surfactants play a major role in stabilizing the interface. When placed at the interface, surfactants reduce the interfacial tension and promote droplet formation. Reduction of the interfacial tension is dependent on the concentration of the surfactant till it reaches the CMC. Even a further addition of surfactant does not produce an appreciable change in the interfacial tension. The formation of a coherent yet flexible layer of surfactants over the newly formed droplets stabilizes them against coalescence. The charges developed over the surface of the droplets that are due to the presence of electrolytes or ions and/or to the chemical nature of the dispersed phase and surfactants in the dispersion medium account for the zeta potential of the droplets imparting repulsive forces on the droplets. Spontaneous emulsification is possible when the surfactant is equally or freely soluble in both immiscible liquids. Schulman recognized in 1940 that instead of a single surfactant, a combination of emulsifiers provides a compact placement of surfactants at the interface, which results in a stable emulsion. Certain polymers such as hydrophilic colloids are also used as emulsifying agents. They stabilize the droplets by multimolecular layer adsorption rather than by lowering the interfacial tension. The suitability of surface-active agents as emulsifying agents is assessed by their hydrophilic–lipophilic balance (HLB) value.

The type of emulsion formed either as oil in water (o/w) or water in oil (w/o) depends on many factors, for example, the nature of the surfactant; the phase volume ratio (oil and water content in the system); the temperature; and the presence of cosurfactants, cosolvents, and electrolytes. The role of the surfactant in the type of emulsion formed can be explained by Bancroft's rule, which states that the phase in which the surfactant is more soluble will become the continuous phase of the emulsion. Surface-active agents with a low HLB (3–6) are lipophilic and favor the formation of w/o type emulsions. The higher HLB (9–12) surfactants facilitate the formation of o/w type emulsions. Figures 1.11 and 1.12 show o/w and w/o emulsions.

Microemulsions or micellar emulsions are defined as single optically isotropic and thermodynamically stable multicomponent fluids composed of oil, water, and surfactant (usually in conjunction with a cosurfactant). The droplets in a microemulsion are 1 to 100 nm in diameter. Microemulsions appear transparent to the eye and exhibit excellent thermodynamic stability.

Microemulsions may exhibit different microstructures like droplet systems, continuous systems, and different aggregations, depending on the nature of the surfactant, composition of the system, temperature, cosurfactant, and cosolutes. Figure 1.13 displays the different types of microemulsions. At low oil content the system shows oil-swollen micelles, and at higher oil concentration the system exists as water-swollen reversed micelles via bicontinuous micelles. The surfactant concentration in microemulsions is usually higher (30%–60%) than in emulsions. Different surfactants stabilize different microstructures by virtue of their aggregation pattern in a particular medium, leading to a system with a minimum free energy and thermodynamic stability. In microemulsions formed by ionic surfactants, the surface tension lowering capacity of the surfactant is insufficient; it requires a cosurfactant for the compact packing of surface-active molecules at the interface to achieve ultralow

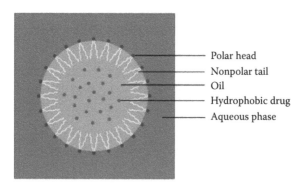

Polar head
Nonpolar tail
Oil
Hydrophobic drug
Aqueous phase

FIGURE 1.11   Oil in water emulsion.

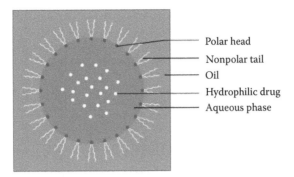

FIGURE 1.12 Water in oil emulsion.

surface tension. Short-chain alcohols and amines are used as cosurfactants in the stabilization of microemulsions. The addition of salt to the system with ionic surfactants promotes closer packing of the surfactants at the interface. Because of the absence of repulsive electrostatic forces, closest packing of the nonionic surfactant molecule is possible, which stabilizes the microemulsion system even in the absence of cosurfactants. The microemulsions formed by nonionic surfactants are more stable toward dilution and salt dependent effects compared to those containing ionic ones.

Self-emulsifying drug delivery systems (SEDDSs) are composed of mixtures of oils and surfactants, ideally isotropic and sometimes containing cosolvents, which emulsify spontaneously to produce fine o/w emulsions when introduced into an aqueous phase under gentle agitation. Liquid SEDDSs can be formulated as solid dosage forms by adsorbing the liquid SEDDSs over inert pellets or tablets. SEDDSs typically produce emulsions with a droplet size between 100 and 300 nm, and self-microemulsifying drug delivery systems form transparent microemulsions with a droplet size of <50 nm. Oil is the most important excipient because it facilitates self-emulsification and increases the fraction of lipophilic drug transported via the intestinal lymphatic system, thereby increasing absorption from the gastrointestinal tract (GIT; Khoo et al., 1998). Long-chain and medium-chain triglyceride oils (modified or hydrolyzed vegetable oils) have been widely explored (Constantinides, 1995). Novel semisynthetic medium-chain triglyceride oils have surfactant properties and are widely replacing the regular medium-chain triglycerides (Khoo et al., 1998). Nonionic surfactants with high HLB values are used in the formulation of SEDDSs (e.g., Tween, Labrasol®, Labrafac CM 10®, and Cremophore®). As in microemulsion systems, SEDDSa require a large quantity of surfactants in the range of 30% to 60% (w/w). Surfactants that are more hydrophilic in nature assist the immediate formation of o/w droplets, rapid spreading of the formulation, or both. Those that are amphiphilic in nature can solubilize large amounts of hydrophobic drug compounds (Shah et al., 1994). Polyalcohols are used as

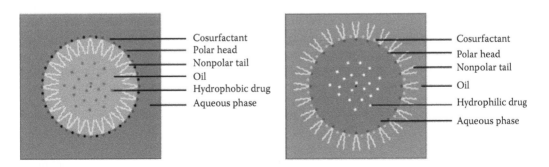

FIGURE 1.13 Oil in water and water in oil type microemulsions.

cosolvents or as cosurfactants, because they play a role in stabilizing newly formed globules in the aqueous environment. These cosolvents, such as poly(ethylene glycol) (PEG), propylene glycol, diethylene glycol monoethyl ether (Transcutol®), propylene carbonate, and tetrahydrofurfuryl alcohol PEG ether (Glycofurol®), may help to dissolve large amounts of hydrophilic surfactants or the hydrophobic drug in the lipid base. The formulation of SEDDSs is similar to that of the microemulsion-based drug delivery system. The prepared SEDDS evaluated transmittance, droplet size, and zeta potential.

1.9 LIPOSOME, NIOSOME, AND POLYMERSOME

Liposomes are concentric bilayered lipid vesicles in which an aqueous volume is enclosed by a membranous lipid bilayer that is mainly composed of natural or synthetic phospholipids. Liposomes may be formed from different surfactants and block copolymers, but phospholipid-based vesicles have been the most widely explored. They have attracted considerable attention because of their capacity to solubilize oil-soluble substances and to encapsulate water-soluble drugs. Bangham and coworkers discovered liposomes in the early 1960s, and since then liposomes have been used to simulate and mimic the behavior of biological membranes. The basic components in a liposome are amphipathic phospholipids, which on contact with the aqueous medium self-assemble into concentric, bilayered vesicles. The amphiphilic nature of phospholipids is due to the presence of a hydrophobic tail comprising fatty acids and a hydrophilic or polar head consisting mainly of phosphoric acid in the ionized form. The most common natural polar phospholipid is phosphatidylcholine (PC). The formation of bilayered structures facilitates a reduction in free energy and thereby renders the bilayered vesicles (liposomes) thermodynamically stable. At lower water content and higher temperature, a variety of lyotropic liquid crystalline phases also exist. Exposure of lipids to temperatures equal to or slightly above the phase transition results in the transformation of the structural integrity of lipids from a tightly packed, rigid form to a relatively loosely bound, flexible, and significantly disordered physical form. At this stage, phospholipid bilayers are capable of enclosing the drug or therapeutic entities within their interior.

Cholesterol is an important constituent in liposomes. It plays a crucial role in maintaining the bilayered structure of liposomes (Papahadjopoulos et al., 1975; Kirby and Gregoriadis, 1980; New, 1989). Because of its amphiphilic nature and relatively small size, cholesterol orients itself within the space adjacent to phospholipids. However, cholesterol is not capable of initiating a bilayered structure; it is quintessential for maintaining the structural integrity and rigidity of the liposomes. By the same mechanism, it also prevents the tilting and distortion of the membrane structure at temperatures exceeding phase-transition temperatures. Liposomes are often classified on the basis of vesicular size and the number of lamellae present. Based on the number of lamellae, they can be classified as unilamellar and multilamellar vesicles, as illustrated in Figure 1.14a and b.

Depending on the nature of the polar head group, phospholipid liposomes may be charged or uncharged. For charged liposomes, colloidal stability is determined largely by the magnitude and range of electrostatic interactions. For zwitterionic phospholipids with a zero net charge, such as PC and phosphatidylethanolamine (PE), phospholipid liposomes repel each other, thereby providing colloidal stability to such systems. Liposomes have been applied to the field of drug delivery in numerous diverse directions. Drug delivery applications of liposomes are discussed elsewhere in the chapter.

Similar to liposomes, nonionic surfactant based vesicles (niosomes) are vesicular formed from the self-assembly of nonionic amphiphiles in aqueous media. This self-assembly into closed bilayers is rarely spontaneous (Lasic, 1990) and usually requires energy input in the form of physical agitation or heat. The bilayer assembly assures minimum contact of the hydrophobic parts of the molecule from the aqueous solvent and the hydrophilic head groups are completely in contact with the same. These structures are analogous to phospholipid vesicles (liposomes) and are able to encapsulate aqueous solutes and serve as drug carriers. The low cost, greater stability, and resultant ease of storage of nonionic surfactants (Florence, 1993) has led to the exploitation

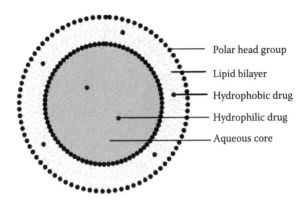

FIGURE 1.14a   Unilamellar liposomal vesicle.

of these compounds as alternatives to phospholipids. Niosome formation is affected by surfactants; additives (usually cholesterol); the nature of the drug; and physical parameters such as the hydration temperature, agitation, and size reduction techniques. Niosome formation is also based on the reduction of interfacial tension as in other vesicular assemblies. Typically, vesicles (closed lamellar structures) are formed in the water-rich phase of the binary phase diagram (Vyas and Uchegbu, 1998).

Polymersomes are polymer-based bilayer vesicles, which are also called nanometer-sized "bags" by scientists. The bilayer structure that is displayed is similar to liposomes and niosomes, which is revealed in Figure 1.15. They can be considered as liposomes but of nonbiological origin. Amphiphilic block copolymers can form various vesicular architectures in solution. They can have different morphologies, such as uniform common vesicles, large polydisperse vesicles, entrapped vesicles, or hollow concentric vesicles. Burke et al. (2001) described the formation of multiple morphologies from six different asymmetric amphiphilic block copolymers: PS-*b*-PAA, PS-*b*-PEO, polybutadiene-*b*-PAA, PS-*b*-poly(4-vinylpyridinium methyl iodide), PS-*b*-(4-vinylpyridinium decyl iodide), and PS-*b*-PMMA-*b*-PAA. These vesicles are inert toward living things and have sizes ranging between 60 and 300 nm. Labeling studies with fluorescent dye describe the block orientation within the vesicles. Scientists have proposed the ratio of the mass of the hydrophilic part to the total mass of the amphiphile as a criterion governing its self-assembly. A ratio of 35% (±10%) is indicative of a balance between the hydrophobic and hydrophilic parts, and polymersomes are formed. Amphiphiles with a higher value form micelles and lower ones form inverted structures. They have also been widely investigated for the delivery of cytotoxic agents (Lecommandoux et al., 2005).

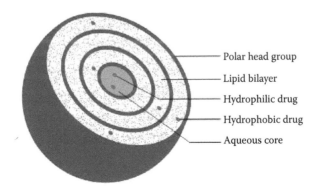

FIGURE 1.14b   Multilamellar liposomal vesicle.

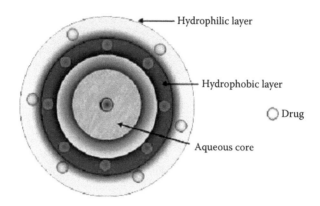

FIGURE 1.15   Drug loaded bilayer polymersome.

1.10   NANOPARTICLES

Nanoparticles are submicron-sized polymeric colloidal particles with a therapeutic agent of interest encapsulated within their polymeric matrix or adsorbed or conjugated on the surface. They may be broadly classified as nanospheres and nanocapsules, as shown in Figure 1.16.

Nanoparticles are mainly based on polymeric materials of natural or synthetic origin. Natural substances (e.g., proteins such as albumin, gelatin, and legumin, and polysaccharides such as starch, alginates, and agarose) have been widely studied. Synthetic hydrophobic polymers from the ester class [poly(lactic acid)] and poly(glycolic acid) copolymers along with ε-caprolactone have been investigated and approved (Couvreur and Vauthier, 1991; Lewis, 1990; Pitt, 1990). The first reported nanoparticles were based on nonbiodegradable polymeric systems such as polyacrylamide, PMMA, and PS. Their nonbiodegradability, immunogenic potential, and safety hazards restrained their applications for biomedical purposes. Soon this was followed by the development of biodegradable systems such as poly(alkyl cyanoacrylate) (PACA; Douglas et al., 1987; Kreuter, 1991), which has opened up a vast number of opportunities to the formulation scientist.

A polymeric nanosphere may be defined as a matrix type, solid colloidal particle in which drugs are dissolved, entrapped, encapsulated, chemically bound, or adsorbed to the constituent polymer matrix (Soppimath et al., 2001). These particles are typically larger than micelles (diameters between 100 and 200 nm) and may also display considerably more polydispersity (Kwon, 1998). Even though elimination may be slowed by the submicron particle size of nanospheres, clearance is still inevitable because of capture by the RES. The hydrophobic surfaces of these particles are highly susceptible to opsonization and clearance by the RES. Hence, it became clear that in order to prolong the circulation of nanoparticles, surfaces must be modified by adsorbing various surfactants (including poloxamine, poloxamer, and Brij detergent) to the particle surface (Mueller and Wallis, 1993) or coupled to a highly hydrophilic moiety like PEG. Although a surfactant coating reduced the total uptake by RES organs over short periods of time, no difference between uncoated and coated particles was found over longer periods, probably because of the desorption of the surfactant (Illum et al., 1986; Douglas et al., 1987).

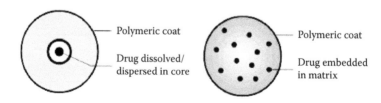

FIGURE 1.16   Nanocapsule and nanosphere.

Polymeric nanocapsules are colloidal-sized, vesicular systems in which the drug is confined to a reservoir or is within a cavity surrounded by a polymer membrane or coating (Soppimath et al., 2001). Frequently, the core is an oily liquid, the surrounding polymer is a single layer of polymer, and the vesicle is referred to as a nanocapsule. These systems have found utility in the encapsulation and delivery of hydrophobic drugs. Polymers used for the formation of nanocapsules have typically included polyester homopolymers such as poly(D,L)-lactide (PLA), PLGA, and polycaprolactone (PCL). In recent years copolymers of PEG and PLA have been used to avoid opsonization of the particles, similar to nanospheres (Ameller et al., 2003).

Nanospheres can be prepared by two general methods, depending on the polymer to be used. If the formed particles require polymerization, this can be achieved by either emulsion polymerization [as for PMMA and poly(ethyl cyanoacrylate)] or interfacial polymerization (as for PACA). For pre-formed polymers such as biodegradable polyesters and their copolymers with PEG, nanosphere preparation can be achieved using emulsification or solvent evaporation, emulsification or solvent diffusion, and salting-out techniques. However, the most popular method is solvent displacement, also referred to as nanoprecipitation (Rodriguez et al., 2004; Reis et al., 2006). The most common method of producing nanocapsules is by the interfacial deposition of preformed polymers (Fessi et al., 1989). In this procedure, a solution of the drug in a water-miscible organic solvent, such as acetone (with or without a lipophilic surfactant), is prepared. To this solution is added an oil that is miscible with the solvent but immiscible with the mixture; this solution is dispersed into the aqueous phase, which frequently contains a hydrophilic surfactant (often poloxamer). Upon moderate agitation, the solvent diffuses into the aqueous phase and the polymer aggregates around the oil droplet. The choice of nanoparticulate system preparation technique depends on the physicochemical properties of both the drug and the polymer. Nanoparticles prepared in this way have been under investigation for diverse biomedical applications, and they are subsequently discussed.

1.11 AEROSOLS AND FOAMS

The term aerosol describes systems in which liquids are dispersed in air, similar in principle to emulsions, stabilized by the presence of a surfactant. For pharmaceutical applications, the size of the particles is usually 0.5 to 5 µm. A fine mist is generated either by breaking up condensed material (liquid), for example, under high pressure, or by condensing gas. The common mechanical means include air nebulizers, spinning disks, ultrasonic nebulizers, and vibrating orifice generators. Aerosol generation by nucleation is via the use of volatile propellants, which on rapid evaporation cause supersaturation and droplet formation. The charged surfactants are expected to improve the stability of the aerosol, but the interaction of surface charges with the low dielectric constant of air makes its presence at the surface unfavorable.

A foam is a dispersion of gas in a liquid or solid. The formation of foam depends on the surface activity of surfactants, polymers, proteins, and colloidal particles to stabilize the interface. Hence, the foamability increases with increasing surfactant concentration up to the CMC, because above the CMC the unimer concentration in the bulk remains nearly constant. The structure and molecular architecture of the foam are known to influence the foamability and its stability. The packing properties at the interface are not excellent for very hydrophilic or very hydrophobic drugs. The surfactant promoting a small spontaneous curvature at the interface is ideal for foams. Nonionic surfactants are the most commonly used. The main advantage with aerosols and foams are their site-specific delivery and multiple dosing of the drug.

1.12 SURFACTANTS AND BLOCK COPOLYMERS IN TARGETED COLLOIDAL DELIVERY SYSTEMS

Colloidal delivery systems have been aimed at improving therapeutics by virtue of their submicron size and targetability. The effectiveness of a drug therapy is often governed by the extent to which temporal and distribution control can be achieved. Temporal control is the ability to manipulate the period of time over which drug release is to take place, the possibility of triggering the release at a specified time during treatment in response to stimuli such as temperature and pH, or both. Distribution control is the ability to direct the delivery system to the desired site of action (Uhrich et al., 1999).

Stimulus-responsive polymers, also referred to as "intelligent," "smart," or "environmentally sensitive" polymers, are systems that exhibit large, sharp changes in response to physical stimuli (such as temperature, solvents, or light) or chemical stimuli (such as reactants, pH, ions in solution, or chemical recognition). Thermosensitive micellar systems can be prepared using block copolymers with one segment exhibiting a lower critical solution temperature. The change in temperature of the environment above it causes a burstlike release of drug that is due to destabilization of the micelles. Examples of such polymers include poly(N-isopropylacrylamide)-containing block copolymers. However, only those micelles located under the skin can be controlled by it (Topp et al., 1997). Metal oxide nanoparticles that can be activated with infrared radiation or the presence of an external alternating magnetic field can be used to circumvent this situation. West et al. (2000) demonstrated the use of gold and iron oxide (γ–Fe_2O_3) particles to trigger the photo- or magnetic-induced transition in thermosensitive polymers. Application of auxiliary agents like channel proteins in block copolymer based systems improves their temporal control. It can allow for the development of delivery systems with pulsatile drug release under the influence of external stimuli such as reduction in the Donnan potential. One of the stimulus-responsive systems is based on pH-induced micellization, and it involves copolymers that exist as unimers at a certain critical pH but undergo micellization when the pH is changed. This occurs whenever two weak basic (or acidic) blocks of different negative logarithm of the equilibrium of association values are associated in the same copolymer. Such systems do not require cosolvents to stabilize the system (Armes et al., 1997, 1999). Scientists have described multistimuli responsive systems. Nowakowska and Zczubiałka (2003) synthesized a series of amphiphilic terpolymers based on sodium 2-acrylamido-2-methyl-1-propane-sulfonate, N-isopropylacrylamide (NIPAAm), and cinnamoyloxyethylmethacrylate. The terpolymers were soluble in water and self-assembled into micelles. They were found to be sensitive to three stimuli: temperature, which is due to the NIPAAm block that imposed a lower critical solution temperature; ultraviolet (UV) light, which is attributable to the presence of the cinnamoyl block; and ionic strength, which at elevated concentration induces the loss of temperature sensitivity. The size of the micelles formed, as indicated by DLS, varied with temperature and UV irradiation.

Distribution control can be exerted by passive targeting or active targeting. A suitable example of passive targeting is the preferential accumulation of colloidal particles in solid tumors by virtue of the enhanced permeation and retention (EPR) effect, which takes advantage of the enhanced vascular permeability of tumor tissue. As a result, high molecular weight substances accumulate inside tumors (Rosler et al., 2001).

Active targeting involves surface functionalization to achieve distribution control. It involves the modification of the hydrophilic chain end with ligands that are recognized specifically by target cells. Figure 1.17 shows the architecture of a typical surface-functionalized micelle. This interaction is more specific and allows for the precise control of targeting to the site of interest. Modification of the crosslinked micelles with appropriate ligands provides a delivery system with sufficient stability, and it also permits active targeting. For example, the aldehyde-functionalized PLA–PEG diblock copolymer allows micellar attachment to appropriate surfaces as well as the introduction of peptide ligands through Schiff base formation (Yamamoto et al., 1999). Functionalization of micellar surfaces with oligopeptides is an important development in targeted colloidal systems. Saccharides have also been used for targeting because they are involved in cell–protein and cell–cell interactions. Kataoka et al. (2001) investigated glucose- and galactose-functionalized PLA–PEG block copolymers. Similarly, Yonese et al. (2001) introduced a lactose ring at one of the ends of a poly(γ-methyl glutamate)–PEG (PMG–PEG) block copolymer.

Several attempts to covalently attach an antibody or ligand to surfactants or polymeric micelles have been described. Poly(L-histidine)–PEG and poly(L-lactic acid)–PEG block copolymer micelles

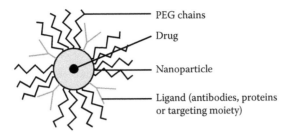

FIGURE 1.17   Surface-functionalized micelle.

carrying folate residue on their surfaces were efficient for the delivery of adriamycin (ADR) to tumor cells *in vitro*, demonstrating the potential for solid tumor treatment (Lukyanov et al., 2004).

The attachment of an anticancer antibody to the micelle surface (creation of immunomicelles) further enhances tumor targeting. Some other examples of ligand-mediated targeting include folate-conjugated PEG-*b*-PCL micelles containing paclitaxel and transferrin-linked PEG-*g*-poly(ethyleneimine) (PEI) micelles containing phosphorothionate oligonucleotide epidermal growth factor (EGF) linked PEG-*g*-PEI loaded with plasmid DNA (Gaucher et al., 2005).

One of the studies used micelles of Pluronics® block copolymers [PEG-*b*-poly(propylene glycol) (PPG)-*b*-PEG] as carriers for central nervous system drug delivery. These micelles were conjugated with either polyclonal antibodies against brain a 2-glycoprotein or insulin as targeting moieties. Both antibody- and insulin-vectorized micelles were able to deliver a drug or a fluorescent probe to the brain *in vivo* (Kabanova and Gendelman, 2007).

The targetability of colloidal carriers is not limited to micelles; concepts have been developed for liposomes, nanoparticles, and other colloidal delivery systems. Liposomes can be explored for drug targeting in the management and chemotherapy of a plethora of fungal, viral, and bacterial diseases such as tuberculosis, leishmaniasis, leprosy, and candidiasis. Because liposomal therapy is site specific, it is capable of achieving high intracellular drug concentrations with reduced distribution to healthy and nonspecific host tissues. Liposomes tagged with sugar moieties, particularly glycolipids with galactose terminals and liposomes consisting of mannosylated phospholipids (Barrat et al., 1986), can be used to target the RES. A number of ligands and targeting moieties such as sialic acid, monosialotetrahexosylganglioside, PEG, and hyaluronic acid (HA) have been explored for prolonging circulatory times along with site-specific drug delivery. Glycolipids and gangliosides such as monosialotetrahexosylganglioside or phosphatidylinositol have been used to prepare sustained-release liposomes. Recently, PEG has been explored extensively for the preparation of stealth liposomes. Liposomes and lipid complexes, particularly surface-modified liposomes such as cationic liposomes, polyplexes, and pH-sensitive liposomes, are being harnessed as nonviral vectors in gene delivery. For example, Holmberg et al. (1994) reported the use of pH-sensitive liposomes conjugated with antibodies specific to glial cells to deliver plasmid DNA, and these liposomes were even more efficacious than simple cationic liposomes. Alving (1982) and Bakker-Wounderberg et al. (1994) observed a significantly high hepatosplenic uptake of liposomal amphotericin B, sodium stilbogluconate with a simultaneous reduction in nonspecific tissue distribution, leading ultimately to minimized toxicity and improved therapeutic benefit. The high localization of liposomes incorporated with radio-opaque materials into the spleen, liver, and similar RES organs makes them valuable tools for the imaging of these organs. Similar to immunomicelles, liposomes tagged with antibodies (immunoliposomes) are effective in tumor targeting.

Several types of targeted nanoparticle preparations have been investigated. Irache et al. (1996) reported the use of lectins such as wheat germ agglutinin, tomato lectin, and bovine serum albumin. Tomato lectin, asparagus pea lectin, *Mycoplasma gallisepticum* lectin, and bovine serum albumin have been covalently coupled as ligands with PS nanoparticles to improve transmucosal absorption.

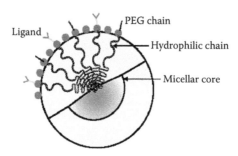

FIGURE 1.18   Targeted stealth nanoparticles.

Premature RES uptake of nanoparticles is a major constraint in achieving site-specific drug target-ing. Fibronectin, lectin, asialoglycoproteins, PEG, sialic acid, and a number of other biomimetic ligands have been used to prevent premature drug uptake by the RES. A schematic depiction is given in Figure 1.18.

A different approach is the use of magnetic nanoparticles prepared by incorporating ferric oxide particles or magnetite particles (10–20 nm) simultaneously with the drug during their preparation. These particles are then injected through the artery supplying the tumor tissue and guided exter-nally with the help of an external magnet to target the nanoparticulate carrier and contents. The attachment of antibodies to the surface of nanoparticles serves as a ligand to facilitate drug delivery to the required sites. Monoclonal antibodies can be tagged to the nanoparticle surface by direct adsorption or by covalent bond formation. The previous section only presents some of the possibili-ties and opportunities presented by surfactant and block copolymer based systems in achieving targeting. However, there are several other concepts and systems that ensure the delivery of thera-peutics to the site of action.

1.13   APPLICATIONS OF COLLOIDAL CARRIERS IN DRUG DELIVERY

1.13.1   MICELLES

The major applications of surfactant or block copolymer micelles are the following:

1. Micelles can be employed to enhance the solubilization of potent but hydrophobic and sparingly soluble drug candidates
2. A low water concentration inside the micellar core retards the degradation of drugs that are prone to hydrolysis
3. Micelles facilitate controlled and sustained drug delivery by virtue of partitioning toward micelles

Apart from these major applications, they also serve as good excipients, act as adjuncts in vaccines, facilitate taste masking, and so forth.

1.13.1.1   Solubilization of Drugs in the Presence of Micelles

The addition of a surfactant above the CMC may strongly increase the solubility of the hydrophobic drug. Polymeric micelles are less toxic for parenteral administration than solubilizing agents cur-rently in use such as polyethoxylated castor oil (Cremophor EL) or polysorbate 80 (Tween 80; Chiappetta and Sosnik, 2007).

Many drugs are moderately stable compounds in aqueous solution and undergo hydrolysis. Their inclusion in micelles may enhance their stability. In fact, an increase in the molecular weight of the block copolymer at a constant composition, for example, in Pluronic (F68 → F88 → F108), results in increased block segregation, increased micellization and solubilization capacity, reduced exposure of the drug to the aqueous environment, and thus a reduced degradation rate.

The most widely investigated copolymers are derivatives of PEO–PPO–PEO block copolymers. Aqueous solutions display a sol–gel transition upon heating (sometimes around 37°C), which makes them suitable for the design of injectable matrices for minimally invasive biomedical applications. Two families are commercially available: the linear and bifunctional PEO–PPO–PEO triblocks or poloxamers (Pluronic) and the branched four-arm counterparts called poloxamine (Tetronic®).

Pluronic formulations have been utilized for enhanced solubilization of poorly water-soluble drugs and a prolonged release profile for many applications (e.g., oral, rectal, topical, ophthalmic, nasal, and injectable preparations). The presence of two tertiary amine groups in poloxamine endows it with temperature and pH sensitivity. A large number of drugs with varied pharmacological effects have been investigated with Pluronic to increase the solubilization and stabilization and achieve delivery-specific requirements, for example, mydriatic tropicamide, immunosuppressant cyclosporins, anticancer doxorubicin, psychotropic propranolol, antibiotic vancomycin, NSAID diclofenac, and corticosteroid triamcinalone (Chiappetta and Sosnik, 2007).

1.13.1.2 Micelles in Intravenous Administration

Intravenous administration of a drug generally results in rapid clearance of the drug from the bloodstream and subsequent accumulation in the RES. Consequently, the bioavailability of drug molecules in tissues other than the RES is negligible. These erratic drug absorption organs may trigger toxicity in RES tissues. Colloidal drug carriers such as liposomes, nanoparticles, and micellar systems can be sought to overcome this problem. The small size and slow disintegration time of PEO-containing block copolymer micelles has attracted a great deal of attention from various researchers. These block copolymeric micelles may prove to be potential drug carriers for site-specific and controlled targeting of cytotoxic drugs in neoplastic disorders. An example of this is PEO–polyaspartate block copolymer conjugates with the anticancer drug ADR. Small micelles formed by copolymer conjugate forms provide sustained release of ADR by prolonging the disintegration time after intravenous administration. The mechanism underlying prolonged disintegration time is biodegradation of aspartate segments, which acts as a rate-limiting step for disintegration.

Most of the cytotoxic drugs unleash severe toxicity that is due to nonspecific drug distribution. The surfactant and block copolymeric micelles reduce host tissue toxicity. It has been reported that intravenous administration of ADR at a concentration of 10 mg/kg in tumor-bearing mice results in a significant reduction of mean survival time because of severe side effects. However, ADR solubilized in block copolymer conjugates reduces the side effects by one to two orders of magnitude (Yokoyama, 1992). Poly(L-histidine)–PEG and poly(L-lactic acid)–PEG block copolymer micelles carrying folate residue on their surface were efficient for the delivery of ADR to tumor cells *in vitro*, demonstrating potential for solid tumor treatment (Lukyanov et al., 2004).

Micelles with blocks made of PEO are sterically stabilized (stealth) and undergo less opsonization and uptake by macrophages of the RES, allowing the micelles to circulate longer in the blood. Physically and chemically stabilized micelles have also been developed to improve their disintegration kinetics. A novel strategy introduced in the last few years is crosslinking of the micellar shell. Terminal hydroxyl groups undergo modification to improve the reactivity and are further reacted with coupling agents or by free radical polymerization reactions. Two different effects can be observed: micellar stabilization can be enhanced and shell permeability can be manipulated, which is critical for drug inclusion and release rates. Wooley et al. (1996) pioneered the concept of chemically crosslinking the corona of amphiphilic diblock copolymer micelles using a broad range of polymers including styrene, isoprene, butadiene, ε-caprolactone, and methyl methacrylate. Monomers that have been used for the preparation of the hydrophilic, water-soluble corona include 4-vinylpyridine, MAA, and 2-di-methylaminoethyl methacrylate. They possess great potential for the development of advanced drug delivery systems (Rosler et al., 2001).

Micelles can also be core crosslinkable micelles, shell crosslinkable micelles, and functionalized site-directed micelles. For core crosslinking, the most suitable method is to modify the hydrophobic chain of the block copolymer with a polymerizable group. Kataoka et al. (2001) reported a number of PLA–PEG copolymers containing a methacryloyl group at the PLA chain. Several authors explored amphiphilic block copolymers containing crosslinkable groups that were laterally distributed along the hydrophobic block.

Superior drug targeting can be achieved by binding pilot molecules such as antibodies or sugars or by introducing a polymer sensitive to variations in temperature or pH (Jones et al., 1999). This strategy allows active targeting with ligands that are selectively recognized by receptors on the surface of the cells of interest. Kataoka et al. (2001) prepared galactose- and glucose-functionalized PLA–PEG block copolymers. Yonese et al. (2001) prepared sugar-substituted PMG–PEO block copolymers. Several attempts to covalently attach an antibody to surfactants or polymeric micelles have been described.

Another important characteristic of micelles, when compared with microspheres or many liposomal formulations, is their small and uniform particle size. In theory, particle sizes can go down to the order of 10 nm for nonloaded polymeric micelles. This size is still large enough to accomplish passive targeting to tumors and inflamed tissues by the so-called EPR effect.

Micelles from certain polymer–lipid conjugates can also be loaded with a number of poorly soluble drugs, and they are quite stable in their ability to preserve their morphology and retain micellized drugs in conditions modeling parenteral administration. The micelles can spontaneously accumulate in tumors via the EPR effect. The attachment of an anticancer antibody to the micelle surface (creation of immunomicelles) further enhances tumor targeting. Some other examples of ligand-mediated targeting include folate-conjugated PEG-b-PCL micelles containing paclitaxel and transferrin-linked PEG-g-PEI micelles containing phosphorothionate oligonucleotide EGF-linked PEG-g-PEI loaded with plasmid DNA (Gaucher et al., 2005).

PEO–poly(L-amino acid) block copolymers have been explored to form polyion complexes with DNA or peptides and undergo micellization. For example, poly(L-lysine) has been used for encapsulating plasmid DNA. Depending on the type of amino acid, PEO-b-PLA block copolymers may have a positive or negative charge at their side chains (Lavasanifar et al., 2002).

Micelles have been explored in drug delivery for specific applications. For example, because of toxicity concerns, many water-insoluble photosensitizers are preferentially required to deliver micelles at pathogenic sites by passive or active targeting (Nostrum, 2004). Hioka et al. (2002) studied the use of Pluronic P123 to solubilize a benzoporphyrin derivative for photodynamic therapy. Leroux et al. (2000) used a poorly water-soluble aluminum chloride phthalocyanine as the photosensitizer and applied random copolymers as the carrier composed of NIPAAm and MAA (typically 3–5 mol %) to create pH sensitivity.

Recently, novel polymeric micelles with smart functions, such as targetability and stimuli sensitivity, have emerged as promising carriers that enhance the efficacy of drugs and genes with minimal side effects. Several scientist groups (e.g., Kataoka et al., Bae et al., Frechet et al., and Katayma et al.) have been working in this direction, in which polymeric micelles respond to intracellular signals (Nishiyama et al., 2005).

Most of the work on polymeric micelles has focused on anticancer drug delivery and gene delivery (Torchilin, 1998), but it can also have other important applications. For example, an early study used micelles of Pluronics block copolymers (PEG-b-PPG-b-PEG) as carriers for central nervous system drug delivery. These micelles were conjugated with either polyclonal antibodies against brain a2-glycoprotein or insulin as targeting moieties. Both antibody- and insulin-vectorized micelles delivered a drug or a fluorescent probe to the brain *in vivo* (Kabanova and Gendelman, 2007).

The use of an external physical stimulus to control micellization or micelle disruption processes has just begun to be exploited (Rapoport, 2007). Jiang et al. (2005) synthesized a block copolymer based on PEO and polymethacrylate attached to pyrene. UV irradiation caused hydrolysis of the ester group of the polymer, leading to micelle dissolution.

Several scientists have investigated micelles prepared from water-soluble polymers conjugated with lipids. The hydrophobic part is represented by a lipid instead of a hydrophobic polymer block. The ability of PEG–PE molecules to form micelles in an aqueous environment was observed as early as 1994. Some of the PEG-conjugated lipids that have been explored are PEG–distearoylphos phatidylethanolamine, PEG–dioleoylphosphatidylethanolamine, and so forth. Drugs (such as paclitaxel) and vitamin K_3 have been loaded in such micelles and explored. These also circulate a long time in nature (Lukyanov et al., 2004).

1.13.2 Liquid Crystalline Phases

Liquid crystals offer several useful properties for drug delivery. The solubilization of a drug in the liquid crystalline phase is similar to that in micelles. Simultaneously, an increase in the viscosity of the system provides more localized effects in parenteral (intramuscular), topical, or oral delivery. The phase transitions of liquid crystals can be achieved either by temperature or by dilution. Systems can be tailored in such a way that transitions can be achieved at body temperature or in contact with body fluids.

1.13.2.1 Oral Cavity

The administration of drugs in the oral cavity is challenging because of the possible ingestion of the formulation and the mechanical stress exerted by masticatory movements. *In situ* thickening of Pluronics with the aqueous system makes it suitable for formulations intended for the oral cavity. Local anesthetics (e.g., lidocaine and prilocaine) used for the treatment of periodontal diseases are investigated for their liquid crystal based delivery system.

1.13.2.2 Topical Delivery

Systems undergoing temperature-induced thickening are found to be more useful in the topical delivery system. Liquid crystal based wound dressings are used for the protection of artificial skin and for the prevention of microbial infections during severe skin burns. Atyabi et al. (2007) investigated cellulose acetate and cellulose nitrate containing liquid crystals as topically applied thermoresponsive barriers using paracetamol and methimazole as model drugs. Liquid crystal based topical formulations containing nimesulide have higher stability, better release and absorption, and subsequent enhanced bioavailability (U.S. Pat. 6,288,121).

1.13.2.3 Parenteral Route of Administration

Liquid crystal based injectables have gained importance as dosage forms because of their salient features of *in situ* thickening. *In situ* thickening results in sustained drug release; hence, liquid crystal based injectables may serve as repository preparations. Pluronics are biocompatible and are being explored further. Boyd and Whittaker (2006) demonstrated the sustained release of a series of model hydrophobic and hydrophilic drugs (paclitaxel, irinotecan, glucose, histidine, and octreotide) from lyotropic liquid crystalline phases based on glycerate surfactants.

1.13.3 Emulsions

Emulsions are extensively used as drug delivery systems, particularly for the topical and oral routes of administration. Generally, o/w emulsions are intended for internal use for obvious reasons. Some of the advantages offered by emulsions in drug delivery are solubilization of hydrophobic drugs, high drug release rates, and enhanced chemical stability or protection of the drugs from hydrolysis in an aqueous environment. Pharmaceutical emulsions contain less amounts of surfactants (phospholipids, lecithins) that are natural in origin and thus nontoxic in nature. Oil components are always selected from oils of natural origin, such as soybean oil, cottonseed oil, coconut oil, corn oil, sesame oil, cod liver oil, olive oil, and linseed oil, or mixtures of different fatty acids in different proportions (Malmsten, 2002).

1.13.3.1 Parenteral Administration

Prior to formulating an emulsion, the components are subjected to screening for purity, as well as purification by chromatographic techniques if required. While formulating emulsions intended for parenteral use, much emphasis should be given to droplet size, iso-osmolarity, and sterility, because droplet sizes >2 μm may cause emboli or thrombosis. Iso-osmolarity of the delivery systems should be ensured by the addition of an electrolyte without affecting the stability of the delivery system. Systemic administration of an emulsion results in rapid plasma clearance and is readily taken up by the RES. This can be overcome by the surface modification of droplets with ethylene oxide containing surfactants or block copolymers. Hydrophobic drugs, nutrients such as vitamin K_1, diagnostic agents, and vaccines intended for parenteral use are formulated as emulsions. Fluorocarbon-based emulsions are used as plasma substitutes. A chlorambucil-loaded parenteral emulsion was prepared by a high-energy ultrasonication method using soybean oil as a triglyceride oil core and egg PC as an emulsifier. A pharmacokinetics and pharmacodynamics study in mice suggested that the chlorambucil parenteral emulsion could be an effective parenteral carrier for chlorambucil delivery in cancer treatment (Ganta et al., 2008). Wang et al. (2008) developed lipid emulsions as parenteral drug delivery systems for morphine and its ester prodrugs such as morphine propionate, morphine enanthate, and morphine decanoate using PE as an emulsifier and squalene as an inner oil phase. They suggested that the combination of a prodrug strategy and lipid emulsions may be practical for improving analgesic therapy with morphine.

1.13.3.2 Oral Administration

Emulsions for oral drug delivery are well suited to mask bitter and obnoxious tastes. The absorption of o/w formulations mainly depends on dissolution of the drug, the partition coefficient of the drug between the emulsion components, the flow of bile juice, and the membrane permeability to the drug in the GIT. Generally, sparingly soluble and poorly bioavailable drugs showing drug uptake variability can be incorporated in o/w emulsions to minimize the variability. Liquid paraffin oral emulsions and castor oil emulsions are commercially available in the market for laxative purposes. Simethione oral emulsions are used for the relief of painful symptoms of flatulence and dyspepsia associated with flatulence. The poor bioavailability of poorly water-soluble cefpodoxime proxetil was greatly improved by o/w submicron emulsions (Nicolaos et al., 2003). Coenzyme Q_{10} (CoQ_{10}) was emulsified with fats and four types of emulsifiers (lecithin, monoglycerides, calcium stearoyl-2-lactate, and diacetyl tartaric acid esters of monoglycerides), and the oral bioavailability of CoQ_{10} was confirmed to be slightly greater than that of a standard commercial CoQ_{10} product (Thanatuksorn et al., 2009).

1.13.3.3 Topical Administration

Emulsions meant for topical use are more popular as cosmetics than as pharmaceuticals. Both o/w and w/o for mutations are used topically; however, o/w-based formulations produce a cooling effect due to the evaporation of external aqueous phase, whereas w/o-based formulations produce a softening effect. The uptake of drug from topically administered formulations mainly depends on partitioning of the drug between the oil and water components, droplet size, concentration of the drug, skin condition, and interaction between the surfactants/components of the delivery system and the proteins present at the stratum corneum. Estrasorb—a topical emulsion containing estradiol—is marketed by Navavax, King Pharmaceuticals for the treatment of hot flushes and night sweats associated with menopause in females.

1.13.4 MICROEMULSIONS

Microemulsions offer a range of advantages for oral as well as topical drug delivery. Currently, they are being explored for nasal and buccal delivery.

1.13.4.1 Oral Administration

In general, microemulsions increase the bioavailability of drugs because of their small droplet sizes when compared with emulsions of the same drug. Microemulsions themselves provide a suitable delivery system for proteins and peptides. Apart from enhancing the bioavailability of proteins, microemulsion-based protein delivery systems provide protection against the enzymatic and chemical degradation of drugs, and consequently enhance the bioavailability of peptides. The o/w microemulsions may offer an alternative delivery system for pH-sensitive drugs susceptible to degradation at the acidic pH values of the stomach. An improvement in the bioavailability of cyclosporine incorporated in microemulsions for oral delivery is disclosed in patents (U.S. Pat 6,306,434, 2001; U.S. Pat 6,638,522, 2003). Oral microemulsions of mometasone furoate are highly effective compared to conventionally available and marketed preparations of the drug in the treatment of erosive–ulcerative oral lichen planus (Aguire et al., 2004). Cliek et al. (2006) tried to develop a stable oral microemulsion system for insulin delivery using lecithin. The paclitaxel containing microemulsion with or without cellulosic polymers improves the oral bioavailability of the drug (U.S. Pat 7,115,565, 2006).

1.13.4.2 Topical Route of Administration

The barrier to topical drug absorption, the stratum corneum, has been partially overcome by the use of surfactant-based delivery systems. Surfactants react with the lipids of the stratum corneum and enhance drug penetration across the skin. Increased solubility and insecticidal activity of azadirachitin A, a natural component from neem oil (U.S. Pat 6,703,034, 2004), is found when it is formulated as a microemulsion-based topical formulation. Willimann et al. (1992) prepared lecithin containing w/o microemulsions for the transdermal administration of scopolamine and broxaterol. They found that the transport rate obtained with the lecithin microemulsion gel was much higher than that obtained with an aqueous solution at the same concentration.

1.13.4.3 Parenteral Route of Administration

The presence of a high content of surfactant in microemulsions may trigger systemic toxicity. Apart from this, in some cases the cosurfactant may not be stable on dilution and causes phase separation leading to emboli formation. Various parameters such as acid–base balance, blood gases, plasma electrolytes, mean arterial pressure, and heart rate should be monitored on parenteral administration of microemulsions, which complicates microemulsion-based parenteral therapy. However, medium-chain triglyceride-based microemulsions are suitable for parenteral use, because the medium-chain triglyceride on spontaneous *in situ* emulsification produces droplet sizes acceptable for intravenous application. The problems associated with the solubility and odor of propofol (anesthetic), and pain during propofol injection have been overcome by the formulation of a microemulsion-based injection with an emulsifier composition of a long-chain polymer surfactant component and a medium-chain fatty acid surfactant component (U.S. Pat 6,623,765, 2003).

1.13.4.4 Block Copolymers in a Microemulsion-Based Drug Delivery System

Microemulsions conventionally prepared with surfactants are not very robust and may change their structure with dilution and temperature change. Barker and Vincent (1984) investigated stable w/o microemulsions with PS–PEO diblock polymers. Attempts were made to compare the stability and drug-releasing properties of emulsions prepared with conventional surfactants and block polymers. It was shown that PCL-*b*-PEO block copolymers are promising nonionic macromolecular surfactants for the stabilization of emulsions because they display stronger adsorption and provide increased long-term stability (Chausson and Fluchère, 2008).

Andrij Pich and Nadine Schiemenz (2006) demonstrated that PEO–PS block copolymers are the best stabilizers for the preparation of polymeric particles compared to conventional surfactants such as sodium dodecyl sulfate or cetyl trimethyl ammonium bromide in emulsions.

Indomethacin was used as a model drug in a micellar system composed of copolymers of poly-(L-lactic acid) and methoxy end-capped PEG, and the release pattern of the drug from the micellar structure in aqueous media was studied (Kim and Kim, 2001).

Scherlund and Malmsten (2000) investigated thermosetting microemulsions and mixed micellar solutions based on poloxamer block polymers (Lutrol® F127 and Lutrol F68) for the delivery of lidocaine and prilocaine intended for induction of periodontal anesthesia. The micellar systems showed better stability and increased drug release compared to conventional emulsion-based formulations.

Opanasopit and Yokoyama (2004) synthesized a PEG–poly(β-benzyl-L-aspartate) block copolymer and studied the suitability of a camptothecin-loaded system for tumor targeting. Camptothecin-loaded nanoparticles prepared from the emulsion of a PLA/PEG–PPG–PEG system were reported by Kunii and Onishi (2007). They found with fairly high and stable entrapment of the drug and more area under the curve in normal rats and significant tumor growth suppression in mice with sarcoma 180 solid tumors.

Kwon and Naito (1997) reported doxorubicin-loaded micelles of PEO-*b*-poly(β-benzyl-L-aspartate) as depot preparation for the sustained release of the drug.

1.13.5 SEDDS AS A DRUG DELIVERY SYSTEM

SEDDSs offers advantages in the drug delivery system by improving the solubility, absorption, bioavailability, sustained release, and stability of the poorly soluble and less bioavailable drugs.

Atef and Belmonte (2008) developed SEDDSs loaded with phenytoin for comparison with a marketed solution of phenytoin (Dilantin). She observed a 2.3-fold increase in the area under the curve accompanied by an increased rate of absorption of phenytoin and a sustained effect of phenytoin in plasma.

Ji-Yeon Hong (2006) investigated the dissolution profile of itraconazole solubilized in a SEDDS composed of Transcutol, Pluronic L64, and tocopherol acetate. The SEDDS reduced the interference of food in drug absorption. It was demonstrated that the absorption of torcetrapib from a SEDDS is not affected in the presence of food (Perlman and Murdane, 2008).

SEDDSs has been shown to improve the stability of the entrapped drug compared to the plain drug solution. Gang Soo Chae and Jin Soo Lee (2005) studied the *in vitro* stability of 1,3-bis-(2-chloroethyl)-1-nitrosourea (BCNU) after its release from a PLGA wafer and evaluated the *in vitro* antitumor activity against 9L gliosarcoma cells. The study concluded that SEDDS is capable of not only stabilizing BCNU but also facilitating the sustained release of BCNU from self-emulsified BCNU-loaded PLGA wafers. The faster dissolution and drug release of CoQ_{10} from a self-nanoemulsifying drug delivery system was demonstrated by Palamakula et al. (2004).

The oral bioavailability of poorly water-soluble drugs such as WIN 54954 (Charman and Charman, 1992), immunosuppressive drugs (Lattuada and Martini, 1998), halofantrine (Shui-Mei Khoo, 1998), diclofenac (Attama and Nzekwe, 2003), and simavastatin (Kang and Lee, 2004) and lipophilic drugs such as cyclosporine, aritonavir, saquinavir (Gursoy and Benita, 2004, ritonavir (Arunothayanun and Pirayavaraporn, 2004), silymarin (Wu and Wang, 2006), 9-nitrocamptothecin (9-NC; Lu and Wang, 2008), and oridonin (Ping Zhang and Ying Liu, 2008) were increased by formulating them as SEDDSs.

1.13.6 LIPOSOMES

1.13.6.1 Intravenous Administration

Liposomes tend to extravasate to the spleen, liver, and other RES organs because of loose junctions between their endothelial cells. Moghimi and Hunter (2001) reported that neutral and smaller liposomes (<100 nm) undergo slow clearance compared to larger and charged liposomes. This behavioral profile makes liposomes ideal candidates for targeting cytotoxic drugs to the RES along with minimization of their lethal toxicities. Anthracyclines have been successfully explored and commercialized in liposomal formulations, for example, an intravenous doxorubicin liposomal formulation (Myocet®). The liposomal drug has a reduced incidence of cardiomyopathy and similar cardiotoxicity in comparison to conventional doxorubicin formulations. Shimizu et al. (2002) found that intravenous injection of TAS-103 (a topoisomerase inhibitor encapsulated in nanometric liposomes) improves survival and increases the life span to 42 days for mice induced with Lewis lung carcinoma compared to 38.6 days for those treated with nonliposomal TAS-103.

Liposomes have been successfully used for intracellular targeting of chemotherapeutic agents. Achieving optimum or inhibitory concentrations of antibiotic intracellularly is a major challenge for the formulation chemist. Wu et al. (2004) investigated the liposomal formulation of a popular cephalosporin—cefoxitin. The major therapeutic constraints of this agent are its short half-life and poor intracellular penetration. When administered intravenously, liposomal cefoxitin was absorbed 6 to 16 times more in target tissues compared to free cefoxitin.

1.13.6.2 Subcutaneous Administration

Oussoren and Storm (2001) and Gregoriadis et al. (2002) used subcutaneously administered liposomes for lymphatic delivery to target therapeutic and diagnostic agents for imaging, vaccination, and so forth. The subcutaneous route was found to be promising for preventing malignancy and metastatic development through the lymphatic system. Liposomal gadolinium was used for imaging (Fujimoto et al., 2000; Misselwitz and Sasche, 1997). Similarly, subcutaneous administration of a nonionic liposomal-encapsulated plasmid for a nucleoprotein of the influenza virus was therapeutically more efficacious compared to free DNA (Perrie et al., 2004).

1.13.6.3 Intraperitoneal Administration

The intraperitoneal route of administration is usually preferred when high localization of drugs in tumors and low plasma drug levels are required simultaneously. Liposomal drug delivery by the intraperitoneal route leads to increased local drug concentrations in the peritoneal cavity. Sadzuka et al. (2000) investigated doxorubicin liposomes for their potential in solid tumor therapy in Ehrlich ascites carcinoma bearing mice. Similarly, intraperitoneal administration of a liposomal formulation of an L-dopa prodrug derivative to rats resulted in a two times higher concentration in rat corpus striatum compared to L-dopa or the free prodrug itself (Di Stefano et al., 2004).

1.13.6.4 Pulmonary Drug Delivery

The salient anatomical and physiological features of lungs make them ideal for drug targeting. The large surface area offered by alveoli (Patton et al., 2004), avoidance of first-pass hepatic metabolism, noninvasiveness of the pulmonary route, accomplishment of local drug delivery in antiasthmatic drugs, a higher rate and extent of penetration, and better bioavailability even in the case of peptide drugs make the lungs a fascinating site for drug targeting. Aerosolized liposomes or liposomal dry powder inhalers with particle sizes not exceeding 3 μm are ideal for pulmonary delivery. Literature citations report the use of liposomal drug delivery in the treatment of various pulmonary disorders ranging from cystic fibrosis to lung cancer and in gene delivery using liposome-based formulations. Modified liposomal systems such as aerosolized liposomes, dry powder inhalers, and liposome DNA complexes have been studied for efficacious lung targeting (Schwarz et al., 1996; Eastman et al., 1997). Drugs such as amphotericin B in liposomes have been successfully commercialized for systemic fungal infections like candidiasis and aspergillosis. Vyas et al. (2004) reported the use of liposomal rifampicin in the treatment of AIDS-related pulmonary tuberculosis. Liposomal rifampicin showed significantly higher retention in the lungs compared to conventional rifampicin formulations. The same scientists also reported the encapsulation of amphotericin B in liposomes for the treatment of aspergillosis. Liposomes were prepared by conjugating them to o-palmitoylated pullulan as well as unconjugated liposome. Ligand-tagged amphotericin B liposomes showed promising uptake compared to nonconjugated ones as well as conventional formulations (Vyas et al., 2005).

Doddoli et al. (2005) reported reduced nephrotoxicity of methotrexate in liposome-entrapped methotrexate in comparison to the free one. Similarly, liposomal 9-NC exhibited good pulmonary tumor concentrations when studied in human cancers xenografted subcutaneously in mice and murine melanoma and human pulmonary metastases.

1.13.6.5   Oral Delivery

Liposomes have attracted a great deal of attention for the oral delivery of drugs, especially for drugs that are acid labile, proteinous, and prone to first-pass metabolism.

1.13.6.6   Transdermal Delivery

The excellent biocompatibility and potential of liposomes as ideal transdermal carriers are attributed to their composition that greatly resembles that of the skin structure, their ability to act as potent permeation enhancers, their elasticity, their rejuvenation ability, as well as their revitalizing action on skin. These attributes render them extremely useful as cosmetic delivery vectors. Liposomal composition, size, charge, and zeta potential are critical parameters that should be controlled to optimize transdermal drug delivery. Manosroi (2004) reported a 10-fold increase in the rate and extent of absorption of amphotericin B incorporated in charged liposomes compared to chargeless, plain amphotericin B liposomes. Similar results were reported by Liu (2004) for acyclovir, where there was a much higher release of acyclovir from positively charged liposomes compared to plain liposomes incorporated with acyclovir. Modified liposomes such as niosomes and transferosomes for topical or transdermal drug delivery have been developed in recent years. Transformable or elastic vesicles called transferosomes are currently attracting research interest for their possible role in transdermal drug delivery. Similarly, a newer concept of liposomes incorporated with ethanol, called "ethosomes," is also rapidly gaining popularity. Godin and Touitou (2003) reported the potential of ethosomes in the drug delivery of highly lipophilic drugs such as sex hormones, contraceptives, cannabinoids, and cationic drugs across the skin. Han et al. (2004) reported the higher diffusion and extent of drug absorption of ADR when delivered by the transfollicular route using liposomal drug delivery assisted with iontophoresis. Wells et al. (2000) used electroporation synergized with cationic liposomes to augment gene transfer in murine breast tumor skin.

1.13.6.7   Ocular Drug Delivery

One of the major constraints of the ocular route of delivery is the very low residence time of a drug in the ocular cavity, leading to a subsequent reduction in the bioavailability of therapeutic moieties. Utilized liposomal drugs in the treatment of dry eye syndrome. Charge and vesicular size are important parameters that affect the biodistribution of liposomes. Law et al. (2000) reported a higher corneal uptake of positively charged liposomal acyclovir.

1.13.6.8   Nasal Delivery

The nasal route has been widely explored and is considered to be prospective for the delivery of various drugs for local and systemic effect. Shahiwala and Misra (2004) reported the improved bioavailability of contraceptives such as leuprolide and levonorgestrol due to the improved mucoadhesion and residence time of liposomal formulations of these drugs studded with chitosan and Carbopol®. Liposomal vaccines are also being tried for their possible delivery by the nasal route. Wang and Nagata (2004) reported intranasal administration of liposomes loaded with plasmid DNA encoding influenza virus hemagglutinin in mice and found it to be better than existing commercial vaccines.

1.13.6.9   Vaginal Delivery

Liposomes and liposome-based gels are gradually becoming popular as delivery systems. Paveli et al. (2001) investigated liposomally entrapped chemotherapeutics for the local treatment of vaginitis.

They also developed liposomal gel formulations with an appreciable degree of mucoadhesion for antifungal drugs such as clotrimazole and metronidazole for vaginal local therapy.

1.13.6.10 Liposomal Targeting in Brain Tumors

Drug targeting to brain tumors or gliomas is a major challenge for the formulation scientist because of the presence of a highly resistant blood–brain barrier. The reason for the poor prognosis of brain tumors is mainly poor localization of cytotoxic drugs in tumors that is attributable to the blood–brain barrier and subsequent nonspecific tissue distribution leading to toxicity. Liposomal cytotoxic drugs are believed to be more biocompatible and nontoxic to healthy host tissues along with efficacious tumor targeting. Nanosized liposomal vesicles (size range = 200–400 nm) tagged with a suitable ligand can be an ideal delivery system for brain tumor targeting. The rationale underlying brain targeting is the overexpression of certain receptors in pathological conditions such as cancer. Ligands capable of identifying such targets can be chemically anchored on the surface of liposomes to facilitate their internalization to such tumors and exert the required therapeutic action. Disialoganglioside (GD2) is overexpressed in neuronal ectodermal cancer cells compared to normal brain cells. A monoclonal antibody such as anti-GD2 linked with liposomes via pegylation can serve as a good targeting moiety. Brignole et al. (2005) reported the use of the Fab' fragment of anti-GD2–PEG liposomes incorporated with antisense oligonucleotide-c-*myb* and found an increase in both survival times and overall life span with intravenous injections of these site-specific liposomes in nude mice with HTLA-230 xenografts.

Similarly, the combination of liposomal anthracyclines with other antineoplastic agents was determined to be a synergistic combination. Caralgia et al. (2006) studied the combination regimen of doxorubicin liposomes (Doxil) with temozolomide in the management of brain tumors and found the combination to be quite effective in a phase II study. The incidence of severe toxicities such as cardiomyopathy and mucosal damage were greatly diminished with liposomal doxorubicin compared to conventional intravenous doxorubicin injections.

1.13.6.11 Breast Cancer

Chemotherapy for breast cancer has only been partially successful because of unwanted and highly toxic side effects unleashed by these chemotherapeutic agents. Liposomal doxorubicin has been better tolerated by breast cancer patients in comparison to conventional dosage forms. The liposomal formulation of existing chemotherapeutic agents is safer and relatively less toxic and has intact therapeutic efficacy (Hofheinz et al., 2005; Gradishar, 2005). Mrozek et al. (2005) reported liposomal doxorubicin in conjunction with docetaxel as being safe and well tolerated in patients with breast cancer.

Liposomes can also be used in tumor targeting by altering the biodistribution and pharmacokinetic profile of a drug. Stover et al. (2005) studied ceramide-loaded liposomes for anticancer action. Ceramide, which is a cytotoxic molecule produced after sphingomyelin metabolism, was loaded in PEGylated liposomes and injected intravenously in mice bearing syngeneic or human xenografts of breast cancer. Administration of this liposomal ceramide led to tumor regression (by six times).

1.13.6.12 Lung Cancer

9-NC is a lipophilic derivative of a naturally occurring anticancer agent, camptothecin. It has been reported to possess a substantial cytotoxic effect against lung cancer but with marked hematological toxicity. Verschraegen et al. (2004) investigated 9-NC incorporated liposomes in aerosolized formulations in patients with primary and metastatic lung cancer and showed there was a significant reduction in hematological toxicities.

Peer and Margalit (2004) investigated HA liposomes encapsulated with doxorubicin. In this study, HA was used as a targeting moiety to CD44 and RHAMM receptors overexpressed in lung cancer. The hydrophilic coat provided by HA also served as a stealthing moiety and facilitated controlled release. HA-conjugated doxorubicin and nonconjugated doxorubicin liposomes were investigated in three mice models, and liposomal and HA liposomal doxorubicin were found to be long circulating with higher accumulation in target lung cancer.

1.13.7 Nanoparticles

1.13.7.1 Drug Delivery Applications of Natural Polymeric Nanoparticles

Alginates, chitosan, gelatin, and albumin have been widely explored as natural polymeric materials for the fabrication of nanoparticles. The biodegradability and biocompatibility of these materials make them appropriate candidates for the formulation of nanoparticles.

1.13.7.2 Alginate Nanoparticles

Pandey et al. (2005) reported an enhancement in the bioavailability of antifungal agents incorporated in chitosan-stabilized alginate particles. The prepared nanoparticles executed a sustained-release profile of these drugs. Similar improved stability was seen in chitosan–alginate nanoparticles containing ovalbumin (Borges et al., 2005).

1.13.7.3 Chitosan Nanoparticles

Chitosan nanoparticles can be useful in stabilizing and protecting labile bioactives and therapeutic entities. Nanoparticles offer protection and thereby improved delivery of enzymes, oligonucleotides, nucleic acids, and so forth. Chitosan nanoparticles rendered hydrophobic by surface treatment were used for the delivery of trypsin (Liu et al., 2005). Mansourie et al. (2006) utilized chitosan nanoparticles targeted to folate receptors to improve gene transfection. Hyung Park et al. (2006) investigated chitosan particulate delivery of doxorubicin and found a significant reduction in the cardiotoxicity of doxorubicin.

1.13.7.4 Gelatin Nanoparticles

Gelatin nanoparticles have been investigated in recent years for drug delivery and site-specific drug targeting. Literature citations report the use of gelatin for the delivery of hydrocortisone (Vandervoort and Ludwig, 2004), paclitaxel (Lu et al., 2004), and chloroquine (Bajpai and Choubey, 2006). PEGylation of gelatin nanoparticles can be used to achieve the sustained release of hydrophilic drugs by prolonging their circulation times in the body (Kaul and Amiji, 2002). Gelatin nanoparticles can also be efficacious in tumor targeting. A higher accumulation of PEGylated nanoparticles was found in the liver and tumors (Kaul and Amiji, 2004) and lysosomes of neuronal dendrites (Coester, 2006), justifying their use in targeting gliomas. However, cardiotoxicity and immunogenicity due to its proteinous nature are major constraints for the use of gelatin in drug delivery.

1.13.7.5 Synthetic Polymeric Nanoparticles

1.13.7.5.1 Polycaprolactone Nanoparticles

Nanoparticles prepared from PCL–PEG blends were found to augment the cytotoxicity of retinoic acid (Jeong et al., 2004). Similarly, chemical modification of the surface of these nanoparticles by anchoring it with folic acid can be effectively used in tumors overexpressing folate receptors and significantly improves drug uptake (Park et al., 2005).

1.13.7.5.2 Polyanhydride Nanoparticles

Pfiefer et al. (2005) used polyanhydride–lactic acid blends for plasmid transfection using firefly luciferase DNA and found improved delivery of firefly DNA entrapped in polyanhydride nanoparticles.

1.13.7.5.3 Poly(Alkyl Cyanoacrylate) Nanoparticles

McCarron (2004) reported PACA nanoparticles to be potentially effective in the treatment of oral candidiasis. The enhanced efficacy of PACA nanoparticles may be attributed to improved bioadhesion with these particles. PACA nanoparticles were also effective in DNA transfection. Nanoparticles increased the accumulation and retention of antisense oligonucleotides in vascular smooth muscle cells (Toub, 2005). These nanoparticles can also be used for insulin delivery in diabetic patients. The major constraints associated with oral delivery of insulin are its acid lability and susceptibility to proteolytic enzyme degradation in the GIT. Insulin also has a short half-life, which necessitates frequent administration and results in poor patient compliance. PACA nanoparticles incorporated with insulin were found to enhance the absorption of insulin along with prolonging the duration of action. Pronounced hypoglycemic effects for a prolonged duration were reported, which may help to render insulin therapy more controlled, sustained, and patient friendly (McCarron, 2004).

1.13.7.5.4 Cyclodextrins in Nanoparticles

The major limitation of nanoparticles is poor drug loading capacity. Poor payload necessitates the use of a large amount of polymeric materials for the fabrication of nanoparticles, and this in turn raises safety and toxicity issues. Various approaches have been sought to alleviate this problem. Drug loading problems can be overcome by cyclodextrins (CDs) by virtue of their solubilization and stabilization properties, as explored by several scientists. Boudad (2001) investigated an antiviral agent, saquinavir, to improve its drug loading by complexing saquinavir with hydroxypropyl-β-CD followed by loading it into PACA nanoparticles. Similarly, chitosan nanoparticles in conjunction with CDs were tried by some workers for solubility enhancement and protection of labile drug components in an aqueous microenvironment. Chitosan nanoparticles of frusemide and triclosan were prepared after complexation with CDs followed by crosslinking of chitosan with sodium tripolyphosphate. The resultant nanoparticles exhibited improved drug loading capacity along with sustained release of the drug (Maestrelli, 2006).

Hede (2005) studied CD-based nanoparticles conjugated with transferrin for the delivery of small interfering RNA to neoplastic cells and found a nanoparticles-based delivery system efficacious in controlling the growth of neoplasm along with freedom from tissue toxicity.

1.13.7.5.5 Amphiphilic Cyclodextrins

Amphiphilic derivatives of modified CDs have been explored over the last few years for the synthesis of nanoparticles. The amphiphilic nature of these derivatized CDs eliminates the use of other polymers or surfactants. These CDs have been investigated as potential substitutes for currently used polymer-based systems. These systems offer some salient advantages:

1. Improved interaction and ready adaptability to biological membranes result in higher biocompatibility and reduced tissue toxicity.
2. These systems serve as excellent carriers for hydrophobic drugs because of the improved degree of interaction with hydrophobic moieties. Loading of lipophilic–hydrophobic bioactives can be further enhanced by modifying the structure of amphiphilic CDs and using more hydrophobic derivatives.
3. Spontaneity in the formation of nanoconstructs makes them very appealing for use in nanotechnology for drug delivery.

1.13.7.5.5.1 Nonionic Amphiphilic Cyclodextrins
Nonionic amphiphilic CDs possess appreciable surface activity. These systems also possess inclusion-forming and complex-forming properties. Bilensoy et al. (2007) hypothesized that the unsubstituted secondary face of these CDs facilitates the entrapment of drugs in the cavity. Neutral CDs can be used for transfection of nucleic acids and may be considered as powerful tools in the field of gene delivery.

1.13.7.5.5.2 Cationic Amphiphilic Cyclodextrins
These CDs mainly comprise hydrophobic thioalkyl chains and a hydrophilic ethylene glycol skeleton. The stability of colloidal carriers formed by CDs may be attributed to the balance existing between these hydrophobic and hydrophilic fragments of the molecule. Cationic CDs are more promising in gene delivery compared to neutral

ones because of the improved efficiency of nucleic acid entrapment. Cationic CDs are strongly complexed with negatively charged nucleic acid due to electrostatic attraction and thus may provide better payload and improved targeting efficacy (Matsumoto, 2004, 67).

1.13.7.5.5.3 Anionic Amphiphilic Cyclodextrins   The presence of a sulfate group in the structure renders this molecule anionic. Anionic amphiphilic CDs have been tried for drugs such as acyclovir. Granger et al. (2000) pioneered the concept of fluorine-containing anionic CDs. Peroche et al. (2005) delineated the synthesis of amphiphilic perfluorohexyl, perfluorooctyl thio β CDs and their alkyl derivatives and studied their possible role in the fabrication of nanoparticles.

1.13.7.5.6   Applications in the Drug Delivery of Cyclodextrin-Based Nanoparticles

1.13.7.5.6.1 Cancer Therapy   The nonselectivity of cytotoxic agents results in lethal and hazardous toxicities to cancer patients. Colloidal delivery systems have been extensively and successfully studied for their efficacy in targeting cancer. Nanoparticles are readily taken up by the leaky vasculature of tumors and can exert controlled and sustained drug delivery to cancerous cells because of EPR. Amphiphilic CD nanoparticles incorporated with tamoxifen were studied by Bilensoy et al. (2007). Tamoxifen is the drug of choice in the management of breast cancer and an adjunct for metastatic breast cancer. However, the drug suffers from the major limitation of nonspecific tissue toxicity. Tamoxifen CD nanoparticles were found to exert a sustained-release effect and were therapeutically as active as the free drug.

Amphiphilic CDs have excellent solubilization and complexation properties. Researchers have exploited these properties to improve the absorption and solubilization of highly hydrophobic molecules such as camptothecin. The very low water solubility of this drug results in poor bioavailability and renders it clinically less acceptable. Different derivatives of amphiphilic CDs have been used successfully to enhance the solubilization of this drug by formulating it in the form of nanoparticles. CDC6 and 6-O-CAPRO-CD nanoparticles have also been found to maintain the therapeutic efficacy of the drug (Cirpanh, 2006, 2007).

1.13.7.5.6.2 Oxygen Delivery   Perfluorinated amphiphilic CDs are more efficient in dissolving oxygen and therefore can be used as efficient vectors for oxygen delivery. These nanoconstructs were more efficient as oxygen carriers because their higher number in colloidal solution permits a greater rate of oxygen dissolution (Skiba et al., 2002).

1.13.8 AEROSOLS

1.13.8.1 Foam Systems

One of the most popular types of aerosol products is foam aerosols. In foam-based aerosols, liquid components may not form a homogeneous phase because of immiscibility. Hence, liquids may be added in the form of biphasic, heterogeneous phases. These biphasic forms tend to form either o/w or w/o emulsions. The liquefied propellant is emulsified and is generally found in the internal phase. Surfactants are added to stabilize the emulsion. During the passage of this emulsion through a foam head on agitating the container, evaporation of the oil phase furnishes copious foam. These copious and rich lather yielding foam aerosols are quite popular in cosmetics and dermatological preparations. Foam-based aerosols have found applications as hair creams, shaving creams, shower preparations, and so forth.

Foam aerosols are dispersed as stable or quick breaking foam, depending on the nature of the ingredient and the formulation. Aerosol emulsions are dispensed as foams and preferred for the application of irritant ingredients or when the application is to be confined to a given site of application.

1.13.8.2 Aqueous Stable Foams

The formulation part of this type of preparation can be broadly divided into two parts: active ingredients (oil waxes, o/w surfactant, and water up to 95–96.5%) and hydrocarbon propellant in a proportion of 3.5% to 5%. Hydrocarbon propellants produce stiffer and dry foams at higher concentrations and wet foams at lower concentrations.

1.13.8.3   Nonaqueous Stable Foams

Glycols are predominantly used to fabricate these systems. Glycol esters are the chief class of emulsifiers in this case along with other routine excipients and drugs.

1.13.8.4   Breaking Foams

These aerosol systems differ from the rest of the foam-based systems in some aspects such as the formation of a quick breaking and collapsible foam. As far as the formulation is concerned, propellant forms the external phase. This system is mainly meant for topical and dermatological applications. The formulation components comprise 46% to 66% ethanol, 0.5% to 5% surfactant, 28% to 42% water, and 3% to 15% hydrocarbon propellant.

Shaving foams typically consist of blends of fatty acids like stearic acid and lauric acid, which serve as the oil phase; various grades of PEGs serving as emulsifiers, cosurfactants, and cosolvents; triethanolamine derivatives such as surfactants; glycerol as a humectant; and so forth. These excipients are in turn emulsified with propellants I, II, III, or IV according to the requirement. Similarly, shower foam-based products generally consist of all of these excipients, usually consisting of isopropyl myristate as the oily phase, macrogels of different grades as cosolvents and cosurfactants, glycerol as the humectant, and alkylglycol ether sulfate as the surfactant. Generally, these excipients constitute 90% (w/w) of the formulation and 10% propellant mixtures.

1.13.8.5   Deodorant Sprays

Apart from the key excipients of foam aerosols, deodorant aerosol sprays comprise cosolvents like ethanol, antibacterials and antimicrobials like phenolic derivatives, and perfume as an organoleptic additive. Phenolic derivatives are usually added to exert germicidal and antimicrobial action to prevent and mask body odor.

1.14   BIOLOGICAL IMPLICATIONS OF SURFACTANT AND BLOCK COPOLYMERS: TOXICITY ASPECTS

Chemical agents used in a pharmacy were evaluated for their toxicity to the biological system and environment, their degradation pathways, and the relative toxicity study of degraded products. The degradation of chemical substances under aerobic and anaerobic conditions of the natural environment was studied to understand their biodegradation profile. The surfactants and polymers used in the drug delivery system were evaluated in the same context and for their suitability for the different routes of administration in biological systems.

1.14.1   Toxicological Studies

Toxicity studies are conducted to assess the systemic exposure achieved in animals and its relationship to quantity level and the time course of the toxicity study. In-vivo mammalian toxicity studies (acute, subacute, long term) are conducted in species such as dog, guinea pig, mouse, and rabbit by different routes of administration such as oral, inhalation, and intravenous. Chemical substances intended for the oral route of administration are evaluated for their LD_{50} value. LD_{50} value is the amount/dose of chemical substance which is lethal to 50% of the experimental animal population. A high LD_{50} value indicates the safer use of the chemical substances.

Acute toxicity studies are carried out in two rodent species (mice and rat) using the same route as intended for humans. If the route of administration is other than parenteral, more than one route is selected in order to ensure the systemic absorption of the substance under investigation.

Animals are observed for 14 days after administration, and minimal lethal dose (MLD) and maximum tolerated Dose (MTD) are calculated for the components under investigation. Mortality is observed for upto 7 days after parenteral administration and upto 14 days after oral administration. During acute toxicity studies, symptoms, signs and mode of death, and macroscopic and microscopic findings are taken into consideration. LD_{10} and LD_{50} are calculated preferably with 95% confidence limits.

The chronic toxicity studies are carried out in two mammalian species, of which one should be nonrodent. These studies are carried out for the duration of 14, 28, 90, and 180 days.

For the products meant for parenteral route of administration, the site of injection is specially examined grossly and microscopically. The reversibility of the adverse effects is also monitored. For the excipients in the use of intramuscular injections, muscle toxicity is evaluated by gross morphological examination of tissue and serum creatinine phosphokinase (CPK) levels, which is considered as the index of muscle tissue injury. In general, lower HLB value copolymers (lipophilic in nature) show greater CPK values (Rodriguez Stephen and Singer Edward, 1996).

Inhalation toxicity studies are to be undertaken in one rodent and one non rodent species for the acute, subacute and chronic use of the chemical under investigation. Gases and vapors are given in whole body exposure chambers, and aerosols are given by nose only method. Exposure time and concentrations of test substance are validated to ensure exposure levels comparable to multiples of intended human exposure. During this study, the observations include effects on respiratory rate, findings of bronchial lavage fluid examination, histological examination of respiratory passages and lung tissue.

Dermal toxicity studies are done in rabbit and rat. The test material is applied on shaved skin covering not less than 10% of the total body surface area. Porous gauze dressing is used to hold liquid material in place. Period of application may vary from 7 to 90 days depending on the clinical duration of use. Local signs (erythema, oedema, and eschar formation), as well as histological examination of sites of application, are observed. Vaginal toxicity study is done in rabbit or dog for a minimum of 7 days and a maximum of 30 days and observed for swelling, closure of introits and histopathology of vaginal wall. Figure 1.19 shows the schematic of toxicity studies.

1.14.2 Degradation of Surfactants, Polymers, and Phospholipids

1.14.2.1 Degradation of Surfactants

Surfactants are used as emulsifiers, wetting agents, solubilizers, and dispersing agents in pharmaceutical formulations, which can influence the biological activity of the drug in the living system.

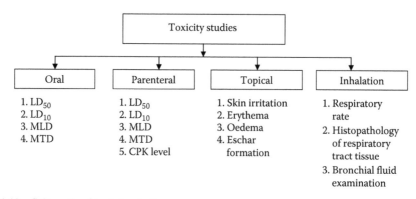

FIGURE 1.19 Schematic of toxicity studies.

Surfactants may influence the desegregation and dissolution of drugs by controlling the rate of precipitation of drugs and increase the membrane permeability by affecting membrane integrity. Surfactants above the CMC increase the saturation solubility of the drug and the peak plasma concentration. However, at higher concentrations, surfactants decrease drug absorption by affecting the chemical potential of the drugs. Complex interactions between drugs and protein may be possible, which leads to alterations in drug-metabolizing enzyme activity (Tadros, 2005).

Many commercially available surfactants degrade under biological conditions. The ability to undergo degradation mainly depends on the hydrophobicity and steric hinderance properties of the surfactants. The speed of degradation of surfactants greatly depends on the alkyl chain length, linearity of the chain, degree of branching, and branch distribution in the main alkyl chain. Linear molecules undergo degradation rapidly when compared to molecules having the same number of carbon atoms. The degradation pathways of anionic and nonionic surfactants have been well studied, but the degradation pathways of cationic and amphoteric surfactants remain limited. Mostly all surfactants undergo ω and β oxidation of the alkyl chains of the surfactants.

1.14.2.2 Biodegradation of Polymers

Biodegradable polymers have attracted a great deal of interest in drug delivery because they can undergo degradation under normal physiological conditions. In general, polymers undergo either chemical or enzymatic degradation in the biological system, varying in their degradation rate. Different polymers have different stabilities toward degradation; this allows the formulator to design the drug delivery system with varying drug release rates. Polymers that show more stability or a slow degradation pattern are considered as suitable candidates for sustained- or controlled-release formulations. Polyanhydride, polyketal, and polyorthoesters are more susceptible to hydrolytic degradation; polyamide, polyurethane, and polycarbonate polymers are found to be stable and inert in a hydrolytic environment.

When drugs are incorporated in polymers to achieve sustained release of the drugs, release of the drugs will predominantly depend on degradation of the polymer. Polymers can be tailored to obtain the desired release rate from the dosage form. Because of the degradation of polymers, there will be a compositional variation in the polymers, which still affects the drug release pattern. In polylactide–polyglycolide block copolymers, on degradation they become enriched with polylactide because of the preferential degradation of polyglycolide blocks.

Like surfactants, the degradation of polymers by the hydrolytic pathway depends not only on the composition of the individual monomer but also on the polymer architecture, polymer crystallinity, pH, and temperature, as well as the presence of other components like drugs in the delivery system.

The pH-dependent degradation of polymers allows the formulator to design the site-specific delivery of a drug, especially in the oral route of administration. Because the GIT offers different pH ranges at various sites (acidic at gastric sites and alkaline at intestinal sites), polymers that are sensitive to pH will degrade at the desired pH and release the drug at the hydrolytic site. In general, hydrolytic degradation is found in acidic and alkaline pH and is limited in neutral pH, which enables the stability of the formulation during storage.

Polymers that are considered to be stable for chemical degradation have been degraded by enzymatic degradation, while naturally occurring polymers such as polysaccharides and polypeptides are more susceptible to enzymatic hydrolysis. Many synthetic polymers such as poly(ether urethane), poly(ε-caprolactone), poly(ester urea), poly(hydroxyl butyrates), and poly(glycolic acid) undergo enzymatic degradation.

Apart from the chemical nature of the polymer, enzymatic degradation is affected by the pH, temperature, presence of enzyme inhibitors, presence of surfactants, other organic compounds, electrolytes, and concentration. Common enzymes that facilitate the enzymatic degradation of polymers are trypsin, chymotrypsin, papain, esterase, carboxy peptidase, leucin amino peptidase, lipase, and elastase.

Although polymers are selected based on the least toxicity of the polymers and their degradation products, PEO polymers encounter problems such as the formation of aldehydes when they are exposed to oxygen and light, which may lead to irritation and certain toxic effects. The degradation of polymers influences other characteristics of the delivery system, for example, gel strength and bioadhesion.

In the body, alginates degrade by acidic hydrolysis into glucuronic and mannuronic units. On intravenous administration of chitosan delivery systems, chitosan nanoparticles accumulate in the liver with minimal concentrations in the heart and lung. However, chitosan-based drug delivery systems have not been found to be toxic. Gelatin is degraded by proteases into amino acids. In the body, PLGA and PLA polymers degrade into the monomers lactic and glycolic acids, which enter the citric acid cycle, where they are metabolized and eliminated as carbon dioxide and water. Glycolic acid may also be excreted through the kidney. In poly(ε-caprolactone), bulk scission of the polymer chain occurs in the RES of macrophages. In the polyanhydride-based drug delivery system, the anhydride bond degrades to form diacid monomers that are eliminated from the body. PACA polymers are hydrolyzed into water-soluble alcohols and poly(cyanoacrylic acid).

1.14.2.3 Biodegradation of Phospholipids

Phospholipids are widely used in drug delivery systems such as liposomes, emulsions, microemulsions, and dispersed lipid particles that are intended for internal use. Phospholipids undergo enzymatic degradation by lipases and phospholipases. As in the enzymatic degradation of polymers, the degradation of phospholipids by enzymes depends on the concentration of enzymes, the nature of the lipid, pH, temperature, and the presence of other substances such as salt, enzyme inhibitors, and cholesterol. Longer and saturated phospholipids show more stability toward enzymatic degradation than branched and unsaturated phospholipids (Malmsten, 2002).

1.15 CONCLUSION

Surfactants have been used for centuries in many medicinal and other applications. In the last 10 years there has been a paradigm shift in the utility and versatility of applications of surfactants, and they have been extensively exploited in the detergent industry and in drug delivery. Their increasing applications in the area of drug delivery development are the outcome of the continued efforts of scientists to identify natural surfactants with less toxicological implications and to develop many more synthetic polymers in an effort to reduce the toxicities of the available surfactants. The drug carriers developed using these surfactants to increase their utility in drug delivery by attaching ligands or block copolymers render them less toxic and increasingly site specific. These efforts have resulted in drug delivery systems with an improved therapeutic index, reduced dose or frequency of dosing and toxicities, and probably better patient compliance and therapeutic benefit. Still, many of these surfactants, polymers, and block copolymers must be applied in the product used in clinical practice for the lack of preclinical acute and long-term toxicological evaluation. The collaborative and continued efforts of scientists from various disciplines are necessary to develop newer polymers suited for the delivery of drugs through various routes of administration without causing any change to the barrier layer except transit, even in long-term use.

ABBREVIATIONS

ADR	Adriamycin
BCNU	1,3-Bis(2-chloroethyl)-1-nitrosourea
CD	Cyclodextrin
CMC	Critical micelle concentration
CMT	Critical micelle temperature
CoQ_{10}	Coenzyme Q_{10}
CPP	Critical packing parameter

DLS	Dynamic light scattering
EGF	Epidermal growth factor
EPR	Enhanced permeation and retention
GD2	Disialoganglioside
GIT	Gastrointestinal tract
HA	Hyaluronic acid or hyaluronan
HLB	Hydrophilic–lipophilic balance
LD_{50}	Lethal dose in 50% population of animals
MAA	Methacrylic acid
9-NC	9-Nitrocamphothecin
NIPAAm	*N*-Isopropylacrylamide
NMR	Nuclear magnetic resonance
o/w	Oil in water
PAA	Poly(acrylic acid)
PACA	Poly(alkyl cyanoacrylate)
PC	Phosphatidylcholine
PCL	Polycaprolactone
PE	Phosphatidylethanolamine
PEG	Poly(ethylene glycol)
PEI	Poly(ethyleneimine)
PEO	Poly(ethylene oxide)
PLA	Poly(D,L)-lactide
PLGA	Poly(D,L-lactide-*co*-glycolide)
PMG	Poly(γ-methyl glutamate)
PMMA	Poly(methyl methacrylate)
PPG	Poly(propylene glycol)
PPO	Poly(propylene oxide)
PS	Polystyrene
RES	Reticuloendothelial system
SEDDSs	Self-emulsifying drug delivery systems
UV	Ultraviolet
w/o	Water in oil

SYMBOLS

n_{w}	Weight average
n_{n}	Number average
Z	Aggregation number
R_{g}	Radius of gyration
R_{h}	Hydrodynamic radius of micelle
R_{c}	Micellar core radius
L	Thickness of corona
G	Free energy
N	Number of repeating units
F	Number of arms
χ	Flory–Huggins parameter
δ	Solubility parameter
M_{w}	Weight-average molecular weight
u	Unimer
m	Micelle
x	Weight fraction of unimers

v Volume of the hydrophobic tails
a Polar head group area
l Length of the hydrophobic chain of the surfactant

REFERENCES

Afsaneh, L., John, S., and Glen, S. K., 2002. Poly(ethylene oxide)-*block*-poly(L-amino acid) micelles for drug delivery. *Adv. Drug Deliv. Rev.*, 54, 169–190.

Aguire, J. M., Bagan, J. V., Rodriguez, C., Jimenez, Y., Martinez, C. R., Diaz de Rojas, F., and Ponte, A., 2004. Efficacy of mometasone furoate microemulsion in the treatment of erosive, ulcerative oral lichen planus: Pilot study. *J. Oral Pathol. Med.*, 33(7), 381–385.

Alexandris, P., 2000. *Amphiphilic Block Copolymers: Self Assembly and Applications*. Amsterdam, Elsevier.

Alving, C. R., 1982. Therapeutic potential of liposomes as carriers in leishmaniasis, malaria and vaccines. In: *Targeting of Drugs*, G. Gregoriadis, J. Senior, and A. Trouet (Eds), p. 285. New York, Plenum Press.

Ameller, T., Marsaud, V., Legrand, P., et al., 2003. Polyester–poly(ethylene glycol) nanoparticles loaded with the pure antiestrogen RU 58668: Physicochemical and opsonization properties. *Pharm. Res.*, 20, 1063–1070.

Anacker, E. W., Geer, R. D., and Eylar E. H., 1971. Dependence of micelle aggregation number on polar head structure. I. Light scattering by aqueous solutions of decylammonium salts and related surfactants. *J. Phys. Chem.*, 75(3), 369–374.

Armes, S. P., Bütün, V., and Billingham, N. C., 1997. Synthesis and aqueous solution properties of novel hydrophilic–hydrophilic block copolymers based on tertiary amine methacrylates. *Chem. Commun.*, 671–672.

Armes, S. P., Lee, A. S., Gast, A. P., and Bütün, V., 1999. Characterizing the structure of pH dependent polyelectrolyte block copolymer micelles. *Macromolecules*, 32, 302–310.

Arnarson, T. and Elworthy, P. H., 1981. Effects of structural variations on non-ionic surfactants on micellar properties and solubilization: Surfactants containing very long hydrocarbon chains. *J. Pharm. Pharmacol.*, 33, 141.

Arunothayanun, P. and Pirayavaraporn, C., 2004. Self-emulsifying drug delivery systems (SEDDS) for ritonavir oral solution. International Conference on AIDS (15th, Bangkok, Thailand). Abstract no. A10030.

Atef, E. and Belmonte, A. A., 2008. Formulation and *in vitro* and *in vivo* characterization of a phenytoin self-emulsifying drug delivery system (SEDDS). *Eur. J. Pharm. Sci.*, 35(4), 257–263.

Attama, A. A. and Nzekwe, I. T., 2003. The use of solid self-emulsifying systems in the delivery of diclofenac. *Int. J. Pharm.*, 262(1–2), 23–28.

Attwood, D. and Florence, A. T., 1983. *Surfactant Systems—Their Chemistry, Pharmacy and Biology*, pp. 72–117. New York, Chapman & Hall.

Atyabi, F., Khodaverdi, E., and Dinarvand, R., 2007. Temperature modulated drug permeation through liquid crystal embedded cellulose membranes. *Int. J. Pharm.*, 339(1–2), 213–221.

Bajpai, A. K. and Choubey, J., 2006. Design of gelatin nanoparticles as swelling controlled delivery system for chloroquin phosphate. *J. Mater. Sci. Mater. Med.*, 17(4), 345–358.

Bakker-Wounderberg, I. A., Storm, G., and Woodle, M. C., 1994. Liposomes in treatment of infections. *J. Drug Target*, 2, 363–371.

Balmbra R. R., Clunie, J. S., Corkill, J. M., et al., 1962. Effect of temperature on the micelle size of a homogeneous non-ionic detergent. *Trans. Faraday Soc.*, 58, 1661.

Barker, M. and Vincent, B., 1984. Micellar solutions and microemulsions containing polystyrene/poly(ethylene oxide) AB block copolymers. *Colloids Surf.*, 8, 297–314.

Barrat, G. M., Tenu, J. P., Yapo, A., and Petit, J. F., 1986. Preparation and characterization of liposomes containing mannosylated phospholipids capable of targetting drugs to macrophages. *Biochim. Biophys. Acta*, 862, 153–164.

Berne, B. J. and Pecora, R., 2000. *Dynamic Light Scattering with Applications to Chemistry, Biology and Physics*. New York, Dover Publications.

Bilensoy, E., Dogan, A. L., Sen, M., and Hincal, A. A., 2007. Complexation behavior of antiestrogen drug tamoxifen citrate with natural and modified cyclodextrins. *J. Inclusion Phenom. Macrocyclic Chem.*, 57, 651–655.

Borges, O., Borchard, G., Verhoef, J. C., et al., 2005. Preparation of coated nanoparticles for a new mucosal vaccine delivery system. *Int. J. Pharm.*, 299(1–2), 268–276.

Boudad, H., Legrand, P., LeBas, G., Cheron, M., Duchene, D., and Ponchel, G., 2001. Combined hydroxypropyl-beta-cyclodextrin and poly(alkylcyanoacrylate) nanoparticles intended for oral administration of saquinavir. *Int. J. Pharm.*, 218, 113–124.

Boyd, B. J., 2005. Colloidal drug delivery, Drug delivery report, 63–70.

Boyd, B. J., 2008. Past and future evolution in colloidal drug delivery systems. *Expert Opin. Drug Deliv.*, 5(1), 69–85.

Boyd, B. J. and Whittaker, D. V., 2006. Lyotropic liquid crystalline phases formed from glycerate surfactants as sustained release drug delivery systems. *Int. J. Pharm.*, 309(1–2), 218–226.

Brignole, C., Marimpietri, D., Pagnan, G., et al., 2005. Neuroblastoma targeting by c-*myb*-selective antisense oligonucleotides entrapped in anti GD2 immunoliposome: Immune cell mediated anti tumour activities. *Cancer Lett.*, 228, 181–186.

Burchard, W., 1983. Static and dynamic light scattering from branched polymers and biopolymers. *Adv. Polym. Sci.*, 48, 1–124.

Burke, S., Shen, H., and Eisenberg, A., 2001. Multiple vesicular morphologies from block copolymers in solution. *Macromol. Symp.*, 175, 273–283.

Caralgia, M., Addeo, R., Costanzo, R., et al., 2006. Phase II study of temozolomide plus pegylated liposomal doxorubicin in the treatment of brain metastases from solid tumours. *Cancer Chemother. Pharmacol.*, 57, 34–39.

Chae, G. S. and Lee, J. S., 2005. Enhancement of the stability of BCNU using self-emulsifying drug delivery systems (SEDDS) and *in vitro* antitumor activity of self-emulsified BCNU-loaded PLGA wafer. *Int. J. Pharm.*, 301(1–2), 6–14.

Charman, S. A. and Charman, W. N., 1992. Self-emulsifying drug delivery systems: Formulation and biopharmaceutic evaluation of an investigational lipophilic compound. *Pharm. Res.*, 9(1), 87–93.

Chausson, M. and Fluchère, A.-S., 2008. Block copolymers of the type poly(caprolactone)-*b*-poly(ethylene oxide) for the preparation and stabilization of nanoemulsions. *Int. J. Pharm.*, 362(1–2), 153–162.

Chiappetta, D. A. and Sosnik, A., 2007. Poly(ethylene oxide)–poly(propylene oxide) block copolymer micelles as drug delivery agents: Improved hydrosolubility, stability and bioavailability of drugs. *Eur. J. Pharm. Biopharm.*, 66, 303–317.

Cilek, A., Celebi, N., and Tirnaksiz, F., 2006. Lecithin based microemulsion of a peptide for oral administration, preparation, characterization and physical stability of the formulation. *Drug Deliv.*, 13(1), 19–24.

Cirpanh, Y., Bilensoy, E., Calis, S., and Hincal, A. A., 2006. Camphothecin inclusion complexes with natural and modified beta cyclodextrins. In: *Proceedings of the 33rd Annual Meeting and Exhibition on Controlled Release Society*, Vienna, July 22–26, pp. 988–989.

Cirpanh, Y., Bilensoy, E., Calis, S., and Hincal, A. A., 2007. Development of camphothecin loaded nanoparticles from amphiphilic beta cyclodextrin derivatives. Paper presented at the Pharmaceutical Sciences World Congress PSWC, Amsterdam, April 22–25.

Coester, C., Nayyar, P., and Samuel, J., 2006. *In vitro* uptake of gelatin nanoparticles by murine dendritic cells and their intracellular localization. *Eur. J. Pharm. Biopharm.*, 62(3), 306–314.

Constantinides, P. P., 1995. Lipid microemulsion for improving drugs dissolution and oral absorption: Physical and biopharmaceutical aspects. *Pharm. Res.*, 12, 1561–1572.

Corkill, J. M., Goodman, J. F., Harrold, S. P., 1964. Thermodynamics of micellization of non-ionic detergents. *Trans. Faraday Soc.*, 60, 202.

Couvreur, P. and Vauthier, C., 1991. Poly alkyl cyanoacrylate nanoparticles as drug carriers: Present state and future perspective. *J. Control. Release*, 17, 187–198.

Daoud, M. and Cotton, J. P., 1982. Star-shaped polymers: A model for the conformation and its concentration dependence. *J. Phys. (Fr)*, 43, 531–538.

Di Stefano, A., Carafa, M., Sozio, P., et al., 2004. Evaluation of rat striatal L-dopa and DA concentration after intraperitoneal administration of L-dopa pro drugs in liposomal formulations. *J. Control. Release*, 99, 293–300.

Doddoli, C., Ghez, O., Barlesi, F., D'journo, B., Robitail, S., Thomas, P., and Clerc, T., 2005. *In vitro* and *in vivo* methotrexate disposition in alveolar macrophages: Comparison of pharmacokinetic parameters of two formulations. *Int. J. Pharm.*, 297, 180–189.

Dominguez, A., Fernandez, A., Gonzalez, N., et al., 1997. Determination of critical micelle concentration of some surfactants by three techniques. *J. Chem. Educ.*, 74(10), 1227–1231.

Douglas, S. J., Davis, S. S., and Illum, L., 1987. Nanoparticles in drug delivery. *Crit. Rev. Ther. Drug Carrier Syst.*, 3, 233–261.

Eastman, S. J., Tousignant, J. D., Lukason, M. J., et al., 1997. Optimization of formulations and conditions for the aerosol delivery of functional cationic lipid: DNA complexes. *Hum. Gene Ther.*, 8, 313–322.

Eisenberg, A. and Yu, K. E., 1998. Bilayer morphologies of self-assembled crew-cut aggregates of amphiphilic PS-*b*-PEO diblock copolymers in solution. *Macromolecules*, 31, 3509–3518.

Fessi, H., Puisieux, F., Devissaguet, J. P., et al., 1989. Nanocapsule formation by interfacial polymer deposition following solvent displacement. *Int. J. Pharm.* 55, R1–R4.

Florence, A. T., 1993. Non-ionic surfactant vesicles preparation and characterisation. In: *Liposome Technology*, G. Gregoradis (Ed.), 2nd ed., Vol. 2, pp. 157–176. Boca Raton, FL, CRC Press.

Forster, S. and Plantenberg, T., 2002. From self-organizing polymers to nanohybrid and biomaterials. *Angew. Chem. Int.*, 41, 688–714.

Fujimoto, Y., Okuhata, Y., Tyngi, S., Namba, Y., and Oku, N., 2000. Magnetic resonance lymphography of profounded lymph nodes with liposomal gadolinium–diethylenetriamine penta acetic acid. *Biol. Pharm. Bull.*, 23, 97–100.

Ganta, S., Paxton, J. W., Baguley, B. C., and Garg, S., 2008. Pharmacokinetics and pharmacodynamics of chlorambucil delivered in parenteral emulsion. *Int. J. Pharm.*, 360(1–2), 115–121.

Gaucher Geneviève, Dufresne Marie-Hélène, Sant Vinayak, P., et al., 2005. Block copolymer micelles: Preparation, characterization and application in drug delivery. *J. Control. Release*, 109, 169–188.

Godin, B. and Touitou, E., 2003. Ethosomes: New prospects in transdermal delivery. *Crit. Rev. Ther. Drug Carrier Syst.*, 20, 63–102.

Gradishar, W. J., 2005. The future of breast cancer: The role of prognostic factors. *Breast Cancer Res. Treat.*, 89, S17–S26.

Granger, C. E., Feliz, C. P., Parrot-Lopez, H., and Langlois, B. R., 2000. Fluorine containing beta cyclodextrin: A new class of ampbiphilic carriers. *Tetrahedron Lett.*, 41, 9257–9260.

Gregoriadis, G., Bacon, A., Caparros-Wanderley, W., and McCormack, B., 2002. A role for liposomes in genetic vaccinatin. *Vaccine*, 20, B1–B9.

Gursoy, N. and Benita, S., 2004. Self-emulsifying drug delivery systems (SEDDS) for improved oral delivery of lipophilic drugs. *Biomed. Pharmacother.*, 58(3), 173–182.

Hadjichristidis, N., Pispas, S., and Floudas, G. A., 2003. *Block Copolymers: Synthetic Strategies, Physical Properties, and Applications.* New York, Wiley.

Han, I., Kim, M., and Kim, J., 2004. Enhanced transfollicular delivery of adriamycin with a liposome and ion-tophoresis. *Exp. Dermatol.*, 13, 86–92.

Hede, K., 2005. Blocking cancer with RNA interference moves towards the clinic. *J. Natl. Cancer Inst.*, 97, 626–628.

Hioka, N., Chowdhary, R. K., Chansarkar, N., et al., 2002. Studies of a benzoporphyrin derivative with Pluronics. *Can. J. Chem.*, 80, 1321–1326.

Hofheinz, R. D., Gnad-Vogt, S. U., Beyer, U., and Hochchaus, A., 2005. Liposomal encapsulated anti cancer drugs. *Anti Cancer Drugs*, 16, 691–707.

Holmberg, E. G., Reuer, Q. R., Geisert, E. E., et al., 1994. Delivery of plasmid DNA to glial cells using pH sensitive liposomes. *Biochem. Biophys. Res. Commun.*, 201(2), 888–893.

Hong, J. Y., Kim, J. K., Song, Y. K., et al., 2006. A new self-emulsifying formulation of itraconazole with improved dissolution and oral absorption. *J. Control. Release*, 110(2), 332–338.

Hurtrez, G., 1992. Study of PS–PEO and PEO–PS–PEO block copolymers (in French). PhD Thesis. University Haute Alsace, France.

Hyung Park, J., Kwon, S., Lee, M., et al., 2006. Self-assembled nanoparticles based on glycol chitosan bearing hydrophobic moieties as carriers for doxorubicin: *In vivo* biodistribution and anti tumour activity. *Biomaterials*, 27(1), 119–126.

Illum, L., Hunneyball, I. M., and Davis, S. S., 1986. The effect of hydrophilic coatings on the uptake of colloidal particles by the liver and by peritoneal macrophages, *Int. J. Pharm.*, 29, 53–65.

Irache, J. M., Durrer, C., Duchene, D., and Ponchel, G., 1996. Bioadhesion of lectin latex conjugates to rat intestinal mucosa. *Pharm. Res.*, 13, 1716.

Israelachvili, J. N., 1992. *Intermolecular and Surface Forces.* London, Harcourt Brace and Company.

Jeong, Y. L., Kang, M. K., Sun, H. S., et al., 2004. All trans retinoic acid release from core shell type nanoparticles of polyepsilon caprolactone poly(ethylene glycol) diblock copolymer. *Int. J. Pharm.*, 273, 95–107.

Jiang, J., Tong, X., and Zhao, Y., 2005. A new design for light-breakable polymer micelles. *J. Am. Chem. Soc.*, 127, 8290–8291.

Jones, M. S., 1999. Effect of pH on the lower critical solution temperatures of random copolymers of *N*-isopropylacrylamide and acrylic acid. *Eur. Polym. J.*, 35(5), 795–801.

Kabanova, A. V. and Gendelman, H. E., 2007. Nanomedicine in the diagnosis and therapy of neurodegenerative disorders. *Progr. Polym. Sci.*, 32, 1054–1082.

Kalyanasundaram, K. and Thomas, J. K. 1977. Environmental effects on vibronic band intensities in pyrene monomer fluorescence and their application in studies of micellar systems. *J. Am. Chem. Soc.*, 99, 2039–2044.

Kang, B. K. and Lee, J. S., 2004. Development of self-microemulsifying drug delivery systems (SMEDDS) for oral bioavailability enhancement of simvastatin in beagle dogs. *Int. J. Pharm.*, 274(1–2), 65–73.

Kaparissides, C., Alexandridou, S., Kotti, K., and Chaitidou, S., 2006. Recent advances in novel drug delivery systems. Available at http://www.azonano.com/oars.asp

Kataoka, K., Harada, A., and Nagasaki, Y., 2001. Block copolymer micelles for drug delivery: Design, characterization and biological significance. *Adv. Drug Deliv. Rev.*, 47, 113–131.

Kaul, G. and Amiji, M., 2002. Long circulating poly ethylene glycol modified gelatin nanoparticles for intracellular delivery. *Pharm. Res.*, 19(7), 1061–1067.

Kaul, G. and Amiji, M., 2004. Biodistribution and targeting potential of poly(ethylene glycol)-modified gelatin nanoparticles in subcutaneous murine tumor model. *J. Drug Targeting.* 12(9–10), 585–591.

Khoo, S. M., Humberstone, A. J., Porter, C. J. H., et al., 1998. Formulation design and bioavailability assessment of lipidic self-emulsifying formulations of halofantrine. *Int. J. Pharm.*, 167(1–2), 155–164.

Kim, S. Y. and Kim, J. H., 2001. Drug-releasing kinetics of MPEG/PLLA block copolymer micelles with different PLLA block lengths. *J. Appl. Polym. Sci.*, 82(10), 2599–2605.

Kirby, C. J. and Gregoriadis, G., 1980. Action of indomethacin on neutral and positively charged monolayers. *Life Sci.*, 27, 2223–2230.

Klevens, H. B., 1950. Solubilization. *Chem. Rev.*, 47, 1.

Kolthoff, I. M. and Stricks, W., 1948. Solubilization of dimethylaminoazobenzene in solutions of detergents. I. The effect of temperature on the solubilization and upon the critical concentration. *J. Phys. Colloid Chem.*, 52, 195.

Kosaric, N., Cairns, W. L., and Gray N. C. C., 1987. Introduction: Biotechnology and the surfactant Industry. In: *Biosurfactants and Biotechnology. Surfactant Science Series.* N. Kosaric, W.L. Cairns, and N. C. C. Gray (Eds), Vol. 25, pp. 1–21. New York, Marcel Dekker.

Kositza, M. J., Bohne, C., Hatton, T. A., et al., 1999. Micellization dynamics of PEO–PPO–PEO block copolymers measured by stopped flow. *Progr. Colloid Polym. Sci.*, 112, 146–151.

Kreuter, J., 1991. Nanoparticles based drug delivery systems. *J. Control. Release*, 16, 169–176.

Kunii, R. and Onishi, H., 2007. Preparation and antitumor characteristics of PLA/(PEG–PPG–PEG) nanoparticles loaded with camptothecin. *Eur. J. Pharm. Biopharm.*, 67(1), 9–17.

Kwon, G. S., 1998. Diblock copolymer nanoparticles for drug delivery. *Crit. Rev. Ther. Drug Carrier Syst.*, 15, 481–512.

Kwon, G. and Naito, M., 1997. Block copolymer micelles for drug delivery: Loading and release of doxorubicin. *J. Control. Release*, 48(2–3), 195–201.

Lasic, D. D., 1990. On the thermodynamic stability of liposomes. *J. Colloid Interface Sci.*, 140, 302–304.

Lattuada, R. and Martini, A., 1998. Smedds® (self micro-emulsifying drug delivery systems) for oral administration of an immunosuppressive drug. *Eur. J. Pharm. Sci.*, 6(1), S67.

Law, S. L., Huang, K. J., and Chiang, C. H., 2000. Acyclovir containing liposomes for potential ocular delivery. Corneal penetration and absorption. *J. Control. Release*, 631(1–2), 135–140.

Lawrence, A. S. C., 1937. Internal solubility in soap micelles. *Trans. Faraday Soc.*, 33(325), 815.

Lecommandoux, S., Rodríguez-Hernández, J., Chécot, F., et al., 2005. Toward "smart" nano-objects by self-assembly of block copolymers in solution. *Progr. Polym. Sci.*, 30(7), 691–724.

Leroux, J.-C., Taillefer, J., Jones, M.-C., and Brasseur, N., 2000. Preparation and characterization of pH-responsive polymeric micelles for the delivery of photosensitizing anticancer drugs. *J. Pharm. Sci.*, 89, 52–62.

Lewis, D. H., 1990. Controlled release of bioactive agents from lactide/glycolide polymers. In: *Biodegradable Polymers as Drug Carrier Systems,* M. Chasin and R. Langer (Eds), pp. 1–43. New York, Marcel Dekker.

Linse, P. 2000. Modelling of self-assembly of block copolymers in selective solvent. In: *Amphiphilic Block Copolymers: Self Assembly and Applications*, B. Lindman, P. Alexandridis (Eds), pp. 13–40. Amsterdam, Elsevier.

Liu, C. G., Desai, K. G., Chen, X. G., and Park, H. J., 2005. Preparation and characterization of nanoparticles containing trypsin based on hydrophobically modified chitosan. *J. Agric. Food Chem.*, 53(5), 1728–1733.

Liu, H., Pan, W. S., Tang, R., and Luo, S. D., 2004. Topical delivery of acyclovir palmitate liposome formulations through rat skin *in vitro*. *Pharmazie*, 59, 203–206.

Lu, J.-L. and Wang, J.-C., 2008. Self-microemulsifying drug delivery system (SMEDDS) improves anticancer effect of oral 9-nitrocamptothecin on human cancer xenografts in nude mice. *Eur. J. Pharm. Biopharm.*, 69(3), 899–907.

Lu, Z., Yeh, T. K., Tsai, M., Au, J. L., and Wientjes, M. G., 2004. Paclitaxel loaded gelatin nanoparticles for intravesical bladder cancer therapy. *Clin. Cancer Res.*, 10(22), 7677–7684.

Lukyanov Anatoly, N. and Torchilin Vladimir, P., 2004. Micelles from lipid derivatives of water-soluble polymers as delivery systems for poorly soluble drugs. *Adv. Drug Deliv. Rev.*, 56, 1273–1289.

Maestrelli, F., Garcia-Fuentes, M., Mura, P., and Alonso, M. J., 2006. A new nanocarrier consisting of chitosan and hydroxypropylcyclodextrin. *Eur. J. Pharm. Biopharm.*, 69, 79–86.

Malmsten, M., 2002. *Surfactants and Polymers in Drug Delivery*. New York, Marcel Dekker.

Manosroi, A., Konganeramit, L., and Manosroi, J., 2004. Stability and transdermal absorption of topical amphotericin B liposome formulations. *Int. J. Pharm.*, 270, 279–286.

Mansourie, S., Cuie, Y., Winnik, F., et al., 2006. Characterization of folate chitosan DNA nanoparticles for gene therapy. *Biomaterials*, 27(9), 2060–2065.

Matsumoto, M., Matsuzawa, Y., Noguchi, S., Sakai, H., and Abe, M., 2004. Structure of Langmuir Blodgett films of amphiphilic cyclodextrin and water soluble benzophenone. *Mol. Cryst. Liq. Cryst.*, 425, 197–204.

McCarron, P. A., Donnelly, R. F., Canning, P. E., et al., 2004. Bioadhesive non drug loaded nanoparticles as modulators of candidial adherence to buccal epithelial cells: A potentially novel prophylaxis for candidosis. *Biomaterials*, 25, 2399–2407.

Misselwitz, B. and Sasche, A., 1997. Interstitial MR lymphography using GD-carrying liposomes. *Acta Radio. Suppl.*, 412, 51–55.

Moffitt, M., Khougaz, K., and Eisenberg, A., 1996. Micellization of ionic block copolymers. *Acc. Chem. Res.*, 29(2), 95–102.

Moghimi, S. M. and Hunter, A. S., 2001. Recognition by macrophages and liver cells of opsonized phospholipid vesicles and phospholipid headgroups. *Pharm. Res.*, 18, 1–8.

Mrozek, E., Rhoades, C. A., Allen, J., Hade, E. M., and Shapiro, C. L., 2005. Phase I trial of liposomal encapsulated doxorubicin (Myocet™, D-99) and weekly docetaxel in advanced breast cancer patients. *Ann. Oncol.*, 16, 1087–1093.

Mueller, R. H. and Wallis, K. H., 1993. Surface modification of i.v. injectable biodegradable nanoparticles with poloxamer polymers and Poloxamine 908. *Int. J. Pharm.*, 89, 25–31.

Mukherjee, P., 1972. Size distribution of small and large micelles. Multiple equilibrium analysis. *J. Phys. Chem.*, 76, 565.

Muller, N. 1972. Kinetics of micelle dissociation by temperature-jump techniques. Reinterpretation. *J. Phys. Chem.*, 76, 3017.

Munk, P. 1996. Equilibrium and nonequilibrium polymer micelles. In: *Solvents and Selforganization of Polymer. NATO ASI Series, Series E: Applied Sciences*, S. E. Webber, P. Munk, and Z. Tuzar (Eds), Vol. 327, pp. 19–32. Dordrecht, Kluwer Academic.

Murray, R. C. and Hartley, G. S., 1935. Equilibrium between micelles and simple ions, with particular reference to the solubility of long-chain salts. *Trans. Faraday Soc.* 31, 185.

Nagarajan, R. 1996. Solubilization of hydrophobic substances by block copolymer micelles in aqueous solution. In: *Solvents and Selforganization of Polymer. NATO ASI Series, Series E: Applied Sciences*, S. E. Webber, P. Munk, and Z. Tuzar (Eds), Vol. 327, pp. 121–165. Dordrecht, Kluwer Academic.

New, R. R. C., 1989. Introduction. In: *Liposomes: A Practical Approach*. R. R. C. New (Eds), p. 1. Oxford, London, OIRL Press.

Nicolaos, G., Crauste-Manciet, Farinotti, R., et al., 2003. Improvement of cefpodoxime proxetil oral absorption in rats by an oil-in-water submicron emulsion. *Int. J. Pharm.*, 263(1–2), 165–171.

Nishiyama, N., Bae, Y., Miyata, K., et al., 2005. Smart polymeric micelles for gene and drug delivery. *Drug Discov. Today: Technol.*, 2(1), 21–26.

Noolandi, J. and Hong, K. M., 1982. Interfacial properties of immiscible homopolymer blends in the presence of block copolymers. *Macromolecules*, 15, 482–492.

Nowakowska, M. and Zczubiałka, K., 2003. Response of micelles formed by smart terpolymers to stimuli studied by dynamic light scattering. *Polymer*, 44, 5269–5274.

Opanasopit, P. and Yokoyama, M., 2004. Block copolymer design for camptothecin incorporation into polymeric micelles for passive tumor targeting. *Pharm. Res.*, 21(11), 2001–2008.

Oussoren, C. and Storm, G., 2001. Liposomes to target the lymphatics by subcutaneous administration. *Adv. Drug Deliv. Rev.*, 50, 143–156.

Palamakula, A., Nutan, M. T. H., and Khan, M. A., 2004. Response surface methodology in optimization and characterization of limonene-based Coenzyme Q10 from self-nanoemulsified capsule dosage form. *AAPS Pharm. Sci. Technol.*, 5(4), 1–8.

Pandey, R., Ahmad, Z., Sharma, S., and Khuller, G. K., 2005. Inhalable alginate nanoparticles as antitubercular drug carriers against experimental tuberculosis. *Int. J. Antimicrob. Agents*, 26(4), 298–303.

Papahadjopoulos, D., Vali, W. J., Jacobson, K., and Poste, G., 1975. Effects of local anaesthetics on membrane properties I changes in the fluidity of phospholipids bilayers. *Biochim. Biophys. Acta*, 394, 504–519.

Park, E. K., Lee, S. B., and Lee, Y. M., 2005. Preparation and characterization of methoxy poly(ethylene glycol)/poly(epsilon-caprolactone) amphiphilic block copolymeric nanospheres for tumour specific folate mediated targeting of anticancer drugs. *Biomaterials*, 26, 1053–1061.

Patton, J. S., Fishburn, C. S., and Weers, J. G., 2004. The lungs as a portal for entry of systemic drug delivery. *Proc. Am. Thorac. Soc.*, 1, 339–344.

Paveli, Z., Skalko-Basnet, N., and Schubert, R., 2001. Liposomal gels for vaginal drug delivery. *Int. J. Pharm.*, 219, 139–149.

Peer, D. and Margalit, R., 2004. Tumour targeted hyaluronan nanoliposomes increase the antitumour activity of liposomal doxorubicin in syngeneic and human xenograft mouse tumour models. *Neoplasia*, 6, 343–353.

Perlman, M. E., Murdande, S. B., Gumkowski, M. J., et al., 2008. Development of a self-emulsifying formulation that reduces the food effect for torcetrapib. *Int. J. Pharm.*, 351(1–2), 15–22.

Peroche, S., Degobert, G., Putaux, J. L., Blanchin, M. G., Fessi, H., and Parrot-Lopez, H., 2005. Synthesis and characterization of novel nanospheres made from amphiphilic perfluoroalkylthio-beta cyclodextrins. *Eur. J. Pharm. Biopharm.*, 60, 123–131.

Perrie, Y., Barralet, J. E., McNeil, S., and Vangala, A., 2004. Surfactant vesicle mediated delivery of DNA vaccines via the subcutaneous route. *Int. J. Pharm.*, 284, 31–41.

Pethica, B. A., 1960. *Proceedings of the 3rd International Congress on Surface Activity*, Cologne. 1, 212.

Pfeifer, B. A., Burdick, J. A., Little, S. R., and Langer, R., 2005. Poly(ester-anhydride) poly(beta-amino ester) micro and nanospheres: DNA encapsulation and cellular transfection. *Int. J. Pharm.*, 304, 210–219.

Pich, A. and Schiemenz, N., 2006. Preparation of poly(3-hydroxybutyrate-*co*-3-hydroxyvalerate) (PHBV) particles in O/W emulsion. *Polymer*, 47(6), 1912–1920.

Ping Zhang and Ying Liu, 2008. Preparation and evaluation of self-microemulsifying drug delivery system of oridonin. *Int. J. Pharm.*, 355(1–2), 269–276.

Pitt, C. G., 1990. Poly(E-caprolactone) and its co polymers, In: *Biodegradable Polymers as Drug Carrier Systems*, M. Chasin and R. Langer (Eds), p. 71. New York, Marcel Dekker.

Price, C., 1982. Colloidal properties of block copolymers. In: *Developments in Block Copolymers*, Vol. 1, p. 39–79. London, Applied Science.

Quintana, J. R., Villacampa, M., and Katime, I., 1996. Block copolymers: Micellization in solution. In: *Polymer Material Encyclopedia*, Vol. 1, pp. 815–821. Boca Raton, FL, CRC Press.

Quirk, R. P., Kinning, D. J., and Fetters, L. J., 1989. *Block Copolymers. Comprehensive Polymer Science*, Vol. 7, pp. 1–26. Oxford, Pergamon Press.

Rapoport, N., 2007. Physical stimuli-responsive polymeric micelles for anti-cancer drug delivery. *Progr. Polym. Sci.*, 32, 962–990.

Reigelman, S., Allawala, N. A., Hrenoff, M. K., et al. 1958. The ultraviolet absorption spectrum as a criterion of the type of solubilization. *J. Colloid Sci.*, 13, 208.

Reis, C. P., Neufeld, R. J., Ribeiro, A. J., et al., 2006. Nanoencapsulation I. Methods for preparation of drug-loaded polymeric nanoparticles. *Nanomedicine*, 2(1), 8–21.

Riess, G., 2003. Micellization of block copolymers. *Prog. Polym. Sci.*, 28, 1107–1170.

Riess, G., Hurtrez, G., and Bahadur, P., 1985. Block copolymers. In: *Encyclopedia of Polymer Science and Engineering*, H. F. Mark and J. I. Kroschwitz (Eds), 2nd ed., Vol. 2, pp. 324–434. New York, Wiley.

Rodrigues, L., Banat Ibrahim, M., Teixeira, J., et al., 2006. Biosurfactants: Potential applications in medicine. *J. Antimicrob. Chemother.*, 57(4), 609–618.

Rodriguez, S. Galindo, Allemann, E., et al. 2004. Physicochemical parameters associated with nanoparticle formation in the salting-out, emulsification–diffusion, and nanoprecipitation methods. *Pharm. Res.*, 21, 1428–1439.

Rodriguez Stephen, C. and Singer Edward, J., 1996. Toxicology of polyoxyalkylene block copolymer. In: *Non Ionic Surfactants, Surfactant Science Series*, V. M. Nace (Ed.), Vol. 60, pp. 211–231. New York, Marcel Dekker.

Rosler, A., Vandermeulen, G. W.M., and Klok, H.-A., 2001. Advanced drug delivery devices via self-assembly of amphiphilic block copolymers. *Adv. Drug Deliv. Rev.*, 53, 95–108.

Sadzuka, Y., Horota, S., and Sonobe, T., 2000. Intraperitoneal administration of doxorubicin encapsulating liposomes against peritoneal dissemination. *Toxicol. Lett.*, 116, 51–59.

Scherlund, M., Malmsten, M., Holmqvist, P., et al., 2000. Thermosetting microemulsions and mixed micellar solutions as drug delivery systems for periodontal anesthesia. *Int. J. Pharm.*, 194(1), 103–116.

Schott, H., 1967. Solubilization of a water-insoluble dye. II. *J. Phys. Chem.*, 71, 3611.

Schwarz, L. A., Johnson, J. L., Black, M., et al., 1996. Delivery of DNA-cationic liposome complexes by small particle aerosol. *Hum. Gene Ther.*, 24, 35–36.

Shah, N. H., Carvajal, M. T., Patel, C. I., et al., 1994. Self-emulsifying drug delivery systems (SEDDS) with polyglycolized glycerides for improving *in vitro* dissolution and oral absorption of lipophilic drugs. *Int. J. Pharm.*, 106, 15–23.

Shahiwala, A. and Misra, A., 2004. Nasal delivery of levonorgestrol for contraception: An experimental study in rats. *Fertil. Steril.*, 81(Suppl. 1), 893–898.

Shimizu, K., Takada, M., Asai, T., Kuromi, K., Baba, K., and Oku, N., 2002. Cancer chemotherapy by liposomal 6-[[2-(dimethyal amino ethyl]amino]-3-hydroxy-7H-indeno[2,1-c] quinolin-7-one dihydrochloride (TAS-103), a novel anticancer gent. *Biol. Pharm. Bull.*, 25, 1385–1387.

Skiba, M., Skiba-Lahiani, M., and Arnaud, P., 2002. Design of nanocapsules based on novel fluorophilic cyclodextrin derivatives and their potential role in oxygen delivery. *J. Inclusion Phenom. Macrocyclic Chem.*, 44, 151–154.

Soppimath, K. S., Aminabhavi, T. M., Kulkarni, A. R., et al., 2001. Biodegradable polymeric nanoparticles as drug delivery devices. *J. Control. Release*, 70(1–2), 1–20.

Stover, T. C., Sharma, A., Robertson, G. P., and Kester, M., 2005. Systemic delivery of liposomal short chain ceramide limits solid tumour growth in murine models of breast adenocarcinoma. *Clin. Cancer Ther.*, 11, 3465–3474.

Tadros Tharwat, F., 2005. *Applied Surfactants: Principles and Applications*, pp. 2–17. Hoboken, NJ, Wiley.

Thanatuksorn, P., Kawai, K., Hayakawa, M., et al., 2009. Improvement of the oral bioavailability of coenzyme Q_{10} by emulsification with fats and emulsifiers used in the food industry. *LWT-Food Sci. Technol.*, 42(1), 385–390.

Topp, M. D. C., Dijkstra, P. J., Talsma, H., et al., 1997. Thermosensitive micelle-forming block copolymers of poly-(ethylene glycol) and poly(*N*-isopropylacrylamide). *Macromolecules*, 30(26), 8518–8520.

Torchilin, V. P., 1998. Liposomes as carriers of contrast agents in *in vivo* diagnostics. In: *Medical Applications of Liposomes*, D. D. Lasic and D. Papahadjapoulos (Eds.), p. 515. Oxford, Elsevier.

Toub, N., Angiari, C., Eboue, D., et al., 2005. Cellular fate of oligonucleotides when delivered by nanocapsules of poly(isobutylcyanoacrylate). *J. Control. Release*, 106, 209–213.

Tuzar, Z. and Kratochvil, P., 1993. Micelles of block and graft copolymers in solution. In: *Surface and Colloid Science*, E. Matijevic (Ed.), Vol. 15, pp. 1–83. New York, Plenum Press.

Uhrich, K. E., Cannizzaro, S. M., et al., 1999. Polymeric systems for controlled drug release. *Chem. Rev.*, 99, 3181–3198.

U.S. Pat. 6288121, Stefano Bader, Helsinn Healthcare Inc., 2001. Nimesulide topical formulations in the form of liquid crystals.

U.S. Pat 6306434, Chung II Hong, Chong Kun Dang Corp. (KR), 2001. Pharmaceutical composition comprising cyclosporine solid state microemulsion.

U.S. Pat. 6623765, Donn, M. Denis, University of Florida, Research Foundation, 2003. Microemulsion and micelle system for solubilising drugs.

U.S. Pat. 6638522, Nirmal Mulye, Pharmasolutions, 2003. Microemulsion concentrate composition of cyclosporine.

U.S. Pat. 6703034, Balraj, S. Parmar, University of Florida, 2004. Neem oil microemulsion without cosurfactants and alcohol and the process to form the same.

U.S. Pat. 7115565, Ping Goa, Pharmacia and Upjohn Company, 2006. Chemotherapeutic microemulsions composition of paclitaxel with improved oral bioavailability.

van Nostrum, C. F., 2004. Polymeric micelles to deliver photosensitizers for photodynamic therapy. *Adv. Drug Deliv. Rev.*, 56, 9–16.

Vandervoort, J. and Ludwig, A., 2004. Preparation and evaluation of drug loaded gelatin nanoparticles for topical ophthalmic use. *Eur. J. Pharm. Biopharm.*, 57(2), 251–261.

Verschraegen, C. F., Gilbert, B. E., Loyer, E., et al., 2004. Clinical evaluation of the delivery and safety of aerosolized liposomal 9-nitro camphothecin in patients with advanced pulmonary malignancies. *Clin. Cancer Res.*, 100, 1449–1458.

Vyas, S. P. and Uchegbu, I. F., 1998. Non-ionic surfactant based vesicles (niosomes) in drug delivery. *Int. J. of Pharm.*, 172, 33–70.

Vyas, S. P., Kannan, M. P., Jain, S., Mishra, V., and Singh, P., 2004. Design of liposomal aerosols for improved delivery of rifampicin to alveolar macrophages. *Int. J. Pharm.*, 269, 37–49.

Vyas, S. P., Quraishi, S., Gupta, S., and Jaganathan, K. S., 2005. Aerosolized liposome based delivery of amphotericin B to alveolar macrophages. *Int. J. Pharm.*, 296, 12–25.

Wang, D., Nagata, L. P., Christopher, M. E., et al., 2004. Intranasal immunization with liposome-encapsulated plasmid DNA encoding influenza virus hemagglutinin elicits mucosal, cellular and humoral immune responses. *J. Clinical Virology.* 31(1), 99–106.

Wang, J.-J., Sung, K. C., Yeh, C.-H., and Fang, J.-Y., 2008. The delivery and antinociceptive effects of morphine and its ester prodrugs from lipid emulsions. *Int. J. Pharm.*, 353(1–2), 95–104.

Wells, J. M., Li, H., Sen, A., Jahreis, G. P., and Hui, S. W., 2000. Electroporation enhanced gene delivery in mammary tumours. *Gene Ther.*, 7, 541–547.

West, J. L., Sershen, S. R., Westcott, S. L., et al., 2000. Temperature-sensitive polymer–nanoshell composites for photothermally modulated drug delivery. *J. Biomed. Mater. Res.*, 51, 293–296.

Willimann, H., Walde, J. I., Luisi, A., Gazzaniga, A., and Dtroppolo, F., 1992. Lecithin organogels as matrix for transdermal transport of drugs. *J. Pharm. Sci.* 81(9), 871–874.

Wooley, K. L., Thurmond, K. B., Kowalewski, T., 1996. Water-soluble needle-like structures: The preparation of shell-cross-linked small particles. *J. Am. Chem. Soc.*, 118, 7239–7240.

Wu, P. C., Tsai, Y. H., Liao, C. C., et al., 2004. Characterization and biodistribution of cefoxitin loaded liposomes. *Int. J. Pharm.*, 271, 31–39.

Wu, W. and Wang, Y., 2006. Enhanced bioavailability of silymarin by self-microemulsifying drug delivery system. *Eur. J. Pharm. Biopharm.*, 63(3), 288–294.

Yamamoto, Y., Nagasaki, Y., Kato, M., and Kataoka, K., 1999. Surface charge modulation of poly(ethylene glycol)–poly(D,L-lactide) block copolymer micelles: conjugation of charged peptides. *Colloids Surf. B*, 16, 135–146.

Yokoyama, M., 1992. Block copolymers as drug carriers. *Crit. Rev. Ther. Drug Carrier Syst.*, 9, 213–248.

Yonese, M., Toyotama, A., Kugimiya, S., et al., 2001. Preparation of a novel aggregate like sugar-ball micelle composed of poly(methylglutamate) and poly(ethyleneglycol) modified by lactose and its molecular recognition by lectin. *Chem. Pharm. Bull.*, 49, 169–172.

Zana, R., 2000. Fluorescence studies of amphiphilic block copolymers in solution. In: *Amphiphilic Block Copolymers: Self Assembly and Applications*, B. Lindman and P. Alexandridis (Eds), pp. 221–252. Amsterdam, Elsevier.

2 Application of Colloidal Properties in Drug Delivery

Swarnlata Saraf

CONTENTS

2.1 INTRODUCTION

In the last few decades a renaissance of colloid science and technology has occurred due to the expanding field of nanotechnology and its application (Rill et al., 2008). A large variety of areas have been covered in this field, including drug delivery systems, biochemical systems, and industrial systems (Gennaro, 2004). The term "colloids" applies broadly to a system containing at least two components in any state of matter, one dispersed in another, in which the dispersed components consist of large molecules or small particles (Bungenberg, 1949). The size of colloidal particles ranges between 1 nm and 1 μm (Hunter and O'Brien, 1989). Generally, the significant properties of colloids that play a role in drug delivery are morphological, optical, physicochemical, kinetic, electrical, and magnetic.

The properties of colloids provide information about the absorption; absorption, distribution, metabolism, and elimination; cell–membrane interaction; and drug molecule behavior in surrounding environments. Such properties play a major role in drug loading capacity and transport mechanisms because the charge and colloidal size range is set for this capacity. Colloidal particles greatly affect the rate of sedimentation, osmotic pressure, and stability and biocompatibility of colloidal drug carriers. Colloidal properties also provide a selection basis for excipients and manufacturing process criteria, such as the viscosity, interfacial tension, and aggregation properties, which results in bioavailability and biodistribution of colloidal carriers in the human body (Yang and Alexandridis, 2000). Particle properties like the Tyndall effect, turbidity, and dynamic light scattering (DLS) can be used as tools to access rational criteria for the development of formulations. The optical property is widely used to observe the size, shape, and structure of colloidal particles. The kinetic properties of colloids deal with the motion of particles with respect to the dispersion medium. These properties mainly include Brownian motion, diffusion, osmosis, sedimentation, and viscosity. They provide detail of the movement of colloidal carriers within the body and provide transportation criteria of colloidal carriers across cell membranes. The physicochemical properties of colloids such as the physical state of colloids, lyophilicity, lyophobicity (Swarbrick and Martin, 1991), drug polymorphs (Wong et al., 2008), and interfacial properties (Richard et al., 2006) are helpful in the selection of colloidal carriers (e. g., liposomes, nanosomes, nanoparticles, etc.) for a particular drug. This zeta potential (Wiacek, 2007) and conductivity (Sripriya et al., 2007) related to the migration of particles yield information about the selection basis for excipients and the interaction of colloidal carriers with cell membranes. The magnetic property of colloids deals with the specific delivery of drugs inside the body, which helps to increase the efficacy of the drug and reduces its toxicity (Vyas and Khar, 2002).

2.2 MORPHOLOGICAL PROPERTIES OF COLLOIDS

The morphological properties of colloidal particles greatly influence the morphological characteristics of the drug delivery system, for example, microparticles and nanoparticles. The morphological properties mainly include the size, shape, and surface area of colloidal particles. Any alteration in the morphological properties of colloids may greatly influence the stability, biocompatibility, and biodistribution of drug carrier systems (Yang and Alexandridis, 2000). Several morphologies of colloids can be studied with atomic force microscopy images as observations of the adsorption behavior of colloids, for example, the periodicity of discrete adsorbed aggregates on the surface of particles.

2.2.1 PARTICLE SIZE OF COLLOIDS

The colloidal size range is approximately 1 nm to 1 μm, and most colloidal systems are heterodisperse (Burgess, 2007). Because of their size, colloidal particles can be separated from molecular particles with relative ease (Swarbrick and Martin, 1991). Particle size is one of the most important parameters that assists in selection of the route of drug delivery. The size of particles reflects the surface area for absorption and its settling in suspensions. To characterize heterodisperse systems, it is necessary to determine the particle size distribution. Various sizes of colloids affect the biodistribution of drug delivery (Figure 2.1). Colloids can be measured as follows:

Martin's diameter (d_m): The length of a line that dissects the image of the colloids.

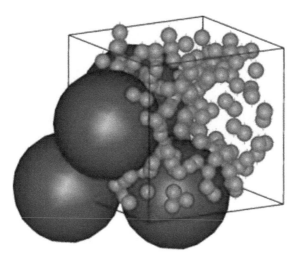

FIGURE 2.1 Texture of colloid.

Feret's diameter (d_f): The distance between two tangents on opposite sides of the colloid, which are parallel to some fixed direction.
Projected area diameter (d_a): The diameter of a circle having the same area as the colloids (Allen, 1975).

Colloidal drug delivery carriers are usually required for the following reasons (Table 2.1 and Figure 2.2):

1. Many therapeutically effective drugs are characterized by poor aqueous solubility and need to be solubilized.
2. Many drugs, such as proteins and peptides, are very fragile and often large in size and need a microenvironment for protection from hydrolysis or enzymatic degradation and absorption through membranes.
3. Colloidal drug delivery carriers increase drug efficacy by reducing its size to nanoparticles and decrease its toxicity.
4. Because of their small size, they are not recognized by normal reticuloendothelial system organs.
5. Targeted delivery of drugs can be achieved by conjugating a specific vector to the carrier.

TABLE 2.1
Various Sizes of Colloidal Carrier Systems

System	Size (nm)
Microspheres	200–1000
Multilamellar vesicles	200–1000
Nanocapsules	50–200
Unilamellar vesicles	25–200
Nanoparticles	25–200
Microemulsions	20–50
Low-density lipoproteins	20–25

Source: Florence, A. T., 1994. *Drug Delivery: Advances and Commercial Opportunities.* Oxford, UK: Connect-Pharma.

6. Because of their smaller size, they can be easily administered intravenously. These particles cannot extravagate except in tissues with a discontinuous capillary endothelium, which are the liver, spleen, and bone marrow. The size of the gap between endothelial cells is approximately 100 nm, which allows only the smallest particles to penetrate into the tissue. Carriers extravagate into solid tumors and into inflamed or infected sites, where the capillary endothelium is defective (Barratt, 2000; Yang and Alexandridis, 2000). The anticancer drug doxorubicin (adriamycin) is active against a wide spectrum of tumors, but it has dose-limiting cardiotoxicity. Encapsulation within liposomes or nanoparticles decreases this toxicity by reducing the amount of drug that reaches the myocardium.

In parenteral drug delivery by means of colloidal emulsions, the particle size is one of the most important characteristics of an emulsion. Large particle size emulsions are clinically unacceptable because of the formation of emboli, and small emulsion droplet sizes promote good physical stability for long circulation times in the body (Kawaguchi et al., 2008).

2.2.2 Shape

The shape adopted by colloidal particles in dispersion is important; the more extended the particle, the greater its specific surface and the greater the opportunity for attractive force to develop between particles of the dispersed phase and the dispersion medium. The rate of sedimentation and osmotic pressure are greatly affected by a change in the shape of colloidal particles (Swarbrick and Martin, 1991). The shape of the colloid derives the surface properties that strongly affect the stability as well as the biocompatibility and biodistribution of colloidal drug carriers. Any alteration in the shape of colloidal particles tends to increase the chances of aggregation (Figure 2.3). The aggregation of particles provides selection criteria for excipients for colloidal drug delivery systems. The shape of colloidal particles also provides a selection basis for the kinetics and mechanism of drug release from carriers into the biological system. Particles produced by dispersion methods have shapes that depend partly on the natural cleavage planes of crystals and partly on any points of weakness within the crystals. Particle shape can be determined by scanning electron microscopy, transmission electron microscopy, electron microscopy, and confocal laser scanning microscopy (Nixon and Mathews, 1974). The surface properties of any colloid can be assessed by its angular or spherical shape, which affects the efficacy of drug delivery. For instance, the molecular shape of phosphatidylcholine (PC) is cylindrical and the aggregation state is lamellar as biomembranes. Cylindrical molecules produce close vesicles, but lysophosphatidylcholine causes hemolysis; hence, it is not used for parenteral emulsions (Kawaguchi et al., 2008).

2.2.3 Surface Area

The surface area of any colloid is derived from its size, which also reflects a greater charge on particles. These charged particles repel each other, overcoming the tendency to aggregate and

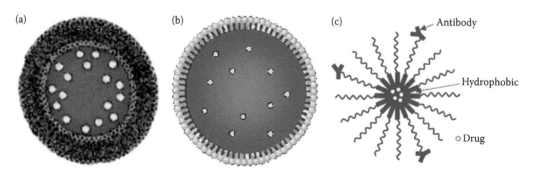

FIGURE 2.2   (a) Liposome; (b) nanoparticles; and (c) specific miceller system.

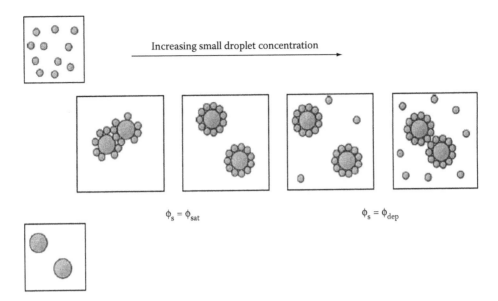

$\phi_s = \phi_{sat}$ $\phi_s = \phi_{dep}$

FIGURE 2.3   Alteration in shape.

remain dispersed. The surface area mainly affects the viscosity, interfacial tension, and aggregation properties. It also provides selection criteria for drug delivery systems. For instance, the combination of lecithin and a hexanoyl oil in water microemulsion decreases emulsion droplet size and increases surface area and stability, which in turn also increases the circulation time of the drug (Kawaguchi et al., 2008). A large surface area is associated with the characteristic size of colloidal particles, which is an intrinsic property of the colloidal system. For example, a typical micellar system for drug delivery containing 0.1 m amphiphile has 40,000 m^2 of interfacial area per liter of solution.

2.3   OPTICAL PROPERTIES OF COLLOIDS

One of the most promising properties of colloids is the optical property, which is widely used to observe the size, shape, and structure of colloidal particles. The different types of optical properties are described in the following subsections.

2.3.1   Tyndall Effect

The Tyndall effect is the most common optical property of colloids. When a beam of light is passed through a colloidal solution, a visible cone resulting from the scattering of light by colloidal particles is formed. This is called the Faraday–Tyndall effect. The smaller the particle size, the greater is the Tyndall effect. The Tyndall effect is attributable to the interaction of particles with light, and it provides selection criteria for the route of administration by knowing the interaction of colloidal particles with cell membranes.

2.3.2   Turbidity

Turbidity is defined as the reduction in the intensity of light as it passes through a colloidal sample. The loss of intensity is due to the scattering of light, and it can be used to determine the average molar mass of lyophilic colloids. Such a knowledge of molar mass helps to access the developability, processability, and quality criteria of any drug delivery system. The molecular weight of colloids may be obtained from the following equation:

$$\frac{Hc}{\tau} = \frac{1}{M} + 2Bc,$$ (2.1)

where τ is the turbidity, c is the concentration of solute (g/cm³), M is the weight-average molecular weight, and B is the interaction constant (Swarbrick and Martin, 1991).

The determination of molecular mass by the turbidity provides a selection basis for absorption, distribution, and interaction of colloidal carriers with cell membranes.

2.3.3 DYNAMIC LIGHT SCATTERING

DLS is one of the properties of colloids known as quasielastic light scattering or photon correlation spectroscopy. The DLS technique utilizes the principle of Brownian motion and can measure particles in the range of 0.6 nm to 6.0 µm (Chabay and Bright,1978; Spragg,1980). This technique reports the particle size as the hydrodynamic radius, which is defined as the radius of a hard sphere that diffuses at the same rate as the particle under examination. The factors that affect the hydrodynamic radius include the shape, density, properties of the suspending fluid, and temperature (Wong et al., 2008). For example, the mean water droplet size is dependent on the water concentration as confirmed by DLS measurement: the higher the water concentration, the larger the droplet size. Water droplet size was determined by DLS measurement at 25°C, confirming that a water volume fraction between 3% and 20% ensured that the size was in the nanoemulsion range (Chiesa et al., 2008).

2.4 KINETIC PROPERTIES OF COLLOIDS

The most interesting subset of colloid properties is the kinetic properties (*The Columbia Electronic Encyclopedia*, 2007). Kinetic properties relate to the motion of particles with respect to the dispersion medium. The motion may be thermally induced (Brownian motion, diffusion, and osmosis), gravitationally induced (sedimentation), or applied externally (viscosity) (Swarbrick and Martin, 1991).

2.4.1 BROWNIAN MOTION

The thermal motion of particles in the colloidal size range (~5 µm) is known as Brownian motion. Robert Brown (1827) was the first to observe the random directional movement of colloidal particles. The particles display a zigzag-type movement, which is the result of random collision with molecules of the suspending medium. Brownian motion increases with an increased particle velocity and a decreased particle size. With increase in the viscosity of the medium, Brownian motion first decreases and then finally stops (Everett, 1987; Shaw, 1992; Swarbrick and Martin, 1991). As a result, the suspension settles down and aggregates. Thus, Brownian motion provides a basis for selection of excipients (e.g., viscosity choice) for colloidal drug delivery systems as well as a basis for carrier–cell membrane interaction (via motion).

2.4.2 DIFFUSION

As a result of Brownian motion, colloidal particles diffuse from a region of higher concentration to one of lower concentration until the concentration is uniform throughout (Diane, 2007). Diffusion is governed by Fick's law, which states that

$$Dq = -DS\frac{dc}{dx}\,dt,$$ (2.2)

where D represents the diffusion coefficient, which is the amount of material diffusing per unit time across a unit area when dc/dx (the concentration gradient) is unity (Swarbrick and Martin, 1991).

Diffusion is inversely related to the particle radius. Because colloidal particles are small, they can diffuse easily through membranes such as a porous plug. The diffusion of colloidal particles is relatively slow compared to that of small molecules or ions. Gravitational forces, which cause particles to sediment, and Brownian motion (diffusion force) oppose one another within the size range at which the Brownian force is stronger than the gravitational force, so particles tend to remain suspended (Diane, 2007). Aggregate transport through the skin is the colloid-induced opening of originally narrow (0.4 nm) gaps between cells in the barrier to pores with diameters above 30 nm. Colloids are incapable of enforcing such widening and simultaneously of a self-adopting size of 20–30 nm without destruction are confined to the skin surface. Hence, colloids within the size range applied to the skin have drug molecules diffusing through it. Colloids only enforce skin penetration and not colloid-enhanced drug permeation. For instance, the transdermal drug delivery system is based on a corresponding colloidal dispersion and therefore relies on simple drug diffusion through the skin (Ceve, 2004).

2.4.3 Osmosis and Osmotic Pressure

Osmosis and the osmotic pressure of colloids greatly influence drug delivery. During osmosis, molecules move from a lower concentration to a higher concentration. An osmotic pressure is generated in colloidal solution when it is separated from its solvent by a barrier that is impermeable to solutes but permeable to solvents. Pure solvent will flow across the membrane and dilute the colloidal dispersion; and because the colloidal solution cannot flow in the opposite direction, a pressure difference (osmotic pressure) will be created between the two compartments (Diane, 2007). This osmotic pressure provides a selection criterion for the absorption of colloidal carriers through the topical administration (Figure 2.4).

For example, a transferosomal formulation of diclofenac in gel form applied to the skin surface is partly dehydrated by water evaporation loss, and then the lipid vesicles feel this osmotic gradient and try to escape complete drying by moving along the gradient (Ceve, 1996).

2.4.4 Sedimentation

Sedimentation is one of the most critical properties of colloids that significantly influences the processing, manufacturing, and market launching of a drug delivery system. Sedimentation is mainly influenced by gravitational force. The velocity (V) of sedimentation of spherical particles with a density ρ in a medium of density ρ^0 and a viscosity η is given by Stokes law:

$$V = \frac{2r^2(\rho - \rho^0)g}{9\eta}. \tag{2.3}$$

Because a colloidal dispersion is within the size range and exhibits Brownian motion, gravity sedimentation will be minimal. In contrast, coarse particles will tend to fall out even if they receive an electric charge like the smaller particles, because gravity will have a greater influence than electrical force (which maintains the dispersion given a constant particle size). The more likely attraction force will overcome the repelling charge, creating large masses. At the same point, mass will precipitate out because of gravitation. At lesser concentrations, the attraction force is insufficient for precipitate particle bonding and groups are light enough that gravitational force will not pull them out of solution. Colloidal systems can be destroyed because particles can aggregate, become larger and noncolloidal in size, and then drop out of the medium. Thus, sedimentation provides information on the concentration of a colloidal dispersion in drug delivery, which reflects the loading and entrapment efficiency of drugs.

For instance, externally applied suspensions for topical use are legion and are designed for dermatologic, cosmetic, and protective purposes. The concentration of the dispersed phase may exceed 20%, and parenteral suspensions may contain from 0.5% to 30% solid particles (Swarbrick and Martin, 1991).

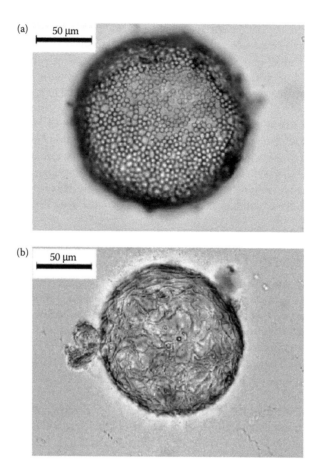

FIGURE 2.4 Swelling of colloidal particles (a) immediately after emulsification and (b) after 20 min.

2.4.5 VISCOSITY

The viscosity of colloid executes by shape of particles to medium, type of colloidal system and molecular weight of particles also affects the viscosity of colloids. The viscosity of a colloidal dispersion is mainly affected by the shapes of particles of the dispersed phase. Spherocolloids form dispersions of relatively low viscosity, whereas systems containing linear particles are more viscous. The relationship between the shape and viscosity reflects the degree of solvation of particles. If a linear colloid is placed in a solvent for which it has a low affinity, it tends to assume a spherical shape and the viscosity falls (Swarbrick and Martin, 1991). For instance, the viscosity of bicontinuous microemulsions is highly dependent on the nature of the cosurfactant with respect to its solubility in the microemulsion phase and chain length. The effect of the structure of the oil phase is not significant in the viscosity of microemulsions. The structure effect of the surfactant is very low in *n*-butanol-based microemulsions. These microemulsions exhibit significantly lower viscosity than *n*-pentanol-based microemulsions (Sripriya et al., 2007).

2.5 PHYSICOCHEMICAL PROPERTIES OF COLLOIDS

The physicochemical property is one of the most interesting subsets of colloidal properties to achieve a characteristic physical appearance and to optimize the therapeutic effects of a medication. Several of these properties are the physical state of colloids, lyophilicity, lyophobicity, and drug polymorphs, as well as the effect of temperature and interfacial properties and the dissociation and erosion property in polymeric colloids. Such properties are commonly used to improve the biocompatibility and biodistribution of drugs in the body. For example, peptide and protein protection from the microenvironment and micellar colloids improve the solubility of biopharmaceutically classified type IV drugs.

2.5.1 Physical State

Colloids are present in different physical states, including microemulsions, nanoemulsions, nanosuspensions, liposomal dispersions, nanoparticles, and low-density lipoproteins (LDLs).

2.5.1.1 Microemulsion

This is a nonpolymeric colloidal system used in drug delivery to solubilize water-insoluble drug substances and to control and sustain release and drug targeting. The stability of parenterally administered emulsion droplets is dependent on the nature of the emulsifying agent. Interaction of emulsions with macrophage cells can be reduced by coating them with hydrophilic polymeric materials. Drug-release rates from emulsions are controlled by the particle size of the dispersed phase and emulsion viscosity (Davis and Hansrani, 1985).

2.5.1.2 Nanosystem

A nanosuspension preparation containing finely divided particles (3 nm to 3 μm) is distributed throughout a fluid or semisolid vehicle. The physicochemical aspects of a nanosystem affect drug loading and release, as well as the surface properties of colloidal properties; for example, in peptide and protein drug delivery systems, copolymers such as poly(lactide glycolic acid) or poly(lactic acid)–poly(ethylene glycol) offer the advantages of manipulating biodegradation kinetics and compatibility with drugs and improving surface properties. Nanosuspensions are generally used for surface-modified drug nanoparticles for site-specific delivery to the brain (Mullar and Kayser, 2001).

2.5.1.3 Lipoproteins

These are natural body transporters of cholesterol, triacycloglycerols, and phospholipids. They include high-density lipoproteins (10 nm), LDLs (23 nm), and very low density lipoproteins (30–100 nm). LDL has been targeted into endothelial cells and Kupffer cells (Mahley and Innerarity, 1982; Nagelkerke et al., 1983; Van Berkel et al., 1985).

2.5.1.4 Liposomes

Liposomes are artificial lipid vesicles consisting of one or more lipid bilayers enclosing a similar number of aqueous compartments. They can be subcategorized into (a) small unilamellar vesicles (25–70 nm) that consist of a single lipid bilayer, (b) large unilamellar vesicles (100–400 nm) that consist of a single lipid bilayer, and (c) multilamellar vesicles (200 nm to several microns, but in colloidal dispersion the size ranges to 5 μm) that consist of two or more concentric bilayers. From the physicochemical viewpoint, lipids with long hydrocarbon chains and a low degree of unsaturation and branching can form tightly packed bilayers with improved liposomal stability and reduced leakage; for example, polymeric-based liposomes prepared by poly(ethylene oxide)–poly(ethylene thylene) block copolymer are tougher and less permeable than common phospholipid liposomes (Gabizon and Martin, 1997; Godfredsen et al., 1983; Lasic and Martin, 1995; Szoka and Paphadjopoulas, 1980; Torchilin, 1998).

2.5.1.5 Gels

Gels are a relatively newer class of colloidal systems created by the entrapment of large amounts of aqueous or hydro-alcoholic liquid in a network of colloidal solid particles. Depending on the nature of the colloidal substances and liquids in the formulation, gels will range in appearance from entirely clear to opaque. When these colloidal drug delivery carriers are given topically in gel form, the amount penetrating the superficial layers may be increased compared with free drugs. The sol–gel transition with a change in temperature attributable to the usual fluctuations of body temperature, pH, physiology, and ionic strength affect drug loading and drug release. Physically self-assembled gels can load drugs in mild conditions with higher drug-loading amounts and faster drug-loading kinetics than cross-linked colloid gels (Kim et al., 1993).

2.5.2 LYOPHILICITY AND LYOPHOBICITY OF COLLOIDS

The lyophilicity and lyophobicity of colloids affect drug delivery to a large extent. Lyophilicity is a solvent-loving property of colloids, in which the dispersed phase is dissolved in the continuous phase. These lyophilic colloids are best treated as a single-phase system; they are thermodynamically stable and form spontaneously when a solute and a solvent are brought together. For example, dispersion of liposomes and nanoparticles in water represent hydrophilic colloidal systems. The viscosities of lyophilic colloids are higher than those of the dispersion medium, and these colloids are coacervates and precipitate out in higher concentrations. Polymer-based colloids offer opportunities to achieve optimal medication effects by their chemical composition and molecular weight. This is readily tailored to accommodate drugs of varying hydrophobicity and specific delivery systems (Figure 2.5).

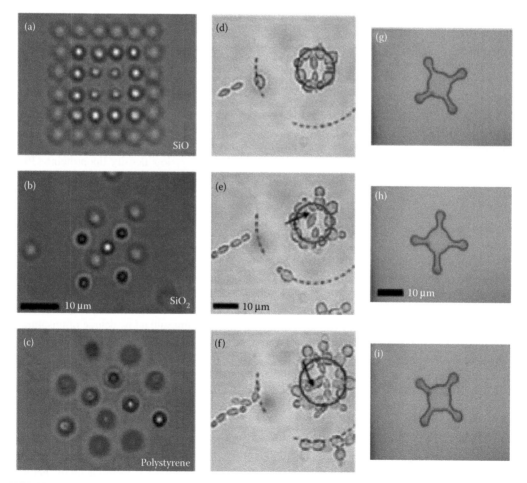

FIGURE 2.5   Micromanipulation of colloid.

In lyophilic or hydrophilic colloids (in higher concentrations), particles are mainly of opposite charges; hence, they neutralize each other. This is attributable to a reduction in the Gibbs free energy (ΔG) on dispersion of a lyophilic colloid. The ΔG is related to the interfacial area (ΔA), interfacial tension (γ), and entropy of the system (ΔS):

$$G = \gamma A - T, \tag{2.4}$$

where T is the absolute temperature.

The strong interaction between solute and solvent usually supplies sufficient energy to break up the dispersed phase. In addition, there is an increase in the entropy of the solute on dispersion, which is generally greater than any decrease in solvent entropy. The γ is negligible if the solute has a high affinity for the solvent; thus, the $\gamma \Delta A$ term will approximate to zero. The shape of macromolecular colloids will vary with the affinity for the solvent. Macromolecules will take on elongated configurations in a solvent for which they have a high affinity and will tend to decrease their total area of contact with a solvent for which they have little affinity by forming compact coils. By contrast, lyophobic colloids are thermodynamically unstable and have a tendency to aggregate. The ΔG increases when a lyophobic material is dispersed throughout a medium. The greater the extent of dispersion, the greater the total surface area exposed and thus the greater the increase in the free energy of the system. When a particle is broken down into smaller particles, work is needed to separate the pieces against the forces of attraction between them (W). The resultant increase in free energy is proportional to the area of the new surface created (A):

$$G = W = 2\gamma A \tag{2.5}$$

Molecules that were originally bulk molecules become surface molecules. In the surface environment, molecules have different configurations and energies than those in the bulk. An increase in free energy arises from the difference between the intermolecular forces experienced by the surface and bulk molecules. Lyophobic colloids are aggregatively unstable and can remain dispersed in a medium only if the surface is treated to cause a strong repulsion between the particles. Such treated colloids are thermodynamically unstable yet are kinetically stable because aggregation can be prevented for long periods. For instance, a microemulsion is a lyophobic colloid and is widely used in topical, parenteral, and ocular drug delivery systems because of its small size.

2.5.3 POLYMORPHS AND CRYSTALLINITY

Polymorphs and the crystallinity of drugs can be modified by colloidal particles to develop stable and efficient drug delivery. If sterile drugs of the thermodynamically stable polymorph are available, direct homogenization for surfactant coating and reduction of colloidal size may be performed to aseptically produce an IV injectable suspension with mean volume weighted particle sizes of <1 μm. However, crystals formed by precipitation may contain inclusion pockets of entrapped impurities such as solvent. Crystals with solvent inclusions may show different physical properties. Different crystallization conditions produce different polymorphs. For example, polymorphs I and II showing different x-ray diffraction profiles of itraconazole were produced in the development of a potential suspension dosage form.

Depending on the pharmacokinetics and pharmacodynamic requirements, amorphous or crystalline injectable suspensions may be developed. A suspension with amorphous solid particles can be stabilized by lyophilization to prepare a stable colloidal system. A crystalline particle containing injectable suspension as a depot formulation may be developed for the controlled release of drugs. Microparticle piroxicam, dantrolene, and flurbiprofen have been reported to achieve rapid dissolution upon injection in animals. There are still numerous methods of medication and optimization to tap the potential in drug delivery as the microemulsion approach for nanoparticle synthesis (Wong et al., 2008).

2.5.4 TEMPERATURE

The colloidal properties of a system are greatly affected by temperature. As the temperature is increased, the viscosity of the colloidal dispersion tends to decrease and affects the stability. At room temperature, a colloidal dispersion becomes stable. For instance, three parenteral emulsions were prepared from three different phospholipids: purified egg yolk lecithin, PC, and phosphatidylethanolamine. Emulsions prepared with PC showed no creaming for a week at room temperature. Thus, the stability of the emulsion prepared with PC was more stable than that with purified egg yolk lecithin (Kawagauchi, 2008).

2.6 ELECTRICAL PROPERTIES OF COLLOIDS

The electrical property of colloids is one of the most valuable properties and plays a critical role in drug delivery. This property depends on, or is affected by, the presence of charge on the surface of the particle. The electrical properties include electrokinetic phenomena and the Donnan membrane equilibrium. Electrokinetic phenomena include the zeta potential and conductivity. In general, dispersed particles possess a charge on their surface. The distribution of ions in the environment of charged particles is explained by the concept of an electrical double layer. When particles move, this double layer shell also moves along with the particle. As the shear plane of the particle is located at the periphery of the tight bound layer, the rate-determining potential is the zeta potential. This potential is needed for the migration of charged particles in the dispersion medium. As the velocity of the particles is high, the rate of migration also increases with the increase in zeta potential. The change in zeta potential is pH dependent, but it settles rather fast. The values of zeta potential are calculated as

$$\zeta = \frac{v}{E} \times \frac{4\pi\eta}{\varepsilon} \times (9 \times 10^4), \tag{2.6}$$

where ζ is the zeta potential (V), v is the velocity of migration (cm sec) in an electrophoresis tube of definite length (cm), η is the viscosity (P), ε is the dielectric constant of the medium, E is the potential gradient (V/cm), and the term v/E is known as the mobility (Swarbrick and Martin, 1991).

For instance, small liposomes and those containing some negatively charged lipids (ganglioside GM_1 or hydrogenated phosphatidylinositol) remain in the bloodstream for longer periods of time. Increased circulation times can also be achieved, to some extent, by the administration of high doses, which saturate the phagocytic system. The major breakthrough, however, is the use of phospholipids substituted with poly(ethylene glycol) chains of molecular weights from 1000 to 5000, as 5–10% of the total lipid. This provides a "cloud" of hydrophilic chains at the particle surface, which repels plasma proteins. Such "sterically stabilized" liposomes possess circulating half-lives of up to 45 h, as opposed to a few hours or even minutes for conventional liposomes (Barratt, 2000).

2.6.1 CONDUCTIVITY

Conductivity is also one of the electrokinetic properties of colloids. An increase in conductivity is normally associated with a corresponding decrease in viscosity. For instance, n-butanol-based microemulsions exhibit higher conductivity compared to n-pentanol-based microemulsions. The viscosity changes are dependent on the cosurfactant, whereas the conductivity changes are mainly dependent on the surfactant (Wiacek, 2007).

The Donnan membrane equilibrium is one of the electrical properties that has a large affect on drug delivery. It is used to enhance the absorption of drugs like sodium salicylate and potassium benzyl penicillin. Sodium carboxymethylcellulose is an anionic polyelectrolyte, and it is not diffusible through semipermeable membranes. However, it enhances drug absorption when it combines with another anionic drug, provided the drug is diffusible. Similarly, ion exchange resins and even sulfate and phosphate ions that do not diffuse readily tend to drive anions from the intestinal tract into the bloodstream (Swarbrick and Martin, 1991).

2.7 MAGNETIC PROPERTY

Generally, magnetic colloidal particles consist of hematite and yttrium oxide particles. The magnetic property reflects both the magnetic behavior of inner magnetic particles and the surface properties of the shell. The idea has already been applied to the production of mixed systems in which the outer shell is polymeric (biodegradable, when the drug delivery application is considered). Magnetically modulated systems like magnetic nanoparticles and magnetic microemulsions contain the active drug moiety that is selectively delocalized by applying an external magnetic field to the specific site, so the drug starts to release specifically at that particular site. The advantages of this property are the release of the drug in the specific area and the reduction of the dose of the drug to 1/10th of the free drug dose. For instance, the emulsion is magnetically responsive to an oil in water type emulsion bearing the anticancer agent methyl chloroethyl-3-cyclohexyl-*1*-nitrosourea, which could be selectively localized by applying an external magnetic field to a specific target site. The magnetic emulsion consists of ethyl oleate-based magnetic fluid as the dispersed phase and casein solution as the continuous phase. The anticancer agent methyl chloroethyl-3-cyclohexyl-*1*-nitrosourea was trapped in the oil dispersed phase. The emulsion showed high retention by a magnetic field *in vitro* (Akimoto and Marimoto, 1983).

2.8 CONCLUSION

We have focused on the influence of colloidal properties in the development of applications in drug delivery. On the basis of their properties, these colloidal drug carriers are particularly useful in the delivery of new drugs and biologicals like peptides, proteins, genes, and oligonucleotides because they can provide protection from degradation in biological fluids and promote their penetration into cells, resulting in the efficacy and safety of the delivery system. The size and shape of colloidal particles are mainly helpful in the selection of drug delivery systems. For delivering the drug into the body, a rational approach needs to be chosen by a formulation scientist, who can strike a balance between the particle size and stability of a system. By altering the surface and charge properties of colloidal particles, the formulation scientist can improve the biocompatibility and biodistribution of colloidal dispersions. An improved understanding of colloidal properties provides an idea of the absorption, distribution, metabolism, and elimination; transport mechanism; and colloidal carriers with cell–membrane interaction. Therefore, it is likely that, on the basis of these properties, a formulation scientist can improve the efficacy of established drugs and new molecules will soon be available.

ABBREVIATIONS

DLS   Dynamic light scattering
LDL   Low-density lipoproteins
PC   Phosphatidylcholine

SYMBOLS

B Interaction constant
c Concentration of solute
d_a Projected area diameter
d_f Feret's diameter
d_m Martin's diameter
dc/dx Concentration gradient

D	Diffusion coefficient
E	Potential gradient
ε	Dielectric constant
ζ	Zeta potential
ΔG	Gibbs free energy
M	Molecular weight
η	Viscosity
ρ	Spherical particles density
ρ^0	Density of medium
τ	Turbidity
T	Absolute temperature
ΔS	Entropy of the system
W	Work
v	Velocity
V	Velocity of migration

REFERENCES

Akimoto, M. and Marimoto, Y., 1983. Use of magnetic emulsion as a novel drug carrier for chemotherapeutic agents. *Biomaterials*, 4, 49.

Allen, T., 1975. Particle size measurement. In *The Powder Technology Series*, J. C. William (Ed.), p. 41. London: Chapman & Hall.

Barratt, G. M., 2000. Therapeutic application of colloidal drug carriers. *Pharm. Sci. Technol. Today*, 3, 163–171.

Bungenberg, D. J. H. G., 1949. A survey of the study objects of the volume. In *Colloid Science: Reversible Systems*, H. R. Kruyt (Ed.), Vol. 2, pp. 1–17. New York: Elsevier.

Burgess, D. J., 2007. Colloids and colloid drug delivery. In *Encyclopedia of Pharmaceutical Technology*, pp. 636–643. London: Informa Healthcare.

Ceve, G., 1996. Transferosomes, liposomes and other lipid suspensions on the permeation enhancement, vesicles penetration and transdermal drug delivery. *Crit. Rev. Ther. Drug Carrier Syst.*, 13, 257–290.

Ceve G., 2004. *Advanced Drug Delivery Reviews*, 56, 675–711.

Chabay, I. and Bright, D. S., 1978. The physical behaviour of macromolecules with biological functions. *J. Colloid Interface Sci.*, 63, 304–309.

Chiesa, M., Garg, J., Kang, Y. T., and Chen, G., 2008. Thermal conductivity and viscosity of water-in-oil nano-emulsions. *Colloids Surf. A*, 326, 67–72.

Davis, S. S. and Hansrani, P. J., 1985. The effect of minor components on the stability of lipid emulsion systems. *Colloid Interface Sci.*, 108, 285–287.

Diane J. B., 2007. *Encyclopedia of Pharmaceutical Technology*, pp. 636–643. USA: Informa Healthcare.

Everett, D. H., 1987. *Basic Principles of Colloid Science*, pp. 76–108. London: Royal Society of Chemistry.

Florence, A. T., 1994. *Drug Delivery: Advances and Commercial Opportunities*. Oxford, UK: Connect-Pharma.

Gabizon, A. and Martin, F., 1997. Polyethylene glycol-coated (pegylated) liposomal doxorubicin. Rationale for use in solid tumours. *Drugs*, 54, 15–21.

Gennaro, A. R., 2004. *Remington Pharmaceutical Science*, 20th edition, pp. 288–315. Baltimore, MD: Lippincott Williams & Wilkins.

Godfredsen, C. F., Van Berkel, Th. J. C., Kruijt, J. K., and Goethais, A., 1983. Cellular localization of stable solid liposomes in the liver of rats. *Biochem. Pharmacol.*, 32, 3389–3396.

Gregor, C., 2004. Lipid vesicles and other colloids as drug carriers on the skin. *Adv. Drug Deliv. Rev.*, 56, 675–711.

Hunter, R. J. and O'Brien, R. W., 1989. Electrokinetic effects. In *Foundations of Colloid Science*, Vol. II, Chap. 13. Oxford, UK: Oxford University Press.

Kawaguchi, E., Shimokawa, K. I., and Ishii, F., 2008. Physicochemical properties of structured phosphatidyl-choline in drug carrier lipid emulsions for drug delivery systems. *Colloids Surf. B*, 62, 130–135.

Kim, C. K., Kim, J. J., Chi, S. C., and Shim, C. K., 1993. Effect of fatty acids and urea on the penetration of ketoprofen through rat skin. *Int. J. Pharm.*, 93, 109–118.

Lasic, D. D. and Martin, F., 1995. *Stealth Liposome.* Boca Raton, FL: CRC Press.

Mahley, R. W. and Innerarity, T. L., 1982. Lipoprotein receptors and cholesterol homeostasis. *Biochim. Biophys. Acta*, 737, 197–222.

Mullar, R. H. and Kayser, O., 2001. Nanosuspensions as particulate drug formulations in therapy: Rationale for development and what we can expect for the future. *Adv. Drug Deliv. Rev.*, 47, 3.

Nagelkerke, J. E., Barto, K. P., and Van Berkel, Th. J. C., 1983. *In vivo* and *in vitro* uptake and degradation of acetylated low density lipoprotein by rat liver endothelial, Kupffer, and parenchymal cells. *J. Biol. Chem.*, 258, 12221–12227.

Nixon, J. R. and Matthews, B. R., 1974. Surface characteristics of gelatin microcapsules by scanning electron microscopy. *J. Pharm. Pharmacol.*, 26, 383.

Richard, V., Virgiliis, A. D., Wolfsheimer, S., Schilling, T., et al., 2006. Interfacial properties of colloidal model systems. *NIC Symp.*, 32, 235–242.

Rill, C., Bauer, M., Bertagnolli, H., and Kickelbick, G., 2008. Microemulsion approach to neodymium, europium and ytterbium oxide/hydrocolloids. Effect of precursors and preparation parameters on particle size and crystallinity. *J. Colloid Interface Sci.*, 325, 179.

Shaw, D. J., 1992. Good introduction to the whole field of colloid science. In *Electrophoresis and Zeta Potential Introduction to Colloid and Surface Chemistry*, 2nd edition, pp. 41–48. London: Butterworths.

Spragg, S. P., 1980. *The Physical Behaviour of Macromolecules with Biological Functions*, pp. 126–140. New York: Wiley.

Sripriya, R., Raja, K. M., Santhosh, G., Chandrasekaran, M., and Noel, M., 2007. The effect of structure of oil phase, surfactant and co-surfactant on the physicochemical and electrochemical properties of bicontinuous microemulsion. *J. Colloid Interface Sci.*, 314, 712–717.

Swarbrick, J. and Martin, A., 1991. Colloids. In *Physical Pharmacy*, 3rd edition, Vol. 545, pp. 471–486. Philadelphia, PA: Lea & Febiger.

Szoka, F. J. and Paphadjopoulas, D., 1980. Comparative properties and methods of preparation of lipid vesicles (liposomes). *Ann. Rev. Biophys., Bioeng.*, 9, 467–508.

The Columbia Electronic Encyclopedia, 6th edition, 2007. New York: Columbia University Press.

Torchilin, V. P., 1998. Polymer-coated long-circulating microparticulate pharmaceuticals. *J. Microencapsul.*, 15, 1–19.

Van Berkel, Th. J. C., Kruijt, J. K., Spanjer, H. H., Nagelkerke, J. F., Harkes, L., and Kempen, H. J. M., 1985. Specific targeting of high density lipoproteins to liver hepatocytes by incorporation of a trigalactoside-terminated cholesterol derivatives. *J. Biol. Chem.*, 260, 2694–2699.

Vyas, S. P. and Khar, R. K., 2002. Liposomes. In *Targeted and Controlled Drug Delivery—Novel Carrier Systems*, 1st edition, p. 478. New Delhi: CBS Publisher and Distributors.

Wiacek, A. E., 2007. Electrokinetic properties of *n*-tetradecane/lecithin solution emulsion. *Colloids Surf. A*, 293, 20–27.

Wong, J., Brugger, A., Khare, A., Chaubal, M., et al., 2008. Suspensions for intravenous (IV) injection: A review of development, preclinical and clinical aspects. *Adv. Drug Deliv. Rev.*, 60, 939–954.

Yang, L. and Alexandridis, P., 2000. Physicochemical aspects of drug delivery and release. *Curr. Opin. Colloid Interface Sci.*, 5, 132–143.

3 Polymeric Nanocapsules for Drug Delivery
An Overview

Sílvia S. Guterres, Fernanda S. Poletto, Letícia M. Colomé, Renata P. Raffin, and Adriana R. Pohlmann

CONTENTS

3.1 INTRODUCTION

The pharmacological response to a drug is directly related to its concentration at the action site in the organism. Usually, drugs are biodistributed to the entire body, implying in high drug concentrations in non-target sites. This is strictly related to the adverse effects of drugs. To circumvent this problem, different strategies have been proposed in the last few decades to control the distribution of a drug within the organism. Examples of these strategies include the chemical modification of molecules to prodrugs and drug entrapment in delivery systems.

Submicron devices are more advantageous as drug carriers compared with conventional formulations due to the possibility of drug targeting and intravenous administrations without any risk of embolization. Oily-core polymeric nanocapsules are a particular class of submicron devices, which are composed of an oil core surrounded by a polymeric wall. Their mean sizes are generally around 200–300 nm with a monomodal and narrow size distribution. There are reports of nanocapsules being proposed for therapeutic applications dating from the 1970s. Over the last 40 years, several academic investigations furnished insights into their properties and supramolecular architecture. This nanoscience permits the tuning of nanocapsule characteristics in order to improve their efficacy as a drug delivery system. This chapter focuses on the raw materials, architecture, preparation methods, and physicochemical characterization of oily-core nanocapsules. The main biomedical applications of these systems are also described according to the therapeutic classes of nanoencapsulated drugs.

3.2 NANOCAPSULE ARCHITECTURE AND RAW MATERIALS

The theoretical model for a polymeric nanocapsule is a vesicle, in which an oily or an aqueous core is surrounded by a thin polymeric wall (Couvreur et al., 2002). These devices are stabilized by surfactants, such as phospholipids, polysorbates, and poloxamers, and by cationic surfactants (Schaffazick et al., 2003a). Different raw materials have been described to prepare the nanocapsules, such as polyesters and polyacrylates as polymers and triglycerides, large-sized alcohols, and mineral oil as oily cores. Numerous bioactive molecules have been loaded into these systems, such as antitumorals, antibiotics, antifungals, antiparasitics, antiinflammatories, hormones, steroids, proteins, and peptides. As a general rule, the type of raw material used to compound nanocapsules can influence their morphological and functional characteristics, which may influence *in vitro* release, *in vivo* response, or both.

3.2.1 OILY CORE

An advantage of oily-core nanocapsules over matrix systems is the higher drug loading capacity, especially when the lipophilic core is a good solvent for the drug. Other advantages are the reduction of burst release, the protection of drugs from degradation, and the reduction of drug side effects (Couvreur et al., 2002). A wide range of oils is suitable for the preparation of nanocapsules, including vegetable or mineral oils and pure compounds such as ethyl oleate (Mosqueira et al., 2000). In some cases, the oily core is the active component, such as the chemical sunscreen octyl methoxycinnamate (Alvarez-Román et al., 2001; Weiss-Angeli et al., 2008). Oil selection criteria include the absence of toxicity, risk of degradation and/or dissolution of the polymer, and high capacity to dissolve the drug (Couvreur et al., 2002; Guterres et al., 2000). The nanocapsule oily core should be compatible with the administration route (Table 3.1).

3.2.2 POLYMERIC WALL

3.2.2.1 Organic Polymers

Synthetic and natural biodegradable polymers are reported to make up the polymeric wall of nanocapsules (Table 3.1). The most common are hydrophobic polyesters, such as poly(lactide) (PLA), poly(lactide-*co*-glycolide) (PLGA), and poly(ε-caprolactone) (PCL). These polymers have been widely used in drug delivery systems because of their biocompatibility and degradability in forming

TABLE 3.1
Examples of Raw Materials, Methods, and Administration Routes for Oily-Core Nanocapsules

Component	Examples	Preparation Method	Drug/Bioactive Molecule	Administration Route	Reference
Oil	Triglycerides	Interfacial deposition of preformed polymers	Nimesulide	Topical	Alves et al., 2007
	Miglyol 812N	In situ interfacial polymerization	Insulin	Oral	Pinto-Alphandary et al., 2003
	Mineral oil	Interfacial deposition of preformed polymers	Indomethacin	—	Pohlmann et al., 2002
	CCT	Interfacial deposition of preformed polymers	Melatonin	Intraperitoneal	Schaffazick et al., 2008
	Octylsalicylate	Emulsification–solvent diffusion	Hinokitiol	Topical	Hwang and Kim, 2008
	Miglyol 810 or ethyl oleate	Interfacial deposition of preformed polymers	—	—	Mosqueira et al., 2000
Monomer[a]/polymer[b]	Isobutylcyanoacrylate (monomer)	In situ interfacial polymerization	Insuline	Oral	Cournarie et al., 2002
	PLA	Interfacial deposition of preformed polymers	Diclofenac	Intravenous and oral	Guterres et al., 1995a,b
	PCL	Interfacial deposition of preformed polymers	Chlorhexidine	Topical	Lboutounne et al., 2002
	PLGA	Interfacial deposition of preformed polymers	Insulin	Oral	Saï et al., 1996
	Eudragit S100	Interfacial deposition of preformed polymers	Melatonin	—	Schaffazick et al., 2005
	Cellulose acetate phthalate	Emulsification–solvent diffusion	Octyl methoxycinnamate (chemical sunscreen)	Topical	Olvera-Martínez et al., 2005
Surfactants	Epikuron 170® and Poloxamer 188	Interfacial deposition of preformed polymers	Halofantrine	Intravenous	Leite et al., 2007
	Tween 80®	In situ interfacial polymerization	Idebenone	—	Palumbo et al., 2002
	CTAC	Emulsification–solvent diffusion	Hinokitiol	Topical	Hwang and Kim, 2008
	Poly(vinyl alcohol)	Emulsification–solvent diffusion	—	—	Quintanar-Guerrero et al., 1998
Organic solvent	Ethanol, acetone, or acetonitrile	In situ interfacial polymerization	Ethosuximide, 5,5-diphenyl hydantoin and carbamazepine	—	Fresta et al., 1996
	Acetone	Interfacial deposition of preformed polymers	Ketoprofen	Intraperitoneal	Matoga et al., 2002
	Ethyl acetate	Emulsification–solvent diffusion	—	—	Quintanar-Guerrero et al., 1996

Note: CTAC—cetyltrimethylammonium chloride.

[a] For chemical methods.
[b] For physicochemical methods.

nontoxic residues. They have been approved by the U.S. Food and Drug Administration in several medicines, including those administered systemically. In addition, the release kinetics of entrapped drugs can be controlled by varying the molecular weight (MW) of the polymers (Cha and Pitt, 1988; Sinha et al., 2004; Wise et al., 1987). The *in vitro* and *in vivo* degradation kinetics and biological behavior of the polyesters have been extensively characterized (Middleton and Tipton, 2000; Sinha et al., 2004). They present slow degradation, which can be catalyzed by lipases, causing minimal immune response. The MW influences the degradation kinetics. The hydrolytic degradation of low MW PLA polymers starts in a few days, whereas it takes much longer for high MW PLA polymers (Andreopoulos et al., 1999). Considering this, PLA, PLGA, and PCL are the first choices for producing nanocapsules for parenteral routes. Other polymers, such as polyacrylates, polyacrylamides, and polyureas, have also been described as alternatives for producing nanocapsules because of their adequate biocompatibility (Montasser et al., 2007; Vauthier et al., 2007).

3.2.2.2 Functionalized Surfaces

In recent years, engineering approaches have been devised to create "smart" nanocapsules. In this context, functionalized-surface nanocapsules present interesting properties such as stealth ability and active targeting via ligand binding to specific cell receptors. This is especially important for some molecules that are considered as "undeliverable" compounds, such as nucleic acids, which need to reach an intracellular target to achieve their therapeutic effect (Behr et al., 1989; Felgner et al., 1987), or highly toxic drugs (Couvreur et al., 2002).

Functionalization strategies are divided into two groups: the ligand is incorporated at the nanocapsule during or after its preparation. In the former case, the polymer is chemically bound to the ligand and this complex material is used as the raw material to produce the nanocapsules. In the latter case, nonfunctionalized nanocapsules are produced earlier and the ligand is attached to the nanocapsule polymeric wall by physicochemical or chemical processes. An example of the first strategy is nanocapsules that are covalently attached to poly(ethylene glycol) (PEGylated), which are prepared using a diblock copolymer as the polymeric wall. The second strategy includes post-insertion of whole antibodies and antibody fragments into the surface of nanocapsules (Figure 3.1).

Bioadhesive properties can be obtained by coating the nanocapsule surface with polymers presenting positive charge, such as the polysaccharide chitosan and its derivatives. The adsorption process of chitosan onto nanocapsules occurs via ionic interactions between chitosan and

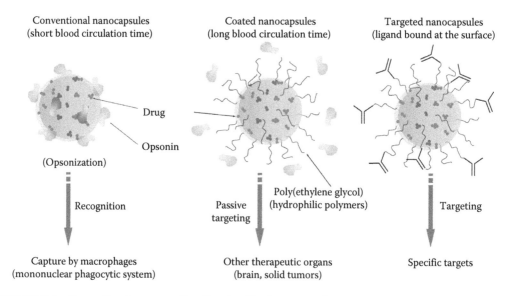

FIGURE 3.1 Conventional versus surface-functionalized nanocapsules.

specimens containing negative charges on the particle surface (such as lecithin), which is reflected by an inversion in the zeta potential signal (Calvo et al., 1997; Prego et al., 2006a). Bioadhesion can increase the residence time of the drug at the site of absorption, which allows improvement in drug penetration (Kriwet et al., 1998). Polyacrylates also have bioadhesive properties, which occur through hydrogen bonding between the carboxylic acid groups of the polymer and the hydroxyl groups of the mucous membrane (Dodou et al., 2005).

Active targeting of specific organs and tissues is based on molecular recognition processes. Examples of relevant targets are the folate receptor or the integrin surface receptor for tumor cells (Couvreur et al., 2002) and mannose and fucose recognizing receptors for macrophages, which are the host cells of the *Leishmania* sp. parasite (Date et al., 2007). A particularly interesting example of surface-functionalization devices is the stealth nanocapsule. The rapid removal of conventional nonfunctionalized-surface nanocapsules from the bloodstream by the mononuclear phagocytic system decreases their effectiveness as site-specific drug delivery devices (Gref et al., 1994; Owens and Peppas, 2006). Macrophages remove the nanocapsules because they can recognize specific opsonin proteins that bind at their surfaces (Frank and Fries, 1991). In this way, the introduction of long hydrophilic polymer chains and nonionic surfactants at the nanocapsule surface can provide slow opsonization because of a steric effect that delays electrostatic and hydrophobic interactions to bind opsonins onto nanoparticle surfaces (Owens and Peppas, 2006). As a consequence, the nanoparticle blood circulation half-life increases (Gref et al., 1994; Kaul and Amiji, 2004). Steric shielding can be obtained using polymers such as polysaccharides, polyacrylamides, poly(vinyl alcohol), poly(N-vinyl-2-pyrrolidone), PEG, PEG block copolymers, and PEG-containing surfactants such as poloxamines, poloxamers, and polysorbates. PEGylation of the particles is the most effective and commonly used strategy to obtain stealth nanoparticles (Owen et al., 2006; Veronese and Pasut, 2005). PEG can be introduced at the surface in two ways: by the adsorption of surfactants (Illum et al., 1987; Moghimi et al., 1993) or by the use of block or branched copolymers, usually with PLA or poly(alkyl cyanoacrylate) (PAC; Peracchia et al., 1999).

Functionalized surfaces provide obvious advantages for nanocapsule performance in comparison with those that are nonfunctionalized. However, formulating nanocapsules with new materials that have not been well characterized may make their preclinical and eventual clinical evaluation difficult. The toxicity profiles of these materials must be established before the nanocapsule clinical trials (Jabr-Milane et al., 2008). Furthermore, it is noteworthy that not only the surface characteristics of nanocapsules but also their size and morphology play a key role in their biological fate (Couvreur and Vauthier, 2006; Gaumet et al., 2008).

3.2.3 SUPRAMOLECULAR STRUCTURE MODELS

Nanocapsule supramolecular organization may differ for each particular combination of raw materials because of the complexity of structures (Mosqueira et al., 2000; Müller et al., 2001). Different organizations can be obtained by selecting specific raw materials. Therefore, supramolecular structure models for nanocapsules can only be proposed in each case after extensive physicochemical investigation.

The theoretical model for oily-core nanocapsules is a vesicle, in which the oil is surrounded by a polymeric wall (Couvreur et al., 2002). However, a careful combination of polymers and oil is required to obtain truly vesicular nanocapsules. For instance, swelling experiments of polyesters (PLA or PCL) immersed in benzyl benzoate demonstrated the dissolution of polymers, indicating that nanocarriers prepared using these materials are nanoemulsions instead of nanocapsules (Guterres et al., 2000). Conversely, PCL was not dissolved by octyl methoxycinnamate, suggesting the presence of a polymeric wall in the nanocapsules composed of this polymer–oil combination (Weiss-Angeli et al., 2008).

Materials structured at the submicron level may present properties different from those at the macroscopic level. In this way, drug release experiments, light scattering analyses, and nuclear magnetic resonance spectroscopy are valuable tools for determining the supramolecular structures of nanocapsules.

Dynamic light scattering (DLS) studies have demonstrated that nanocapsules prepared with PCL, mineral oil, sorbitan monostearate, and polysorbate 80 present low interactions with water because of the presence of sorbitan monostearate dispersed in the polymer (Pohlmann et al., 2002). In contrast, sorbitan monostearate was dispersed in the oil phase of nanocapsules prepared with the same composition but using caprylic/capric triglyceride (CCT) as the oil core instead of mineral oil (Cruz et al., 2006a; Müller et al., 2001). Hence, the organization of the wall and core domains depends on the nature of the polymer, oil, and surfactants. Insights concerning the supramolecular structure of nanocapsules using nuclear magnetic resonance spectroscopy have also been reported (Hoffmann and Mayer, 2000; Mayer et al., 2002). The nanocapsule polymer wall structure composed of poly(n-butyl cyanoacrylate) did not significantly differ from the structure of the solid compact polymer (bulk) used as the reference. In contrast, the triglyceride (core) and Pluronic® F68 (surfactant) presented high mobility and were temporarily adsorbed on the nanocapsule wall. Another study investigated nanocapsule particle–water interface interactions using fluorescent dyes chemically bound to poly(methyl methacrylate), which constituted the polymeric wall (Jäger et al., 2007). The chemical environment-sensitive fluorescent dyes showed that the particle–water interface was composed of oil, surfactant, polymer, and water.

The nature of the surfactant used to stabilize the nanocapsules can also influence their morphology. PLGA capsules containing the ultrasound contrast agent perfluorooctyl bromide (PFOB) as the oily core were prepared with different surfactants (Pisani et al., 2008). Sodium taurocholate generated capsules with a corn morphology in which a hemisphere of liquid PFOB coexisted with a hemisphere of solid polymer, whereas sodium cholate provided liquid PFOB perfectly encapsulated within a PLGA wall. The use of poly(vinyl alcohol) resulted in the coexistence of both morphologies in the same suspension.

In vitro drug release profiles can be useful in determining drug localization in the nanocapsule structure. A comparative study conducted with nanocapsules, nanospheres, and nanoemulsions showed similar indomethacin release behavior, while its ethyl ester derivative showed different release profiles (Cruz et al., 2006b). Kinetic data indicated that indomethacin was adsorbed on the nanocarriers, whereas the indomethacin ethyl ester was entrapped within the nanocarriers.

3.3 PREPARATION METHODS

3.3.1 PHYSICOCHEMICAL METHODS

3.3.1.1 Solvent Displacement-Based Methods

The interfacial deposition of preformed polymers was the first method described (Fessi et al., 1989) for the preparation of nanocapsules, and it is a combination of the spontaneous emulsification of oily droplets and the simultaneous precipitation of polymer onto the water–oil interface. Spontaneous emulsification occurs because of the initial nonequilibrium states of two miscible liquids when they are brought into close contact with each other. Depending on the specific conditions used to carry out spontaneous emulsification, nanodroplets can also be formed (Bouchemal et al., 2004). This process increases the entropy of the system and, as a consequence, decreases its Gibbs free energy. The motive force is the interface tension gradient induced by the diffusion of solutes between the two liquid phases (Marangoni effect). Katz and coworkers (Ganachaud and Katz, 2005; Vitale and Katz, 2003) provided new insights into the physical phenomenon behind spontaneous emulsification. They named this phenomenon the ouzo effect due to a beverage with the same name, where an ethanol extract of anise seeds becomes instantaneously cloudy when diluted with water. In the ouzo effect, the addition of water to a solution of oil in a totally water-soluble solvent causes supersaturation of the oil in the mixture, which leads to its nucleation in small droplets. The mean droplet diameter is exclusively a function of the oil/solvent ratio at a given temperature.

Considering the above, a fundamental requirement for the interfacial deposition of preformed polymers is the high miscibility between the organic and aqueous phases. The most commonly used solvents for preparing nanocapsules by this method are acetone and ethanol. In addition, the organic solvent must be a good solvent for the polymer, but water must not (Fessi et al., 1989). The interfacial tension between oil and the aqueous phase seems to have a greater effect on nanocapsule size than oil viscosity (Mosqueira et al., 2000). A decrease in interfacial tension values causes a reduction in nanocapsule sizes. The advantages of the interfacial deposition of preformed polymers are its simplicity and robustness at small scale and high reproducibility among different batches (Fessi et al., 1989; Teixeira et al., 2005).

3.3.1.2 High-Shear Techniques

High-shear techniques involve the emulsification of an organic phase with an aqueous phase by rotor/stator devices or ultrasound generators. The most common high-shear technique of preparing nanocapsules is emulsification–solvent diffusion, which involves the formation of a conventional oil-in-water emulsion within a partially water-soluble solvent (Quintanar-Guerrero et al., 1996). Water and solvent are previously saturated with each other. The subsequent addition of a second aqueous phase to the system makes the solvent diffuse into the external phase, resulting in the formation of nanocapsules.

Several preparative variables can affect nanocapsule size, such as the type and concentration of the stabilizers, stirring rate of the primary emulsion, internal/external phase ratio of the primary emulsion, polymer concentration in the organic phase, pH, and viscosity of the external phase (Quintanar-Guerrero et al., 1996). Examples of solvents that are used to prepare nanocapsules are ethyl acetate, propylene carbonate, and benzyl alcohol (Quintanar-Guerrero et al., 1998). The sizes of nanocapsules prepared by emulsification–solvent diffusion are a function of thermodynamic parameters such as the mutual diffusion coefficients of water and solvent and the solvent–polymer interaction parameters (Choi et al., 2002). In addition, the sizes of nanocapsules prepared by this technique are related to the chemical composition of the organic phase and the size of the primary emulsion droplets by a simple geometrical relationship (Moinard-Chécot et al., 2008). Therefore, most of the properties of the nanocapsules are decided at the emulsification step. A modification of this technique involving the use of ethanol as cosolvent in the organic phase was proposed to produce controlled-size poly(hydroxybutyrate-*co*-hydroxyvalerate) nanocapsules containing CCT or mineral oil as the oily core (Poletto et al., 2008a). In this study, the control of nanocapsule diameters was achieved by adjusting the surface tension of the organic phase.

The advantages of emulsification–solvent diffusion involve high yields of encapsulation, high reproducibility, control of particle size, and easy scale-up (Moinard-Chécot et al., 2008). However, a large amount of water has to be removed to concentrate the suspensions.

3.3.2 CHEMICAL METHODS

3.3.2.1 *In Situ* Interfacial Polymerization

In situ polymerization to obtain nanocapsules consists of the polymerization of a monomer at the interface between the organic and the aqueous phase of an emulsion. The monomers can be added into the aqueous phase or organic phase. In the former case, the monomers must have good solubility in the external phase, but the polymer must be immiscible in both phases (Zhang et al., 2008). The drug is dissolved in the polymerization medium either before the addition of the monomer or at the end of the reaction. Hence, drugs are encapsulated within or adsorbed onto the particles (Couvreur et al., 1990). The polymerization reaction occurs because of the presence of initiators and specific process conditions, such as pH (Soppimath et al., 2001). The morphology of the polymeric wall is a function of the monomer's initial concentration and reaction time (Anton et al., 2008; Turos et al., 2007). With the addition of monomers in the organic phase, the solvent serves as a vehicle for the monomer. This organic phase is emulsified with an aqueous phase containing water and a hydrophilic surfactant. Nanodroplets are formed to give a milky suspension, and the organic solvent is then removed under reduced pressure. The polymer is synthesized by changing some process parameters (temperature or pH), and it segregates toward the water–oil interface, thereby generating the nanocapsules. The polymerization must be very fast in order to allow efficient formation of the polymer envelope around the oil droplet and thus achieve an effective encapsulation of drugs (Couvreur et al., 2002; Gallardo et al., 1993). Alkyl cyanoacrylates are commonly employed as monomers because of their very fast polymerization rates when they come into contact with water (Gallardo et al., 1993).

Covalent bonds may be created between the active molecule and the polymer during polymerization. Thus, their potential mutual reactivity must be taken into account to obtain the designed nanocapsules (Anton et al., 2008). In addition, nanocapsule formation and structure may be highly affected by the nature of the reaction initiator in chemical methods (Han et al., 2003).

3.4 PHYSICOCHEMICAL CHARACTERIZATION

3.4.1 MORPHOLOGY

Different microscopy techniques can be used to observe the nanocapsule morphology and structure. Scanning electron microscopy (SEM) allows observation of the sample after drying and coating with a thin layer of gold or platinum. This method gives information about the size, shape, and surface aspects of nanocapsules. However, particles smaller than 100 nm may be difficult to observe (Gaumet et al., 2008). Freeze–fracture SEM has also been helpful in visualizing different organizations of lipophilic surfactant in nanocapsule suspensions, which can simultaneously form vesicles, micelles, bilayers, or monolayers, depending on the concentration (Mosqueira et al., 2000). This technique can also be applied to identify the presence and morphology of nanocapsules on animal tissues (Lboutounne et al., 2002) or in a gel matrix after storage (Milão et al., 2003).

Transmission electron microscopy (TEM) is a technique in which a beam of electrons is transmitted through the sample. The sample is dried and stained with contrast agents, such as uranyl acetate or phosphotungstic acid, and then analyzed. TEM provides information about the size, shape, and integrity of nanocapsules and permits observation of their vesicular structure (Mosqueira et al., 2000). In addition, TEM performed after freeze–fracture has given helpful information about the polymeric wall and the core, allowing the wall thickness to be estimated (Fresta et al., 1996). There have been other studies to determine the thickness of the nanocapsule polymeric wall. The wall thickness of nanocapsules prepared with PCL and Miglyol® 812 was estimated to be about 2.0 nm (Guinebretière et al., 2002). Moreover, the polymeric wall can also be characterized using neutron scattering techniques. A wall thickness of about 10 nm for nanocapsules composed of PLA and triglycerides was determined by small-angle neutron scattering (Rübe et al., 2005).

Atomic force microscopy is a useful tool for determining the simultaneous characterization of particle shape, surface structure, and interparticle organization by tridimensional images. In some cases, atomic force microscopy may differentiate nanocapsules according to their composition (e.g., presence vs. absence of PEG as coating; de Assis et al., 2008). The drawback of this technique is the complexity of the sample preparation process before the microscopy analysis.

3.4.2 MEAN AVERAGE SIZE AND PARTICLE SIZE DISTRIBUTION

Many tools based on different physical principles are currently available to measure the size of submicron particles. DLS, also called photon correlation spectroscopy, is based on the interaction of the particle with an incident light beam. The intensity of scattered light detected at a fixed angle provides the mean size, size distribution, and polydispersity index (PI) of the sample. Because the calculation model is based on the equivalent sphere principle, the presence of aggregates greatly increases the mean size. In addition, parameters such as the viscosity or pH of the suspension medium, temperature, concentration, and particle sedimentation may influence the data. Microscopic methods (SEM and TEM) are also described to evaluate the nanocapsule mean size. Particle sizing with microscopic techniques requires image treatment of a large number of particles, thus making them subjective and time consuming. Furthermore, nanocapsules may shrink during the drying step, leading to an underestimation of their actual diameter (Gaumet et al., 2008).

Size distribution is a parameter as important as mean particle size in nanotechnology. Size distribution can be monomodal (one population) or multimodal (several populations) and monodisperse (narrow distribution) or polydisperse (broad distribution). Generally, the size distribution of nanocapsules is evaluated on the basis of the PI, which takes into account the mean particle size, the refractive index of the solvent, the measurement angle, and the variance of the distribution (Koppel, 1972). However, a linear correlation between the PI value and the true polydispersity of a sample cannot be drawn (Gaumet et al., 2008). None of the methods available to determine nanoparticle size and distribution are fully satisfactory and a combination of at least two methods, one of which should be a microscopic method, is highly recommended.

3.4.3 ZETA POTENTIAL

The zeta potential is the electric potential at the interfacial double layer between the dispersion medium and the stationary layer of fluid attached to the dispersed particle. This parameter can be determined by electrophoresis, in which an electric field is applied across the dispersion. Particles migrate toward the electrode of opposite charge with a velocity proportional to the magnitude of the zeta potential (Delgado et al., 2005). The most common experimental electrophoretic technique used to determine the zeta potential of nanocapsules is electrophoretic light scattering.

The zeta potential is influenced by the charge of the different components of nanocapsules, especially surfactants located at the interface with the dispersion medium, as well as the composition of this dispersion medium (Couvreur et al., 2002). Lecithin, poloxamers, and polymers are the major components that affect this parameter. For instance, the polysaccharide chitosan gives a positive zeta potential to nanocapsules (Prego et al., 2006a). By contrast, many polymers, especially α-hydroxyacids (such as PLA) and lecithin, contribute a negative charge to the surface, which is reflected in the zeta potential value. Nonionic surfactants, such as poloxamer and poly(vinyl alcohol), tend to reduce the absolute zeta potential value (Mosqueira et al., 2000). A zeta potential of 30 mV (positive or negative) is indicative of the adequate stability behavior of nanocapsules that is attributable to the charge repulsion between particles, which may be sufficient to prevent their aggregation (Couvreur et al., 2002). However, surface-coated nanocapsules can be stable despite a zeta potential close to zero because of the steric effect of surfactants or PEG copolymers.

Zeta potential measurements may also be used to investigate whether a biologically active compound is encapsulated within the nanocapsule oily core or adsorbed onto the particle surface (Aboubakar et al., 1999) and to confirm nanocapsule coating by a specific material (Preetz et al., 2008). Zeta potential values may be generally associated with the pH values of the bulk (Sussman et al., 2007).

3.4.4 DRUG CONTENT

The total content of a drug in the nanocapsule suspension is determined using high-performance liquid chromatography or another analytical technique after dissolving or extracting the drug from the system using an adequate solvent. Free drug (nonencapsulated) is usually determined in the ultrafiltrate after separation of the nanocapsules by the ultrafiltration–centrifugation technique. The amount of encapsulated drug (associated within or adsorbed onto the nanocapsules) is calculated from the difference between the total and the free drug concentrations determined in the nanocapsule suspension (after dissolution) and in the ultrafiltrate, respectively. The amount of drug encapsulated in the nanocapsule expressed as the percentage of the total amount of drug in the suspension is commonly called the "encapsulation efficiency." Aggregates of the pure drug (stabilized by surfactants) are retained with the nanocapsules in the upper compartment.

Although the encapsulation efficiency of lipophilic drugs is generally related to drug solubility in the oily core (Fresta et al., 1996), hydrophilic compounds such as peptides have also been successfully encapsulated in oily-core nanocapsules prepared by chemical methods. One explanation for this may be the extremely rapid polymerization of the polymer occurring at the surface of the oil droplet, which limits the diffusion of the peptide toward the aqueous phase and therefore leads to higher entrapment in the nanocapsules (Aboubakar et al., 1999).

3.4.5 *In Vitro* Drug Release

The composition of nanocapsules influences the mechanism of drug encapsulation (adsorption on the polymer or entrapment within the core; Couvreur et al., 2002). The localization of drug in the nanocapsule structure affects its release kinetics (Cruz et al., 2006a). Therefore, *in vitro* drug release experiments can give valuable information about the interactions between the drug and the nanocapsule. This concept was illustrated by a comparative study carried out with nanoencapsulated 3-methoxyxanthone and xanthone (Teixeira et al., 2005). Similar kinetics were observed for 3-methoxyxanthone released from nanocapsules and nanoemulsions (without polymer), indicating that drug release was mainly governed by the partition between the oily core and the external aqueous medium. In contrast, the release of xanthone from nanocapsules was significantly slower than that from nanoemulsions, suggesting an interaction of the drug with the polymer.

The release of drugs from nanocapsules depends on drug desorption, drug diffusion, or polymer erosion (Lopes et al., 2000; Soppimath et al., 2001). Similar *in vitro* rapid release kinetics were observed for benzathine penicillin G from PLGA nanocapsules and from nanoemulsions without polymer (Santos-Magalhães et al., 2000). Although experimental data indicate that the polymeric wall is an important factor for drug release kinetics (Pisani et al., 2006; Poletto et al., 2008b), hydrophilic drugs and peptides are generally adsorbed on the polymer surface rather than encapsulated within the oily core. In this case, the drug does not need to cross the polymeric wall to be released, and the release profile from nanocapsules and nanoemulsions may be quite similar. Conversely, the controlled release of 4-nitroanisole from different types of PLA nanoparticles (oily-core nanocapsules and matrix nanospheres) was explained by a Fickian diffusion mechanism (Romero-Cano and Vincent, 2002), in which the exact particle morphology (presence of an oily core, concentration of polymer, and localization of drug in the particle) influenced the release rate kinetics.

To determine the drug release profiles from nanocapsules, the drug may be separated from the nanostructures using ultracentrifugation, ultrafiltration–centrifugation, or dialysis techniques. However, these methods are limited to the determination of the drug partition coefficient between nanoparticles and the continuous phase, and an experimental sink condition is not achieved (Washington, 1990).

3.4.6 Physicochemical Stability of Nanocapsule Suspensions

The stability of nanocapsules can be evaluated in terms of macroscopic aspects, free and total drug contents, zeta potential, pH, and mean particle size as a function of time (Guterres et al., 1995a; Schaffazick et al., 2007). Previous reports demonstrated that the reduction in the pH values of nanocapsule suspensions is related to a partial hydrolysis of the polymer during storage (Calvo et al., 1996a; Guterres et al., 1995a). Nanocapsules may aggregate because of attractive forces among particles. Steric stabilization and electrostatic stabilization are needed to overcome these attractive interactions. Steric stabilization is based on the osmotic stress created by encroaching steric layers of bulk polymers and surfactants at the surface of nanocapsules, whereas electrostatic stabilization provides charge repulsion caused by charged species onto the particle surface. Generally, steric stabilization alone is sufficient to prevent irreversible aggregation. However, attractive forces among particles may still remain to cause reversible flocculation in some cases. Steric stabilization combined with electrostatic stabilization may circumvent flocculation due to the additional repulsive contribution (Kesisoglou et al., 2007; Vauthier et al., 2008). Aggregation of nanocapsules during storage can be monitored by DLS. Nevertheless, this technique requires a dilution of suspensions before the measurement, and reversible flocculation may not be detected. Multiple light scattering analyses have been carried out recently to evaluate the physicochemical instability phenomena of nanodevices, such as precipitation, creaming, and aggregation (Daoud-Mahammed et al., 2007; Lemarchand et al., 2003). This technique permits analysis of the sample without further dilution. Variations in zeta potential values during storage can also be indicative of aggregation.

The reduction in the drug content of nanocapsules can occur because of its chemical degradation after storage, which is caused by external agents such as oxygen, temperature, and ultraviolet irradiation. The degradation products can be identified and quantified by chromatography or other analytical techniques (Müller et al., 2004). Drug nanocrystals stabilized by surfactants can also be concomitantly formed with nanocapsules (Calvo et al., 1996a; Guterres et al., 1995a). Drug oversaturation of nanocapsules causes the formation of nanocrystals, which agglomerate and precipitate during storage (Pohlmann et al., 2008). The consequence is the reduction of drug content within the nanocapsules. Although ultrafiltration–centrifugation cannot separate nanocapsules and drug nanocrystals, their simultaneous presence in suspension can be detected by static light scattering analysis.

3.5 APPLICATIONS IN THERAPEUTICS

3.5.1 ANTIINFLAMMATORIES

Nonsteroidal antiinflammatory drugs (NSAIDs), such as diclofenac and indomethacin, are known to cause gastrointestinal side effects such as irritation and mucosal damage, which are due to local contact between the mucosa and solid drug particles and by an indirect effect (inhibition of prostaglandins and prostacyclin) (Reynolds, 1993). Thus, these drugs are excellent models to evaluate the potentiality of nanoencapsulation to protect biological mucosae against the ulcerative effect of NSAIDs (Friedrich et al., 2008). One of the first works to consider this issue focused on nanocapsules containing diclofenac (free acid) and indomethacin prepared by the deposition of PLA (Guterres et al., 1995b). After oral administration, a significant reduction of gastrointestinal toxicity in rats was observed for the nanoencapsulated drugs. The protective effect of nanocapsules was attributed to a reduction in the local irritation caused by the direct contact of these drugs.

Indomethacin-loaded nanocapsules prepared by the interfacial polymerization of isobutyl cyanoacrylate exhibited a 10-fold increase in antiinflammatory activity compared to free indomethacin (Gursoy et al., 1989). Nanoencapsulation of indomethacin using PLA and poly(isobutyl cyanoacrylate) (PIBC) as polymers induced a protective effect for the jejunal tissue compared to the ulcerative effect of a commercial indomethacin solution (Ammoury et al., 1991). No difference was observed in the pharmacokinetic parameters.

Indomethacin was also encapsulated in PLA by nanoprecipitation using benzyl benzoate as the oil core (Ammoury et al., 1993). The drug release from nanocapsules occurred within minutes, indicating that the nanocapsules were not able to retain the drug in the oil core. Regarding gastrointestinal tolerance, these nanocapsules significantly reduced the ulceration caused by indomethacin in solution. A subsequent work (Guterres et al., 2000) verified that benzyl benzoate completely dissolves both PLA and PCL, showing that nanocapsules prepared with this lipophilic phase and these polyesters do not have a polymer wall. In the same work, the authors used Miglyol 810 as the oil core, which is a nonsolvent for PLA and PCL. The results indicated that the polymer wall acted as a barrier and the oil remained in the spray-dried powder, in contrast to the results observed for the benzyl benzoate formulation (Guterres et al., 2000). In addition, nanocapsules prepared using Miglyol 810 reduced the indomethacin gastrointestinal side effects after oral administration in rats (Raffin et al., 2003).

Despite the advantage of protecting the gastrointestinal mucosae, some authors (Guterres et al., 1995a; Saez et al., 2000; Schaffazick et al., 2003b) stated that the nanocapsules in aqueous suspensions are unstable formulations because of the risk of ingredient degradation, microbiological growth, or both. To improve the stability of nanocapsule formulations, the spray-drying technique was applied to dry the suspensions and produce solid dosage forms (Müller et al., 2001). The production of a nanocapsule spray-dried powder involves droplet formation from the atomized suspension, followed by their solidification driven by water evaporation. This one-step process is easy to scale-up and equipment is readily available, so the processing costs can be reduced (Masters, 1991). This strategy was used for the first time to dry diclofenac-loaded nanocapsules using colloidal silica as the drying adjuvant (Müller et al., 2000, 2001). Biological evaluation of the powder demonstrated the maintenance of gastrointestinal tolerance observed for diclofenac-loaded nanocapsule suspensions (Guterres et al., 2001). A pharmacokinetic study in rats demonstrated that the drug was completely absorbed after oral administration of diclofenac-loaded nanocapsule powder (reconstituted in water and administered by gavage), presenting a higher half-life in plasma than the sodium diclofenac aqueous solution.

An organic–inorganic system based on polymeric nanocapsules prepared by the spray-drying technique has also been described. The system consists of microagglomerates containing the drug dispersed in the core (Aerosil® 200) and polymeric nanocapsules as the coating material (Beck et al., 2004, 2006). Nanoparticle-coated microparticles were prepared using Eudragit® S100 nanocapsules as the coating material of a core composed of diclofenac and silicon dioxide (Beck et al., 2005). This powder showed a protective effect against mucosal diclofenac damage in rats, indicating that this coating strategy presents a potential use for the oral administration of drugs.

The efficient control of drug release from dexamethasone-loaded nanoparticle-coated microparticles was shown by an *in vitro* drug transport study across Caco-2 cell monolayers (Beck et al., 2007). In accordance with the *in vitro* drug release studies at pH 7.4, nanocapsule-coated microparticles presented lower permeability coefficient values across this human intestinal cell line compared to the free dexamethasone solution. Furthermore, cytotoxicity studies showed that the nanoparticle-coated microparticles were nontoxic to membranes of Caco-2 cells. This recent study reinforced that nanocapsule-coated microparticles represent a promising platform for the development of controlled oral NSAID delivery systems.

In a subsequent study, silica xerogel was used as the core material instead of Aerosil 200 to prepare nanocapsule-coated microagglomerates encapsulating sodium diclofenac as the hydrophilic drug model (da Fonseca et al., 2008). The new system showed gastroresistance, and it was efficient in reducing burst release and in sustaining the drug dissolution profile.

An indomethacin derivative, indomethacin ethyl ester, was used as a drug model in order to determine the nanocapsule architecture and drug release mechanism (Pohlmann et al., 2004). The strategy was based on the interfacial alkaline hydrolysis of indomethacin ester simulating a sink condition of release. Indomethacin ethyl ester loaded nanocapsules were prepared, varying the PCL concentration to evaluate the influence of the polymeric wall in drug release (Cruz et al., 2006b). The increase in polymer concentration enhanced the drug release sustained half-lives. The antiedematogenic activity of indomethacin ethyl ester loaded nanocapsules in rats showed a significant pharmacological effect in comparison with an indomethacin ethyl ester loaded nanoemulsion (Cruz et al., 2006a). The pharmacokinetics of the nanocapsule formulation were evaluated in rats (Cattani et al., 2008), demonstrating a fast *in vivo* release of ester from the nanocapsules and its conversion to indomethacin independently of the administration route (intravenous or oral). Hence, indomethacin was the entity responsible for the antiedematogenic activity after indomethacin ethyl ester-loaded nanocapsule dosing.

In addition to the oral route, polymeric nanocapsules were evaluated by alternative routes of administration. Indomethacin-loaded nanocapsules were administered rectally (Fawaz et al., 1996) and showed 100% bioavailability. Further, the nanocapsule formulation was effective in protecting the rectal mucosa against the local toxic effects of indomethacin.

Indomethacin *in vitro* corneal penetration was evaluated using nanocapsules as drug carriers (Calvo et al., 1996b). The transcorneal flux of the drug through isolated rabbit cornea showed a considerable increase of 4–5 times the penetration rate of the nanoencapsulated drug compared to commercial eyedrops. In addition, PCL nanocapsules containing indomethacin were coated with chitosan, poly(L-lysine), or both in order to combine the features of nanocapsules with the advantages of a cationic mucoadhesive coating (Calvo et al., 1997). Chitosan-coated nanocapsules provided an optimal corneal penetration of indomethacin and displayed good ocular tolerance.

NSAIDs, apart from their classical peripheral site of action, display a central analgesic effect (Matoga et al., 2002). In this sense, indomethacin-loaded nanocapsules were tested in glioma cell lines, because this drug presents an antiproliferative effect because of the arrest of cell cycle progression (Bernardi et al., 2008). Indomethacin-loaded nanocapsules were at least 2 times more cytotoxic than indomethacin in solution for glioma cell lines, indicating that nanoencapsulation improved the indomethacin effect without the undesirable side effects of conventional chemotherapy.

Other NSAIDs have been the focus of nanoencapsulation studies. The pharmacokinetics of ketoprofen in solution and in nanocapsules were evaluated in plasma and cerebrospinal fluid (Matoga et al., 2002). Nanocapsules were prepared with PLA and injected intraperitoneally in rats. The extent of absorption was similar for both solution and nanocapsule suspension, but with different plasma profiles. Ketoprofen administered in solution was rapidly absorbed, and for the nanoencapsulated formulation two peaks of concentrations were noted after administration. The second and lower peak was attributed to a progressive release of the drug from nanocapsules during the elimination phase.

Nimesulide is an NSAID that selectively inhibits cyclooxygenase-2. A semisolid topical formulation containing nimesulide-loaded nanocapsules was evaluated using Franz diffusion cells and a tape-stripping technique in order to investigate whether nanoencapsulation is able to modify the drug distribution in the different strata of full-thickness human skin (Alves et al., 2007). Gels containing nimesulide-loaded nanocapsules were able to promote drug penetration in the stratum corneum compared to a conventional formulation, and they allow higher penetration in the skin compared to the nimesulide-loaded nanospheres and nimesulide-loaded nanoemulsion.

3.5.2 ANTICANCER DRUGS

Conventional chemotherapeutics are often limited by the inadequate delivery of therapeutic concentrations to tumor target tissue. Therefore, it is important to develop new nanotechnologies for targeted delivery to tumors at both the cellular and tissue levels, thereby improving the therapeutic index of the carried anticancer molecules. Strategies for developing new efficient targeted nanocarriers of anticancer drugs may result from the combined knowledge of cancer physiopathology features, drug characteristics, and *in vivo* fate and behavior of nanocarriers (Couvreur and Vauthier, 2006).

Nanotechnological devices can be crucial approaches for the stability of the carried drug. Gemcitabine, for example, is an anticancer drug that suffers from rapid plasmatic metabolization. To overcome this problem, both physical and chemical protection of gemcitabine were developed (Stella et al., 2007). Lipophilic derivatives of gemcitabine [4-(N)-valeroylgemcitabine, 4-(N)-lauroylgemcitabine, and 4-(N)-stearoylgemcitabine] were synthesized and incorporated into poly[aminopoly(ethylene glycol)cyanoacrylate-co-hexadecyl cyanoacrylate] nanocapsules by nanoprecipitation of the copolymer. The cytotoxicity assay showed that 4-(N)-stearoylgemcitabine was more toxic than gemcitabine on two cell lines (human cervix carcinoma cell line KB3-1 cells, and human breast adenocarcinoma MCF-7 cells). Moreover, the incorporation of 4-(N)-stearoylgemcitabine in nanocapsules did not change its half-maximal inhibitory concentration values, showing that the cytotoxic activity of 4-(N)-stearoylgemcitabine was not modified by the nanoencapsulation.

Besides protecting the carried chemotherapeutic, nanoparticles can also be used to avoid cellular mechanisms such as multidrug resistance, because they allow entry into the cancer cells and act as an intracellular anticancer drug reservoir. Taking this into account, an approach was proposed using lipid nanocapsules (LNCs) consisting of Labrafac®, lecithin and PEG-660 hydroxystearate (Lamprecht and Benoit, 2006). Etoposide-loaded LNCs reverted multidrug resistance and reduced the cell growth in glioma cell lines, showing higher efficiency than the drug solution. The mechanism of action proposed for etoposide LNC was cell uptake followed by sustained drug release in combination with intracellular P-glycoprotein inhibition, ensuring a higher anticancer drug concentration inside the cancer cells.

The importance of developing new formulations for carrying anticancer drugs emerges from the failure of established therapies. Tamoxifen is an antiestrogenic molecule, which is the endocrine therapy of choice for the treatment of estrogen receptor (ER) positive breast cancer. Unfortunately, tamoxifen resistance often occurs, and in that case a blockade of human prostaglandin E_2 synthesis by aromatase inhibitors is sometimes of benefit. However, no response to aromatase inhibitors can occur. Thus, despite encouraging improvements in breast cancer treatment, the prognosis of metastatic disease is dramatic, stressing the need for new drugs and new administration strategies (Maillard et al., 2005).

With this in mind, biodegradable PLA and PEG–PLA nanocapsules containing either 4-hydroxytamoxifen (4-HT) or RU 58668 (RU), both antiestrogens, were prepared using the interfacial deposition of preformed polymer following solvent displacement (Renoir et al., 2006). The small sizes of the particles (233–246 nm) were compatible with their extravasations through the discontinuous endothelium of tumor vasculature. This allowed their accumulation in MCF-7 cell xenografts and led to prolonged exposure of the tumor to antiestrogen in athymic nude mice bearing established xenografts. In these tumors and in MCF-7/Ras xenografts, RU- and 4-HT-loaded nanocapsules inhibited tumor growth. In addition, RU-loaded nanocapsules promoted ER-α subtype loss in the tumor cells, according to the immunohistochemistry assay. The same *in vivo* tumor models had already been used by the authors to evaluate the antitumoral activity of RU encapsulated within PEG–PLA nanoparticles (Ameller et al., 2003). RU-loaded PEG–PLA nanocapsules were more active than free RU or RU entrapped with PEGylated nanospheres at an equivalent dose.

Approaches of coating nanocapsules with PEG have been widely described in anticancer therapy. At the tissue level, upon intravenous injection, colloids without this coating are opsonized and rapidly cleared from the bloodstream by the normal reticuloendothelial defense mechanism, irrespective of particle composition (Nguyen et al., 2008). Hence, in order to increase the circulation time in the bloodstream and to enhance the probability of the molecule to extravasate in tumor tissues, a great deal of work has been devoted to developing so-called stealth particles, which are "invisible" to macrophages (Figure 3.1). For this, PEG (MW = 1000 to 5000 Da) placed at the nanoparticle surfaces reduces the opsonization process, thus enhancing the blood circulation time (Couvreur and Vauthier, 2006). This approach is the so-called passive targeting approach and subsequent drug accumulation in the tumor interstitium is due to the known enhanced permeability and retention effect, as a result of the gaps of the discontinuous endothelium of cancer cells, which are richly vascularized (Lozano et al., 2008; Maillard et al., 2005).

Another system, based on the incorporation of RU in PLA nanocapsules and in PEG–PLA nanocapsules, potentiated the RU effect of increasing the population of MCF-7 cells in sub-G1 by blocking cell cycle progression (Maillard et al., 2005). In addition, the nanocapsules enhanced the activity of the free drug, inducing MCF-7 apoptosis and supporting the notion that the incorporation of RU within the nanocapsules increased their antitumor activity. However, RU-loaded nanocapsules were not able to inhibit the E_2-induced tumor growth rate through intravenous administration in nude mice bearing MCF-7 cell tumors.

Cases of intrinsic and acquired resistance, when the insensitive ER tumor cells point out the limitation of hydroxytamoxifen, highlight the need for new active molecules with broader therapeutic scopes. In this context, a potentially cytotoxic moiety (the organometallic group ferrocene) was added to the competitive bioligand hydroxytamoxifen scaffold (Nguyen et al., 2008). This attachment enhanced hydroxytamoxifen cytotoxicity, increasing the lipophilicity of the compound to facilitate its passage through the cellular membrane. Two organometallic triphenylethylene compounds, 1,1-di(4′-hydroxyphenyl)-2-ferrocenylbut-1-ene (Fc-diOH) and 1,2-di(4′-hydroxyphenyl)-1-[4″-(2″-ferrocenyl-2″-oxoethoxy)phenyl]but-1-ene (DFO), which present strong antiproliferative activity in breast cancer cells but are insoluble in biological fluids, were synthesized and incorporated in PEG–PLA nanocapsules. The influence of the encapsulated drugs on the cell cycle and apoptosis was studied by flow cytometry analyses. Whether free or encapsulated in the nanocapsules, Fc-diOH arrested the cell cycle in the S-phase. However, free DFO had no significant effect on MCF-7 cells, whereas nanoencapsulated DFO slightly increased the number of cells in the S-phase.

One group recently incorporated Fc-diOH in LNCs (Allard et al., 2008) with high drug loading capacity because of a larger oily core in their structure. The cytostatic activity of Fc-diOH was conserved after its encapsulation in LNCs, which were taken up by glioma cells. Fc-diOH-loaded LNCs were very effective on 9L-glioma cells, showing low toxicity levels when in contact with healthy cells. In addition, Fc-diOH LNC treatment significantly reduced both tumor mass and volume evolution after 9L-cell implantation into rats, indicating the *in vivo* efficacy of this kind of organometallic compound.

A nanosystem based on oil-encapsulating poly(ethylene oxide)-*b*-poly(propylene oxide)-*b*-poly(ethylene oxide) (PEO–PPO–PEO)/PEG nanocapsules conjugated with folic acid was synthesized using PEO–PPO–PEO and amine-functionalized six-arm branched PEG (MW = 20,000 Da; Bae et al., 2007). The shell encapsulating an oil phase was developed as a target-specific carrier for a water-insoluble drug, paclitaxel. Folate-mediated targeting significantly enhanced the cellular uptake and apoptotic effect on KB cells overexpressing folate receptors. The anticancer effect of paclitaxel-loaded nanocapsules was comparable to that of a clinically available formulation of paclitaxel (Taxol®). However, the cytotoxicity of Taxol was mainly caused by the toxicity of the Cremophor® EL vehicle rather than the drug. In contrast, the folate-conjugated nanocapsules exhibited far greater cytotoxicity against KB cells at a lower dosage than Taxol (Bae et al., 2007).

Besides paclitaxel, docetaxel belongs to the taxane class that is characterized by its hydrophobic character, resulting in the necessity of using solubilizers for its intravenous administration. These solubilizers, however, are responsible for severe side effects, which limit the amount of drug that can be safely administered. To overcome these problems, an alternative formulation based on chitosan colloidal carriers (nanocapsules) was prepared by the solvent displacement technique (Lozano et al., 2008). Chitosan nanocapsules were rapidly internalized by human tumor cells. Docetaxel-loaded chitosan carriers had an effect on cell proliferation, which was significantly greater than that of free docetaxel. Another work demonstrated that the encapsulation of docetaxel within LNCs dramatically increased the drug biological half-life, providing substantial accumulation at the tumoral site in mice bearing subcutaneously implanted C26 colon adenocarcinoma (Khalid et al., 2006).

Paclitaxel-loaded LNCs were used to elucidate whether LNCs were able to improve anticancer hydrophobic drug bioavailability and overcome multidrug resistance (Garcion et al., 2006). The results revealed an interaction between LNCs and efflux pumps, which results in an inhibition of multidrug resistance in rat glioma cells both in culture and in cell implants in animals. Paclitaxel-loaded LNCs were also more efficient than Taxol.

Nanocapsules with an external layer made up of PLA, PLA grafted with PEG (PLA–PEG), and PLA coated with poloxamer 188 (PLA–polox) have been proposed to incorporate photosensitizers for tumor tissue in photodynamic therapy (Bourdon et al., 2002). The cellular uptake, localization, and phototoxicity of *m*-tetra(hydroxyphenyl)chlorine (mTHPC) encapsulated in submicron colloidal carriers were studied in macrophage-like J774 cells and HT29 human adenocarcinoma cells. Cellular uptake by J774 was reduced with mTHPC encapsulated within surface-modified nanocapsules (PLA–PEG and PLA–polox) compared to naked PLA, indicating a possible limitation of the clearance of such carriers by the reticuloendothelial system. A specific punctate fluorescence pattern was revealed with PLA–PEG and PLA–polox nanocapsules, in contrast to a more diffuse distribution with solution, indicating that photodamage targeting could be different.

The same formulations were studied to evaluate the biodistribution of mTHPC in nude mice bearing HT29 human tumors (Bourdon et al., 2002). Compared to PLA nanocapsules, incorporation of mTHPC in surface-modified nanocapsules resulted in strong modifications of drug biodistribution and tumoral retention with an increase in drug levels. Reduced liver uptake was observed, indicating that surface-modified nanocapsules are effective in limiting reticuloendothelial system uptake and are potential carriers for enhancing the therapeutic ratio of lipophilic photosensitizers.

3.5.3 Hormones, Proteins, and Peptides

Hormones are substances that present low therapeutic doses and require chronic administration, so they are suitable molecules to nanoencapsulate. Melatonin was associated with Eudragit S100 nanocapsules in order to improve its protective properties against lipid peroxidation induced by ascorbyl free radicals using liposomes and microsomes as substrates (Schaffazick et al., 2005). The antioxidant capacity of melatonin was significantly increased when it was nanoencapsulated. The *in vivo* acute antioxidant capacity (lipid peroxidation, total antioxidant reactivity, and free radical levels in the brain and liver of mice) showed that lipid peroxidation significantly decreased in the cortex and in the hippocampus when melatonin-loaded nanocapsules were administered. In contrast, a melatonin aqueous solution did not exert any significant activity against lipid peroxidation (Schaffazick et al., 2008).

Peptide drugs are poorly absorbed after oral administration because of their susceptibility to enzymatic degradation and their low permeability across the intestinal epithelium. Keeping these important biopharmaceutical limitations in mind, many pharmaceutical scientists have taken the challenge of designing new delivery strategies intended to enhance the oral absorption of these macromolecules (Prego et al., 2006b).

A challenge in ocular drug delivery is to enhance the permeation of macromolecules across the cornea. PCL nanocapsules containing cyclosporin A, an immunosuppressive drug, were developed for ocular delivery in order to reduce its systemic side effects (Calvo et al., 1996b). The nanocapsules promoted the penetration of cyclosporin A to a very high degree.

Novel drug delivery systems for insulin administration, avoiding injectable formulations, have been the focus of much research for over 20 years. As a hypothesis, oral formulations containing insulin provide the peptide directly to the liver by hepatic portal circulation. This is a major advantage because this pathway mimics the physiological traffic of insulin when it is secreted by the pancreas of healthy individuals (Saffran et al., 1997). However, mucosal routes are extremely challenging for the administration of peptides and proteins because these generally hydrophilic macromolecules are unable to overcome mucosal barriers by themselves and are degraded before reaching the bloodstream (Couvreur and Vautier, 2006).

Insulin-loaded nanocapsules have been studied since 1988 (Damgé et al., 1988), when it was proved that PAC nanocapsules preserve the therapeutic effect of insulin in rats when administered orally, prolonging its effect. Insulin-loaded nanocapsules controlled glycemia for at least 13 days in streptozotocin-induced diabetic rats (Michel et al., 1991). *In vitro* nanocapsules protect insulin against proteolysis from pepsin, chymotrypsin, and trypsin (Michel et al., 1991). In addition, nanocapsules administered orally induce several beneficial persistent effects in both normal and diabetic dogs (Damgé et al., 1995). This formulation was tested as a prophylactic strategy to prevent diabetes in nonobese diabetic mice via oral administration. In humans, this form of prophylactic insulin administration was claimed to be less constraining than insulin injections (Saï et al., 1996). The early administration of insulin nanocapsules reduced diabetes and insulitis in the nonobese diabetic mouse model that mimics human Type 1 diabetes.

More recently, PIBC nanocapsules containing Texas Red® labeled and gold labeled insulin were studied (Pinto-Alphandary et al., 2003). Insulin was located inside nanocapsules that were observed on both sides of the jejunum. In the lumen, the environment was suitable to protect insulin from degradation. Nanocapsules were absorbed by portions of the M-cell-free epithelium and were highly degraded in M-cell-containing epithelium.

After oral administration, insulin from PIBC nanocapsules was very quickly but heterogeneously absorbed. Furthermore, nanocapsules allowed the delivery of noticeable levels of insulin into the blood of diabetic rats (Cournarie et al., 2002). Nevertheless, high levels of plasma insulin were necessary to produce efficient hypoglycemic activity, and they were not reached after oral administration of nanocapsules.

The activity of insulin-loaded PAC nanocapsules was evaluated after subcutaneous administration, showing a reduction in blood glucose in diabetic rats, which was delayed by the nanoencapsulation (Watnasirichaikul et al., 2002). The formulation of insulin-loaded nanocapsules dispersed in a water-in-oil microemulsion showed a significant increase in the oral bioavailability of insulin in diabetic rats (Watnasirichaikul et al., 2002).

Salmon calcitonin (model peptide) was encapsulated in chitosan–PEG nanocapsules prepared by the solvent displacement technique (Prego et al., 2006a). The PEGylation of the chitosan coating facilitated the retention of peptide in the nanocapsules. The hypocalcemic effect after the oral administration of nanocapsules was significantly higher than the nanoemulsion and free peptide aqueous solution. The permeation study using Caco-2 cells indicated that nanocapsules penetrated the tissue by a transcellular pathway and were randomly distributed. The mucoadhesive character of chitosan nanocapsules was the determinant for interaction with intestinal mucosae, facilitating the intestinal absorption of salmon calcitonin (Prego et al., 2006a).

3.5.4 ANTIFUNGALS, ANTIBIOTICS, AND ANTIPARASITICS

The encapsulation of antifungal agents in nanoparticulate carriers was proposed with the objective of modifying the pharmacokinetics of drugs, resulting in more efficient treatments with fewer side effects (de Assis et al., 2008). In this way, fluconazole was radiolabeled and encapsulated in PLA–PEG nanocapsules, demonstrating a fast release of radioactivity. The PEG layer around the nanocapsules probably reduced the amount of drug released by impairing protein binding at the nanocapsule surface. Another antifungal, griseofulvin, is rarely used because of its high lipophilicity, which makes both formulation and delivery difficult (Zili et al., 2005). Griseofulvin was very rapidly released from PCL nanocapsules, which was probably due to the dissolution of the polymer by the oily phase (benzyl benzoate). Griseofulvin nanocapsules showed a higher dissolution rate, which indicated that lower doses of this molecule can be used for oral applications, thus reducing its side effects (Zili et al., 2005).

Several antiseptics can be incorporated in hand-washing agents. However, their frequent use induces contact dermatitis and allergies. To improve and sustain antimicrobial activity, chlorhexidine-loaded nanocapsules were tested *in vitro* against some microorganisms (Lboutounne et al., 2002). The activity of the nanocapsules was also tested *ex vivo* in porcine ear skin. The encapsulation maintained the chlorhexidine effect, sustaining the *ex vivo* topical antimicrobial activity against *Staphylococcus epidermidis*. Chlorhexidine was incorporated in a hydrophilic gel and tested as a hand-rub gel against resident skin flora. This product had immediate antibacterial effect, explained by the rapid desorption of chlorhexidine from the nanocapsule wall, subsequent diffusion within bacteria, and sustained antibacterial effect. This immediate effect was a consequence of the slow release of chlorhexidine from the nanocapsule core against further bacterial colonization (Nhung et al., 2007).

Injectable formulations of benzathine penicillin G were developed in the 1950s and consisted of intramuscular depot formulations because of the low solubility of the drug. New formulations (nanoemulsion and PLGA nanocapsules containing benzathine penicillin G) were developed that were stable over 120 days when stored at 4°C, exhibiting *in vitro* antimicrobial activity against *S. pyogenes* (Santos-Magalhães et al., 2000).

The emergence of chloroquine resistance in *Plasmodium falciparum* has increased the search for new antimalarial drugs, such as halofantrine hydrochloride, one of the most active antimalarial drugs against *P. falciparum in vitro* but presenting serious cardiotoxicity (Mosqueira et al., 2004). Considering that the cardiac epithelium is continuous, stealth nanocapsules containing halofantrine have been developed to reduce drug side effects and to increase drug circulation (Mosqueira et al., 2004). No signs of toxicity or abnormal behavior were observed after intravenous administration of halofantrine-loaded nanocapsules in mice, and the maximum tolerated dose was higher than that of the free drug in solution. Poloxamer-coated nanocapsules provided a fast effect, whereas PEG-coated nanocapsules provided a more sustained effect because of their long blood circulation. PEG-coated nanocapsules provided a reduced halofantrine cardiotoxic profile when compared to the free drug, showing that drug distribution seemed to be modified by these nanocarriers (Leite et al., 2007).

The potential of colloidal drug carriers in the targeted and controlled delivery of antileishmanial compounds has also received much attention. Leishmania are obligate intracellular parasites in mammals that live exclusively in the cells of the mononuclear phagocyte system (Cauchetier et al., 2003a). Taking into consideration that conventional nanocapsules undergo phagocytosis by macrophages after opsonization (Stolnik et al., 1995), atovaquone-loaded nanocapsules were prepared by the interfacial deposition of preformed polymer using different polyesters (Cauchetier et al., 2003b). Atovaquone was released from PLA nanocapsules by diffusional transport of the drug through the polymer, associated with the first stage of polymer degradation. *In vivo,* nanocapsules were significantly more effective than the free drug in the treatment of mice with visceral leishmaniasis. The dose–response data indicated that livers were cleared of parasites if the nanocapsule preparation was administered in three doses, whereas the maximum suppression possible with the free drug is about 61%, whatever the dose (Cauchetier et al., 2003a).

3.5.5 OTHER DRUGS

Nanocapsules can be used to increase the accessibility of drugs to the receptors localized in specific areas. They can serve as vehicles for use in the treatment of ophthalmic pathologies, because increased corneal penetration and prolonged therapeutic response have been achieved for some drugs (Losa et al., 1993). Several nanocapsule formulations containing metipranolol were tested for drug release and the ability to prevent conjunctival absorption. The nanoencapsulation was capable of reducing bradycardia, showing lower systemic toxicity. Another drug used as eyedrops, pilocarpine, was encapsulated in PIBC nanocapsules incorporated in a Pluronic F127 gel (Desai and Blanchard, 2000). The formulation increased the contact time of the drug with the absorbing tissue in the eye and improved ocular bioavailability.

Many authors have focused on different drugs used to control drug release and improve drug bioavailability and stability (Calvo et al., 1996c; Fresta et al., 1996; Ourique et al., 2008). Antiepileptic drugs (5,5-diphenyl hydrotoin, carbamazepine, and ethosuximide) were encapsulated in poly(ethyl-2-cyanoacrylate) nanocapsules and zero order release was achieved, providing controlled drug release (Fresta et al., 1996).

Idebenone is a lipophilic benzoquinone electron carrier, which behaves as an antioxidant free radical scavenging molecule, that is active in central nervous system disorders (Palumbo et al., 2002). Idebenone-loaded poly(ethyl cyanoacrylate) nanocapsules showed a greater effectiveness for the antioxidant effect *in vitro,* under different stress conditions, toward human fibroblasts than the free drug.

Tretinoin is the active form of a metabolic product of vitamin A, which is indicated in the topical treatment of different skin diseases such as acne vulgaris, ichtyose, and psoriasis. However, this drug presents some drawbacks such as poor solubility, high chemical instability and photoinstability (which gives inactive metabolites), and irritation of the treated area. Nanocapsules containing tretinoin were developed, which were aimed at reducing drug side effects and drug photoinstability (Ourique et al., 2008). Tretinoin-loaded nanocapsules improved tretinoin photostability, independently of the type of oil core used.

Spironolactone is a specific steroid antagonist that is used as a potassium-sparing diuretic in premature infants to reduce lung congestion, but no liquid formulations are available because of its low solubility (Blouza et al., 2006). Different parameters were tested in order to obtain an optimized formulation of nanocapsules. The release of spironolactone from nanocapsules was rapid and complete in a simulated gastric fluid. Therefore, recourse to spironolactone nanoencapsulation should enhance its oral bioavailability and probably its efficiency. Concerning drug therapies based on nanocapsules, Table 3.2 gives an overview of the systems carrying different molecules and their main results.

TABLE 3.2

Oily-Core Polymeric Nanocapsules Proposed for Therapeutic Goals

Therapeutic Class	Drug	Nanocapsule Composition	Main Result	Reference
Antiinflammatory	Indomethacin and diclofenac	PLA and Miglyol 810	Reduction of gastrointestinal toxicity	Guterres et al., 1995a,b
	Indomethacin	PIBC and benzyl benzoate	10-fold increase in antiinflammatory activity	Gursoy et al., 1989
		PLA, PIBC, and benzyl benzoate	Protective effect on jejunal tissue	Ammoury et al., 1991
		PLA and benzyl benzoate	Reduction of ulceration in intestine and rectum	Ammoury et al., 1993; Fawaz et al., 1996
		PCL and Miglyol 840	Increases corneal penetration rate 4–5 times	Calvo et al., 1996c
		PCL coated with chitosan or poly(L-lysine) and Miglyol 840	Optimal corneal penetration and good ocular tolerance	Calvo et al., 1997
		PCL and CCT	Twice as cytotoxic for glioma cell lines	Bernardi et al., 2008
	Indomethacin ethyl ester	PCL and Miglyol 810	Increase in antiedemadogenic activity	Cruz et al., 2006b
	Ketoprofen	PLA and benzyl benzoate	Progressive release of the drug	Matoga et al., 2002
	Nimesulide	PCL and CCT	Higher penetration in the skin	Alves et al., 2007
Diuretic	Spironolactone	PCL and Labrafac hydro	Fast and complete release	Blouza et al., 2006
Hormone	Melatonin	PCL and Miglyol 810	Protective properties against lipid peroxidation induced by ascorbyl free radical in cortex and in hippocampus	Schaffazick et al., 2005, 2008
	Insulin	PAC and Miglyol	Preserves therapeutic effect orally in rats and dogs	Damgé et al., 1988, 1995
		PIBC and Miglyol	Glycemia for at least 13 days in rats	Michel et al., 1991
		PIBC and Miglyol	Reduced diabetes and insulitis in mice	Saï et al., 1996
		PIBC and Miglyol 812N	Nanocapsules at both sides of the jejunum	Couarnie et al., 2002; Pinto-Alphandary et al., 2003
		PAC and Capmul® MCM		Watnasirichaikul et al., 2002
Peptide	Cyclosporine A	PCL and Miglyol 840	Promotes penetration of cyclosporine A across cornea	Calvo et al., 1996b
	Salmon calcitonin	Chitosan–PEG and Miglyol 812	Higher hypocalcemic effect and penetration by transcellular pathway	Prego et al., 2006a,b
Antifungal	Fluconazole	PLA–PEG and CCT	Controlled release	de Assis et al., 2008
	Griseofulvin	PCL and benzyl benzoate	Higher dissolution rate	Zili et al., 2005
Antibiotics	Chlorhexidine	PCL and Labrafac hydrophile WL 1219	Sustained effect against *Staphylococcus epidermidis*	Lboutounne et al., 2002; Nhung et al., 2007

continued

TABLE 3.2 (continued)
Oily-Core Polymeric Nanocapsules Proposed for Therapeutic Goals

Therapeutic Class	Drug	Nanocapsule Composition	Main Result	Reference
	Benzathine penicillin G	PLGA and sunflower oil and benzyl benzoate	Stability over 120 days and activity against *Streptococcus pyogenes*	Santos-Magalhães et al., 2000
Antimalarial	Halofantrine	PLA, PLA–PEG, and Miglyol 810	No signs of toxicity after IV administration	Mosqueira et al., 2004
		PCL and Miglyol 810	Reduced halofantrine cardiotoxic	Leite et al., 2007
Antileishmanial	Atovaquone	PLA, PCL, PLGA, and benzyl benzoate	More effective in visceral leishmaniasis in rats	Cauchetier et al., 2003a
Antiepileptic	5,5-Diphenyl hydrotoin, carbamazepine, and ethosuximide	PEC and Miglyol 812	Controlled drug release	Fresta et al., 1996
Antioxidant	Idebenone	PEC and Miglyol 812	Greater *in vitro* antioxidant activity	Palumbo et al., 2002
Vitamin	Tretinoin	PCL and Miglyol 810	Improved tretinoin photostability	Ourique et al., 2008
Anticancer drug	Gencitabine lipofilic derivate	Poly(H$_2$NPEGCA-*co*-HDCA) and Miglyol 812N	Higher cytoxicity than gencitabine, not modified by nanocapsule incorporation	Stella et al., 2007
	Etoposide	PEG–HS and Labrafac	Reverts multidrug resistance, reduce tumor growth	Lamprecht and Benoit, 2006
	4-HT and RU	PLA, PLA–PEG, and Miglyol 810	Block cell cycle progression in tumor cells, increase antitumor activity, promote loss ERα in tumor cells	Ameller et al., 2003; Maillard et al., 2005; Renoir et al., 2006
	Organometallic tamoxifen derivatives	PLA–PEG and Miglyol 810	Increase number of cells in S-phase	Nguyen et al., 2008
	Organometallic tamoxifen derivative	PEG–HS and Labrafac	Reduce tumor mass *in vivo*	Allard et al., 2008
	Paclitaxel	PEO-PPO–PEO/PEG folic acid conjugated and Lipiodol®	Enhance cellular uptake and apoptotic effect on cells overexpressing folate receptors	Bae et al., 2007
	Docetaxel	Chitosan and Miglyol 812	Rapidly internalized by human tumor cells, promotes reduction of cell proliferation	Lozano et al., 2008
	Docetaxel	PEG–HS and tricaprylin	Increase drug biological half-life and accumulate at tumoral site *in vivo*	Khalid et al., 2006
	Paclitaxel	PEG–HS and Labrafac	Inhibition of multidrug resistance *in vivo* and *in vitro*	Garcion et al., 2006
Photosensitizer	Tetra(hydroxyphenyl)chlorin	PLA, PLA–PEG, PLA–polox, and Miglyol 812N	Reduce cellular uptake and clearance with surface-modified nanocapsules	Bourdon et al., 2000, 2002

Note: HS—hydroxystearate; PEC—poly(ethyl cyanoacrylate); poly(H$_2$NPEGCA-*co*-HDCA)—poly[aminopoly(ethylene glycol)cyanoacrylate-*co*-hexadecyl cyanoacrylate].

CONCLUSION

Investigations concerning polymeric nanocapsules are generally a multidisciplinary study involving engineering, chemistry, pharmacy, and pharmacology. Challenges have been presented during the past 40 years about the physicochemical evaluation of these formulations. The complexity of these systems is a consequence of both submicron-size and soft-matter properties. However, successful applications using complementary techniques to elucidate the nanocapsule supramolecular structure have been reported since the 1970s. The current knowledge concerning the supramolecular organization of nanocapsules allowed the tuning of their characteristics for specificities of use, such as the administration route, drug release time, and drug target site in the organism. Major reports in the literature about the therapeutic applications of nanocapsules describe the delivery of antiinflammatory drugs, chemotherapeutic agents, hormones, peptides, and other agents, with increases of drug efficacy and reduction of drug toxicity. Considering this, nanocapsule science and technology open new and interesting perspectives for human health.

ABBREVIATIONS

CCT	Caprylic/capric triglyceride
DFO	1,2-Di(4′-hydroxyphenyl)-1-[4″-(2″-ferrocenyl-2″-oxoethoxy)phenyl]but-1-ene
DLS	Dynamic light scattering
E_2	Type of human prostaglandin
ER	Estrogen receptor
Fc-diOH	1,1-di(4′-hydroxyphenyl)-2-ferrocenylbut-1-ene
4-HT	4-Hydroxytamoxifen
KB3-1	Human cervix carcinoma cell line
LNC	Lipid nanocapsules
MCF-7	Human breast adenocarcinoma cell line
mTHPC	Meta-*tetra*(hydroxyphenyl)chlorine
MW	Molecular weight
NSAID	Nonsteroidal antiinflammatory drug
PAC	Poly(alkyl cyanoacrylate)
PCL	Poly(ε-caprolactone)
PEG	Poly(ethylene glycol)
PEO	Poly(ethylene oxide)
PFOB	Perfluorooctyl bromide
PIBC	Poly(isobutyl cyanoacrylate)
PI	Polydispersity index
PLA	Polylactide
PLGA	Poly(lactide-*co*-glycolide)
PPO	Poly(propylene oxide)
RU	RU 58668 steroidal antiestrogen
SEM	Scanning electron microscopy
TEM	Transmission electron microscopy

TRADEMARKS

Aerosil® (Evonik Degussa GmbH, Frankfurt am Main, Germany)
Capmul® (Abitec Corporation, Columbus, Ohio, USA)
Cremophor® (BASF Corporation, Florham Park, New Jersey, USA)
Eudragit® (Evonik Degussa GmbH, Frankfurt am Main, Germany)
Epikuron® (Degussa GmbH, Hamburg, Germany)

Labrafac® (Gattefossé, Saint-Priest, France)
Lipiodol® (Guerbet S.A., Roissy Charles-de-Gaulle Cedex, France)
Miglyol® (SASOL Germany GmbH, Witten, Germany)
Pluronic® (BASF Corporation, Florham Park, New Jersey, USA)
Taxol® (Bristol-Myers Squibb, New York, USA)
Texas Red® (Molecular Probes, Leiden, The Netherlands)
Tween® (ICI Americas Inc., London, England)

REFERENCES

Aboubakar, M., F. Puisieux, P. Couvreur, M. Deyme, and C. Vauthier, 1999. Study of the mechanism of insulin encapsulation in poly(isobutylcyanoacrylate) nanocapsules obtained by interfacial polymerization. *Journal of Biomedical Materials Research* 47 (4): 568–576.

Allard, E., C. Passirani, E. Garcion, P. Pigeon, A. Vessieres, G. Jaouen, and J. P. Benoit, 2008. Lipid nanocapsules loaded with an organometallic tamoxifen derivative as a novel drug-carrier system for experimental malignant gliomas. *Journal of Controlled Release* 130 (2): 146–153.

Alvarez-Román, R., G. Barre, R. H. Guy, and H. Fessi, 2001. Biodegradable polymer nanocapsules containing a sunscreen agent: Preparation and photoprotection. *European Journal of Pharmaceutics and Biopharmaceutics* 52 (2): 191–195.

Alves, M. P., A. L. Scarrone, M. Santos, A. R. Pohlmann, and S. S. Guterres, 2007. Human skin penetration and distribution of nimesulide from hydrophilic gels containing nanocarriers. *International Journal of Pharmaceutics* 341 (1–2): 215–220.

Ameller, T., W. Marsaud, P. Legrand, R. Gref, and J. M. Renoir, 2003. *In vitro* and *in vivo* biologic evaluation of long-circulating biodegradable drug carriers loaded with the pure antiestrogen RU 58668. *International Journal of Cancer* 106 (3): 446–454.

Ammoury, N., H. Fessi, J. P. Devissaguet, M. Dubrasquet, and S. Benita, 1991. Jejunal-absorption, pharmacological activity, and pharmacokinetic evaluation of indomethacin-loaded poly(D,L-lactide) and poly(isobutylcyanoacrylate) nanocapsules in rats. *Pharmaceutical Research* 8 (1): 101–105.

Ammoury, N., M. Dubrasquet, H. Fessi, J. Ph. Devissaguet, F. Puisieux, and S. Benita, 1993. Indomethacin-loaded poly(D,L-lactide) nanocapsules: Protection from gastrointestinal ulcerations and anti-inflammatory activity evaluation in rats. *Clinical Materials* 13 (1–4): 121–130.

Andreopoulos, A. G., E. Hatzi, and M. Doxastakis, 1999. Synthesis and properties of poly(lactic acid). *Journal of Materials Science—Materials in Medicine* 10 (1): 29–33.

Anton, N., J. P. Benoit, and P. Saulnier, 2008. Design and production of nanoparticles formulated from nano-emulsion templates—A review. *Journal of Controlled Release* 128 (3): 185–199.

Bae, K. H., Y. Lee, and T. G. Park, 2007. Oil-encapsulating PEO–PPO–PEO/PEG shell cross-linked nanocapsules for target-specific delivery of paclitaxel. *Biomacromolecules* 8 (2): 650–656.

Beck, R. C. R., A. R. Pohlmann, E. V. Benvenutti, T. D. Costa, and S. S. Guterres, 2005. Nanostructure-coated diclofenac-loaded microparticles: Preparation, morphological characterization, *in vitro* release and *in vivo* gastrointestinal tolerance. *Journal of the Brazilian Chemical Society* 16 (6A): 1233–1240.

Beck, R. C. R., S. E. Haas, S. S. Guterres, M. I. Re, E. V. Benvenutti, and A. R. Pohlmann, 2006. Nanoparticle-coated organic–inorganic microparticles: Experimental design and gastrointestinal tolerance evaluation. *Quimica Nova* 29 (5): 990–996.

Beck, R. C. R., A. R. Pohlmann, and S. S. Guterres, 2004. Nanoparticle-coated microparticles: Preparation and characterization. *Journal of Microencapsulation* 21 (5): 499–512.

Beck, R. C. R., A. R. Pohlmann, C. Hoffmeister, M. R. Gallas, E. Collnot, U. F. Schaefer, S. S. Guterres, and C. M. Lehr, 2007. Dexamethasone-loaded nanoparticle-coated microparticles: Correlation between *in vitro* drug release and drug transport across Caco-2 cell monolayers. *European Journal of Pharmaceutics and Biopharmaceutics* 67 (1): 18–30.

Behr, J. P., B. Demeneix, J. P. Loeffler, and J. P. Mutul, 1989. Efficient gene-transfer into mammalian primary endocrine-cells with lipopolyamine-coated DNA. *Proceedings of the National Academy of Sciences of the United States of America* 86 (18): 6982–6986.

Bernardi, A., R. L. Frozza, E. Jager, F. Figueiro, L. Bavaresco, C. Salbego, A. R. Pohlmann, S. S. Guterres, and A. M. O. Battastini, 2008. Selective cytotoxicity of indomethacin and indomethacin ethyl ester-loaded nanocapsules against glioma cell lines: An *in vitro* study. *European Journal of Pharmacology* 586 (1–3): 24–34.

Blouza, I. L., C. Charcosset, S. Sfar, and H. Fessi, 2006. Preparation and characterization of spironolactone-loaded nanocapsules for paediatric use. *International Journal of Pharmaceutics* 325 (1–2): 124–131.

Bouchemal, K., S. Briançon, E. Perrier, and H. Fessi, 2004. Nano-emulsion formulation using spontaneous emulsification: Solvent, oil and surfactant optimisation. *International Journal of Pharmaceutics* 280 (1–2): 241–251.

Bourdon, O., I. Laville, D. Carrez, A. Croisy, P. Fedel, A. Kasselouri, P. Prognon, P. Legrand, and J. Blais, 2002. Biodistribution of *meta-tetra*(hydroxyphenyl)chlorin incorporated into surface-modified nanocapsules in tumor-bearing mice. *Photochemical & Photobiological Sciences* 1 (9): 709–714.

Bourdon, O., V. Mosqueira, P. Legrand, and J. Blais, 2000. A comparative study of the cellular uptake, localization and phototoxicity of *meta-tetra*(hydroxyphenyl) chlorin encapsulated in surface-modified submicronic oil/water carriers in HT29 tumor cells. *Journal of Photochemistry and Photobiology B: Biology* 55 (2–3): 164–171.

Calvo, P., M. J. Alonso, J. L. VilaJato, and J. R. Robinson, 1996c. Improved ocular bioavailability of indomethacin by novel ocular drug carriers. *Journal of Pharmacy and Pharmacology* 48 (11): 1147–1152.

Calvo, P., A. Sanchez, J. Martinez, M. I. Lopez, M. Calonge, J. C. Pastor, and M. J. Alonso, 1996b. Polyester nanocapsules as new topical ocular delivery systems for cyclosporin A. *Pharmaceutical Research* 13 (2): 311–315.

Calvo, P., J. L. VilaJato, and M. J. Alonso, 1996a. Comparative *in vitro* evaluation of several colloidal systems, nanoparticles, nanocapsules, and nanoemulsions, as ocular drug carriers. *Journal of Pharmaceutical Sciences* 85 (5): 530–536.

Calvo, P., J. P. VilaJato, and M. J. Alonso, 1997. Evaluation of cationic polymer-coated nanocapsules as ocular drug carriers. *International Journal of Pharmaceutics* 153 (1): 41–50.

Cattani, V. B., A. R. Pohlmann, and T. D. Costa, 2008. Pharmacokinetic evaluation of indomethacin ethyl ester-loaded nanoencapsules. *International Journal of Pharmaceutics* 363 (1–2): 214–216.

Cauchetier, E., M. Deniau, H. Fessi, A. Astier, and M. Paul, 2003b. Atovaquone-loaded nanocapsules: Influence of the nature of the polymer on their *in vitro* characteristics. *International Journal of Pharmaceutics* 250 (1): 273–281.

Cauchetier, E., M. Paul, D. Rlvollett, H. Fessit, A. Astier, and M. Deniau, 2003a. Therapeutic evaluation of free and nanocapsule-encapsulated atovaquone in the treatment of murine visceral leishmaniasis. *Annals of Tropical Medicine and Parasitology* 97 (3): 259–268.

Cha, Y. and C. G. Pitt, 1988. A one-week subdermal delivery system for l-methadone based on biodegradable microcapsules. *Journal of Controlled Release* 7 (1): 69–78.

Choi, S. W., H. Y. Kwon, W. S. Kim, and J. H. Kim, 2002. Thermodynamic parameters on poly(D,L-lactide-*co*-glycolide) particle size in emulsification–diffusion process. *Colloids and Surfaces A: Physicochemical and Engineering Aspects* 201 (1–3): 283–289.

Cournarie, F., D. Auchere, D. Chevenne, B. Lacour, M. Seiller, and C. Vauthier, 2002. Absorption and efficiency of insulin after oral administration of insulin-loaded nanocapsules in diabetic rats. *International Journal of Pharmaceutics* 242 (1–2): 325–328.

Couvreur, P., G. Barratt, E. Fattal, P. Legrand, and C. Vauthier, 2002. Nanocapsule technology: A review. *Critical Reviews in Therapeutic Drug Carrier Systems* 19 (2): 99–134.

Couvreur, P., L. Roblot-Treupel, M. F. Poupon, F. Brasseur, and F. Puisieux, 1990. Nanoparticles as micro-carriers for anticancer drugs. *Advanced Drug Delivery Reviews* 5 (3): 209–230.

Couvreur, P. and C. Vauthier, 2006. Nanotechnology: Intelligent design to treat complex disease. *Pharmaceutical Research* 23 (7): 1417–1450.

Cruz, L., S. R. Schaffazick, T. Dalla Costa, L. U. Soares, G. Mezzalira, N. P. da Silveira, E. E. S. Schapoval, A. R. Pohlmann, and S. S. Guterres, 2006a. Physico-chemical characterization and *in vivo* evaluation of indomethacin ethyl ester-loaded nanocapsules by PCS, TEM, SAXS, interfacial alkaline hydrolysis and antiedematogenic activity. *Journal of Nanoscience and Nanotechnology* 6 (9–10): 3154–3162.

Cruz, L., L. U. Soares, T. D. Costa, G. Mezzalira, N. P. da Silveira, S. S. Guterres, and A. R. Pohlmann, 2006b. Diffusion and mathematical modeling of release profiles from nanocarriers. *International Journal of Pharmaceutics* 313 (1–2): 198–205.

da Fonseca, L. S., R. P. Silveira, A. M. Deboni, E. V. Benvenutti, T. M. H. Costa, S. S. Guterres, and A. R. Pohlmann, 2008. Nanocapsule–xerogel microparticles containing sodium diclofenac: A new strategy to control the release of drugs. *International Journal of Pharmaceutics* 358 (1–2): 292–295.

Damgé, C., D. Hillairebuys, R. Puech, A. Hoeltzel, C. Michel, and G. Ribes, 1995. Effects of orally-administered insulin nanocapsules in normal and diabetic dogs. *Diabetes Nutrition & Metabolism* 8 (1): 3–9.

Damgé, C., C. Michel, M. Aprahamian, and P. Couvreur, 1988. New approach for oral-administration of insulin with polyalkylcyanoacrylate nanocapsules as drug carrier. *Diabetes* 37 (2): 246–251.

Date, A. A., M. D. Joshi, and V. B. Patravale, 2007. Parasitic diseases: Liposomes and polymeric nanoparticles versus lipid nanoparticles. *Advanced Drug Delivery Reviews* 59: 505–521.

de Assis, D. N., V. C. F. Mosqueira, J. M. C. Vilela, M. S. Andrade, and V. N. Cardoso, 2008. Release profiles and morphological characterization by atomic force microscopy and photon correlation spectroscopy of (99 m) technetium–fluconazole nanocapsules. *International Journal of Pharmaceutics* 349 (1–2): 152–160.

Delgado, A. V., E. Gonzalez-Caballero, R. J. Hunter, L. K. Koopal, and J. Lyklema, 2005. Measurement and interpretation of electrokinetic phenomena (IUPAC technical report). *Pure and Applied Chemistry* 77 (10): 1753–1805.

Desai, S. D. and J. Blanchard, 2000. Pluronic® F127-based ocular delivery system containing biodegradable polyisobutylcyanoacrylate nanocapsules of pilocarpine. *Drug Delivery* 7 (4): 201–207.

Dodou, D., P. Breedveld, and P. A. Wieringa, 2005. Mucoadhesives in the gastrointestinal tract: Revisiting the literature for novel applications. *European Journal of Pharmaceutics and Biopharmaceutics* 60 (1): 1–16.

Daoud-Mahammed, S., P. Couvreur, and R. Gref, 2007. Novel self-assembling nanogels: Stability and lyophilisation studies. *International Journal of Pharmaceutics* 332 (1–2): 185–191.

Fawaz, F., F. Bonini, M. Guyot, A. M. Lagueny, H. Fessi, and J. P. Devissaguet, 1996. Disposition and protective effect against irritation after intravenous and rectal administration of indomethacin loaded nanocapsules to rabbits. *International Journal of Pharmaceutics* 133 (1–2): 107–115.

Felgner, P. L., T. R. Gadek, M. Holm, R. Roman, H. W. Chan, M. Wenz, J. P. Northrop, G. M. Ringold, and M. Danielsen, 1987. Lipofection—A highly efficient, lipid-mediated DNA-transfection procedure. *Proceedings of the National Academy of Sciences of the United States of America* 84 (21): 7413–7417.

Fessi, H., F. Puisieux, J. P. Devissaguet, N. Ammoury, and S. Benita, 1989. Nanocapsule formation by interfacial polymer deposition following solvent displacement. *International Journal of Pharmaceutics* 55 (1): R1–R4.

Frank, M. M. and L. F. Fries, 1991. The role of complement in inflammation and phagocytosis. *Immunology Today* 12 (9): 322–326.

Fresta, M., G. Cavallaro, G. Giammona, E. Wehrli, and G. Puglisi, 1996. Preparation and characterization of polyethyl-2-cyanoacrylate nanocapsules containing antiepileptic drugs. *Biomaterials* 17 (8): 751–758.

Friedrich, R. B., M. C. Fontana, R. C. R. Beck, A. R. Pohlmann, and S. S. Guterres, 2008. Development and physicochemical characterization of dexamethasone-loaded polymeric nanocapsule suspensions. *Química Nova* 31 (5): 1131–1136.

Gallardo, M., G. Couarraze, B. Denizot, L. Treupel, P. Couvreur, and F. Puisieux, 1993. Study of the mechanisms of formation of nanoparticles and nanocapsules of polyisobutyl-2-cyanoacrylate. *International Journal of Pharmaceutics* 100 (1–3): 55–64.

Ganachaud, F. and J. L. Katz, 2005. Nanoparticles and nanocapsules created using the Ouzo effect: Spontaneous emulsification as an alternative to ultrasonic and high-shear devices. *Chemphyschem* 6 (2): 209–216.

Garcion, E., A. Lamprecht, B. Heurtault, A. Paillard, A. Aubert-Pouessel, B. Denizot, P. Menei, and J. P. Benoit, 2006. A new generation of anticancer, drug-loaded, colloidal vectors reverses multidrug resistance in glioma and reduces tumor progression in rats. *Molecular Cancer Therapeutics* 5 (7): 1710–1722.

Gaumet, M., A. Vargas, R. Gurny, and F. Delie, 2008. Nanoparticles for drug delivery: The need for precision in reporting particle size parameters. *European Journal of Pharmaceutics and Biopharmaceutics* 69 (1): 1–9.

Gref, R., Y. Minamitake, M. T. Peracchia, V. Trubetskoy, V. Torchilin, and R. Langer, 1994. Biodegradable long-circulating polymeric nanospheres. *Science* 263 (5153): 1600–1603.

Guinebretière, S., S. Briançon, H. Fessi, V. S. Teodorescu, and M. G. Blanchin, 2002. Nanocapsules of biodegradable polymers: Preparation and characterization by direct high resolution electron microscopy. *Materials Science and Engineering C: Biomimetic and Supramolecular Systems* 21 (1–2): 137–142.

Gursoy, A., L. Eroglu, S. Ulutin, M. Tasyurek, H. Fessi, F. Puisieux, and J. P. Devissaguet, 1989. Evaluation of indomethacin nanocapsules for their physical stability and inhibitory activity on inflammation and platelet-aggregation. *International Journal of Pharmaceutics* 52 (2): 101–108.

Guterres, S. S., H. Fessi, G. Barratt, J. P. Devissaguet, and F. Puisieux, 1995a. Poly(D,L-lactide) nanocapsules containing diclofenac. 1. Formulation and stability study. *International Journal of Pharmaceutics* 113 (1): 57–63.

Guterres, S. S., H. Fessi, G. Barratt, F. Puisieux, and J. P. Devissaguet, 1995b. Poly(D,L-lactide) nanocapsules containing nonsteroidal antiinflammatory drugs—Gastrointestinal tolerance following intravenous and oral-administration. *Pharmaceutical Research* 12 (10): 1545–1547.

Guterres, S. S., C. R. Muller, C. B. Michalowski, A. R. Pohlmann, and T. Dalla Costa, 2001. Gastro-intestinal tolerance following oral administration of spray-dried diclofenac-loaded nanocapsules and nanospheres. *STP Pharma Sciences* 11 (3): 229–233.

Guterres, S. S., V. Weiss, L. D. Freitas, and A. R. Pohlmann, 2000. Influence of benzyl benzoate as oil core on the physicochemical properties of spray-dried powders from polymeric nanocapsules containing indomethacin. *Drug Delivery* 7 (4): 195–199.

Han, M., E. Lee, and E. Kim, 2003. Preparation and optical properties of polystyrene nanocapsules containing photochromophores. *Optical Materials* 21 (1–3): 579–583.

Hoffmann, D. and C. Mayer, 2000. Cross polarization induced by temporary adsorption: NMR investigations on nanocapsule dispersions. *Journal of Chemical Physics* 112 (9): 4242–4250.

Hwang, S. L. and J. C. Kim, 2008. *In vivo* hair growth promotion effects of cosmetic preparations containing hinokitiol-loaded poly(epsilon-caprolactone) nanocapsules. *Journal of Microencapsulation* 25 (5): 351–356.

Illum, L., L. O. Jacobsen, R. H. Muller, E. Mak, and S. S. Davis, 1987. Surface characteristics and the interaction of colloidal particles with mouse peritoneal-macrophages. *Biomaterials* 8 (2): 113–117.

Jabr-Milane, L., L. van Vlerken, H. Devalapally, D. Shenoy, S. Komareddy, M. Bhavsar, and M. Amiji, 2008. Multi-functional nanocarriers for targeted delivery of drugs and genes. *Journal of Controlled Release* 130 (2): 121–128.

Jäger, A., V. Stefani, S. S. Guterres, and A. R. Pohlmann, 2007. Physico-chemical characterization of nanocapsule polymeric wall using fluorescent benzazole probes. *International Journal of Pharmaceutics* 338 (1–2): 297–305.

Kaul, G. and M. Amiji, 2004. Biodistribution and targeting potential of poly(ethylene glycol)-modified gelatin nanoparticles in subcutaneous murine tumor model. *Journal of Drug Targeting* 12 (9–10): 585–591.

Kesisoglou, F., S. Panmai, and Y. H. Wu, 2007. Nanosizing-oral formulation development and biopharmaceutical evaluation. *Advanced Drug Delivery Reviews* 59: 631–644.

Khalid, M. N., P. Simard, D. Hoarau, A. Dragomir, and J. C. Leroux, 2006. Long circulating poly(ethylene glycol)-decorated lipid nanocapsules deliver docetaxel to solid tumors. *Pharmaceutical Research* 23 (4): 752–758.

Koppel, D. E., 1972. Analysis of macromolecular polydispersity in intensity correlation spectroscopy—Method of cumulants. *Journal of Chemical Physics* 57 (11): 4814–4820.

Kriwet, B., E. Walter, and T. Kissel, 1998. Synthesis of bioadhesive poly(acrylic acid) nano- and microparticles using an inverse emulsion polymerization method for the entrapment of hydrophilic drug candidates. *Journal of Controlled Release* 56 (1–3): 149–158.

Lamprecht, A. and J. P. Benoit, 2006. Etoposide nanocarriers suppress glioma cell growth by intracellular drug delivery and simultaneous P-glycoprotein inhibition. *Journal of Controlled Release* 112 (2): 208–213.

Lboutounne, H., J. F. Chaulet, C. Ploton, F. Falson, and F. Pirot, 2002. Sustained *ex vivo* skin antiseptic activity of chlorhexidine in poly(epsilon-caprolactone) nanocapsule encapsulated form and as a digluconate. *Journal of Controlled Release* 82 (2–3): 319–334.

Leite, E. A., A. Grabe-Guimaraes, H. N. Guimaraes, G. L. L. Machado-Coelho, G. Barratt, and V. C. F. Mosqueira, 2007. Cardiotoxicity reduction induced by halofantrine entrapped in nanocapsule devices. *Life Sciences* 80 (14): 1327–1334.

Lemarchand, C., P. Couvreur, M. Besnard, D. Costantini, and R. Gref, 2003. Novel polyester–polysaccharide nanoparticles. *Pharmaceutical Research* 20 (8): 1284–1292.

Lopes, E., A. R. Pohlmann, V. Bassani, and S. S. Guterres, 2000. Polymeric colloidal systems containing ethionamide: Preparation and physico-chemical characterization. *Pharmazie* 55 (7): 527–530.

Losa, C., L. Marchalheussler, F. Orallo, J. L. V. Jato, and M. J. Alonso, 1993. Design of new formulations for topical ocular administration—Polymeric nanocapsules containing metipranolol. *Pharmaceutical Research* 10 (1): 80–87.

Lozano, M. V., D. Torrecilla, D. Torres, A. Vidal, F. Dominguez, and M. J. Alonso, 2008. Highly efficient system to deliver taxanes into tumor cells: Docetaxel-loaded chitosan oligomer colloidal carriers. *Biomacromolecules* 9 (8): 2186–2193.

Maillard, S., T. Ameller, J. Gauduchon, A. Gougelet, F. Gouilleux, P. Legrand, V. Marsaud, E. Fattal, B. Sola, and J. M. Renoir, 2005. Innovative drug delivery nanosystems improve the anti-tumor activity *in vitro* and *in vivo* of anti-estrogens in human breast cancer and multiple myeloma. *Journal of Steroid Biochemistry and Molecular Biology* 94 (1–3): 111–121.

Masters, K., 1991. *Spray Drying Handbook*, 1st edition. Harlow, England: Longman Scientific and Technical.

Matoga, M., F. Pehourcq, F. Lagrange, F. Fawaz, and B. Bannwarth, 2002. Influence of a polymeric formulation of ketoprofen on its diffusion into cerebrospinal fluid in rats. *Journal of Pharmaceutical and Biomedical Analysis* 27 (6): 881–888.

Mayer, C., D. Hoffmann, and M. Wohlgemuth, 2002. Structural analysis of nanocapsules by nuclear magnetic resonance. *International Journal of Pharmaceutics* 242 (1–2): 37–46.

Michel, C., M. Aprahamian, L. Defontaine, P. Couvreur, and C. Damgé, 1991. The effect of site of administration in the gastrointestinal-tract on the absorption of insulin from nanocapsules in diabetic rats. *Journal of Pharmacy and Pharmacology* 43 (1): 1–5.

Middleton, J. C. and A. J. Tipton, 2000. Synthetic biodegradable polymers as orthopedic devices. *Biomaterials* 21 (23): 2335–2346.

Milão, D., M. T. Knorst, W. Richter, and S. S. Guterres, 2003. Hydrophilic gel containing nanocapsules of diclofenac: Development, stability study and physico-chemical characterization. *Pharmazie* 58 (5): 325–329.

Moghimi, S. M., I. S. Muir, L. Illum, S. S. Davis, and V. Kolb-Bachofen, 1993. Coating particles with a block co-polymer (poloxamine-908) suppresses opsonization but permits the activity of dysopsonins in the serum. *Biochimica et Biophysica Acta* 1179 (2): 157–165.

Moinard-Chécot, D., Y. Chevalier, S. Briancon, L. Beney, and H. Fessi, 2008. Mechanism of nanocapsules formation by the emulsion–diffusion process. *Journal of Colloid and Interface Science* 317 (2): 458–468.

Montasser, I., S. Briançon, and H. Fessi, 2007. The effect of monomers on the formulation of polymeric nanocapsules based on polyureas and polyamides. *International Journal of Pharmaceutics* 335 (1–2): 176–179.

Mosqueira, V. C. F., P. Legrand, H. Pinto-Alphandary, F. Puisieux, and G. Barratt, 2000. Poly(D,L-lactide) nanocapsules prepared by a solvent displacement process: Influence of the composition on physico-chemical and structural properties. *Journal of Pharmaceutical Sciences* 89 (5): 614–626.

Mosqueira, V. C. F., P. M. Loiseau, C. Bories, P. Legrand, J. P. Devissaguet, and G. Barratt, 2004. Efficacy and pharmacokinetics of intravenous nanocapsule formulations of halofantrine in *Plasmodium berghei*-infected mice. *Antimicrobial Agents and Chemotherapy* 48 (4): 1222–1228.

Müller, C. R., V. L. Bassani, A. R. Pohlmann, C. B. Michalowski, P. R. Petrovick, and S. S. Guterres, 2000. Preparation and characterization of spray-dried polymeric nanocapsules. *Drug Development and Industrial Pharmacy* 26 (3): 343–347.

Müller, C. R., S. E. Haas, V. L. Bassani, S. S. Guterres, H. Fessi, M. Peralba, and A. R. Pohlmann, 2004. Degradation and stabilization of diclofenac in polymeric nanocapsules. *Quimica Nova* 27 (4): 555–560.

Müller, C. R., S. R. Schaffazick, A. R. Pohlmann, L. D. Freitas, N. P. da Silveira, T. D. Costa, and S. S. Guterres, 2001. Spray-dried diclofenac-loaded poly(epsilon-caprolactone) nanocapsules and nanospheres. Preparation and physicochemical characterization. *Pharmazie* 56 (11): 864–867.

Nguyen, A., V. Marsaud, C. Bouclier, S. Top, A. Vessieres, P. Pigeon, R. Gref, P. Legrand, G. Jaouen, and J. M. Renoir, 2008. Nanoparticles loaded with ferrocenyl tamoxifen derivatives for breast cancer treatment. *International Journal of Pharmaceutics* 347 (1–2): 128–135.

Nhung, D. T. T., A. M. Freydiere, H. Constant, F. Falson, and F. Pirot, 2007. Sustained antibacterial effect of a hand rub gel incorporating chlorhexdine-loaded nanocapsules (Nanochlorex(R)). *International Journal of Pharmaceutics* 334 (1–2): 166–172.

Olvera-Martínez, B. I., J. Cazares-Delgadillo, S. B. Calderilla-Faardo, R. Villalobos-Garcia, A. Ganem-Quintanar, and D. Quintanar-Guerrero, 2005. Preparation of polymeric nanocapsules containing octyl methoxycinnamate by the emulsification–diffusion technique: Penetration across the stratum corneum. *Journal of Pharmaceutical Sciences* 94 (7): 1552–1559.

Ourique, A. F., A. R. Pohlmann, S. S. Guterres, and R. C. R. Beck, 2008. Tretinoin-loaded nanocapsules: Preparation, physicochemical characterization, and photostability study. *International Journal of Pharmaceutics* 352 (1–2): 1–4.

Owens, D. E. and N. A. Peppas, 2006. Opsonization, biodistribution, and pharmacokinetics of polymeric nanoparticles. *International Journal of Pharmaceutics* 307 (1): 93–102.

Palumbo, M., A. Russo, V. Cardile, M. Renis, D. Paolino, G. Puglisi, and M. Fresta, 2002. Improved antioxidant effect of idebenone-loaded polyethyl-2-cyanoacrylate nanocapsules tested on human fibroblasts. *Pharmaceutical Research* 19 (1): 71–78.

Peracchia, M. T., S. Harnisch, H. Pinto-Alphandary, A. Gulik, J. C. Dedieu, D. Desmaele, J. d'Angelo, R. H. Muller, and P. Couvreur, 1999. Visualization of *in vitro* protein-rejecting properties of PEGylated Stealth® polycyanoacrylate nanoparticles. *Biomaterials* 20 (14): 1269–1275.

Pinto-Alphandary, H., M. Aboubakar, D. Jaillard, P. Couvreur, and C. Vauthier, 2003. Visualization of insulin-loaded nanocapsules: *In vitro* and *in vivo* studies after oral administration to rats. *Pharmaceutical Research* 20 (7): 1071–1084.

Pisani, E., N. Tsapis, J. Paris, V. Nicolas, L. Cattel, and E. Fattal, 2006. Polymeric nano/microcapsules of liquid perfluorocarbons for ultrasonic imaging: Physical characterization. *Langmuir* 22 (9): 4397–4402.

Pisani, E., E. Fattal, J. Paris, C. Ringard, V. Rosilio, and N. Tsapis, 2008. Surfactant dependent morphology of polymeric capsules of perfluorooctyl bromide: Influence of polymer adsorption at the dichloromethane–water interface. *Journal of Colloid and Interface Science* 326 (1): 66–71.

Pohlmann, A. R., G. Mezzalira, C. D. Venturini, L. Cruz, A. Bernardi, E. Jager, A. M. O. Battastini, N. P. da Silveira, and S. S. Guterres, 2008. Determining the simultaneous presence of drug nanocrystals in drug-loaded polymeric nanocapsule aqueous suspensions: A relation between light scattering and drug content. *International Journal of Pharmaceutics* 359 (1–2): 288–293.

Pohlmann, A. R., L. U. Soares, L. Cruz, N. P. da Silveira, and S. S. Guterres, 2004. Alkaline hydrolysis as a tool to determine the association form of indomethacin in nanocapsules prepared with poly(ε-caprolactone). *Current Drug Delivery* 1 (2): 103–110.

Pohlmann, A. R., V. Weiss, O. Mertins, N. P. da Silveira, and S. S. Guterres, 2002. Spray-dried indomethacin-loaded polyester nanocapsules and nanospheres: Development, stability evaluation and nanostructure models. *European Journal of Pharmaceutical Sciences* 16 (4–5): 305–312.

Poletto, F. S., L. A. Fiel, B. Donida, M. I. Re, S. S. Guterres, and A. R. Pohlmann, 2008a. Controlling the size of poly(hydroxybutyrate-*co*-hydroxyvalerate) nanoparticles prepared by emulsification–diffusion technique using ethanol as surface agent. *Colloids and Surfaces A: Physicochemical and Engineering Aspects* 324 (1–3): 105–112.

Poletto, F. S., E. Jaeger, L. Cruz, A. R. Pohlmann, and S. S. Guterres, 2008b. The effect of polymeric wall on the permeability of drug-loaded nanocapsules. *Materials Science and Engineering C: Biomimetic and Supramolecular Systems* 28 (4): 472–478.

Preetz, C., A. Rübe, I. Reiche, G. Hause, and K. Mäder, 2008. Preparation and characterization of biocompatible oil-loaded polyelectrolyte nanocapsules. *Nanomedicine: Nanotechnology, Biology, and Medicine* 4 (2): 106–114.

Prego, C., M. Fabre, D. Torres, and M. J. Alonso, 2006b. Efficacy and mechanism of action of chitosan nanocapsules for oral peptide delivery. *Pharmaceutical Research* 23 (3): 549–556.

Prego, C., D. Torres, E. Fernandez-Megia, R. Novoa-Carballal, E. Quinoa, and M. J. Alonso, 2006a. Chitosan-PEG nanocapsules as new carriers for oral peptide delivery—Effect of chitosan PEGylation degree. *Journal of Controlled Release* 111 (3): 299–308.

Quintanar-Guerrero, D., E. Allemann, E. Doelker, and H. Fessi, 1998. Preparation and characterization of nanocapsules from preformed polymers by a new process based on emulsification–diffusion technique. *Pharmaceutical Research* 15 (7): 1056–1062.

Quintanar-Guerrero, D., H. Fessi, E. Allemann, and E. Doelker, 1996. Influence of stabilizing agents and preparative variables on the formation of poly(D,L-lactic acid) nanoparticles by an emulsification–diffusion technique. *International Journal of Pharmaceutics* 143 (2): 133–141.

Raffin, R. P., E. S. Obach, G. Mezzalira, A. R. Pohlmann, and S. S. Guterres, 2003. Dried nanocapsules containing indomethacin: Formulation study and gastrointestinal tolerance evaluation in rats. *Acta Farmaceutica Bonaerense* 22 (2): 163–172.

Renoir, J. M., B. Stella, T. Ameller, E. Connault, P. Opolon, and V. Marsaud, 2006. Improved anti-tumoral capacity of mixed and pure anti-oestrogens in breast cancer cell xenografts after their administration by entrapment in colloidal nanosystems. *Journal of Steroid Biochemistry and Molecular Biology* 102 (1–5): 114–127.

Reynolds, J. E. F., 1993. *Martindale: The Extra Pharmacopoeia*, 30th edition. London: Pharmaceutical Press.

Romero-Cano, M. S. and B. Vincent, 2002. Controlled release of 4-nitroanisole from poly(lactic acid) nanoparticles. *Journal of Controlled Release* 82 (1): 127–135.

Rübe, A., G. Hause, K. Mader, and J. Kohlbrecher, 2005. Core–shell structure of Miglyol/poly(D,L-lactide)/Poloxamer nanocapsules studied by small-angle neutron scattering. *Journal of Controlled Release* 107 (2): 244–252.

Saez, A., M. Guzman, J. Molpeceres, and M. R. Aberturas, 2000. Freeze-drying of polycaprolactone and poly(D,L-lactic-glycolic) nanoparticles induce minor particle size changes affecting the oral pharmacokinetics of loaded drugs. *European Journal of Pharmaceutics and Biopharmaceutics* 50 (3): 379–387.

Saffran, M., B. Pansky, G. C. Budd, and F. E. Williams, 1997. Insulin and the gastrointestinal tract. *Journal of Controlled Release* 46 (1–2): 89–98.

Saï, P., C. Damgé, A. S. Rivereau, A. Hoeltzel, and E. Gouin, 1996. Prophylactic oral administration of metabolically active insulin entrapped in isobutylcyanoacrylate nanocapsules reduces the incidence of diabetes in nonobese diabetic mice. *Journal of Autoimmunity* 9 (6): 713–722.

Santos-Magalhães, N. S., A. Pontes, V. M. W. Pereira, and M. N. P. Caetano, 2000. Colloidal carriers for benzathine penicillin G: Nanoemulsions and nanocapsules. *International Journal of Pharmaceutics* 208 (1–2): 71–80.

Schaffazick, S. R., A. R. Pohlmann, T. Dalla-Costa, and S. S. Guterres, 2003b. Freeze-drying polymeric colloidal suspensions: Nanocapsules, nanospheres and nanodispersion. A comparative study. *European Journal of Pharmaceutics and Biopharmaceutics* 56 (3): 501–505.

Schaffazick, S. R., A. R. Pohlmann, C. A. S. de Cordova, T. B. Creczynski-Pasa, and S. S. Guterres, 2005. Protective properties of melatonin-loaded nanoparticles against lipid peroxidation. *International Journal of Pharmaceutics* 289 (1–2): 209–213.

Schaffazick, S. R., A. R. Pohlmann, and S. S. Guterres, 2007. Nanocapsules, nanoemulsion and nanodispersion containing melatonin: Preparation, characterization and stability evaluation. *Pharmazie* 62 (5): 354–360.

Schaffazick, S. R., I. R. Siqueira, A. S. Badejo, D. S. Jornada, A. R. Pohlmann, C. A. Netto, and S. S. Guterres, 2008. Incorporation in polymeric nanocapsules improves the antioxidant effect of melatonin against lipid peroxidation in mice brain and liver. *European Journal of Pharmaceutics and Biopharmaceutics* 69 (1): 64–71.

Schaffazick, S. R., S. S. U. Guterres, L. D. Freitas, and A. R. Pohlmann, 2003a. Physicochemical characterization and stability of the polymeric nanoparticle systems for drug administration. *Quimica Nova* 26 (5): 726–737.

Sinha, V. R., K. Bansal, R. Kaushik, R. Kumria, and A. Trehan, 2004. Poly-epsilon-caprolactone microspheres and nanospheres: An overview. *International Journal of Pharmaceutics* 278 (1): 1–23.

Soppimath, K. S., T. M. Aminabhavi, A. R. Kulkarni, and W. E. Rudzinski, 2001. Biodegradable polymeric nanoparticles as drug delivery devices. *Journal of Controlled Release* 70 (1–2): 1–20.

Stella, B., S. Arpicco, F. Rocco, V. Marsaud, J. M. Renoir, L. Cattel, and P. Couvreur, 2007. Encapsulation of gemcitabine lipophilic derivatives into polycyanoacrylate nanospheres and nanocapsules. *International Journal of Pharmaceutics* 344 (1–2): 71–77.

Stolnik, S., L. Illum, and S. S. Davis, 1995. Long circulating microparticulate drug carriers. *Advanced Drug Delivery Reviews* 16 (2–3): 195–214.

Sussman, E. M., M. B. Clarke, and V. P. Shastri, 2007. Single-step process to produce surface-functionalized polymeric nanoparticles. *Langmuir* 23: 12275–12279.

Teixeira, M., M. J. Alonso, M. M. M. Pinto, and C. M. Barbosa, 2005. Development and characterization of PLGA nanospheres and nanocapsules containing xanthone and 3-methoxyxanthone. *European Journal of Pharmaceutics and Biopharmaceutics* 59 (3): 491–500.

Turos, E., J. Y. Shim, Y. Wang, K. Greenhalgh, G. S. K. Reddy, S. Dickey, and D. V. Lim, 2007. Antibiotic-conjugated polyacrylate nanoparticles: New opportunities for development of anti-MRSA agents. *Bioorganic and Medicinal Chemistry Letters* 17 (1): 53–56.

Vauthier, C., B. Cabane, and D. Labarre, 2008. How to concentrate nanoparticles and avoid aggregation? *European Journal of Pharmaceutics and Biopharmaceutics* 69 (2): 466–475.

Vauthier, C., D. Labarre, and G. Ponchel, 2007. Design aspects of poly(alkylcyanoacrylate) nanoparticles for drug delivery. *Journal of Drug Targeting* 15 (10): 641–663.

Veronese, F. M. and G. Pasut, 2005. PEGylation, successful approach to drug delivery. *Drug Discovery Today* 10 (21–24): 1451–1458.

Vitale, S. A. and J. L. Katz, 2003. Liquid droplet dispersions formed by homogeneous liquid–liquid nucleation: "The ouzo effect." *Langmuir* 19 (10): 4105–4110.

Washington, C., 1990. Drug release from microdisperse systems—A critical review. *International Journal of Pharmaceutics* 58 (1): 1–12.

Watnasirichaikul, S., T. Rades, I. G. Tucker, and N. M. Davies, 2002. *In-vitro* release and oral bioactivity of insulin in diabetic rats using nanocapsules dispersed in biocompatible microemulsion. *Journal of Pharmacy and Pharmacology* 54 (4): 473–480.

Weiss-Angeli, V., F. S. Poletto, L. R. Zancan, F. Baldasso, A. R. Pohlmann, and S. S. Guterres, 2008. Nanocapsules of octyl methoxycinnamate containing quercetin delayed the photodegradation of both components under ultraviolet A radiation. *Journal of Biomedical Nanotechnology* 4 (1): 80–89.

Wise, D. L., D. J. Trantolo, R. T. Marino, and J. P. Kitchell, 1987. Opportunities and challenges in design of implantable biodegradable polymeric systems for drug delivery of antimicrobial agents and vaccines. *Advanced Drug Delivery Reviews* 1 (1): 19–40.

Zhang, Y., S. Y. Zhu, L. C. Yin, F. Qian, C. Tang, and C. H. Yin, 2008. Preparation, characterization and biocompatibility of poly(ethylene glycol)-poly(*n*-butyl cyanoacrylate) nanocapsules with oil core via miniemulsion polymerization. *European Polymer Journal* 44 (6): 1654–1661.

Zili, Z., S. Sfar, and H. Fessi, 2005. Preparation and characterization of poly-e-caprolactone nanoparticles containing griseofulvin. *International Journal of Pharmaceutics* 294 (1–2): 261–267.

4 Poly(Alkyl Cyanoacrylate) Nanoparticles for Drug Delivery and Vaccine Development

Anja Graf, Karen Krauel-Göllner, and Thomas Rades

CONTENTS

4.1 INTRODUCTION

Alkyl cyanoacrylates have been used as surgical materials, in particular as biodegradable tissue adhesives, since the 1960s (Leonard et al., 1966). Since their discovery, the U.S. Food and Drug Administration has approved two monomers, butyl cyanoacrylate (2002) and octyl cyanoacrylate (1998), for human use, which are available on the market as liquid bandaids (Indermil®, LiquiBand®, and Dermabond®). Poly(alkyl cyanoacrylate) (PACA) nanoparticles emerged in drug delivery around 30 years ago (Couvreur et al., 1979). They have since been explored with increasing interest as colloidal drug delivery systems and for targeting of bioactives, including low molecular weight drugs, peptides, proteins, and nucleic acids. The extensive interest in PACA nanoparticles as drug carriers is due to the biocompatibility and biodegradability of the polymer, the ease of preparation of the particles, and their ability to entrap bioactives, including subunit antigens. Many modern drugs are proteins that are susceptible to chemical and enzymatic degradation in the physiological environment and exhibit low bioavailability. They thus require chemical and enzymatic protection, bioavailability enhancement, and even targeting to certain delivery sites. Encapsulation of proteins into PACA nanoparticles may be one way of overcoming some of the challenges in effective protein delivery. However, although PACA nanoparticles have been considered as promising polymeric colloidal drug delivery systems for some time, no product has entered the market. Initial progress has been made in the clinical development of a PACA formulation in cancer therapy. In 2006 BioAlliance Pharma announced clinical phase II/III trials for Transdrug®, doxorubicin (DOX)-loaded poly(isohexyl cyanoacrylate) (PiHCA) nanoparticles suitable for intra-arterial, intravenous (i.v.), or oral administration. However, because of acute pulmonary damage, phase II trials of Transdrug were suspended in July 2008.

Various methods have been used to prepare PACA nanoparticles including polymerization in a continuous aqueous phase and interfacial polymerization of submicron emulsions or microemulsions. The preparation of PACA nanoparticles on the basis of various templates involving these techniques will be discussed in this chapter with a particular emphasis on biocompatible microemulsions as polymerization templates. For further details on the design aspects of PACA nanoparticles including surface property modifications, which are critical in controlling the *in vivo* fate of the delivery system, the reader is referred to an excellent review by Vauthier et al. (2007). In general, we will introduce methods for the pharmaceutical characterization of PACA nanoparticles and application of these nanoparticles as second- and third-generation colloidal carriers, that is, carriers without surface modification (passive body distribution) and with surface modifications (for drug targeting; Couvreur et al., 2002).

4.2 PREPARATION OF POLY(ALKYL CYANOACRYLATE) NANOPARTICLES ON THE BASIS OF DIFFERENT POLYMERIZATION TEMPLATES

PACA nanoparticles can be differentiated into nanospheres and nanocapsules and can contain a drug either in the core of the particle (encapsulation) or adsorbed on the outside of the nanoparticle (sorption; Krauel et al., 2004). The types of PACA nanoparticles and differences in entrapment are illustrated in Figure 4.1. The type of template will determine the preparation technique and type of particles formed. PACA nanoparticles can be prepared by polymerization of alkyl cyanoacrylate monomers or from preformed polymers. The two polymerization techniques most commonly applied for the preparation of PACA nanoparticles are polymerization in an aqueous acidic phase (Couvreur et al., 1979) and interfacial polymerization of (submicron) emulsions (Al Khouri Fallouh et al., 1986; El-Samaligy et al., 1986; Weiss et al., 2007) or microemulsions (Gasco and Trotta, 1986). Polymerization in an aqueous acidic medium is also referred to in the literature as micellar polymerization and emulsion or dispersion polymerization. The term micellar polymerization is based on the theory that polymerization occurs around micelles formed by surfactant molecules added to the acidic phase as stabilizers (Couvreur, 1988). However, Fitch and Tsai (1970) have

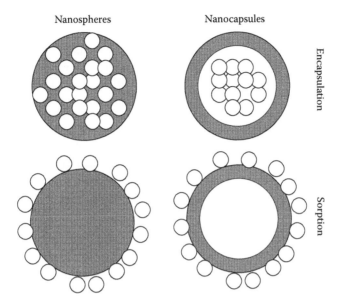

Nanospheres   Nanocapsules

Encapsulation

Sorption

FIGURE 4.1 Schematic representation of types of PACA nanoparticles and methods of drug entrapment.

shown that polymerization does not depend on the presence of these stabilizers. The differentiating factor between dispersion and emulsion polymerization is the type of stabilizer used, which in the former is dextran (Behan et al., 2001) and in the latter a surfactant, usually a poloxamer (Reddy and Murthy, 2004). Emulsion polymerization is potentially misleading because the polymerization medium is not an emulsion but an aqueous solution.

Alkyl cyanoacrylates are able to polymerize very rapidly by free-radical, base-catalyzed anionic and zwitterionic polymerization mechanisms. Neither high-energy radiation nor chemical initiators, which can affect the stability of the entrapped drug, are required for the polymerization process. Polymerization can be initiated by hydroxyl ions from the dissociation of water in a polymerization medium. Thus, typical initiators of anionic polymerization are water, anions, and weak bases including alcohols and amino acids. The excellent tissue adhesiveness of PACA is attributable to polymerization initiated by amino acids of proteins in the skin. When the bioactive itself can act as a nucleophilic polymerization initiator, copolymerization of the bioactive into the polymer network may occur (Grangier et al., 1991; Weiss et al., 2007). Proteins and peptides are nucleophilic, and it is thus possible that these molecules might intervene in the polymerization process (Leonard et al., 1966). This can result in a loss of bioactivity of the protein (Gibaud et al., 1998; Grangier et al., 1991) or it may enhance stabilization and increase entrapment, as shown by the use of copolymerized peptide particles (Hillery et al., 1996a, 1996b; Liang et al., 2008). However, a large excess of alcohol seems to prevent the copolymerization of peptides with the monomer, leaving, for instance, insulin unmodified, so it was still recognized by the insulin receptor of hepatocytes (Aboubakar et al., 1999a; Roques et al., 1992).

Preparation of PACA nanoparticles was first established by Couvreur et al. (1979) by the addition of the monomer to an acidic aqueous phase containing surfactant. This polymerization mechanism leads to the formation of nanospheres in a three-step process. After polymerization has been initiated by nucleophiles, particularly after contact of the monomer with water, oligomeric chains form and are terminated by protons. Hence, an acidic pH is used to regulate the polymerization reaction. Upon termination the oligomeric chains aggregate and swell with the remaining monomer. This is followed by an *in situ* reinitiation of terminated oligomeric chains by live chains leading to further polymerization until an equilibrium molecular weight is reached (Behan et al., 2001). The nanospheres are stabilized against agglomeration by the surfactant in the polymerization medium.

Polymerization of alkyl cyanoacrylates at the interface of oil-in-water (O/W) submicron dispersions was first reported by Al Khouri Fallouh et al. (1986). For this interfacial polymerization technique, the monomer, drug, and oil are dissolved in a water-miscible organic solvent; this mixture is then injected into a surfactant-containing aqueous phase. The polymerization occurs at the oil–water interface that forms following the diffusion of the water-miscible solvent into the aqueous phase. The organic solvent is subsequently removed under reduced pressure. Surfactants are added to the aqueous phase to prevent aggregation of the nanocapsules upon storage (Al Khouri Fallouh et al., 1986). The type of nanoparticles formed by this technique depends on the type of water-miscible solvent used (Puglisi et al., 1995). When a protic water-miscible solvent, such as ethanol, isopropanol, or butanol, is used, the resulting nanoparticles will be a mixture of oily-cored nanocapsules and nanospheres. Gallardo et al. (1993) identified that an oil/ethanol ratio of 2% will yield nanocapsules alone. Two studies proposed that, in addition to interfacial polymerization, the protic solvent can initiate the polymerization in the organic phase leading to precipitation of the preformed polymer in a large excess of nonsolvent at the oil–water interface (Chouinard et al., 1994) and assembly of polymerization nuclei in the aqueous medium in the absence of oil droplets (Puglisi et al., 1995). A fragmentation process of the polymeric interfacial film was also suggested (Gallardo et al., 1993). This method is mainly useful for the encapsulation of oil-soluble substances. However, because the polymerization rate is extremely high, highly water-soluble molecules such as insulin have also been encapsulated into oily-core nanocapsules (Aboubakar et al., 1999a; Damgé et al., 1988).

El-Samaligy et al. (1986) applied interfacial polymerization of alkyl cyanoacrylate monomers to water-in-oil (W/O) submicron emulsions, which led to the preparation of nanocapsules with an aqueous core. Here an aqueous solution was emulsified in a surfactant-containing organic phase. Ultrasonication was used to reduce the emulsion droplets to submicron size. A washing process was required to remove potentially toxic solvents. However, the toxicity issue can be overcome by using (W/O) submicron emulsions prepared from biocompatible compounds (Vranckx et al., 1996).

Interfacial polymerization techniques in general are attractive because the preparation is simple and easy to scale-up (Al Khouri Fallouh et al., 1986). However, using submicron emulsions as templates is disadvantageous because of the high energy input required either by ultrasonication (El-Samaligy et al., 1986) or by vigorous stirring (Vranckx et al., 1996), and the difficulty of obtaining nanocapsules with a low polydispersity, that is, a narrow particle size distribution (El-Samaligy et al., 1986), unless high surfactant concentrations are used (Vranckx et al., 1996). Formation of nanocapsules without the need of high-energy input can be achieved using microemulsions as templates (Gasco and Trotta, 1986; Watnasirichaikul et al., 2000). The advantages of using a microemulsion template instead of a coarse or submicron emulsion template for the preparation of nanoparticles are enhanced physical stability of the system, minimal energy input required for the template formation, and a small and monodisperse droplet size (Watnasirichaikul et al., 2000). Particle size distribution is an important factor in that a monodisperse distribution would be advantageous for uniform administration of the drug. In addition, particle morphology has implications on how the drug is entrapped and in turn on the release profile. Therefore, particle size and surface properties are important because these will have implications on the drug entrapment, bioadhesiveness of the vehicles, amount and extent of particle uptake, and particle and drug biodistribution (Pitaksuteepong et al., 2002). Characterization of nanoparticles is necessary to ensure that particle formation has occurred and appropriate delivery vehicles have been formulated. Common techniques to characterize nanoparticles are described later.

PACA nanocapsules with an aqueous core were first prepared by Gasco and Trotta (1986) using interfacial polymerization of W/O microemulsions. However, the microemulsion system they used contained organic solvents that subsequently had to be removed. Sophisticated purification processes followed by redispersion can lead to aggregation of the nanoparticles and difficulties in the scale-up of this method. This disadvantage was overcome by using Watnasirichaikul et al.'s (2000) biocompatible microemulsion system. When biocompatible oils and surfactants are used, there is no need to isolate the nanoparticles from the polymerization template. In addition, coadministration of nanoparticles dispersed in the microemulsion template can allow the exploitation of the permeability enhancing effects of microemulsions, which are beneficial for oral absorption (Damgé et al., 1987; Ritschel, 1991). Watnasirichaikul et al. (2000) developed a simple one-step process by which PACA nanocapsules could be readily prepared from a biocompatible W/O droplet-type microemulsion via interfacial polymerization of ethyl-2-cyanoacrylate as the monomer, which can be scaled up easily. This is unlike Gasco and Trotta's (1986) proposition of nanoparticle formation around the water-droplet swollen micelles, which would result in nanoparticles with a core–shell structure and the size of the swollen micelle (~10–20 nm), Watnasirichaikul et al. (2000) hypothesized that the mechanism of nanoparticle formation based on a W/O droplet-type microemulsion takes place at the water–oil interface of micelle clusters with a concomitant structural collapse resulting in approximately 250-nm nanocapsules.

Further, it was found that the use of microemulsions as templates for PACA nanoparticles is not limited to W/O microemulsions. Krauel et al. (2005, 2006) demonstrated that all types of microemulsions, including W/O, bicontinuous, O/W, and water-free microemulsions, can be used as templates to form PACA nanoparticles. This was surprising because, according to the interfacial polymerization theory, particle formation takes place at the interface around the water droplets. However, interfacial polymerization in a bicontinuous microemulsion was also reported previously (Holtzscherer et al., 1987). As a possible mechanism for the formation of particles, Krauel et al. proposed that polymerization is initiated upon contact with hydroxyl ions or other nucleophiles and the resulting polymer forms spherical particles in an attempt to minimize its interfacial area by minimizing its surface–volume ratio. Subsequently, the interface, if not already present in the microemulsion template, is created *in situ* by the forming polymer. Therefore, the expression "interfacial polymerization" still holds true even for bicontinuous and solution-type microemulsions. From the studies performed on microemulsions as PACA templates, it appears that neither a droplet structure nor an interface is a prerequisite for particle formation by interfacial polymerization. However, entrapment of fluorescently labeled ovalbumin [fluorescein isothiocyanate–ovalbumin (FITC-OVA); Krauel et al., 2005] and insulin (Graf et al., 2009) was found to be dependent on the microemulsion template and the amount of monomer used for polymerization. In water-free microemulsions it is likely that other components with nucleophilic characters could act as a polymerization initiator.

PACA nanocapsules with an aqueous core efficiently entrap insulin (Watnasirichaikul et al., 2000), OVA (Pitaksuteepong et al., 2002), salmon calcitonin (Vranckx et al., 1996), and oligonucleotides (Lambert et al., 2000b).

In addition to oral delivery of proteins and peptides with PACA nanoparticles, the use of these as delivery systems for subunit antigens has received considerable attention (Kreuter, 1988). The areas of oral and vaccine drug delivery using PACA nanoparticles are further elucidated below.

4.3 CHARACTERIZATION OF POLY(ALKYL CYANOACRYLATE) NANOPARTICLES

It is necessary to thoroughly characterize not only the nanoparticles but also the templates (e.g., microemulsions) used to prepare them to be able to explain the polymerization process and the characteristics of the nanoparticles. However, the focus of this section will be on the characterization of the nanoparticles. For a detailed description of methods to characterize microemulsion templates, the reader is referred to Alany et al. (2008).

Of major interest in the characterization of nanoparticles is their size, because it can give useful information about whether the polymerization process was successful and insight into the polymerization mechanism. It was also shown that particle size can influence body distribution (Jani et al., 1990). In addition to the size, the polydispersity index (a measure of particle size distribution) should be determined. A monomodal and narrow size distribution should be the aim of nanoparticle preparation to ensure uniform distribution behavior of the particles upon administration. Further, the surface charge of the nanoparticles is of interest for the entrapment of drugs by adsorption.

The morphology of the PACA nanoparticles is interesting, which can be visualized by various electron microscopic techniques. Ultracentrifugation was employed in the early days of PACA nanoparticle research to differentiate between nanospheres and nanocapsules (Rollot et al., 1986), and matrix-assisted laser desorption ionization time-of-flight mass spectrometry (MALDI-TOF-MS) was used recently to investigate the chemical composition of nanoparticles (Bootz et al., 2005). Another valuable technique with which several aspects of nanoparticle characterization can be covered in one set of experiments is nuclear magnetic resonance (NMR; Mayer et al., 2002). In the following sections of the chapter we will highlight some advantages and disadvantages of techniques employed in the characterization of nanoparticles. However, it is not the aim of this chapter to give a thorough explanation of each technique and the reader is referred to specific textbooks for this. In addition, issues around drug loading and release are also discussed.

4.3.1 PARTICLE SIZE MEASUREMENT

4.3.1.1 Photon Correlation Spectroscopy

Photon correlation spectroscopy (PCS), also known as quasielastic light scattering or dynamic light scattering, takes advantage of the high spatial coherence of monochromatic light sources (laser) to analyze the intensity fluctuation of scattered light from particle samples dispersed in solutions. By employing the Stokes–Einstein equation, which is applicable to spheres in Brownian motion, the diffusion coefficient of the particles is computed and derived to average hydrodynamic particle diameters and particle size distributions using an autocorrelation function. The method is noninvasive and the measurement time is on the order of minutes with a measurement range of ~5–5000 nm. Latex particles with standardized sizes can be used to calibrate the PCS instrument. Mixtures of various standardized size nanoparticles and consequent measurements allow for control of whether the chosen measurement settings are able to separate nanoparticles of different size populations. Cross-correlation, where dual laser beams cross over in the sample container and generate two similar signal patterns, or scanning of the sample for the optimum measurement concentration minimizes the problems of multiscattering. Nanoparticles have to be measured suspended in a liquid; and because a hydrodynamic diameter is measured, particles appear to be slightly bigger when compared to their respective electron microscopy (EM) image (Bootz et al., 2004). This can be of specific interest when measuring the size of coated particles where different surface coatings or stabilizers may produce a different hydrodynamic radius of the particles that is due to varying interactions of these compounds with water molecules. Another point of interest can be the aggregation of nanoparticles over time in a certain dispersion medium. The dispersion medium should always be filtered to avoid contamination with dust particles that could obscure results before carrying out any PCS measurements.

4.3.1.2 Microscopy

The typical dimensions of nanoparticles are below the diffraction limit of visible light and are thus outside the range for optical microscopy. Therefore, EM must be used to visualize nanoparticles. EM-based techniques for particle size determination are based on direct observation of the nanoparticles, and the measured diameter of the particles will depend on the projected image of the particles. The recent development of particle-counting software and even the ability to calculate particle size distributions with the software has reintroduced EM as a technique to size nanoparticles. The major limitation of conventional electron microscopes, such as transmission and scanning microscopes, is that they have to be operated under vacuum conditions. This means that liquid samples cannot be introduced into the sample chamber and sample preparation (dehydration, cryofixation, embedding, freeze-fracturing, etc.) is necessary, which can lead to sample alteration and creation of artifacts. Improvements in particle preparation have been made, however, for instance, the development of plunge-freezing techniques to achieve faster freezing rates and thereby more homogeneous freezing throughout the sample.

One possibility to image nanoparticles under more natural conditions is the use of environmental scanning EM (ESEM). In an ESEM the sample chamber is separated from the column and can be operated at almost all atmospheric conditions (10–50 Torr), which allows the observation of non-dried samples. In addition, the gas ionization in the ESEM chamber eliminates charging artifacts and materials therefore do not have to be coated with a conducting layer. The image contrast, however, becomes increasingly poor with increasing humidity of the sample and drifting of the sample can be problematic, as is loss of resolution in the range of ~10 to ~100 nm.

A further microscopic technique to analyze the size of nanoparticles is atomic force microscopy (AFM). In AFM an oscillating cantilever scans the sample surface and electrostatic forces (down to 10^{-12} N) are measured between the tip and the sample surface. With AFM we can obtain three-dimensional surface profiles through force measurements with a height resolution of ~0.5 nm. However, the measurement of liquid samples has the drawback that particles may float and even stick to the cantilever because of smearing effects and changes in the cantilever oscillating properties as the tip gets heavier. Operating the AFM in noncontact mode can overcome these issues (Balnois et al., 2007; Reddy and Murthy, 2004). As force measurements are the basis of AFM, this technique certainly has its value in confirming coating of nanoparticles with drug molecules, stabilizers, and so forth. Routine particle sizing of nanoparticles should be carried out with PCS, however, because results can be obtained within minutes and are easy to interpret.

4.3.2 ZETA POTENTIAL

The zeta potential of nanoparticles is the potential at the slipping plane of the electric double layer and represents the "effective" surface charge of the particles. Simple and rapid measurements can be carried out by measuring the electrophoretic mobility of the particles in a certain electrolyte solution. Using Equation 4.1, the electrophoretic mobility can then be used to calculate the zeta potential:

$$\zeta = \frac{\mu_E 4\pi\eta}{\varepsilon}, \tag{4.1}$$

where ζ is the zeta potential, μ_E is the electrophoretic mobility, η is the viscosity of the medium, and ε is the dielectric constant.

Because the presence of an electrolyte is generally necessary, the result of the zeta potential measurement always has to be interpreted in relation to the type of electrolyte that was used in the measurements (Reddy and Murthy, 2004). Sodium or potassium chloride are commonly used as electrolytes. The zeta potential gives information about the potential stability of a colloidal system, and a large negative or positive potential (>±30 mV) will lead to greater stability as particles will electrostatically repel each other. When measuring the zeta potential of PACA nanoparticles in media of different pH values, Arias et al. (2001) found that the zeta potential decreased to neutral values when lowering the pH because the acrylic acid groups at the end of the polymer chains are less likely to dissociate at low pH. Successful decoration of the particle surface (see sections on brain and tumor delivery) can be verified easily with zeta potential measurements, because the surface charge of the particles will change according to the attached molecule used for the coating (Kreuter et al., 1995). The adsorption of drug molecules can also be confirmed with this method, which Aboubakar et al. (2000) demonstrated; entrapping insulin via adsorption to the poly(isobutyl cyanoacrylate) (PiBCA) nanoparticle surface led to a decrease in zeta potential from −19 to −10 mV (Aboubakar et al., 2000).

4.3.3 Morphology

4.3.3.1 Electron Microscopy

The most common methods used to study the morphology of nanoparticles are freeze-fracture transmission EM (FF-TEM), SEM, their variations cryo-TEM and cryo-SEM, and AFM. Sometimes TEM is also used for samples prepared by negative staining. For FF-TEM a replica of the actual sample is prepared by initially sandwiching the sample on a grid between two copper support disks. The sandwich is then snap-frozen by immersion in liquid propane (−180°C) and afterward loaded into a double-replica device immersed in liquid nitrogen (−196°C). For fracturing, the double-replica device is mounted onto the sample stage of a freeze-etch device. The fractured surfaces are then shadowed with platinum at a 45° angle, which is immediately followed by carbon at a 90° angle. The replicas are washed with solvents to remove any residual sample material and mounted onto copper grids for viewing. FF-TEM was used very early on in the investigation of polymeric nanoparticles (O'Hagan et al., 1989). The sample preparation for FF-TEM is rather invasive and only a replica is investigated under the microscope. The creation of artifacts is likely, and interpretation of micrographs has to be performed with great care. Typical micrographs of PACA nanoparticles can be found in various publications (Couvreur et al., 1996; Krauel et al., 2007). Whether nanoparticles are really fractured may be questionable, and O'Hagan et al. (1989) stated that the spherical shapes in the replica are often imprints of nanoparticles rather than fractured particles. Fracturing of these colloidal systems seems very unlikely and interpretations of a shadow, which can be seen in images obtained by Watnasirichaikul et al. (2002a, 2002b) as the capsule wall, remain speculative. Even in pure microemulsion samples one can find shapes that resemble nanoparticles, which again makes interpretation of FF-TEM micrographs difficult (Krauel et al., 2007). An improvement in sample preparation has been the development of cryo-TEM, where the sample is prepared by immersion into liquid propane followed by observation on a cryostage of the TEM. Plunge freezing in liquid propane causes a faster freezing rate and may reduce the amount of artifacts that are created during sample preparation.

Similar to cryo-TEM, cryo-SEM investigates the nanoparticle sample in the frozen state, which makes separation from the preparation template unnecessary and minimizes the amount of artifacts. To prepare nanoparticles for cryo-SEM, the samples are filled into brass rivets and plunge frozen in liquid propane. Samples are then stored in liquid nitrogen and transferred into the cryostage of an SEM. Samples can also be fractured on the cryostage with a knife before being coated with platinum and viewed in the frozen stage at ∼−140°C. Examples of cryo-SEM images can be found in the work of Krauel et al. (2007).

When utilizing cryo-SEM, it is not necessary to separate the nanoparticles from the microemulsion matrix. The liquid dispersion can be frozen and is viewed in the frozen state. Using this preparative technique, samples are viewed close to their natural state and the actual particles are visualized rather than their impression as with FF-TEM.

SEM has become an invaluable technique in the characterization of nanoparticles and a standard in publications dealing with these systems (Bootz et al., 2004, 2005; Chouinard et al., 1991; Couvreur et al., 1995). Sample preparation is usually not very invasive and results can be obtained quickly, for example, within 30 min. For normal SEM, separation of the particles from preparation templates, however, is necessary because only solid samples can be viewed with this technique. Separation can be achieved by centrifugation, followed by repeated washing with a solvent. The samples are then sputter coated with gold and palladium before viewing under the microscope. Alternatively, the separated nanoparticles can be dispersed in ethanol, a drop of the dispersion is then placed on the sample holder, and the ethanol is allowed to evaporate prior to coating (Krauel et al., 2007). The latter method has proven successful in preventing aggregation of PACA nanoparticles (Figure 4.2).

For TEM, samples have to contain contrast, which is often introduced to biological samples by negative staining. This preparation method is suitable for visualizing suspensions of small particle

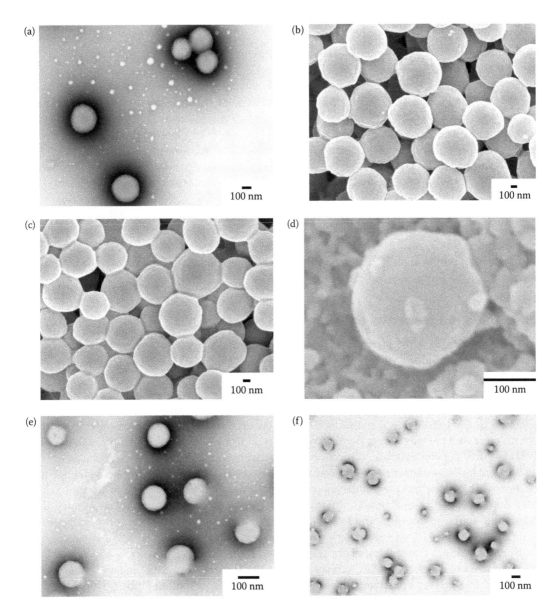

FIGURE 4.2 (a) TEM micrograph of PBCA nanoparticles prepared from a W/O microemulsion template, (b) SEM micrograph of PBCA nanoparticles prepared from a bicontinuous microemulsion template, (c) SEM micrograph of PBCA nanoparticles prepared from an O/W microemulsion template, (d) SEM micrograph of PECA nanoparticles prepared from a W/O microemulsion template, (e) TEM micrograph of PECA nanoparticles prepared from a bicontinuous microemulsion template, and (f) TEM micrograph of PECA nanoparticles prepared from an O/W microemulsion template. TEM samples were negatively stained with 1% phosphotungstic acid. Scale bar = 100 nm.

size, such as nanoparticles. An aqueous dispersion of the nanoparticles is adsorbed onto glow-discharged, carbon-coated copper grids. After blotting off excess sample, the sample grid is stained with a solution of uranyl acetate, ammonium molybdate, or phosphotungstic acid and then dried and viewed.

Examples of SEM and negatively stained TEM micrographs of PACA nanoparticles are provided in Figure 4.2.

4.3.4 Density

4.3.4.1 Ultracentrifugation

The different techniques in preparation of nanoparticles by either polymerization in an aqueous medium or emulsion polymerization result in two different types of nanoparticles: nanospheres and nanocapsules. Rollot et al. (1986) confirmed the existence of the different particles by ultracentrifugation. They assumed a gravity of 1.01 for the solid polymer nanospheres and 0.94 for the oil-filled nanocapsules. Ultracentrifugation of the particle suspensions in 7.5% poly(ethylene glycol) (PEG) separated the two particle fractions, confirming the existence of the two nanoparticle subtypes. The authors confirmed their findings by observation of FF replicas of the nanoparticles under TEM. The nanospheres showed a solid, granulated internal structure, whereas the nanocapsules showed a polymeric envelope surrounding an oily core.

Chouinard et al. (1994) also used centrifugation in a density gradient of colloidal silica to determine the density of nanocapsules and reported the density to be between that of emulsion droplets and nanospheres. It is interesting that doubling of the monomer used to prepare the particles led to nanocapsules with a slightly higher density, which may be caused by particles with a thicker polymer wall.

4.3.5 Chemical Composition of Nanoparticles

Useful methods found in the literature to study the chemical composition of nanoparticles include Fourier transform infrared spectroscopy (FT-IR; Reddy and Murthy, 2004), MALDI-TOF-MS, and size exclusion chromatography (SEC; Bootz et al., 2005).

For FT-IR, nanoparticles have to be separated from the preparation matrix and then freeze-dried to give a solid particles. Characteristic stretches of the polymer structure, for example, CH (~ 2950 cm^{-1}), CN (~ 2200 cm^{-1}), and CO (~ 1750 cm^{-1}), can then be observed in the spectrum and confirm successful polymerization (Reddy and Murthy, 2004). FT-IR is also a versatile technique to investigate whether the polymer was copolymerized with an entrapped drug. Aboubakar et al. (1999b) used FT-IR to show that insulin was not copolymerized with PiBCA by interfacial polymerization when ethanol was present during the polymerization process. The authors attributed this to the nucleophilic ethanol acting as a more efficient polymerization initiator and preventing the nucleophilic amine groups of insulin from interaction with the monomer.

SEC and MALDI-TOF-MS can determine the chemical composition of the polymer nanoparticles by establishing the molecular weight (Bootz et al., 2005). SEC is frequently used to study the molecular mass of polymers. In SEC, polymer chains of different size, shape, and molecular mass are retained in a column containing a porous gel and separated according to their hydrodynamic volume, where the bulkiest molecules are eluted first and the smallest last. Detection occurs via the refractive index or with an ultraviolet (UV)/visible detector. For MALDI-TOF-MS, the sample molecules are cocrystallized with an UV-absorbing matrix, desorbed, and ionized by short UV laser pulses. The ions are then accelerated by a defined voltage into the field-free drift tube. Here, ions of different mass/charge ratios are separated by their TOF and reach the detector. The square of the flight time of the singly charged ions is directly proportional to the molecular mass and can be calculated from the spectrum. Using SEC and MALDI-TOF-MS the authors were able to show the existence of oligomeric poly(butyl cyanoacrylate) (PBCA) chains, as well as two further species with an additional or missing formaldehyde molecule (Bootz et al., 2005). These findings confirm that degradation of PACA nanoparticles occurs to a certain degree in a reverse Knoevenagel reaction because formaldehyde is a decomposition product of PACA in this reaction.

The use of x-ray diffraction to investigate the chemical composition of PACA nanoparticles was reported by Reddy and Murthy (2004). They found that it does not lead to conclusive results other than the polymer particles were amorphous, which is hardly surprising, especially given that the particles were freeze-dried prior to the measurements.

Another technique commonly applied in the characterization of solid-state dosage forms is differential scanning calorimetry. Using x-ray diffraction for particle characterization, Reddy and Murthy (2004) reported the existence of a glass transition temperature (T_g) for a PACA nanoparticle powder in the thermogram, which again would be expected from the freeze-dried material. However, they also showed that the T_g of the polymer changed upon the addition of a drug for entrapment. This leads to the assumption of an intimate mixture of the drug and the polymer; otherwise, two separate T_gs would be detected and the T_g of the polymer would not change if the polymer and drug substance were not mixed.

4.3.6 NUCLEAR MAGNETIC RESONANCE

The characterization of PACA nanoparticles with NMR spectroscopy is a convenient way to gain information about dynamic processes on the molecular level for each individual component and about the structure of the nanoparticles in a single set of completely nondestructive measurements. With NMR, it is possible to detect each component in the particle formulation from characteristic peaks when compared with an NMR spectrum of the pure component, to derive component interactions from peak shifts, and to determine the microstructure of the nanoparticle dispersion from the characteristic diffusive behavior of each component. For instance, if a liquid component is encapsulated in the nanoparticle, its diffusion will be limited by the particle wall and thus be much slower than for the pure component. If the component is attached to the particle or even copolymerized, it will show the same diffusive behavior as the particle itself. However, if under consideration of viscosity, no change in self-diffusion is found in comparison to the pure component, the component is located in the dispersion surrounding the particle. Such NMR experiments using PBCA nanocapsules as a model system were recently shown by Mayer et al. (2002). In this study, nanocapsules were prepared by interfacial polymerization in an acidic organic phase consisting of monomer, triglyceride oil, and hydrochloric acid in ethanol upon injection into an aqueous phase of block copolymer surfactant and water. To fully characterize the nanocapsules the authors used two NMR instruments and four different sets of experiments including ^{13}C direct excitation, (^{1}H)-^{13}C cross-polarization, relaxation of ^{1}H nuclei in the rotating frame, and ^{1}H pulsed field gradient echo experiments. All spectra were Fourier transformed and results were fitted to a numeric simulation algorithm based on a finite grid approximation.

Direct excitation is a very sensitive technique to detect components in the liquid and dissolved state. When compared to reference spectra of the individual components of the nanocapsule dispersion, no monomer peak was found with this method, implying that all of the monomer must have polymerized into a solid state.

For components in the solid state, (^{1}H)-^{13}C cross-polarization can be efficiently used, which is based on a transfer of magnetization between these nuclei. Spectra clearly showed the polymer in the solid state, associating with the triglyceride oil and surfactant in a time-dependent manner. Cross-polarization between the polymer, triglyceride, and surfactant is initiated by a temporary adsorption of these liquid molecules to the solid capsule surface (Hoffmann and Mayer, 2000). During this short period of immobilization, ^{1}H nuclei transfer their polarization (magnetization) to ^{13}C nuclei. After desorption, the molecules gain high mobility again, resulting in a narrow line in the corresponding spectrum, whereas solid-state signals are usually broad and less well resolved.

The phenomenon of spin–lattice relaxation in the rotating frame of ^{1}H nuclei led to monoexponential decay in direct excitation measurements. In the cross-polarization experiments a biexponential relaxation was found and could be attributed to the interaction between the solid and the liquid components.

FT pulse gradient spin echo (FT-PGSE) experiments, which are also a very convenient tool in the characterization of microemulsions (as described in detail in Alany et al., 2008), can give valuable information on the location of the liquid components. The sensitivity of this technique is limited to ^{1}H nuclei. Rotational mobility is measured with the previous techniques to distinguish between the solid and the liquid state, but PGSE measures the lateral mobility to detect the location of liquid components (inside or surrounding the particle). Briefly, in PGSE experiments the magnetic field is superimposed twice causing dephasing and refocusing of the spins, which leads to a decay in echo intensity depending on the self-diffusion (Brownian motion) of the species or hindrance by the capsule wall. Analysis of the PGSE experiments revealed rapidly decreasing water and ethanol resonances and a much slower decrease for the surfactant and triglyceride oil. The diffusion coefficients for water were clearly on the order of the magnitude of pure water, thus forming the particle surrounding the continuous phase. However, the oil showed hindered diffusion but equivalent to the diffusion of the nanoparticles. From the diffusion coefficient, the particles were calculated to be 120 nm. Therefore, the oil is located in the nanocapsule core and diffusion in the long-range order is dominated by the diffusion of the nanocapsules. Self-diffusion of the surfactant was unhindered and the coefficient related to about 10-nm particles, thus referring to micelles in the continuous aqueous phase. A rapid exchange of components inside and outside the capsules was not completely ruled out.

As indicated, NMR measurements not only give valuable insight into the microstructure of the particles and location of the individual components, but they can also be used to determine the particle size and wall permeability.

With NMR, the particle size can be measured by analyzing the NMR line shape or from the PGSE decay for the particles under observation of diffusion in the long-range order and the encapsulated components in the short-range order.

Particle size measurements with self-diffusion NMR are based on the following relationships (Equations 4.2 and 4.3), in which the correlation time (τ) of the rotational diffusion of a spherical particle and the self-diffusion coefficient (D) is determined by its radius (r) in a liquid corresponding to its Brownian motion and depending on the η of the medium, the Boltzmann constant (k_B), and the temperature (T):

$$\tau = \frac{4\pi\eta r^3}{3k_B T} \tag{4.2}$$

and

$$D = \frac{k_B T}{6\pi\eta r} \tag{4.3}$$

In Mayer et al.'s (2002) study, a particle size below 100 nm could only be estimated from the line shape analysis because of difficulties resulting from poor spectral resolution. Fitting the PGSE data to a numerical simulation yielded a mean particle size of 120 nm, which was in agreement with optical particle size measurements.

For the capsule permeability, a free and an encapsulated fraction could be assigned from the PGSE resonances of ethanol. The encapsulated fraction, however, decreased when increasing the observation period. This led to the conclusion that ethanol can permeate freely through the nanocapsule wall. In a subsequent study of PGSE NMR on PBCA nanocapsules (Wohlgemuth and Mayer, 2003), a higher exchange of benzene used as a tracer molecule was found for the PBCA nanocapsules prepared by the authors in comparison to the original nanocapsules prepared by Al Khouri Fallouh et al. (1986). Carrying out permeability studies on PACA nanoparticles with time-resolved PGSE NMR will thus allow long-term release data to be obtained.

Although NMR is a versatile technique to characterize PACA nanoparticles, to our knowledge the technique has not yet established itself as a standard technique to be commonly used in the characterization of such drug delivery systems. For instance, in a recent study on poly(ethyl cyanoacrylate) (PECA) nanocapsules, direct excitation NMR experiments were only used to show that polymerization was complete (Bogdan et al., 2008). Particle size and morphology were determined using PCS and TEM, respectively.

Possible reasons for the less frequent use of NMR in comparison to the more common techniques introduced above might be a lack of availability of this specialized equipment and the expertise to analyze and model the data. Considering that its application to PACA nanocapsules was only published recently, we can assume that the technique will become more popular in the near future.

In general, depending on how thorough of an investigation we want on the nature of PACA nanoparticles, a combination of techniques has to be applied to get meaningful results. The choice of techniques also depends on what the focus of the study is. Researchers may focus on the entrapment, release, or distribution profile of the particles of choice rather than the size or morphology. Finally, as already mentioned, the availability of equipment often determines the particle characteristics to be investigated.

4.3.7 Drug Loading

Drug loading and entrapment efficiency (EE) are two important properties of nanoparticles. A variety of drugs, including hydrophilic or lipophilic drugs, peptides, proteins, and nucleic acid derivatives, have been successfully associated with PACA nanoparticles (Table 4.1). This is a result of the different types of PACA nanoparticles and the available templates that can be used to prepare these particles. Drugs to be entrapped in PACA nanoparticles are generally dissolved in the polymerization medium; hence, EE will greatly depend on the drug solubility in the polymerization medium. Therefore, nanospheres favor the incorporation of water-soluble drugs whereas entrapment in nanocapsules depends on the component forming the core of the particle. For nanospheres, however, high drug loadings of poorly water-soluble drugs have been achieved by the addition of cyclodextrins to the polymerization medium (Boudad et al., 2001; Duchêne et al., 1999; Monza da Silveira et al., 1998). The use of different structure types of biocompatible microemulsion templates has resulted in nanoparticles with similar properties (Krauel et al., 2005, 2006; Graf et al., 2008b). This enables the formulator to select the microemulsion template according to the solubilization behavior of the drug for optimal loading and release without changing the nanoparticle properties. It has been reported that the EE of lipophilic drugs correlates with their solubility in the contained oil (Fresta et al., 1996).

In addition to solubility, the polymerization kinetics and molecular weight of the drug are important factors to be considered for the entrapment of drugs in PACA nanoparticles. In general, the encapsulation efficiency depends on the partition of the drug between the oil and aqueous phase and its molecular weight. This is particularly important when entrapping peptide drugs because these are normally not oil but water soluble. In contrast, the shorter the alkyl chain length of the alkyl cyanoacrylate is, the faster the polymerization (and the degradation); thus, a higher mechanical entrapment is possible (Arias et al., 2007). Conversely, the longer the alkyl chain is, the higher the degree of interdigitation of the polymer (El-Samaligy et al., 1986). These will have implications on drug entrapment and release from the nanoparticles. Entrapment of hydrophilic drugs into nanocapsules prepared from O/W templates was found as a result of the very fast polymerization rate of alkyl cyanoacrylate monomers (Aboubakar et al., 1999a; Arias et al., 2007; Damgé et al., 1988; Lowe and Temple, 1994). Drug molecules with a small molecular weight have a faster diffusion coefficient and may thus escape the forming polymer network, particularly if they are soluble in the continuous phase of the polymerization template, leading to low entrapment (Aboubakar et al., 1999a; Graf et al., 2008a, 2008b; Pitaksuteepong et al., 2002). Entrapment is generally believed to be higher in a droplet-type template in which the molecule is confined within the droplet around which the polymerization occurs (Krauel et al., 2005). However, encapsulation of low molecular weight hydrophilic and lipophilic molecules was compromised in those templates as well (Damgé et al., 1997a; Hillaireau et al., 2007). Pitaksuteepong et al. (2002) found that ionic interactions play an important role in the entrapment of low molecular weight compounds. Formation of ion pairs of low and high molecular weight polycations was recently suggested as a successful strategy in overcoming this problem (Hillaireau et al., 2007). In that regard, the pH at which the polymerization is performed and possible implications on the charge of the drug at this pH should be considered, because entrapment may be hampered by electrorepulsive forces between the particle and the drug.

TABLE 4.1

Summary of *In Vivo* and *In Vitro* Studies on PACA Nanoparticles as Drug and Vaccine Delivery Systems

Polymer	Bioactive	Preparation Method or Template	Size (nm)	EE (%)	Species	Observations	Reference
				In Vitro and *In Vivo* Studies			
PMCA, PECA, PBCA	Fluroescein, DOX	Interfacial polymerization or W/O emulsion	500–1500	55–74	*In vitro*	Drug release PMCA > PECA > PBCA, increasing monomer concentration led to retardation of drug release	(El-Samaligy et al., 1986)
PECA	ETS, DPH, CBZ	Interfacial polymerization or O/W emulsion	100–400	≤12	*In vitro*	Controlled drug release by diffusion from oil core through intact polymer wall	(Fresta et al., 1996)
PECA	FITC-OVA	Interfacial polymerization or W/O microemulsion	300–400	97	Dendritic cells	Morphological changes indicative of dendritic cell maturation	(Pitaksuteepong et al., 2004)
PECA	FITC-OVA	Interfacial polymerization or microemulsions	250	8–92	*In vitro*	Nanoparticle morphology determined more by dynamics within template than its microstructure	(Krauel et al., 2005)
PECA, BSA, chitosan	Ganciclovir	Interfacial polymerization or W/O emulsion	300	65	Rabbit	Improved ocular drug concentration	(El-Samaligy et al., 1996)
PECA	Insulin (bovine)	Aqueous acidic medium	<500	85	Rat	Significant glucose reduction with absorption enhancer	(Radwan and Aboul-Enein, 2002)
PECA	Insulin (Humulin®)	Interfacial polymerization or W/O microemulsion	150	86	*In vitro*	Interfacial polymerization in microemulsion convenient for peptide entrapment in PACA	(Watnasirichaikul et al., 2000)
PECA, PBCA	Insulin (Humulin)	Interfacial polymerization or microemulsion	200–400	10–50	*In vitro*	Decisive release factors different *in vitro* as compared to *in vivo*	(Graf et al., 2009)
PECA, PBCA	Insulin (Humulin)	Interfacial polymerization or microemulsions	160–400	10–20	*In vitro*	Insulin interference with polymerization, moderate entrapment due to insulin escaping	(Graf et al., 2008b)
PECA, PBCA	Insulin (Humulin)	Interfacial polymerization or microemulsions	150–370	9–18	*In vitro*	Valuable characterization with self-diffusion NMR, promising microemulsion templates	(Graf et al., 2008a)

PECA, PBCA	LCMV$_{33-41}$	Interfacial polymerization or W/O microemulsion	200	≥90	In vitro	Functionalized copolymerization	(Liang et al., 2008)
PBCA	Avarol	Alcohol medium	—	—	Rat	Eight- to ninefold increase in oral bioavailability	(Beck et al., 1994)
PBCA	BSA	Aqueous acidic medium	100	70–80	Rat	Significantly increased immune response	(O'Hagan et al., 1989)
PBCA	Insulin (porcine, crystalline, Zn)	Aqueous acidic medium	255	79	Rat	57.2% Relative pharmacological bioavailability*	(Zhang et al., 2001)
PBCA	LHRH	Aqueous acidic medium	100	Copolymerized	Rat	Significant serum testosterone reduction	(Hillery et al., 1996a, 1996b.)
PBCA	Methotrexate	Aqueous acidic medium	178–318	42–87	In vitro	Variation in polymerization technique results in particles with different properties suitable for different drug delivery routes	(Reddy and Murthy, 2004)
PBCA	Calcitonin (salmon)	Interfacial polymerization or W/O emulsion	50–300	30–50	Rat	Hypocalcemic response in the presence of deoxycholic acid as absorption enhancer, 40–45% bioavailability*	(Vranckx et al., 1996)
PBCA, PHCA	HIV-2	Aqueous acidic medium	443–961	—	Mouse	Necessity of using vaccines with two or more different adjuvants to induce the required immune response	(Stieneker et al., 1995)
PiBCA	ODNs	Interfacial polymerization or W/O emulsion	350	70	Fetal calf serum	Efficient ODN protection from degradation by nucleases upon encapsulation	(Lambert et al., 2000b)
PiBCA	CDV, AZT-TP, ODN with PEI, chitosan	Interfacial polymerization or W/O microemulsion	20–250	90	In vitro	Presence of cationic polymers in nanocapsule led to successful encapsulation of AZT-TP and ODN	(Hillaireau et al., 2007)
PiBCA	Darodipine	Interfacial polymerization or O/W emulsion	—	—	Rat	Reduction in blood pressure	(Hubert et al., 1991)
PiBCA	Indomethacin	Interfacial polymerization or O/W emulsion	—	—	Rat	Protection from ulceration, sustained bioactivity of thromboxane blood levels over 24 h	(Ammoury et al., 1991)

continued

TABLE 4.1 (continued)
Summary of *In Vivo* and *In Vitro* Studies on PACA Nanoparticles as Drug and Vaccine Delivery Systems

Polymer	Bioactive	Preparation Method or Template	Size (nm)	EE (%)	Species	Observations	Reference
PiBCA	Insulin (bovine)	Nanoprecipitation interfacial polymerization or O/W emulsion	312 or 249	99, 90	*In vitro*	Competitive polymerization initiation by ethanol	(Aboubakar et al., 1999)
PiBCA	Insulin (Zn, semicrystalline)	Interfacial polymerization or O/W emulsion	250–300	≥98	Rat/CHO cells	*In vitro* binding of insulin to receptor	(Roques et al., 1992)
PiBCA	Insulin (Humulin, insulin (porcine)	Interfacial polymerization or O/W emulsion	220	55	Rat	Strong hypoglycemic response for up to 20 days	(Damgé et al., 1988)
PiBCA	Insulin (bovine, Texas Red® labeled)	Interfacial polymerization or O/W emulsion	265	90	Rat	Nanocapsules protect insulin from degradation and are significantly involved in absorption mechanism	(Aboubakar et al., 2000)
PiBCA	Insulin (human, Zn), insulin (^{125}I labeled)	Aqueous acidic medium	145	80	Rat	50% Decrease of fasted glycemia for 10–13 days, increased or prolonged uptake in GIT, liver, and blood	(Damgé et al., 1997b)
PiBCA	Insulin (Humulin)	Interfacial polymerization or W/O microemulsion	200	>80	Rat	Significant reduction of glycemia for 60 h over control when dispersed in microemulsion	(Watnasirichaikul et al., 2002)
PiBCA	Insulin (Velosulin®)	Interfacial polymerization or O/W emulsion	297	98	Rat	Intensive and prolonged after ileal administration	(Michel et al., 1991)
PiBCA	Insulin	Interfacial polymerization or O/W emulsion	<300	—	Rat	Long-lasting strong hypoglycemic response, reduction of hyperglycemic peak	(Damgé et al., 1990)
PiBCA	Insulin	Aqueous acidic medium	300	40	Rat	Prolonged hypoglycemic effect s.c. but not orally	(Couvreur et al., 1980)
PiBCA	Insulin (bovine)	Aqueous acidic medium	<500	65–95	Rat	Significant reduction of glycemia for more than 8 h, pharmacological availability* of 37.6%	(Radwan, 2001)
PiBCA	Insulin (Humalog®)	Interfacial polymerization or O/W emulsion	400	90	Rat	High variability in the concentration of insulin in plasma, no decrease of glycemia	(Cournarie et al., 2002)
PiBCA	Insulin (Humulin)	Aqueous acidic medium	85	59	Rat	Significant reduction of glycemia over control	(Mesiha et al., 2005)

Polymer	Drug	Method	Size	Loading	Model	Effect	Reference
PiBCA	Insulin (human, Zn)	Interfacial polymerization or O/W emulsion	280	>98	NOD mouse	Reduced incidence and delayed onset of diabetes, reduced lymphocytic inflammation of endogenous islets	(Saï et al., 1996)
PiBCA	Insulin/calcitonin (human, ^{125}I labeled)	Interfacial polymerization or O/W emulsion	305	90–95	Rat	No significant overall enhanced peptide absorption	(Lowe and Temple, 1994)
PiBCA	Lipiodol	Interfacial polymerization or O/W emulsion	165	—	Dog	Accelerated, intensified, and prolonged passage of iodine through intestinal mucosa	(Damgé et al., 1987)
PiBCA	ODNs	Interfacial polymerization or O/W emulsion	290	81	Mouse	Inhibition of Ewing sarcoma related tumor after intratumoral injection of cumulative dose	(Lambert et al., 2000a)
PiBCA	Octreotide	Interfacial polymerization or O/W emulsion	260	60	Rat	Significant reduction of prolactin secretion	(Damgé et al., 1997a)
PiBCA	Pilocarpine	Interfacial polymerization or O/W emulsion	370–460	13.5	Rabbit	Significantly increased and prolonged miosis implying improved ocular bioavailability	(Desai and Blanchard, 2000)
PiBCA, PiHCA	GRF	Aqueous acidic medium	146–163	74–83	In vitro	Co-polymerization	(Grangier et al., 1991)
PiBCA, PiHCA	rh-G-CSF	Aqueous acidic medium or nanoprecipitation	200	66 or 90	Mouse	Copolymerization, loss of bioactivity	(Gibaud et al., 1998)
PiBCA, PiHCA	Saquinavir (HPβCD complex)	Aqueous acidic medium	250–350	45–50 µg/mg	In vitro	400-Fold increased apparent solubility and 20-fold increased drug loading by HPβCD inclusion complex	(Boudad et al., 2001)
PHCA, PiHCA	ODNs	Aqueous acidic medium or DEAE dextran	142–624	—	Vero cells, mouse	Protection from degradation, significantly enhanced cellular uptake, improved antisense treatments for tumor, and antiviral therapy, liver targeting	(Zimmer, 1999)
Surface Modifications							
PEG-PECA	Acyclovir	Aqueous acidic medium	200	2 (w/w)	Rabbit	25-fold increase of drug levels in aqueous humor	(Fresta et al., 2001)
PEG-/PS80-PBCA	Dalargin	Aqueous acidic medium	100	40	Mouse	Significant analgesia due to successful brain targeting	(Das and Lin, 2005)

continued

TABLE 4.1 (continued)
Summary of *In Vivo* and *In Vitro* Studies on PACA Nanoparticles as Drug and Vaccine Delivery Systems

Polymer	Bioactive	Preparation Method or Template	Size (nm)	EE (%)	Species	Observations	Reference
PEG-PHDCA	DOX	Nanoprecipitation	140–205	70–85	Rat	Encapsulated DOX unable to elicit improved therapeutic response in the gliosarcoma	(Brigger et al., 2004)
PEG-PHDCA	rHuTNF-α	Double-emulsion method	150	60	Cells, mouse	Protective effect of human serum albumin or BSA during preparation, increased antitumor potency	(Li, et al., 2001a; Li, et al., 2001b)
Polysaccharide–PiBCA	Heparin	Aqueous acidic redox radical polymerization	93–59000	Copolymerized	In vitro	Antithrombotic activity preserved	(Chauvierre et al., 2003)
PS80-PBCA	Dalargin	Aqueous acidic medium	230	30	Mouse	Phagocytic uptake of nanoparticles by the brain–blood vessel endothelial cells, significant analgesic effect	(Kreuter et al., 1995)
PS80-PBCA	DOX	Aqueous acidic medium	270	80	Rat	High DOX brain concentration, reduced toxicity	(Gulyaev et al., 1999; Gelperina et al., 2002)
PS80-PBCA	Loperamide	Aqueous acidic medium	290	47	Mouse	Endocytotic uptake, long and significant analgesic effect	(Alyautdin et al., 1997)
PS80-PBCA	Probenecid, MRZ 2/576	Aqueous acidic medium	251	25–100	Mouse	Prolonged duration of anticonvulsive activity in brain	(Friese et al., 2000)
PS80-PBCA	Tacrine	Aqueous acidic medium	36	15 (w/w)	Rat	Significantly increased tacrine concentration in brain	(Wilson et al., 2008)

Note: AZT-TP, azidothymidine-triphosphate; BSA, bovine serum albumin; CBZ, carbamazepine; CDV, cidofovir; CHO, Chinese hamster ovary; DEAE, diethylaminoethyl; DPH, 5,5-diphenyl hydantoin; ETS, ethosuximide; GRF, growth hormone releasing factor; HIV-2, human immunodeficiency virus 2; HPβCD, hydroxylpropyl-β-cyclodextrin; LCMV, lymphocytic choriomeningitis virus glycoprotein; LHRH, luteinizing hormone releasing hormone; NOD, nonobese diabetic; ODN, oligodeoxynucleotide; PEI, poly(ethyleneimine); rh-G-CSF, recombinant human-granulocyte colony stimulating factor.

Data referring to loading instead of EE indicated by w/w and µg/mg, respectively.

* Pharmacological availability/bioavailability determined based on the extent of hypoglycemic/hypocalcemic response relative to subcutaneous (Vranckx et al., 1996; Radwan, 2001; Zhang et al., 2001) injection of insulin or salmon calcitonin, respectively.

Moreover, a study by Cournarie et al. (2004) on entrapment of insulin in PACA nanoparticles identified that the origin of the monomer with varying synthesis by-products, the common addition of stabilizers or polymerization inhibitors, the pH, the preservatives of the insulin solution, and even the association state of the peptide in solution are key parameters that influence EE. Because of these parameters, it is difficult to compare individual studies. EE can generally be improved by increasing the monomer concentration (El-Samaligy et al., 1986; Krauel et al., 2005).

Monomers are highly reactive and polymerization can be initiated by any kind of nucleophilic group, so one the possibility of copolymerization must be considered, which may compromise the activity of the drug and lead to an uncontrollable polymerization and the formation of polymer aggregates instead of nanoparticles (Behan and Birkinshaw, 2001; Damgé et al., 1997b; Grangier et al., 1991). Aboubakar et al. (1999a) suggested that interference of the drug in the polymerization can be avoided by adding the drug after the polymerization has been initiated or to preformed polymers. The disadvantage of this approach, however, is that the drug is likely to be adsorbed onto the particle surface rather than encapsulated in the core, providing little protection for peptides from enzymatic degradation in the gastrointestinal tract (GIT).

The EE refers to the amount of drug entrapped in the nanoparticles, and it is usually expressed as the percentage of the total amount of drug added to the polymerization template. In contrast, drug loading refers to the amount of drug as a percentage of the polymer content. Nanocapsules with a thin polymer wall confining a central cavity with a high amount of solubilized drug are advantageous in greatly increasing the drug loading when compared to nanospheres, where the drug is distributed throughout a polymer network (Couvreur et al., 2002). Alternatively, drug loading can be expressed as the percentage of drug by weight in the final formulation (Couvreur et al., 2002).

Drug loading and EE can be determined by either a direct or an indirect method. In both methods the nanoparticles are separated from the polymerization template by centrifugation, SEC, or ultrafiltration. The remaining amount of unentrapped drug is determined in the polymerization template in the indirect method, and the amount of entrapped drug is determined after washing and lysis of the nanoparticles with acetonitrile (Damgé et al., 1997b) or tetrahydrofuran (Yang et al., 2000) in the direct method. Quantitative analysis of the drug is usually performed with high-performance liquid chromatography. Both methods can be combined for characterization of nanoparticles.

4.3.8 Drug Release

Once the drug delivery system has reached the site of action, the drug is released. Throughout the literature, degradation of PACA appears to be the driving factor for drug release. For instance, for insulin and human growth hormone, polymer degradation and drug release occur in parallel (Aboubakar et al., 2000; Grangier et al., 1991). This also correlated with the biological effect of insulin, and results indicated that its duration depends on the insulin/polymer ratio (Damgé et al., 1997b). PACA nanoparticles are biodegradable, and hydrolysis of the alkyl side chain has been identified as the major degradation pathway in vivo (Lenaerts et al., 1984). Intestinal and serum esterases hydrolyze the polymer to alkyl alcohol and poly(cyanoacrylic acid), which are both water soluble and eliminated from the body via kidney filtration. The degradation rate of PACA nanoparticles and therefore the drug release depend on the alkyl side chain length, and an increase in length leads to a decrease in the degradation rate (Leonard et al., 1966) because of the higher interdigitation of the longer alkyl cyanoacrylate homologues (El-Samaligy et al., 1986). For low molecular weight molecules, however, diffusion through the polymer rather than degradation can be assumed to be the predominant release mechanism as shown by NMR (Wohlgemuth and Mayer, 2003).

Drug release from nanospheres can be described by matrix kinetics, whereas reservoir-type kinetics are applicable to nanocapsules (Vauthier et al., 2007), often described by zero-order release kinetics (Watnasirichaikul et al., 2002a, 2002b). The use of high monomer concentrations to improve drug entrapment also leads to an increase in wall thickness of nanocapsules (El-Samaligy et al., 1986; Watnasirichaikul et al., 2002a, 2002b) and thus a slower release profile.

In the previous sections we have looked at the most important characteristics of PACA nanoparticles and methods to determine these. A thorough characterization of the delivery systems is a prerequisite to evaluate their application in drug and vaccine delivery, which is the topic of the next section.

4.4   APPLICATIONS OF POLY(ALKYL CYANOACRYLATE) NANOPARTICLES IN DRUG DELIVERY AND VACCINE DEVELOPMENT

4.4.1   Oral Drug Delivery

PACA nanoparticles have a huge potential for biomedical and pharmaceutical applications. Their ability to protect proteins from enzymatic degradation (Lowe and Temple, 1994) and to facilitate absorption from the gut lumen (Damgé et al., 1987) provides the opportunity for safe and effective oral delivery, which is the most convenient and desirable route for drug therapy. Despite their non-swellable and anionic character, PACA nanoparticles exhibit bioadhesive tendencies, providing intimate contact with mucosal sites (Ponchel and Irache, 1998). Mucosal adhesion slows down the transit of the carrier and the drug, thereby enhancing drug absorption and translocation. Translocation through the intestinal barrier has been visualized using Texas Red® labeled insulin PACA nanoparticles (Aboubakar et al., 2000). Among others, uptake of PACA nanoparticles via Peyer's patches appears to be the major pathway described in the literature (Aprahamian et al., 1987; Damgé et al., 1990; Michel et al., 1991; Vauthier et al., 2003b). Research has focused on oral delivery of peptides and proteins because they are important in the treatment of chronic metabolic diseases. Oral delivery of proteins and peptides is a problem that is attributable to several unfavorable physicochemical properties of these molecules, especially taking into account the physiology of the GIT. The presence of digestive enzymes and the barrier of epithelial transport are physiological obstacles that significantly hinder absorption. As a result, the oral bioavailability of proteins and peptides is extremely low, usually less than 1%.

One of the first peptide drugs to be delivered with PACA nanoparticles was insulin, which is the most studied model peptide in the context of particulates for oral drug delivery. The first attempt was made in 1980 by Couvreur et al. (1980), but they were unable to achieve a hypoglycemic effect after oral administration because insulin was adsorbed to the nanoparticle surface. Upon encapsulation of insulin into the oily core of PiBCA nanocapsules, however, this group achieved a prolonged hypoglycemic response for up to 20 days in a dose-dependent manner (Damgé et al., 1988). The particles were prepared by interfacial emulsion polymerization from an O/W emulsion and yielded an EE of only 55%. The prolonged effect was assigned to "a progressive arrival of nanocapsules from the stomach to the gut" and delayed absorption via paracellular translocation in the ileum (Aprahamian et al., 1987; Michel et al., 1991). A study by Lowe and Temple (1994) found maximum plasma levels of insulin and human calcitonin administered in PiBCA nanoparticles as early as 5–15 min after administration with a rapid decrease thereafter and no overall significant absorption enhancement. The authors could not confirm the long-term effect of insulin reported by Damgé et al. (1988), but they suggested that it may have been the result of a local effect of insulin that directly affected intestinal cells. A few years later, Damgé et al. (1997b) studied PiBCA nanospheres prepared by dispersion polymerization for oral delivery of insulin. To avoid copolymerization, insulin was added after polymerization had been initiated, yielding an entrapment of 80%. The authors were again able to achieve a long-term hypoglycemic effect for 10–13 days when nanospheres were administered in an oily medium containing surfactants. However, the hypoglycemic effect lasted only for two days in an aqueous medium containing surfactants. Radwan (2001) performed a similar study on PiBCA nanospheres and the absorption enhancement of insulin in the presence of surfactants. Cholic acid had a better absorption enhancing effect than deoxycholic acid in an aqueous medium. A hypoglycemic effect with the nanospheres was maintained for less than 24 h and the authors suggested that this was due to an initial burst release of insulin in an aqueous medium enhanced by the presence of surfactants from the nanosphere surface. Thus, nanocapsules should be the preferred delivery system for peptides. Following on from the study by Damgé et al. (1988) on oral delivery of insulin, Cournarie et al. (2002) recently attempted to characterize the absorption of insulin in the blood. PiBCA nanocapsules were prepared in the same manner as Damgé et al. but with another type of insulin, and this resulted in an EE of 90%. Although significant levels of insulin were detected in rat plasma 30–60 min after intragastric administration of insulin-loaded nanocapsules, the levels were very erratic between animals and lacked a hypoglycemic response. Parenteral injection of insulin indicated that a high insulin concentration was necessary to achieve a hypoglycemic response, and a different response between normal and diabetic rats was found. The authors concluded that the high insulin concentration required to elicit a hypoglycemic response was not achieved orally and that diabetic rats developed insulin resistance.

The variability in EE and pharmacodynamic responses found from PACA nanoparticles prepared by the same method presumably led to an interesting study by the same group (Cournarie et al., 2004) on the parameters affecting the formulation of insulin-loaded nanocapsules, which was mentioned in Section 4.3.7.

A pharmaceutically acceptable and easily scaled up formulation for oral delivery of insulin was developed by Watnasirichaikul et al. (2002a, 2002b). Biocompatible W/O microemulsions were used as templates for PiBCA nanocapsules that are suitable for oral administration as a dispersion to facilitate the absorption of insulin. The EE of 200-nm nanocapsules was 80%. Insulin retained its bioactivity in the insulin-loaded nanocapsules dispersed in the microemulsion as shown by subcutaneous (s.c.) administration. Intragastric encapsulation of insulin in PACA nanocapsules resulted in a significantly greater hypoglycemic response than unencapsulated insulin in a microemulsion or in solution. The parenteral hypoglycemic response could be modulated, depending on the monomer concentration. The polymer wall of the nanocapsules increased with increasing monomer concentration, thus controlling the release (Watnasirichaikul et al., 2002a, 2002b). The authors did not examine intragastric formulations prepared with different monomer concentrations. However, a significant hypoglycemic response over controls was achieved for nearly 60 h following administration. Following this study, insulin-loaded PACA nanoparticles dispersed in microemulsions with different microstructures were recently examined (Graf et al., 2009). Microemulsions were polymerized with different monomer types (ethyl cyanoacrylate and isobutyl cyanoacrylate) and different amounts of monomer. An increasing monomer concentration led to a higher EE, but an increasing aqueous fraction, depending on the microemulsion structure type, decreased the entrapment of insulin. Insulin loading showed an opposite trend. Nanoparticles were spherical and 200–400 nm in size without any significant difference among different microemulsion templates and types and amounts of monomer. Confirming the results of Watnasirichaikul et al. (2002a, 2002b), *in vitro* release could be controlled by the monomer concentration. Formulations prepared with 1200 mg of monomer were used for *in vivo* experiments and a consistent and significant hypoglycemic effect over controls was found for up to 36 h, depending on the type of monomer. Release of insulin could be controlled by the monomer concentration *in vitro*, but the degradation kinetics of the monomer type took effect on the *in vivo* pharmacodynamic response. Similar to the study by Cournarie et al. (2002), insulin plasma levels were not detectable. These findings support the hypothesis by Lowe and Temple (1994) of local intestinal effects of insulin. Another recent study on oral absorption of insulin PiBCA nanospheres with an average size of only 85 nm and 59% EE resulted in a significant reduction in blood glucose over controls. Plasma insulin levels were not determined. The aim of the study was to show better bioavailability that was due to the smaller particle size, but an enhanced hypoglycemic effect was only found for up to 60 h and no comparison to other particle sizes was made (Mesiha et al., 2005).

Insulin-loaded PACA nanoparticles have also been pursued as a prophylactic strategy to prevent diabetes instead of the common treatment approach. Sai et al. (1996) found that oral administration of insulin encapsulated in PiBCA nanocapsules prepared by interfacial polymerization in an O/W emulsion reduced the incidence of diabetes in a nonobese diabetic mice model, which mimics human type 1 diabetes mellitus. The authors suggested that in humans this form of prophylactic insulin administration would be more convenient, comfortable, and less constraining than parenteral administration by making use of the long-term effect observed previously with these nanoparticles by Damgè et al. (1995, 1988). However, diabetes could not be prevented in all mice and the question remained as to what extent and for how long the incidence and the progression of diabetes could be reduced.

Peptides other than insulin, such as calcitonin (Lowe and Temple, 1994; Vranckx et al., 1996) and octreotide (Damgé et al., 1997a), have also been studied in context with PACA nanoparticles. Although calcitonin (Lowe and Temple, 1994) entrapped in PiBCA nanocapsules was protected from degradation after duodenal administration, maximum plasma levels reached 15–30 min after administration were not significant from controls. Vranckx et al. (1996) achieved an absolute bioavailability of 40% after oral administration of salmon calcitonin in PBCA nanocapsules. The increase in bioavailability was attributable to coadministration of deoxycholic acid as a penetration enhancer. For octreotide (Damgé et al., 1997a), encapsulation in PACA nanoparticles resulted in a significant therapeutic effect after peroral administration to estrogen-treated rats with increased plasma levels over controls.

Apart from proteins and peptides, PACA nanoparticles have also been used as a delivery system for other drugs including vincamine (Maincent et al., 1986), mitoxantrone (Beck et al., 1994), and darodipine (Hubert et al., 1991). Administration of these drugs with PACA nanoparticles resulted in increased bioavailability. To avoid gastric side effects, encapsulation of indomethacin in PACA nanocapsules was used as an oral delivery method and resulted in a significant reduction in ulceration (Ammoury et al., 1991). Mucosal protection was assigned to the particles preventing the drug from having direct contact with the mucosa in combination with sustained release from the nanoparticles.

Given the many interesting properties that PACA nanoparticles offer for oral delivery of drugs and the numerous studies involving PACA nanoparticles as delivery systems, it is remarkable that they have not yet entered the market. A question about the clinical relevance of PACA nanoparticles arises; for instance, Damgé et al. (2007) have recently moved away from PACA to polycaprolactone nanoparticles for oral delivery of insulin.

4.4.2 Ocular Delivery

The bioadhesive properties inherent to PACA offer the possibility for drug delivery to the eye. Drugs delivered to the eye are usually cleared rapidly because of tear turnover and lacrimal drainage. Poor permeability of the corneal epithelium also imposes a problem to ocular drug delivery. PACA nanoparticles have been investigated as sustained release delivery systems to maintain an efficient drug concentration in the ocular tissue, thereby reducing the number of administrations in the treatment of glaucoma and cytomegalovirus infection. PACA nanoparticles containing betaxolol were the first preparation developed for the treatment of glaucoma, yet it did not yield a significant therapeutic effect (Marchal-Heussler et al., 1992). Pilocarpine-loaded PACA nanoparticles administered in a Pluronic® gel were more promising and significantly increased the bioavailability of the drug (Desai and Blanchard, 2000).

For the treatment of cytomegalovirus infection, ganciclovir encapsulated in PECA nanocapsules provided a controlled sustained release of the drug and maintained a therapeutic level for up to 10 days (El-Samaligy et al., 1996). The drug is accumulated specifically in the retina and the vitreous humor. Although this would be desirable in the treatment of this viral retinitis, toxic side effects including lens opacification and humor turbidity would have to be eliminated. Ocular tolerability has recently been improved by coating the nanoparticles with PEG (Fresta et al., 2001). Coating with PEG has also improved the bioavailability of tamoxifen upon intraocular administration in the treatment of autoimmune uveoretinitis (de Kozak et al., 2004).

For further reading on PACA nanoparticles in ocular delivery, the reader is referred to reviews by Ludwig (2005) and Bu et al. (2007).

4.4.3 DRUG DELIVERY ACROSS THE BLOOD–BRAIN BARRIER

Attempts to deliver drugs into the systemic circulation are the challenge of every formulation scientist, and there have been numerous attempts to improve the uptake of compounds. One of the most difficult challenges faced in this context is delivery across the blood–brain barrier (BBB). The brain is protected by capillary endothelial cells lining the microvessels, which contain much tighter junctions than in peripheral vessels (Wolberg and Lippoldt, 2002). This prevents the entry of most drug molecules, peptides, antibodies, and so forth.

There have been various attempts to overcome the BBB, including the osmotic opening of the BBB (Wilson et al., 2008), the use of biologically active agents (e.g., histamine, serotonin, and bradykinin), and the intracerebral application of drugs. However, manipulating the BBB by these means is generally very invasive and may pose the risk of causing cerebral edema. Polymeric nanoparticles have gained increased interest as possible delivery systems for the BBB, and special credit has to be given to Jörg Kreuter and his collaborators for the great advances made in the use of PACA nanoparticles coated with polysorbate 80 (PS80) as delivery systems (Kreuter et al., 1995). Examples of drugs that have been successfully delivered across the BBB are the hexapeptide dalargin (Kreuter et al., 1995); the dipeptide kytorphin (Schroeder et al., 2000); loperamide (Alyautdin et al., 1997); tubocurarine (Alyautdin et al., 1998); the N-methyl-D-aspartic acid receptor antagonist MRZ 2/576 (Friese et al., 2000); DOX (Gulyaev et al., 1999); and recently tacrine, a cholinesterase inhibitor used in the treatment of Alzheimer's disease (Wilson et al., 2008).

To deliver the drug tacrine across the BBB, Wilson et al. (2008) prepared PBCA nanoparticles by polymerization in an aqueous phase based on the method described by Kreuter (1994). The drug was added to the aqueous medium before polymerization, so that entrapment of the drug occurred by encapsulation into the polymer matrix and particles were coated with 1% PS80 after formation. A freeze-dried preparation of the nanoparticles was resuspended in phosphate-buffered saline (PBS) for i.v. application into the tail vein of rats at a dose of 1 mg/kg body weight. However, it is unclear from the publication whether this dose was related to the amount of tacrine or nanoparticles. One hour after application the animals were sacrificed and the brain, liver, lungs, spleen, and kidneys were removed and analyzed for drug content by high-performance liquid chromatography. When bound to nanoparticles, tacrine showed increased accumulation in the liver, spleen, and lungs compared to the free drug applied in PBS. When the drug was encapsulated in PS80-coated nanoparticles, lower amounts were found in the liver, spleen, and lungs and the amount of tacrine found in the brain was increased compared to the formulations containing the free drug in PBS or the drug bound to uncoated nanoparticles. At the same time an increased amount of tacrine was found in the kidneys when delivered with PS80-coated particles. The amount of tacrine found in the kidneys was also high for the free drug and the drug bound to uncoated particles, so this could be a point of concern for possible kidney damage and should be investigated more thoroughly. The study by Wilson et al. (2008) showed, however, that delivery of tacrine across the BBB was possible. It is now necessary to investigate further any possible toxic effects of the drug that are due to the higher availability in the brain and other organs, especially after long-term application. Any effects against Alzheimer's characteristics (e.g., amyloid-rich plaques and neuronal degeneration) must be shown in further studies.

Recent interest in BBB delivery with PACA nanoparticles has focused on the delivery of anticancer drugs to treat brain tumors, because effectiveness was shown for the drug DOX against various tumors (Gelperina et al., 2002) and the ability of this drug to cross the BBB when attached to PACA nanoparticles (Kreuter et al., 1995). Possible reasons for the accumulation of particles in solid tumors are the discontinuous vascularization of the endothelium, allowing extravasation of the particles and reduced lymphatic drainage, which supports accumulation. Apart from being an anatomical barrier, the BBB possesses a P-glycoprotein (P-gp) efflux pump for antineoplastic drugs, which can also be found in other tumors on the cellular level and is the main reason for multidrug resistance (MDR) in chemotherapy. DOX is a P-gp substrate that results in low uptake of the free drug into tumors. Couvreur and Vauthier (2006) established an explanation of how entrapment of DOX to PACA nanoparticles can overcome MDR, and a more concise explanation is given in Section 4.4.4.

Without going into further detail about the numerous studies that have been carried out and to summarize, PACA nanoparticles are able to transport various drugs efficiently across the BBB. The coating of the nanoparticles with PS80 seems to be vital for efficient uptake of the particles (Gelperina et al., 2002). Other common molecules to coat nanoparticles are PEGs. Coating with PEG efficiently prevents opsonization of particles by plasma proteins and thereby extends their circulation time. For the delivery across the BBB, Brigger et al. (2004) demonstrated that coating of poly(hexadecyl cyanoacrylate) (PHDCA) nanoparticles with PEG proved to be counterproductive for brain delivery of DOX. DOX associated with PEG-PHDCA particles did not show an increased uptake across the BBB or into intracerebrally implanted gliosarcoma in rats but instead accumulated in the lungs and spleen. The authors attributed this to the increased plasma protein interaction that is due to the positive charge of the resulting PEG-PHDCA particles.

When binding a drug to a colloidal carrier, the distribution profile and its toxicity may change. Various studies have already proved a decrease in toxicity of DOX, more specifically reduced cardiotoxicity, when attached to PACA nanoparticles (Vauthier et al., 2003a). When looking at the delivery of a substance across the BBB, however, toxicity studies have to be evaluated even more carefully and a safe dose established, which should not be automatically concluded from toxicity studies aimed at other organs. In a study involving the delivery of DOX bound to PS80-coated PBCA nanoparticles to implanted glioblastoma, Gelperina et al. (2002) established that the maximum tolerated dose for DOX formulations was close to 7.5 mg/kg body weight in rats (~20 mg/kg particles), which consequently allowed three doses at 2.5 mg/kg and three doses at 1.5 mg/kg on various days after tumor implantation. At doses considerably exceeding the necessary amount to achieve a therapeutic level of DOX (100–400 mg/kg related to the amount of particles), behavioral and peripheral reactions in the animals were observed, but those were transitory and reversible after 10–15 min.

A recent approach in brain tumor therapy is the combination of the delivery of antisense oligonucleotides (AONs) with PACA nanoparticles across the BBB and the simultaneous application of a cancer vaccine. Schneider et al. (2008) utilized PS80-coated PBCA nanoparticles prepared by aqueous polymerization for the delivery of AONs against transforming growth factor-β_2 (TGF-β_2), which is produced in glioblastoma cells. AONs are short oligonucleotides of deoxyribonucleic acid (DNA) that selectively bind to complimentary messenger ribonucleic acid inside the cytoplasma and thus block specific protein production, for instance, TGF-β_2. In contrast, TGF-β_2, which is produced by glioblastoma cells, inhibits the immune system of patients and therefore renders the therapy with Newcastle disease virus modified tumor cells inefficient. Schneider et al. (2008) first investigated the delivery of a reporter gene across the BBB with PS80-coated PBCA nanoparticles in rats. After this approach was successful, the group next attached an FITC-decorated AON encoding for the blockage of TGF-β_2 to the nanoparticles and showed that delivery of the AON was also successful and consequently reduced the production of TGF-β_2 in glioblastoma cells. The next step was to combine the application of the Newcastle disease virus vaccine and the AON-bearing nanoparticles to rats with implanted glioblastomas. This approach led to an 18% improvement in the survival rate of the animals, which the authors regarded as significant considering the high malignancy of these types of tumors. They attributed the improved survival rate to the reduction in TGF-β_2 by the AON and an increase in the amount of CD25[+] T-lymphocytes, the latter of which is a sign of an improved immune response.

This section highlighted that delivery of drugs across the BBB is possible, at least in animal studies. In particular, the treatment of brain tumors with PS80-coated PACA seems to be a feasible approach.

4.4.4 DRUG DELIVERY INTO TUMORS

Since the first publication in the 1980s on the preparation and use of PACA nanoparticles as colloidal drug delivery systems, delivery of drugs into tumors has been one of the most extensively investigated areas (Couvreur et al., 1979) and has mainly focused on DOX as a model drug. Generally, when seeking to deliver drugs into tumors, decreasing the high toxicity of the drug is the aim. High doses of the drug are commonly administered because biodistribution of the drug into untargeted healthy organs leads to low uptake levels in tumors. DOX is a prime example of this because it has the adverse effect of cardiotoxicity when delivered in the free form. However, tumors are heterogeneous systems and delivery of drugs has to overcome various obstacles. These include the physiological barrier that a tumor presents, resulting in drug resistance on the noncellular level; poor vascularization of the tumor; and increased interstitial pressure, which may lead to an outward fluid flow. The properties of the interstitium can also prevent uptake of the drug or the drug itself can exhibit unfavorable properties for tumor uptake, such as those due to its surface charge and conformation. Changes in the enzymatic activity of malignant cells, for example, topoisomerase activity, apoptosis regulation, and the P-gp efflux pump, present obstacles on the cellular level of the tumor.

After i.v. injection, colloidal particles are opsonized by plasma proteins and rapidly cleared from the blood by uptake into the mononuclear phagocyte system. Therefore, biodistribution of PACA nanoparticles often shows high counts of particles accumulated in the liver. Targeting of hepatic cancers or metastasis in the liver thus seems intuitive. Chiannilkulchai et al. (1990) showed that DOX bound to PACA nanoparticles was superior in the reduction of hepatic metastasis numbers in a murine reticulosarcoma model (M5076) compared to the free drug. Further extensive research in this area over the past few years resulted in the development of Transdrug by BioAlliance Pharma. Until recently, this product was in phase II clinical trials for the treatment of primary hepatocellular carcinoma, which has a survival rate of only 5%. The formulation is a freeze-dried preparation of DOX entrapped into PiHCA nanoparticles with cyclodextrin as a stabilizer; application occurs via the intra-arterial, i.v., or oral route. As mentioned earlier, the drug safety monitoring board of the company decided to suspend the trials because of adverse pulmonary events.

One specific advantage of PACA nanoparticles for the delivery of DOX and other anthracyclines is their ability to overcome MDR on the cellular level. MDR on the cellular level is caused by the active drug efflux from the tumor cells by the P-gp pump. Couvreur and Vauthier's (2006) excellent review explains that the mechanism of how PACA nanoparticles can overcome this effect is by adhesion of the nanoparticles into tumor cells and subsequent formation of the ion pair poly(cyanoacrylic acid) (a degradation product of PACA) and DOX. This ion pair is then able to cross into the cell because it is no longer a substrate of the P-gp efflux pump.

Other approaches to target tumors with PACA nanoparticles are the decoration of particles with PEG to prevent opsonization. Biodistribution studies of PEG-PHDCA nanoparticles showed that these particles circulated longer in the blood, showed reduced liver uptake (Peracchia et al., 1999), and increased brain accumulation (Calvo et al., 2001). These particles also exhibited an increased accumulation in implanted gliomas (Gelperina et al., 2002). A recent study by Brigger et al. (2004), however, which used the same particles loaded with DOX, found that uptake of DOX into implanted glioblastomas was not improved when compared to the free drug, which demonstrates the importance of remembering that incorporation of a drug into colloidal carriers may alter their distribution profile.

The fact that human cancer cells often express folic acid binding proteins on their surface has been utilized in the investigation of PEG-coated PACA nanoparticles conjugated with folic acid. These particles showed a higher affinity to the folate receptor on human cancer cells than free folate, which was attributed to the existence of a multivalent folic acid form on the surface of the nanoparticles (Stella et al., 2000). A cancer cell line overexpressing the folate receptor on their surface also confirmed the enhanced receptor-mediated endocytosis of folic acid decorated PACA nanoparticles as shown by confocal microscopy.

Another new perspective is the decoration of PACA nanoparticles with polysaccharides, where the properties of the sugar can influence the surface charge or hydrophobicity of the particles as desired. Examples of polysaccharides used in the coating of PACA nanoparticles are orosmucoid (Olivier et al., 1995), dextran (Chauvierre et al., 2003), diethylaminoethyl dextran (Zimmer, 1999), and chitosan (Chauvierre et al., 2003). More information on polysaccharide coating of colloidal carriers can be found in the review by Lemarchand et al. (2004).

Further investigations utilizing PEG-coated PACA nanoparticles in tumor delivery have been carried out by entrapping tumor necrosis factor-α (TNF-α) into these systems. TNF-α shows direct cytotoxicity against various kinds of tumor cells, activation of an immune antitumor response, and induction of hemorrhagic necrosis in an animal model of some transplanted tumors (Li et al., 2001a). The plasma half-life of these agents is very short however, so high doses are required that lead to toxic side effects. Li et al. (2001b) demonstrated the attachment of recombinant human TNF-α (rHuTNF-α) to PEG-PHDCA nanoparticles, and coating with albumin improved the stability of rHuTNF-α. However, cumulative release was only 47% after 14 days, which may affect the effectiveness of this system *in vivo*. A consecutive *in vivo* study with these particles showed delayed blood clearance and higher accumulation of rHuTNF-α in S-180 tumor nodules in mice, resulting in a marked reduction of the tumor size when compared to free rHuTNF-α. Moreover, the formulation was well tolerated by the animals, demonstrating the reduced toxicity of rHuTNF-α when bound to nanoparticles. This may open a new option of tumor treatment with PACA nanoparticles.

AONs present another measure in the treatment of tumors. Because they are short DNA fragments, AONs face the problem of low bioavailability that is due to fast degradation by exo- and endonucleases, and their negative charge also lowers their uptake into cells. Lambert et al. (2000b) showed that the encapsulation of AONs into the aqueous core of PiHCA nanoparticles by interfacial polymerization in a W/O emulsion is possible and that improved stability of AONs can be achieved via this method. *In vivo* experiments with these particles utilized encapsulated phosphorothioate AONs against EWS Fli-1 chimeric ribonucleic acid and showed an inhibition of tumor growth in experimental Ewing sarcoma after intratumoral application compared to free AON (Lambert et al., 2000a). Finding ways that allow for more convenient administration of AONs with PACA nanoparticles (e.g., i.v. or s.c. by coating with PEGs) may develop these colloidal carriers as an interesting alternative in tumor treatment.

Arias et al. (2007) recently demonstrated successful encapsulation of Ftorafur, an anticancer drug used for the treatment of various cancers such as brain tumors and liver cancer, into PECA and PBCA nanoparticles. The authors deliberately chose these monomers for chronic treatment because their fast degradation rate and accumulation of degradation products should be low. Release of the encapsulated drug was complete after 80 min. The *in vivo* activity of these systems, however, still needs to be demonstrated.

4.4.5 Vaccine Delivery

Immunization is the most cost-effective method for controlling and eradicating microbial infections (WHO/UNICEF, 1996). Hurdles in developing suitable vaccines lie in their delivery to the desired site, stability, and cost. A specific problem in this context are subunit vaccines, which have a better safety profile but are poorly immunogenic. However, modern vaccine delivery systems are expected to offer more than delivery to a certain site, better stability, and low cost. In general, the system should be able to stimulate an immunogenic effect at the most desired site, ideally a mucosal and systemic effect, and the induced antibody levels should generate a primed cell population that can rapidly respond to a renewed contact with the pathogen. The need for booster injections should also be avoided with an effective delivery system. Furthermore, it should be possible to administer several antigens with one application and the antigen formulation needs to be easily adjustable to new strains developed by the original pathogen. Approved adjuvants for human use, which help to address these problems to a certain degree, have only been aluminum-based salts and gels until recently. However, these adjuvants show difficulties in achieving a cell-mediated immune response and may lead to adverse local immunoglobulin E stimulation. Considering the outlined obstacles, it is not surprising that the application of "Ehrlich's magical bullet" (i.e., nanoparticles or microparticles) has also found its way into the field of vaccine delivery. Thus far only a liposomal product based on viral membrane proteins (virosome) has been approved for the use in hepatitis A and as an influenza vaccine. An immunostimulating complex (ISCOMS)-based vaccine against equine influenza has been commercialized for some time, and ISCOMSs have also been tested in trials for human use but have not gained approval thus far (Barr and Mitchell, 1996).

Because of their outlined features (e.g., ease of manufacture, good stability, high loading efficiency, controllable release, and low toxicity), PACA nanoparticles have also been investigated for use as vaccine delivery systems (O'Hagan et al., 1989) with very limited success, although the early studies showed promising results. Various reasons can be given for their failure as vaccine delivery systems. Many studies investigating the immunogenicity of nanoparticles used OVA or bovine serum albumin as the model antigen with promising initial immune responses (O'Hagan et al., 1989; Pitaksuteepong et al., 2004). However, albumins are very stable proteins and do not necessarily reflect the greater sensitivity of "real" antigens such as DNA and protein fragments, for which generally only poor immunogenicity is found. An increase in the dose of the nanoparticles and the antigen is often not possible because toxic effects on the immunocompetent cells are a limiting factor. This is especially the case in *in vitro* studies, where the cells are in prolonged contact with the nanoparticles and their degradation products, because they cannot be cleared by the systemic circulation. Brayden (2001) also recently commented on the comparability of animal studies to human studies in the oral application of antigens in particles. It is widely understood that after oral administration particles are taken up by the microfold (M) cells of Peyer's patches in the GIT, which would make them desirable antigen delivery systems because of their immunogenic potential. Brayden considers that the amount of particles actually taken up by the M cells is very low overall and that the amount of M cells in human Peyer's patches accounts for only 5% of the cells compared to 40% in rabbits and 10% in mice. Success in creating an immune response in rodents or rabbits may therefore not be easily transferable to human studies.

Another issue of PACA nanoparticles may be their reasonably short degradation time, which does not allow for an extended release of the antigen, and no advantage may be gained over the application of the antigen in a soluble form. Furthermore, unmodified PACA nanoparticles have a negative surface charge that may prevent (at least to some extent) their attachment to dendritic cells because these also carry a negative surface charge. The more promising results of ISCOMSs as vaccine delivery systems are due to the adjuvant Quil A as part of their structure. Our group therefore investigated whether the immunogenicity of PECA nanoparticles could be improved by the incorporation of Quil A (Krauel, 2005). Using the same assay as for ISCOMS, the Quil A bearing PECA nanoparticles failed to induce an immune response in mouse dendritic cells despite the high entrapment efficiencies of Quil A. ISCOMSs always proved to be the more promising immune-stimulating particles (Demana et al., 2004). One possible reason for the failure of PECA nanoparticles to stimulate an immune response may have been the encapsulation of Quil A into the core of the nanoparticles so that Quil A was not able to interact with the dendritic cells or only interacted in its soluble form because the particles would have degraded within several hours and released Quil A. However, ISCOMSs and their cagelike structure allow Quil A moieties to interact with the cells while still being part of the particle structure.

In our opinion, PACA nanoparticles do not seem to be a promising vaccine delivery system because of the various issues that we have highlighted. The high EE of Quil A, however, may be an incentive for other investigators to try this approach for other, more promising polymer particles.

4.5 TOXICITY

The fact that cyanoacrylates (CAs) are commonly used as adhesives (Superglue®) immediately sparks the concern that these monomers and the polymer must be highly toxic. The toxicity profile of either of them is reasonably low, and the recent U.S. Food and Drug Administration approval of octyl cyanoacrylate (Dermabond®) as an external wound closure agent is promising. An increasing interest in the use of CAs for surgical wound repair (Quinn et al., 1997), embolotherapy (Oowaki et al., 2000), and so forth has raised new concerns about the safety profile of CAs. Earlier studies did show that *in vivo* CAs cause tissue inflammation and necrosis (Toriumi et al., 1991). In another study where CAs were used for the repair of a nonruptured aneurysm, several patients suffered from arterial occlusive lesions. Other effects found in various toxicity studies were occlusion of intracranial arteries (Kawamura et al., 1998), calcification of the dura (Agarwal et al., 1998), and tissue histotoxicity (Kawamura et al., 1998). These effects must be taken into consideration when using CAs in the clinical setting and certainly limit their application. The utilization of longer alkyl chain homologues reduces the toxic effects (Kante et al., 1982), and the C8 homologue octyl cyanoacrylate shows a greatly improved safety profile when compared to the C1 or C2 homologues of CAs. Clinical papers generally claim the production of the degradation product formaldehyde as the main reason for the toxic effects of CAs. This degradation via the reverse Knoevnenagel route only takes place to a very minor extent and the main degradation path is via ester hydrolysis (Lenaerts et al., 1984), which results in the formation of poly(cyanoacrylic acid) and an alkyl alcohol, for example, ethanol for ethyl cyanoacrylate. A major factor in the cause of toxicity, however, is the degradation rate in relation to the alkyl chain length. The degradation rate of the shorter homologues is reasonably short (within 3 h) compared to the longer chain CAs (~14 days for octyl cyanoacrylate), which leads to fast accumulation of degradation products and the need for rapid clearance for the short chain CAs (Müller et al., 1992). This is a major issue when CAs are used for external wound closure, where the area is poorly vascularized and drainage of degradation products is slow. CAs with a longer chain are thus better tolerated. Apart from the degradation products, the polymerization reaction itself can also lead to tissue damage because the reaction is exothermic, producing a reasonable amount of heat. Because the polymerization rate of the longer chain CAs is also slower compared to their shorter homologues, reduced heat production is another reason for the better tolerability of the longer chain CAs.

When using PACAs in the form of nanoparticles, the effect of heat production does not play a role in the toxicity of these systems, because only the polymerized nanoparticles are applied. Investigations regarding the toxicity of PACA nanoparticles have been conducted in conjunction with their discovery as colloidal drug delivery systems. Some of the important findings are summarized below. The details on the studies can be found in the original publications.

In Kante et al.'s (1982) early *in vivo* study in mice, the lethal dose for 50% of subjects was established as 196 mg/kg for PiBCA and 230 mg/kg for PBCA. After s.c. injection of 3 mg of PBCA nanoparticles, no histotoxicity was visible in these animals, not even after two further consecutive injections. Poly(methyl cyanoacrylate) (PMCA) and PBCA nanoparticles also showed no mutagenicity. Morphological changes in macrophages and cell death in hepatocytes were seen *in vitro*, however. Fifteen milligrams of PBCA nanoparticles in 100 mL of incubation medium caused mice hepatocyte death after 2 h. At a concentration of 1% in a culture medium, PMCA nanoparticles showed a damaging effect on mice macrophages. The macrophages were completely perforated when viewed under SEM. These first toxicological results suggested a certain toxicity of the various polymers *in vitro*. Kante et al. (1982) hypothesized that the toxicity may be caused by the degradation products rather than by the polymer itself. A reliable interpretation of these results is problematic because the *in vitro* setting does not allow the drainage of degradation products that would be possible *in vivo*. The observed perforation of the macrophages has to be viewed with skepticism, because the invasive sample preparation of fixing with glutaraldehyde and lyophilization may have been responsible for the damaging effects as well.

A later study by Lherm et al. (1992) investigated the cytotoxicity of various alkyl cyanoacrylates and their degradation products on L929 fibroblasts. They found that the effect on cell death was in the order PiHCA < PMCA < PECA = PiBCA. It was interesting that the nanoparticles displayed greater cytotoxicity than their degradation products alone, which the authors attributed to the ability of the particles to adhere to the cells. Internalization of the particles, however, did not seem to be a means of cytotoxicity but rather the release of degradation products once the particles had attached to the cells. The finding that the PiHCA nanoparticles showed the lowest toxicity supports the hypothesis that the longer alkyl chain homologues exhibit lower toxicity that is due to their slower degradation rate, although some inconsistency seems to exist because PiBCA and PECA appeared to be more toxic than PMCA in Lherm et al.'s (1992) study. *In vitro* incubation of fibroblasts with PECA nanoparticles for 72 h caused an initial decrease in the cell number followed by a slight increase. This means that the cells were able to recover, which was confirmed in later studies (Simeonova et al., 2003). The same study investigated the influence of particle size on the toxicity of the polymer PiHCA. They found that smaller particles (50 nm) showed higher cytotoxicity than the larger particles (200 nm). This was explained by the effect of the larger surface area of the smaller particles forcing them to degrade faster and leading to a higher concentration of degradation products in the cell culture medium.

Because PACA is rapidly taken up mainly by the liver as a part of the reticuloendothelial system, it is important to study the toxic effects of these colloidal systems on the liver. Fernandez-Urrusono et al. (1995) investigated the influence of PiBCA and PiHCA nanoparticles with and without ampicillin in *in vivo* and *ex vivo* models of rat liver cells. After application of an accumulated dose of 200 mg of nanoparticles per kilogram for 14 days, hepatocytes were isolated and tested for release of α 1-acid glycoprotein (AGP, a positive acute-phase protein), albumin (a negative acute-phase protein), and fructose metabolism, all of which are indicators for possible tissue damage of liver cells. Application of the PACA nanoparticles caused a 40–50% inhibition of albumin release and a twofold increase in AGP release when compared to the controls (saline solution and polymerization medium), which shows the occurrence of an inflammatory process. Glucose production was significantly inhibited by 40%, indicating some form of liver damage. Similar modifications were caused by the drug-loaded nanoparticles. All effects were reversible 15 days after the treatment was stopped. To test the influence of Kupffer cells (KCs) and PiBCA nanoparticles on the protein synthesis of hepatocytes, KCs incubated with PiBCA nanoparticles were cocultured with hepatocytes. PiBCA nanoparticle cultured KCs induced the protein synthesis (AGP) of hepatocytes only marginally. *In vivo* studies found this stimulation of AGP synthesis of the hepatocytes by KCs to be caused by the release of inflammatory cytokines from KCs after phagocytosis of PACA nanoparticles. In the present *in vitro* study, however, AGP synthesis was also stimulated in hepatocytes alone after incubation with PACA nanoparticles. Therefore, for communication between KCs and hepatocytes, the presence of a physiological system such as circulating blood cells, serum hormones, or both may be required.

Carrying on from their established results, the same group investigated the effects of PACA nanoparticles on liver macrophages (Fernandez-Urrusuno et al., 1996). Macrophages phagocytose toxic or antigenic material, so impairment of their function could be potentially dangerous. Furthermore, repeated large doses of colloidal particles are toxic to liver macrophages (Ellens et al., 1982). Fernandez-Urrusuno et al. (1996) therefore used a colloidal carbon clearance assay to investigate the scavenging ability of macrophages in the liver and spleen after application of PiBCA and PiHCA nanoparticles and their degradation products. Histological analysis and organ weight were also studied. After application of nanoparticles the ability of liver macrophages to clear carbon particles was temporarily diminished in a dose-dependent manner. Cell recovery was complete after 24 h. The degradation products did not induce an effect on the macrophages. The weight of the liver remained normal but an increase in the weight of the spleen was observed, which may indicate the occurrence of granulopoiesis and erythropoiesis that is attributable to an inflammatory process.

Macrophage recovery after exposure to PACA nanoparticles was also demonstrated in a more recent study by Simeonova et al. (2003). Peritoneal mice cells composed of a mixture of macrophages, monocytes, and polymorphonuclear leukocytes were harvested 3–120 h after intraperitoneal injection of 10 or 200 mg of PBCA nanoparticles per kilogram or the equivalent degradation products. Upon *in vitro* mixing of the harvested cells with sheep red blood cells, the phagocytic index and ingestion capacity of the peritoneal cells was tested. As shown in earlier studies, the activity of the macrophagic cells dropped initially upon incubation with the nanoparticles and the degradation products, but this effect was only transitory and cells were able to recover completely after five days.

Considering the results highlighted from various toxicity studies, PACA nanoparticles seem to be a colloidal delivery system with a low toxicity profile. As mentioned, a clinical trial with PiHCA particles was suspended because of adverse pulmonary events and we hope that further information about these toxic effects will be published by BioAlliance. Long-term *in vivo* mutagenicity testing of these compounds would be another interesting point that needs to be addressed. The incorporation of drugs can change the distribution profile of the nanoparticles or the drug, so toxicity testing will have to be carried out for every new PACA nanoparticle system that is being developed. The same must apply to the various surface and polymer modifications that have been reported in the literature.

4.6 FUTURE OF POLY(ALKYL CYANOACRYLATE) NANOPARTICLES

The future of PACA nanoparticles in drug delivery is uncertain. Approximately 30 years of research into the usefulness of these systems for drug and vaccine delivery have not yet yielded the successful market introduction of nanomedicines based on this technology. Although numerous studies have demonstrated the benefit of these systems compared to, for example, drugs in solution, a major area of research in the future should be the direct comparison of PACA nanoparticles to other colloidal delivery systems that are based on polymers and lipids.

The most promising areas of delivery for these nanoparticles seems to be delivery into tumors and across the BBB. Other routes, such as oral or ocular delivery and especially the use of these particles as vaccine delivery systems, appear to be less likely to substantially improve drug delivery.

ABBREVIATIONS

AFM	Atomic force microscopy
AGP	α 1-Acid glycoprotein
AON	Antisense oligonucleotide
BBB	Blood–brain barrier
CA	Cyanoacrylate
DNA	Deoxyribonucleic acid
DOX	Doxorubicin
EE	Entrapment efficiency
EM	Electron microscopy
ESEM	Environmental scanning electron microscopy
FF	Freeze-fracture
FF-TEM	Freeze-fracture transmission electron microscopy
FITC	Fluorescein isothiocyanate
FT-IR	Fourier transform infrared spectroscopy
FT-PGSE	Fourier transform pulse gradient spin echo
GIT	Gastrointestinal tract
i.v.	Intravenous
ISCOMS	Immunostimulating complex
KCs	Kupffer cells
M cells	Microfold cells
MALDI-TOF-MS	Matrix-assisted laser desorption ionization time-of-flight mass spectrometry

MDR	Multidrug resistance
NMR	Nuclear magnetic resonance
O/W	Oil-in-water
OVA	Ovalbumin
PACA	Poly(alkyl cyanoacrylate)
PBCA	Poly(butyl cyanoacrylate)
PBS	Phosphate-buffered saline
PCS	Photon correlation spectroscopy
PECA	Poly(ethyl cyanoacrylate)
PEG	Poly(ethylene glycol)
P-gp	P-glycoprotein
PGSE	Pulse gradient spin echo
PHDCA	Poly(hexyldecyl cyanoacrylate)
PiBCA	Poly(isobutyl cyanoacrylate)
PiHCA	Poly(isohexyl cyanoacrylate)
PMCA	Poly(methyl cyanoacrylate)
PS80	Polysorbate 80
rHu	Recombinant human
s.c.	Subcutaneous
SEC	Size exclusion chromatography
SEM	Scanning electron microscopy
TEM	Transmission electron microscopy
TGF-β_2	Transforming growth factor-β_2
TNF-α	Tumor necrosis factor-α
UV	Ultraviolet
W/O	Water-in-oil

SYMBOLS

D	Diffusion coefficient
ε	Dielectric constant
η	Dynamic viscosity
k_B	Boltzmann constant
μ_E	Electrophoretic mobility
r	Radius
τ	Correlation time
T_g	Glass transition temperature
ζ	Zeta potential

REFERENCES

Aboubakar, M., P. Couvreur, H. Pinto-Alphandary, B. Gouritin, B. Lacour, R. Farinotti, F. Puisieux, and C. Vauthier, 2000. Insulin-loaded nanocapsules for oral administration: *In vitro* and *in vivo* investigation. *Drug Development Research* 49(2): 109–117.

Aboubakar, M., F. Puisieux, P. Couvreur, M. Deyme, and C. Vauthier, 1999a. Study of the mechanism of insulin encapsulation in poly(isobutylcyanoacrylate) nanocapsules obtained by interfacial polymerization. *Journal of Biomedical Materials Research* 47(4): 568–576.

Aboubakar, M., F. Puisieux, P. Couvreur, and C. Vauthier, 1999b. Physico-chemical characterization of insulin-loaded poly(isobutylcyanoacrylate) nanocapsules obtained by interfacial polymerization. *International Journal of Pharmaceutics* 183(1): 63–66.

Agarwal, A., A. Varma, and C. Sarkar, 1998. Histopathological changes following the use of biological and synthetic glue for dural grafts. An experimental study. *British Journal of Neurosurgery* 12: 213–216.

Alany, R. G., G. M. M. El Maghraby, K. Krauel-Goellner, and A. Graf, 2008. Microemulsion systems and their potential as drug carriers. In *Structure, Interactions and Dynamic Processes in Microemulsions*. M. Fanun (Ed.). Taylor&Francis/CRC Press: Boca Raton, Florida.

Al Khouri Fallouh, N., L. Roblot-Treupel, H. Fessi, J. P. Devissaguet, and F. Puisieux, 1986. Development of a new process for the manufacture of polyisobutylcyanoacrylate nanocapsules. *International Journal of Pharmaceutics* 28(2–3): 125–132.

Alyautdin, R. N., V. E. Petrov, K. Langer, A. Berthold, D. A. Kharkevich, and J. Kreuter, 1997. Delivery of loperamide across the blood–brain barrier with polysorbate 80-coated polybutylcyanoacrylate nanoparticles. *Pharmaceutical Research* 14: 325–328.

Alyautdin, R. N., E. B. Tezikov, P. Ramge, D. A. Kharkevich, D. J. Begley, and J. Kreuter, 1998. Significant entry of tubocurarine into the brain of rats with polysorbate 80-coated poly(butylcyanoacrylate) nanoparticles: An *in situ* brain perfusion study. *Journal of Microencapsulation* 15(1): 67–74.

Ammoury, N., H. Fessi, J. P. Devissaguet, M. Dubrasquet, and S. Benita, 1991. Jejunal absorption, pharmacological activity, and pharmacokinetic evaluation of indomethacin-loaded poly(*d,l*-lactide) and poly(isobutyl-cyanoacrylate) nanocapsules in rats. *Pharmaceutical Research* 8(1): 101–105.

Aprahamian, M., C. Michel, W. Humbert, J. P. Devissaguet, and C. Damgé, 1987. Transmucosal passage of polyalkylcyanoacrylate nanocapsules as a new drug carrier in the small intestine. *Biology of the Cell* 61(1–2): 69–76.

Arias, J. L., V. Gallardo, S. A. Gomez-Lopera, R. C. Plaza, and A. V. Delgado, 2001. Synthesis and characterization of poly(ethyl-2-cyanoacrylate) nanoparticles with a magnetic core. *Journal of Controlled Release* 77(3): 309–321.

Arias, J. L., V. Gallardo, M. A. Ruiz, and A. V. Delgado, 2007. Ftorafur loading and controlled release from poly(ethyl-2-cyanoacrylate) and poly(butylcyanoacrylate) nanospheres. *International Journal of Pharmaceutics* 337(1–2): 282–290.

Balnois, E., G. Papastavrou, and K. J. Wilkinson, 2007. Force microscopy and force measurements of environmental colloids. In *Environmental Colloids and Particles: Behaviour, Structure and Characterization*, K. J. Wlkinson and J. R. Lead (Eds). Chichester, U.K.: Wiley.

Barr, I. G. and G. F. Mitchell, 1996. ISCOMs (immunostimulating complexes): The first decade. *Immunology and Cell Biology* 74(1): 8–25.

Beck, P. H., J. Kreuter, W. E. G. Muller, and W. Schatton, 1994. Improved peroral delivery of avarol with polybutylcyanoacrylate nanoparticles. *European Journal of Pharmaceutics and Biopharmaceutics* 40(3): 134–137.

Behan, N. and C. Birkinshaw, 2001. Preparation of poly(butyl cyanoacrylate) nanoparticles by aqueous dispersion polymerisation in the presence of insulin. *Macromolecular Rapid Communications* 22(1): 41–43.

Behan, N., C. Birkinshaw, and N. Clarke, 2001. Poly *n*-butyl cyanoacrylate nanoparticles: A mechanistic study of polymerisation and particle formation. *Biomaterials* 22(11): 1335–1344.

Bogdan, M., A. Nan, C. V. L. Pop, L. Barbu-Tudoran, and I. Ardelean, 2008. Preparation and NMR characterization of polyethyl-2-cyanoacrylate nanocapsules. *Applied Magnetic Resonance* 34(1): 111–119.

Bootz, A., T. Russ, F. Gores, M. Karas, and J. Kreuter, 2005. Molecular weights of poly(butyl cyanoacrylate) nanoparticles determined by mass spectrometry and size exclusion chromatography. *European Journal of Pharmaceutics and Biopharmaceutics* 60(3): 391–399.

Bootz, A., V. Vogel, D. Schubert, and J. Kreuter, 2004. Comparison of scanning electron microscopy, dynamic light scattering and analytical ultracentrifugation for the sizing of poly(butylcyanoacrylate) nanoparticles. *European Journal of Pharmaceutics and Biopharmaceutics* 57(3): 369–375.

Boudad, H., P. Legrand, G. Lebas, M. Cheron, D. Duchêne, and G. Ponchel, 2001. Combined hydroxypropyl-[beta]-cyclodextrin and poly(alkylcyanoacrylate) nanoparticles intended for oral administration of saquinavir. *International Journal of Pharmaceutics* 218(1–2): 113–124.

Brayden, D. J., 2001. Oral vaccination in man using antigens in particles: Current status. *European Journal of Pharmaceutical Sciences* 14(3): 183–189.

Brigger, I., J. Morizet, L. Laudani, G. Aubert, M. Appel, V. Velasco, M.-J. Terrier-Lacombe, D. Desmaele, et al., 2004. Negative preclinical results with stealth nanospheres-encapsulated doxorubicin in an orthotopic murine brain tumor model. *Journal of Controlled Release* 100(1): 29–40.

Bu, H.-Z., H. J. Gukasyan, L. Goulet, X.-J. Lou, C. Xiang, and T. Koudriakova, 2007. Ocular disposition, pharmacokinetics, efficacy and safety of nanoparticle-formulated ophthalmic drugs. *Current Drug Metabolism* 8(2): 91–107.

Calvo, P., B. Gouritin, H. Chacun, D. Desmaele, J. D'Angelo, J. P. Noel, D. Georgin, et al., 2001. Long-circulating PEGylated polycyanoacrylate nanoparticles as new drug carrier for brain delivery. *Pharmaceutical Research* 18(8): 1157–1166.

Chauvierre, C., D. Labarre, P. Couvreur, and C. Vauthier, 2003. Novel polysaccharide-decorated poly(isobutyl cyanoacrylate) nanoparticles. *Pharmaceutical Research* 20(11): 1786–1793.

Chiannilkulchai, N., N. Ammoury, J. Caillou, J. Ph. Devissaguet, and P. Couvreur, 1990. Hepatic tissue distribution of doxorubicin-loaded nanoparticles after i.v. administration in reticulosarcoma M 5076 metastasis-bearing mice. *Cancer Chemotherapy and Pharmacology* 26(2): 122–126.

Chouinard, F., S. Buczkowaski, and V. Lenaerts, 1994. Poly(alkylcyanoacrylate) nanocapsules: Physicochemical characterization and mechanism of formation. *Pharmaceutical Research* 11(6): 869–874.

Chouinard, F., F. W. K. Kan, J.-C. Leroux, C. Foucher, and V. Lenaerts, 1991. Preparation and purification of polyisohexylcyanoacrylate nanocapsules. *International Journal of Pharmaceutics* 72(2): 211–217.

Cournarie, F., D. Auchere, D. Chevenne, B. Lacour, M. Seiller, and C. Vauthier, 2002. Absorption and efficiency of insulin after oral administration of insulin-loaded nanocapsules in diabetic rats. *International Journal of Pharmaceutics* 242(1–2): 325–328.

Cournarie, F., M. Cheron, M. Besnard, and C. Vauthier, 2004. Evidence for restrictive parameters in formulation of insulin-loaded nanocapsules. *European Journal of Pharmaceutics and Biopharmaceutics* 57(2): 171–179.

Couvreur, P., 1988. Polyalkylcyanoacrylates as colloidal drug carriers. *Critical Reviews in Therapeutic Drug Carrier Systems* 5(1): 1–20.

Couvreur, P., G. Barratt, E. Fattal, P. Legrand, and C. Vauthier, 2002. Nanocapsule technology: A review. *Critical Reviews in Therapeutic Drug Carrier Systems* 19(2): 99–134.

Couvreur, P., G. Couarraze, J. Devissaguet, and F. Puisieux, 1996. Nanoparticles: Preparation and characterisation. In *Microencapsulation: Methods and Industrial Applications*, S. Benita (Ed.). New York: Marcel Dekker.

Couvreur, P., C. Dubernet, and F. Puisieux, 1995. Controlled drug delivery with nanoparticles: Current possibilities and future trends. *European Journal of Pharmaceutics and Biopharmaceutics* 41(1): 2–13.

Couvreur, P., B. Kante, M. Roland, P. Guiot, P. Bauduin, and P. Speiser, 1979. Polycyanoacrylate nanocapsules as potential lysosomotropic carriers—Preparation, morphological and sorptive properties. *Journal of Pharmacy and Pharmacology* 31(5): 331–332.

Couvreur, P., V. Lenaerts, B. Kante, M. Roland, and P. Speiser, 1980. Oral and parenteral administration of insulin associated to hydrolysable nanoparticles. *Acta Pharmaceutica Technologica* 26: 220–222.

Couvreur, P. and C. Vauthier, 2006. Nanotechnology: Intelligent design to treat complex disease. *Pharmaceutical Research* 23(7): 1417–1450.

Damgé, C., M. Aprahamian, G. Balboni, A. Hoeltzel, V. Andrieu, and J. P. Devissaguet, 1987. Polyalkylcyanoacrylate nanocapsules increase the intestinal absorption of a lipophilic drug. *International Journal of Pharmaceutics* 36(2–3): 121–125.

Damgé, C., D. Hillaire-Buys, R. Puech, A. Hoeltzel, C. Michel, and G. Ribes, 1995. Effects of orally administered insulin nanocapsules in normal and diabetic dogs. *Diabetes, Nutrition and Metabolism—Clinical and Experimental* 8(1): 3–9.

Damgé, C., P. Maincent, and N. Ubrich, 2007. Oral delivery of insulin associated to polymeric nanoparticles in diabetic rats. *Journal of Controlled Release* 117(2): 163–170.

Damgé, C., C. Michel, M. Aprahamian, and P. Couvreur, 1988. New approach for oral administration of insulin with polyalkylcyanoacrylate nanocapsules as drug carrier. *Diabetes* 37(2): 246–251.

Damgé, C., C. Michel, M. Aprahamian, P. Couvreur, and J. P. Devissaguet, 1990. Nanocapsules as carriers for oral peptide delivery. *Journal of Controlled Release* 13(2–3): 233–239.

Damgé, C., J. Vonderscher, P. Marbach, and M. Pinget, 1997a. Poly(alkyl cyanoacrylate) nanocapsules as a delivery system in the rat for octreotide, a long-acting somatostatin analogue. *Journal of Pharmacy and Pharmacology* 49(10): 949–954.

Damgé, C., H. Vranckx, P. Balschmidt, and P. Couvreur, 1997b. Poly(alkyl cyanoacrylate) nanospheres for oral administration of insulin. *Journal of Pharmaceutical Sciences* 86(12): 1403–1409.

Das, D. and S. Lin, 2005. Double-coated poly (butylcynanoacrylate) nanoparticulate delivery systems for brain targeting of dalargin via oral administration. *Journal of Pharmaceutical Sciences* 94(6): 1343–1353.

de Kozak, Y., K. Andrieux, H. Villarroya, C. Klein, B. Thillaye-Goldenberg, M.-C. Naud, E. Garcia, and P. Couvreur, 2004. Intraocular injection of tamoxifen-loaded nanoparticles: A new treatment of experimental autoimmune uveoretinitis. *European Journal of Immunology* 34(12): 3702–3712.

Demana, P. H., C. Fehske, K. White, T. Rades, and S. Hook, 2004. Effect of incorporation of the adjuvant Quil A on structure and immune stimulatory capacity of liposomes. *Immunology and Cell Biology* 82(5): 547–554.

Desai, S. D. and J. Blanchard, 2000. Pluronic® F127-based ocular delivery system containing biodegradable polyisobutylcyanoacrylate nanocapsules of pilocarpine. *Drug Delivery* 7(4): 201–207.

Duchêne, D., G. Ponchel, and D. Wouessidjewe, 1999. Cyclodextrins in targeting: Application to nanoparticles. *Advanced Drug Delivery Reviews* 36(1): 29–40.

Ellens, H., E. Mayhew, and Y. M. Rustum, 1982. Reversible depression of the reticuloendothelial system by liposomes. *Biochimica Biophysica Acta* 714: 479–485.

El-Samaligy, M. S., P. Rohdewald, and H. A. Mahmoud, 1986. Polyalkyl cyanoacrylate nanocapsules. *Journal of Pharmacy and Pharmacology* 38(3): 216–218.

El-Samaligy, M. S., Y. Rojanasakul, J. F. Charlton, G. W. Weinstein, and J. K. Lim, 1996. Ocular disposition of nanoencapsulated acyclovir and ganciclovir via intravitreal injection in rabbit's eye. *Drug Delivery: Journal of Delivery and Targeting of Therapeutic Agents* 3(2): 93–97.

Fernandez-Urrusuno, R., E. Fattal, D. Porquet, J. Feger, and P. Couvreur, 1995. Evaluation of liver toxicological effects induced by polyalkylcyanoacrylate nanoparticles. *Toxicology and Applied Pharmacology* 130(2): 272–279.

Fernandez-Urrusuno, R., E. Fattal, J. M. Rodrigues, J. Feger, P. Bedossa, and P. Couvreur, 1996. Effect of polymeric nanoparticle administration on the clearance activity of the mononuclear phagocyte system in mice. *Journal of Biomedical Materials Research* 31: 401–408.

Fitch, R. M. and C.-H. Tsai, 1970. Polymer colloids: Particle formation in nonmicellar systems. *Journal of Polymer Science Part B: Polymer Letters* 8(10): 703–710.

Fresta, M., G. Cavallaro, G. Giammona, E. Wehrli, and G. Puglisi, 1996. Preparation and characterization of polyethyl-2-cyanoacrylate nanocapsules containing antiepileptic drugs. *Biomaterials* 17(8): 751–758.

Fresta, M., G. Fontana, C. Bucolo, G. Cavallaro, G. Giammona, and G. Puglisi, 2001. Ocular tolerability and *in vivo* bioavailability of poly(ethylene glycol) (PEG)-coated polyethyl-2-cyanoacrylate nanosphere-encapsulated acyclovir. *Journal of Pharmaceutical Sciences* 90(3): 288–297.

Friese, A., E. Seiller, G. Quack, B. Lorenz, and J. Kreuter, 2000. Increase of the anticonvulsive activity of a novel NMDA receptor antagonist using poly(butylcyanoacrylate) nanoparticles as a parenteral controlled release system. *European Journal of Pharmaceutics and Biopharmaceutics* 49(2): 103–109.

Gallardo, M., G. Couarraze, B. Denizot, L. Treupel, P. Couvreur, and F. Puisieux, 1993. Study of the mechanisms of formation of nanoparticles and nanocapsules of polyisobutyl-2-cyanoacrylate. *International Journal of Pharmaceutics* 100(1–3): 55–64.

Gasco, M. R. and M. Trotta, 1986. Nanoparticles from microemulsions. *International Journal of Pharmaceutics* 29(2–3): 267–268.

Gelperina, S. E., A. S. Khalansky, I. N. Skidan, Z. S. Smirnova, A. I. Bobruskin, S. E. Severin, B. Turowski, et al., 2002. Toxicological studies of doxorubicin bound to polysorbate 80-coated poly(butyl cyanoacrylate) nanoparticles in healthy rats and rats with intracranial glioblastoma. *Toxicology Letters* 126(2): 131–141.

Gibaud, S., C. Rousseau, C. Weingarten, R. Favier, L. Douay, J. P. Andreux, and P. Couvreur, 1998. Polyalkylcyanoacrylate nanoparticles as carriers for granulocyte-colony stimulating factor (G-CSF). *Journal of Controlled Release* 52(1–2): 131–139.

Graf, A., E. Ablinger, S. Peters, A. Zimmer, S. Hook, and T. Rades, 2008a. Microemulsions containing lecithin and sugar-based surfactants: Nanoparticle templates for delivery of proteins and peptides. *International Journal of Pharmaceutics* 350(1–2): 351–360.

Graf, A., K. S. Jack, A. K. Whittaker, S. M. Hook, and T. Rades, 2008b. Protein delivery using nanoparticles based on microemulsions with different structure-types. *European Journal of Pharmaceutical Sciences* 33(4–5): 434–444.

Graf, A., T. Rades, and S. M. Hook, 2009. Oral insulin delivery using nanoparticles based on microemulsions with different structure-types: Optimisation and *in vivo* evaluation. *European Journal of Pharmaceutical Sciences* 37(1): 53–61.

Grangier, J. L., M. Puygrenier, J. C. Gautier, and P. Couvreur, 1991. Nanoparticles as carriers for growth hormone releasing factor. *Journal of Controlled Release* 15(1): 3–13.

Gulyaev, A. E., S. E. Gelperina, A. S. Skidan, A. S. Antropov, G. Y. Kivman, and J. Kreuter, 1999. Significant transport of doxorubicin into the brain with polysorbate 80-coated nanoparticles. *Pharmaceutical Research* 16: 1564–1569.

Hillaireau, H., T. Le Doan, H. Chacun, J. Janin, and P. Couvreur, 2007. Encapsulation of mono- and oligo-nucleotides into aqueous-core nanocapsules in presence of various water-soluble polymers. *International Journal of Pharmaceutics* 331(2): 148–152.

Hillery, A. M., I. Toth, and A. T. Florence. 1996a. Biological activity of luteinizing hormone releasing hormone after oral dosing with a nanoparticulate delivery system: Co-polymerised peptide particles. *Pharmaceutical Science* 2:281–283.

Hillery, A. M., I. Toth, and A. T. Florence. 1996b. Co-polymerised peptide particles II: Oral uptake of a novel copolymeric nanoparticulate delivery system for peptides. *Journal of Controlled Release* 42(1):65–73.

Hoffmann, D. and C. Mayer, 2000. Cross polarization induced by temporary adsorption: NMR investigations on nanocapsule dispersions. *Journal of Chemical Physics* 112(9): 4242–4250.

Holtzscherer, C., J. P. Durand, and F. Candau, 1987. Polymerization of acrylamide in nonionic microemulsions—Characterization of the microlattices and polymers formed. *Colloid and Polymer Science* 265(12): 1067–1074.

Hubert, B., J. Atkinson, M. Guerret, M. Hoffman, J. P. Devissaguet, and P. Maincent, 1991. The preparation and acute antihypertensive effects of a nanocapsular form of darodipine, a dihydropyridine calcium entry blocker. *Pharmaceutical Research* 8(6): 734–738.

Jani, P., G. W. Halbert, J. Langridge, and A. T. Florence, 1990. Nanoparticle uptake by the rat gastrointestinal mucosa: Quantitation and particle size dependency. *Journal of Pharmacy and Pharmacology* 42: 821–826.

Kante, B., P. Couvreur, G. Dubois-Krack, C. De Meester, P. Guiot, M. Roland, M. Mercier, and P. Speiser, 1982. Toxicity of polyalkylcyanoacrylate nanoparticles I: Free nanoparticles. *Journal of Pharmceutical Sciences* 71(7): 786–790.

Kawamura, S., H. Hadeishi, A. Suzuki, and N. Yasui, 1998. Arterial occlusive lesions following wrapping and coating and unruptured aneurysms. *Neurologia Medico-Chirurfica* 38(1): 12–18.

Krauel, K., 2005. Formulation and characterisation of nanoparticles based on biocompatible microemulsions. PhD dissertaion, University of Otago, Dunedin, School of Pharmacy.

Krauel, K., N. M. Davies, S. Hook, and T. Rades, 2005. Using different structure types of microemulsions for the preparation of poly(alkylcyanoacrylate) nanoparticles by interfacial polymerization. *Journal of Controlled Release* 106(1–2): 76–87.

Krauel, K., L. Girvan, S. Hook, and T. Rades, 2007. Characterisation of colloidal drug delivery systems from the naked eye to cryo-FESEM. *Micron* 38(8): 796–803.

Krauel, K., A. Graf, S. M. Hook, N. M. Davies, and T. Rades, 2006. Preparation of poly (alkylcyanoacrylate) nanoparticles by polymerization of water-free microemulsions. *Journal of Microencapsulation* 23(5): 499–512.

Krauel, K., T. Pitaksuteepong, N. M. Davies, and T. Rades, 2004. Entrapment of bioactive molecules in poly (alkylcyanoacrylate) nanoparticles. *American Journal of Drug Delivery* 2(4): 251–259.

Kreuter, J., 1988. Possibilities of using nanoparticles as carriers for drugs and vaccines. *Journal of Microencapsulation* 5(2): 115–127.

Kreuter, J. (Ed.), 1994. Nanoparticles. In *Colloidal Drug Delivery Systems*. New York: Marcel Dekker.

Kreuter, J., R. N. Alyautdin, D. A. Kharkevich, and A. A. Ivanov, 1995. Passage of peptides through the blood–brain barrier with colloidal polymer particles (nanoparticles). *Brain Research* 674: 171–174.

Lambert, G., E. Fattal, H. Pinto-Alphandary, A. Gulik, and P. Couvreur, 2000b. Polyisobutylcyanoacrylate nanocapsules containing an aqueous core as a novel colloidal carrier for the delivery of oligonucleotides. *Pharmaceutical Research* 17(6): 707–714.

Lambert, G., J. R. Bertrand, E. Fattal, F. Subra, H. Pinto-Alphandary, C. Malvy, C. Auclair, and P. Couvreur, 2000a. EWS Fli-1 antisense nanocapsules inhibits Ewing sarcoma related tumor in mice. *Biochemical and Biophysical Research Communications* 279: 401–406.

Lemarchand, C., R. Gref, and P. Couvreur, 2004. Polysaccharide-decorated nanoparticles. *European Journal of Pharmaceutics and Biopharmaceutics* 58(2): 327–341.

Lenaerts, V., P. Couvreur, D. Christiaens-Leyh, E. Joiris, M. Roland, B. Rollman, and P. Speiser, 1984. Degradation of poly (isobutyl cyanoacrylate) nanoparticles. *Biomaterials* 5(2): 65–68.

Leonard, F., R. K. Kulkarni, G. Brandes, J. Nelson, and J. J. Cameron, 1966. Synthesis and degradation of poly (alkyl α-cyanoacrylates). *Journal of Applied Polymer Science* 10(2): 259–272.

Lherm, C., R. H. Müller, F. Puisieux, and P. Couvreur, 1992. Alkylcyanoacrylate drug carriers: II. Cytotoxicity of cyanoacrylate nanoparticles with different alkyl chain length. *International Journal of Pharmaceutics* 84: 13–22.

Li, Y.-P., Y.-Y. Pei, Z.-H. Zhou, X.-Y. Zhang, Z.-H. Gu, J. Ding, J.-J. Zhou, and X.-J. Gao, 2001a. PEGylated polycyanoacrylate nanoparticles as tumor necrosis factor-[alpha] carriers. *Journal of Controlled Release* 71(3): 287–296.

Li, Y.-P., Y.-Y. Pei, Z.-H. Zhou, X.-Y. Zhang, Z.-H. Gu, J. Ding, J.-J. Zhou, X.-J. Gao, and J.-H. Zhu, 2001b. Stealth polycyanoacrylate nanoparticles as tumor necrosis factor-alpha carriers: Pharmacokinetics and anti-tumor effects. *Biological and Pharmaceutical Bulletin* 24(6): 662–665.

Liang, M., N. M. Davies, and I. Toth, 2008. Increasing entrapment of peptides within poly(alkyl cyanoacrylate) nanoparticles prepared from water-in-oil microemulsions by copolymerization. *International Journal of Pharmaceutics* 362(1–2): 141–146.

Lowe, P. J. and C. S. Temple, 1994. Calcitonin and insulin in isobutylcyanoacrylate nanocapsules—Protection against proteases and effect on intestinal absorption in rats. *Journal of Pharmacy and Pharmacology* 46(7): 547–552.

Ludwig, A., 2005. The use of mucoadhesive polymers in ocular drug delivery. *Advanced Drug Delivery Reviews* 57(11): 1595–1639.

Maincent, P., R. Le Verge, P. Sado, P. Couvreur, and J. P. Devissaguet, 1986. Disposition kinetics and oral bioavailability of vincamine-loaded polyalkyl cyanoacrylate nanoparticles. *Journal of Pharmaceutical Sciences* 75(10): 955–958.

Marchal-Heussler, L., H. Fessi, J. P. Devissaguet, M. Hoffman, and P. Maincent, 1992. Colloidal drug delivery systems for the eye. A comparison of the efficacy of three different polymers: Polyisobutylcyanoacrylate, polylactic-*co*-glycolic acid, poly-epsilon-caprolactone. *S.T.P. Pharma Sciences* 2(1): 98–104.

Mayer, C., D. Hoffmann, and M. Wohlgemuth, 2002. Structural analysis of nanocapsules by nuclear magnetic resonance. *International Journal of Pharmaceutics* 242(1–2): 37–46.

Mesiha, M. S., M. B. Sidhom, and B. Fasipe, 2005. Oral and subcutaneous absorption of insulin poly-(isobutylcyanoacrylate) nanoparticles. *International Journal of Pharmaceutics* 288(2): 289–293.

Michel, C., M. Aprahamian, L. Defontaine, P. Couvreur, and C. Damgé, 1991. The effect of site of administration in the gastrointestinal tract on the absorption of insulin from nanocapsules in diabetic rats. *Journal of Pharmacy and Pharmacology* 43(1): 1–5.

Monza da Silveira, A., G. Ponchel, F. Puisieux, and D. Duchene, 1998. Combined poly(isobutylcyanoacrylate) and cyclodextrins nanoparticles for enhancing the encapsulation of lipophilic drugs. *Pharmaceutical Research* 15(7): 1051.

Müller, R. H., C. Lherm, J. Herbort, T. Blunk, and P. Couvreur, 1992. Alkylcyanoacrylate drug carriers: I. Physicochemical characterisation of nanoparticles with different alkyl chain length. *International Journal of Pharmaceutics* 84(1): 1–11.

O'Hagan, D. T., K. J. Palin, and S. S. Davis, 1989. Poly (butyl-2-cyanoacrylate) particles as adjuvants for oral immunisation. *Vaccine* 7: 213–216.

Olivier, J.-C., C. Vauthier, M. Taverna, D. Ferrier, and P. Couvreur, 1995. Preparation and characterization of biodegradable poly(isobutylcyano acrylate) nanoparticles with the surface modified by adsorption of proteins. *Colloids Surfaces B: Biointerfaces* 4: 349–356.

Oowaki, H., S. Matsuda, and N. Sakai, 2000. Non-adhesive cyanoacrylate as an embolic material for endovascular neurosurgery. *Biomaterials* 21: 1039–1046.

Peracchia, M. T., E. Fattal, D. Desmaele, M. Besnard, J. P. Noel, J. M. Gomis, M. Appel, J. d'Angelo, and P. Couvreur, 1999. Stealth(R) PEGylated polycyanoacrylate nanoparticles for intravenous administration and splenic targeting. *Journal of Controlled Release* 60(1): 121–128.

Pitaksuteepong, T., N. M. Davies, M. Baird, and T. Rades, 2004. Uptake of antigen encapsulated in poly (ethylcyanoacrylate) nanoparticles by D1-dendritic cells. *Pharmazie* 59(2): 134–142.

Pitaksuteepong, T., N. M. Davies, I. G. Tucker, and T. Rades, 2002. Factors influencing the entrapment of hydrophilic compounds in nanocapsules prepared by interfacial polymerisation of water-in-oil microemulsions. *European Journal of Pharmaceutics and Biopharmaceutics* 53(3): 335–342.

Ponchel, G. and J.-M. Irache, 1998. Specific and non-specific bioadhesive particulate systems for oral delivery to the gastrointestinal tract. *Advanced Drug Delivery Reviews* 34(2–3): 191–219.

Puglisi, G., M. Fresta, G. Giammona, and C. A. Ventura, 1995. Influence of the preparation conditions on poly-(ethylcyanoacrylate) nanocapsule formation. *International Journal of Pharmaceutics* 125(2): 283–287.

Quinn, J., J. Maw, K. Ramotar, G. Wenckebach, and G. Wells, 1997. Octylcyanoacrylate tissue adhesive versus suture wound repair in a contaminated wound model. *Surgery* 122(1): 69–72.

Radwan, M. A., 2001. Enhancement of absorption of insulin-loaded polyisobutylcyanoacrylate nanospheres by sodium cholate after oral and subcutaneous administration in diabetic rats. *Drug Development and Industrial Pharmacy* 27(9): 981–989.

Radwan, M. A. and H. Y. Aboul-Enein. 2002. The effect of oral absorption enhancers on the in vivo performance of insulin-loaded poly(ethylcyanoacrylate) nanospheres in diabetic rats. *Journal of Microencapsulation* 19(2): 225–235.

Reddy, L. H. and R. R. Murthy, 2004. Influence of polymerization technique and experimental variables on the particle properties and release kinetics of methotrexate from poly(butylcyanoacrylate) nanoparticles. *Acta Pharmaceutica* 54(2): 103–118.

Ritschel, W. A., 1991. Microemulsions for improved peptide absorption from the gastrointestinal-tract. *Methods and Findings in Experimental and Clinical Pharmacology* 13(3): 205–220.

Rollot, M., P. Couvreur, L. Roblot-Treupel, and F. Puisieux, 1986. Physicochemical and morphological characterization of polyisobutyl cyanoacrylate nanocapsules. *Journal of Pharmaceutical Sciences* 75(4): 361–364.

Roques, M., C. Damgé, C. Michel, C. Staedel, G. Cremel, and P. Hubert, 1992. Encapsulation of insulin for oral administration preserves interaction of the hormone with its receptor in vitro. *Diabetes* 41(4): 451–456.

Saï, P., Ch. Damgé, A. S. Rivereau, A. Hoeltzel, and E. Gouin, 1996. Prophylactic oral administration of meta-bolically active insulin entrapped in isobutylcyanoacrylate nanocapsules reduces the incidence of diabe-tes in nonobese diabetic mice. *Journal of Autoimmunity* 9(6): 713–721.

Schneider, T., A. Becker, K. Ringe, A. Reinhold, R. Firsching, and B. A. Sabel, 2008. Brain tumor therapy by combined vaccination and antisense oligonucleotide delivery with nanoparticles. *Journal of Neuroimmunology* 195(1–2): 21–27.

Schroeder, U., H. Schroeder, and B. A. Sabel, 2000. Body distribution of 3HH-labelled dalargin bound to poly(butyl cyanoacrylate) nanoparticles after I.V. injections to mice. *Life Sciences* 66(6): 495–502.

Simeonova, M., M. Antcheva, and K. Chorbadjiev, 2003. Study on the effect of polybutyl-2-cyanoacrylate nanoparticles and their metabolites on the phagocytic activity of peritoneal exudate cells of mice. *Biomaterials* 24(2): 313–320.

Stella, B., S. Arpicco, M. T. Peracchia, D. Desmaele, J. Hoebeke, M. Renoir, J. D'Angelo, L. Cattel, and P. Couvreur, 2000. Design of folic acid-conjugated nanoparticles for drug targeting. *Journal of Pharmaceutical Sciences* 89(11): 1452–1464.

Stieneker, F., G. Kersten, L. van Bloois, D. J. A. Crommelin, S. L. Hem, J. Löwer, and J. Kreuter. 1995. Comparison of 24 different adjuvants for inactivated HIV-2 split whole virus as antigen in mice. Induction of titres of binding antibodies and toxicity of the formulations. *Vaccine* 13(1): 45–53.

Toriumi, D. M., W. F. Raslan, M. Friedmann, and M. E. Tardy, 1991. Variable histotoxicity of histoacryl when used in a subcutaneous site: An experimental study. *Laryngoscope* 101: 339–343.

Vauthier, C., C. Dubernet, C. Chauvierre, I. Brigger, and P. Couvreur, 2003a. Drug delivery to resistant tumors: The potential of poly(alkyl cyanoacrylate) nanoparticles. *Journal of Controlled Release* 93(2): 151–160.

Vauthier, C., C. Dubernet, E. Fattal, H. Pinto-Alphandary, and P. Couvreur, 2003b. Poly(alkylcyanoacrylates) as biodegradable materials for biomedical applications. *Advanced Drug Delivery Reviews* 55(4): 519–548.

Vauthier, C., D. Labarre, and G. Ponchel, 2007. Design aspects of poly(alkylcyanoacrylate) nanoparticles for drug delivery. *Journal of Drug Targeting* 15(10): 641–663.

Vranckx, H., M. Demoustier, and M. Deleers, 1996. A new nanocapsule formulation with hydrophilic core: Application to the oral administration of salmon calcitonin in rats. *European Journal of Pharmaceutics and Biopharmaceutics* 42(5): 345–347.

Watnasirichaikul, S., N. M. Davies, T. Rades, and I. G. Tucker, 2000. Preparation of biodegradable insulin nanocapsules from biocompatible microemulsions. *Pharmaceutical Research* 17(6): 684–689.

Watnasirichaikul, S., T. Rades, I. G. Tucker, and N. M. Davies, 2002a. Effects of formulation variables on characteristics of poly(ethylcyanoacrylate) nanocapsules prepared from w/o microemulsions. *International Journal of Pharmaceutics* 235(1–2): 237–246.

Watnasirichaikul, S., T. Rades, I. G. Tucker, and N. M. Davies, 2002b. *In-vitro* release and oral bioactivity of insulin in diabetic rats using nanocapsules dispersed in biocompatible microemulsion. *Journal of Pharmacy and Pharmacology* 54(4): 473–480.

Weiss, C. K., U. Ziener, and K. Landfester, 2007. A route to nonfunctionalized and functionalized poly(n-butylcyanoacrylate) nanoparticles: Preparation in miniemulsion. *Macromolecules* 40(4): 928–938.

WHO/UNICEF (Ed.), 1996. *State of the World's Vaccines and Immunization.* Geneva: WHO/UNICEF.

Wilson, B., M. K. Samanta, K. Santhi, K. P. S. Kumar, N. Paramakrishnan, and B. Suresh, 2008. Targeted delivery of tacrine into the brain with polysorbate 80-coated poly(n-butylcyanoacrylate) nanoparticles. *European Journal of Pharmaceutics and Biopharmaceutics* 70(1): 75–84.

Wohlgemuth, M. and C. Mayer, 2003. Pulsed field gradient NMR on polybutylcyanoacrylate nanocapsules. *Journal of Colloid and Interface Science* 260(2): 324–331.

Wolberg, H. and A. Lippoldt, 2002. Tight junctions of the blood–brain-barrier: Development, composition and regulation. *Vascular Pharmacology* 38: 323–337.

Yang, S. C., H. X. Ge, Y. Hu, X. Q. Jiang, and C. Z. Yang, 2000. Doxorubicin-loaded poly(butylcyanoacrylate) nanoparticles produced by emulsifier-free emulsion polymerization. *Journal of Applied Polymer Science* 78(3): 517–526.

Zhang, Q., Z. Shen, and T. Nagai. 2001. Prolonged hypoglycemic effect of insulin-loaded polybutylcyanoacry-late nanoparticles after pulmonary administration to normal rats. *International Journal of Pharmaceutics* 218(1-2) :75–80.

Zimmer, A., 1999. Antisense oligonucleotide delivery with polyhexylcyanoacrylate nanoparticles as carriers. *A Companion to Methods in Enzymology* 18: 286–295.

5 Stimuli-Sensitive Polymer Gels for Dermal and Transdermal Drug Delivery and Their Application in the Development of Smart Textile Materials

Witold Musial and Vanja Kokol

CONTENTS

5.1 INTRODUCTION

The main reason to apply an active substance (or a drug) on the skin is for *in situ* therapy. Mineral substances such as the famous terra sigillata were used during the Roman Empire to cure superficial skin conditions. Similarly, different compositions containing, *inter alia*, hot or warm vegetable mass, which is suspected to relieve pain, have been utilized for centuries in Asia, Africa, and Europe

(Ling, 1998). The topical delivery of compounds to the surface and deeper layers of the skin and the transdermal delivery of pharmacologically active substances is an issue of great interest for both professionals and consumers. The cosmetic industry is performing continuous research on sunscreens, insect repellents, and beauty products. The pharmaceutical sector studies topical antiinfectives and emollients for the skin surface. Antibiotics, antiinflammatory drugs, corticosteroids, and fluorouracil are intended to act in the inner skin tissues. Skin delivery systems are being evaluated to maintain the concentration of drugs, such as nicotine, antihypertensives, scopolamine, nitroglycerine, testosterone, and antiarthritics, in the central compartment. During the last four decades, some drug forms have been developed for delivery of active substances to systemic circulation. Since the 1960s, transdermal drug delivery has been developed systematically (Scheuplein, 1978), and the skin as an application site for drug delivery systems has been recognized.

5.2 SKIN STRUCTURE, COMPOSITION, AND BIOFUNCTIONALITY

Skin is a tissue, which some consider as an organ; its barrier function is recognized as the most important facet. The skin protects the body against a number of egzogenic factors with different natures, such as physical, chemical, or biological. The skin maintains conditions inside the human body—the proper temperature range, amounts of water and minerals, and even body shape.

There are three main skin layers, and all are important for the function of the skin as a barrier. The uppermost layer, the stratum corneum (SC), is characterized by a thickness of only 10–20 µm, but it is considered as the main barrier within the skin structures. This layer has a big impact on drug delivery to the deeper skin layers. It consists mainly of cornified unviable cells with a high level of keratin. The unviable cells (40-µm length, 0.5-µm thickness) are embedded in a firm, strict SC layer and connected by lipids and desmosome–intercellular connections. The cells are consequently desquamated in the life cycle of the skin epidermis, which lasts about four weeks. Below the level of cornified and desquamating skin cells, the viable epidermis maintains its activity. Rows of cells develop there from the stratum basale to form finally the unviable keratinocytes, which form the SC. The total thickness of the epidermis (both SC and viable epidermis) is considered to be approximately 100 µm (Kendall et al., 2007).

The human dermis thickness is characterized in the range of 500–1000 µm. The dermis layer provides support for the epidermis in terms of nutritive, immune, and mechanical aspects through a thin papillary layer, which is adjacent to the epidermis. The dermal vascular system comprises a matrix of mainly the proteins collagen and elastin. There are also mucopolysaccharides in the collagen–elastin hydrogel. The extensive vascular network present (~300 µm) below the skin surface plays a role in the nutrition of skin tissue and in thermoregulation of the human body. The blood flow rate in the skin tissue is assessed as 0.05 mL/min, which is important for the theoretical prediction of the amounts of drug delivered to the systemic circulation. The dense flat meshwork of the lymphatic system exists inside the dermis. The lymphatic system is important for intradermal removal of large molecules, and blood flow determines the clearance of relatively small molecules. Nerve connections are present inside the dermis tissue, including nerve fibers for temperature, pain, and pressure (Slivka et al., 1993).

The dermis also possesses other structures that should be considered in terms of drug transport: accrine and apocrine sweat glands with sweat ducts, hair follicles with sebaceous glands, and self-standing sebaceous glands. Hair follicles with sebaceous glands are widely distributed over the body surface with some exceptions, like the internal parts of the feet and palms, as well as the lips. Another area is the specific skin appendage, the nails; the area of applied therapeutic components is rather narrow, including mainly antifungal components.

5.2.1 SKIN AS A BARRIER

As mentioned, the SC layer is considered to be crucial for the barrier functions of the skin. The diffusion rate of pharmacologically active substances through the dermis matrix is about 1000-fold higher than through the SC. The barrier properties are related to the high density and low hydration of the SC. The keratinization process contributes to the barrier function of the SC. Inside a keratinocyte there is approximately 70% keratin and approximately 20% lipids. The cornified cells are practically insoluble in most solutions applied on the skin (Suhonen et al., 1999). The explanation for this is the internal composition of cellular proteins. The major protein in keratinocytes, keratin, is extensively cross-linked by intermolecular disulfide bridges, creating keratin filaments (Sun and Green, 1978). In addition, the cornified cell envelope, consisting of loricrin and involucrin, is extremely resistant to water solutions (Yaffe et al., 1993). Desmosomes are found between the cornified keratinocytes in the intercellular area, which provide epidermal cell cohesion and communication. The area between the corneocytes (0.1 μm) is filled by the so-called intercellular lipid lamellae. This lipid-rich environment is composed mainly of cholesterol and cholesterol derivatives; several classes of ceramoides are identified by the numbers 1–8; in class 6 there are still three subtypes, fatty acids, and other components (Wertz et al., 1985). After removal of skin lipids from the surface and partially from the intercellular area, the drug diffusion rate may increase up to 1000-fold. Chemical species, including drugs, may diffuse through the keratinocytes or between them through the intercellular lipid lamellae. According to the traditional point of view, hydrophilics diffuse intracellularly whereas hydrophobics permeate through the intercellular region. Newer reports characterize the possibility for hydrophilic drug transport by the intercellular route, with hydration of the intercellular lipid lamellae and reversible degeneration of the desmosomes (Elias and Menon, 1991).

5.2.2 Biotransformations in the Skin Layers

Metabolism is possible within the dermis and viable epidermis. These processes influence the diffusion rate of a drug through the skin layers (Hikima et al., 2005). According to contemporary results, the main enzymatic activity of the skin is maintained in the epidermis. The main processes of oxidation, reduction, and hydrolysis influence the distribution of a drug in the skin (Ahmad and Mukhtar, 2004). Oxidation is activated by 17-β-hydroxysteroid dehydrogenase, which completes the functions of oxidase and monoamineoxidase. This enzyme assists in degradation of testosterone to androsten-3,17-dion, cortisol to cortisone, and 17-β-estradiol to estron. The reductive enzymes are ketoreductase and 5-α-reductase. They play a main role in the pathway of reduction of progesterone to pregnadiol, estron to 17-β-estradiol, and cortisol to allodihydrocortisol. The most active hydrolyzing enzymes are esterase and arylester-O-dealkylase, which take part in biodegradation of 17,21-cortisol to cortisol, 17-valereate-β-metazone to metazone, metronidazole esters to metronidazole, and peroxybenzoate to benzoate.

The prodrugs formulated according to the different metabolic pathways in the epidermis and dermis should pass easily through the SC lipophilic barrier. Then they are degraded by the activity of enzymes. The degraded form is still biologically active; however, it is more hydrophilic and may be distributed to the dermis structures or to the systemic circulation. The enzymatic activity of the skin is also important for the degradation of some drugs, which develop their activity in the nondeep layers of the skin, although with time they are metabolized to biologically inactive components, which are eliminated by the main circulation. This is true for benzoate peroxide. However, most of the known topically applied components do not undergo metabolic degradation in the skin. This is considered as an advantage when comparing the intragastrointestinal and hepatic degradation of biologically active components. In this case the transdermal route enables the elimination of the so-called first pass effect.

5.2.3 Biotransformations on the Skin Surface

The SC matrix develops under the influence of pH gradients, sodium ions, and enzymes: synthetases, reductases, hydrolases, and lipases (Feingold, 1991). The surface of the horny layer is covered by a thin amorphous film that contributes to the SC structure and function. In newborns there is an almost neutral skin surface (pH 6.6) that change within days or weeks into an acidic surface film (pH 5.9), leading to activation of pH-dependent hydrolytic enzymes such as β-glucocerebrosidase (GlcCer'ase) and SC secretory phospholipase A2 (Behne et al., 2003; Fluhr et al., 2004). Because SC acidification in adults is required for normal permeability barrier homeostasis and lipid processing occurs via acidic pH-dependent enzymes, the skin surface pH is modulated by microbial harvest, eccrine and sebaceous gland secretions, and endogenous catabolic pathways. The acidification of the horny layer is necessary for barrier function. Exposure of it to a neutral buffer or blocking of acidification increases the pH, leading to activation of proteases that digest desmoglein 1, a major constituent of corneosomes. Metabolization of desmoglein 1 coherently decreases corneocytes, enforces horny layer permeability, and alters the SC integrity and cohesion.

The lipids of the skin's SC comprise a distinctive mixture of ceramides, cholesterol, and free fatty acids, which appear to provide the barrier against excess water loss and limit the ingress of xenobiotics. A number of lipid catabolic enzymes, including sphingomyelinase, phospholipase A, triacylglycerol hydrolase, and steroid sulfatase, have been localized in sites where these transformations occur, as well as within epidermal lamellar bodies. This suggests that, although alternative delivery pathways may exist, most of these lipid hydrolytic activities in the SC result from the secretion of this organelle's contents.

A recent study showed that the SC pH changes alter the barrier recovery kinetics because of a decrease of β-GlcCer'ase and acidic sphingomyelinase catalytic activity, which is consequently attributable to pH-induced, sustained serine protease activity (Hachem et al., 2005). The hydrolysis of glucosylceramides to ceramides by GlcCer'ase activity in the intercellular membrane domains within the SC was critical for epidermal homeostasis and permeability barrier function (Takagi et al., 1999).

The activity of a β-nicotinamide adenine dinucleotide disodium salt dependent prostaglandin E2–9-ketoreductase was also demonstrated in human skin, the increase of which may be partially due to the increased generation of β-nicotinamide adenine dinucleotide disodium salt by the tissues and partially due to the alteration of the prostaglandin E2–9-ketoreductase by the excessive proliferation of the tissues (Ziboh et al., 1977). Some of the functions of dermal fibroblasts are documented by the presence of collagenases secreted by 72-kDa dermal fibroblasts, such as matrix metalloproteinase-2 (MMP-2), and 92-kDa keratinocytes (MMP-9; Auger et al., 2000). The synthesis of the collagen matrix, that is, the formation of cross-links between the chains of Type I collagen, is dependent on the production of prolyl-4-hydroxylase by dermal fibroblasts (Cvetkovska et al., 2008; Kivirikko and Myllyharju, 1998). It is also recognized that skin has the enzymatic machinery to produce vitamin D3. This compound and its analogs have been developed for treating psoriasis, a hyperproliferative disease (Holick, 2003).

Important enzymatic activity was also revealed in sweat glands. Enzymes, such as alkaline phosphatase, acid phosphatase, adenosine triphosphatase, cholinesterase, and pseudocholin esterase, were determined in isolated sweat gland tissue for normal subjects and for patients with cystic fibrosis. Isocitric, glucose-6-phosphate, lactic, and malic dehydrogenases were also determined as a possible occurrence of an energy metabolism disorder related to the secretion of sweat in cystic fibrosis (Gibbs and Reimer, 1964).

5.2.4 Skin Flora

Human skin flora are defined as the microbes present on the surface of healthy skin, within the SC, in the duct (infundibulum) of sebaceous glands, and in hair follicles. The skin flora are composed of the resident flora that are continuously present on the skin and thus may be regularly sampled, the transient flora that are only sampled in low frequency or density, and the temporary resident flora that transiently grow on the skin without leading to infection. The resident skin flora make up the following genera: micrococci with coagulase-negative staphylococci, *Peptococcus* spp. and *Micrococcus* spp., diptheroids with corynebacteria and *Brevibacterium* spp., propionibacteria, and

Gram-negative rods. The relevance of resident skin flora for healthy skin is that it generates an ecological system that protects the skin from pathogens. Thus, *Staphylococcus epidermidis*, *Propionibacterium acnes*, corynebacteria, and *Pityrosporum ovale* produce lipases and esterases that break triglycerides into free fatty acids, leading to a low skin surface pH and thereby unfavorable growing conditions for skin pathogens. *S. epidermidis* and *Propionibacterium acnes* are known to produce antibiotics that may interfere with pathogenic organisms.

5.3 DERMAL AND TRANSDERMAL DRUG DELIVERY

5.3.1 Transcutaneous Drug Delivery Routes

The alternative for transcellular and intercellular transport is transfollicular transport, which may involve both pilosebaceous units and sweat glands. The fractional area of pilosebaceous units is approximately 1/1000 of the area of the skin surface; however, in the areas of higher hair density the fractional area may increase up to 1/100. The mean number of pilosebaceous units is estimated to be around 750 cm² of the skin. The dimension of the hair follicles is assessed as 50–100 μm, whereas the diameter of adjacent sebaceous glands is estimated to be in the range of 200–2000 μm. There are different opinions on the importance of transfollicular transport. This route presently cannot be considered as main one for known substances, although minoxidil was successfully applied in this area in some studies (Reddy et al., 2006). The high reservoir and permeability barrier function in appendage-free skin may support the thesis on transappendageal transport (Hueber et al., 1992, 1994). The presence of a specific sebum in the sebaceous glands plays an important role in the possible transport of active substances, and a respective method for this transport route was proposed (Musial and Kubis, 2003, 2006). The passage of hydrophilic drugs is postulated for the sweat glands, although no concise data have been presented up to now. It must be emphasized that transfollicular transport of drugs encounters many obstacles according to the low surface area of the appendages and high lipophilicity in the area of hair follicles. However, there are many research efforts being performed on this subject to provide better pharmacotherapy for associated dermatological conditions.

5.3.2 Transdermal Drug Transport

The diffusion rate of chemical species increases with their hydrophobicity, which is understandable considering the lipophilic conditions of the SC and the associated solubility of lipophilic drugs in this medium. Basic research confirmed that, with the increase of lipophilicity, the permeation of homologous alcohols increased in parallel as a straight or almost straight line dependency (Le and Lippold, 1995). The parameter that describes the lipophilicity of a component is the log of the partition coefficient of octanol–water (log P). The most appropriate value of log P to obtain a reasonable drug diffusion rate into the skin is between 1 and 2. With the increase of this value to more than 2, however, the absorbed dose of drug in an *in vivo* experiment decreases as the drug molecule, after passing the SC, has to move through the deeper layers of the skin, which are characterized as a hydrophilic environment. The solubility and partitioning between a more and less soluble environment may be described by using the energy that is needed to dissolve the drug in the vehicle applied in the product and the energy that is needed to dissolve the drug in the SC.

Charge also plays an important role in drug delivery to the skin layers. Nonionized species are preferred, which increase the bioavailability compared to the salt. Molecules with a molecular mass of up to 1–2 kDa are diffusible. Higher differences in drug transport are observed for molecules with a molecular mass over 0.4 kDa; below this value, the diffusion rate is similar. For molecules with a very high molecular mass of over 2 kDa, such as proteins, enzymes, and insulin, the delivery was not proved with contemporary technology (Magnusson et al., 2004). The physicochemical requirements for the candidate drug, which could be applied in a transdermal therapeutic system (TTS), are given in Table 5.1.

TABLE 5.1
Physicochemical Requirements of a Drugs Applied
in the Transdermal System

Parameter	Value
Molecular weight	<500 Da
Melting point	<150°C
Log P	1–2
Solubility in lipids and water	>1mg/mL

5.3.3 Drug Diffusion

A drug may be applied on the skin in different forms; it may be dissolved or dispersed in the form of a suspension or emulsion. The penetration of the drug into the skin is possible only in the case of molecular dispersion; thus, the drug in suspension needs to dissolve in the vehicle, sweat, or sebum, and then it can diffuse into the SC and the deeper layers of the skin. Dissolution is often a slow process, so this stage may limit the rate of the overall process (Anissimov and Roberts, 1999). However, the absorption rate is usually slower than the dissolution, so the solute is present in the saturated concentration. The diffusion rate of the drug through the skin is proportional to its concentration in the vehicle. The diffusion described by Fick's law proceeds according to the concentration gradient. The rate of drug penetration is dependent on the solubility of the drug in the layers of the skin, which is actually the partition coefficient between two environments. The diffusion of the drug that is released from the polymer matrix and permeates across the SC may be expressed in terms of Fick's law of diffusion:

$$\frac{dM}{dt} = \frac{D\Delta CK}{h},$$

where dM/dt is the steady-state flux across the SC, D is the diffusion coefficient or diffusivity of drug molecules, ΔC is the drug concentration gradient across the SC, K is the partition coefficient of the drug between the skin and the vehicle, and h is the thickness of the SC.

The active substance diffuses into the muscles and tissues supporting the skin and finally into the systemic circulation. The pharmacological activity of the drug can be manifested and the drug may be degraded according to the biodegradation by the skin enzymes.

The diffusion of a solute from the vehicle into the SC depends on the solubility of the component in the vehicle and in the SC, which can be described as a partition coefficient. Because it is a constant value, the amount of drug that is absorbed by the skin increases with the amount of dissolved drug in the vehicle. Saturated solutions provide the maximal thermodynamic activity of the drug on the border of the considered phases and the highest concentration and rate of penetration in the SC. In addition, the lipophilicity and solubility and the tortuosity of the route in the intercellular areas prolong the time it takes for a drug to pass through the skin, so the so-called lag time is observed. For many substances the lag time is minutes or hours. In additional extensive drug binding, for example, by proteins, the lag time may be days (Li et al., 1998). The proposed reason for this phenomenon is binding of the drug with the strongly ionic groups of the keratin. In some therapeutic systems or ointments the drug is also present in the systemic circulation after its removal from the skin surface. This is the case for corticosteroids, verapamil, griseofulvin, fusidic acid, and amino acids, which were found in the circulation several days after topical application (Barbero and Frasch, 2006).

The viable epidermis is considered as a hydrophilic layer, so the diffusion of a solute from the SC to the viable epidermis is connected to the change from lipophilic to hydrophilic at pH 7.4. The distribution in this area depends on the partition coefficient between both phases. The hydrophilic substances can easily diffuse into the viable epidermis, whereas the lipophilic molecules can be deposited in the SC. For strongly lipophilic components, this will be the limiting stage of diffusion through the skin.

5.3.4   VEHICLE COMPOSITION AND DRUG ABSORPTION

Biologically active components are applied on the skin mostly in semisolid preparations (ointments, creams, gels), liquids (such as aqueous or oil solutions), and powders. Emplastra are also used for transdermal application. The influence of the drug form on the absorption, although very important, is sometimes considered as overestimated. The traditional drug form is likely to influence the metabolism of the released drug, its binding with the proteins, or its distribution in the skin layers. As mentioned earlier, the main factor influencing the drug absorption rate is diffusion of the molecule through the SC. In this pharmacokinetic stage, the drug form plays the most important role. The vehicle type determines the penetration of the drug through the SC and the bioavailability of the topically applied drug (Imanidis et al., 1998). To increase drug absorption, several technological processes are performed. The increase of drug solubility in the vehicle will increase the drug absorption according to the increase in the gradient of the drug concentration on the phase border. The increase of the thermodynamic activity of the drug may be obtained by applying very concentrated or saturated solutions of the drug in the vehicle that is going to evaporate during its presence on the skin. A preparation with a proper partition coefficient of the vehicle/SC should be considered; the systems with low drug solubility in the vehicle, compared to the skin lipids, are believed to be beneficial. The permeability of the SC may also be influenced by the vehicular composition. Sorption promoters are applied for this, which reversibly change the characteristics of the SC and enable the increased diffusion of the drug in this layer. In last few decades, microcarriers and nanocarriers have been proposed for facilitated transport of drug molecules to the surface layers of the SC.

The vehicle applied topically on the skin influences the concentration of the drug in the SC. However, considering the systemic drug concentration, the best results are observed in a lipophilic substance applied on the skin in a hydrophilic vehicle. The same drug applied on the skin in a lipophilic vehicle, at the same concentration, will exhibit lower concentrations in the SC and decrease its bioavailability. In a hydrophilic drug, the very lipophilic vehicle should enable the increase in concentration of this active molecule in the SC, if the partition coefficient of the vehicle/SC is higher than 1. In a hydrophilic vehicle, with a hydrophilic drug, the drug diffusion to the SC as well as to the deeper skin layers and systemic circulation has the most pessimistic outcome.

Dermatological preparations are usually deposited on the skin surface in the 10–30 μm thick layer. The thickness of this layer is another physical factor that influences drug diffusion to the SC. The amount of the drug is dependent on the thickness of the drug form on the skin. A thin layer of the vehicle is beneficial; when the vehicle components evaporate, the concentration of drug increases, with a consequent increase in absorption. The amount of drug that diffuses into the skin depends on the skin area exposed to the preparation. The high areas of body skin surface will enhance the possibility of side effects, whereas using transdermal systems with a defined surface will maintain proper drug amounts diffusing through the skin. Thus, the surface is the important factor that determines the amount of the absorbed drug (Chowhan and Pritchard, 1975). According to these data, the release experiments and *in vivo* data should be treated with great care, because the initial release experiments may be treated only as a preliminary indication of drug behavior on the skin in terms of pharmacokinetics (Csóka et al., 2005). In occlusive preparation, the water content in the thin layer over the SC will enhance drug deposition in the epidermis, and eventually the transdermal transport, which is unknown for the actual topical application (Iordanskii et al., 2000). In advanced dermal or transdermal systems, the micro- or nanoparticles (often liposomes) are embedded in the hydrogel matrix (Jenning et al., 2000).

The polymers applied in the dermal and transdermal drug forms are listed in Table 5.2.

TABLE 5.2
Polymers Applied in the Formulations of Matrices for Dermal and Transdermal Drug Delivery

Natural polymers	Acacia gum
	Alginic acid salts
	Cargeenans
	Chitosan and derivates
	Collagen
	Tragacanth
	Xanthan gum
Synthetic polymers	Polyacrylates and derivates
	Copolymers of poly(methacrylates)
	Poly(vinyl alcohol)
	Poly(vinyl pyrrolidone)
Cellulose and derivates	Carboxymethylcellulose salt
	Hydroxyethylcellulose
	Hydroxypropylcellulose
	Hydroxypropylmethylcellulose
	Microcrystalline cellulose
Others	Poloxamers
	Polyoxyethylates

5.3.5 TRANSDERMAL THERAPEUTIC SYSTEMS

The most developed drug forms applied on the skin are the TTSs (Table 5.3). Nitroglycerine, isosorbide, estradiole, testosterone, nicotine, scopolamine, and fentanyl are applied in the form of TTSs. According to the maintenance of the determined surface, thickness, drug concentration, viscosity, and matrix composition of the drug, delivery at a determined rate is possible. The drug is incorporated in the matrix or reservoir with a membrane, which controls the release process. In the case of a TTS, the system is supposed to limit the flux of the drug to the skin surface. For this kind of device, only drugs that quickly diffuse through the skin and exhibit the activity in low concentrations may be used. A model drug form should control drug absorption for many drugs, which is

TABLE 5.3
Examples of Clinical Studies of Drugs Applied in Transdermal Systems

Active Agent	TTS Description	Reference
Fentanyl	Transdermal patches	Schneider et al., 2008
Nimodipine	Menthol-based TTS	Krishnaiah and Bhaskar, 2004
Nicorandil	Limonene-based TTS	Krishnaiah et al., 2005
Scopolamine	Patch	Gil et al., 2005
Tulobuterol	TTS	Horiguchi et al., 2004
Glyceryl trinitrate	Systemic transdermal treatment	Colak et al., 2003
Buprenorphine	TTS	Böhme, 2002
Estradiol	TTS	Matsumoto et al., 2000
Physostigmine	TTS	Möller et al., 1999

possible only by the reversible modification of the skin barrier conditions (Böhme, 2002; Colak et al., 2003; Gil et al., 2005; Horiguchi et al., 2004; Krishnaiah and Bhaskar, 2004; Krishnaiah et al., 2005; Matsumoto et al., 2000; Möller et al., 1999; Schneider et al., 2008).

5.4 HYDROGELS FOR TOPICAL APPLICATION

Hydrogels are polymeric networks with the capability of swelling in water or biological fluids, according to the hydrophilicity of the polymer. After swelling, this three-dimensional network may retain a large volume of fluids in the swollen state. The absorption of aqueous fluids is attributable to the hydrophilic groups, which have the potential to create van der Waals, ionic, and other bonds. Among the functional groups, the most well-known and intentionally applied ones to obtain water binding in the product are, *inter alia*, hydroxyl groups, carboxylic groups, and sulfate groups (Peppas and Khare, 1993). The permeability, mechanical properties, surface properties, and biocompatibility are affected by the estimated water volume stored in the polymeric network. Hydrophilic networks with a high content of water are considered to have physical properties similar to that of living tissue, that is, consistency, rheological characteristics, and low interfacial tension on the phase border between the hydrogel and water or biological fluids (Blanco et al., 1996). Pharmacologically active substances may diffuse through the polymeric matrix, enabling the saturation of the hydrogel with drug molecules. Consequently, after loading, drug release is permitted and the loaded hydrogel can be used as a drug delivery system. The main mechanism generally considered for drug delivery by hydrogels is diffusion, and water-soluble drugs and proteins are mostly used in this drug form. However, in dermal and transdermal drug delivery, an additional important factor controls drug release: the partition coefficient of the molecule between the SC and hydrogel.

Hydrogels can be designed to change properties, for example, swelling and collapse or solution–gel transitions, in response to externally applied triggers, such as temperature, ionic strength, solvent polarity, electric/magnetic field, and light or small (bio)molecules (Ulijn et al., 2007). The so-called smart biomaterials change properties in response to selective biological recognition events. When exposed to a biological target (nutrient, growth factor, receptor, antibody, enzyme, or whole cell), molecular recognition events trigger changes in molecular interactions that translate into macroscopic responses such as swelling and collapse or solution–gel transitions. Hydrogel transitions may be used directly as optical readouts for biosensing linked to the release of actives for drug delivery or instigate biochemical signaling events that control or direct cellular behavior. Accordingly, bioresponsive hydrogels have gained significant interest for application in diagnostics, drug delivery, and tissue regeneration or wound healing. The modifications in polymer composition, amount of water, cross-linking density, and morphological forms may influence the diffusion of the drug in the polymeric matrix; these are used to control the drug delivery rate to the target (Andrianov and Payne, 1998). In the main classification of hydrogels, there are pH-, temperature-, enzyme-, and electric-sensitive hydrogels (Kost and Langer, 2001). The pH-sensitive hydrogels may be ionic or neutral, according to the functional group composition, amount, and degree of ionization. Cationic functional groups are included in cationic hydrogels, and positively charged functionals are included in anionic hydrogels. The swelling of gels is driven by water–polymer thermodynamic mixing contributions and elastic polymer contributions in the neutral gels. For ionic hydrogels, ionic interactions between charged polymers and free ions must also be considered, because the ionizable functional groups render the polymers more hydrophobic and the water uptake is high (Kudela et al., 1987). Anionic polymeric networks containing carboxylic or sulfonic acid groups ionize when the pH of the external swelling medium rises above the negative logarithm of the equilibrium constant for association of the ionic functional group, whereas the cationic hydrogels ionize at pH values below that same constant of the cationic group (e.g., amine group) and swelling occurs. The diffusion of water-soluble drugs follows free volume theory, which suggests a pore-type mechanism. In water-insoluble drugs, the diffusion seems to follow a partition or solution–diffusion mechanism (Varshosaz and Falamarzian, 2001). The electrical impulse may induce a response in hydrogel-containing ionizable groups and may influence the release of the drug. There are several mechanisms proposed for selective and controlled transport of proteins and neutral solutes across hydrogel membranes. The swelling of a membrane may be induced electrically and chemically, affecting the effective pore size and permeability. The solute flux within a membrane may also be increased electrophoretically or using electro-osmotic augmentation. Electrostatic partitioning of charged solutes into charged membranes is used as well (Murdan, 2003). The release rates of hydrocortisone, benzoic acid, and lidocaine hydrochloride from chitosan hydrogels were assessed in response to different currents as a function of time (Ramanathan and Block, 2001). To enhance skin permeation of ascorbyl palmitate, it was encapsulated in liposomes and formulated into a liposomal hydrogel by dispersing the liposome into a poloxamer hydrogel matrix. An electric current was applied to improve the skin permeation of ascorbyl palmitate. The system was evaluated as a helpful drug delivery system to enhance skin permeation of the drug. The combined use of a negative lipogel with cathodal electric assistance was found to be promising in enhancing the skin delivery of ascorbyl palmitate (Lee et al., 2007). Thermosensitive polymers possess the unique property of swelling and deswelling according to changes in the environmental temperature. Hydrophobic groups (e.g., methyl, ethyl, and propyl functional groups) interact when the temperature increases or decreases; consequently, the water is exposed in the interchain areas and the polymer particle collapses. Thermosensitive polymers are often characterized by a lower critical solution temperature (CST), upper CST, or volume phase transition temperature (VPPT); above or below it the water is exposed to the environment. Poly(N-isopropylacrylamide) [poly(NIPAM)] is one of the most widely studied thermosensitive polymers with a lower CST of around 32°C, which is close to human body temperature (Figure 5.1). Essentially, discrete colloidal gel particles, as a result of their very high surface area to volume ratio compared to bulk gels, have a much faster response to external stimuli such as temperature or pH. The reversible shrinking and swelling as a function of external stimuli provides a novel drug release system. Copolymers of poly(ethylene oxide) and poly(propylene oxide) are characterized by a phase change from sol to gel in the range of internal body temperature, and they are applied in injectable implants, *inter alia*.

Temperature-sensitive hydrogels are classified into three groups: negatively thermosensitive with a characteristic lower CST, positively thermosensitive with a characteristic upper CST, and thermally reversible gels (Qui and Park, 2001). The on–off release profile of drugs from the thermosensitive matrices was explained by the formation of a dense, less permeable surface layer of gel. The barrier was formed by the fast temperature change, which was due to the faster collapse of the gel surface than the interior of the matrix. This surface shrinking process was found to be regulated by the length of the methacrylate alkyl side chain, that is, the hydrophobicity of the comonomers (Okano et al., 1990).

Although skin possesses numerous microorganisms with the potential for hydrogel biodegradation, enzyme-sensitive hydrogels are mainly used in drug forms applied to the colon or skin wound dressings. This target is achieved because of the presence of pH-sensitive monomers and azo cross-linking agents

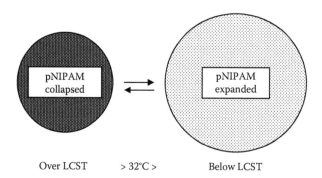

FIGURE 5.1   The scheme of the volume phase transition around the lower CST of poly(NIPAM) microgel.

in the hydrogel structure. During the passage of the hydrogel matrix through the gastrointestinal tract, the hydrogel swelling increases with a pH increase in the surrounding environment of the colon. In the colon, the hydrogel cross-links are accessible to the azoreductases produced by the intestinal bacterial flora, so the entrapped drug is released from the degraded polymeric net (Yang et al., 2002).

The phase transition may occur in the presence of various ions. Some of the polysaccharides are classified as ion sensitive (Bhardwaj et al., 2000; Guo et al., 1998). K-carrageenan forms strict gels in the presence of potassium ions, whereas i-carrageenan forms elastic gels in the presence of calcium ions. Gellan gum, an anionic polysaccharide, undergoes a phase transition under the influence of both mono- and divalent ions. Similar phase changes were observed for pectins and alginic acid. In the more advanced systems, the nano- and microparticles of polymers are applied to obtain pulsatile release (Kikuchi and Okano, 2002).

5.4.1 Hydrogels for Dermal Application

The basic dermal therapeutic systems contain a flexible backing layer, an adhesive-controlled release matrix that controls the drug release, and a removable protection layer. Dermal therapeutic systems must not retain the water excreted by the skin and sweat glands, because water deposition may result in damage of the skin microstructure, positively influence the growth of dermal pathogens, and increase uncontrolled drug permeation (Bucks et al., 1991; Hurkmans et al., 1985). To prepare dermal therapeutic systems, artificial silk was selected as a backing layer because it has good water vapor permeability, compatibility with the coating process, and cohesion with the matrices. Two adhesive hydrophilic copolymers of dimethylaminoethyl methacrylate and neutral methacrylic esters mixed with a nonadhesive hydrophobic copolymer of ethyl acrylate and methyl methacrylate, which was supplied in suspension in two different concentrations, were used to prepare dermal therapeutic systems. The water vapor permeability and adhesion properties of the prepared systems were evaluated. Adding 10–30% (w/w) of both adhesive and nonadhesive polymers permits the formulation of patches with high water vapor permeability and good adhesive properties (Minghetti et al., 1997). Capsaicin has a broad range of topical applications, including rheumatoid arthritis, osteoarthritis, diabetic neuropathy, and postherpetic neuralgia (Fusco and Giacovazzo, 1997). *In vitro* and *in vivo* skin absorption of capsaicin and nonivamide from hydrogels and various commercialized creams of capsaicin were compared with hydrogels. The incorporation of nonionic pluronic F-127 polymer into the hydrogels resulted in a retarded release of capsaicin, whereas the *in vitro* capsaicin permeation showed higher levels in cationic chitosan and anionic carboxymethylcellulose hydrogels than in cream bases. The permeation of nonivamide was retarded at the later stage of *in vitro* application (Fang et al., 2001, 2002). Topical application was proposed to bypass the side effects of chlorfeniramine used perorally as an antihistamine. According to the researchers, the percutaneous absorption of chlorfeniramine across the skin seems to be possible through topical application when using hydrogels with cellulose derivatives. Polymer preparations with various viscosities present different release profiles (Tas et al., 2003). The release of caffeine, crystal violet, and phenol red in porous ionic thermosensitive poly(NIPAM) derivative hydrogels was investigated. The release rate of caffeine from the hydrogels was not affected by the ionicity of the hydrogels, but the crystal violet strongly interacted with the anionic hydrogel and revealed a very fast release rate. The authors suggest that crystal violet is adsorbed on the thin skin layer of the cationic hydrogel because of charge repulsion and is released rapidly (Lee and Chiu, 2002). Thermosensitivity and ion sensitivity are both being considered in the intensive studies of the newly proposed poly(NIPAM) thermosensitive polymers (Musial et al., 2008; Figure 5.2).

A number of hydrogel preparations were analyzed for enhanced delivery of griseofulvin to obtain effective antifungal drug concentrations. Of the preparations containing the main vehicle component (carbopol and essential oils, propylene glycol, and N-methyl-2-pyrrolidone), the hydrogel with the N-methyl-2-pyrrolidone enhancer was the most effective (Shishu, 2006). Hydrogels are also used as carriers for nanostructured particles, like liposomes, for optimal rheology and release rate.

FIGURE 5.2 The scanning electron microscopy image of modified poly(NIPAM) obtained during surfactant free emulsion polymerization. (The image was obtained in the Electron Microscopy Center of the University of Maribor.)

The dispersion of liposomes in the hydrogel was assessed to evaluate the optimal preparation for clindamycin release on the skin surface (Arnardottir et al., 1996). The deposition of vitamin E is of great interest, because it influences the condition of the skin. A system for vitamin E delivery with antioxidative properties was evaluated, applying methacrylated dextran; it was copolymerized with aminoethyl methacrylate and subsequently esterified with *trans*-ferulic acid (Cassano et al., 2009).

5.4.2 Hydrogels for Transdermal Application

The skin permeation of flurbiprofen from sodium salt of carboxymethylcellulose hydrogels approximated that from an aqueous solution, suggesting the feasibility of sodium salt of carboxymethylcellulose hydrogels in further *in vivo* or clinical situations because of its excellent release of the drug and bioadhesive properties. D-Limonene, a cyclic monoterpene, had the highest ability to enhance the flux of flurbiprofen. However, phospholipids as retarders reduced the skin absorption of flurbiprofen (Fang et al., 2003). The release rates of the drugs dibucaine hydrochloride, theophylline, and sodium benzoate with different charges from poly(NIPAM) hydrogels were studied. The release rate was temperature dependent for these types of drug. When the temperature was lower than the phase transition temperature, the release rate was higher at lower temperatures and increased as the temperature rose (Makino et al., 2000). A thermally reversible gel of Pluronic F127 was evaluated as a vehicle for the percutaneous administration of indomethacin. *In vivo* studies suggest that a 20% (w/w) aqueous gel may be of practical use as a base for topical administration of the drug. Poloxamer 407 gel was found to be suitable for transdermal delivery of insulin. The combination of chemical enhancers and iontophoresis resulted in synergistic enhancement of insulin permeation (Miyazaki et al., 1995; Pillai and Panchagnula, 2003). Systemic methotrexate toxicity disables its wider use, and transdermal delivery is expected to protect patients from side effects. Transdermal permeation of methotrexate loaded into a polyacrylamide-based hydrogel patch across mice skin was studied *in vitro* after pretreatment with terpenes and ethanol, alone or in combination with iontophoresis; a binary mixture of menthol and ethanol in combination with square wave iontophoresis gave the highest drug levels in systemic circulation (Prasad et al., 2007). Capsaicin, which has been used for years as a topical rubefaciens, also possesses a number of activities that could be beneficial after drug delivery to the central compartment. Researchers proved that application of modified β-cyclodextrin enables the transport of capsaicin (Zi et al., 2008). The therapeutic level of clonazepam, a drug applied in epilepsy and psychiatric conditions, should be maintained in specified range. The evaluated hydrogel with poly(acrylic acid) and diethyleneglycol monoethyl ether as the penetration enhancer was efficient in increasing the drug flux through rat skin (Mura et al., 2000). The lipid spheres were embedded into the poly(acrylic acid) matrix to obtain controlled release of triamcynolone acetonide acetate, which is used as a potent antiinflammatory component (Liu et al., 2008). Because of its short elimination half-life and the low oral bioavailability of nalbuphine (an opioid analgesic), frequent injections are needed (Lo et al., 1987). The blood nalbuphine concentration would be better maintained by newly synthesized nalbuphine prodrugs. In last few years, topical forms of the drug were proposed with methylcellulose hydrogel as the carrier (Huang et al., 2005). The transdermal system with L-dopa, an anti-Parkinson's agent, was evaluated, applying an alcoholic hydrogel. The system is intended for elderly dementive patients, who fail to receive regular peroral or injectable therapy (Sudo et al., 1998).

5.5 HYDROGEL-FUNCTIONALIZED TEXTILE MATERIALS

As textiles become more functional, stimuli-responsive polymeric hydrogels can also find application in the creation of smart textiles. The environmentally responsive fabrics based on smart polymer modification can be tailored to respond to a variety of stimuli such as temperature and pH. They can also be used in cosmetic and nutrient or drug delivery fabrics, which can release at body temperature and are reusable with repeated rinsing. Because of these attributes, smart textiles may keep us warm in a cold environment or cool in a hot environment; guard against bacterial attack; and provide us with considerable convenience, support, and even pleasure in our daily activities. The phenomenon of the temperature and humidity swelling and deswelling properties of surface-modified textile materials, for example, the competition between drying and swelling processes that occur at higher temperatures, can be exploited to develop stimuli-sensitive textiles. In addition, environmentally responsive fabrics can enhance the protective functions of the skin's keratinous layer, reduce skin irritation, and improve the skin's barrier properties by providing breathable, antistatic, and antistain characteristics. Hydrogel-based materials can be used in various applications including nonwoven films, diapers, skin treatments, prosthetic devices, excipients, and the like.

Hydrogel-based textile materials placed near the skin can also be used to absorb body fluids such as wound exudate. There is a growing interest in the development of wound dressings that possess functionality beyond providing physical protection and an optimal moisture environment for the wound. In gel-based wound dressings, the gel adhering on the suitable solid material (e.g., cotton) has to be amenable to bind to the skin surface and simultaneously promote healing by maintaining a moist wound environment while absorbing excess exudates and controllably releasing a variety of therapeutic agents (Miyata, 1981). An advanced dressing material has to be hydrophilic; easy to prepare on fabrics, such as polyester, cotton, and nylon, with excellent mechanical integrity when hydrated; and stable to a variety of chemicals and sterilization methodologies.

Dual stimuli-responsive hydrogels (pH and temperature) prepared by combining thermoresponsive polymers have recently been the most widely studied, because these two factors have a physiological significance and versatile systems, mainly for biomedical applications. Recent attempts have also been made to develop dual stimuli-responsive textiles by modifying their surface with polymeric systems in various forms and combinations. To increase the surface area per unit mass of hydrogels and thus improve the chemical reactivity and response times (Qian and Hinestroza, 2004), the feasibility of stimuli-responsive hydrogels prepared in the micro- and nanometer size range was

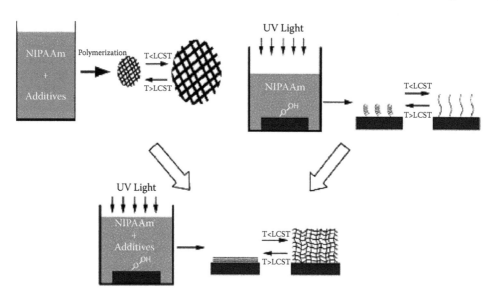

FIGURE 5.3 Schematic illustration of the polymerization and cross-linking process onto plasma-activated synthetic fabric. (From Chen, K.-S., et al., 2002. *Mater. Sci. Eng.* 20: 203–208. With permission.)

realized as the surface-modifying system for different textile materials. Among the polymers used for this purpose, poly(NIPAM) has attracted considerable attention (Chen et al., 2002; Liu and Hu, 2005; Serra, 2002). Poly(NiPAM) gels were also grafted onto plasma-activated poly(ethylene terephthalate) film and a polypropylene nonwoven fabric surface (Chen et al., 2002) by using suitable additives (ammonium peroxodisulfate as an initiator, *N,N,N′,N′-tetra*-methylethylene-diamine as a promoter, and *N,N′*-methylenebisacrylamide as a cross-linking agent; Figure 5.3). In addition, the γ-preirradiation-induced grafting of poly(NIPAM) on the surface of cotton cellulose fabric was introduced (Jianqin et al., 1999), showing that the grafting reaction was induced mainly by trapped radicals located in the interphase between the crystal and amorphous regions. Poly(NIPAM) and polyurethane (PU) were simultaneously grafted onto nonwoven cellulose fabrics by ammonium persulfate initiated copolymerization (Hu et al., 2006a), showing that the water absorption of the grafted fabrics is responsive to both temperature and pH when coated as a thin or thick layer of hydrogel. The incorporation of PU suppressed the syneresis of poly(NIPAM) hydrogel without affecting the phase transition temperature. The temperature and pH dual-responsive poly(NIPAM)/ PU copolymer hydrogel-grafted fabrics have great potential as smart wound dressings and cosmetic materials for skin care. Similarly, a surface-modifying system based on chitosan and poly(NIPAM) was formulated as microhydrogels and incorporated to a previously aminized cotton fabric surface (Glampedaki et al., 2008). A cellulose-supported chitosan-modified temperature-responsive poly(NIPAM)/PU hydrogel was introduced that could be used in preparation of a facial mask (Hu et al., 2006b), which not only improves the handling and skin affinity but also renders the facial mask antibacterial.

Protein-based hydrogels formed from animal or vegetable protein polymers such as gelatins and soybean proteins (Battista, 1983) were also introduced as a suitable raw material for producing novel ultrafine fibers or monofils, which were extruded through suitable spinnerets (as in producing viscose rayon and cellulose acetate filaments). These materials can be fashioned into many conventional textile forms and are suitable for a variety of medical and apparel uses, as well as sutures for surgery. A highly absorbent keratin solid fiber or powder capable of forming a hydrogel upon the addition of water was introduced by partially oxidizing hair keratin disulfide bonds to sulfonic acid residues and then reacting with a cation (Van Dyke et al., 2001). The hydrogel may be useful as a therapeutic for skin or as an excipient.

Controlled delivery of drugs in response to environments has the potential of targeting therapies and personalized treatments. A protocol to synthesize hydrogels with cross-links composed of different enzyme-cleavable peptide sequences (e.g., disease-associated wound proteases; Law et al., 2006; Plunkett et al., 2005) or temperature-responsive polymers (Kumashiro et al., 2004; Thornton et al., 2005) has been proposed as a controllable and targeted delivery mechanism and as skin wound dressing materials. This technique allows the release of drug molecules, which depends on both enzyme selectivity and changes in body temperature. In a similar way, a bioresponsive hydrogel was explored in tissue engineering to regenerate damaged or diseased tissues, where the peptide sequences are cleavable by MMPs to form a gel into which cells can infiltrate (Lutolf and Hubbell, 2005; Lutolf et al., 2003). MMPs are a family of enzymes that have many roles, including the breakdown of extracellular matrix molecules during tissue remodeling and disease. Such a material may have far-reaching applications for functionally targeted drug delivery. Accordingly, a polyester-supported polymer dressing was modified by ion exchange to prepare sodium, silver, or doxycycline salts; these formulations inhibited neutrophil proteases, elastase, and collagenase-2 (MMP-8) under the same conditions (Vachon and Yager, 2006). Chitosan–collagen hydrogel-coated textile scaffolds were promising for tissue-engineering applications, allowing favorable hepatocyte attachment, spheroid formation, and maintenance of function (Risbud et al., 2003). Using biodegradable hydrogel technologies and nonwoven fabrics, textile-based scaffold biomaterials were developed for engineering new tissues and organs for human body repair (Chu et al., 2003). In addition, a new kind of plaster was introduced that employed a hydrogel [made from insoluble and nontoxic polymers such as hydroxyethyl acrylate, acrylamide, and poly(ethylene oxide)] on a piece of textile material to help cool down a child's fever (http://www.newton.dep.anl.gov/askasci/eng99/eng99302.htm). The cooling effect that can last for 2 h arises from the absorbed water, which has high heat capacity. The yarn is also able to absorb sweat by introduction of superabsorbent hydrogel-based material inside the yarn and eventually releases a perfume, which was previously absorbed inside the hydrogel (Sannino et al., 2006).

ABBREVIATIONS

CST	Critical solution temperature
GlcCer'ase	β-Glucocerebrosidase
MMP	Matrix metalloproteinase
poly(NIPAM)	Poly(N-isopropylacrylamide)
PU	Polyurethane
SC	Stratum corneum
TTS	Transdermal therapeutic system

SYMBOLS

ΔC	Drug concentration gradient across the SC
dM/dt	Steady-state flux across the SC
D	Diffusion coefficient of drug molecules
h	Thickness of the SC
K	Partition coefficient of the drug between skin and the vehicle
$\log P$	Log of partition coefficient octanol/water

REFERENCES

Ahmad, N. and Mukhtar, H., 2004. Cytochrome P450: A target for drug development for skin diseases. *J. Invest. Dermatol.* 123: 417–425.

Andrianov, A. K. and Payne, L. G., 1998. Polymeric carriers for oral uptake delivery of microparticulates. *Adv. Drug Deliv. Rev.* 34: 155–170.

Anissimov, Y. G. and Roberts, M. S., 1999. Molecular size as the main determinant of solute maximum flux across the skin. *J. Pharm. Sci.* 122: 993–999.

Arnardottir, H. B., Sveinsson, S. J., and Kristmundsdottir, T., 1996. The release of clindamycin phosphate from a suspension of different types of liposomes and selected topical dosages forms. *Int. J. Pharm.* 134: 71–77.

Auger, F. A., Pouliot, R., Tremblay, N., Guignard, R., Noël, P., Juhasz, J., Germain, L., and Goulet, F., 2000. Multistep production of bioengineered skin substitutes: Sequential modulation of culture conditions. *In Vitro Cell. Dev. Biol. Anim.* 36: 96–103.

Barbero, A. M. and Frasch, H. F., 2006. Transcellular route of diffusion through stratum corneum: Results from finite element models. *J. Pharm. Sci.* 95: 2186–2194.

Battista, O. A., 1983. Protein polymer hydrogels. U.S. Patent 4416814.

Behne, M. J., Barry, N. P., Hanson, K. M., Aronchik, I., Clegg, R. W., Gratton, E., Feingold, K., Holleran, W. M., Elias, P. M., and Mauro, T. M., 2003. Neonatal development of the stratum corneum pH gradient: Localization and mechanisms leading to emergence of optimal barrier function. *J. Invest. Dermatol.* 120: 998–1006.

Bhardwaj, T. R., Kanwar, M., Lal, R., and Gupta, A., 2000. Natural gums and modified natural gums as sustained-release carriers. *Drug Dev. Ind. Pharm.* 26: 1025–1038.

Blanco, M. D., Garcia, O., Trigo, R. M., Teijon, J. M., and Katime, I., 1996. 5-Fluorouracil release from copolymeric hydrogels of itaconic acid monoester. 1. Acrylamide-*co*-monomethyl itaconate. *Biomaterials* 17: 1061–1067.

Böhme, K., 2002. Buprenorphine in a transdermal therapeutic system—A new option. *Clin. Rheumatol.* 21: 13–16.

Bucks, D., Guy, R., and Malbach, H., 1991. Effect of occlusion. In: R. L. Bronaugh and H. I. Malbach (Eds.), *In Vitro Percutaneous Absorption: Principles, Fundamentals, and Applications*, pp. 85–114. CRC Press, Boca Raton, FL.

Cassano, R., Trombino, S., Muzzalupo, R., Tavano, L., and Picci, N., 2009. A novel dextran hydrogel linking *trans*-ferulic acid for the stabilization and transdermal delivery of vitamin E. *Eur. J. Pharm. Biopharm.* 72: 232–238.

Chen, K.-S., Tsai, J.-C., Chou, C.-W., Yang, M.-R., and Yang, J.-M., 2002. Effects of additives on the photo-induced grafting polymerization of *n*-isopropylacrylamide gel onto PET film and PP nonwoven fabric surface. *Mater. Sci. Eng.* 20: 203–208.

Chowhan, Z. T. and Pritchard, R., 1975. Release of corticoids from oleaginous ointment bases containing drug in suspension. *J. Pharm. Sci.* 64: 754–759.

Chu, C. C., Zhang, X. Z., Mathew, A., and Van Buskirk, R., 2003. National Textile Center Research Briefs— Materials Competency, June 1.

Colak, T., Ipek, T., Urkaya, N., Kanik, A., and Dirlik, M., 2003. A randomised study comparing systemic trans-dermal treatment and local application of glyceryl trinitrate ointment in the management of chronic anal fissure. *Eur. J. Surg. Suppl.* 588: 18–22.

Csóka, I., Csányi, E., Zapantis, G., Nagyb, E., Fehér-Kiss, A., Horváth, G., Blazsó, G., and Eros, I., 2005. *In vitro* and *in vivo* percutaneous absorption of topical dosage forms: Case studies. *Int. J. Pharm.* 291: 11–19.

Cvetkovska, B., Islam, N., Goulet, F., and Germain, L., 2008. Identification of functional markers in a self-assembled skin substitute *in vitro*. *In Vitro Cell Dev. Biol. Anim.* 44: 444–450.

Elias, P. M. and Menon, G. K., 1991. Structural and lipid biochemical correlates of the epidermal permeability barrier. *Adv. Lipid Res.* 24: 1–26.

Fang, J.-Y., Hwang, T.-L., and Leu J.-L., 2003. *In vitro* and *in vivo* evaluations of the efficacy and safety of skin permeation enhancers using flurbiprofen as a model drug. *Int. J. Pharm.* 255: 153–166.

Fang, J.-Y., Leu, Y.-L., Wang, Y.-Y., and Tsai, Y.-H., 2002. *In vitro* topical application and *in vivo* pharmacodynamic evaluation of nonivamide hydrogels using Wistar rat as an animal model. *Eur. J. Pharm. Sci.* 15: 417–423.

Feingold, K. R., 1991. The regulation and role of epidermal lipid synthesis. *Adv. Lipid Res.* 24: 57–82.

Fluhr, J. W., Behne, M. J., Brown, B. E., Moskowitz, D. G., Selden, C., Mao-Qiang, M., Mauro, T. M., Elisian, P. M., and Feingold, K. R., 2004. Stratum corneum acidification in neonatal skin: Secretory phospholipase A(2) and the sodium/hydrogen antiporter-1 acidify neonatal rat stratum corneum. *J. Invest. Dermatol.* 122: 320–329.

Fusco, M. M. and Giacovazzo, M., 1997. Peppers and pain—The promise of capsaicin. *Drugs* 53: 909–914.

Gibbs, G. E. and Reimer, K., 1964. Quantitative microdetermination of enzymes in the sweat gland. *J. Pediatr.* 65: 540–541.

Gil, A., Nachum, Z., Dachir, S., Chapman, S., Levy, A., Shupak, A., Adir, Y., and Tal, D., 2005. Scopolamine patch to prevent seasickness: Clinical response vs. plasma concentration in sailors. *Aviat. Space Environ. Med.* 76: 766–770.

Glampedaki, P., Tourrette, A., Warmoeskerken, M. M. C. G., and Jocić, D., 2008. Preparation and properties of stimuli-responsive surface modifying systems applicable to textiles. AUTEX 2008 World Textile Conference, June 24–26, Città Studi, Biella, Italy, Poster 124 (CD-ROM).

Guo, J.-H., Skinner, G. W., Harcum, W. W., and Barnum, P. E., 1998. Pharmaceutical applications of naturally occurring water-soluble polymers. *Pharm. Sci. Technol. Today* 1: 254–261.

Hachem, J. P., Man, M. Q., Crumrine, D., Uchida, Y., Brown, B. E., Rogiers, V., Roseeuw, D., Feingold K. R., and Elias, P. M., 2005. Sustained serine proteases activity by prolonged increase in pH leads to degradation of lipid processing enzymes and profound alterations of barrier function and stratum corneum integrity. *J. Invest. Dermatol.* 125: 510–520.

Hikima, T., Tojo, K., and Maibach, H. I., 2005. Skin metabolism in transdermal therapeutic systems. *Skin Pharmacol. Physiol.* 18: 153–159.

Holick, M. F., 2003. A millennium perspective. *J. Cell. Biochem.* 88: 296–307.

Horiguchi, T., Kondo, R., Miyazaki, J., Fukumokto, K., and Torigoe, H., 2004. Clinical evaluation of a transdermal therapeutic system of the β_2-agonist tulobuterol in patients with mild or moderate persistent bronchial asthma. *Arzneimittel-Forsch* 54: 280–285.

Hu, J., Liu, W., and Liu, B., 2006b. Fabric-supported chitosan modified temperature responsive PNIPAAm/PU hydrogel and the use thereof in preparation of facial mask. U.S. Patent 0286152.

Hu, J.-L., Liu, B.-H., and Liu, W.-G., 2006a. Temperature/pH dual sensitive *N*-isopropylacrylamide/polyurethane copolymer hydrogel-grafted fabrics. *Text. Res. J.* 76: 853–860.

Huang, J.-F., Sung, K. C., Hu, O. Y., Wang, J.-J., Lin, Y.-H., and Fang, J.-Y., 2005. The effects of electrically assisted methods on transdermal delivery of nalbuphine benzoate and sebacoyl dinalbuphine ester from solutions and hydrogels. *Int. J. Pharm.* 297: 162–171.

Hueber, F., Schaefer, H., and Wepierre, J., 1994. Role of transepidermal and transfollicular routes in percutaneous absorption of steroids: *In vitro* studies on human skin. *Skin Pharmacol.* 7: 237–244.

Hueber, F., Wepierre, J., and Schaefer, H., 1992. Role of transepidermal transfollicular routes in percutaneous absorption of hydrocortisone and testosterone: *In vivo* study in the hairless rat. *Skin Pharmacol.* 5: 99–107.

Hurkmans, J. F. G., Bodde, H. E., Van Driel, L. M. J., Van Doorne, H., and Junginger, H. E., 1985. Skin irritation caused by transdermal drug delivery systems during long-term (5 days) application. *Br. J. Dermatol.* 112: 461–467.

Imanidis, G., Helbing-Strausak, S., Imboden, R., and Leuenberger, H., 1998. Vehicle dependent *in situ* modification of membrane controlled drug release. *J. Control. Release* 51: 23–34.

Iordanskii, A. L., Feldstein, M. M., Markin, V. S., Hadgraft, J., and Plate, N. A., 2000. Modeling of the drug delivery from a hydrophilic transdermal therapeutic system across polymer membrane. *Eur. J. Pharm. Biopharm.* 49: 287–293.

Jenning, V., Schafer-Korting, M., and Gohla, S., 2000. Vitamin A-loaded solid lipid nanoparticles for topical use: Drug release properties. *J. Control. Release* 66: 115–126.

Jianqin, L., Maolin, Z., and Hongfei, H., 1999. Pre-irradiation grafting of temperature sensitive hydrogel on cotton cellulose fabric. *Radiat. Phys. Chem.* 55: 55–59.

Kendall, M. A. F., Chong, Y. F., and Cock, A., 2007. The mechanical properties of the skin epidermis in relation to targeted gene and drug delivery. *Biomaterials* 33: 4968–4977.

Kikuchi, A. and Okano, T., 2002. Pulsatile drug release control using hydrogels. *Adv. Drug Deliv. Rev.* 54: 53–77.

Kivirikko, K. I. and Myllyharju, J., 1998. Prolyl 4-hydroxylases and their protein disulfide isomerase subunit. *Matrix Biol.* 16: 357–368.

Kost, J. and Langer, R., 2001. Responsive polymeric delivery systems. *Adv. Drug Deliv. Rev.* 46: 125–148.

Krishnaiah, Y. S. and Bhaskar, P., 2004. Studies on the transdermal delivery of nimodipine from a menthol-based TTS in human volunteers. *Curr. Drug Deliv.* 1: 93–102.

Krishnaiah, Y. S., Chandrasekhar, D. V., Rama, B., Jayaram, B., Satyanarayana, V., and Al-Saidan, S. M., 2005. *In vivo* evaluation of limonene-based transdermal therapeutic system of nicorandil in healthy human volunteers. *Skin Pharmacol. Physiol.* 18: 263–272.

Kudela, V., 1987. Hydrogels. In: H. F. Mark, N. Bikales, C. H. Overberger, G. Menges, and J. I. Kroschwitz (Eds.), *Encyclopedia of Polymer Science and Technology*, Vol. 7, pp. 783–806. Wiley, New York.

Kumashiro, T., Ooya, T., and Yui, N., 2004. Dextran hydrogels containing poly(*N*-isopropylacrylamide) as grafts and cross-linkers exhibiting enzymatic regulation in a specific temperature range. *Macromol. Rapid Commun.* 25: 867–872.

Law, B., Weissleder, R., and Tung, C.-H., 2006. Peptide-based biomaterials for protease-enhanced drug delivery. *Biomacromolecules* 7: 1261–1265.

Le, V. H. and Lippold, B. C., 1995. Influence of physicochemical properties of homologous esters of nicotinic acid on skin permeability and maximum flux. *Int. J. Pharm.* 124: 285–292.

Lee, S., Lee, J., and Choi, Y. W., 2007. Skin permeation enhancement of ascorbyl palmitate by liposomal hydrogel (lipogel) formulation and electrical assistance. *Biol. Pharm. Bull.* 30: 393–396.

Li, S. K., Suh, W., Parikh, H. H., Ghanem, A.-H., Mehta, S. C., Peck, K. D., and Higuchi, W. I., 1998. Lag time data for characterizing the pore pathway of intact and chemically pretreated human epidermal membrane. *Int. J. Pharm.* 170: 93–108.

Ling, M. R., 1998. Extemporaneous compounding: The end of the road? *Dermatol. Clin.* 16: 321–327.

Liu, B. and Hu, Y., 2005. Sensitive hydrogels to textiles: A review of Chinese and Japanese investigations. *Fibres Text. East. Eur.* 13: 45–49.

Liu, W., Hu, M., Liu, W., Xue, C., Xu, H., and Yang, X.-L., 2008. Investigation of the carbopol gel of solid lipid nanoparticles for the transdermal iontophoretic delivery of triamcinolone acetonide acetate. *Int. J. Pharm.* 364: 135–141.

Lo, M. W., Schary, W. L., and Whitney, Jr., C. C., 1987. The disposition and bioavailability of intravenous and oral nalbuphine in healthy volunteers. *J. Clin. Pharmacol.* 27: 866–873.

Lutolf, M. P. and Hubbell, J. A., 2005. Synthetic biomaterials as instructive extracellular microenvironments for morphogenesis in tissue engineering. *Nat. Biotechnol.* 23: 47–55.

Lutolf, M. P., Lauer-Fields, J. L., Schmoekel, H. G., Metters, A. T., Weber, F. E., Fields, G. B., and Hubbell, J. A., 2003. Synthetic matrix metalloproteinase-sensitive hydrogels for the conduction of tissue regeneration: Engineering cell-invasion characteristics. *Proc. Natl. Acad. Sci. USA* 100: 5413–5418.

Magnusson, B. M., Anissimov, Y. G., Cross, S. E., and Roberts, M. S., 2004. Molecular size as the main determinant of solute maximum flux across the skin. *J. Invest. Dermatol.* 122: 993–999.

Makino, K., Hiyoshi, J., and Ohshima, H., 2000. Effects of thermosensitivity of poly(*N*-isopropylacrylamide) hydrogel upon the duration of a lag phase at the beginning of drug release from the hydrogel. *Colloid Surf. B* 20: 341–346.

Matsumoto, S., Sugiyama, T., Hanai, T., Ohnishi, N., Park, Y. C., and Kurita, T., 2000. A study of the clinical effect of estradiol transdermal therapeutic system alone on pollakisuria and urinary incontinence in postmenopausal woman. *Nippon Hinyokika Gakkai Zasshi* 91: 501–505.

Minghetti, P., Cilurzo, F., Liberti, V., and Montanari, L., 1997. Dermal therapeutic systems permeable to water vapour. *Int. J. Pharm.* 158: 165–172.

Miyazaki, S., Tobiyama, T., Takada, M., and Attwood, D., 1995. Percutaneous absorption of indomethacin from pluronic F127 gels in rats. *J. Pharm. Pharmacol.* 47: 455–457.

Möller, H.-J., Hampel, H., Hegerl, U., Schmitt, W., and Walter, K., 1999. Double-blind, randomized, placebo-controlled clinical trial on the efficacy and tolerability of a physostigmine patch in patients with senile dementia of the Alzheimer type. *Pharmacopsychiatry* 32: 99–106.

Mura, P., Faucci, M. T., Bramanti, G., and Corti, P., 2000. Evaluation of transcutol as a clonazepam transdermal permeation enhancer from hydrophilic gel formulations. *Eur. J. Pharm. Sci.* 9: 365–372.

Murdan, S., 2003. Electro-responsive drug delivery from hydrogels. *J. Control. Release* 92: 1–17.

Musial, W. and Kubis, A. A., 2003. Preliminary assessment of alginic acid as a factor buffering triethanolamine interacting with artificial skin sebum. *Eur. J. Pharm. Biopharm.* 55: 237–240.

Musial, W. and Kubis, A. A., 2006. Preliminary evaluation of interactions between selected alcohol amines and model skin sebum components. *Chem. Pharm. Bull.* 54: 1076–1081.

Musial, W., Voncina, B., and Kokol, V., 2008. The influence on the temperature on the conductivity of polyNIPAM microgels for controlled release of drugs. *Farm. Vest.* 59: 289–290.

Miyata, T., 1981. Collagen skin dressing. U.S. Patent 4294241.

Okano, T., Bae, Y. H., Jacobs, H., and Kim, S. W., 1990. Thermally on–off switching polymers for drug permeation and release. *J. Control. Release* 11: 255–261.

Peppas, N. A. and Khare, A. R., 1993. Preparation, structure and diffusional behavior of hydrogels in controlled release. *Adv. Drug Deliv. Rev.* 11: 1–35.

Pillai, O. and Panchagnula, R., 2003. Transdermal delivery of insulin from poloxamer gel: *Ex vivo* and *in vivo* skin permeation studies. *J. Control. Release* 89: 127–140.

Plunkett, K. N., Berkowski, K. R., and Moore, J. S., 2005. Chymotrypsin responsive hydrogel: Application of a disulfide exchange protocol for the preparation of methacrylamide containing peptides. *Biomacromolecules* 6: 632–637.

Prasad, R., Koula, V., Ananda, S., and Khar, R. K., 2007. Effect of DC/mDC iontophoresis and terpenes on transdermal permeation of methotrexate: *In vitro* study. *Int. J. Pharm.* 333: 70–78.

Qian, L. and Hinestroza, J. P., 2004. Application of nanotechnology for high performance textiles. *J. Text. Appar. Technol. Manag.* 4: 1–7.

Qui, Y. and Park, K., 2001. Environment-sensitive hydrogels for drug delivery. *Adv. Drug Deliv. Rev.* 53: 321–339.

Ramanathan, S. and Block, L. H., 2001. The use of chitosan gels as matrices for electrically-modulated drug delivery. *J. Control. Release* 70: 109–123.

Reddy, M. S., Mutalik, S., and Rao, G. V., 2006. Preparation and evaluation of minoxidil gels for topical application in alopecia. *Indian J. Pharm. Sci.* 68: 432–436.

Risbud, M. V., Karamuk, E., Schlosser, V., and Mayer, J., 2003. Hydrogel-coated textile scaffolds as candidate in liver tissue engineering: II. Evaluation of spheroid formation and viability of hepatocytes. *Biomater. Sci. Polym. Ed.* 14: 719–731.

Sannino, A., Maffezzoli, A., and Pollini, M., 2006. Natural or synthetic yarns with high absorption property obtained by introduction of superabsorbent hydrogel. World International Property Organization Document WO/2006/126233.

Scheuplein, R. J., 1978. Permeability of the skin: A review of major concepts. *Curr. Probl. Dermatol.* 7: 172–186.

Schneider, S., Ait-M-Bark, Z., Schummer, C., Lemmer, P., Yegles, M., Appenzeller, B., and Wennig, R., 2008. Determination of fentanyl in sweat and hair of a patient using transdermal patches. *J. Anal. Toxicol.* 32: 260–264.

Serra, M., 2002. Adaptable skin—Hydrogel gives wetsuit protection. *Smart Mater. Bull.* 7: 7–8.

Shishu, A. N., 2006. Preparation of hydrogels of griseofulvin for dermal application. *Int. J. Pharm.* 326: 20–24.

Slivka, S. R., Landeen, L. K., Zeigler, F., Zimber, M. P., and Bartel, R. L., 1993. Characterization, barrier function and drug metabolism of an *in vitro* skin model. *J. Invest. Dermatol.* 100: 40–46.

Sudo, J.-I., Iwase, H., Terui, J., Kakuno, K., Soyama, M., Takayama, K., and Nagai, T., 1998. Transdermal absorption of Image-Dopa from hydrogel in rats. *Eur. J. Pharm. Sci.* 7: 67–71.

Suhonen, T. M., Bouwstra, J. A., and Urtti, A., 1999. Chemical enhancement of percutaneous absorption in relation to stratum corneum structural alterations. *J. Control. Release* 59: 149–161.

Sun, T. T. and Green, H., 1978. Keratin filaments of cultured human epidermal cells. Formation of intermolecular disulfide bonds during terminal differentiation. *J. Biol. Chem.* 253: 2053–2060.

Takagi, Y., Kriehuber, E., Imokawa, G., Elias, P. M., and Holleran, W. M., 1999. β-Glucocerebrosidase activity in mammalian stratum corneum. *J. Lipid Res.* 40: 861–869.

Tas, C., Ozkan, Y., Savaser, A., and Baykara, T., 2003. *In vitro* release. Studies of chlorpheniramine maleate from gels prepared by different cellulose derivatives. *Il Farmaco* 58: 605–611.

Thornton, P. D., McConnell, G., and Ulijn, R. V., 2005. Enzyme responsive polymer hydrogel beads. *Chem. Commun.* 47: 5913–5915.

Ulijn, R. V., Bibi, N., Jayawarna, V., Thornton, P. D., Todd, S. J., Mart, R. J., Smith, A. M., and Gough, J. E., 2007. Bioresponsive hydrogels. *Mater. Today* 10: 40–48.

Vachon, D. J. and Yager, D. R., 2006. Novel sulfonated hydrogel composite with the ability to inhibit proteases and bacterial growth. *Biomed. Mater. Res. A* 76: 35–43.

Van Dyke, M. E., Timmons, S. F., Blanchard, C. R., Siller-Jackson, A. J., and Smith R. A., 2001. Absorbent keratin wound dressing. U.S. Patent 6270793.

Varshosaz, J. and Falamarzian, M., 2001. Drug diffusion mechanism through pH-sensitive hydrophobic/polyelectrolyte hydrogel membranes. *Eur. J. Pharm. Biopharm.* 51: 235–240.

Wertz, P. W., Miethke, M. C., Long, S. A., Strauss, J. S., and Downing, D. T., 1985. The composition of the ceramides from human stratum corneum and from comedones. *J. Invest. Dermatol.* 84: 410–412.

Yaffe, M. B., Murthy, S., and Eckert, R. L., 1993. Evidence that involucrin is a covalently linked constituent of highly purified keratinocyte cornified cell envelopes. *J. Invest. Dermatol.* 100: 3–9.

Yang, L., Chu, J. S., and Fix, J. A., 2002. Colon-specific drug delivery: New approaches and *in vitro/in vivo* evaluation. *Int. J. Pharm.* 235: 1–15.

Zi, P., Yang, X., Kuang, H., Yang, Y., and Yu, L., 2008. Effect of HPβCD on solubility and transdermal delivery of capsaicin through rat skin. *Int. J. Pharm.* 358: 151–158.

Ziboh, V. A., Lord, J. T., and Penneys, N. S., 1977. Alterations of prostaglandin E2–9-ketoreductase activity in proliferating skin. *J. Lipid Res.* 18: 37–43.

6 Micelles
The Multifunctional Nanocarrier for Colloidal Drug Delivery

Chandana Mohanty, Sarbari Acharya, and Sanjeeb K. Sahoo

CONTENTS

6.1 INTRODUCTION

Numerous pharmacological active compounds have been discovered, isolated, and synthesized in the last few decades. Many of these promising compounds did not advance beyond the laboratory bench to the clinical setting because of difficulties in finding a suitable delivery system. The colloidal drug delivery system (DDS) provides a crucial approach for problematic drug candidates and gives a solution to the hurdles involved in conventional drug delivery (Boyd, 2008). Colloidal science is often defined as the science of materials with a length scale below 1 μm, mostly in the sub-100-nm regime. Particle size is often the primary property of concern in a colloidal system and its final application. The huge surface area that results from dividing a mass of materials down to colloidal dimension offers an important advantage in terms of tailoring the surface properties and subsequent modification of particle behavior. In terms of drug delivery, flexibility in tailoring the internal structure and surface has led to the adoption of colloidal drug carriers as a platform for a wide range of existing products as a means to achieve new delivery modes for existing drugs and to improve their therapeutic profile. In order to improve the specific delivery of drugs with a low therapeutic index, several drug carriers such as liposomes (Krauze et al., 2006), microparticles (Guglielmini, 2008), nanoparticles (Sahoo et al., 2002), drug–polymer conjugates (Sahoo et al., 2007), and polymeric micelles (Sutton et al., 2007) have been developed. In recent years, polymeric micelles have been the object of growing scientific attention for the solubilization and tumor-targeted delivery of chemotherapeutic agents after systemic administration (Rijcken et al., 2007). They are becoming a powerful nanomedicine platform for therapeutic application because of their small size (10–100 nm), good *in vivo* biocompatibility, stability, and successful use in pharmaceuticals to solubilize water-insoluble drugs. They are often compared to naturally occurring carriers such as viruses or lipoproteins (Jones and Leroux, 1999; Sutton et al., 2007). The nanoscopic dimension, stealth properties induced by the hydrophilic polymeric brush on the micellar surface, capacity for stabilized encapsulation of hydrophobic drugs offered by the hydrophobic and rigid micellar core, and possibility for the chemical manipulation of the core–shell structure have made polymeric micelles one of the most promising carriers for drug delivering, targeting, and imaging.

The combination of micelles with several useful properties in one particle can significantly enhance the efficacy of many therapeutic and diagnostic protocols. Micelles can possess a combination of the following abilities: long circulation and targeting the site of the disease via both nonspecific and specific mechanisms, such as the enhanced permeability and retention (EPR) effect and ligand-mediated recognition; responding to local stimuli, which are characteristic of the pathological site, for example, releasing an entrapped drug or deleting a protective coating under the slightly acidic conditions in side tumor tissue; and enhanced intracellular delivery of an entrapped drug. In addition, these carriers can be supplied with contrast moieties to follow their real-time biodistribution and tumor accumulation and with certain moieties to provide more exotic properties (e.g., magnetic). This multifunctional nature of polymeric micelles appears to fulfill several tasks required for an ideal carrier capable of selective drug delivery at different levels. This chapter considers the current status and possible future directions in the emerging area of multifunctional micelles with primary attention for colloidal drug delivery with some novel properties for drug loading, targeting to the specific site via intracellular trafficking, and so forth.

6.2   DRUG DELIVERY PRINCIPLE AND CHALLENGE

Drug delivery is the method or process of administering a pharmaceutical compound to achieve a therapeutic effect in humans or animals. The insight behind drug delivery should be to provide a constant concentrations of drugs and to bring the compound with pharmaceutical activity directly to the site of need in order to enhance the effectiveness of action (Duncan, 2007). The problems and challenges experienced by free drug delivery are instability, poor solubility, toxicity, nonspecificity, and inability to cross the blood–brain barrier. The problem experienced by free drugs can be overcome by the use of a DDS. In DDSs, to direct the active substance to its site of action, their biodistribution needs to be modified by entrapping it in particulate drug carriers such as nanoparticles, micelles, or liposomes (Nakayama and Okano, 2005). The preferred delivery principle for patients includes noninvasive peroral, topical, transmucosal routes, as well as inhalation. The delivery of drugs to pathological tissue can be achieved primarily in two basic ways: passive targeting and active targeting. Passive targeting approaches make use of the anatomical and functional differences between the normal and tumor vasculature (Bardelmeijer et al., 2002). Most cancerous cells have leaky vasculature, allowing the nanocarriers to extravasate and localize in the tissue interstitial spaces. Angiogenic blood vessels, unlike the tight blood vessels in most normal tissues, have gaps as large as 600–800 nm between adjacent endothelial cells. Carriers can extravasate through these gaps into the tumor interstitial space, in a size-dependent manner. Because tumors have impaired lymphatic drainage (Sahoo et al., 2002), the carriers concentrate in the tumor and increase in tumor drug concentration (10-fold or higher) can be achieved relative to administration of the same dose of free drug. This process is known as EPR (Maeda et al., 2000; Northfelt et al., 1996; Sahoo et al., 2007). Drug delivery through nanocarrier systems to pathological sites does not involve only wrapping up drugs in a new formulation for different routes of delivery; the focal point is on targeted therapy. New techniques of targeted drug delivery have tremendous significance in the field of therapy. Various radioimmunopharmaceuticals, immunotoxins, and immunoconjugates are already in the market; immunoliposomes, immunopoly-mers, and antibody-directed enzyme prodrug therapies are in clinical development. In the field of drug delivery and cancer therapy this technique is actively engaged in improving the stability (LaVan et al., 2003), solubility (Maeda et al., 2000), absorption, and therapeutic action of the drug within the target tissue (Maeda et al., 2000), which permits long-term release of the drugs.

Nanocarriers' size creates the potential for crossing various biological barriers within the body, especially the potential to cross the blood–brain barrier, and may open new ways for drug delivery into the brain. In addition, nanocarriers also allow access into the cell and various cellular compartments including the nucleus. When a drug is associated with a carrier, the drug clearance decreases (half-life increases), the volume of distribution decreases, and the area under the time versus concentration curve increases. In addition to these advantages, the DDS has some unresolved problems such as in polymer–drug conjugates or liposome systems; the drug is inactive (not bio-available) and the failure to release the drug from the carrier in a timely manner may result in a reduced therapeutic effect relative to the free drug. In contrast, rapid release of the drug from the carrier may result in therapeutic effects that are similar to those seen for administration of the free drug (Cabanes et al., 1998). Associating a therapeutic molecule with a carrier may, however, result in the generation of immune reactions (Pichler et al., 2006). Most DDSs use nontoxic, biodegrad-able ingredients, so toxicities associated with the carrier molecules are likely to be mild. Perhaps the most common side effect is a hypersensitivity reaction after intravenous administration, which is possibly attributable to complement activation (Szebeni et al., 2002). This can be avoided by slowing the rate of infusion of the product or by patient premedication. Hypersensitivity reactions and side effects often fail to appear on repeated administration of the DDS relative to the free drug; for instance, the cardiotoxicity of doxorubicin (DOX) is reduced when a DDS is used, because of reduc-tions in the peak cardiac levels of the drug. Thus, recognition of these challenges is leading to establishing new approaches that will help to make these goals increasingly practicable.

6.3 COMPOSITION OF MICELLAR STRUCTURE

In recent years, polymeric micelles have been investigated as potential carriers for poorly water-soluble drugs. Efforts have been made for the preparation, characterization, and pharmaceutical application of polymeric micelles. Surfactant molecules (e.g., cetyl trimethyl ammonium bromide, sodium dodecyl sulfate, Triton X-100) self-aggregate into supermolecular structures when dissolved in water or oil. The simplest aggregate of these surfactant molecules is called a micelle (Tanford, 1973). Micelles made of nonionic surfactants are widely used as adjuvants and drug carrier systems in many areas of pharmaceutical technology and controlled drug delivery (Bardelmeijer et al., 2002; Malik et al., 1975; Tanford, 1973). An important criterion in the formation of these micelles is the critical micelle concentration (CMC). The CMC is the concentration at which amphiphilic polymers in aqueous solution begin to form micelles (i.e., self-aggregate) while coexisting in the equilibrium with individual polymer chains or unimers. At CMC and slightly above it, the micelles form loose aggregates and contain some water in the core. With further increase in the amphiphilic polymer concentration, the unimer–micelle equilibrium shifts toward micelle formation (Kwon and Kim, 2002). Surfactant micelles are formed only above the CMC and rapidly break apart upon dilution, which can result in premature leakage of the drug and its precipitation *in situ*. These limitations of surfactant micelles as drug delivery carriers triggered the search for micelles of significantly enhanced stability and solubilizing power (Marie-Helene, 2005). The use of polymer-based micelles has gained much attention recently because of the high diversity of polymers, their biocompatibility and biodegradability, as well as the multiplicity of functional groups they display. These polymeric micelles often provide better kinetics and thermodynamic stability than conventional surfactants.

Polymeric micelles composed of block copolymers have been investigated as drug delivery vehicles because of their characteristic advantages, such as small size, thermodynamic stability, and solubilization of hydrophobic molecules (Liyan et al., 2007). Polymeric micelles of block

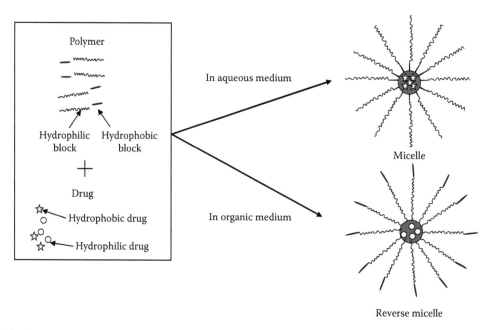

FIGURE 6.1 Formation of micelles and reverse micelles.

copolymers have been reported primarily for di- or triblock copolymers. Various types of drugs can be loaded into the hydrophobic core of polymeric micelles [e.g., hydrophobic low molecular weight (MW) drugs, cisplatin, and deoxyribonucleic acid (DNA)] by chemical conjugation or physical entrapment utilizing various interactions such as hydrophobic interactions, ionic interactions, and hydrogen bonding (Yokoyama et al., 1998). The hydrophobic core serves as a reservoir from which the drug is released slowly over extended periods of time. For example, the water solubility of pacli-taxel (PTX) can be increased by several orders of magnitude from 0.0015 to 2 mg/mL through micelle incorporation (Wang et al., 2008b).

In the late 1960s, micelles drew much attention as drug carriers because of their easily controlled properties and good pharmacological characteristics. Micelles are formed when amphiphiles are placed in water. Polymer micelles are composed of amphiphilic macromolecules that have distinct hydrophobic and hydrophilic block domains; and the structure of the copolymers usually is a diblock, triblock, or graft copolymer. At the appropriate ratio of block lengths, these copolymers spontaneously form spherical particles in water: the hydrophobic blocks form the "core," and the hydrophilic blocks form the surrounding "corona." The inner micellar core is created by cohesive interactions of the hydrophobic portions of the block copolymers in aqueous media (i.e., hydrophobic interactions), whereas the outer hydrophilic portions surround the inner hydrophobic core as a hydrated shell (Kakizawa and Kataoka, 2002) that serves as a stabilizing interface between the hydrophobic core and the external aqueous environment (Figure 6.1). The size of polymeric micelles ranges from ~10 to ~100 nm, and usually the size distribution is narrow. This topology is similar to that of surfactant micelles. Another type of micelle is the reverse micelle. Reverse micelles are nanometer-sized (1–10 nm) water droplets dispersed in organic media obtained by the action of surfactants. Surfactant molecules organize with the polar part on the inner side and are able to solu-bilize water and the apolar part in contact with the organic solvent (Figure 6.1). Proteins and other hydrophilic drugs can be solubilized in the water pool of reverse micelles (Khmelnitsky et al., 1992). Micelles are formed by the competition of two forces: the hydrophobic interaction between the tails that provides the driving force for aggregation and the electrostatic or steric repulsion between the head groups that limits the size that a micelle can attain.

Many types of copolymers have been used for micelle formation, but the requirements of bio-compatibility and often biodegradability have limited the choice of copolymers in clinical applications. Broad interest in biodegradable and biocompatible polymers has led to the synthesis of an increasingly wide range of amphiphilic diblock copolymers (Lavasanifar et al., 2002). Generally, amphiphilic block copolymers of A–B type, where A represents a hydrophilic block and B represents a hydrophobic block, are used to design polymeric micelles and vesicles. Other examples include A–B–A triblock copolymers and graft copolymers. As a drug carrier, it is preferable that the hydrophobic B (inner micelle core block) comprises a biodegradable polymer and the hydrophilic A (outer micelle shell block) comprises a polymer that is capable of interacting with plasma proteins and cell membranes (Rijcken et al., 2007).

For the hydrophilic segment, the most commonly used polymer is poly(ethylene glycol) (PEG) because it is a nontoxic polymer with U.S. Food and Drug Administration approval. Its exclusive physicochemical properties (high water solubility, high flexibility, and large exclusion volume) and good stealth properties make it an ideal choice for the hydrophilic segment of the micelles. Other hydrophilic polymers such as poly(N-vinyl pyrrolidone), poly(N-isopropylacrylamide), or poly(acrylic acid) (PAA) have also been used to form the micelle corona layer (Benahmed et al., 2001). For the hydrophobic segments, the most common materials are hydrophobic polyesters such as poly(lactic acid) (PLA), poly(ε-caprolactone) (PCL), and poly(trimethylene carbonate); but other materials, such as polyethers, polypeptides, or poly(β-aminoester), have been used (Sutton et al., 2007). Polyesters and polyamides can undergo hydrolytic and enzyme-catalyzed degradations, respectively, and are considered biodegradable. The choice of the core-forming segment is the major determinant for vital properties of polymeric micelles such as stability, drug-loading capacity, and drug-release profile and explains why so many core-forming hydrophobic polymers have been used for the development of polymeric micelles.

Thus, polymeric micelles are nanodelivery systems formed through self-assembly of amphiphilic block copolymers in an aqueous environment. The nanoscopic dimension, stealth properties induced by the hydrophilic polymeric brush on the micellar surface, capacity for stabilized encapsulation of hydrophobic drugs offered by the hydrophobic and rigid micellar core, and the possibility for the chemical manipulation of the core–shell structure have made polymeric micelles one of the best potential carriers for drug targeting.

6.4 MICELLAR SYSTEM AS DRUG DELIVERY SYSTEM

Most systemically administered drugs exert their biological effects not only at their target sites but also at nontarget sites, which often results in undesired side effects that hamper their therapeutic potential. The goal of all sophisticated DDSs is to deploy intact medications to specifically targeted parts of the body through a medium that can control the therapy's administration by means of either a physiological or chemical trigger. The current methods of drug delivery exhibit specific problems that scientists are attempting to address. For example, the potencies and therapeutic effects of many drugs are limited or otherwise reduced because of their partial degradation that occurs before they reach a desired target in the body. Further, slow release medications deliver drugs continuously, rather than providing relief and protection from adverse events solely when necessary (Torchilin, 2005). Injectable medications could be made less expensive and administered more easily if they could simply be dosed orally. For this purpose, model drug carriers should preferably have a high drug-loading capacity, adequate stability in the bloodstream, long circulating properties, selective accumulation at the site of action, along with a suitable drug-release profile and good biocompatibility. To achieve this goal, researchers are turning to advances in the world of nanotechnology. During the past decade, polymeric micelles have been shown to be effective in enhancing drug targeting specificity, lowering systemic drug toxicity, improving treatment absorption rates, and providing protection of pharmaceuticals against biochemical degradation and hence have emerged as a competent drug delivery vehicle (Abdullah et al., 2007).

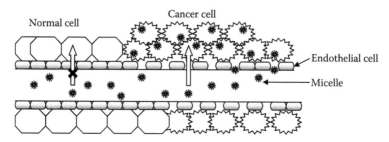

FIGURE 6.2 Because of the tight junction in normal vasculature, drug-loaded micelles are not able to extravasate whereas the leaky tumor vasculature creates a high local concentration and increased accumulation leading to the EPR effect.

The application of polymeric micelles as DDSs was pioneered by H. Ringsdorf's group in 1984 and subsequently used by Kataoka in the early 1990s through the development of DOX-conjugated block copolymer micelles (Yokoyama et al., 1992). Polymeric micelles are nanodelivery systems formed through self-assembly of amphiphilic block copolymers in an aqueous environment. Micelles as drug carriers are able to provide a series of unbeatable advantages: they can solubilize poorly soluble drugs by their hydrophobic core, resulting in the increase of drug stability and bioavailability. The nanoscopic dimensions, stealth properties induced by the hydrophilic polymeric brush on the micellar surface, capacity for stabilized encapsulation of hydrophobic drugs offered by the hydrophobic and rigid micellar core, and the possibility for the chemical manipulation of the core–shell structure have made polymeric micelles one of the most promising carriers for drug targeting (Kim et al., 2005). To date, three generations of polymeric micellar delivery systems, that is, polymeric micelles for passive, active, and multifunctional drug targeting, have arisen from research efforts, with each subsequent generation displaying greater specificity for the diseased tissue.

The purpose of ideal DDSs in cancer chemotherapy is to achieve selective delivery of anticancer agents to cancer tissues at effective concentrations for an appropriate duration of time, to reduce the adverse effects of the drug and simultaneously enhance the antitumor effect. Two major mechanisms can be distinguished for addressing the desired sites for drug release: passive and active targeting. An example of passive targeting is the preferential accumulation of chemotherapeutic agents in solid tumors as a result of the enhanced vascular permeability of tumor tissues compared with healthy tissues. Polymeric micelles increase the accumulation of drugs in tumor tissues utilizing the EPR effect (Figure 6.2). The EPR effect was proposed by Maeda et al. in 1986 and is attributed to two factors. First, the angiogenic tumor vasculature, as well as the blood vessels in tumor tissues, have higher permeability compared to normal ones because of its discontinuous endothelium. Second, lymphatic drainage is not fully developed in tumors (Maeda, 2001). These features lead to colloidal particles (polymeric micelles) extravasating through the "leaky" endothelial layer in the tumor and other inflamed tissues and subsequently being retained there (Maeda, 2001). A strategy that could allow active targeting involves the surface functionalization of micelles with ligands that are selectively recognized by receptors on the surface of the cells of interest. Ligand–receptor interactions can be highly selective, so this could allow a more precise targeting of the site of interest.

A key issue for prolonged circulation is to reduce the rate of nonspecific recognition and uptake by the reticuloendothelial system (RES). It has been demonstrated that grafting hydrophilic polymers (e.g., PEG and poloxamer) on the surface of particles is effective to oppose opsonization and subsequent uptake by the RES cells of the liver, spleen, and bone marrow. Polymeric micelles are made of amphiphilic copolymers such as PEG–PCL. The PEG coating helps the micelles escape the RES system (Owens and Peppas, 2006).

One important issue determining the effectiveness of a micellar drug carrier is the ability to control the time over which drug release takes place. Fast release may lead to premature loss of drug, causing systemic side effects and low concentration of the drug at the target site; slow release may reduce the efficacy of the drug at the site of action and increase drug resistance in cells. This challenge has motivated the development of new micellar systems that are designed to release their loaded drug in a controlled manner upon reaching the target site. The release of physically entrapped drugs from polymeric micelles is controlled by diffusion of the drug through the micellar core and the partition coefficient of the drug over the micellar core and the aqueous phase, provided that the micelles remain intact. Other factors influencing drug release are the length of the core-forming polymer segment, the affinity between the drug and the core (i.e., partition coefficient between the core and the aqueous phase), and the amount of loaded drug. Changes in acidity is a particularly useful environmental stimulus to exploit the development of drug carriers because of the numerous pH gradients that exist in both normal and pathophysiological states (Gillies and Frechet, 2004). For example, it is well documented that the extracellular pH of tumors is slightly more acidic than normal tissues, with a mean pH of 7.0 in comparison with 7.4 for the blood and normal tissues. Polymeric micelles that are responsive to these pH gradients can be designed to release their payload selectively in tumor tissues or within tumor cells. Release of the anticancer drug DOX could be modulated using pH, and it was found that the toxicity of the DOX-loaded mixed micelles to tumor cells *in vitro* was strongly dependent on pH with toxicity comparable to free DOX near the triggering pH (Yoo et al., 2002).

Presently, the development of integrated cancer nanomedicines, which consist of drugs that exploit cancer-specific molecular targets combined with effective carriers like micelles for tumor-targeted drug delivery, has shown significant promise in expanding therapeutic indices for chemotherapy. The cytotoxic effect of β-lapachone, a novel, plant-derived anticancer drug, is significantly enhanced by reduced nicotinamide adenine dinucleotide (phosphate); quinone oxidoreductase 1, a flavoprotein, overexpressed in a variety of human cancers, including those of the lung, prostate, pancreas, and breast. To exploit the numerous advantages of the hydrophobic drug, which is comparatively less ineffective in its native form, it was loaded in polymer micelles to develop β-lapachone-containing micelles for a quinone oxidoreductase 1 specific therapy (Blanco et al., 2007). Another hydrophobic drug, triamterene, which is used clinically to treat high blood pressure and fluid retention caused by heart disease and various other conditions, was loaded directly into micellar suspensions as was done previously with hydrophobic dyes (Kim et al., 2005). Hydrocamptothecin, a novel camptothecin congener widely used in China, has shown more powerful antitumor activity to lung, ovarian, breast, pancreas, and stomach cancers than camptothecin. Micelles using methoxy-PEG (MPEG)–PCL as a carrier material and hydrocamptothecin as a model drug were prepared to enhance the therapeutic efficacy of the drug (Shi et al., 2005). NK911 is a DOX-encapsulated polymeric micellar system. The preclinical pharmacological study revealed that the plasma area under the curve and intratumor area under the curve of NK911 was 15- to 30-fold higher than those of DOX (Hamaguchi, 2003). Cisplatin is an important class of antitumor agents that is widely used for the treatment of many malignancies, including testicular, ovarian, bladder, head and neck, small-cell, and non-small-cell lung cancers. Polymeric micelles incorporating cisplatin were prepared through a polymer–metal complex formation between the cisplatin polymeric micelles and PEG–PGA block copolymers, and their utility as a tumor-targeted DDS was investigated (Nishiyama et al., 2003). A novel polymeric micelle formulation of PTX was prepared with the purpose of improving *in vitro* release as well as prolonging the blood circulation time of PTX in comparison to a current PTX formulation.

Thus, the need for research into DDSs extends beyond ways to administer new pharmaceuticals. The safety and efficacy of the current treatments may be improved if their delivery rate, biodegradation, and site-specific targeting can be predicted, monitored, and controlled. This is mostly done through nanoparticular systems that act as effectual drug delivery vehicles.

6.5 FUNCTIONALIZATION OF POLYMERIC MICELLES

Polymeric micelles are particularly attractive because of their ability to deliver large payloads of a variety of drugs [e.g., small molecules, proteins, and DNA/ribonucleic acid (RNA) therapeutics],

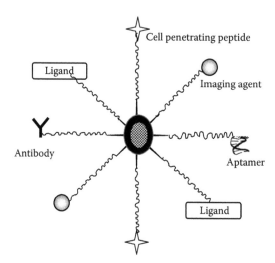

FIGURE 6.3 Multifunctional micelles.

their improved *in vivo* stability compared to other colloidal carriers (e.g., liposomes), and their nanoscopic size that allows for passive accumulation in diseased tissues such as solid tumors by the EPR effect (Torchilin, 2005). Typical chemotherapeutic agents have low water solubility, short blood half-lives, narrow therapeutic indices, and high systemic toxicity, which lead to patient morbidity and mortality while compromising the desirable therapeutic outcome of the drugs. Polymer micelles have been shown to increase the aqueous solubility of chemotherapeutic agents and prolong their *in vivo* half-lives with lessened systemic toxicity (Torchilin, 2007). Although stealth micelles allow for passive accumulation inside tumors with leaky vasculature, the majority of these nanoparticles are still cleared by the RES, resulting in short half-lives and unwanted micelle deposition in the liver and spleen. Development of multifunctional micelles through conjugation of targeting ligands on the micelle surface can lessen these problems by increasing particle and drug exposure to the tumor (Sutton et al., 2007). Thus, using appropriate surface functionality, polymeric micelles are further decorated with cell-targeting groups and permeation enhancers that can actively target diseased cells and aid in cellular entry, resulting in improved cell-specific delivery (Figure 6.3).

Functionalization of the outer surface of polymeric micelles to modify their physiochemical and biochemical properties is of great value from the standpoint of designing micellar carrier systems for receptor-mediated drug delivery. This can be accomplished in a regulated fashion by constructing micelles from a variety of end-functionalized block copolymers (Nagasaki et al., 2001). Polymeric micelles having sugars and peptides on their periphery have been prepared to explore their utility in the field of drug and gene delivery. Furthermore, functional groups on the nanoparticle surface may be conjugated with metal-chelating systems for radiolabeling, thus providing as effective tools for imaging and radiotherapy. Targeting ligands are conjugated to the corona of the micelle to induce specific targeting and uptake of the micelle by tumor cells. These ligands tend to fall into the categories of small organic molecules, carbohydrates, antibodies, and aptamers (Apts; Sutton et al., 2007). Regardless of the targeting moiety, the principle outcome is essentially the same, mainly improved tumor cell recognition, improved tumor cell uptake, and reduced recognition at nonspecific sites.

6.5.1 Folic Acid as Ligand

Folate targeting is a widely used active targeting ligand in micelles because the receptor recognizing the folic acid is commonly overexpressed on a wide variety of tumor types such as ovarian, colorectal, breast, and nasopharyngeal carcinomas. Folate coupling to the surface of micelles via PEG results in a greater intracellular uptake of the micelles in folate receptor (FR)-positive KB tumor cells over uptake of the nanocarriers lacking the targeting ligand (Yoo and Park, 2004). To overcome multidrug resistance (MDR) existing in tumor chemotherapy, polymeric micelles encoded with folic acid on the micelle surface were prepared with the encapsulation of a potent MDR modulator, FG020326. The micelles were fabricated from diblock copolymers of PEG and biodegradable PCL with folate attached to the distal ends of PEG chains. The *in vitro* rhodamine 123 efflux experiment using MDR KB (v200) cells revealed that, when cells were pretreated with folate attached and FG020326-loaded micelles, the P-glycoprotein drug efflux function was significantly inhibited (Yang et al., 2008). Recently, a new folate-conjugated polymeric nanoparticle, known as shell cross-linked nanoparticles, was developed for tumor targeting via both FR-mediated endocytosis and the EPR effect. To address the nontargeting nature of PTX, cholesterol-grafted poly(*N*-isopropylacrylamide-*co*-*N*,*N*-dimethylacrylamide-*co*-undecenoic acid was synthesized. Folate was subsequently attached to the hydrophilic segment of the polymer in order to target FRs overexpressing cancer cells. *In vitro* cytotoxicity assays performed against KB cells provided conclusive evidence that the cellular uptake of micelles surface functionalized with folate was indeed enhanced because of a receptor-assisted endocytosis process. This novel polymeric design thus has the potential to be a useful PTX vehicle for the treatment of FR-positive cancers (Seow et al., 2007). The anticancer drug DOX and the contrast agent for magnetic resonance imaging (MRI), superparamagnetic iron oxide (SPIO), were accommodated in the core of micelles self-assembled from an amphiphilic block copolymer of PEG and PCL with a targeting ligand (folate) attached to the distal ends of PEG (folate–PEG–PCL). The *in vitro* tumor cell targeting efficacy of these folate-functionalized and DOX/SPIO-loaded micelles (folate–SPIO–DOX micelles) was evaluated by observing the cellular uptake of micelles by human hepatic carcinoma cells (Bel 7402 cells). The folate–SPIO–DOX multifunctional polymeric micelles had better targeting to the hepatic carcinoma cells *in vitro* than their nontargeting counterparts (Chen et al., 2008).

6.5.2 PROTEIN AS LIGAND

A peptide analog conjugated to a micelle can also be used as a targeting moiety. Novel poly(ethylene oxide)-*b*-PCL (PEO-*b*-PCL) based copolymers modified with RGD ligands on PEO and pendent functional groups on PCL, that is, GRGDS–PEO-*b*-poly(α-benzylcarboxylate-ε-caprolactone) and GRGDS–PEO-*b*-poly(α-carboxyl-ε-caprolactone), were synthesized. Chemical conjugation of DOX to the poly(α-carboxyl-ε-caprolactone) core produced GRGDS–PEO-*b*-PCL–DOX micellar conjugates. The GRGDS-modified micelles showed enhanced cellular internalization through endocytosis, increased intracellular DOX release, nuclear localization, and improved cytotoxicity against metastatic B16F10 cells compared to their unmodified counterparts (Xiong et al., 2008).

A large number of peptide receptors are overexpressed in many tumor cells that can act as ligands for specific targeting. One such receptor is $\alpha_v\beta_3$ integrins. Multifunctional polymeric micelles with cancer-targeting capability via $\alpha_v\beta_3$ integrins, controlled drug delivery, and efficient MRI contrast characteristics have come under much focus today. DOX and a cluster of SPIO nanoparticles were loaded successfully inside a micellar core. The presence of RGD on the micelle surface resulted in cancer-targeted delivery to $\alpha_v\beta_3$-expressing tumor cells. *In vitro* MRI and cytotoxicity studies demonstrated the ultrasensitive MRI imaging and $\alpha_v\beta_3$-specific cytotoxic response of these multifunctional polymeric micelles (Nasongkla et al., 2006). Reactive micelles based on a diblock copolymer of poly(ethyl ethylene phosphate) and PCL can be used as attractive tools for targeted intracellular drug delivery. The micelles were surface conjugated with galactosamine to target the asialoglycoprotein receptor of HepG2 cells. PTX-loaded micelles with galactose ligands exhibited comparable activity to free PTX in inhibiting HepG2 cell proliferation, in contrast to the poor inhibition activity of micelles without galactose ligands, particularly at lower PTX doses (Wang et al., 2008a).

6.5.3 ANTIBODY AS LIGAND

Another promising class of tumor-targeting ligand is cancer-specific monoclonal antibodies. These large and high-affinity ligands have the advantage of being able to be tailored to bind specifically to a large variety of targets such as cancer cell specific antigens. These were among the first ligands used for micellar targeting; Kabanov et al. (1989) used them to target haloperidol-loaded pluronic micelles for psychiatric treatment. Brain-specific antibody conjugation increased the neuroleptic action of the loaded micelles by 5-fold over nontargeted micelles and 20-fold over the free drug (Kabanov et al., 1989). Torchilin et al. (2003) developed these ligands to target micelles to lung cancer cells. Diacyllipid–PEG-conjugated polymer micelles were functionalized with one of two antibodies, either an anticancer monoclonal antibody (2C5) or an antimyosin monoclonal antibody (2G4; Torchilin et al., 2003). Intravenous administration of tumor-specific 2C5 immunomicelles loaded with a sparingly soluble anticancer agent, taxol, into experimental mice bearing Lewis lung carcinoma resulted in an increased accumulation of PTX in the tumor compared with free PTX or PTX in nontargeted micelles as well as enhanced tumor growth inhibition. The major benefit of using antibodies as targeting ligands is their high binding affinity. Antiepidermal growth factor receptor antibody [antiepithelial growth factor receptor (EGFR) antibody] was conjugated with a block copolymer micelle based on PEG and PCL for active targeting to EGFR overexpressing cancer cells. The cytotoxicity of DOX–anti-EGFR micelles to induce apoptosis in RKO human colorectal cancer cells was significantly better than that of free DOX or DOX micelles. These results demonstrated that the presence of the anti-EGFR antibody on the DOX micelle surface (DOX–anti-EGFR micelle) amplified the internalization of the DOX micelle and nuclear accumulation of DOX (Noh et al., 2008).

6.5.4 APTAMER AS LIGAND

Besides antibodies, Aptamers (Apts) are also a potential class of ligands that are used to functionalize polymeric micelles. Aptamers are RNA or DNA oligonucleotides that fold by intramolecular interaction into unique three-dimensional confirmations capable of binding to a target antigen. Farokhzad et al. used an RNA Apt for the prostate-specific membrane antigen (PSMA) to target PEG–PLA micelles to prostate tumors (Farokhzad et al., 2004, 2006; Torchilin et al., 2003). These nanoparticles showed specific binding to PSMA-expressing cancer cells, with the Apt inducing a 77-fold increase in binding versus the control group; they were subsequently loaded with docetaxel and examined in prostate cancer treatment (Farokhzad et al., 2006). A unique strategy to deliver cisplatin to prostate cancer cells is by constructing platinum(IV)-encapsulated PSMA targeted micelles of poly(D,L-lactic-*co*-glycolic acid)–PEG. By using poly(D,L-lactic-*co*-glycolic acid)-*b*-PEG micelles with PSMA targeting Apts on the surface, a lethal dose of cisplatin was delivered specifically to prostate cancer cells. A comparison between the cytotoxic activities of platinum(IV)-encapsulated micelles with the PSMA Apt on the surface and the nontargeted platinum(IV)-encapsulated micelles against human prostate PSMA-overexpressing LNCaP and PSMA⁻ PC3 cancer cells revealed significant differences between the two (Dhar et al., 2008).

6.5.5 OTHER MOLECULES AS LIGANDS

Other examples of functionalized micelles include micelles formed from PEO-*b*-poly(methylidene malonate 2.1.2) block copolymers bearing a primary amino group at the PEO chain; then the reactive amino endgroup was subsequently modified by reaction with functional ligands such as mannose and fluorescein. Another amphiphilic, biotinylated poly[*N*-isopropylacrylamide-*co*-*N*-(3-dimethylamino propyl) methacrylamide]-*b*-PCL block copolymer was synthesized that had thermoresponsive properties. These surface-modified micellar formulations showed great promise as DDSs (Heath et al., 2007). An asymmetric PEG-*b*-poly-[2-(*N,N*-dimethylamino) ethyl methacrylate] copolymer presenting a thiol group at the end of the PEP chain was synthesized to form stimuli-responsive micellar networks. Thiol groups are highly appealing because they react almost exclusively and quantitatively with maleimides under physiological conditions, thereby facilitating biotin functionalization on the copolymer. The biotinylated copolymers self-assemble with an oligonucleotide in aqueous media to form polyion complex micelles with biotin groups at their outer surface. These micelles are capable of molecular recognition toward streptavidin. Alternatively, thiol-decorated (nonderivatized) micelles showed improved mucoadhesion properties through the formation of disulfide bonds with mucin (Gaucher et al., 2005). PAA-coated core–shell gold and magnetite nanoparticles prepared by surface-initiated atomic transfer radical polymerization (grafting from method) are applied in a wide range of fields, including DDSs, diagnostics, gene analysis, proteomics, affinity purifications, quantum dots, and the like. To functionalize magnetic nanoparticles for bioconjugation and subsequent delivery and imaging studies, phospholipid–PEG molecules with primary amines were incorporated into the micelle structure. For cellular delivery, the cell-penetrating peptide Tat custom modified with a linker (6-aminohexonic acid) at the C-terminus followed by a reactive with cysteine was used.

Thus, superfunctionalized polymeric micelle mounting genes, drugs, and others in the inner core were developed and nanogene therapy was created, which achieved the diagnosis and treatment necessary for the required time and position with minimum side effects.

6.6 STEALTH POLYMERIC MICELLES AS DRUG DELIVERY SYSTEMS

The long desired goal of targeting drugs to specific sites in the body (where the pharmacological action is desired) sparing other tissues has been actively pursued for many years. Targeting drugs to specific sites and maintaining pharmacologically relevant drug levels at the site is what makes nanosystems the burgeoning magic wands. Advanced DDSs try to adjust the site and the rate of drug release depending on the physiological conditions of the patient, progression of the illness, or the circadian rhythms of the patient. Because they are different from classical preprogrammed controlled release dosage forms, these nanocarriers aim to provide the drug release profile that is best for the needs of each patient. However, substantial challenges still exist in terms of biological barriers hindering the nanocarriers to exhibit their properties. Since their discovery in the early 1980s, polymeric micelles have been the subject of several studies as delivery systems that can potentially improve the therapeutic performance and modify the toxicity profile of encapsulated drugs by changing their pharmacokinetic characteristics. In spite of these advantages, a major issue that limits the systemic application of micellar nanocarriers is the nonspecific uptake by the RES. The body normally treats the micelles as foreign particles, so they become easily opsonized and get removed from the circulation before completion of their function. Thus, one of the most important and growing areas of biomedical research includes increasing the longevity of these circulating micelles (Vasir and Labhasetwar, 2005).

Plasma proteins, known as opsonins, bind to these circulating micelles and eliminate them from the circulation within seconds to minutes through the RES (Owens and Peppas, 2006). Stealth technologies mean hiding these micelles from the immune system, thus avoiding an immune reaction. Imparting stealth shielding on the surface of these DDSs (micelles) prevents opsonins from recognizing these particles, thereby limiting phagocytosis by the RES cells and increasing the systemic circulation time from minutes to hours or even days (Owens and Peppas, 2006). A key issue for prolonged circulation is to reduce the rate of nonspecific recognition and uptake by the RES. A great example to escape the RES is exhibited in nature by bacteria. Van Oss showed in 1978 that many bacteria have a highly hydrophilic hydrated surface layer of protein, polysaccharide, and glycoprotein that reduces interactions with blood components and inhibits phagocytosis in the human body. Keeping this in mind researchers have shown that grafting of hydrophilic polymers on the surface of particles is effective to oppose opsonization and subsequent uptake by the RES cells of the liver, spleen, and bone marrow.

Among several strategies to impart particles with stealth shielding, the most prominent are surface modification with polysaccharides, polyacrylamide, and poly(vinyl alcohol). Surface modification with PEG and PEG copolymers proved to be most effective, fueling its widespread use. Amphiphilic block copolymers, such as poloxamers and poloxamines, consisting of blocks of hydrophilic PEG (or PEO) and hydrophobic poly(propylene oxide), are additional forms of PEG derivatives, which are often employed for modification by surface adsorption or entrapment (Veronese and Pasut, 2005). PEG (also known as PEO when MW > 20 kDa) modification has emerged as a common strategy to ensure such stealth shielding and long circulation of therapeutics or delivery devices. The hydrophilic shells of micelles usually consist of PEG. PEG is a linear polyether terminated with one to two hydroxyl groups. As a protecting polymer, PEG provides a very attractive combination of properties: excellent solubility in aqueous solutions, high flexibility of its polymer chain, very low toxicity, immunogenicity, and antigenicity, as well as lack of accumulation in RES cells and minimal influence on specific biological properties of modified pharmaceuticals (Gref et al., 2000). It prevents the interaction between the hydrophobic micellar core and biological membranes, reducing their uptake by RES cells and also prevent the adsorption of plasma proteins onto nanoparticle surfaces. The mechanisms by which PEG prevents opsonization include shielding of the surface charge and antigenic epitopes; increased surface hydrophilicity; enhanced repulsive interaction between polymer-coated nanocarriers and blood components; and formation of a polymeric layer over the particle surface, which is impermeable to large molecules of opsonins, even at relatively low polymer concentrations (Kommareddy et al., 2005). PEG conjugation also increases the MW and hydrodynamic volume of micelles, resulting in decreased blood clearance by renal filtration. PEG-b-poly(D, L- lactide) micelles showed proper circulation times (25% of the injected dose was still detected in the blood after 24 h), but polymeric micelles are generally cleared from the systemic circulation of experimental animals within the first 8–10 h after intravenous administration (Moghimi et al., 2001).

The protective (stealth) action of PEG is mainly attributable to the formation of a dense, hydrophilic cloud of long flexible chains on the surface of the colloidal particle that reduces the hydrophobic interactions with the RES. The tethered and chemically anchored PEG chains can undergo spatial conformations, thus preventing the opsonization of particles by the macrophages of the RES, which leads to preferential accumulation in the liver and spleen. PEG surface modification therefore enhances the circulation time of molecules and colloidal particles in the blood (Veronese and Pasut, 2005). The term "steric stabilization" has been introduced to describe the phenomenon of polymer-mediated protection, and the mechanism of steric hindrance by the PEG-modified surface has been thoroughly examined. The water molecules form a structured shell through hydrogen bonding to the ether oxygen molecules of PEG. The tightly bound water forms a hydrated film around the particle and repel the protein interactions (Gref et al., 1999). In addition, PEG surface modification may also increase the hydrodynamic size of the particle and decrease its clearance, a process that is dependent on the molecular size as well as particle volume. Ultimately, this helps in greatly increasing the circulation half-life of the particles (Hamidi et al., 2006).

Currently, clinical data on three polymer micelle systems, SP1049C, NK911, and Genexol-PM, have been reported that are stealth micelle formulations; that is, they all have stabilizing PEG coronas to minimize opsonization of the micelles and maximize blood circulation times (Danson et al., 2004; Kim et al., 2004; Matsumura et al., 2004). SP1049C is formulated as DOX-encapsulated pluronic micelles, NK911 as DOX-encapsulated micelles, and Genexol-PM as a PTX-encapsulated PEG–PLA micelle formulation. Amphiphilic polycations with a stealth cationic nature have also been designed and synthesized by the PEGylation of a polycationic amphiphile via a novel pH-responsible benzoic imine linker. The linkage is stable in aqueous solution at physiological pH but cleaves in slightly acidic conditions such as the extracellular environment of solid tumors and endosomes. The polymeric micelle formed from the amphiphilic stealth polycation contains a pH-switchable cationic surface driven by the reversible detachment and reattachment of the shielding PEG chains due to the cleavage and formation process of the imine linkage. At physiological pH the micellar surface is shielded by the PEG corona, leading to lower cytotoxicity and less hemolysis; in a mild acidic condition, such as in endosomes or solid tumors, the deshielding of the PEG chains exposes the positive charge on the micellar surface and retains the membrane disrupting ability (Gu et al., 2008).

Thus, the properties that have made polymeric micelles one of the most promising carriers for drug targeting consists of the nanoscopic dimension, stealth properties induced by the hydrophilic polymeric brush on the micellar surface, capacity for stabilized encapsulation of hydrophobic drugs offered by the hydrophobic and rigid micellar core, and the possibility for the chemical manipulation of the core–shell structure.

6.7   MICELLES AS DIAGNOSTIC AGENT

Medical diagnostic imaging modalities currently utilized in a certain quantity of reporter groups include γ-scintigraphy, MRI, computed tomography (CT), and ultrasonography. In any of these modalities, clinical diagnostic imaging requires a certain signal intensity from an area of interest that is achieved in order to differentiate structures under observation from surrounding tissues. To reach the required criteria of a contrast agent, it was a natural progression to use microparticulate carriers, such as liposomes, and radiation (x-rays) with the aid of computers that are able to selectively carry multiple reporter moieties for the efficient delivery of contrast agents into the required areas (Torchilin, 1997, 2000, 2002; Torchilin et al., 1999). Among those carriers, micelles (amphiphilic compounds formed from colloidal particles with a hydrophobic core and a hydrophilic corona) have recently drawn much attention because of their easily controlled properties and good pharmacological characteristics. Chelated paramagnetic metals, such as gadolinium (Gd), manganese (Mn), or dysprosium, are of major interest for the design of magnetic resonance positive (T1) contrast agents. Mixed micelles obtained from monoolein and taurocholate with Mn-mesoproporphyrin were shown to be a potential oral hepatobiliary imaging agent for T1-weighted MRI (Torchilin, 1997). Since chelated metal ions possess a hydrophilic character, it can be ideally incorporated in micellar formulation. Several amphiphilic chelating probes, where a hydrophilic chelating residue is covalently linked to a hydrophobic (lipid) chain, have been developed earlier for the liposome membrane incorporation studies, such as diethylene triamine pentaacetic acid (DTPA) conjugate with phosphatidyl ethanolamine (DTPAYPE; Schmiedl et al., 1995), DTPAY stearylamine (DTPAYSA; Grant et al., 1989), and amphiphilic acylated paramagnetic complexes of Mn and Gd (Mulder et al., 2007). The lipid part of such an amphiphilic chelate molecule can be anchored into the micelle's hydrophobic core, whereas a more hydrophilic chelating group is localized inside the hydrophilic shell of the micelle. The amphiphilic chelating probes (paramagnetic GdYDTPAYPE and radioactive [111]InYDTPAYSA) were incorporated into PEGYPE micelles and used *in vivo* for MR and γ-scintigraphy imaging. The main feature that makes PEGY lipid micelles attractive for diagnostic imaging applications is their small size, which facilitates better penetration to the targets to be visualized. In one study MPEG–iodolysine micelles were synthesized and injected into rats via the tail vein at a dose of 170 mg iodine/kg. Three animals were used, and tissue enhancement was followed on serial CT scans. In rats these MPEG–iodolysine micelles were shown to be a long-lived blood-pool contrast agent useful for CT (Torchilin et al., 1999). Because of their small size and surface properties imparted by the PEG corona, the micellar particulates can move with ease from the injection site along the lymphatics to the systemic circulation with the lymph flow. Their action is based on the visualization of lymph flowing through different elements of the lymphatics, whereas the action of other lymphotropic contrast media is based primarily on their active uptake by the nodal macrophages. Mulder et al. (2007) recently reported that, when micelles were conjugated with macrophage scavenger receptor specific antibodies, the macrophages in the abdominal aortas of atherosclerotic (apoE-KO) mice can be effectively and specifically detected by molecular MRI and optical methods upon administration of a PEGylated micellar contrasting agent. The efficacy of micelles as a contrast medium can be further increased by increasing the quantity of carrier-associated reporter metal (such as Gd or ^{111}In), thus enhancing the signal intensity. To solve this task, the use of amphiphilic chelating polymers was suggested (Trubetskoy et al., 1996). These modality polymers represent a family of soluble single-terminus lipid-modified polymers containing multiple chelating groups attached to the polylysine chain, which are suitable for incorporation into the hydrophobic surroundings (such as a hydrophobic core of corresponding micelles). A pathway was developed for the synthesis of the amphiphilic polychelator *N*,(DTPAY polylysyl)glutarylYPE, which sharply increases the number of chelated metal atoms attached to a single lipid anchor (Trubetskoy et al., 1996). Micelles formed by self-assembled amphiphilic polymers (such as PEGYPE) can easily incorporate such amphiphilic polylysine-based chelates carrying multiple diagnostically important metal ions such as Gd and [111]In (Unger et al., 1994). In addition, in MRI contrast agents, it is especially important that chelated metal atoms are directly exposed to the aqueous environment, which enhances the relaxivity of the paramagnetic ions and leads to the enhancement of the micelle contrast properties. CT represents an imaging modality with high spatial and temporal resolution, which uses x-ray absorbing heavy elements, such as iodine, as contrast agents. Diagnostic CT imaging requires an iodine concentration of millimoles per milliliter of tissue (Kong et al., 2007), so that large doses of a low MW CT contrast agent (iodine-containing organic molecules) are normally administered to patients. The selective enhancement of blood upon such administration is brief because of rapid extravasation and clearance. Currently suggested particulate contrast agents possess a relatively large particle size (between 0.25 and 3.5 mm) and are actively cleared by phagocytosis (Torchilin, 2005). The synthesis and *in vivo* properties of a block copolymer of MPEG and triiodobenzoic acid substituted poly-L-lysine have been described (Leander, 1996; Torchilin, 2007), which easily micellizes in the solution and form stable and heavily iodine-loaded particles (up to 35% of iodine by weight) with a size of 50–70 nm. These are in practice and act efficiently as contrasting agents. When the micellar iodine-containing CT contrast agent was injected intravenously into rats and rabbits, a fourfold enhancement of the x-ray signal in the blood pool was visually observed in both animal species for a period of 2 h following the injection (Leander, 1996; Torchilin, 2007). Thus, such technology may provide an innovative viewpoint and shed light on solving the imaging issue.

6.8 FUTURE PROSPECTS AND CONCLUSION

Polymeric micelles possess an excellent ability to solubilize poorly water-soluble drugs and increase their bioavailability. Because of their small size, they demonstrate a very efficient spontaneous accumulation of drugs taking place via the EPR effect in pathological areas by exploiting the leaky vasculature. Micelles of different composition can be used for a broad variety of drugs, mostly poorly soluble anticancer drugs. In the context of drug targeting, modifications are needed in designing polymeric micellar systems that can provide optimum disposition of the incorporated drug in the biological system, that is, within the reach of its molecular targets in cancerous tissue and away from nontarget healthy cells. Taking advantage of the chemical flexibility of the core–shell structure has resulted in the development of multifunctional polymeric micelles that bear multiple targeting functionalities on an individual carrier and are expected to achieve superior selectivity for diseased tissue (Haxton and Burt, 2008). By varying the micelle composition and the sizes of hydrophilic and hydrophobic blocks of the micelle-forming material we can easily control the properties of micelles, such as size, loading capacity, and longevity in the blood. Biodegradable micelles, in which the controlled degradation of hydrophilic blocks can facilitate drug liberation from the hydrophobic core, can also become a subject of interest. Another interesting option is provided by stimuli-responsive micelles, the degradation and subsequent drug release of which should proceed at abnormal pH values and temperatures characteristic for many pathological zones. Similarly, the design and preparation of pH-sensitive micelles is a new and exciting field of research to improve the selective delivery of therapeutic molecules using physiological triggers. One of the major challenges is the relatively narrow pH range in which the micellar carrier must both retain the drug over prolonged periods and then release it relatively rapidly. This challenge has been met by many different approaches including incorporation of titratable groups into the copolymer backbone such that the solubility of the polymer is altered by protonation or deprotonation events and by incorporation of pH-sensitive linkages that are designed to undergo hydrolysis under distinct physiological conditions either to directly release the drug or to alter the polymer structure to disrupt or break apart the micelles.

Metal-based drugs have recently been widely used in the treatment of cancer. Significant side effects and drug resistance, however, have limited its clinical applications. Advances in biocoordination chemistry are crucial for improving the design of compounds to reduce toxic side effects and understand their mechanisms of action. Haxton and Burt (2008) demonstrated that platinum–polymer complex micelles, conjugated to cisplatin (metal-based drugs) analogs, have improved specificity for tumor tissue, thereby reducing side effects and drug resistance. However, little is known about the interaction of polymeric micelles with plasma and cellular components and drug delivery in order to design micelles to efficiently deliver a drug to its site of action. Until now, only a limited number of polymeric micellar formulations have been tested as DDSs. Therefore, they still need systematic investigation on the micellar composition, structure, and zeta potential on the pharmacokinetics, biodistribution, and activity of the carried drug. The presence of a variety of colloidal particles in the bloodstream ranging from soluble proteins to lipoproteins to blood cells and a variety of unforeseen interactions between the micelles and biological components are also problematic in tracing micelles in physiological environments. Recently, a Förster resonance energy transfer imaging method was applied to reveal potential interactions of polymeric micelles constructed by a particular diblock copolymer with biological components, which in turn affected the drug release behaviors of unmodified polymeric micelles under *in vitro* (Asokan and Cho, 2005) and *in vivo* conditions (Chen et al., 2008). Such technologies may provide an innovative viewpoint and shed some light on solving the stability issue. *In vivo* biodistribution and excretion of polymeric micelles after intravenous administration, which can greatly affect the pharmacokinetics of encapsulated drugs, remains another challenge in the field of drug delivery.

The future of micelle formulation research should center on more scientific understanding of the application of practical systems. Another direction should focus on identifying intact unmodified micelles that have reached their intended target sites using various microscopic techniques, and the investigators should be able to identify micelles from other biological components. Interaction of micelles with biological components that can influence micelle stability will be another area to be explored. Such micelles need to be dissociated by certain signals or in the local environment after finishing their assigned role. Additional concerns might be linked to the drug-loading capacity of the micelles to reasonably meet the required dose size of an active drug. Handling less hydrophobic drugs will be another aspect to consider. Most of the new drug entities that are in the discovery stage are discarded because of dissatisfaction with the solubility and poor bioavailability. Micelle formulations in the pharmaceutical industry have not been well adopted, but the micelle approach is a versatile option that should be considered. For example, oral formulations based on micelle technology are one topic to be investigated. As long as micelle stability in the blood is clarified, designing intelligent micelles to maximize drug efficacy will be a novel challenge in the area. Thus, micelles that are easy to prepare and load with the drug or diagnostic moiety with a combination of properties should lead to the successful practical application of micellar drugs in the foreseeable future.

ACKNOWLEDGMENT

The third author (S.K.S.) thanks the Department of Biotechnology, Government of India, for providing Grants BT/04(SBIRI)/48/2006-PID and BT/PR7968/MED/14/1206/2006.

ABBREVIATIONS

CMC	Critical micelle concentration
CT	Computed tomography
DDS	Drug delivery system
DNA	Deoxyribonucleic acid

DOX	Doxorubicin
DTPA	Diethylene triamine pentaacetic acid
DTPAYPE	Diethylene triamine pentaacetic acid conjugate with phosphatidyl ethanolamine
DTPAYSA	Diethylene triamine pentaacetic acid conjugate with phosphatidyl stearylamine
EGFR	Epithelial growth factor receptor
EPR	Enhanced permeability and retention
FR	Folate receptor
MDR	Multidrug resistance
MPEG	methoxy-poly(ethylene glycol)
MRI	Magnetic resonance imaging
MW	Molecular weight
PAA	Poly(acrylic acid)
PCL	Poly(ε-caprolactone)
PEG	Poly(ethylene glycol)
PEO	Poly(ethylene oxide)
PLA	Poly(lactic acid)
PSMA	Prostate-specific membrane antigen
PTX	Paclitaxel
RES	Reticuloendothelial system
RNA	Ribonucleic acid
SPIO	Superparamagnetic iron oxide

REFERENCES

Abdullah, M., X. X. Bing, H. M. Alibad, and L. Afsaneh, 2007. Polymeric micelles for drug targeting. *J Drug Target* 15(9): 553–584.

Asokan, A. and M. J. Cho, 2005. Cytosolic delivery of macromolecules 4. Head group-dependent membrane permeabilization by pH-sensitive detergents. *J Control Release* 106(1–2): 146–153.

Bardelmeijer, H. A., M. Ouwehand, M. M. Malingre, J. H. Schellens, J. H. Beijnen, and O. van Tellingen, 2002. Entrapment by cremophor EL decreases the absorption of paclitaxel from the gut. *Cancer Chemother Pharmacol* 49(2): 119–125.

Benahmed, A., M. Ranger, and J. C. Leroux, 2001. Novel polymeric micelles based on the amphiphilic diblock copolymer poly(Nvinyl-2-pyrrolidone)-block-poly(dllactide). *Pharmaceutical Research* 18: 323–328.

Blanco, E., E. A. Bey, Y. Dong, B. D. Weinberg, D. M. Sutton, D. A. Boothman, and J. Gao, 2007. Beta-lapachone-containing PEG–PLA polymer micelles as novel nanotherapeutics against NQO1-overexpressing tumor cells. *J Control Release* 122(3): 365–374.

Boyd, B. J., 2008. Past and future evolution in colloidal drug delivery systems. *Expert Opin Drug Deliv* 5(1): 69–85.

Cabanes, A., K. E. Briggs, P. C. Gokhale, J. A. Treat, and A. Rahman, 1998. Comparative *in vivo* studies with paclitaxel and liposome-encapsulated paclitaxel. *Int J Oncol* 12(5): 1035–1040.

Chen, H., S. Kim, W. He, H. Wang, P. S. Low, K. Park, and J. X. Cheng, 2008. Fast release of lipophilic agents from circulating PEG–PDLLA micelles revealed by *in vivo* Forster resonance energy transfer imaging. *Langmuir* 24(10): 5213–5217.

Danson, S., D. Ferry, V. Alakhov, J. Margison, D. Kerr, D. Jowle, M. Brampton, G. Halbert, and M. Ranson, 2004. Phase I dose escalation and pharmacokinetic study of pluronic polymer-bound doxorubicin (SP1049C) in patients with advanced cancer. *Br J Cancer* 90(11): 2085–2091.

Dhar, S., F. X. Gu, R. Langer, O. C. Farokhzad, and S. J. Lippard, 2008. Targeted delivery of cisplatin to prostate cancer cells by aptamer functionalized Pt(IV) prodrug–PLGA–PEG nanoparticles. *Proc Natl Acad Sci USA* 105(45): 17356–17361.

Duncan, R., 2007. Designing polymer conjugates as lysosomotropic nanomedicines. *Biochem Soc Trans* 35(Pt 1): 56–60.

Farokhzad, O. C., J. Cheng, B. A. Teply, I. Sherifi, S. Jon, P. W. Kantoff, J. P. Richie, and R. Langer, 2006. Targeted nanoparticle–aptamer bioconjugates for cancer chemotherapy *in vivo*. *Proc Natl Acad Sci USA* 103(16): 6315–6320.

Farokhzad, O. C., S. Jon, A. Khademhosseini, T. N. Tran, D. A. Lavan, and R. Langer, 2004. Nanoparticle–aptamer bioconjugates: A new approach for targeting prostate cancer cells. *Cancer Res* 64(21): 7668–7672.

Gaucher, G., M. H. Dufresne, V. P. Sant, N. Kang, D. Maysinger, and J. C. Leroux, 2005. Block copolymer micelles: Preparation, characterization and application in drug delivery. *J Control Release* 109(1–3): 169–188.

Gillies, E. R. and J. M. J. Frechet, 2004. Development of acid-sensitive copolymer micelles for drug delivery. *Pure Appl Chem* 76: 1295–1307.

Grant, C. W., S. Karlik, and E. Florio, 1989. A liposomal MRI contrast agent: Phosphatidylethanolamine-DTPA. *Magn Reson Med* 11(2): 236–243.

Gref, R., M. Luck, P. Quellec, M. Marchand, E. Dellacherie, S. Harnisch, T. Blunk, and R. H. Muller, 2000. "Stealth" corona–core nanoparticles surface modified by polyethylene glycol (PEG): Influences of the corona (PEG chain length and surface density) and of the core composition on phagocytic uptake and plasma protein adsorption. *Colloids Surf B Biointerfaces* 18(3–4): 301–313.

Gref, R., A. Domb, P. Quellec, T. Blunk, R. H. Muller, J. M. Verbavatz, and R. Langer, 1999. The controlled intravenous delivery of drugs using PEG-coated sterically stabilized nanospheres. *Adv Drug Deliv Rev* 16(2–3): 215–233.

Gu, J., W. P. Cheng, J. Liu, S. Y. Lo, D. Smith, X. Qu, and Z. Yang, 2008. pH-Triggered reversible "stealth" polycationic micelles. *Biomacromolecules* 9(1): 255–262.

Guglielmini, G., 2008. Nanostructured novel carrier for topical application. *Clin Dermatol* 26(4): 341–346.

Hamaguchi, T., Y. Matsumura, K. Shirao, Y. Shimada, Y. Yamada, K. Muro, T. Okusaka, H. Ueno, M. Ikeda, and N. Watanabe, 2003. Phase I study of novel drug delivery system, NK911, a polymer micelle encapsulated doxorubicin. *Proc Am Soc Clin Oncol* 22(1): 1–3.

Hamidi, M., A. Azadi, and P. Rafiei, 2006. Pharmacokinetic consequences of pegylation. *Drug Deliv* 13(6): 399–409.

Haxton, K. J. and H. M. Burt, 2008. Polymeric drug delivery of platinum-based anticancer agents. *J Pharm Sci* 1(1): 1–22.

Heath, F., P. Haria, and C. Alexander, 2007. Varying polymer architecture to deliver drugs. *AAPS J* 9(2): E235–E240.

Jones, M. and J. Leroux, 1999. Polymeric micelles—A new generation of colloidal drug carriers. *Eur J Pharm Biopharm* 48(2): 101–111.

Kabanov, A. V., V. P. Chekhonin, V. Yu. Alakhov, E. V. Batrakova, A. S. Lebedev, N. S. Melik-Nubarov, S. A. Arzhakov, et al., 1989. The neuroleptic activity of haloperidol increases after its solubilization in surfactant micelles. Micelles as microcontainers for drug targeting. *FEBS Lett* 258(2): 343–355.

Kakizawa, Y. and K. Kataoka, 2002. Block copolymer micelles for delivery of gene and related compounds. *Adv Drug Deliv Rev* 54(2): 203–222.

Khmelnitsky, Y. L., A. K. Gladilin, V. L. Roubailo, K. Martinek, and A. V. Levashov, 1992. Reversed micelles of polymeric surfactants in nonpolar organic solvents. A new microheterogeneous medium for enzymatic reactions. *Eur J Biochem* 206(3): 737–745.

Kim, T. Y., D. W. Kim, J. Y. Chung, S. G. Shin, S. C. Kim, D. S. Heo, N. K. Kim, and Y. J. Bang, 2004. Phase I and pharmacokinetic study of Genexol-PM, a cremophor-free, polymeric micelle-formulated paclitaxel, in patients with advanced malignancies. *Clin Cancer Res* 10(11): 3708–3716.

Kim, Y., P. Dalhaimer, D. A. Christian, and D. E. Discher, 2005. Polymeric worm micelles as nano-carriers for drug delivery. *Nanotechnology* 16(Suppl): S1–S8.

Kommareddy, S., S. B. Tiwari, and M. M. Amiji, 2005. Long-circulating polymeric nanovectors for tumor-selective gene delivery. *Technol Cancer Res Treat* 4(6): 615–625.

Kong, W. H., W. J. Lee, Z. Y. Cui, K. H. Bae, T. G. Park, J. H. Kim, K. Park, and S. W. Seo, 2007. Nanoparticulate carrier containing water-insoluble iodinated oil as a multifunctional contrast agent for computed tomography imaging. *Biomaterials* 28(36): 5555–5561.

Krauze, M. T., J. Forsayeth, J. W. Park, and K. S. Bankiewicz, 2006. Real-time imaging and quantification of brain delivery of liposomes. *Pharm Res* 23(11): 2493–2504.

Kwon, S. Y. and M. W. Kim, 2002. Topological transition in aqueous nonionic micellar solutions. *Phys Rev Lett* 89(25): 258–302.

LaVan, D. A., T. McGuire, and R. Langer, 2003. Small-scale systems for *in vivo* drug delivery. *Nat Biotechnol* 21(10): 1184–1191.

Lavasanifar, A., J. Samuel, and G. S. Kwon, 2002. Poly(ethylene oxide)-*block*-poly(L-amino acid) micelles for drug delivery. *Adv Drug Deliv Rev* 54(2): 169–190.

Leander, P., 1996. A new liposomal contrast medium for CT of the liver. An imaging study in a rabbit tumour model. *Acta Radiol* 37(1): 63–68.

Liyan, Q., Z. Cheng, J. Yi, and Z. Kangjie, 2007. Polymeric micelles as nanocarriers for drug delivery. *Expert Opin Ther Patents* 17(7): 819–830.

Maeda, H., 2001. SMANCS and polymer-conjugated macromolecular drugs: Advantages in cancer chemotherapy. *Adv Drug Deliv Rev* 46(1–3): 169–185.

Maeda, H., J. Wu, T. Sawa, Y. Matsumura, and K. Hori, 2000. Tumor vascular permeability and the EPR effect in macromolecular therapeutics: A review. *J Control Release* 65(1–2): 271–284.

Malik, S. N., D. H. Canaham, and M. W. Gouda, 1975. Effect of surfactants on absorption through membranes III: Effects of dioctyl sodium sulfosuccinate and poloxalene on absorption of a poorly absorbable drug, phenolsulfonphthalein, in rats. *J Pharm Sci* 64(6): 987–990.

Marie-Helene, D., G. A. Marc, and L. Jean-Christophe, 2005. Thiol-functionalized polymeric micelles: From molecular recognition to improved mucoadhesion. *Bioconjug Chem* 16(4): 1027–1033.

Matsumura, Y., T. Hamaguchi, T. Ura, K. Muro, Y. Yamada, Y. Shimada, K. Shirao, et al., 2004. Phase I clinical trial and pharmacokinetic evaluation of NK911, a micelle-encapsulated doxorubicin. *Br J Cancer* 91(10): 1775–1781.

Moghimi, S. M., A. C. Hunter, and J. C. Murray, 2001. Long-circulating and target-specific nanoparticles: Theory to practice. *Pharmacol Rev* 53(2): 283–318.

Mulder, W. J., G. J. Strijkers, K. C. Briley-Saboe, J. C. Frias, J. G. Aguinaldo, E. Vucic, V. Amirbekian, et al., 2007. Molecular imaging of macrophages in atherosclerotic plaques using bimodal PEG-micelles. *Magn Reson Med* 58(6): 1164–1170.

Nagasaki, Y., K. Yasugi, Y. Yamamoto, A. Harada, and K. Kataoka, 2001. Sugar-installed block copolymer micelles: Their preparation and specific interaction with lectin molecules. *Biomacromolecules* 2(4): 1067–1070.

Nakayama, M. and T. Okano, 2005. Drug delivery systems using nano-sized drug carriers. *Gan To Kagaku Ryoho* 32(7): 935–940.

Nasongkla, N., E. Bey, J. Ren, H. Ai, C. Khemtong, J. S. Guthi, S. F. Chin, A. D. Sherry, D. A. Boothman, and J. Gao, 2006. Multifunctional polymeric micelles as cancer-targeted, MRI-ultrasensitive drug delivery systems. *Nano Lett* 6(11): 2427–2430.

Nishiyama, N., S. Okazaki, H. Cabral, M. Miyamoto, Y. Kato, Y. Sugiyama, K. Nishio, Y. Matsumura, and K. Kataoka, 2003. Novel cisplatin-incorporated polymeric micelles can eradicate solid tumors in mice. *Cancer Res* 63(24): 8977–8983.

Noh, T., Y. H. Kook, C. Park, H. Youn, H. Kim, E. K. Choi, H. J. Park, and C. Kim, 2008. Block copolymer micelles conjugated with anti-EGFR antibody for targeted delivery of anticancer drug. *J Polym Sci Part A: Polym Chem* 46(22): 7321–7331.

Northfelt, D. W., F. J. Martin, P. Working, P. A. Volberding, J. Russell, M. Newman, M. A. Amantea, and L. D. Kaplan, 1996. Doxorubicin encapsulated in liposomes containing surface-bound polyethylene glycol: Pharmacokinetics, tumor localization, and safety in patients with AIDS-related Kaposi's sarcoma. *J Clin Pharmacol* 36(1): 55–63.

Owens, 3rd, D. E. and N. A. Peppas, 2006. Opsonization, biodistribution, and pharmacokinetics of polymeric nanoparticles. *Int J Pharm* 307(1): 93–102.

Pichler, W. J., A. Beeler, M. Keller, M. Lerch, S. Posadas, D. Schmid, Z. Spanou, A. Zawodniak, and B. Gerber, 2006. Pharmacological interaction of drugs with immune receptors: The p-i concept. *Allergol Int* 55(1): 17–25.

Rijcken, C. J., O. Soga, W. E. Hennink, and C. F. van Nostrum, 2007. Triggered destabilisation of polymeric micelles and vesicles by changing polymers polarity: An attractive tool for drug delivery. *J Control Release* 120(3): 131–148.

Sahoo, S. K., J. Panyam, S. Prabha, and V. Labhasetwar, 2002. Residual polyvinyl alcohol associated with poly(D,L-lactide-co-glycolide) nanoparticles affects their physical properties and cellular uptake. *J Control Release* 82(1): 105–114.

Sahoo, S. K., S. Parveen, and J. J. Panda, 2007. The present and future of nanotechnology in human health care. *Nanomed Nanotechnol Biol Med* 3(1): 20–31.

Schmiedl, U. P., J. A. Nelson, L. Teng, F. Starr, R. Malek, and R. J. Ho, 1995. Magnetic resonance imaging of the hepatobiliary system: Intestinal absorption studies of manganese mesoporphyrin. *Acad Radiol* 2(11): 994–1001.

Seow, W. Y., J. M. Xue, and Y. Y. Yang, 2007. Targeted and intracellular delivery of paclitaxel using multi-functional polymeric micelles. *Biomaterials* 28(9): 1730–1740.

Shi, B., C. Fang, M. Y. Xian, Y. Zhang, S. Fu, and Y. Y. Pei, 2005. Stealth MePEG–PCL micelles: Effects of polymer composition on micelle physicochemical characteristics, *in vitro* drug release, *in vivo* pharmacokinetics in rats and biodistribution in S180 tumor bearing mice. *Colloid Polym Sci* 283(9): 954–967.

Sutton, D., N. Nasongkla, E. Blanco, and J. Gao, 2007. Functionalized micellar systems for cancer targeted drug delivery. *Pharm Res* 24(6): 1029–1046.

Szebeni, J., L. Baranyi, S. Savay, J. Milosevits, R. Bunger, P. Laverman, J. M. Metselaar, et al., 2002. Role of complement activation in hypersensitivity reactions to doxil and hynic PEG liposomes: Experimental and clinical studies. *J Liposome Res* 12(1–2): 165–172.

Tanford, C., 1973. *The Hydrophobic Effect: Formation of Micelles and Biological Membranes.* New York: Wiley.

Torchilin, V. P., 1997. Pharmacokinetic considerations in the development of labeled liposomes and micelles for diagnostic imaging. *Q J Nucl Med* 41(2): 141–153.

Torchilin, V. P., 2000. Polymeric contrast agents for medical imaging. *Curr Pharm Biotechnol* 1(2): 183–215.

Torchilin, V. P., 2002. PEG-based micelles as carriers of contrast agents for different imaging modalities. *Adv Drug Deliv Rev* 54(2): 235–252.

Torchilin, V. P., 2005. Block copolymer micelles as a solution for drug delivery problems. *Expert Opin Ther Patents* 15(1): 63–75.

Torchilin, V. P., 2007. Micellar nanocarriers: Pharmaceutical perspectives. *Pharm Res* 24(1): 1–16.

Torchilin, V. P., M. D. Frank-Kamenetsky, and G. L. Wolf, 1999. CT visualization of blood pool in rats by using long-circulating, iodine-containing micelles. *Acad Radiol* 6(1): 61–65.

Torchilin, V. P., A. N. Lukyanov, Z. Gao, and B. Papahadjopoulos-Sternberg, 2003. Immunomicelles: Targeted pharmaceutical carriers for poorly soluble drugs. *Proc Natl Acad Sci USA* 100(10): 6039–6044.

Trubetskoy, V. S., M. D. Frank-Kamenetsky, K. R. Whiteman, G. L. Wolf, and V. P. Torchilin, 1996. Stable polymeric micelles: Lymphangiographic contrast media for gamma scintigraphy and magnetic resonance imaging. *Acad Radiol* 3(3): 232–238.

Unger, E., D. Shen, T. Fritz, G. L. Wu, B. Kulik, T. New, T. Matsunaga, and R. Ramaswami, 1994. Liposomes bearing membrane-bound complexes of manganese as magnetic resonance contrast agents. *Invest Radiol* 29(Suppl 2): S168–S169.

Vasir, J. K. and V. Labhasetwar, 2005. Targeted drug delivery in cancer therapy. *Technol Cancer Res Treat* 4(4): 363–374.

Veronese, F. M. and G. Pasut, 2005. PEGylation, successful approach to drug delivery. *Drug Discov Today* 10(21): 1451–1458.

Wang, Y., Y. Li, L. Zhang, and X. Fang, 2008b. Pharmacokinetics and biodistribution of paclitaxel-loaded pluronic P105 polymeric micelles. *Arch Pharm Res* 31(4): 530–538.

Wang, Y. C., X. Q. Liu, T. M. Sun, M. H. Xiong, and J. Wang, 2008a. Functionalized micelles from block copolymer of polyphosphoester and poly(epsilon-caprolactone) for receptor-mediated drug delivery. *J Control Release* 128(1): 32–40.

Xiong, X. B., A. Mahmud, H. Uludag, and A. Lavasanifar, 2008. Multifunctional polymeric micelles for enhanced intracellular delivery of doxorubicin to metastatic cancer cells. *Pharm Res* 25(11): 2555–2566.

Yang, X., W. Deng, L. Fu, E. Blanco, J. Gao, D. Quan, and X. Shuai, 2008. Folate-functionalized polymeric micelles for tumor targeted delivery of a potent multidrug-resistance modulator FG020326. *J Biomed Mater Res A* 86(1): 48–60.

Yokoyama, M., G. S. Kwon, T. Okano, Y. Sakurai, T. Seto, and K. Kataoka, 1992. Preparation of micelle-forming polymer–drug conjugates. *Bioconjug Chem* 3(4): 295–301.

Yokoyama, M., A. Satoh, Y. Sakurai, T. Okano, Y. Matsumura, T. Kakizoe, and K. Kataoka, 1998. Incorporation of water-insoluble anticancer drug into polymeric micelles and control of their particle size. *J Control Release* 55(2–3): 219–229.

Yoo, H. S., E. A. Lee, and T. G. Park, 2002. Doxorubicin-conjugated biodegradable polymeric micelles having acid-cleavable linkages. *J Control Release* 82(1): 17–27.

Yoo, H. S. and T. G. Park, 2004. Folate-receptor-targeted delivery of doxorubicin nano-aggregates stabilized by doxorubicin–PEG–folate conjugate. *J Control Release* 100(2): 247–256.

7 Multiple Emulsions
An Overview and Pharmaceutical Applications

Jatin Khurana, Somnath Singh, and Alekha K. Dash

CONTENTS

7.1   INTRODUCTION

Multiple emulsions are complex polydispersed systems in which water-in-oil (W/O) and oil-in-water (O/W) emulsions exist simultaneously. They are thermodynamically unstable systems and need special and careful consideration for their stability. Stabilization of multiple emulsions requires both hydrophilic and hydrophobic surfactants in the formulation. The ratio between these surfactants also plays an important role in their stability. The stability of multiple emulsions has been further enhanced by reduction of the droplet size and use of polymeric amphiphiles and complex adducts in the formulation. The two commonly used multiple emulsions in practice are water-in-oil-in-water (W/O/W) and oil-in-water-in-oil (O/W/O). The W/O/W emulsion has been widely used in pharmaceutical applications and has been extensively studied and reported. In addition to pharmaceutical applications, multiple emulsions have shown potential use in the agricultural, cosmetic, food, separation sciences, and nutraceutical industries.

Several methods have been reported for the preparation of multiple emulsions. The double emulsification technique is the most commonly employed method of preparation compared to the one-step technique, which is sometimes called the phase inversion method (Matsumuto et al., 1985). Some of the more recent methods of preparation include the membrane emulsification technique (Okochi and Nakano, 1997) and microchannel emulsification methods (Sugiura et al., 2004). The membrane method uses a microporous glass membrane with a particular pore size as the emulsifying tool. The microchannel method is a very novel approach of producing monodispersed multiple emulsions. Various formulation variables including the method of preparation, emulsifiers, lipophilic phase, phase volume ratio of various phases, temperature, and shear or agitation used during preparation can affect the stability of multiple emulsions (Florence and Whitehill, 1981; Matsumoto et al., 1976). The rheological properties of these emulsions are affected by the nature of the external phase, phase volume ratio, and particle size distribution of the dispersed phase (Jiao and Burgess, 2003; Khan et al., 2006). Various *in vitro* methods have been used for the quality assurance of multiple emulsions in pharmaceutical practice. These include macroscopic and microscopic examinations, entrapment efficiency studies, rheological analysis, zeta potential measurements, phase separation, and *in vitro* drug release characteristics (Khan et al., 2006). The difference in the nuclear magnetic resonance signals of a water proton in simple emulsions versus multiple emulsions has been utilized as an analytical tool in quality control (Khopade and Jain, 2001).

Pharmaceutical applications of multiple emulsions include sustaining and controlling the release of the active component, transport and targeting of the active moiety, masking the taste of medicaments, enhancing bioavailability, immobilization of enzymes, lymphatic delivery, transdermal delivery, preparation of microparticulate and nanoparticulate drug delivery systems, drug intoxication, and red blood cell substitutes (Jiao and Burgess, 2008).

7.2   METHODS OF PREPARATION

Emulsions are thermodynamically unstable systems because of the excess surface free energy created during the preparation of smaller droplets from the different dispersed phases. The excess surface free energy is a result of cohesive forces being stronger than the adhesive forces. Multiple emulsions are even more unstable because of their complex nature and because they have two different interfaces giving rise to multiple-fold free energy. Therefore, unlike simple emulsions, they require two different sets of surfactants or emulsifiers for their stability. The order of their use depends on the type of emulsion to be prepared. For instance, while preparing W/O/W-type multiple emulsions, the first set of emulsifiers must be hydrophobic and the other should be hydrophilic and vice versa for preparing an O/W/O-type emulsion. The common formulation techniques used for preparing multiple emulsions are discussed below.

7.2.1 Phase Inversion (One-Step Emulsification)

The first multiple emulsion that was prepared utilized only one set of emulsifiers and an inversion process (Matsumuto and Yonezawa, 1976). This process includes formation of a lamellar structure from a concentrated simple emulsion that can be redispersed in the external phase to give a multiple emulsion. It is a single-step process. However, the resultant multiple emulsions are highly unstable because it is really difficult to control the migration of emulsifiers between the two phases, which often results in destabilization of the emulsion. Another disadvantage of the phase inversion technique is a wide inner droplet size distribution that gives rise to a high degree of polydispersity. Therefore, this technique cannot be used to prepare uniform size multiple emulsions (Hou and Papadopoulos, 1997).

Methods have been reported recently for the preparation of multiple emulsions via the inversion technique using two or more surfactants. An aqueous phase containing poly(ethylene glycol) and polyoxyethylene (Tween) 20 was added to an oil phase (comprising 1-octanol, hydroxypropyl cellulose, and sorbitan monooleate over a magnetic stirrer) that resulted in the formation of an O/W/O emulsion. The oil phase diffused into the inner aqueous phase across the hydroxypropyl cellulose to yield O/W/O-type multiple emulsions (Oh et al., 2004).

7.2.2 Double Emulsification (Two-Step Emulsification)

This is the most common process employed for the preparation of multiple emulsions of both W/O/W and O/W/O types. It is very simple yet reproducible and gives a high yield. This process involves two steps. The first step involves preparation of a simple primary W/O or O/W emulsion using a suitable emulsifier. The primary emulsion obtained in the first step is then reemulsified into an excess of the final continuous phase to which a second emulsifier has already been added. A schematic representation of this process is depicted in Figure 7.1.

The first step generally requires a high shear homogenization and more recently sonication to obtain the primary emulsion. The second step requires gentle agitation to prevent fracture of the inner globules leading to the formation of a simple emulsion (Fukushima et al., 1983). Thus, in the second step the external phase is added with stirring instead of sonication or homogenization.

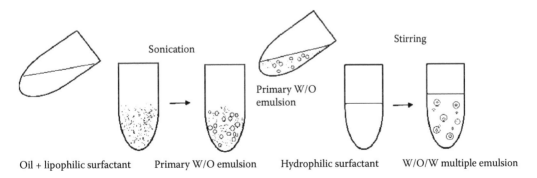

FIGURE 7.1 Preparation of a W/O/W multiple emulsion via a two-step emulsification process.

However, in some recent studies a modified two-step method is used. This involves dividing each step into two substeps, a preemulsification via sonication followed by stirring (Okochi and Nakano, 1996, 2000). Studies have shown that production parameters such as temperature and pressure influence the droplet size and the encapsulation efficiency in W/O/W-type multiple emulsions (Lindenstruth and Muller, 2004).

7.2.3 MEMBRANE EMULSIFICATION TECHNIQUE

The membrane emulsification technique is a highly effective method of producing narrow droplet size distributions (Higashi et al., 1995). A schematic representation of such a process is provided in Figure 7.2. It makes use of microporous membranes with a defined pore size as an emulsifying tool. The first membrane used for this purpose was a glass membrane known as Shirasu porous glass (Nakashima et al., 1991). Membranes made up of other materials such as ceramic, metallic, and polymers have been reported recently for producing multiple emulsions. Various microengineered devices such as parallel arrays of microgrooves formed by engraving on a single-crystal silicon substrate have been used for the preparation of multiple emulsions (Vladisavljevic and Williams, 2005). A low-shear pump is generally employed to recirculate the continuous phase along the membrane. This ensures regular droplet detachment from the pores (Nakashima et al., 1991). The same process can be completed by agitation using a stirrer in the vessel (Kosvintsev et al., 2008). This process involves the drop by drop creation of individual droplets by pressurizing the primary emulsion to pass through a set pore size into the final continuous phase, thus forming monosized globules (Nakashima et al., 1991; Okochi and Nakano, 1997). This process is more suited for large-scale production of multiple emulsion products.

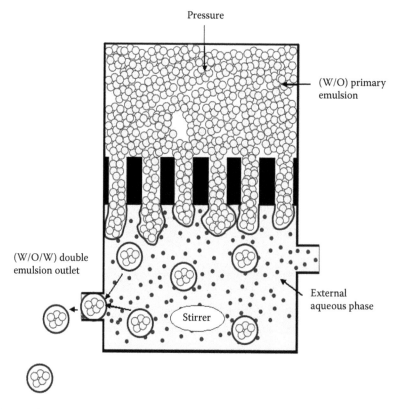

FIGURE 7.2 Membrane emulsification process.

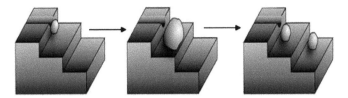

FIGURE 7.3   Schematic representation of droplet formation via microchannel emulsification process.

7.2.4   MICROCHANNEL EMULSIFICATION

This is a more recent method employed for the preparation of multiple emulsions (Sugiura et al., 2004). In this process the primary emulsion is passed through a channel onto a flat surface called a terrace. As more of the dispersed phase is pushed through the channel, the droplet grows in size. The expansion is followed by a spontaneous detachment as the droplet reaches the edge of the terrace and falls into a deeper well. A W/O/W-type emulsion was successfully prepared through this technique using decane, ethyl oleate, and medium chain triglyceride as the intermediate oil phase (Sugiura et al., 2004). Figure 7.3 depicts a channel process of making multiple emulsions.

A further modification of the microchannel emulsification technique is the use of microfluidic devices, which utilize a T-junction or microcapillary devices as shown in Figure 7.4. These devices are generally made of quartz glass (Nisisako et al., 2005), poly(dimethylsiloxane) (Yi et al., 2004), and polyurethane (Nie et al., 2006). They are highly efficient and can form multiple emulsions in a single step while maintaining not only the inner and outer droplet size but also the number of inner droplets encapsulated in the larger droplet.

7.3   STABILITY

7.3.1   MECHANISM

As previously mentioned, multiple emulsions are thermodynamically unstable systems with abundant surface free energy because of the multitude of interfaces presented by the droplets of the dispersed phases. The excess free energy is also a result of cohesive forces between the molecules

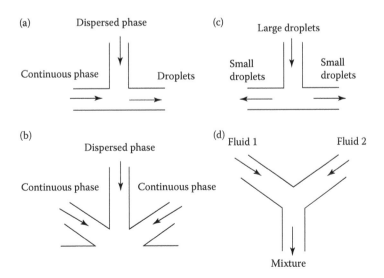

FIGURE 7.4   Different types of T-junctions used in microfluidic devices.

of the individual phase empowering the adhesive forces between the two phases (Jantzen and Robinson, 2002; Martin et al., 1993). The interfacial free energy associated with the interface between two immiscible liquids can be represented mathematically as

$$\Delta G = \gamma \Delta A,$$

where G is the interfacial free energy, γ is the interfacial tension, and A is the total interfacial area of the dispersed phase. Clearly an increase in interfacial area is directly proportional to an increase in interfacial free energy and thus a thermodynamically unstable system.

One of the ways to make these systems thermodynamically more stable is to reduce the interfacial tension. This can be achieved by the addition of surfactants or emulsifiers. Use of a certain kind of surfactant, either hydrophilic or hydrophobic, depending on the type of emulsion to be prepared, is well known for the production of simple emulsions. Multiple emulsions are more complex because W/O and O/W exist simultaneously in the system. Hence, two different sets of surfactants are required to impart stability to multiple emulsions. These surfactant molecules usually get adsorbed at the interfacial layer and prevent droplets from coalescing. They help to reduce the interfacial tension by forming an interfacial film. Low interfacial tension favors the formation of smaller droplets and a narrow size distribution and imparts greater kinetic stability. The long-term stability of an emulsion is dependent on the strength of this film (Jao and Burgess, 2008). For the film to be an effective barrier, it should remain intact at all points in time when forced between two droplets. The film should have a certain degree of surface elasticity, which will help it reform rapidly once it is broken. Thus, interfacial elasticity and interfacial strength can be used to predict the stability of multiple emulsions (Opawale and Burgess, 1998).

In an attempt to impart greater stability, conventionally used monomeric surfactants have been replaced by polymeric emulsifiers. The gain in free energy is greater for polymeric amphiphiles compared to absorbed monomeric surfactants (Ortega-Vinuesa et al., 1996). The use of polymeric emulsifiers also helps to control and sustain the release of the active ingredient from the formulation. Smaller monomeric surfactants tend to migrate from the interface toward one of the two phases, depending on their affinity with the two phases, and in the process disrupting the hydrophilic–hydrophobic balance of the system. Another mechanism of instability is the transport of water or diffusion of solubilized materials across the oily membrane, depending on the osmotic pressure and concentration gradient across the membrane. Polymeric emulsifiers also have instabilities. The common destabilization mechanisms of multiple emulsions include flocculation, aggregation, and coalescence. Flocculation is a typical instability in emulsions stabilized by polymeric emulsions. It can be achieved by either depletion or bridging.

The polymeric emulsifiers generally used for preparation of multiple emulsions may be categorized as synthetic polymers and natural polymers, which include hydrocolloids and proteins. Low molecular weight surfactants and proteins differ in their mechanism of stabilization of the emulsion systems. Surfactants rapidly diffuse to stabilize any disturbance at the interface, dragging the fluid into the interlamellar space between the droplets and hence keeping them separated. Such a mechanism is known as the Gibbs–Marangeni mechanism. In contrast, proteins, after being adsorbed at the interface, tend to unfold, presenting their hydrophobic groups to the hydrophobic phase. In this process, they develop strong interactions with the other neighboring protein molecules and form an interconnected viscoelastic gel structure at the interface. This viscoelastic gel prevents breakage of the film and thus stabilizes the emulsion.

7.3.1.1 Steric and Depletion Stabilization

As the polymer gets adsorbed on the surface of the droplet, the polymer chains cover the entire droplet surface area. When two such droplets covered with polymeric chains come closer to each other, there is an increase in the local concentration and this results in an osmotic pressure difference between the external solution and the overlap zone. The osmotic pressure is much lower in the overlap zone compared to the external regions. This results in an instant flow of the external continuous phase in the overlap zone, thus separating the particles (Babak and Stebe, 2002). In addition, the overlap results in compression of the polymer molecules and causes a change of configuration and a loss of entropy. Hence, droplets must segregate in order to regain the lost freedom of movement.

7.3.1.2 Electrostatic Repulsion Stabilization

This is a common and well-established mechanism of stabilization. The amphiphilic molecules present at the droplet surface contain either positively or negatively charged chains. This leads to the development of static charges at the surface of the droplets. These surface static charges form electrostatic repulsive forces that tend to keep the particles separated.

The use of synthetic polymers is generally not recommended, paving the way for naturally occurring biopolymers. These polymers largely include hydrocolloids and proteins. Hydrocolloids are high molecular weight biopolymers that are generally used as viscosity building agents. They include various polysaccharides and gums of mainly plant origin that are modified by enzymatic action. Proteins have a broad history of being used as emulsifiers. They are macromolecular amphiphiles and impart stability to multiple emulsions. Proteins such as bovine serum albumin (BSA; Dickinson et al., 1994), gelatin (Zhang et al., 1992), whey proteins (Cornec et al., 2000), human serum albumin, and casein have been extensively evaluated since the early 1990s for their emulsifying properties. The amphiphilic nature of a protein molecule makes it a suitable candidate to be used as an emulsifying agent. The utilization of proteins as emulsifying agents has been meticulously evaluated independently and in combination with other monomeric surfactants. Studies have been performed to understand the mechanism, if any, by which they complement or compete with each other to get adsorbed at the interface. However, it is still debatable whether the adsorption is competitive or complementary (Kerstens et al., 2006; Wilde et al., 1993).

Certain protein molecules like BSA form interfacial complexes with the nonionic surfactant that imparts stability to the multiple emulsions and helps in slowing down the release of the entrapped solute. Studies have shown that BSA can serve as a replacement for monomeric surfactants in the inner phase in providing good stabilization for W/O/W emulsions (Garti and Aserin, 1996). However, when BSA was used in combination with monomeric surfactants, the emulsion that was formed had better stability compared to BSA used alone. A delay in the release of the active moiety was also observed (Garti et al., 1994). The use of lysozyme was also investigated with other monomeric surfactants in the internal and external phases. Lysozyme is a small protein and the droplets tend to coalesce (Silletti et al., 2007).

Recent studies evaluated sodium caseinate with polyglycerol poly(ricinoleic acid) for stabilizing multiple emulsions. When increasing the caseinate concentration from 0.03% to 1%, there was a significant reduction in the mean droplet size. The reduction in droplet size was attributed to loss of flocculation upon increasing the caseinate concentration (Su et al., 2006). Hence, studies concluded that it is not possible to completely replace monomeric surfactants by proteins. However, when present along with monomeric surfactants, proteins certainly improve the stability and adjourn the release of the solute from the internal phase (Dickinson et al., 1994).

7.3.2 Effect of Formulation Variables on Stability

7.3.2.1 Phase Volume

The aqueous to oily phase ratio is an important parameter that influences the stability of multiple emulsions. The phase volume of the primary emulsion has no significant effect on the final product, but the phase volume ratio of the secondary phase is very important. Thus, a varied range of phase proportions may be used while preparing the primary emulsion. Stable multiple droplets are produced at low volume fractions only (Matsumoto et al., 1976). However, some studies also reported that they obtained stable multiple emulsions at very high phase volume ratios (70–90%;

DeLuca et al., 1990). Another important consideration while preparing multiple emulsions is the order of phase addition. To prepare a stable multiple emulsion, the dispersed phase should be added slowly into the continuous phase to prevent rupture of the inner globules.

7.3.2.2 Emulsifiers

The use of emulsifiers is the most common way to impart stability to multiple emulsions. The choice of the emulsifier or sets of emulsifiers depends on the type of emulsion to be prepared (Magdassi et al., 1985; Safwat et al., 1994). For the preparation of multiple emulsions, generally a combination of two or more emulsifiers is used. In most of the earlier studies, the focus was to find a blend of monomeric hydrophilic and hydrophobic surfactants in the most appropriate ratio required for optimal stabilization. Matsumoto and coworkers came up with the empirical weight ratio rule of 10 for the internal hydrophobic surfactant to the external hydrophobic surfactant (Matsumoto et al., 1976, 1985). However, current studies involve calculation of the effective hydrophilic–lipophilic balance (HLB) value of emulsifiers for optimal stability (Magdassi and Garti, 1999). All the available emulsifiers are classified as either hydrophilic or hydrophobic on the basis of their HLB value. Emulsifiers with low HLB (2–8) are considered lipophilic whereas ones with high HLB (6–16) are hydrophilic. Optimum concentrations of primary and secondary emulsifiers are required to formulate a stable multiple emulsion. If present in excess, the primary emulsifier can reach the surface of the secondary micelle, resulting in instability. The use of polymeric emulsifiers is preferred over synthetic ionic surfactants because of toxicity constraints. Polymeric emulsifiers also offer other advantages like better yield, improved stability, and the ability to control and sustain the release of the entrapped molecule (Florence et al., 1982; Matsumoto et al., 1976; Okochi et al., 1996).

7.3.2.3 Oils Used

A wide variety of oils have been tried and tested for the preparation of multiple emulsions. Different oils are used for different purposes in a multiple emulsion. Some of them are used for medicinal or nutritional value whereas some are used to act as a membrane to deliberately impede the release of the active solute from the inner phase. Others may be used to impart a suitable viscosity or consistency required for the formulation, or they may even form a part of the emulsifier system.

These oils may be chemically categorized as fixed oils, volatile oils, simple esters, hydrocarbons, terpenoid derivatives, and so forth. The choice of oil also depends on the route of administration. Nonbiodegradable mineral oils and castor oils are generally used in preparations of emulsions for oral administration. They also provide a local laxative effect. Some of the oils used for nutritional value include fish liver oil, olive oil, and other fixed oils of vegetable origin like arachis, cottonseed, and maize. Studies have shown a better yield and better stability with mineral oils compared to vegetable oils (Davis et al., 1983; Florence et al., 1982; Sinha et al., 2002).

The rheology of multiple emulsions is significantly affected by the nature of the oil. In addition to the rheology of the emulsion system, the nature of the oil also affects the diffusion of the solute across the oily membrane. The drug release profile can be altered by using oils of varying permeability, density, and viscosity (Omotosho et al., 1986). The release can be decreased by increasing the fat content of the oil. In recent studies, the release of the drug was delayed using a semicrystalline oil phase (Weiss et al., 2005).

7.3.2.4 Applied Shear

Shear, which is the force applied while preparing the emulsion, is an important parameter affecting the stability of multiple emulsions. Varying shear force is applied to the emulsions prepared via the two-step emulsification process. The first step, preparation of the primary emulsion, is a high shear rate step, whereas the second involves gentle stirring. A high shear applied in the second step can lead to breakage of the inner globules and thus results in the formation of a simple emulsion instead of a multiple emulsion and hence instability. Similarly, a low shear rate during the first step leads to the formation of a highly unstable emulsion system with bigger droplets and a high tendency toward coalescence. Therefore, an optimal shear rate is vital for the production of a stable multiple emulsion.

Shear time is another important parameter for preparing stable multiple emulsions. A high shear employed for longer periods can lead to incorporation of air within the emulsion. Excessive frothing caused by air entrapment causes displacement of the surfactant from the interface. Such an emulsion system is highly unstable (Florence et al., 1982; Matsumoto et al., 1976; Okochi et al., 1996; Yan et al., 1992).

7.3.2.5   Temperature and Pressure

Temperature and pressure are important parameters that affect the size of the droplets. Precise control of these parameters is critical to obtain reproducible results in terms of yield and droplet size. It is reported that a relatively colder temperature (5–10°C) during the second step of emulsification favors the production of a stable multiple emulsion (Geiger et al., 1998; Khopade et al., 1997).

Accelerated stability studies have shown that adverse conditions lead to instability. Therefore, it can be postulated that drastic variations in temperature during storage and transport can lead to instability (Clausse et al., 1999).

7.4   EVALUATION AND QUALITY CONTROL

7.4.1   Macroscopic Examination

A few of the instabilities in multiple emulsions can be easily viewed by the naked eye. Such instabilities include breaking of the emulsion, creaming, sedimentation, caking, aggregation, flocculation, and deflocculation. Changes in volume, homogeneity, color, consistency, and spreadability are some of the macroscopic properties that can be observed easily. These observations can help in ensuring that the emulsion is formed.

Dilution of the emulsion with the continuous or external phase is one of the macroscopic ways to validate the type of emulsion formed (Florence et al., 1982). Conductivity measurements are also useful in determining the type of emulsion that is formed. Only a W/O/W type of emulsion allows the passage of electric current through it. Conductivity tests can also serve to analyze the release of electrolytes entrapped in the emulsion globules.

7.4.2   Microscopic Examination

A light microscope or an electron microscope make it possible to capture direct images of the double emulsion globules. A group of microscopic techniques have been employed for the verification of the multiplicity of multiple emulsions (DeLuca et al., 2000; Nakhare et al., 1995a, 1995b).

7.4.2.1   Photomicrography

The use of photomicrography dates back to 1978, when Kavaliunas and Frank used it for the first time to determine the structure of a water, *p*-xylene, and nonylphenol diethylene glycol ether emulsion system (Kavaliunas and Frank, 1978). Photomicrography has been used not only to validate the structure of the double emulsion but also to visually look for instabilities. An increase in droplet size as a result of the coalescence of emulsion globules and other destabilizations induced by shear force can also be observed microscopically (Oliveiri et al., 2003). The role of different components can be better understood by microscopic visualization of their assembly in the system.

Photomicroscopy has also been used to determine the droplet size (Driscol et al., 2006; Pallandre et al., 2007). The existence of a complex four-phase W/O/W/O system was detected by the use of this technique (Goubault et al., 2001). Other studies have made use of different techniques for observing double emulsions with photomicrography (Florence et al., 1981). Cournarie and coworkers used immersion microscopy for the evaluation of emulsion droplets. Some studies have also made use of phase contrast microscopy to investigate emulsion properties and instabilities (Doucet et al., 1998; Pays et al., 2001, 2002). Even very thin oil films present between the internal and external aqueous phase can be detected using phase contrast and dark field microscopy (Florence et al., 1985).

7.4.2.2 Videomicrography

Various events occurring in the multiple emulsion can be directly observed in the actual sequential order by using an immersion microscope equipped with a video camera (Cheng et al., 2007). The use of sequential microscopic imaging was first employed in 1981 when Florence and Whitehill used the technique to view a double emulsion (Florence and Whitehill, 1981). Video micrography has been of great help in understanding the mechanisms of instability. Various studies have employed video enhanced confocal microscopy to study the stability of double emulsions loaded with fluorescent markers (Puppo et al., 2007).

7.4.2.3 Capillary Microscopy

This type of microscopy offers an advantage over the other microscopic techniques because the dynamic changes of a single emulsion globule can be analyzed right from its preparation. Unlike the other microscopic techniques, which involve analyzing the globules in the bulk system, capillary microscopy involves formation of individual double emulsion globules in an extremely thin capillary. The diameter of the capillaries range from 150 to 200 μm. This technique has been used by various groups in investigational studies for the observation of dynamic behavior (Hou and Papadopulos, 1997; Villa et al., 2003; Wen and Papadopulos, 2000).

7.4.2.4 Electron Microscopy

Miniature and extremely fine details of the emulsion system can be captured by electron microscopy. The major drawbacks of the optical microscopic systems are their inability to observe fine droplets smaller than a few microns (Zheng et al., 1991, 1993) and their inability to observe water droplets entrapped inside oil droplets (Matsumoto et al., 1980). Electron microscopy has been used in various studies to overcome both these drawbacks, and it has become possible to view and capture micrographs of droplets as small as 0.02 μm in diameter (DiStefano et al., 1983). The size of both inner and outer globules can be measured by using freeze etching electron microscopy (Davis and Burbace, 1977; Ursica et al., 2005).

7.4.3 Number of Globules

Another method for characterization of a multiple emulsion is to count the number of globules present per cubic millimeter. A higher number of globules can be correlated with high stability (Kita et al., 1977). A hemocytometer can be used for the calculation of the number of globules per cubic millimeter. It consists of numerous small squares, which are filled by the multiple emulsions after appropriate dilution. The number of globules in each square is then counted. Generally, 5 groups of 16 squares each are used for the calculation. The following equation is then used to mathematically calculate the number of globules per cubic millimeter:

$$\text{Number of globules/mm}^3 = \frac{\text{number of globules in each square} \times \text{dilution} \times 4000}{\text{number of small squares counted}}.$$

7.4.4 Zeta Potential

As previously discussed, the droplets carry surface static charges that are responsible for development of electrostatic repulsion between the particles. These surface charges also give rise to a potential that can be analyzed using a potentiometer. The instrument is typically a microelectrophoresis cell with platinum–iridium electrodes (Nakhare et al., 1997). It measures the zeta potential by determining the electrophoretic mobility of the globules when a fixed or variable voltage is applied. The zeta potential can be calculated using the following equation:

$$\zeta = \frac{4\pi\eta\mu}{\varepsilon E}$$

where ζ is the zeta potential (mV), η is the viscosity of the dispersion medium (P), μ is the migration velocity (cm/s), ε is the dielectric constant of the dispersion medium, and E is the potential gradient (voltage applied/distance between electrodes).

7.4.5 ENTRAPMENT EFFICIENCY

The entrapment efficiency is one of the most commonly used parameters that is employed for the evaluation of multiple emulsions. The quality of the multiple emulsions that are formed depends on how efficiently the active ingredient is entrapped in the innermost phase of the globules. Some studies have revealed that the entrapment efficiency depends on the ratio of the lipophilic and hydrophilic surfactants used in the respective phases. A ratio of more than 10 is said to yield emulsions with 90% or more entrapment efficiency (Matsumoto et al., 1976).

To measure the entrapment efficiency or yield of the multiple emulsion, an internal tracer is generally added, which is then analyzed to evaluate the yield (Davis et al., 1983). This internal tracer is mostly an impermeable marker incorporated in the inner phase. The amount or percentage of drug entrapped is calculated indirectly by estimating the amount of unentrapped drug present in the outermost continuous phase and then applying weight balance to determine the amount of entrapped marker (Magdassi et al., 1984; Matsumoto and Kang, 1988). The various markers used for this purpose include glucose (Matsumoto et al., 1976), hydrogen ions (Martin et al., 1993), electrolytes (Lee et al., 2001), new coccine (Ohwaki et al., 1992, 1993), various dyes (like sulfane blue, polytartrazane), and some radioactive tracers (D_2O; Sela et al., 1995). It can be mathematically expressed as the following:

$$\text{Entrapment efficiency} = \frac{(\text{total amount of drug} - \text{free (unentrapped) drug}) \times 100}{\text{total amount of drug}}.$$

The various processes used to separate the marker include dialysis, centrifugation, filtration, and conductivity measurement (Yan et al., 2006). Studies have shown that certain additives can effectively improve the yield of an emulsion. These additives include sodium salts of alkyl sulfonates and alkyl carbonates and some other osmotic agents like sorbitol, sodium carbonate, and Tweens. These additives should be added in the inner aqueous phase (Ohwaki et al., 1992). Another study showed similar results upon addition of sodium chloride to the inner phase; however, inverse results were observed when sodium chloride was added in the outer continuous phase in a W/O/W emulsion containing tryptophan (Hino et al., 2001).

7.4.6 PERCENTAGE OF PHASE SEPARATION

The coalescence of globules results in separation of one of the phases from the emulsion and thus causes instability in the emulsion. The percentage of phase separation may be defined as the percentage of the volume of the separated phase from the total volume of the emulsion. The percentage of phase separation is determined by physically examining the separated phase in 20 mL of the freshly prepared emulsion that is allowed to stand in a 25-mL graduated glass cylinder for an extended period of time at room temperature or any other specified conditions. The percentage of phase separation (B) at any particular time interval can be calculated as the following:

$$B(\%) = 100 \times \frac{V_{sep}/20}{(V_1 + V_2)/(V_1 + V_2 + V_0)},$$

where V_1 represents the volume of the inner aqueous phase, V_2 represents the volume of the continuous phase, V_0 represents the volume of the middle phase (Nakhare et al., 1997), and V_{sep} represents the volume of the separated phase observed periodically.

7.4.7 IN VITRO DRUG RELEASE

An appropriate release profile from the emulsion system is another parameter that is used to evaluate the quality of the emulsion. The drug is generally incorporated in the innermost phase, and the intermediate phase acts as a physical barrier regulating the release of the drug from the globules.

Dialysis is the most common method employed to determine the drug release from the emulsion. Five to ten milliliters of the emulsion is placed in a dialysis tube covered by a cellophane membrane at both ends that is then placed in 200–250 mL of suitable dissolution media (phosphate buffered saline, assay buffer, etc.) in a rotating basket United States Pharmacopoeia Type I dissolution apparatus. A sink condition is maintained throughout the duration of the study. Samples are withdrawn at different time intervals, and the volume of the dissolution media is kept constant by replacing it with an equal volume of fresh media (Khopade et al., 1998, 1999; Roy et al., 2006).

7.5 PHARMACEUTICAL APPLICATIONS

7.5.1 CONTROLLED RELEASE DRUG DELIVERY

To be used as a controlled release drug delivery system, the drug is incorporated in the inner globules of multiple emulsions. Many researchers have expansively evaluated multiple emulsions, especially W/O/W types, as controlled release formulations. The categories of drugs incorporated in multiple emulsions vary from analgesics to anticancer agents. Multiple emulsion systems are also known to prolong the effect of drugs that have extremely short half-lives. The middle oil barrier of the W/O/W emulsion generally sustains the release of the entrapped drug in the inner globules. The release rate generally depends on the diffusion of the drug across the oily membrane. Multiple emulsions have been investigated for both oral and intravenous administration. Table 7.1 lists several drugs incorporated in multiple emulsions to achieve a prolonged and sustained delivery rate.

7.5.2 MANUFACTURING PARTICULATE PRODUCTS

One of the major and novel pharmaceutical applications of multiple emulsions is in the manufacture of particulate products (Martín-Banderas et al., 2006). This includes micro- and nanoparticulate systems. Multiple emulsions are used as an intermediate step for a wide range of particulate drug delivery systems ranging from solid–lipid nano- and microparticles (Nakashima et al., 2000), gel microbeads (Kobayashi et al., 2001; Nakashima et al., 1991), polymeric microspheres and microcapsules (Liu et al., 2005), core–shell microparticles (Bandera et al., 2006; Nie et al., 2005; Utada et al., 2005), polymerosomes (Lorenceau et al., 2005), and colloidosomes (Nakashima et al., 1991). These delivery systems act as alternate delivery vehicles to the more conventional liposomes and microencapsulated beads. Table 7.2 provides a list of particulate drug delivery systems prepared by intermediate multiple emulsion techniques. Solid–lipid microcarriers containing irinotecan hydrochloride have been prepared using Shirasu porous glass membranes. The entrapment efficiency of the freeze-dried product was found to be 90% (Nakashima et al., 2000). Particulate products are generally obtained by one of the following techniques:

1. Cross-linking the monomer in the outer aqueous phase
2. Heat denaturation of proteins
3. Solvent evaporation
4. Precipitation by the addition of salt or alcohol

TABLE 7.1
Multiple Emulsions Used for Controlled Delivery of Drugs

Category	Drug
Analgesic and antipyretic agent	Pentazocine (Mishra and Pandit, 1989, 1990)
	Morphine (Wang et al., 2008b)
	Diclofenac diethylamine (Vasiljevic et al., 2006)
	Paracetamol (Khan, 2004)
Anticancer agents	Cisplatin (Xiao et al., 2004)
	Cytarabine (Kim et al., 1995)
	Doxorubicin (Lin et al., 1992)
	Epirubicin (Higashi et al., 1999; Tanaka et al., 2008)
	Etoposide (Tian et al., 2007)
	5-Fluorouracil (Fukushima et al., 1983; Omotosho et al., 1986, 1989)
	Mercaptopurine (Khopade and Jain, 1999a, 2000)
	Methotrexate (Karasulu et al., 2007)
	Paclitaxel (Fan et al., 2004)
	Tegafur (Oh et al., 1998)
	Vancomycin (Okochi and Nakano, 2000)
Antimalarials	Chloroquine (Vaziri and Warburton, 1994)
Antitubercular agents	Rifampicin (Khopade and Jain, 1999b; Nakhare and Vyas, 1995a, 1995b, 1997)
	Isoniazid (Khopade and Jain, 1998)
Antiasthmatic agents	Salbutamol (Pandit et al., 1987)
	Theophylline (Jourquin and Kauffmann, 1998)
Antibiotics	Cefadroxil (Miyakawa et al., 1993)
	Cephardine (Miyakawa et al., 1993)
	Nitroimidazole (Ozer et al., 2007)
	Nitrofurantoin (Rudnic et al., 2006)
Antihistaminic	Chlorpheniramine (Ghosh et al., 1997)
Proteins	Insulin (Liu et al., 2005)
Peptides	Salmon calcitonin (Dogru et al., 2000)
Miscellaneous	Dexamethasone (Suitthimeathegorn et al., 2007)
	Prednisolone (Safwat et al., 1994)
	Testosterone (Leichtnam et al., 2006)
	Antipyrene, 4-aminoantipyrene (Miyakawa et al., 1993)
	Pilocarpine (Saettone et al., 2000)
	Phenylephrine (Sasaki et al., 1996)

Various polymers used to prepare nano- and microspheres include eudragit, poly(lactic acid), poly(glycolic acid), polycaprolactone, poly(lactide-*co*-glycolide), alginic acid, and chitosan. Table 7.3 lists a few drugs that are encapsulated in multiple emulsions and formulated into particulate systems.

7.5.3 Drug Targeting

One of the reasons for the side effects of various drugs, which are administered via the conventional route, is that drugs are generally distributed at multiple sites in the body other than the desired site of action. Therefore, one of the ways to reduce toxicity is to make the delivery system site specific. In addition to the reduction in the side effects, drug targeting also concentrates the drug at the diseased tissue, thus reducing the total amount of drug required for the therapeutic effect. Multiple emulsions have been successfully investigated for drug targeting for a variety of drugs and more so

TABLE 7.2
Particulate Products Manufactured Using Multiple Emulsions

Products	Examples
Chitosan and glyceryl monooleate nanoparticles	Paclitaxel (Trickler et al., 2008)
Poly(D,L-lactide-*co*-glycolide)/ montmorillonite nanoparticles	Paclitaxel (Dong and Feng, 2005), 5-fluorouracil (Bozkir and Saka, 2005), streptomycin (Pandey and Khuller, 2007)
Solid–lipid nanoparticles	Insulin (Gallarate et al., 2008)
Chitosan microparticle	Astaxanthin (Higuera-Ciapara et al., 2003), ranitidine (Zhou et al., 2005)
Solid–lipid microcarriers	W/O microcarrier (Nakashima et al., 2000)
Microspheres	Nonsteriodal antiinflammatory drugs (Sam et al., 2008), naltrexone (Dinarvand et al., 2005) antiacne drugs (Castro and Ferreira, 2008)
Gel microbeads	Calcium pectinate beads (Kobayashi et al., 2001) Calcium alginate beads (Nakashima et al., 1991)
Inorganic microparticles	Silica particles (Nakashima et al., 1991)
Microparticles	Polystyrene microparticles of indomethacin (Tamilvanan and Sa, 2008), diclofenac sodium (Nahla et al., 2008)
Polymeric microspheres	Polylactide spheres (Liu et al., 2005)
Polymeric core–shell microcapsules	Poly(tripropylene glycol diacrylate) particles (Martín-Banderas et al., 2006; Utada et al., 2005)
Polymersomes	Poly(normal-butyl acrylate)–poly(acrylic acid) polymerosomes (Lorenceau et al., 2005)

for highly potent drugs like anticancer agents at the desired tumor sites. Various organs and tissues that have been investigated for the targeted delivery of multiple emulsions include the lymphatic tissue, liver, lungs, and brain (Nakhare et al., 1994).

Multiple emulsions can also be used for not targeting some organs. A W/O/W multiple emulsion of diclofenac sodium, containing poloxamer 403, was found to target reticuloendothelial system rich organs (e.g., spleen, liver) inversely in comparison to the identical emulsion but not containing poloxamer 403 (Talegaonkar and Vyas, 2005). In contrast, a nanoemulsion of the antimalarial drug primaquine is taken up preferentially by the liver, at least 45% more than the conventional dosage form (Singh and Vingkar, 2007). Table 7.4 lists a few drugs that were effectively targeted to the respective tissue or organ by multiple emulsions.

TABLE 7.3
Multiple Emulsions Used in Formulation of Particulate Drug Delivery Systems

Albumin (Shah et al., 1987)
Ciprofloxacin (Jeong et al., 2008)
Diclofenac sodium (Nahla et al., 2008)
Leuprolide acetate (Luan and Bodmeier, 2006)
Low molecular weight heparin (Rewat et al., 2008)
Lysozyme (Adachi et al., 2003)
Lysozyme (Wang et al., 2008a)
Paclitaxel (Trickler et al., 2008)
Pseudoephedrine (Bodmeier et al., 1992)
Retinol (Lee et al., 2001)
Salbutamol sulfate (Jaspart et al., 2007)
Sulfadiazine (Shah et al., 1987)
Theophylline, propranolol, acetaminophen, tacrine (Lee et al., 2000)

TABLE 7.4
Multiple Emulsions Used in Drug Targeting

Target Tissue or Organ	Investigated Drug
Lymphatic system	5-Fluorouracil (Takahashi et al., 1977), iodohippuric acid (Hashida et al., 1977), bleomycin (Hirnle 1997), isoniazid (Khopade and Jain, 1998)
Tumor	Bleomycin (Takahashi et al., 1977), paclitaxel (Trickler et al., 2008)
Brain	Rifampicin (Khopade and Jain, 1999b)
Liver	5-Fluorouracil (Omotosho et al., 1989), epirubicin (Hino et al., 2001)
Lungs	Rifampicin (Nakhare and Vyas, 1997)
Inflammatory tissue	Diclofenac sodium (Talegaonkar and Vyas, 2005)
Intranasal and oral	Ovalbumin (Shahiwala and Amiji, 2008)
Vaginal	Nitroimidazole (Ozer et al., 2007)
Skin	Glycolic acid (Yener and Baitokova, 2006)

7.5.4 TASTE MASKING

Another application of multiple emulsions is for taste masking of orally administered bitter drugs. Water-soluble bitter drugs can be efficiently incorporated in the inner aqueous phase of a W/O/W multiple emulsion. Various edible oils have been investigated for this purpose. Chlorpromazine, an antipsychotic drug (Garti, 1997), and chloroquine, an antimalarial agent (Vaziri and Warburton, 1994), are some of the drugs for which taste masking was effectively achieved by formulating them as multiple emulsions.

7.5.5 BIOAVAILABILITY ENHANCEMENT

Multiple emulsions can be used for enhancing the oral absorption of drugs that are not effectively absorbed through the gastrointestinal (GI) tract. Some of the more polar water-soluble drugs like isoniazid are ionized at gastric pH and thus are not absorbed well from the GI tract. Isoniazid absorption has been improved by delivering them in a multiple emulsion formulation (Khopade et al., 1998). Multiple emulsions have also been examined to prevent environmental and enzymatic degradation of proteins and peptide drugs such as insulin (Degim and Celebi, 2007; Gallarate et al., 2008; Shen et al., 2008) and vancomycin (Shively et al., 1997) via the peroral route. Multiple emulsions impart lipophilic characteristics to highly polar drugs, hence enhancing their absorption through the lymphatic system. This approach not only improves its absorption through the GI tract but also avoids first pass metabolism of the drug by escaping the hepatic portal system. Other drugs that have shown enhanced bioavailability in a multiple emulsion include the antifungal agents griseofulvin (Ahmed et al., 2008), nitrofurantoin (Rudnic et al., 2008), and pyrenetetrasulfonic acid tetrasodium salt (Adachi et al., 2003).

7.5.6 DETOXIFICATION

Acute drug intoxications resulting from accidental overdose are a common occurrence reported in the public health sector. There are only a few detoxification methods available, which include active charcoal, ipecac-induced vomiting, and gastric lavage. Reactive detoxifying emulsion technology offers an effective treatment option for drug overdoses. Multiple emulsions have been effectively used for the treatment of drug overdosing and detoxification. The first multiple emulsion used for overdose treatment was proposed in 1978 by Chiang et al. It works on the principle of the solvent extraction technique and pH-partition theory. Different parts of a multiple emulsion can have different pH values, which can affect the ionization and partition of the drug. Detoxifying emulsions are necessarily W/O/W-type emulsions (Hamoudeh et al., 2006). A weakly acidic drug overdose like barbiturates has been treated by a W/O/W emulsion containing an inner aqueous phase with a basic buffer. In the acidic pH of the stomach, barbiturates remain in the unionized state and are partitioned through the oil phase into the inner basic aqueous phase. In this basic pH, barbiturates ionize back and stay in this phase without any partition into the oil phase. Morimoto et al. (1979) reported the use of multiple emulsions for the treatment of quinine sulfate overdoses. Detoxification of blood using multiple emulsions was also reported by Volkel et al. (1982, 1984).

7.5.7 ENZYME IMMOBILIZATION

Most enzymes used in industrial and pharmaceutical processes are relatively unstable, and the cost of their isolation and purification is still high. It is also technically very challenging and expensive to recover an active enzyme from the reaction mixture at the end of a process. Therefore, enzyme immobilization on various carriers has been developed. The number of methods available for this purpose has increased significantly over the past two decades (Kennedy and Cabral, 1987).

Multiple emulsions have also been utilized in enzyme immobilization in place of solid-membrane or conventional methods (May and Li, 1972; Scheper et al., 1987). This enzyme immobilization via the multiple emulsion technique involves the entrapment of the enzymes, which catalyzes the process in the interior phase of the emulsion. Alcohol dehydrogenase has been immobilized by the multiple emulsion method and used in the conversion of alcohol to acetaldehyde (May and Landgraff, 1976). The process of conversion of ketoisocaproate to L-leucine was carried out by using enzyme immobilization for L-leucine dehydrogenase by multiple emulsions (Kajiwara et al., 1990). Scheper and coworkers (1984) reported the enzymatic production of L-phenyl alanine, an amino acid from immobilized chymotrypsin. Immobilized lipase for the hydrolysis of fatty acids was also described by Iso et al. (1987).

7.5.8 TOPICAL APPLICATION

Farahmand et al. (2006) showed that multiple emulsions release the active drug more slowly than the solutions. Raynal and coworkers (1993) developed a topical W/O/W emulsion containing sodium lactate in the inner phase, spironolactone in the intermediate phase, and chlorhexidine gluconate in the outer aqueous phase. Other studies entailed comparisons of the release profiles of metronidazole from W/O, O/W, and W/O/W emulsions (Ferreira et al., 1994, 1995a, 1995b; Ozer et al., 2007). Bonina and coworkers (1992) developed a multiple emulsion containing testosterone, caffeine, and water for topical use. Ferreira et al. (1994, 1995a, 1995b) devolved a multiple emulsion for metronidazole and glucose and compared its percutaneous absorption from different formulations. Multiple emulsions were also shown to enhance the stability of rapidly degrading active drugs (Gallarate et al., 2008). They were also used to enhance the stability of ascorbic acid (Farahmand et al., 2006). Laugel et al. (1996, 1998) developed W/O/W multiple emulsions for both dihydralazine and hydrocortisone. A multiple emulsion of benzalkonium chloride for vaginal use was investigated by Tedajo et al. (2002). The antibacterial spectrum of this formulation was further enhanced by incorporation of chlorhexidine digluconate to the internal phase (Tedajo et al., 2005).

A recent topical application of multiple emulsions is formulating an emulsion that releases the active drug entity under shear. Studies have shown that thickening of the external aqueous phase accelerates the release of the active drug under shear (Muguet et al., 1999, 2001). A recent development in topically used multiple emulsions is a shear-sensitive thermogelling emulsion. The emulsion contains a thermoresponsive hydrogel in the external aqueous phase. This results in the gelling of the emulsion only at skin temperature and hence an increased delivery of the active moiety when applied topically (Guillot et al., 2009; Kanl et al., 2005; Oliveri et al., 2003).

7.6   PATENTS ON MULTIPLE EMULSIONS

Preparation of a stable multiple emulsion has been a challenge, and various patents exist on the technology involved in its preparation and potential pharmaceutical applications. Generally, water-insoluble topically active compounds such as hair conditioners, hair dyes, skin care products, or topical medicaments are used as emulsions where nonadherence to the applied surface is a great challenge. A patent claims to find the solution of this problem by using the medicaments in a multiple emulsion formulation, W/O/W (W1-O-W2), which comprises a primary W/O (W1/O) emulsion as the internal phase and an external aqueous phase (Herb et al., 1997). The various components of a primary emulsion are a first topically active compound; a surfactant or a mixture of surfactants; an oil phase consisting of a silicone compound, a hydrocarbon compound, or a mixture thereof; and water. The external aqueous phase contains the second topically active compound and an emulsifier capable of forming stabilizing liquid crystals, for example, dicetyldimonium chloride, distearyldimonium chloride, dipalmitylamine, cetyl alcohol, stearyl alcohol, steareth-2, steareth-21, dioctylsodium sulfosuccinate, phosphatidylserine, phosphatidylcholine, and mixtures thereof. When this multiple emulsion is applied, the second topically active medicament is delivered that can be rinsed from or allowed to remain on the skin or hair. After evaporation of the oil phase or after rupture of the primary emulsion by friction (e.g., rubbing), the first topically effective compound remains on the skin or hair to perform its intended function.

However, emulsions stabilized via the formation of a "lamellar liquid crystal phase" produce less foam or lather, and a limited number of surfactants can be used. This problem is solved by a patent where multiple emulsions consist of isotropic surfactants that are stable for more than 8 weeks at room temperature. The only limitations are that the external surfactant phase must not comprise an amido group containing an anionic surfactant and the oil must not be greater than 50% unsaturated and not be a volatile silicone (Naser and Vasudevan, 2001).

There is another patent on the method of preparation that involves first preparing an O/W microemulsion that is destabilized by dilution with sufficient water to produce a W/O/W multiple emulsion (Gaonkar, 1994). This method is unique in the sense that a no-mixing step of the primary and external aqueous phase and use of a lipophilic emulsifier are required. A simple mixture of oil, water, and a polar protic solvent having at least one alkyl group and a hydrophilic emulsifier results in a submicron-sized division of the oil phase, forming the O/W microemulsion. Then, a multiple emulsion is obtained by diluting the microemulsion with water.

The standard process of preparation of multiple emulsions involves two reactors: the first for preparing the W/O inverse emulsion and the second for preparing the external aqueous phase. The inverse emulsion from the first reactor is added to the second reactor containing the external phase. This process works well if the inverse emulsion is relatively nonviscous, that is, if there is no significant difference in the viscosity of inverse emulsion and external phase. However, it suffers from the drawback of using two reactors, necessitating the installation of specific means to ensure that at least one reactor is cleaned between two production runs. The major problem encountered is attributable to the large viscosity difference between the inverse emulsion and external aqueous phase, which drastically decreases the efficacy of stirring. Consequently, stirring becomes necessary to generate more energy, sufficient to produce a satisfactory droplet size. However, such a stirring technique may shear the droplets of the inverse emulsion, resulting in the high risk of the release of the active material encapsulated in the inner aqueous phase of the multiple emulsions.

This problem is solved by the claim in a recent patent that uses heat-thickening polymers in the external aqueous phase, which results in an increase in its viscosity, thus decreasing the viscosity difference from the inverse emulsion (Mercier and Vallier, 2008). Consequently, the stirring methods required could be less powerful and the problem of release of the active material encapsulated in the inner aqueous phase would also be minimized. This method could be applied to the preparation of multiple emulsions having broad viscosity ranges and where there is a risk of leaking the active medicaments encapsulated in the inner aqueous phase.

Multiple emulsions find uses in formulating two incompatible active constituents together where one of the components could be incorporated in the inverse emulsion and the other in the continuous phase, for which the inverse emulsion would serve as the internal phase. When such multiple emulsion based formulations are administered, both active components would come into contact and exhibit the intended therapeutic effect. However, multiple emulsions are thermodynamically extremely unstable systems and there is a great risk of mixing the two incompatible active constituents even before the actual use. The known methods for remedying these instability problems work well while preparing the formulation; however, they fail to ensure the stability of the multiple emulsions during storage. This problem is overcome by a patent that claims to disperse the granules, which are obtained by drying an inverse emulsion dispersed in an outer aqueous phase that can be stored in water without compromising its stability, in water to form a multiple emulsion (Lannibois-Drean et al., 2006).

One patent specifically claims to control the release of the active principle encapsulated in a multiple emulsion (Bibette et al., 2003). The active principle encapsulated in the drops of the internal phase of multiple emulsions is released via the process of coalescence that the patent claims to oppose or induce by virtue of a specific concentration of surfactants in the external aqueous phase. If the concentration of the surfactant is beyond a critical concentration threshold, destabilization of the multiple emulsions is induced with the effect of releasing the active principle. If the concentration of hydrophilic surface-active agent present in the external aqueous phase is below this concentration threshold, then no coalescence is observed for several months.

7.7 CONCLUSIONS

In addition to its application in other fields of science, multiple emulsions have shown promising and potential applications in pharmaceutical research and development over the last two decades. Their major applications in this field include sustained and controlled drug delivery and targeting. Their potential use in imaging and diagnosis has also generated some interest in the recent past. This field will definitely experience a major growth in the future. One of the unique advantages of this multiple emulsion is the facile and inexpensive method of their preparation. Novel methods have also been introduced recently to produce monodispersed multiple emulsions on a commercial scale. The unique challenge facing the pharmaceutical scientists today is long-term instability for this multiple emulsion drug delivery system. Use of amphiphilic macromolecules instead of low molecular weight surfactants has shown some improvement in their stability. Steric stabilization, mechanical stabilization, and depletion stabilization are some of the techniques used to enhance the stability of multiple emulsions. Currently, the ideal multiple emulsions are most likely to be prepared by a membrane separation technique using polymeric amphiphiles as emulsifiers with viscosity enhancing agents and electrolytes to control the osmotic pressure. This area of stabilization is still in an embryonic stage and needs more extensive work for drug delivery applications. The future trend in multiple emulsion formulations may see a replacement of the interior emulsion with thermodynamically stable microemulsions. Microemulsions, thermodynamic stability, and nanostructures will certainly enhance the stability of multiple emulsions. Over the last few years O/W/O multiple emulsions have gained more interest for the sustained delivery of hydrophobic drugs in micro- and nanoencapsulation formulations. In the future, this technique will be used more frequently for the delivery and targeting of many hydrophobic drugs.

ABBREVIATIONS

BSA   Bovine serum albumin
GI   Gastrointestinal
HLB   Hydrophilic–lipophilic balance
O/W   Oil-in-water

O/W/O   Oil-in-water-in-oil
W/O   Water-in-oil
W/O/W   Water-in-oil-in-water

REFERENCES

Adachi, S., Imaoka, H., Hasegawa, Y., and Matsuno, R., 2003. Preparation of water-in-oil-in-water (W/O/W) type microcapsules by a single-droplet-drying method and change in encapsulation efficiency of a hydrophilic substance during storage. *Bioscience, Biotechnology, and Biochemistry* 67(6): 1376–1381.

Ahmed, I. S., Aboul-Einien, M. H., Mohamed, O. H., and Farid, S. F., 2008. Relative bioavailability of griseofulvin lyophilized dry emulsion tablet vs. immediate release tablet: A single-dose, randomized, open-label, six-period, crossover study in healthy adult volunteers in the fasted and fed states. *European Journal of Pharmaceutical Sciences* 35(3): 219–225.

Babak, V. G. and Stebe, M. J., 2002. Highly concentrated emulsions: Physicochemical principles of formulation. *Journal of Dispersion Science and Technology* 23(1): 1–22.

Bibette, J., Ficheux, M.-F., Leal Calderon, F., and Bonnakdar, L., 2003. Method for releasing an active principle contained a multiple emulsion. U.S. Patent 6627603.

Bodmeier, R., Wang, J., and Bhagwatwar, H., 1992. Process and formulation variables in the preparation of wax microparticles by a melt dispersion technique. II. W/O/W multiple emulsion technique for water-soluble drugs. *Journal of Microencapsulation* 9(1): 99–107.

Bonina, F., Bader, S., Montenegro, L., Scrofani, C., and Visca, M., 1992. Three phase emulsions for controlled delivery in the cosmetic field. *International Journal of Cosmetic Science* 14(2): 65–74.

Bozkir, A. and Saka, O. M., 2005. Formulation and investigation of 5-FU nanoparticles with factorial design-based studies. *Il Farmaco* 60(10): 840–846.

Castro, G. A. and Ferreira, L. A. M., 2008. Novel vesicular and particulate drug delivery systems for topical treatment of acne. *Expert Opinion on Drug Delivery* 5(6): 665–679.

Cheng, J., Chen, J. F., Zhao, M., Luo, Q., Wen, L. X., and Papadopoulos, K. D., 2007. Transport of ions through the oil phase of W(1)/O/W(2) double emulsions. *Journal of Colloid and Interface Science* 305(1): 175–182.

Chiang, C. W., Fuller, G. C., Frankenfeld, J. W., and Rhodes, C. T., 1978. Potential of liquid membranes for drug overdose treatment: *In vitro* studies. *Journal of Pharmaceutical Sciences* 67(1): 63–66.

Clausse, D., Pezron, I., and Komunjer, L., 1999. Stability of W/O and W/O/W emulsions as a result of partial solidification. *Colloids and Surfaces A: Physicochemical and Engineering Aspects* 152(1–2): 23–29.

Cournarie, F., Savelli, M. P., Rosilio, V., Bretez, F., Vauthier, C., Grossiord, J. L., and Seiller, M., 2004. Insulin-loaded W/O/W multiple emulsions: comparison of the performances of systems prepared with medium-chain-triglycerides and fish oil. *European Journal of Pharmaceutics and Biopharmaceutics* 58(3): 477–482.

Cornec, M., Cho, D., and Narsimhan, G., 2000. Adsorption and exchange of whey proteins onto spread lipid monolayers. In: *Emulsions, Foams, and Thin Films*, K. L. Mittal and P. Kumar (Eds.). Boca Raton, FL: CRC Press.

Davis, S. S. and Burbace, A. S., 1977. Electron micrography of water-in-oil-in-water emulsions. *Journal of Colloid and Interface Science* 62(2): 361–363.

Davis, S. S. and Walker, I., 1983. Measurement of the yield of multiple emulsion droplets by a fluorescent tracer technique. *International Journal of Pharmaceutics* 17(2–3): 203–213.

Degim, I. T. and Celebi, N., 2007. Controlled delivery of peptides and proteins. *Current Pharmaceutical Design* 13(1): 99–117.

DeLuca, M., Grossiord, J. L., Medard, J. M., Vaution, C., and Sellier, M., 1990. A stable W/O/W multiple emulsion. *Cosmetics and Toiletries* 105(11): 65–69.

DeLuca, M., Grossiord, J. L., and Sellier, M., 2000. A very stable W/O/W multiple emulsion. *International Journal of Cosmetic Science* 22(2): 157–161.

DiStefano, F. V., Shaffer, O. M., El-Aasser, M. S., and Vanderhoff, J. W., 1983. Multiple oil-in-water-in-oil emulsions of extremely fine droplet size. *Journal of Colloid and Interface Science* 92(1): 269–272.

Dickinsone, E., 1994. Protein-stabilized emulsions. *Journal of Food Engineering* 22(1–4): 59–74.

Dinarvand, R., Moghadam, S. H., Sheikhi, A., and Atyabi, F., 2005. Effect of surfactant HLB and different formulation variables on the properties of poly-D,L-lactide microspheres of naltrexone prepared by double emulsion technique. *Journal of Microencapsulation* 22(2): 139–151.

Dogru, S. T., Calis, S., and Oner, F., 2000. Oral multiple w/o/w emulsion formulation of a peptide salmon calcitonin: *In vitro–in vivo* evaluation. *Journal of Clinical Pharmacy and Therapeutics* 25(6): 435–443.

Dong, Y. and Feng, S. S., 2005. Poly(D,L-lactide-*co*-glycolide)/montmorillonite nanoparticles for oral delivery of anticancer drugs. *Biomaterials* 26(30): 6068–6076.

Doucet, O., Ferrero, L., Garcia, N., and Zastrow, L., 1998. O/W emulsion and W/O/W multiple emulsion: Physical characterization and skin pharmacokinetic comparison in the delivery process of caffeine. *International Journal of Cosmetic Science* 20(5): 283–295.

Driscoll, D. F., Nehne, J., Peters, H., Klutsch, K., Bistrian, B. R., and Wilhelm, N., 2006. Physical assessments of lipid injectable emulsions via microscopy: A comparison to methods proposed in the United States Pharmacopeia chapter <729>. *International Journal of Pharmaceutical Compounding* 10(4): 309–315.

Fan, T., Takayama, K., Hattori, Y., and Maitani, Y., 2004. Formulation optimization of paclitaxel carried by PEGylated emulsions based on artificial neural network. *Pharmaceutical Research* 21(9): 1692–1697.

Farahmand, S., Tajerzadeh, H., and Farboud, E. S., 2006. Formulation and evaluation of a vitamin C multiple emulsion. *Pharmaceutical Development and Technology* 11(2): 255–261.

Ferreira, L. A. M., Seiller, M., Grossiord, J. L., Marty, J. P., and Wepierre, J., 1994. *In vitro* percutaneous absorption of metronidazole and glucose: Comparison of o/w, w/o/w and w/o systems. *International Journal of Pharmaceutics* 109(3): 251–259.

Ferreira, L. A. M., Seiller, M., Grossiord, J. L., Marty, J. P., and Wepierre, J., 1995a. Vehicle influence on *in vitro* release of glucose: w/o, w/o/w and o/w systems compared. *Journal of Controlled Release* 33(3): 349–356.

Ferreira, L. A. M., Seiller, M., Grossiord, J. L., Marty, J. P., and Wepierre, J., 1995b. *In vitro* percutaneous absorption of metronidazole and glucose: Comparison of o/w, w/o/w and w/o systems. *International Journal of Pharmaceutics* 121(2): 169–179.

Florence, A. T., Law, T. K., and Whateley, T. L., 1985. Nonaqueous foam structures from osmotically swollen w/o/w emulsion droplets. *Journal of Colloid and Interface Science* 107(2): 584–588.

Florence, A. T. and Whitehill, D., 1981. Some features of breakdown in water-in-oil-in-water multiple emulsions. *Journal of Colloid Interface Science* 79(1): 243–256.

Florence, A. T. and Whitehill, D., 1982. The formulation and stability of multiple emulsions. *International Journal of Pharmaceutics* 11(4): 277–308.

Fukushima, S., Juni, K., and Nakano, M., 1983. Preparation of and drug release from W/O/W type double emulsions containing anticancer agents. *Chemical and Pharmaceutical Bulletin (Tokyo)* 31(11): 4048–4056.

Gallarate, M., Trotta, M., Battaglia, L., and Chirio, D., 2008. Preparation of solid lipid nanoparticles from W/O/W emulsions: Preliminary studies on insulin encapsulation. *Journal of Microencapsulation* 26(5): 394–402.

Gaonkar, A. G., 1994. Method for preparing a multiple emulsion. U.S. Patent 5322704.

Garti, N., 1997. Double emulsions—Scope, limitations and new achievements. *Colloids and Surfaces A: Physicochemical and Engineering Aspects* 123–124: 233–246.

Garti, N. and Aserin, A., 1996. Double emulsions stabilized by macromolecular surfactants. *Advances in Colloid and Interface Science* 65(31): 37–69.

Garti, N., Aserin, A., and Cohen, Y., 1994. Mechanistic considerations on the release of electrolytes from multiple emulsions stabilized by BSA and nonionic surfactants. *Journal of Controlled Release* 29(1–2): 41–51.

Geiger, S., Tokgoz, S., Fructus, A., Jager-Lezer, N., Seiller, M., Lacombe, C., and Grossiord, J. L., 1998. Kinetics of swelling–breakdown of a W/O/W multiple emulsion: Possible mechanisms for the lipophilic surfactant effect. *Journal of Controlled Release* 52(1–2): 99–107.

Ghosh, L. K., Ghosh, N. C., Thakur, R. S., Pal, M., and Gupta, B. K., 1997. Design and evaluation of controlled-release W/O/W multiple-emulsion oral liquid delivery system of chlorpheniramine maleate. *Drug Development and Industrial Pharmacy* 23(11): 1131–1134.

Goubault, C., Pays, K., Olea, D., Gorria, P., Bibette, J., Schmitt, V., and Leal-Calderon, F., 2001. Shear rupturing of complex fluids: Application to the preparation of quasi-monodisperse water-in-oil-in-water double emulsions. *Langmuir* 17(17): 5184–5188.

Guillot, S., Tomsic, M., Sagalowicz, L., Leser, M. E., and Glatter, O., 2009. Internally self-assembled particles entrapped in thermoreversible hydrogels. *Journal of Colloid and Interface Science* 330(1): 175–179.

Hamoudeh, M., Seiller, M., Chauvierre, C., Auchere, D., Lacour, B., Pareau, D., Stambouli, M., and Grossiord, J. L., 2006. Formulation of stable detoxifying w/o/w reactive multiple emulsions: *In vitro* evaluation. *Journal of Drug Delivery Science and Technology* 16(3): 223–228.

Hashida, M., Takahashi, Y., Muranishi, S., and Sezaki, H., 1977. An application of water-in-oil and gelatin-microsphere-in-oil emulsions to specific delivery of anticancer agent into stomach lymphatics. *Journal of Pharmacokinetics and Pharmacodynamics* 5(3): 241–255.

Herb, C. A., Chen, L. B., Chung, J., Long, M. A., Sun, W. M., Newell, G. P., Evans, T. A., Kamis, K., and Brucks, R. M., 1997. Water-in-oil-in-water compositions. U.S. Patent 5656280.

Higashi, S., Shimizu, M., Nakashima, T., Iwata, K., Uchiyama, F., Tateno, S., Tamura, S., and Setoguchi, T., 1995. Arterial-injection chemotherapy for hepatocellular carcinoma using monodispersed poppy-seed oil microdroplets containing fine aqueous vesicles of epirubicin. Initial medical application of a membrane-emulsification technique. *Cancer* 75(6): 1245–1254.

Higashi, S., Tabata, N., Kondo, K. H., Maeda, Y., Shimizu, M., Nakashima, T., and Setoguchi, T., 1999. Size of lipid microdroplets effects results of hepatic arterial chemotherapy with an anticancer agent in water-in-oil-in-water emulsion to hepatocellular carcinoma. *Journal of Pharmacology and Experimental Therapeutics* 289(2): 816–819.

Higuera-Ciapara, I., Felix-Valenzuela, L., Goycoolea, F. M., and Argüelles-Monal, W., 2003. Microencapsulation of astaxanthin in a chitosan matrix. *Carbohydrate Polymers* 56(1): 41–45.

Hino, T., Shimabayashi, S., Tanaka, M., Nakano, M., and Okochi, H., 2001. Improvement of encapsulation efficiency of water-in-oil-in-water emulsion with hypertonic inner aqueous phase. *Journal of Microencapsulation* 18(1): 19–28.

Hirnle, P., 1997. Liposomes for drug targeting in the lymphatic system. *Hybridoma* 16(1): 127–132.

Hou, W. and Papadopoulos, K. Y., 1997. $W_1/O/W_2$ and $O_1/W/O_2$ globules stabilized with Span 80 and Tween 80. *Colloids and Surfaces A: Physicochemical and Engineering Aspects* 125(2–3): 181–187.

Iso, M., Shirahase, T., Hanamura, S., Urushiyama, S., and Omi, S., 1989. Immobilization of enzyme by micro-encapsulation and application of the encapsulated enzyme in the catalysis. *Journal of Microencapsulation* 6(2): 165–176.

Jantzen, G. M. and Robinson, J. R., 2002. Sustained- and controlled-release drug-delivery systems. In: *Modern Pharmaceutics*, 4th edition, S. Gilbert, G. S. Banker, and C. Rhodes (Eds.). Jersey, U. K.: Informa Healthcare.

Jaspart, S., Bertholet, P., Piel, G., Dogné, J. M., Delattre, L., and Evrard, B., 2007. Solid lipid microparticles as a sustained release system for pulmonary drug delivery. *European Journal of Pharmaceutics and Biopharmaceutics* 65(1): 47–56.

Jeong, Y. I., Na, H. S., Seo, D. H., Kim, D. G., Lee, H. C., Jang, M. K., Na, S. K., Roh, S. H., Kim, S. I., and Nah, J. W., 2008. Ciprofloxacin-encapsulated poly(DL-lactide-*co*-glycolide) nanoparticles and its anti-bacterial activity. *International Journal of Pharmaceutics* 352(1–2): 317–323.

Jiao, J. and Burgess, D., 2003. Rheology and stability of water-in-oil-in water multiple emulsions containing Span 83 and Tween 80. *AAPS PharmSci* 5(1): E7.

Jiao, J. and Burgess, D., 2008. Multiple emulsion stability: Pressure balance and interfacial film strength. In: *Multiple Emulsions: Technology and Applications*, A. Aserin (Ed.), pp. 1–27. New York: Wiley.

Jourquin, G. and Kauffmann, J. M., 1998. Fluorimetric determination of theophylline in serum by inhibition of bovine alkaline phosphatase in AOT based water/in oil microemulsion. *Journal of Pharmaceutical and Biomedical Analysis* 18(4–5): 585–596.

Kajiwara, S., Maeda, H., and Suzuki, H., 1990. Enzyme immobilization by entrapment in a polymer gel matrix. U.S. Patent 4978619.

Kan, P., Lin, X. Z., Hsieh, M. F., and Chang, K. Y., 2005. Thermogelling emulsions for vascular embolization and sustained release of drugs. *Journal of Biomedical Materials Research Part B: Applied Biomaterials* 75(1): 185–192.

Karasulu, Y., Karabulut, B., Goker, E., Gneri, T., and Gabor, F., 2007. Controlled release of methotrexate from w/o microemulsion and its *in vitro* antitumor activity. *Drug Delivery* 14(4): 225–233.

Kavaliunas, D. R. and Frank, S. G., 1978. Liquid crystal stabilization of multiple emulsions. *Journal of Colloid and Interface Science* 66(3): 586–588.

Kennedy, J. F. and Cabral, J. M. S. 1987. Enzyme immobilization. In: *Biotechnology*, J. H. Rehm and G. Reed (Eds.), Vol. 7, pp. 347–404. Weinheim, Germany: VCH.

Kerstens, S., Murray, B. S., and Dickinson, E., 2006. Microstructure of beta-lactoglobulin-stabilized emulsions containing non-ionic surfactant and excess free protein: Influence of heating. *Journal of Colloid and Interface Science* 296(1): 332–341.

Khan, A. Y., 2004. Development and characterization of (w/o/w) multiple emulsion based system. Master's Thesis, Jamia Hamdard University, New Delhi.

Khan, A. Y., Talegaonkar, S., Iqbal, Z., Ahmed, F. J., and Khar, R. K., 2006. Multiple emulsions: An overview. *Current Drug Delivery* 3(4): 429–443.

Khopade, A. J. and Jain, N. K., 1997. Stabilized multiple emulsions with uni/oligo-droplet internal phase. *Pharmazie* 52(77): 562–565.

Khopade, A. J. and Jain, N. K., 1998. A stable multiple emulsion system bearing isoniazid: Preparation and characterization. *Drug Development and Industrial Pharmacy* 24(3): 289–293.

Khopade, A. J. and Jain, N. K., 1999a. Long-circulating multiple-emulsion system for improved delivery of an anticancer agent. *Drug Delivery* 6(2): 107–110.

Khopade, A. J. and Jain, N. K., 1999b. Multiple emulsions containing rifampicin. *Pharmazie* 54(12): 915–919.

Khopade, A. J. and Jain, N. K., 2000. Concanavalin-A conjugated fine-multiple emulsion loaded with 6-mercaptopurine. *Drug Delivery* 7(2): 105–112.

Khopade, A. J. and Jain, N. K., 2001. Multiple emulsions as drug delivery systems. In: *Advances in Controlled and Novel Drug Delivery*, N. K. Jain (Ed.), p. 381. New Delhi: CBS Publishers & Distributors.

Kim, C. K., Kim, S. C., Shin, H. J., Kim, K. M., Oh, K. H., Lee, Y. B., and Oh, I. J., 1995. Preparation and characterization of cytarabine-loaded w/o/w multiple emulsions. *International Journal of Pharmaceutics* 124(1): 61–67.

Kita, Y., Matsumoto, S., and Yonezawa, D., 1977. Viscometric method for estimating the stability of W/O/W-type multiple-phase emulsions. *Journal of Colloid and Interface Science* 62(1): 87–94.

Kobayashi, I., Nakajima, M., Nabetani, H., Kikuchi, Y., Shohno, A., and Satoh, K., 2001. Preparation of micron-scale monodisperse oil-in-water microspheres by microchannel emulsification. *Journal of the American Oil Chemists' Society* 78(8): 797–802.

Kosvintsev, S. R., Gasparini, G., and Holdich, R. G., 2008. Membrane emulsification: Droplet size and uniformity in the absence of surface shear. *Journal of Membrane Science* 313(1–2): 182–189.

Lannibois-Drean, H., Morvan, M., and Taisne, L., 2006. Granules obtained by drying a multiple emulsion. U.S. Patent 7101931.

Laugel, C., Baillet, A., Youenang Piemi, M. P., Marty, J. P., and Ferrier, D., 1998. Oil–water–oil multiple emulsions for prolonged delivery of hydrocortisone after topical application: Comparison with simple emulsions. *International Journal of Pharmaceutics* 160(1): 109–117.

Laugel, C., Chaminade, P., Baillet, A., Seiller, M., and Ferrier, D., 1996. Moisturizing substances entrapped in W/O/W emulsions: Analytical methodology for formulation, stability and release studies. *Journal of Controlled Release* 38(1): 59–67.

Lee, J., Park, T. G., and Choi, H., 2000. Effect of formulation and processing variables on the characteristics of microspheres for water-soluble drugs prepared by w/o/o double emulsion solvent diffusion method. *International Journal of Pharmaceutics* 196(1): 75–83.

Lee, M.-H., Oh, S. G., Moon, S. K., and Bae, S.-Y., 2001. Preparation of silica particles encapsulating retinol using o/w/o multiple emulsions. *Journal of Colloid and Interface Science* 240(1): 83–89.

Leichtnam, M. L., Rolland, H., Wüthrich, P., and Guy, R. H., 2006. Testosterone hormone replacement therapy: State-of-the-art and emerging technologies. *Pharmaceutical Research* 23(6): 1117–1132.

Lin, S. Y., Wu, W. H., and Lui, W. Y., 1992. *In vitro* release, pharmacokinetic and tissue distribution studies of doxorubicin hydrochloride (Adriamycin HCl) encapsulated in lipiodolized w/o emulsions and w/o/w multiple emulsions. *Pharmazie* 47(6): 439–443.

Lindenstruth, K. and Muller, B. W., 2004. W/O/W multiple emulsions with diclofenac sodium. *European Journal of Pharmaceutics and Biopharmaceutics* 58(3): 621–627.

Liu, R., Huang, S. S., Wan, Y. H., Ma, G. H., and Su, Z. G., 2005. Preparation of insulin-loaded PLA/PLGA microcapsules by a novel membrane emulsification method and its release in vitro. *Colloids and Surfaces B: Biointerfaces* 51(1): 30–38.

Lorenceau, E., Utada, A. S., Link, D. R., Cristobal, G., Joanicot, M., and Weitz, D. A., 2005. Generation of polymerosomes from double-emulsions. *Langmuir* 21(20): 9183–9186.

Luan, X. and Bodmeier, R., 2006. *In situ* forming microparticle system for controlled delivery of leuprolide acetate: Influence of the formulation and processing parameters. *European Journal of Pharmaceutical Sciences* 27(2–3): 143–149.

Magdassi, S., Frenkel, M., and Garti, N., 1985. Correlation between nature of emulsifier and multiple emulsion stability. *Drug Development and Industrial Pharmacy* 11(4): 791–798.

Magdassi, S., Frenkel, M., Garti, N., and Kasan, R., 1984. Multiple emulsions II: HLB shift caused by emulsifier migration to external interface. *Journal of Colloid and Interface Science* 97(2): 374–379.

Magdassi, S. and Garti, N. 1999. Multiple emulsion. In: *Novel Cosmetic Delivery Systems*, S. Magdassi and E. Touitou (Eds.). Boca Raton, FL: CRC Press.

Martin, A. N., Bustamante, P., and Chun, A. H. C., 1993. *Physical Pharmacy*, 4th Edition, pp. 486–496. Baltimore, MD: Lippincott/Williams & Wilkins.

Martín-Banderas, L., Rodríguez-Gil, A., Cebolla, Á., Chávez, S., Berdún-Álvarez, T., Fernandez Garcia, J. M., Flores-Mosquera, M., and Gañán-Calvo, A. M., 2006. Towards high-throughput production of uniformly encoded microparticles. *Advanced Materials* 18(5): 559–564.

Matsumoto, S., Inoue1, T., Kohda, M., and Ikura, K., 1980. Water permeability of oil layers in W/O/W emulsions under osmotic pressure gradients. *Journal of Colloid and Interface Science* 77(2): 555–563.

Matsumoto, S. and Kang, W. W., 1988. An attempt at measuring saccharose flux permeating through the oil layer in W/O/W emulsions on the basis of hydrolytic activity of invertase. *Agricultural and Biological Chemistry* 52(11): 2689–2694.

Matsumoto, S., Kita, Y., and Yonezawa, D., 1976. An attempt at preparing water-in-oil-in-water multiple-phase emulsions. *Journal of Colloid and Interface Science* 57(2): 353–361.

Matsumoto, S., Koh, Y., and Michiure, A., 1985. Preparation of w/o/w emulsions in an edible form on the basis of phase inversion technique. *Journal of Dispersion Science and Technology* 6(5): 507–521.

May, S. W. and Landgraff, L. M., 1976. Cofactor recycling in liquid membrane–enzyme systems. *Biochemical and Biophysical Research Communications* 68(3): 786–792.

May, S. W. and Li, N. N., 1972. The immobilization of urease using liquid-surfactant membranes. *Biochemical and Biophysical Research Communications* 47(5): 1179–1185.

Mercier, J.-M. and Vallier, E., 2008. Method for preparing a water/oil/water multiple emulsion. U.S. Patent 7319117.

Mishra, B. and Pandit, J. K., 1989. Prolonged release of pentazocine from multiple O/W/O emulsions. *Drug Development and Industrial Pharmacy* 15(8): 1217–1230.

Mishra, B. and Pandit, J. K., 1990. Prolonged tissue levels of pentazocine from multiple W/O/W emulsions in mice. *Drug Development and Industrial Pharmacy* 16(6): 1073–1078.

Miyakawa, T., Zhang, W., Uchida, T., Kim, N. S., and Goto, S., 1993. *In vivo* release of water-soluble drugs from stabilized water-in-oil-in-water (W/O/W) type multiple emulsions following intravenous administrations using rats. *Biological and Pharmaceutical Bulletin* 16(3): 268–272.

Morimoto, Y., Sugibayashi, K., Yamaguchi, Y., and Kato, Y., 1979. Detoxification capacity of a multiple (w/o/w) emulsion for the treatment of drug overdose: Drug extraction into the emulsion in the gastro-intestinal tract of rabbits. *Chemical and Pharmaceutical Bulletin (Tokyo)* 27(12): 3188–3192.

Muguet, V., Seiller, M., Barratt, G., Clausse, D., Marty, J. P., and Grossiord, J. L., 1999. W/O/W multiple emulsions submitted to a linear shear flow: Correlation between fragmentation and release. *Journal of Colloid and Interface Science* 218(1): 335–337.

Muguet, V., Seiller, M., Barratt, G., Ozer, O., Marty, J. P., and Grossiord, J. L., 2001. Formulation of shear rate sensitive multiple emulsions. *Journal Controlled Release* 70(1–2): 37–49.

Nahla, S., Barakat, N. S., and Ahmad, A. A. E., 2008. Diclofenac sodium loaded-cellulose acetate butyrate: Effect of processing variables on microparticles properties, drug release kinetics and ulcerogenic activity. *Journal of Microencapsulation* 25(1): 31–35.

Nakashima, T., Shimizu, M., and Kukizaki, M., 1991. *Membrane emulsification: Operation manual*, 1st edition. Miyazaki, Japan: Industrial Research Institute of Miyazaki Prefecture.

Nakashima, T., Shimizu, M., and Kukizaki, M., 2000. Particle control of emulsion by membrane emulsification and its applications. *Advanced Drug Delivery Review* 45(1): 47–56.

Nakhare, S. and Vyas, S. P., 1994. Prolonged release of diclofenac sodium from multiple w/o/w emulsion systems. *Pharmazie* 49(11): 842–845.

Nakhare, S. and Vyas, S. P., 1995a. Prolonged release of rifampicin from multiple W/O/W emulsion systems. *Journal of Microencapsulation* 12(4): 409–415.

Nakhare, S. and Vyas, S. P., 1995b. Prolonged release multiple emulsion based system bearing rifampicin: *In vitro* characterisation. *Drug Development and Industrial Pharmacy* 21(7): 869–878.

Nakhare, S. and Vyas, S. P., 1997. Multiple emulsion based systems for prolonged delivery of rifampicin: *In vitro* and *in vivo* characterization. *Pharmazie* 52(3): 224–226.

Naser, M. S. and Vasudevan, T., 2001. Stable multiple emulsion composition. U.S. Patent 6290943.

Nie, L., Bassi, S. D., and Maningat, C. C., 2006. Composite materials and extruded profiles containing mill feed. U.S. Patent 0228535.

Nisisako, T., Okushima, S., and Torii, T., 2005. Controlled formulation of monodisperse double emulsions in a multiple-phase microfluidic system. *Soft Matter* 1(1): 23–27.

Oh, C., Park, J.-H., Shin, S., and Oh, S.-G., 2004. O/W/O multiple emulsions via one-step emulsification process. *Journal of Dispersion Science and Technology* 25(1): 53–62.

Oh, I., Kang, Y. G., Lee, Y. B., Shin, S. C., and Kim, C. K., 1998. Prolonged release of tegafur from S/O/W multiple emulsion. *Drug Development and Industrial Pharmacy* 24(10): 889–894.

Ohwaki, T., Machida, R., Ozawa, H., Kawashima, Y., Hino, T., Takeuchi, H., and Niwa, T., 1993. Improvement of the stability of water-in-oil-in-water multiple emulsions by the addition of surfactants in the internal aqueous phase of the emulsions. *International Journal of Pharmaceutics* 93(1–3): 61–74.

Ohwaki, T., Nitta, K., Ozawa, H., Kawashima, Y., Hino, T., Takeuchi, H., and Niwa, T., 1992. Improvement of the formation percentage of water-in-oil-in-water multiple emulsion by the addition of surfactants in the internal aqueous phase of the emulsion. *International Journal of Pharmaceutics* 85(1–3): 19–26.

Okochi, H. and Nakano, M., 1996. Basic studies on formulation, method of preparation and characterization of water-in-oil-in-water type multiple emulsions containing vancomycin. *Chemical and Pharmaceutical Bulletin (Tokyo)* 44(1): 180–186.

Okochi, H. and Nakano, M., 1997. Comparative study of two preparation methods of w/o/w emulsions: Stirring and membrane emulsification. *Chemical and Pharmaceutical Bulletin (Tokyo)* 45(8): 1323–1326.

Okochi, H. and Nakano, M., 2000. Preparation and evaluation of w/o/w type emulsions containing vancomycin. *Advanced Drug Delivery Review* 45(1): 5–26.

Olivieri, L., Seiller, M., Bromberg, L., Ron, E., Couvreur, P., and Grossiord, J. L., 2003. Study of the breakup under shear of a new thermally reversible water-in-oil-in-water (W/O/W) multiple emulsion. *Pharmaceutical Research* 18(5): 689–693.

Omotosho, J. A., Whateley, T. L., and Florence, A. T., 1989. Release of 5-fluorouracil from intramuscular w/o/w multiple emulsions. *Biopharmaceutics and Drug Disposition* 10(3): 257–268.

Omotosho, J. A., Whateley, T. L., Law, T. K., and Florence, A. T., 1986. The nature of the oil phase and the release of solutes from multiple (w/o/w) emulsions. *Journal of Pharmacy and Pharmacology* 38(12): 865–870.

Opawale, F. O. and Burgess, D. J., 1998. Influence of interfacial rheological properties of mixed emulsifier films on the stability of water-in-oil-in-water emulsions. *Journal of Pharmacy and Pharmacology* 50(9): 965–973.

Ortega-Vinuesa, J. L., Martin-Rodriguez, A., and Hidalgo-Álvarez, R., 1996. Colloidal stability of polymer colloids with different interfacial properties: Mechanisms. *Journal of Colloid and Interface Science* 184(1): 259–267.

Ozer, O., Ozyazici, M., Tedajo, M., Taner, M. S., and Köseoglu, K., 2007. W/O/W multiple emulsions containing nitroimidazole derivates for vaginal delivery. *Drug Delivery* 14(3): 139–145.

Pallandre, S., Decker, E. A., and McClements, D. J., 2007. Improvement of stability of oil-in-water emulsions containing caseinate-coated droplets by addition of sodium alginate. *Journal of Food Science* 72(9): E518–E524.

Pandey, R. and Khuller, G. K., 2007. Nanoparticle-based oral drug delivery system for an injectable antibiotic—Streptomycin. Evaluation in a murine tuberculosis model. *Chemotherapy* 53(6): 437–441.

Pandit, J. K., Mishra, B., and Chand, B., 1987. Drug release from multiple w/o/w emulsions. *Indian Journal of Pharmaceutical Sciences* 49(3): 103–105.

Pays, K., Giermanska-Kahn, J., Pouligny, B., Bibette, J., and Leal-Calderon, F., 2001. Coalescence in surfactant-stabilized double emulsions. *Langmuir* 17(25): 7758–7769.

Pays, K., Giermanska-Kahn, J., Pouligny, B., Bibette, J., and Leal-Calderon, F., 2002. Double emulsions: How does release occur? *Journal of Controlled Release* 79(1–3): 193–205.

Puppo, M. C., Beaumal, V., Chapleau, N., Speroni, F., de Lamballerie, M., Añón, M. C., and Anton, M., 2007. Physicochemical and rheological properties of soybean protein emulsions processed with a combined temperature/high-pressure treatment. *Food Hydrocolloids* 22(6): 1079–1089.

Raynal, S., Grossiord, J. L., Seiller, M., and Clausse, D., 1993. A topical W/O/W multiple emulsion containing several active substances: Formulation, characterization and study of release. *Journal of Controlled Release* 26(2): 129–140.

Rewat, A., Majumdar, Q. H., and Ashan, F., 2008. Inhalable large porous microspheres of low molecular weight heparin: *In vitro* and *in vivo* evaluation. *Journal of Controlled Release* 128(3): 224–232.

Roy, S., Dasgupta, A., and Das, P. K., 2006. Tailoring of horseradish peroxidase activity in cationic water-in-oil microemulsions. *Langmuir* 22(10): 4567–4573.

Rudnic, E. M., Isbister, J. D., Treacy, Jr., D. J., and Wassink, S. E., 2006. Multiple-delayed released antibiotic product, use and formulation thereof. U.S. Patent 7025989.

Saettone, M. F., Giannaccini, B., and Monti, D., 2000. Ophthalmic emulsions and suspensions. In: *Pharmaceutical Emulsions and Suspensions*, N. Françoise and M.-M. Gilberte (Eds.). Boca Raton, FL: CRC Press.

Safwat, S. M., Kassem, M. A., Attia, M. A., and El-Mahdy, M., 1994. The formulation–performance relationship of multiple emulsions and ocular activity. *Journal of Controlled Release* 32(3): 259–268.

Sam, T. M., Gayathri, D. S., Prasanth, V. V., and Vinod, B., 2008. NSAIDs as microspheres. *Internet Journal of Pharmacology* 6(1): http://www.ispub.com/journal/the_internet_journal_of_pharmacology/volume_6_number_1_32/article/nsaids_as_microspheres.html

Sasaki, H., Yamamura, K., Nishida, K., Nakamura, J., and Ichikawa, M., 1996. Delivery of drugs to the eye by topical application. *Progress in Retinal and Eye Research* 15(2): 583–620.

Scheper, T., Halwachs, W., and Schügerl, K., 1984. Production of L-amino acid by continuous enzymatic hydrolysis of DL-amino acid methyl ester by the liquid membrane technique. *Chemical Engineering Journal* 29(2): B31–B37.

Scheper, T., Makryaleas, K., Nowottny, C., Likidis, Z., Tsikas, D., and Schügerl, K., 1987. Liquid surfactant membrane emulsions. A new technique for enzyme immobilization. *Annals of the New York Academy of Sciences* 501: 165–170.

Sela, Y., Magdassi, S., and Garti, N. 1995. Release of markers from the inner water phase of W/O/W emulsions stabilized by silicone based polymeric surfactants. *Journal of Controlled Release* 33(1): 1–12.

Shah, M. V., de Gennaro, M. D., and Suryakasuma, H., 1987. An evaluation of albumin microcapsules prepared using a multiple emulsion technique. *Journal of Microencapsulation* 4(3): 223–238.

Shahiwala, A. and Amiji, M. M., 2008. Enhanced mucosal and systemic immune response with squalane oil-containing multiple emulsions upon intranasal and oral administration in mice. *Journal of Drug Targeting* 16(4): 302–310.

Shen, B., Pei, F. X., Duan, H., Chen, J., and Mu, J. X., 2008. Preparation and *in vitro* activity of controlled release microspheres incorporating bFGF. *Chinese Journal of Traumatology* 11(1): 22–27.

Shively, M. L., 1997. Multiple emulsions for the delivery of proteins. *Pharmaceutical Biotechnology* 10: 199–211.

Silletti, E., Vingerhoeds, M. H., Norde, W., and van Aken, G. A., 2007. Complex formation in mixtures of lysozyme-stabilized emulsions and human saliva. *Journal of Colloid and Interface Science* 313(2): 485–493.

Singh, K. K. and Vingkar, S. K., 2007. Formulation, antimalarial activity and biodistribution of oral lipid nano-emulsion of primaquine. *International Journal of Pharmaceutics* 347(1–2): 138–143.

Sinha, V. R. and Kumar, A., 2002. Multiple emulsions: An overview of formulation, characterization, stability and applications. *Indian Journal of Pharmaceutical Sciences* 64(3): 191–199.

Su, J., Flanagan, J., Hemar, Y., and Singh, H., 2006. Synergistic effects of polyglycerol ester of polyricinoleic acid and sodium caseinate on the stabilisation of water-oil-water emulsions. *Food Hydrocolloids* 20(2–3): 261–268.

Sugiuraa, S., Nakajima, M., Yamamotoa, K., Iwamotoa, S., Odad, T., Satakee, M., and Sekif, M., 2004. Preparation characteristics of water-in-oil-in-water multiple emulsions using microchannel emulsification. *Journal of Colloid and Interface Science* 270(1): 221–228.

Suitthimeathegorna, O., Turtonb, J. A., Mizuuchia, O., and Florencea, A. T., 2007. Intramuscular absorption and biodistribution of dexamethasone from non-aqueous emulsions in the rat. *International Journal of Pharmaceutics* 331(2): 204–210.

Takahashi, T., Kono, K., and Yamaguchi, T., 1977. Enhancement of the cancer chemotherapeutic effect by anti-cancer agents in the form of fat emulsion. *Tohoku Journal of Experimental Medicine* 123(3): 235–246.

Talegaonkar, S. and Vyas, S. P., 2005. Inverse targeting of diclofenac sodium to reticuloendothelial system-rich organs by sphere-in-oil-in-water (s/o/w) multiple emulsions containing poloxamer 403. *Journal of Drug Targeting* 13(3): 173–178.

Tamilvanan, S. and Sa, B., 2008. *In vitro* and *in vivo* evaluation of single-unit commercial immediate-and sustained-release capsules compared with multiple-unit polystyrene microparticles dosage forms of indomethacin. *PDA Journal of Pharmaceutical Science and Technology* 62(3): 177–190.

Tanaka, T., Ikeda, M., Okusaka, T., Ueno, H., Morizane, C., Ogura, T., Hagihara, A., and Iwasa, S., 2008. A phase II trial of transcatheter arterial infusion chemotherapy with an epirubicin–lipiodol emulsion for advanced hepatocellular carcinoma refractory to transcatheter arterial embolization. *Cancer Chemotherapy and Pharmacology* 61(4): 683–688.

Tedajo, G. M., Bouttier, S., Fourniat, J., Grossiord, J. L., Marty, J. P., and Seiller, M., 2002. *In vitro* microbicidal activity of W/O/W multiple emulsion for vaginal administration. *International Journal of Antimicrobial Agents* 20(1): 50–56.

Tedajo, G. M., Bouttier, S., Grossiord, J. L., Marty, J. P., Seiller, M., and Fourniat, J., 2005. Release of antiseptics from the aqueous compartments of a w/o/w multiple emulsion. *International Journal of Pharmaceutics* 288(1): 63–72.

Tian, L., He, H., and Tang, X., 2007. Stability and degradation kinetics of etoposide-loaded parenteral lipid emulsion. *Journal of Pharmaceutical Sciences* 96(7): 1763–1775.

Trickler, W. J., Nagvekar, A. A., and Dash, A. K., 2008. A novel manoparticle formulation for sustained paclitaxel delivery. *AAPS Pharmaceutical Science and Technology* 9(2): 486–493.

Ursica, L., Tita, D., Palici, I., Tita, B., and Vlaia, V., 2005. Particle size analysis of some water/oil/water multiple emulsions. *Journal of Pharmaceutical and Biomedical Analysis* 37(5): 931–936.

Utada, A. S., Lorenceau, E., Link, D. R., Kaplan, P. D., Stone, H. A., and Weitz, D. A., 2005. Monodisperse double emulsions generated from a microcapillary device. *Science* 308(5721): 537–541.

Vasiljevic, D., Parojcic, J., Primorac, M., and Vuleta, G., 2006. An investigation into the characteristics and drug release properties of multiple W/O/W emulsion systems containing low concentration of lipophilic polymeric emulsifier. *International Journal of Pharmaceutics* 309(1–2): 171–177.

Vaziri, A. and Warburton, B., 1994. Slow release of chloroquine phosphate from multiple taste-masked W/O/W multiple emulsions. *Journal of Microencapsulation* 11(6): 641–648.

Villa, C. H., Lawson, L. B., Li, Y., and Papadopoulos, K. D., 2003. Internal coalescence as a mechanism of instability in water-in-oil-in-water double-emulsion globules. *Langmuir* 19(2): 244–249.

Vladisavljevic, G. T. and Williams, R. A., 2005. Recent developments in manufacturing emulsions and particulate products using membranes. *Advances in Colloid and Interface Science* 113(1): 1–20.

Volkel, W., Bosse, J., Poppe, W., Halwachs, W., and Schger, K., 1984. Development and design of a liquid membrane enzyme reactor for the detoxification of blood. *Chemical Engineering Communications* 30(1–2): 55–66.

Volkel, W., Poppe, W., Halwachs, W., and Schugerl, K., 1982. Extraction of free phenols from blood by a liquid membrane enzyme reactor. *Journal of Membrane Science* 11(3): 333–347.

Wang, A. H., Chen, X. G., Liu, C. S., Meng, X. H., Yu, L. J., and Wang, H., 2008a. Preparation and characteristics of chitosan microspheres in different acetylation as drug carrier system. *Journal of Microencapsulation* 24(1): 1–10.

Wang, J. J., Hung, C. F., Yeh, C. H., and Fang, J. U., 2008b. The release and analgesic activities of morphine and its ester prodrug, morphine propionate, formulated by water-in-oil nanoemulsions. *Journal of Drug Targeting* 16(4): 294–301.

Weiss, J., Scherze, I., and Muschiolik, G., 2005. Polysaccharide gel with multiple emulsion. *Food Hydrocolloids* 19(3): 605–615.

Wen, L. and Papadopoulos, K. D., 2000. Visualization of water transport in $w_1/o/w_2$ emulsions. *Colloids and Surfaces A: Physicochemical and Engineering Aspects* 174(1–2): 159–167.

Wilde, P. J., Clark, D. C., and Marion, D., 1993. Influence of competitive adsorption of a lysopalmitoylphosphatidylcholine on the functional properties of puroindoline, a lipid-binding protein isolated from wheat flour. *Journal of Agricultural and Food Chemistry* 41(10): 1570–1576.

Xiao, C., Qi, X., Maitani, Y., and Nagai, T., 2004. Sustained release of cisplatin from multivesicular liposomes: Potentiation of antitumor efficacy against S180 murine carcinoma. *Journal of Pharmaceutical Sciences* 93(7): 1718–1724.

Yan, L., Thompson, K. E., and Valsaraj, K. T., 2006. A numerical study on the coalescence of emulsion droplets in a constricted capillary tube. *Journal of Colloid and Interface Science* 298(2): 832–844.

Yan, N., Zhang, M., and Ni, P., 1992. A study of the stability of W/O/W multiple emulsions. *Journal of Microencapsulation* 9(2): 143–151.

Yener, G. and Baitokova, A., 2006. Development of a w/o/w emulsion for chemical peeling applications containing glycolic acid. *Journal of Cosmetic Science* 57(6): 487–494.

Yi, G. R., Manoharan, V. N., Michel, E., Elsesser, M. T., Yang, D., and Pine, J., 2004. Colloidal clusters of silica or polymer microspheres. *Advanced Materials* 16(14): 1204–1208.

Zhang, W., Miyakawa, T., Uchida, T., and Goto, S., 1992. Preparation of stable W/O/W type multiple emulsion containing water-soluble drugs and *in vitro* evaluation of its drug-releasing properties. *Journal of the Pharmaceutical Society of Japan* 112(1): 73–80.

Zheng, S., Beissinger, R. L., and Wasan, D. T., 1991. The stabilization of hemoglobin multiple emulsion for use as a red blood cell substitute. *Journal of Colloid and Interface Science* 144(1): 72–85.

Zheng, S., Zheng, Y., Beissinger, R. L., Wasan, D. T., and McCormick, D. T., 1993. Hemoglobin multiple emulsion as an oxygen delivery system. *Biochimica et Biophysica Acta—General Subjects* 1158(1): 65–94.

Zhou, H. Y., Chen, X. G., Liu, C. S., Meng, X. H., Yu, L. J., Liu, X. Y., and Liu, N., 2005. Chitosan/cellulose acetate microspheres preparation and ranitidine release. *In vitro. Pharmaceutical Development and Technology* 10(2): 219–225.

8 Pharmaceutical and Biotechnological Applications of Multiple Emulsions

Rita Cortesi and Elisabetta Esposito

CONTENTS

8.1 INTRODUCTION

Multiple emulsions are complex systems in which dispersed droplets contain smaller droplets inside. They are more complex systems, termed "emulsions of emulsions," in which the droplets of the dispersed phase contain even smaller dispersed droplets. Each dispersed globule in the double emulsion forms a vesicular structure with single or multiple aqueous compartments separated from the aqueous phase by a layer of oil phase compartments (Aserin, 2008; Davis, 1981; Garti and Aserin, 1996; Khan et al., 2006). Multiple emulsions, which have ternary, quaternary, or more complex structures, have been studied since their first description in 1925 (Seifriz, 1925). However, it is only in the past 20 years that they have been studied in more detail.

The simplest multiple emulsions, sometimes called "double emulsions," are actually ternary systems because they have either a water-in-oil-in-water (W/O/W) or an oil-in-water-in-oil (O/W/O) structure in which the dispersed droplets contain smaller droplets of a different phase. A schematic representation of W/O/W and O/W/O double emulsion droplets are provided in Figure 8.1.

Multiple emulsions are widely used to encapsulate active ingredients in a large number of applications, including drug delivery (Davis et al., 1987; Lutz and Aserin, 2008; Nakano, 2000), foods (Lobato-Calleros et al., 2006; Weiss et al., 2005), cosmetics (Akhtar and Yazan, 2008; Yener and Baitokova, 2006), chemical separations (Chakraborty et al., 2006), and syntheses of microspheres

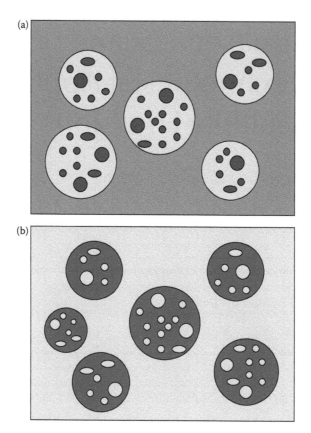

FIGURE 8.1 Schematic representation of (a) W/O/W and (b) O/W/O double emulsion droplets.

and microcapsules (Bocanegra et al., 2005; Couvreur et al., 1997; Esposito et al., 1996a, 1997; Koo et al., 2006; Lorenceau et al., 2005; Nie et al., 2005; Rizkalla et al., 2006; Utada et al., 2005; Zoldesi and Imhof, 2005). These systems have some advantages, such as the protection of the entrapped substances (Rizkalla et al., 2006; Zoldesi and Imhof, 2005) and the incorporation of several actives in different compartments (Koo et al., 2006). The pharmaceutical applications of multiple emulsions include vaccine adjuvants (Jiao and Burgess, 2008; Jiao et al., 2002), red blood cell substitutes (Zheng et al., 1993), lymphatic drug-targeting vehicles (Charman and Stella, 1992; Shively, 2002; Silva-Cunha et al., 1997), prolonged drug delivery systems (Kajita et al., 2000; Lutz and Aserin, 2008; Omotosho et al., 1986; Oza and Frank, 1989; Vaziri and Warburton, 1994), and sorbent reservoirs in drug overdose treatment (Jiao et al., 2002; Khan et al., 2006). Moreover, among pharmaceutical applications of multiple emulsions, their use in the biotechnological field must be emphasized (Bozkir and Hayta, 2004; Bozkir and Saka, 2008; Bozkir et al., 2004; Goto et al., 1995; Kim et al., 2006; McClements et al., 2007; Silva-Cunha et al., 1998). In particular, technological advances in drug development and biological sciences are allowing for the rapid development of new diagnostic methods and drugs based on biological molecules, including proteins and nucleic acids (Bozkir and Saka, 2008; Couvreur et al., 1997; Shively, 2002). In addition, multiple emulsions have been used as the basis of liposome-like lipid vesicles (Nounou et al., 2008) and as a mean for the preparation of injectable microspheres (Couvreur et al., 1997; Dhanaraju et al., 2003).

However, despite their potential usefulness, applications of multiple emulsions have been limited because of thermodynamic instability and their complex structures (Jiao and Burgess, 2003; Law et al., 1986; Muschiolik et al., 2006).

8.2 APPLICATIONS OF MULTIPLE EMULSIONS IN PHARMACEUTICAL AND BIOTECHNOLOGICAL FIELDS

As previously stated, multiple emulsions are complex polydispersed systems where O/W and W/O emulsions coexist simultaneously and are stabilized by lipophilic and hydrophilic surfactants, respectively. The W/O/W-type multiple emulsions have greater potential application in the pharmaceutical field both as a tool to obtain novel delivery systems and as a simple delivery system than O/W/O-type multiple emulsions.

Because of the presence of the external aqueous phase, W/O/W emulsions are also characterized by a lower viscosity than W/O emulsions, thus resulting in the best handling and use of the formulation (i.e., parenteral formulations; Aserin, 2008; Khan et al., 2006). These emulsions may also be used to separate two incompatible hydrophilic substances in the inner and outer aqueous phases by the middle oil phase (i.e., vancomycin and prednisolone).

Because of their physical characteristics, multiple emulsions can be used as carriers for both hydrophilic and lipophilic drugs, depending on the type of application, route of administration, and feasibility of formulation. For therapeutic purposes, multiple emulsions can be administered through oral, topical, and parenteral routes, namely, intravenous, intraperitoneal, intramuscular, and subcutaneous.

In pharmaceuticals the basic pathway for absorption of a drug from a multiple emulsion occurs through the intestinal lymphatic system. The multiple emulsion system can be absorbed directly through intestinal macrophages or Peyer's patches or from mesenteric lymph ducts in the form of chylomicrons and lipoproteins. Because of the intestinal lymphatic absorption of the oils present in multiple emulsions, they are also used for modulating drug absorption kinetics, that is, prolonged or sustained delivery.

Among the pharmaceutical potentials of multiple emulsions, we mention the following three:

1. *Prolonged delivery:* Multiple emulsions (W/O/W) have been investigated for controlling release of different categories of drugs, especially those with short half-lives. Partitioning of drugs from the water to oil phase and then to the external aqueous phase for release in W/O/W multiple emulsions leads to the prolonged delivery of the drug. Table 8.1 summarizes drug-containing multiple emulsions that enable prolonged delivery.
2. *Bioavailability enhancement:* Multiple emulsions are able to increase the bioavailability of many drugs by passing the hepatic first pass metabolism or by protecting them within the gastrointestinal tract physiological and/or ionic or enzymatic environment where these molecules, such as proteins and peptides, are otherwise degraded (Charman and Stella, 1992). Table 8.2 outlines drug-containing multiple emulsions that enhance bioavailability.
3. *Drug targeting:* Multiple emulsions have shown great potential for targeting cytotoxic drugs (highly toxic drugs) to lymphatics. In particular, because of the uptake by the reticuloendothelial system, multiple emulsions have been used as lymphotropic carriers for drug targeting to several organs (i.e., liver and brain; Johnson et al., 2006; Shiau, 1981; Porter and Charman, 2001; Thompson et al., 1989). Table 8.3 gives an overview of the potentials in drug targeting of drug-containing multiple emulsions.

This chapter emphasizes the use of multiple emulsions in the pharmaceutical and biotechnological fields. In the wide range of potential pharmaceutical applications, as a reservoir phase inside droplets of another phase, the use of multiple emulsions to obtain taste masking, transdermal delivery (Farahmand et al., 2006; Laugel et al., 1996, 1998), prolonged drug delivery (Mishra and Pandit, 1990; Tirnaksiz and Kalsin, 2005), sorbent reservoirs in drug overdose treatment (Völkel et al., 1982, 1984), and increase of bioavailability of the drug (Kajita et al., 2000) are considered in particular. In addition, the biotechnological applications of multiple emulsions are reviewed, such as protein delivery (Cournarie et al., 2004a, 2004b; Dogru et al., 2000), enzyme immobilization

TABLE 8.1
Drugs in Multiple Emulsions Enabling Prolonged Delivery

Drug	Type of Study	Reference
Diclofenac sodium	*In vitro* and *in vivo* evaluation in rabbits	Roy and Gupta, 1993; Lindenstruth and Müller, 2004
Paracetamol	*In vitro* characterization	Aserin, 2008
6-Mercaptopurine	*In vitro* and *in vivo* in mice	Khopade and Jain, 2000
Methotrexate	*In vitro* analysis	Omotosho et al., 1986
Cytarabine	*In vitro* analysis	Kim et al., 1995
Doxorubicin	*In vitro* and *in vivo* in rats	Lin et al., 1992
5-Fluorouracil	*In vitro* and *in vivo* in rats	Omotosho et al., 1989
Chloroquine	*In vitro* release	Omotosho et al., 1990
Rifampicin	*In vitro* release	Nakhare and Vyas, 1995
Theophylline	*In vitro* release	Cole and Whateley, 1997
Nitrofunrantoin	*In vivo* study in healthy volunteers	Onyeji and Adesegun, 1995
Vancomycin	*In vitro* and *in vivo* release	Okochi and Nakano, 1997
Insulin	*In vitro* and *in vivo* evaluation in rats	Silva-Cunha et al., 1998
sCT	*In vitro* and *in vivo* evaluation in rats	Dogru et al., 2000
Iodine-131	*In vivo* in rabbits	Davis et al., 1987
Prednisolone	*In vivo* in rabbit eye	Safwat et al., 1994
Naltrexone	*In vitro* release	Brodin et al., 1978
Pentazocine	*In vitro* and *in vivo* in mice	Mishra and Pandit, 1990

TABLE 8.2
Drugs in Multiple Emulsions that Enhance Their Bioavailability

Drug	Type of Study	Reference
Insulin	*In vivo* and *in vitro* evaluation	Cournarie et al., 2004a, 2004b; Silva-Cunha et al., 1997
Griseofulvin	*In vivo* in healthy human volunteers	Onyeji et al., 1991
Vancomycin	*In vivo* in male rats	Shively, 2002
Nitrofurantoin	*In vivo* in healthy human volunteers	Onyeji and Adesegun, 1995

TABLE 8.3
Potentials of Drug Targeting of some Drugs in Multiple Emulsions

Drug	Tissue Target	Type of Study	Reference
5-Fluorouracil	Lymphatic system	*In vivo* in rats	Khopade and Jain, 2008
Bleomycin	Lymphatic system	*In vivo* in rats	Yashioka et al., 1982
Isoniazid	Lymphatic system	*In vitro* release analysis	Khopade and Jain, 1998
Rifampicin	Brain	*In vivo* in male rats	Khopade et al., 1996
Rifampicin	Lungs	*In vitro* and *in vivo*	Nakhare and Vyas, 1997
Epirubicin	Liver	*In vitro* and *in vivo* in rats	Vladisavljević et al., 2006; Hino et al., 2000

(Shively, 2002; Silva-Cunha et al., 1998), adjuvant vaccines (Verma and Jaiswal, 1997), or a gene library (Bernath et al., 2004).

8.2.1 TASTE MASKING

Humans can physiologically detect four kinds of taste: sweet, salt, sour, and bitter. Several pharmaceutical drugs have unpleasant tastes, and most of them are bitter. The major consequence of unpleasant taste is insufficient compliance from the patients and especially from infants, children, and elderly. Bitterness is detected by taste buds at the back of the tongue. To avoid this sensation, it is possible to biologically influence the taste buds by numbing them with an anesthetizing agent or cooling them with menthol. Modifying the apparent taste of the drug by changing its formulation is another possibility.

Multiple emulsions have been proposed as for potential taste masking of bitter drugs (Garti et al., 1983). In particular, hydrophilic bitter drugs have been included in the inner aqueous phase of W/O/W multiple emulsions obtaining a significant taste masking, such as in the case of the bitter antipsychotic drug chlorpromazine (Garti et al., 1983) and the antimalarial drug chloroquine (Rao and Bader, 1993; Vaziri and Warburton, 1994).

Chloroquine is widely available, cheap, well tolerated, and well absorbed orally. However, the unpleasant taste is a problem in oral administration, particularly in children. In particular for chloroquine, "taste masked and controlled release" formulations, such as multiple emulsions, were designed to release the drug through the oil phase in the presence of gastrointestinal fluid. Multiple emulsions containing chloroquine have been prepared by dissolving the drug in the inner aqueous phase of a W/O/W emulsion with good self-stability (Chu et al., 2007; Higashi et al., 1995; Hou and Papadopoulos, 1997). Multiple W/O/W emulsions of chloroquine phosphate showed good stability due to interfacial polymerization or complexion between molecules. Prolonged storage (4 months) of the emulsion resulted in a negligible loss of chloroquine phosphate. This result was ascribed to the diffusion of chloroquine phosphate from the internal globules and not as a consequence of the instability of the W/O/W emulsion. Release assessments showed faster rates for W/O/W emulsions characterized by smaller internal aqueous globules and therefore increased interfacial area.

8.2.2 TOPICAL AND TRANSDERMAL DELIVERY

Emulsions are the most useful systems for the topical applications of drugs because of the presence of both oil and water phases. With respect to simple emulsions, W/O/W multiple emulsions provide the advantage of having an external aqueous phase where any W/O-type emulsion can be incorporated in water, thus overcoming the shortcomings of W/O emulsions; but they also obtain the slow and controlled release of drugs. Multiple emulsion based formulations can be used for different purposes, such as nutritive, moisturizing, and protective in cosmetics (Yazan et al., 1993). Long-term stability multiple emulsions can overcome the stability problems of emulsions (Chen et al., 1999; De Luca et al., 2000; Muguet et al., 2001). For instance, to increase the stability of multiple emulsions, the synergistic interaction between the low hydrophilic–lipophilic balance emulsifier and the high hydrophilic–lipophilic balance surfactant was investigated. These two surfactants produced long-term stability in multiple emulsions due to the very low interfacial tension at the oil–water interface (Vasudevan and Naser, 2002).

As a three phase system, multiple emulsions are very interesting as delivery systems for topical applications because of their ability to obtain slow release of their contents in comparison to solutions (Kundu, 1990). However, only a few studies on topical applications of multiple emulsions have been reported in the literature. Bonina et al. (1992) developed a multiple emulsion containing testosterone, caffeine, and tritiated water for topical use. The effect of perfluoro-polymethyl-lisopropyl-ether (Fomblin HC01) on percutaneous absorption was also investigated. Fomblin did not affect the flux of caffeine, but it decreased the percutaneous absorption of testosterone and increased water permeation. Another topical W/O/W emulsion prepared by Raynal and colleagues (1993) contained an active substance in each of the phases: a moisturizing agent (sodium lactate) in the inner phase, an antiacne agent (spironolactone) in the oily phase, and an antibacterial agent (chlorhexidine digluconate) in the outer phase. The study of the release of these substances demonstrated that two simultaneous release mechanisms could exist, such as a breakdown of the oily membrane followed by the expulsion of the encapsulated substance and a diffusion of this substance through the oily membrane. Ferreira and coworkers (1994, 1995a, 1995b) developed a W/O/W multiple emulsion containing metronidazole and glucose and compared their percutaneous release with W/O and O/W emulsions. They found that percutaneous absorption of metronidazole was similar from W/O/W and O/W emulsions, whereas it was lower from the W/O emulsion. The glucose percutaneous absorption was in the following order: O/W > W/O/W > W/O. Multiple emulsions have been employed to increase the stability of many drugs for topical purposes, such as ascorbic acid (Farahmand et al., 2006) as well as dihydralazine and hydrocortisone (Laugel et al., 1996, 1998). Among W/O/W multiple emulsions, a successful preparation for vaginal application against three microbial strains (*Escherichia coli*, *Staphylococcus aureus*, and *Candida albicans*; Tedajo et al., 2002) was developed. These multiple emulsions contained lactic acid within the internal aqueous phase, octadecylamine in the oily phase, and benzalkonium chloride in the external aqueous phase. The incorporation of chlorhexidine digluconate within the internal aqueous phase of the multiple emulsions led to an enhancement of the antimicrobial spectrum of this formulation (Tedajo et al., 2005).

8.2.3 Prolonged or Controlled Drug Delivery

Multiple W/O/W emulsion systems can be utilized as potential, controlled, and prolonged release dosage forms. The potential of using multiple emulsions as controlled release systems has been described by several research groups (Aserin, 2008; Lindenstruth and Müller, 2004; Okochi and Nakano, 1997; Omotosho et al., 1986, 1990; Roy and Gupta, 1993). Many categories of drugs have been considered in particular (see Table 8.1). The journey of drugs from the internal to external phase across the liquid membrane can occur in many ways. The release kinetics from multiple emulsion systems are particularly affected by various factors such as droplet size, pH, phase volume, and viscosity. There are several possible mechanisms for drug release across the liquid oil phase (Florence and Whitehill, 1982). Rather than a single mechanism, a combination of various mechanisms is responsible for drug release; hence, the exact release mechanism remains unclear. Plausible mechanisms for the drug release include a diffusion mechanism, micellar transport, thinning of the oil phase, rupture of the oil phase, facilitated diffusion (i.e., carrier-mediated transport), photo-osmotic transport, and solubilization of the internal phase in the oil membrane.

Prolonged delivery after oral or parenteral administration has been investigated (Dogru et al., 2000; Mishra and Pandit, 1990; Okochi and Nakano, 1997; Silva-Cunha et al., 1998). In parenteral delivery, multiple emulsions act as reservoirs in the blood. In topical applications, stable W/O/W multiple emulsions were developed by Tirnaksiz and Kalsin (2005). This multiple emulsion contains poloxamine 908 as a hydrophilic surfactant and cetyl dimethicone copolyol as a lipophilic surfactant. Caffeine was used as a water-soluble model. The investigations suggested that poloxamine 908 could be used as a hydrophilic surfactant for the formulation of W/O/W multiple emulsions. The concentration of poloxamine 908 was a particularly important parameter in preparing stable multiple emulsions. They concluded that the concentration of surfactant affected the release rate and caffeine could be transported out by molecular diffusion and through a reverse micellar mechanism controlled by the viscosity of the system.

8.2.4 Delivery of Protein

It is well known that enzymes and proteins are characterized by a high chemical and physical sensitivity toward the environment. The physical instability of conventional systems remains a major factor limiting their wider application. A well-recognized structure providing protection of proteins is a system that behaves as a membrane solid (i.e., microcapsule) or liquid (i.e., multiple emulsion). Multiple emulsions are systems in which a true liquid phase is separately maintained from an external aqueous phase. Multiple emulsions therefore provide an alternative technique to encapsulate protein and other materials that would otherwise be metabolized, rapidly cleared, or toxic to the patient. This may be especially important for bioactive molecules that cannot be appropriately stabilized in the solid state.

Multiple emulsions have been utilized for parenteral and oral administration (Brodin et al., 1978). Although parenteral administration is currently the most widely used route of administration for proteins, there are situations in which the protection or microenvironment provided by multiple emulsions may be desirable, such as the increase of the circulation half-life of the drug. The preparation of a hemoglobin-containing multiple emulsion as a blood substitute is one such example (Davis et al., 1984; Zheng et al., 1993). Hemoglobin is quickly phagocytosed by reticuloendothelial cells throughout the body, and thus a delivery system able to carry oxygen for a longer period of time is desirable. With this view, hemoglobin was incorporated in the inner phase of the emulsion. A stable hemoglobin multiple emulsion simulating red blood cell properties has been reported, in which gases (O_2 and CO_2) are exchanged with hemoglobin. However, there are no reports for multiple emulsions investigated for this purpose in the last decade. The challenges in this are the artificiality, immune response, and questionable *in vivo* compatibility of chemicals (Borwanker et al., 1988; Zheng et al., 1991).

The W/O/W multiple emulsions are systems of potential interest in the oral administration of insulin. Although a single oral administration of an insulin-loaded W/O/W multiple emulsion to diabetic rats led to the significant decrease of blood glucose levels (Silva-Cunha et al., 1998), repeated administrations displayed unpleasant side effects such as diarrhea and steatosis. These unwanted effects were attributed to the high oil concentration used for their preparation. Cournarie et al. (2004a) focused their attention on the reduction of the oil concentration in the formulation of these systems and on the encapsulation of two different insulins. The physical properties and stability of multiple emulsions over long periods of time were assessed by conductivity measurements, granulometric analysis, and microscopic analysis. The encapsulation in the inner aqueous phase of two insulins (umulin and humalog) had non-negligible effects on the formation and stability of W/O/W multiple emulsions. Both insulins improved the formation of multiple emulsions. Circular dichroism studies and surface tension measurements evidenced the contribution of the insulin conformation and surface properties in the multiple emulsion formation and stability. Cournarie et al. (2004b) demonstrated in another study that a stable unloaded multiple emulsion with low fish oil content could be obtained after reducing the viscosity of the primary emulsion and presented almost similar characteristics to those of a multiple emulsion prepared from medium-chain triglycerides. However, incorporation of insulin had opposite effects on the formation and stability of multiple emulsions, which can be explained by the differences in the interaction of insulin at the oil–surfactant–water interface, depending on the nature of the oil. Obviously, medium-chain triglycerides allow the formation of a more stable insulin multiple emulsion than fish oil. However, to combine the advantages of both oils (i.e., small globule formation for medium-chain triglycerides and beneficial therapeutic effects for fish oil), the mixing of the two oils in the preparation of multiple emulsions can be considered.

Among the proposed oral delivery of protein, another example is typified by the administration of calcitonin. Salmon calcitonin (sCT) is a polypeptide hormone consisting of 32 amino acid residues, which can be successfully used for the treatment of osteoporosis, Paget disease, and hypercalcemia. Only nasal and parenteral preparations of sCT are currently available. Oral sCT is poorly bioavailable and is susceptible to enzymatic degradation in the gastrointestinal tract; thus, a W/O/W multiple emulsion formulation was designed for its oral application by Dogru et al. (2000). They placed sCT in the inner water phase and included aprotinin in the outer water phase of this system to investigate the influence of protease inhibitors in the presence of sCT. The effectiveness of the formulation was evaluated *in vitro* and *in vivo* by using a rat model. They found that the incorporation of sCT in the inner aqueous phase of this multiple emulsion appears to protect the peptide from enzymatic degradation. In addition, sCT was protected from the protease inhibitor present in the outer aqueous phase, allowing W/O/W emulsion formulations to be promising carrier systems for peptide protein drugs (Dogru et al., 2000).

Shahiwala and Amiji (2008) developed and evaluated squalane oil containing W/O/W multiple emulsion for mucosal administration of ovalbumin as a model candidate vaccine in BALB/c mice. Controlled and optimized ovalbumin-containing W/O/W emulsion and chitosan-modified W/O/W emulsion formulations were administered intranasally and orally. They found that the ovalbumin-containing W/O/W emulsions resulted in higher immunoglobulin G (IgG) and immunoglobulin A (IgA) responses compared to the aqueous solution. In addition, significant IgG and IgA responses were observed after the second immunization dose using the emulsions with both routes of administration. Intranasal vaccination was more effective in generating the systemic ovalbumin-specific IgG response than the mucosal ovalbumin-specific IgA response. In contrast, oral immunizations showed much higher systemic IgG and mucosal IgA responses compared to the nasally treated groups. The results of this study show that squalane oil containing W/O/W multiple emulsion formulations can significantly enhance the local and systemic immune responses, especially after oral administration, and thus may be proposed as a better alternative in mucosal delivery of prophylactic and therapeutic vaccines.

8.2.5 Enzyme Immobilization

The enzyme immobilization technique involves the entrapment of enzymes. The utility of natural enzymes in industrial processes is limited by their tendency to denature and become inactivated when exposed to organic solvents. However, the use of enzymes in organic media has attracted many researchers interested in reaching effective biocatalysis (Dordick, 1989; Rao et al., 1998; Zaks and Klivanov, 1988). Biocatalysis in organic media has enabled the conversion of hydrophobic compounds and produced favorable shifts in reaction equilibria. To obtain an effective use of enzymatic catalysis, it is necessary to modify the enzyme to protect it from a hazardous environment in organic media. To overcome these drawbacks, a surfactant-coated enzyme immobilized in poly(ethylene glycol) (PEG) microcapsules was developed for the reuse of an oil-soluble enzyme in organic media. This immobilization method is very easy, and all oil-soluble enzymes can be entrapped in the microcapsules without loss of the enzyme. The esterification rate of the surfactant-coated lipase immobilized in these microcapsules was 30-fold higher than that of the powder lipase. In addition, more than 90% of the enzymatic activity of the encapsulated lipases was maintained after recycling them six times (Goto et al., 1995).

The use of multiple emulsions for enzyme immobilization has been reported in the literature since 1972 when hydrocarbon-based multiple emulsions were used to entrap the enzyme urease for the treatment of kidney diseases (May and Li, 1972). Multiple emulsions have become a new tool in the field of biotechnology for the immobilization of many enzymes, proteins, and amino acids (Okochi and Nakano, 1997; Shively, 2002; Silva-Cunha et al., 1998). For instance, immobilized alcohol dehydrogenase was used to convert alcohol to acetaldehyde at the industrial level (May and Landgraff, 1976). Makryaleas and coworkers (1985) carried out the conversion of ketoisocaproate to L-leucine by the immobilization of L-leucine dehydrogenase. Moreover, the enzymatic conversion of highly lipophilic, water-insoluble substrates, such as steroids (Laugel et al., 1998; Safwat et al., 1994), was carried out. In addition, the enzymatic production of L-phenylalanine from immobilized chymotrypsin (Scheper, 1990) and the production of hydrolyzed fatty acids by immobilized lipase enzyme (Iso et al., 1989) were also reported.

8.2.6 Adjuvant Vaccines

Vaccination is one of the most successful achievements of medical science. Routine administration of vaccines is very effective in preventing a number of infectious diseases. The principle at the basis of a vaccine is to mimic an infection as the natural specific mechanism of the host against the pathogen while the host remains free of the disease that normally results from a natural infection. The success of vaccination relies on the induction of a long-lasting immunological memory.

Vaccines are pharmaceuticals that traditionally consist of live attenuated pathogens, whole inactivated organisms, inactivated toxins, or toxoids that are able to prevent infectious diseases by immunization. Unfortunately, modern vaccines (i.e., biosynthetic, recombinants, or others) are often poorly immunogenic, so there is a need in developing potent and safe adjuvants.

Adjuvants are defined as any material able to increase the humoral and cellular responses against an antigen. The success of an immunization depends on the nature of the protective components, their presentation form, the presence of adjuvants, and the route of administration. Many compounds with adjuvant activity are known, but only a few of them are applied routinely in human and veterinary vaccines.

The properties of multiple emulsions, such as low viscosity, easy injection, and a stable formulation, appear to offer significant improvement as delivery vehicles for immunization. The use of a W/O/W multiple emulsion as a new form of adjuvant for antigen was initially reported by Herbert (1965), who incorporated an antigen into the inner water phase to produce a high and prolonged antibody response. These emulsions elicited a better immune response compared to the sole antigen. In addition, these antigenic multiple emulsions could be stored without evidence of breakdown for several months at 56°C and room temperature. Other researchers developed a multiple emulsion vaccine against *Pasteurella multocida* infection in cattle (Verma and Jaiswal, 1997). They found that the vaccine was able to provide protection against the infection activating both humoral and cell-mediated immune responses and that this multiple emulsion based vaccine could be successfully used in the control of hemorrhagic septicemia. Multiple W/O/W emulsion formulations containing influenza virus surface antigen hemagglutinin were recently prepared and characterized *in vitro* and *in vivo* (Bozkir and Hayta, 2004; Bozkir et al., 2004). Results suggested that multiple emulsions carrying the influenza antigen have advantages over conventional preparations and can be effectively used as one of the vaccine delivery systems with adjuvant properties.

8.2.7 OTHER USES

The liquid membrane system was efficaciously used for drug overdosage treatment as early as 1978 (Chiang et al., 1978; Morimoto et al., 1979). This system could be employed for overdosage treatment by utilizing the difference in the pH within the compartments of multiple emulsions. For instance, multiple emulsions have been utilized for acidic drug overdosage treatment, such as barbiturates or quinine sulfate. In these emulsions, the inner aqueous phase of the emulsion has the basic buffer; when the emulsion is taken orally, the acidic pH of the stomach acts as an external aqueous phase. In the acidic phase barbiturate remains mainly in the nonionized form, which transfers through the oil membrane into the inner aqueous phase and becomes ionized. An ionized drug has less affinity to cross the oil membrane and thereby gets entrapped. Thus, entrapping excess drug in multiple emulsions leads to the treatment of overdosage.

Multiple emulsions were recently described for a gene library (Bernath et al., 2004). In this study W/O/W multiple emulsions were explored as compartmentalization systems. The W/O/W emulsions allow the creation of an external aqueous phase without the alteration of the aqueous droplets embedded in the primary W/O emulsion. The aqueous droplets of the primary W/O emulsion acts as cell-like compartments; thus, genes can be transcribed and translated in the aqueous droplets (Griffiths and Tawfik, 2003). Subsequent conversion of the primary W/O emulsion into a W/O/W (double) emulsion makes the emulsion amenable to sorting by flow cytometry without compromising the integrity of the inner aqueous droplets within the oil phase.

Detoxification of blood by multiple emulsions has also been reported (Völkel et al., 1982, 1984).

8.3 METHODS OF MICRO- AND NANOENCAPSULATION INVOLVING MULTIPLE EMULSIONS

Multiple emulsions have been recognized as an intermediate step in the formulation of microspheres, nanospheres, nanoparticles, microcapsules, and so forth. They have been extensively used as intermediate steps for encapsulating many drugs (Adachi et al., 2003; Dhanaraju et al., 2003; Erden and Celebi, 1996; Kar and Choudhury, 2007; Lee et al., 2001; Meng et al., 2003; Ogawa et al., 1988; Shah et al., 1987). The basic technique for microencapsulation involves a two-step emulsification method. In addition, microparticles can be obtained by gelation of the external phase using different methods, such as crosslinking by γ-radiation, heat denaturation, nonsolvent addition and cross-linking, and addition of salts or alcohols. Moreover, in the solvent evaporation method, multiple emulsion formation is an intermediate step that gives rise to the final product. A novel technique of O/W/O double emulsion solvent diffusion was used by Lee et al. (2001) and Bodmeier et al. (1992) to encapsulate hydrophilic compounds, such as theophylline, propranolol, acetaminophen, and tacrine. Microparticles of pseudoephedrine were successfully developed using a multiple emulsion melt dispersion technique with encapsulation efficiencies of more than 80% (Bodmeier et al., 1992). Microencapsulation of different categories of drugs, such as enzymes, hormones, peptides, and synthetic drugs, has also been reported (Couvreur, 1997; Esposito, 1996a, 1996b, 1997). Different polymers [ethylcellulose, eudragit, poly(lactic acid), poly(glycolic acid), etc.] have been employed on the basis of the nature of the drug and intended properties of microcapsules.

Esposito et al. (1996a, 1997) described the production and characterization of biodegradable microparticles containing tetracycline, which was designed for periodontal disease therapies. The influence of production parameters on microparticle characteristics and antibiotic release modality was studied. Microparticles were made by using different preparation procedures and different polyesters: poly(L-lactide) (L-PLA), poly(D,L-lactide) (DL-PLA), and 50:50 poly(D,L-lactide-*co*-glycolide) (DL-PLG). A double emulsion preparation method together with a concentrated salt solution as the external phase gave the best result in terms of the efficacy of tetracycline incorporation. Because the morphology of microparticles reasonably influences the release of the encapsulated drug, the external and internal structure of DL-PLG microparticles was investigated by optical and scanning electron microscopy. The structure of the W/O/W double emulsion was analyzed during the evaporation process by optical microscopy at 5 min after the formation of the secondary emulsion (Figure 8.2a) and after 3 h (Figure 8.2b) and immediately before particle isolation (Figure 8.2c). It is evident that at the beginning of the evaporation process (Figure 8.2a) the

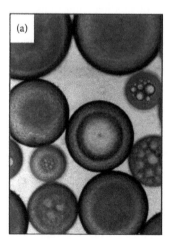  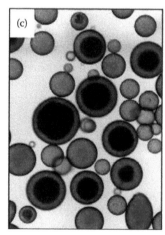

FIGURE 8.2 Structure of W/O/W double emulsion analyzed during the evaporation process. Optical microphotographs taken (a) after 5 min from the formation of the secondary emulsion, (b) after 3 h, and (c) immediately before particle isolation.

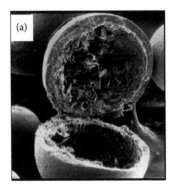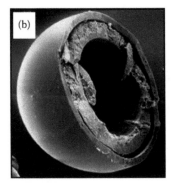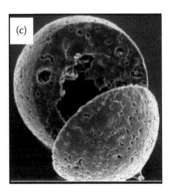

FIGURE 8.3 Scanning electron microscopy microphotographs of (a) DL-PLGA, (b) DL-PLA, and (c) L-PLA microparticles obtained by the W/O/W double emulsion method.

primary W/O emulsion droplets, representing the microparticle precursors, show a few or single inner aqueous droplets. These morphologies represent the respective partial or complete coalescence of the inner aqueous phase. At the end of the evaporation process, the microparticle structure still reflects this situation, displaying a microcapsular structure. Tetracycline crystals are clearly evident in the microcapsular core (Figure 8.2b). The microcapsular structure was confirmed by scanning electron microscopy analysis. Figure 8.3 provides microphotographs of DL-PLG, DL-PLA, and L-PLA microparticles. The DL-PLG microparticles display a smooth surface with only few pores, whereas a rather different morphology was found for microparticles made of DL- and L-PLA.

A double-walled microcapsular system is evident in DL-PLA (Figure 8.3b). The DL-PLA microparticles comprise an external thin and compact layer and an inner thicker layer characterized by a porous matrix. In contrast, L-PLA particles are matrix-type microspheres with high porosity (Figure 8.3c).

In vitro release experiments demonstrated that tetracycline is slowly and appropriately released from microparticles. The release kinetics were influenced by the type of polymer utilized for microparticle production. *In vitro* experiments simulating *in vivo* conditions evidenced that after 30 days only DL-PLG microparticles showed significant changes in their morphology, whereas L-PLA and DL-PLA were almost intact after the same period of time.

The double emulsion solvent evaporation method based on a W/O/W emulsion has been popular to encapsulate various proteins within poly(D,L-lactic-*co*-glycolic acid) (PLGA) microspheres (Cleland and Jones, 1996). Nevertheless, this method elicits drawbacks related to protein stability during the encapsulation process. In particular, protein molecules dissolved in an inner water phase irreversibly aggregate to a great extent as they are continuously exposed onto an interface between the water and oil phases (Maa and Hsu, 1997; Maa et al., 1998). It is conceivable that unfolded protein molecules at the interface form undissociable aggregates (Bam et al., 1998).

For the sustained release formulation of recombinant human growth hormone (rhGH), dissociable rhGH aggregates were microencapsulated by Kim and Park (2001) within PLGA microparticles. The rhGH aggregates were microencapsulated within the PLGA polymer phase by extracting ethyl acetate into an aqueous phase presaturated with ethyl acetate. The released rhGH species were mostly monomeric and had a correct conformation. Monomethoxy-PEG–*b*–DL-PLA microspheres containing bovine hemoglobin (BHb, a model protein) were prepared by four solvent removal methods (Meng et al., 2004). The BHb encapsulated by the W/O/W double emulsion solvent diffusion method with ethyl acetate as the organic solvent displayed bioactivity near to that of native BHb. The efficiency of BHb entrapment achieved by this method was much higher than that by other methods (~90% vs. 30%). Another work by Mok et al. (2007) described the microencapsulation of naked adenovirus (ADV) or PEGylated ADV (PEG-ADV) within PLGA microspheres using a W/O/W double emulsion and solvent evaporation method. Green fluorescent protein transfection efficiencies into HeLa cells were quantified, and the relative extent of the immune response for ADV and PEG-ADV encapsulated within PLGA microspheres was analyzed using macrophage cells. PEG-ADV can be more safely microencapsulated within PLGA microspheres than naked ADV, because of their enhanced physical stability under the harsh formulation conditions and acidic microenvironmental conditions of the microsphere, thereby increasing gene transfection efficiency.

Moreover, multiple emulsions have been used to prepare three kinds of hepatitis B surface antigen–PLGA microspheres. These microspheres showed greater antibody response in mice in comparison to the conventional aluminum-adjuvant vaccine and thus hold promise for controlled delivery of a vaccine (Feng et al., 2006). Tetanus toxoid loaded poly(lactic acid) particles were prepared by the double emulsion technique, and they elicited high and sustained antibody titers after intramuscular immunization (Katare and Panda, 2006).

For the use of a double emulsion in nanoparticle production, Pandey and Khuller (2007) developed PLG nanoparticles encapsulating streptomycin by the multiple emulsion technique and administered them orally to mice for biodistribution and chemotherapeutic studies. There was a 21-fold increase in the relative bioavailability of PLG-encapsulated streptomycin compared with the intramuscular free drug. Further, the nanoparticle formulation did not result in nephrotoxicity assessed biochemically (Pandey et al., 2003). Nanoparticles based on the double emulsion method (W/O/W) were also investigated by Zambaux et al. (1998) using methylene chloride as an organic solvent and poly(vinyl alcohol) (PVA) or human serum albumin as a surfactant. It appeared that the higher the surfactant concentration in the external aqueous phase was, the smaller the particles, the lower the polydispersity index, and the higher the residual amount of surfactant.

Sahoo et al. (2002) formulated PLGA nanoparticles by a multiple emulsion solvent evaporation technique using PVA as an emulsifier and bovine serum albumin as a model protein. They studied the parameters that influence the amount of residual PVA associated with PLGA nanoparticles and its effect on the physical properties and cellular uptake of nanoparticles. They found that residual PVA influenced different pharmaceutical properties of the nanoparticles, such as the particle size, zeta potential, polydispersity index, surface hydrophobicity, protein loading, and *in vitro* release of the encapsulated protein.

Finally, Prabha and Labhasetwar (2004) prepared nanoparticle-containing plasmid deoxyribonucleic acid that carries a wild *p53* gene using a multiple emulsion solvent evaporation technique. This formulation resulted in sustained antiproliferative activity against breast cancer cells that would normally require repeated delivery of the gene. This technique was recently utilized for efficiently encapsulating deoxyribonucleic acid for preparing polymeric nanoparticles with intended characteristics and hence can be a potential tool in the era of nanotechnology in the future (Rizkalla et al., 2006).

8.4   CONCLUSIONS

This chapter demonstrated the use of multiple emulsions for drug delivery and drug targeting.

Multiple emulsions, especially of the W/O/W type, have the possibility of numerous applications in both the pharmaceutical and biotechnological fields. In addition, multiple emulsions are equally suitable for hydrophilic and lipophilic drugs. The availability of newer emulsion technologies has greatly assisted the exploration of liquid membrane systems for both oral and parenteral applications. With an inherent versatility in carrier design, they have shown good results for bioavailability enhancement, prolonged release, taste masking, detoxification, enzyme immobilization, vaccine adjuvants, or gene libraries. They are also a well-stabilized method of production for both micro- and nanoparticles.

Many applications need to be realized, even if the stability of multiple emulsions need to be fully understood and approaches to stabilize multiple emulsions fully rationalized. For instance, the addition of complexing, gelling, and polymeric agents can be considered to obtain more stable systems. Moreover, some specific studies concerning the acquisition of more detailed information about the emulsion have to be investigated.

Combining these various potentials, we expect that there will be a future increase in the use of multiple emulsions for both pharmaceutical and biotechnological applications.

ACKNOWLEDGMENT

We acknowledge the Regione Emilia Romagna (Spinner Project) of Italy for its financial support.

ABBREVIATIONS

ADV	Adenovirus
BHb	Bovine hemoglobin
DL-PLA	Poly(D,L-lactide)
DL-PLG	50:50 Poly(D,L-lactide-*co*-glycolide)
IgA	Immunoglobulin A
IgG	Immunoglobulin G
L-PLA	Poly(L-lactide)
O/W	Oil-in-water
O/W/O	Oil-in-water-in-oil
PEG	Poly(ethylene glycol)
PEG-ADV	PEGylated adenovirus
PLGA	Poly(D,L-lactic-*co*-glycolic acid)
PVA	Poly(vinyl alcohol)
rhGH	Recombinant human growth hormone
sCT	Salmon calcitonin
W/O	Water-in-oil
W/O/W	Water-in-oil-in-water

REFERENCES

Adachi, S., Imaoka, H., Hasegawa, Y., and Matsuno, R. 2003. Preparation of a water-in-oil-in-water (w/o/w) type microcapsules by a single-droplet-drying method and change in encapsulation efficiency of a hydrophilic substance during storage. *Biosci. Biotechnol. Biochem.* 67: 1376–1381.

Akhtar, N. and Yazan, Y. 2008. Formulation and *in-vivo* evaluation of a cosmetic multiple emulsion containing vitamin C and wheat protein. *Pak. J. Pharm. Sci.* 21: 45–50.

Aserin, A. 2008. *Multiple Emulsions: Technology and Applications.* New York: Wiley.

Bam, N. B., Cleland, J. L., Yang, J., Manning, M. C., Carpenter, J. F., Kelley, R. F., and Randolph, T. W. 1998. Tween protects recombinant human growth hormone against agitation-induced damage via hydrophobic interactions. *J. Pharm. Sci.* 87: 1554–1559.

Bernath, K., Hai, M., Mastrobattista, E., Griffiths, A. D., Magdassi, S., and Tawfik, D. S. 2004. *In vitro* compartmentalization by double emulsions: Sorting and gene enrichment by fluorescence activated cell sorting. *Anal. Biochem.* 325: 151–157.

Bocanegra, R., Sampedro, J. L., Ganan-Calvo, A., and Marquez, M. 2005. Monodisperse structured multi-vesicle microencapsulation using flow-focusing and controlled disturbance. *J. Microencapsul.* 22: 745–759.

Bodmeier, R., Wang, J., and Bhagwatwar, H. J. 1992. Process and formulation variables in the preparation of wax microparticles by a melt dispersion technique. I. Oil-in-water technique for water-insoluble drugs. *J. Microencapsul.* 9: 89–98.

Bonina, F., Bader, S., Montenegro, L., Scrofani, C., and Visca, M. 1992. Three phase emulsions for controlled delivery in the cosmetic field. *Int. J. Cosmet. Sci.* 14: 65–74.

Borwanker, C. M., Pfeifer, S. B., Zheng, S., Beissinger, R. L., Wasan, D. T., Sehgal, L. R., and Rosen, A. L. 1988. Formulation and characterization of a multiple emulsion for use as a red blood cell substitute. *Biotechnol. Progr.* 4: 210–217.

Bozkir, A. and Hayta, G. 2004. Preparation and evaluation of multiple emulsions water-in-oil-in-water (W/O/W) as delivery system for influenza virus antigens. *J. Drug Target.* 12: 157–164.

Bozkir, A., Hayta, G., and Saka, O. M. 2004. Comparison of biodegradable nanoparticles and multiple emulsions water-in-oil-in-water containing influenza virus antigen on the *in vivo* immune response in rats. *Pharmazie* 59: 723–725.

Bozkir, A. and Saka, O. M. 2008. Multiple emulsions: Delivery system for antigens. In *Multiple Emulsions: Technology and Applications*, A. Aserin (Ed.), pp. 293–306. New York: Wiley.

Brodin, A. F., Kavaliunas, D. R., and Frank, S. G. 1978. Prolonged drug release from multiple emulsions. *Acta Pharm. Suec.* 15: 1–2.

Chakraborty, M., Ivanova-Mitseva, P., and Bart, H. J. 2006. Selective separation of toluene from *n*-heptane via emulsion liquid membranes containing substituted cyclodextrins as carrier. *Sep. Sci. Technol.* 41: 3539–3552.

Charman, W. N. and Stella, V. J. 1992. *Lymphatic Transports of Drugs*. Boca Raton, FL: CRC Press.

Chen, C. C., Tu, Y. Y., and Chang, H. M. 1999. Efficiency and protective effect of encapsulation of milk immunoglobulin G in multiple emulsion. *J. Agric. Food Chem.* 47: 407–410.

Chiang, C., Fuller, G. C., Frankenfeld, J. W., and Rhodes, C. T. 1978. Investigations of the potential of liquid membranes for the treatment of drug overdose—In vitro studies. *J. Pharm. Sci.* 67: 63–66.

Chu, L. Y., Utada, A. S., Shah, R. K., Kim, J. W., and Weitz, D. A. 2007. Controllable monodisperse multiple emulsions. *Angew. Chem. Int. Ed.* 46: 8970–8974.

Cleland, J. L. and Jones, A. J. S. 1996. Stable formulations of recombinant human growth hormone and interferon-γ for microencapsulation in biodegradable microspheres. *Pharm. Res.* 13: 1464–1475.

Cole, M. L. and Whateley, T. L. 1997. Release rate profiles of theophylline and insulin from stable multiple w/o/w emulsions. *J. Control. Release* 49: 51–58.

Cournarie, F., Rosilio, V., Cheron, M., Vauthier, C., Lacour, B., Grossiord J. L., and Seiller, M. 2004a. Improved formulation of w/o/w multiple emulsion for insulin encapsulation. Influence of the chemical structure of insulin. *Colloid Polym. Sci.* 282: 562–568.

Cournarie, F., Savelli, M. P., Rosilio, V., Bretez, F., Vauthier, C., Grossiord, J. L., and Seiller, M. 2004b. Insulin-loaded W/O/W multiple emulsions: Comparison of the performances of systems prepared with medium-chain-triglycerides and fish oil. *Eur. J. Pharm. Biopharm.* 58: 477–482.

Couvreur, P., Blanco-Prieto, M. J., Puisieux, F., Roques, B., and Fattal, E. 1997. Multiple emulsion technology for the design of microspheres containing peptides and oligopeptides. *Adv. Drug Deliv. Rev.* 28: 85–96.

Davis, S. S. 1981. Liquid membranes and multiple emulsions. *Chem. Ind.* 3: 683–687.

Davis, S. S., Illum, L., and Walker, I. M. 1987. The *in vivo* evaluation of emulsion formulations administered intramuscularly. *Int. J. Pharm.* 38: 133–137.

Davis, T., Asher, W., and Wallace, H. 1984. Artificial red cells with crosslinked hemoglobin membranes. *Appl. Biochem. Biotechnol.* 10: 123–132.

De Luca, M., Grossiord, J. L., and Seiller, M. 2000. A very stable W/O/W multiple emulsion. *Int. J. Cosmet. Sci.* 22: 157–161.

Dhanaraju, M. D., Vema, K., Jayakumar, R., and Vamsadhara, C. 2003. Preparation and characterization of injectable microspheres of contraceptive hormones. *Int. J. Pharm.* 268: 23–29.

Dogru, S. T., Calis, S., and Oner, F. 2000. Oral multiple w/o/w emulsion formulation of a peptide salmon calcitonin: *In vitro–in vivo* evaluation. *J. Clin. Pharm. Ther.* 25: 435–443.

Dordick, J. S. 1989. Enzymatic catalysis in monophasic organic solvents. *Enzyme Microb. Technol.* 11: 194–211.

Erden, N. and Celebi, N. 1996. Factors influencing release of salbutamol sulphate from poly(lactide-*co*-glycolide) microspheres prepared by water-in-oil-in-water emulsion technique. *Int. J. Pharm.* 137: 57–66.

Esposito, E., Cortesi, R., Bortolotti, F., Menegatti, E., and Nastruzzi, C. 1996a. Production and characterization of biodegradable microparticles for the controlled delivery of proteinase inhibitors. *Int. J. Pharm.* 129: 263–273.

Esposito, E., Cortesi, R., Cervellati, F., Menegatti, E., and Nastruzzi, C. 1997. Biodegradable microparticles for sustained delivery of tetracycline to the periodontal pocket: Formulatory and drug release studies. *J. Microencapsul.* 14: 175–187.

Esposito, E., Cortesi, R., Menegatti, E., and Nastruzzi, C. 1996b. Preparation of tetracycline-containing acrylic polymer microparticles by an in-liquid drying process from an oil-in-oil system. *Pharm. Sci.* 2: 215–221.

Farahmand, S., Tajerzadeh, H., and Farboud, E. S. 2006. Formulation and evaluation of vitamin C multiple emulsion. *Pharm. Dev. Technol.* 11: 255–261.

Feng, L., Qi, X. R., Zhou, X. J., Maitani, Y., Wang, S. C., Jiang, Y., and Nagai, T. 2006. Pharmaceutical and immunological evaluation of a single-dose hepatitis B vaccine using PLGA microspheres. *J. Control. Release* 112: 35–42.

Ferreira, L. A. M., Doucet, J., Seiller, M., Grossiord, J. L., Marty, J. P., and Wepierre, J. 1995b. *In vitro* percutaneous absorption of metronidazole and glucose: Comparison of o/w, w/o/w and w/o systems. *Int. J. Pharm.* 121: 169–179.

Ferreira, L. A. M., Seiller, M., Grossiord, J. L., Marty, J. P., and Wepierre, J. 1994. Vehicle influence on *in vitro* release of metronidazole: Role of W/O/W multiple emulsion. *Int. J. Pharm.* 109: 251–259.

Ferreira, L. A. M., Seiller, M., Grossiord, J. L., Marty, J. P., and Wepierre, J. 1995a. Vehicle influence on *in vitro* release of glucose: W/O, W/O/W and O/W systems compared. *J. Control. Release* 33: 349–356.

Florence, A. T. and Whitehill, D. 1982. The formulation and stability of multiple emulsions. *Int. J. Pharm.* 11: 277–308.

Garti, N. and Aserin, A. 1996. Double emulsions stabilized by macromolecular surfactants. *Adv. Colloid Interface Sci.* 65: 37–69.

Garti, N., Frenkel, M., and Shwartz, R. 1983. Multiple emulsions: II. Proposed technique to overcome unpleasant taste of drugs. *J. Display Sci. Technol.* 4: 237–252.

Goto, M., Miyata, M., Kamiya, N., and Nakashio, F. 1995. Novel surfactant-coated enzymes immobilized in poly(ethylene glycol) microcapsules. *Biotechnol. Tech.* 9: 81–84.

Griffiths, A. D. and Tawfik, D. S. 2003. Directed evolution of an extremely fast phosphotriesterase by *in vitro* compartmentalisation. *EMBO J.* 22: 24–35.

Herbert, W. J. 1965. Multiple emulsions: A new form of mineral oil antigen adjuvant. *Lancet* 2: 771–774.

Higashi, H., Shimizu, M., Nakashima, T., Iwata, K., Uchiyama, F., Tateno, S., and Setoguchi, T. 1995. Arterial injection chemotherapy for hepatocellular carcinoma using monodispersed poppy-seed oil microdroplets containing fine aqueous vesicles of epirubicin. *Cancer* 75: 1245–1254.

Hino, T., Kwashima, Y., and Shimabayashi, S. 2000. Basic study for stabilization of w/o/w emulsion and its application to transcatheter arterial embolization therapy. *Adv. Drug Deliv. Rev.* 45: 27–45.

Hou, W. and Papadopoulos, K. Y. 1997. $W_1/O/W_2$ and $O_1/W/O_2$ globules stabilized with Span 80 and Tween 80. *Colloids Surf. A* 125: 181–187.

Iso, M., Shirahase, T., Hanamura, S., Urushiyama, S., and Omi, S. 1989. Application of encapsulated enzyme as a continuous packed-bed reactor. *J. Microencapsul.* 6: 285–299.

Jiao, J. and Burgess, D. J. 2003. Rheology and stability of water-in-oil-in-water multiple emulsions containing Span 83 and Tween 80. *AAPS PharmSci* 5: E7.

Jiao, J. and Burgess, D. J. 2008. Multiple emulsion stability: Pressure balance and interfacial film strength. In *Multiple Emulsions: Technology and Applications*, A. Aserin (Ed.). New York: Wiley.

Jiao, J., Rhodes, D. G., and Burgess, D. J. 2002. Multiple emulsion stability: Pressure balance and interfacial film strength. *J. Colloid Interface Sci.* 250: 444–450.

Johnson, L. R., Barret, K. E., Gishan, F. K., Merchant, J. L., Said, H. M., and Wood, J. D. 2006. *Physiology of the Gastrointestinal Tract*. New York: Academic Press.

Kajita, M., Morishita, M., Takayama, K., Chiba, Y., Tokiwa, S., and Nagai, T. 2000. Enhanced enteral bioavailability of vancomycin using water-in-oil-in-water multiple emulsion incorporating highly purified unsaturated fatty acid. *J. Pharm. Sci.* 89: 1243–1252.

Kar, M. and Choudhury, P. K. 2007. Formulation and evaluation of ethyl cellulose microspheres prepared by the multiple emulsion technique. *Pharmazie* 62: 122–125.

Katare, Y. K. and Panda, A. K. 2006. Influences of excipients on *in vitro* release and *in vivo* performance of tetanus toxoid loaded polymer particles. *Eur. J. Pharm. Sci.* 28: 179–188.

Khan, A. Y., Talegaonkar, S., Iqbal, Z., Ahmed, F. J., and Khar, R. K. 2006. Multiple emulsions: An overview. *Curr. Drug Deliv.* 3: 429–443.

Khopade, A. J. and Jain, N. K. 1998. A stable multiple emulsion system bearing isoniazid: Preparation and characterization. *Drug Dev. Ind. Pharm.* 24: 289–293.

Khopade, A. J. and Jain, N. K. 2000. Concanavalin—A conjugated fine-multiple emulsion loaded with 6-mercaptopurine. *Drug Deliv.* 7: 105–112.

Khopade, A. J. and Jain, N. K 2008. Surface-modified fine multiple emulsions for anticancer drug delivery. In *Multiple Emulsions: Technology and Applications*, A. Aserin (Ed.), pp. 235–256. New York: Wiley.

Khopade, A. J., Mahadik, K. R., and Jain, N. K. 1996. Enhanced brain uptake of rifampicine from W/O/W emulsions via nasal route. *Indian J. Pharm. Sci.* 58: 83–85.

Kim, C.-K., Kim, S.-C., Shin, H.-J., Kim, K. M., Oh, K.-H., Lee, Y.-B., and Oh, I.-J. 1995. Preparation and characterization of cytarabine-loaded w/o/w multiple emulsions. *Int. J. Pharm.* 124: 61–67.

Kim, H. J., Decker, E. A., and McClements, D. J. 2006. Preparation of multiple emulsions based on thermodynamic incompatibility of heat-denatured whey protein and pectin solutions. *Food Hydrocolloids* 20: 586–595.

Kim, H. K. and Park, T. G. 2001. Microencapsulation of dissociable human growth hormone aggregates within poly(D,L-lactic-*co*-glycolic acid) microparticles for sustained release. *Int. J. Pharm.* 229: 107–116.

Koo, H. Y., Chang, S. T., Choi, W. S., Park, J. H., Kim, D. Y., and Velev, O. D. 2006. Emulsion-based synthesis of reversibly swellable, magnetic nanoparticle-embedded polymer microcapsules. *Chem. Mater.* 18: 3308–3313.

Kundu, S. C. 1990. *Preparation and evaluation of multiple emulsions as controlled release topical drug delivery systems.* U.M.I., Ann Arbor, MI.

Laugel, C., Baillet, A., Youenang Piemi, M. P., Marty, J. P., and Ferrier, D. 1998. Oil-water-oil multiple emulsions for prolonged delivery of hydrocortisone after topical application: Comparison with simple emulsions. *Int. J. Pharm.* 160: 109–117.

Laugel, C., Chaminade, P., Baillet, A., Seiller, M., and Ferrier, B. 1996. Moisturizing substances entrapped in W/O/W emulsions: Analytical methodology for formulation, stability and release studies. *J. Control. Release* 38: 59–67.

Law, K., Whateley, T. L., and Florence, A. T. 1986. Stabilisation of W/O/W multiple emulsions by interfacial complexation of macromolecules and nonionic surfactants. *J. Control. Release* 3: 279–290.

Lee, M. H., Oha, S. G., Moona, S. K., and Baeb, S. Y. 2001. Preparation of silica particles encapsulating retinol using o/w/o multiple emulsions. *J. Colloid Interface Sci.* 240: 83–89.

Lin, S.-Y., Wu, W.-H., and Lui, W.-Y. 1992. *In vitro* release, pharmacokinetic and tissue distribution studies of doxorubicin hydrochloride (adriamycin HCl) encapsulated in lipiodolized w/o and w/o/w multiple emulsions. *Pharmazie* 47: 439–443.

Lindenstruth, K. and Müller, B. W. 2004. W/O/W multiple emulsions with diclofenac sodium. *Eur. J. Pharm. Biopharm.* 58: 621–627.

Lobato-Calleros, C., Rodriguez, E., Sandoval-Castilla, O., Vernon-Carter, E. J., and Alvarez-Ramirez, J. 2006. Reduced-fat white fresh cheese-like products obtained from $W_1/O/W_2$ multiple emulsions: Viscoelastic and high-resolution image analyses. *Food Res. Int.* 39: 678–685.

Lorenceau, E., Utada, A. S., Link, D. R., Cristobal, G., Joanicot, M., and Weitz, D. A. 2005. Generation of polymerosomes from double-emulsions. *Langmuir* 21: 9183–9186.

Lutz, R. and Aserin, A. 2008. Multiple emulsions stabilized by biopolymers. In *Multiple Emulsions: Technology and Applications*, A. Aserin (Ed.). New York: Wiley.

Maa, Y. F. and Hsu, C. C. 1997. Effect of primary emulsions on microspheres size and protein-loading in the double emulsion process. *J. Microencapsul.* 14: 225–241.

Maa, Y. F., Nguyen, P. A., and Hsu, C. C. 1998. Spray-drying of air–liquid interface sensitive recombinant human growth hormone. *J. Pharm. Sci.* 87: 152–159.

Makryaleas, K., Scheper, T., Schugent, K., and Kurla, M. R. 1985. Enzymatic production of L-amino acid with continuous coenzyme regeneration by liquid membrane technique. *Ger. Chem. Eng.* 8: 345–350.

May, S. W. and Landgraff, L. M. 1976. Cofactor recycling in liquid membrane–enzyme systems. *Biochem. Biophys. Res. Commun.* 68: 786–792.

May, S. W. and Li, N. N. 1972. The immobilization of urease using liquid-surfactant membranes. *Biochem. Biophys. Res. Commun.* 47: 1179–1185.

McClements, D. J., Decker, E. A., and Weiss, J. 2007. Emulsion-based delivery systems for lipophilic bioactive components. *J. Food Sci.* 72: R109–R124.

Meng, F. T., Ma, G. H., Liu, Y. D., Qiu, W., and Su, Z. G. 2004. Microencapsulation of bovine hemoglobin with high bio-activity and high entrapment efficiency using a W/O/W double emulsion technique. *Colloids Surf. B: Biointerfaces* 33: 177–183.

Meng, F. T., Ma, G. H., Qiu, W., and Su, Z. G. 2003. W/O/W double emulsion technique using ethyl acetate as organic solvent: Effects of its diffusion rate on the characteristics of microparticles. *J. Control. Release* 91: 407–416.

Mishra, B. and Pandit J. K. 1990. Prolonged tissue levels of pentazocine from multiple w/o/w emulsions in mice. *Drug Dev. Ind. Pharm.* 16: 1073–1078.

Mok, H., Park, J. W., and Park, T. G. 2007. Microencapsulation of PEGylated adenovirus within PLGA microspheres for enhanced stability and gene transfection efficiency. *Pharm. Res.* 24: 2263–2269.

Morimoto, Y., Sugibayashi, K., Yamaguchi, Y., and Kato, Y. 1979. Detoxification capacity of a multiple (w/o/w) emulsion for the treatment of drug overdose: Drug extraction into the emulsion in the gastro-intestinal tract of rabbits. *Chem. Pharm. Bull.* 27: 3188–3192.

Muguet, V., Seiller, M., Barratt, G., Ozer, O., Marty, J. P., and Grossior, J. L. 2001. Formulation of shear rate sensitive multiple emulsions. *J. Control. Release* 70: 37–49.

Muschiolik, G., Scherze, I., Preissler, P., Weiss, J., Knoth, A. and Fechner, A. 2006. Multiple emulsions—Preparation and stability. In *IUFoST: 13th World Congress on Food Science & Technology: Food Is Life*, pp. 123–137. http://iufost.edpsciences.org/index.php?option=article&access=standard&Itemid=129&url=/articles/iufost/pdf/2006/01/iufost06000043.pdf

Nakano, M. 2000. Places of emulsions in drug delivery. *Adv. Drug Deliv. Rev.* 45: 1–4.

Nakhare, S. and Vyas, S. P. 1995. Prolonged release multiple emulsion based system bearing rifampicin: *In vitro* characterisation. *Drug Dev. Ind. Pharm.* 21: 869–878.

Nakhare, S. and Vyas, S. P. 1997. Prolonged release of diclofenac sodium from multiple w/o/w emulsion systems. *Pharmazie* 52: 224–226.

Nie, Z., Xu, S., Seo, M., Lewis, P. C., and Kumacheva, E. 2005. Polymer particles with various shapes and morphologies produced in continuous microfluidic reactors. *J. Am. Chem. Soc.* 127: 8058–8063.

Nounou, M. M., El-Khordagui, L. K., Khalafallah, N. A., and Khalil, S. A. 2008. Liposomal formulation for dermal and transdermal drug delivery: Past, present and future. *Recent Pat. Drug Deliv. Formul.* 2: 9–18.

Ogawa, Y., Yamamoto, M., Okada, H., Yashiki, T., and Shimamoto, T. 1988. A new technique to efficiently entrap leuprolide acetate into microcapsules of polylactic acid or copoly(lactic/glycolic) acid. *Chem. Pharm. Bull.* 36: 1095–1103.

Okochi, H. and Nakano, M. 1997. Pharmacokinetics of vancomycin after intravenous administration of a w/o/w emulsion to rats. *Drug Deliv.* 4: 167–172.

Omotosho, J. A., Florence, A. T., and Whateley, T. L. 1990. Absorption and lymphatic uptake of 5-fluorouracil in the rat following oral administration of w/o/w multiple emulsions. *Int. J. Pharm.* 61: 51–56.

Omotosho, J. A., Whateley, T. L., and Florence, A. T. 1989. Methotrexate transport from the internal phase of multiple W/O/W emulsions. *J. Microencapsul.* 6: 183–192.

Omotosho, J. A., Whateley, T. L., Law, T., and Florence, A. T. 1986. The nature of the oil phase and the release of solutes from multiple w/o/w emulsions, *J. Pharm. Pharmacol.* 38: 865–870.

Onyeji, C. O. and Adesegun, S. A. 1995. Influence of viscosity on nitrofurantoin absorption from w/o/w emulsions. *Indian J. Pharm. Sci.* 57: 166–169.

Onyeji, C. O., Omotosho, J. A., and Ogunbona, F. A. 1991. Increased gastro-intestinal absorption of griseofulvin from w/o/w emulsions. *Indian J. Pharm. Sci.* 53: 256–258.

Oza, K. P. and Frank, S. G. 1989. Multiple emulsions stabilized by colloidal microcrystalline cellulose, *J. Display Sci. Technol.* 10: 163–185.

Pandey, R. and Khuller, G. K. 2007. Nanoparticle-based oral drug delivery system for an injectable antibiotic—Streptomycin. *Chemotherapy* 53: 437–441.

Pandey, R., Sharma, A., Zahoor, A., Sharma, S., Khuller, G. K., and Prasad, B. 2003. Poly(DL-lactide-*co*-glycolide) nanoparticle-based inhalable sustained drug delivery system for experimental tuberculosis. *J. Antimicrob. Chemother.* 52: 981–986.

Porter, C. J. H. and Charman, W. N. 2001. Intestinal lymphatic drug transport: An update. *Adv. Drug Deliv. Rev.* 50: 61–80.

Prabha, S. and Labhasetwar, V. 2004. Nanoparticle-mediated *wt-p53* gene delivery results in sustained antiproliferative activity in breast cancer cells. *Mol. Pharm.* 1: 211–219.

Rao, M. B., Tanksale, A. M., Ghatge, M. S., and Deshpande, V. V. 1998. Molecular and biotechnological aspects of microbial proteases. *Microbiol. Mol. Biol. Rev.* 62: 597–635.

Rao, M. Y. and Bader, F. 1993. Masking the taste of chloroquine by multiple emulsion. *East. Pharm.* 123: 11–14.

Raynal, S., Grossiord, J. L., Seiller, M., and Clausse, D. 1993. A topical W/O/W multiple emulsion containing several active substances: Formulation, characterization and study of release. *J. Control. Release* 26: 129–140.

Rizkalla, N., Range, C., Lacasse, F. X., and Hildgen, P. 2006. Effect of various formulation parameters on the properties of polymeric nanoparticles prepared by multiple emulsion method. *J. Microencapsul.* 23: 39–57.

Roy, S. and Gupta, B. K. 1993. *In vitro–in vivo* correlation of indomethacin release from prolonged release W/O/W multiple emulsion system. *Drug Dev. Ind. Pharm.* 19: 1965–1980.

Safwat, S. M., Kassem, M. A., Attia, M. A., and El-Mahdy, M. 1994. The formulation–performance relationship of multiple emulsions and ocular activity. *J. Control. Release* 32: 259–268.

Sahoo, S. K., Panyam, J., Prabha, S., and Labhasetwar, V. 2002. Residual polyvinyl alcohol associated with poly(D,L-lactide-*co*-glycolide) nanoparticles affects their physical properties and cellular uptake. *J. Control. Release* 82: 105–114.

Scheper, T. 1990. Enzyme immobilization in liquid surfactant membrane emulsions. *Adv. Drug Deliv. Rev.* 4: 209–231.

Seifriz, W. 1925. Studies in emulsions. III. Double reversal of oil emulsion occasioned by the same electrolyte. *J. Phys. Chem.* 29: 738–749.

Shah, M. V., De Gennaro, D., and Suryakasuma, H. 1987. An evaluation of albumin microcapsules prepared using a multiple emulsion technique. *J. Microencapsul.* 4: 223–238.

Shahiwala, A. and Amiji, M. M. 2008. Enhanced mucosal and systemic immune response with squalane oil-containing multiple emulsions upon intranasal and oral administration in mice. *J. Drug Target.* 16: 302–310.

Shiau, Y. F. 1981. Mechanisms of intestinal fat absorption. *Am. J. Physiol. Gastrointest. Liver Physiol.* 240: G1–G9.

Shively, M. L. 2002. Multiple emulsions for the delivery of proteins. In *Protein Delivery*, L. M. Sanders and R. W. Hendren (Eds.), pp. 199–211. New York: Springer.

Silva-Cunha, A., Cheron, M., Grossiord, J. L., Puisieux, F., and Seiller, M. 1998. W/O/W multiple emulsions of insulin containing a protease inhibitor and an absorption enhancer: Biological activity after oral administration to normal and diabetic rats. *Int. J. Pharm.* 169: 33–44.

Silva-Cunha, A., Grossiord, J. L., Puisieux, F., and Seiller, M. 1997. W/O/W multiple emulsions of insulin containing a protease inhibitor and an absorption enhancer: Preparation, characterization and determination of stability towards proteases *in vitro*. *Int. J. Pharm.* 158: 79–89.

Tedajo, G. M., Bouttier, S., Fourniat, J., Grossiord, J. L., Marty, J. P., and Seller, M. 2005. Release of antiseptics from the aqueous compartments of a w/o/w multiple emulsion. *Int. J. Pharm.* 288: 63–72.

Tedajo, G. M., Bouttier, S., Grossiord, J. L., Marty, J. P., Seiller, M., and Fourniat, J. 2002. *In vitro* microbicidal activity of W/O/W multiple emulsion for vaginal administration. *Int. J. Antimicrob. Agents* 20: 50–57.

Thompson, A. B. R., Keelan, M., Garg, M. L., and Clandinin, M. T. 1989. Intestinal aspects of lipid absorption: In review. *Can. J. Physiol. Pharmacol.* 67: 179–191.

Tirnaksiz, F. and Kalsin, O. 2005. A topical w/o/w multiple emulsions prepared with Tetronic 908 as a hydrophilic surfactant: Formulation, characterization and release study. *J. Pharm. Pharmaceut. Sci.* 8: 299–315.

Utada, A. S., Lorenceau, E., Link, D. R., Kaplan, P. D., Stone, H. A., and Weitz, D. A. 2005. Monodisperse double emulsions generated from a microcapillary device. *Science* 308: 537–541.

Vasudevan, T. V. and Naser, M. S. 2002. Some aspects of stability of multiple emulsions in personal cleansing systems. *J. Colloid Interface Sci.* 256: 208–215.

Vaziri, A. and Warburton, B. 1994. Slow release of chloroquine phosphate from multiple taste-masked w/o/w multiple emulsions. *J. Microencapsul.* 11: 641–648.

Verma, R. and Jaiswal, T. N. 1997. Protection, humoral and cell-mediated immune responses in calves immunized with multiple emulsion haemorrhagic septicaemia vaccine. *Vaccine* 15: 1254–1260.

Vladisavljević, G. T., Shimizu, M., and Nakashima, T. 2006. Production of multiple emulsions for drug delivery systems by repeated SPG membrane homogenization: Influence of mean pore size, interfacial tension and continuous phase viscosity. *J. Membr. Sci.* 284: 373–383.

Völkel, W., Bosse, J., Poppe, W., Halwachs, W., and Schügerl, K. 1984. Development and design of a liquid membrane enzyme reactor for the detoxification of blood. *Chem. Eng. Commun.* 30: 55–66.

Völkel, W., Poppe, W., Halwachs, W., and Schügerl, K. 1982. Extraction of free phenol from blood: A liquid membrane enzyme reactor. *J. Membr. Sci.* 11: 333–347.

Weiss, J., Scherze, I., and Muschiolik, G. 2005. Polysaccharide gel with multiple emulsions. *Food Hydrocolloids* 19: 605–615.

Yashioka, T., Ikeuchi, K., Hashida, M., Muranishi, L., and Sezaki, H. 1982. Prolonged release of bleomycin from parenteral gelatin sphere-in-oil-in-water multiple emulsion. *Chem. Pharm. Bull.* 30: 1408–1415.

Yazan, Y., Seiler, M., and Puisieux, F. 1993. Multiple emulsions. *Boll. Chim. Farm.* 132: 187–196.

Yener, G. and Baitokova, A. 2006. Development of a w/o/w emulsion for chemical peeling applications containing glycolic acid. *J. Cosmet. Sci.* 57: 487–494.

Zaks, A. and Klibanov, A. M. 1988. Enzymatic catalysis in non-aqueous solvents. *J. Biol. Chem.* 263: 3194–3197.

Zambaux, M. F., Bonneaux, F., Gref, R., Maincent, P., Dellacherie, E., Alonso, M. J., Labrude, P., and Vigneron, C. 1998. Influence of experimental parameters on the characteristics of poly(lactic acid) nanoparticles prepared by a double emulsion method. *J. Control. Release* 50: 31–40.

Zheng, S., Beissinger, R. L., and Wasan, D. T. 1991. The stabilization of hemoglobin multiple emulsion for use as a red blood cell substitute. *J. Coll. Interf. Sci.* 144: 72–85.

Zheng, S., Zheng, Y., Beissinger, R. L., Wasan, D.T, and McCormick, D. L. 1993. Hemoglobin multiple emulsion as an oxygen delivery system. *Biochem. Biophys. Acta* 1158: 65–74.

Zoldesi, C. I. and Imhof, A. 2005. Synthesis of monodisperse colloidal spheres, capsules, and microballoons by emulsion templating. *Adv. Mater.* 17: 924–928.

9 Nanoemulsions as Drug Delivery Systems

Figen Tirnaksiz, Seyda Akkus, and Nevin Celebi

CONTENTS

9.1 INTRODUCTION

Nanoemulsions (NEs) can be defined as extremely small droplet emulsions. They are termed miniemulsions (Constantinides et al., 2008; Solans et al., 2005), ultrafine emulsions (Guglielmini, 2008), submicron emulsions (Benita, 1999; Klang and Benita, 1998), and translucent emulsions (Fernandez et al., 2004; Tadros et al., 2004). These systems have also been called microemulsions (MEs; Koo et al., 2005). There are two types of NEs: thermodynamically stable systems (classic MEs) and metastable systems. Both types can be classified as NEs. The distinction between a thermodynamically stable system and a metastable system is not always made clear (Sarker, 2005). Unlike classic MEs, the stability of metastable NEs depends on the method of preparation (Wang et al., 2007). NEs may possess high kinetic stability and optical transparency resembling MEs (Porras et al., 2004). Despite their metastability, NEs can persist over many months or years because of the presence of stabilizing surfactant micelles (Chiesa et al., 2008).

The structures in NEs are much smaller than visible wavelengths, so many NEs appear applicably transparent (Chiesa et al., 2008). The average droplet size of NEs ranges from 20 to 500 nm (Guglielmini, 2008). NEs are transparent or translucent with a bluish coloration (Salager and Marquez, 2003). They are very sensitive systems by nature. Because they are transparent and usually very fluid, the slightest sign of destabilization easily appears. Their very small droplet size causes a large reduction in gravity force, and Brownian motion may be sufficient for overcoming gravity (Tadros et al., 2004). Because of their small droplet size, Brownian motion prevents sedimentation or creaming, thus offering increased physical stability (Fernandez et al., 2004). They may have high kinetic stability because their small droplet size makes them stable against sedimentation and creaming (Usón et al., 2004). The small droplet size also prevents any flocculation of the

221

droplets. Weak flocculation can be prevented, and this enables the system to remain dispersed with no separation (Tadros et al., 2004). In contrast to MEs, NEs can be diluted with water without changing the droplet size distribution (Fernandez et al., 2004). An NE provides ultralow interfacial tensions and large interfacial areas between the aqueous and oily phases.

NEs are thermodynamically unstable, which may lead to aggregation, flocculation, coalescence, and eventual phase separation. The physical stability of NEs can be changed on admixture of drugs; this is the main factor limiting wider use of the system for drug delivery (Sznitowska et al., 2001b). One of the prime physical properties of NEs is the size of the droplet distribution. NEs should display a narrow particle size distribution, and a stable NE system should retain submicron size; any changes in the particle size distribution over time are indicative of poor physical stability.

NEs require stabilization with emulsifying agents that do not result in the formation of lyotropic liquid crystalline phases. To make a stable emulsion reproducibly, a large number of factors must be controlled. These include selection of an appropriate composition, controlling the order of addition of the components, and applying the shear in a manner that effectively ruptures the droplets. The continuous phase should have a significant excess of surfactant to enable the new surface area of the nanometer droplets to be rapidly coated during emulsification. The main stability problem of NEs is "Ostwald ripening." The Ostwald ripening process is the solubility of the oil phase in the water phase such that oil molecules are transformed from small droplets to big droplets (Sonneville-Auburn et al., 2004). As a result, the droplets become larger. Suppression of Ostwald ripening can be achieved by choosing a very insoluble liquid for the dispersed phase so that this process does not occur rapidly, despite very high Laplace pressure (Chiesa et al., 2008). The coalescence rates of NEs are closely related to the molecular weight of the surfactant adsorbed at the oil–water interface. The longer the molecule is, the less unstable the emulsion will be (Sing et al., 1999).

9.2 PREPARATION OF NANOEMULSIONS

Preparation of NEs requires special application techniques in many cases. The energy input, generally from mechanical devices or from the chemical potential of the components, should be measured (Solans et al., 2005). Several types of oils and surfactants have been used and tested for formulation of NEs (Table 9.1). For delivery of drugs that contain NEs, minimization of surfactants is required to reduce possible side effects and the amount of the internal phase components often needs to be increased to maximize the concentration of lipid-soluble drugs in the final system. In summary, an optimum amount of surfactants and internal phase components should be used for the formulation of NEs (Amani et al., 2008; Nazzal and Khan, 2002; Yuan et al., 2008a).

These systems can be prepared by different mechanical techniques (Antonietti and Landfester, 2002; Constantinides et al., 2008; Table 9.1). Expensive equipment is required as well as high concentrations of emulsifiers (Tadros et al., 2004). In contrast to equilibrium lyotropic liquid crystalline phases called MEs, which spontaneously form when the addition of surfactant effectively causes the surface tension to vanish, NEs are metastable dispersions of submicron droplets that have significant surface tension (which form only when extreme shear is strongly applied to fragment the droplets) and are kinetically inhibited against recombining by repulsive interfacial stabilization due to the surfactant (Graves et al., 2005).

The emulsification of the aqueous and oil phases includes two consecutive steps: first, deformation and disruption of inner phase droplets and, second, the stabilization of newly formed interfaces for all techniques (Antonietti and Landfester, 2002).

9.2.1 HIGH-ENERGY EMULSIFICATION TECHNIQUE

One of the preparation methods for NEs is a high-energy technique that includes ultrasonication, microfluidization, and high-pressure homogenization. Preparation of NEs with high-energy

TABLE 9.1
Some NE Formulations

Drug	Oil	Aqueous Phase	Surfactant	Preparation Method	Reference
Aspirin	Soybean oil	Water	Polysorbate 80	Microfluidization	Subramanian et al., 2008
β-Carotene	MCT oil	Water	Polysorbate 20, polysorbate 40, polysorbate 60, polysorbate 80	High-pressure homogenization	Yuan et al., 2008a, 2008b
Flurbiprofen	Isopropyl myristate, soybean oil, coconut oil	Water	Egg lecithin	Ultrasonication	Fang et al., 2004
Quercetin, methylquercetin	Octyldodecanol	Water	Lipoid® E-80 (Lipoid KG, Ludwigshafen, Germany)	Spontaneous emulsification	Fasolo et al., 2007
—	Soybean oil	Water	Triton® X-100 (Rohm & Haas Co.)	Spontaneous emulsification	Hamouda et al., 2001
Several drugs	Soybean oil	Water + glycerol	Lipoid E-80 (Lipoid KG, Ludwigshafen, Germany)	High-pressure homogenization	Sznitowska et al., 2001a
Pilocarpine	Soybean oil	Water + glycerol	Lipoid E-80 (Lipoid KG, Ludwigshafen, Germany)	High-pressure homogenization	Sznitowska et al., 2000
Benzophenone-3	Capric–caprilic triglyceride, coconut oil	Water	Sorbitan stearate, polyoxyethylene-2-sorbitan monooleate, Laureth-7	Spontaneous emulsification	Fernandez et al., 2000
Octyl methoxy-cinnamate	MCT oil	Water + glycerol	Lipoid E-80 (Lipoid KG, Ludwigshafen, Germany), Pluronic® F-68 (BASF, U.S.)	High-pressure homogenization	Zeevi et al., 1994
—	Capric–caprilic triglyceride	Water	Hydroxypropyl methylcellulose	High-pressure homogenization	Schulz and Daniels, 2000
Benzathine penicillin G	Capric–caprilic triglyceride, soybean oil	Buffer solution (pH 7.4)	Epicuron® 200 (Lucas Meyer Hamburg, Germany), Synperonic® F-68 (Unichema Chemie BV, U.K.)	Spontaneous emulsification	Santos-Magalhães et al., 2000

continued

TABLE 9.1 (continued)
Some NE Formulations

Drug	Oil	Aqueous Phase	Surfactant	Preparation Method	Reference
Diazepam	MCT oil	Water + glycerol	Lipoid E-80 (Lipoid KG, Ludwigshafen, Germany), Synperonic F-68 (Unichema Chemie BV, U.K.)	High-pressure homogenization	Sznitowska et al., 2001a
Diazepam	Soybean oil	Water + glycerol	Pluronic F-68 (BASF, U.S.), egg yolk phospholipids	High-pressure homogenization	Levy and Benita, 1989
Diazepam	Soybean oil	Water + glycerol	Pluronic F-68 (BASF, U.S.), purified fractionated egg yolk phospholipids	High-pressure homogenization	Levy et al., 1994
Diazepam	Capric–caprilic triglyceride, soybean oil	Water	Lipoid E-80 (Lipoid KG, Ludwigshafen, Germany), Tyloxapol (Sigma–Aldrich, U.S.)	High-pressure homogenization	Schwarz et al., 1995
All-trans-retinol acetate	Soybean oil	—	Polyoxyl 35 castor oil, mono- or diglycerides of caprylic acid (cosurfactant)	SNEDDS	Taha et al., 2004
—	Isopropyl myristate	Water	Polyoxyl 35 castor oil, hydrogenated castor oil	Spontaneous emulsification	Usón et al., 2004
—	Decane	Water	Sorbitan monolaurate, sorbitan monooleate, polysorbate 20, polysorbate 80	Spontaneous emulsification	Porras et al., 2004
Cefpodoxime-proxetil	MCT oil	Water	Lipoid E-80 (Lipoid KG, Ludwigshafen, Germany), Imwitor® 742 (Sasol Germany GmbH; cosolvent)	High-pressure homogenization	Nicolaos et al., 2003

Drug	Oil	Aqueous phase	Surfactant/components	Method	Reference
Cefpodoxime-proxetil	MCT oil, soybean oil	Water	Lipoid E-80, Lipoid S40, Lipoid S75 (Lipoid KG, Ludwigshafen, Germany), polysorbate 20, polysorbate 80, polysorbate 85, Imwitor 742 (Sasol Germany GmbH; cosolvent)	High-pressure homogenization	Crauste-Manciet et al., 1998
Ubiquinone	Lemon oil	—	Polyoxyl 35 castor oil, mono- or diglycerides of caprylic acid (cosurfactant)	SNEDDS	Nazzal and Khan, 2002
Oligonucleotides	MCT oil	Water + glycerol	Lipoid E-80 (Lipoid KG, Ludwigshafen, Germany), SynperonicF-68 (Unichema Chemie BV, U.K.)	Microfluidization	Teixeira et al., 2001a, 2001b
Plasmid DNA	Olive oil	Water	Sorbitan monooleate, polysorbate 20	Spontaneous emulsification	Wu et al., 2001a
Inulin	Olive oil	Water	Sorbitan monooleate, polysorbate 20	Spontaneous emulsification	Wu et al., 2001b
Atenolol, danazol, metoprolol	Propylene glycol monolaurate	Water	Pluronic P104, Pluronic® L62, Pluronic L81 (BASF, U.S.)	Spontaneous emulsification	Brüsewitz et al., 2007
Cefpodoxime-proxetil	Propylene glycol monocaprylate		Polyoxyl 35 castor oil, Akoline® MCM (AarhusKarlshamn, Sweden; cosurfactant)	SNEDDS	Date and Nagarsenker, 2007
β-Laktamase	Propylene glycol laurate, propylene glycol monocaprylate	—	Polyoxyl 35 castor oil, polyoxyl 40 hydrogenated castor oil, diethylene glycol monoethyl ether (cosurfactant), propylene glycol (cosurfactant)	SNEDDS	Rao et al., 2008a, 2008c

continued

TABLE 9.1 (continued)
Some NE Formulations

Drug	Oil	Aqueous Phase	Surfactant	Preparation Method	Reference
Ramipril	Propylene glycol monocaprylic ester	Buffer solution (pH 5)	Caprylo caproyl macrogol-8-glyceride, polysorbate 80, diethylene glycol monoethyl ether (cosurfactant), polyglyceryl-6-dioleate (cosurfactant)	Spontaneous emulsification	Shafiq et al., 2007
Paclitaxel	Glycolysed ethoxylated glycerides	Water	Caprylo caproyl macrogol-8-glyceride	Spontaneous emulsification	Khandavilli and Panchagnula, 2007
Egg ceramides	Capric–caprilic triglyceride, tripalmitin	Water	Phospholipon® 90G (American Lecithin Co., U.S.), Solutol® HS 15 (BASF, U.S.)	Ultrasonication	Hatziantoniou et al., 2007
Foscan®	Capric–caprilic triglyceride	Water	Epikuron® 170 (Cargill Europe BVBA, Belgium), poloxamer 188	Spontaneous emulsification	Primo et al., 2007
Oligonucleotides	Capric–caprilic triglyceride	Water + glycerol	Poloxamer188, Lipoid E-80 (Lipoid KG, Ludwigshafen, Germany)	High-pressure homogenization	Hagigit et al., 2008
Paclitaxel + ceramide	Pine-nut oil	Water	Lipoid E-80, Lipoid PE (Lipoid KG, Ludwigshafen, Germany)	Ultrasonication	Desai et al., 2008
Morphine HCl	Soybean oil, sesame oil	Water	Sorbitan monooleate, Polysorbate 80, Brij® 98 (ICI Americas, Inc.), plurol diisostearique	Ultrasonication	Wang et al., 2008a
Risperidone	Mono- or diglycerides of caprylic acid	Water	Polysorbate 80, diethylene glycol monoethyl ether + propylene glycol (cosurfactant)	Spontaneous emulsification	Kumar et al., 2008

Drug	Oil	Aqueous phase	Surfactant	Method	Reference
Camptothecin	Coconut oil + perfluorocarbon	Water	Phospholipon® 80H (American Lecithin Co., U.S.), phosphatidylethanolamine, Pluronic F68 (BASF, U.S.), cholesterol	Ultrasonication	Fang et al., 2008
δ-Tocopherol	Canola oil	Water	Polysorbate 80	Microfluidization	Kotyla et al., 2008
α-Tocopherol	MCT oil	Water + glycerol	Pluronic F68 (BASF, U.S.), Lipoid E-80 (Lipoid KG, Ludwigshafen, Germany)	High-pressure homogenization	Ezra et al., 1996
Camphor + menthol + methyl salicylate	Soybean oil	Water + propylene glycol	Soybean lecithin, polysorbate 80, poloxamer407	High-pressure homogenization	Mou et al., 2008
Aceclofenac	Labrafil + triacetin	Water	Polysorbate 80, diethylene glycol monoethyl ether (cosurfactant)	Spontaneous emulsification	Shakeel et al., 2007
Ramipril	Propylene glycol monocaprylic ester	Water	Polyoxyl 35 castor oil, monoethyl ether of diethylene glycol (cosurfactant)	Spontaneous emulsification	Shafiq-un-Nabi et al., 2007
Ceramides	Octyldodecanol	Water + glycerol	Polysorbate 80, Lipid E-80 (Lipoid KG, Ludwigshafen, Germany)	High-pressure homogenization	Yilmaz and Borchert, 2005, 2006
DNA	MCT oil	Water + glycerol	Poloxamer407, Lipid E-80 (Lipoid KG, Ludwigshafen, Germany)	High-pressure homogenization	Bivas-Benita et al., 2004
—	MCT oil	Water + glycerol	Poloxamer188	High-pressure homogenization	Tamilvanan et al., 2005
—	Capric–caprilic triglyceride	Water	Sorbitan monooleate, sorbitan trioleate, polysorbate 20, polysorbate 80, Pluronic F68 (BASF, U.S.), Lipoid S75 (Lipoid KG, Ludwigshafen, Germany)	Spontaneous emulsification	Bouchemal et al., 2004

continued

TABLE 9.1 (continued)
Some NE Formulations

Drug	Oil	Aqueous Phase	Surfactant	Preparation Method	Reference
—	Paraffin oil	Water	Ceteareth-25	EIP method	Fernandez et al., 2004
—	Paraffin oil	Water	PPG-20 Tocophereth 50	High-speed stirring at the Θ point	Kim et al., 2004
Novel schistosomicidal drug	MCT oil	Water + glycerol	Poloxamer188, sorbitan monooleate, polysorbate 80, Lipoid S75 (Lipoid KG, Ludwigshafen, Germany)	Ultrasonication	Araújo et al., 2007
Carbamazepine	Castor oil + MCT oil	Water + glycerol	Soybean lecithin, polyoxyl 35 castor oil	Spontaneous emulsification	Kelmann et al., 2007
Curcumin	MCT oil	Water	Polysorbate 20	High-pressure homogenization	Wang et al., 2008b
Primaquine	Capric–caprilic triglyceride	Water + glycerol + sorbitol	Egg lecithin, soybean lecithin, poloxamer188	High-pressure homogenization	Singh and Vingkar, 2008
Sevoflurane	Perfluorooctyl bromide	NaCl solution (0.9%)	Fluorinated surfactant (F13M5)	Microfluidization	Fast et al., 2008
Cheliensisin	MCT oil	Water	Soybean lecithin	High-pressure homogenization	Zhao et al., 2008
α-, δ-, γ-Tocopherols	Soybean oil	Water	Polysorbate 80, phosphatidylcholine	Microfluidization	Kuo et al., 2008
Celecoxib	Propylene glycol monocaprylic ester, triacetin	Water	Ceteareth-25, 2-(2-ethoxyethoxy)ethanol (cosurfactant)	Spontaneous emulsification	Shakeel et al., 2008

Drug	Oil	Aqueous phase	Surfactant	Method	Reference
—	Soybean oil	Tri-n-butyl phosphate solution (20%)	Triton X-100 (Rohm & Haas Co.)	High shear stirring	Hamouda et al., 1999
Econazole nitrate, micanozole nitrate	MCT oil	Water + glycerol	Pluronic F68 (BASF, U.S.), Lipoid E-80 (Lipoid KG, Ludwigshafen, Germany)	High-pressure homogenization	Piemi et al., 1999
Indomethacin	MCT oil	Water + glycerol	Lauroamphodiacetate, Lipoid E-80 (Lipoid KG, Ludwigshafen, Germany)	High-pressure homogenization	Muchtar et al., 1997
Indomethacin	Soybean oil, propylene glycol dicaprylate/dicaprate	Water	Lecithin, Synperonic F68 (Unichema Chemie BV, U.K.)	Spontaneous emulsification	Calvo et al., 1996
Physostigmine salicylate	Soybean oil	Water + glycerol	Pluronic F68 (BASF, U.S.), crude phospholipids	High-pressure homogenization	Rubinstein et al., 1991
—	MCT oil	Water + glycerol	Pluronic F68 (BASF, U.S.), Lipoid E-80 (Lipoid KG, Ludwigshafen, Germany)	High-pressure homogenization	Klang et al., 1994
Piroxicam	MCT oil	Water + glycerol	Pluronic F68 (BASF, U.S.), Lipoid E-80 (Lipoid KG, Ludwigshafen, Germany)	High-pressure homogenization	Klang et al., 1999
Steroidal and nonsteroidal antiinflammatory drugs	Capric–caprilic triglyceride, paraffin liquid	Water	Lipoid E-80 (Lipoid KG, Ludwigshafen, Germany), polysorbate 80, Cremophore® EL-620 (BASF, U.S.), Tyloxapol® (Sigma–Aldrich, U.S.)	High-pressure homogenization	Friedman et al., 1995

Note: EIP—emulsion inversion point; HCl—hydrogen chloride; MCT—medium chain trigliceride; PPG—poly(propylene glycol).

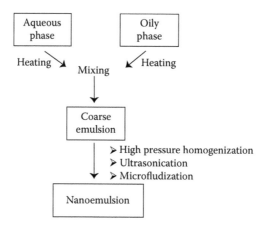

FIGURE 9.1 Schematic representation of the high-energy emulsification technique.

techniques depends on the formation of nanometer-sized droplets in the presence of a surfactant or a surfactant mixture with high energy input.

Ultrasonication has been used especially for small quantities of production, and it is very efficient in reducing droplet size (Fang et al., 2008).

High-pressure homogenization is preferred over the other methods that are not sufficient to obtain nanometer-sized droplets. This method ensures better effectiveness and more homogeneous droplet size distributions and has the possibility of preparing an NE with different types and amounts of surfactants, oils, and water-soluble materials (Schulz and Daniels, 2000). In this technique, the aqueous and oily phases are heated together and dispersed; then, the first emulsification is carried out using a high shear mixer, resulting in a coarse emulsion. This raw emulsion is passed through a high-pressure homogenizer several times to obtain a homogeneous dispersion of small inner phase droplets (Figure 9.1). During preparation, several parameters such as the homogenization pressure, number of homogenization cycles, and homogenization temperature can be changed; these affected the droplet size of the NE, which is very important for the physical stability of the system (Schulz and Daniels, 2000; Yuan et al., 2008b).

9.2.2 Low-Energy Emulsification Techniques

The high-energy methods cannot be used in some cases, especially for labile molecules. In this case, one of the low-energy emulsification techniques such as the spontaneous or self-emulsification method or the phase-inversion temperature method should be chosen.

Phase-inversion methods depend on the spontaneous curling of the surfactant film between the oil and water phases by covering the inner phase, changing factors such as the temperature and electrolyte concentration, or increasing the volume fraction of the dispersed phase (Tadros et al., 2004) (Figure 9.2).

The most popular of these is the phase-inversion temperature method of Shinoda and Saito, which is the most widely used method in the industry (Förster and Tesmann, 1991; Shinoda and Saito, 1968; Solans et al., 2005). For polyoxyethylene-type nonionic surfactants, the solubility changes and molecules become lipophilic with increasing temperature. At low temperature the surfactant monolayer causes oil-swollen droplets [oil-in-water (O/W) emulsion], and at high-temperature water-swollen droplets [water-in-oil (W/O) emulsion] form. This can be achieved by changing the temperature of the system, forcing a transition from an O/W emulsion at low temperatures to a W/O emulsion at higher temperatures (transitional phase inversion). During cooling of the emulsion, the system crosses a point of zero spontaneous curvature and minimal surface tension, promoting the formation of finely dispersed droplets (Fernandez et al., 2004; Solans et al., 2005).

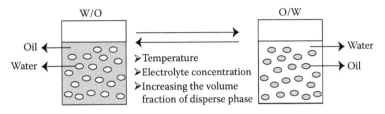

FIGURE 9.2   Schematic representation of the phase-inversion method.

The emulsion inversion point method also provides a transition in the spontaneous radius of curvature obtained by changing the water volume fraction. Initially, water droplets are formed in a continuous oil phase by adding water. Increasing the water volume fraction changes the spontaneous curvature of the surfactant. The system starts as a W/O ME; after that water globules merge together and then decompose into smaller oil globules upon further increasing the water content (Fernandez et al., 2004).

Bouchemal et al. (2004) demonstrated a different spontaneous emulsification method that has three steps. In the first step, the homogeneous organic solution was prepared with oily materials and a lipophilic surfactant in a water-miscible solvent. In the second step, this solution was injected in the aqueous phase composed of water and a hydrophilic surfactant under magnetic stirring. The O/W emulsion was formed instantaneously by diffusion of the organic solvent in the external aqueous phase, leading to the formation of nanodroplets. In the third step, the water-miscible solvent was removed by evaporation under reduced pressure (Figure 9.3). This method has been also used by several other research groups (Akkuş and Tırnaksız, 2006; Bouchemal et al., 2004; Chaix et al., 2003; Fasolo et al., 2007; Kelmann et al., 2007; Primo et al., 2007; Santos-Magalhães et al., 2000).

In another study, Wang et al. (2007) developed a new self-emulsifying alcohol-free NE using the low-energy emulsification technique. All appropriate components were mixed to generate a concentrate and then a certain amount of concentrate was injected into a very much larger volume of water under gentle stirring to achieve the final emulsion. This is termed as a "crash dilution" method by the authors.

In the above-defined methods, the amount of surfactant used in the system is not more than 10%. However, some NEs have been prepared with a high ratio of surfactant in the system (Wu et al., 2001a, 2001b).

Kim et al. (2004) used a novel nanoemulsification method of high-speed stirring of a complex of an oil and water phase; the emulsifier, poly(propylene glycol)-20 Tocophereth-50 a block copolymer amphiphile, is effective in the formation of stable NEs with mean droplet size ranging from 204 to

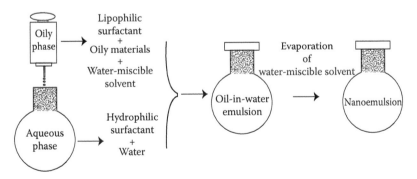

FIGURE 9.3   Schematic representation of the solvent evaporation technique. (From Akkuş, Ş. and Tırnaksız, F. 2006. Formulations of O/W emulsions prepared using spontaneous emulsification technique. Poster presented at Skin and Formulation—2nd Symposium of Association de Pharmacie Galenique Industrielle, 9–10 October, Versailles, France. With permission.)

499 nm. This surfactant causes a flocculated high viscosity complex above the specific temperature at which the oil phase and water phase of the emulsion appear. This complex, which can be dispersed in a stable system with a small particle size, is formed by an appropriate shear rate.

9.3 NANOEMULSIONS AS DRUG DELIVERY CARRIERS

The pharmaceutical field uses NEs as drug delivery systems for parenteral, oral, ocular, and dermal administration for therapeutic needs. Especially during the last few decades, NEs have been designed to deliver and target drugs by various routes of administration (Tamilvanan, 2004; Usón et al., 2004).

9.3.1 PARENTERAL APPLICATION

The O/W-type NEs called lipid emulsions or fat emulsions have long been used in parenteral delivery for nutritional purposes and as drug carriers for lipophilic drugs because of their insufficient water solubility. These systems can solubilize in high quantities of lipophilic drugs in the hydrophobic component of the oil inner phase (Kelmann et al., 2007; Levy and Benita, 1989; Levy et al., 1994). The composition of most of the commercial intravenous (IV) emulsions often contains soybean oil, phospholipids (lecithin) or poloxamers, nonelectrolytic compounds (glycerol or xylitol), and electrolytes. One of the most popular emulsifier combinations is the mixture of phospholipids with poloxamers. The *in vivo* disposition of fat emulsions depends on particle properties such as the size, zeta potential, and composition of the oil phase. The particle size of the droplets should be below 1 μm, and it generally ranges between 100 and 500 nm (Klang and Benita, 1998).

The O/W-type submicron cationic NEs are potent delivery systems for deoxyribonucleic acid (DNA) vectors, oligonucleotides, and plasmid transfection in the field of gene therapy (Tamilvanan et al., 2005; Teixeira et al., 1999, 2001a, 2001b).

Cationic NEs containing a schistosomicidal drug (BphEA) have a higher encapsulation level of drugs and good stability. The *in vitro* activity of BphEA has poor solubility in water, which is included in the cationic NE, and is higher than with the free drug. The positively charged NE probably binds with the anionic groups on the worm surface and facilitates the delivery of BphEA. After IV administration the NE is probably taken up by reticuloendothelial system organs such as the liver, where the schistosomes are located (Araújo et al., 2007).

Zhao et al. (2008) designed lyophilized negatively charged NEs containing the antitumor drug GC-51. The drug was delivered to solid tumors via parenteral administration. They showed that their emulsion system overcame the *in vitro* stability problem of the drug, sustained *in vitro* drug release, improved antitumor efficacy, and produced certain therapeutic effects. The *in vivo* therapeutic efficacy measured in the pulmonary metastasis of a colon cancer bearing mice model was enhanced after administration of a lyophilized NE containing the drug. The efficacy increased from 22.78 to 41.4% compared with injection of a drug solution (Zhao et al., 2008).

Halogenated volatile anesthetics have been used in lipid emulsions for more than 40 years. Fluorinated volatile anesthetics, such as sevoflurane, do not mix well with classic fat emulsions. To solve the significant limitation of clinical use, an NE containing sevoflurane was prepared with perfluorooctyl bromide and a semifluorinated surfactant (F13M5). This new NE system increased the solubility of the drug almost sixfold, and a stable NE was prepared containing up to 30% sevoflurane compared with intralipid (Fast et al., 2008).

NEs also have the potential to serve as parenteral prolonged-release systems for drugs that have a short half-life. Wang et al. (2008a) demonstrated that W/O NEs could be used as sustained-release drug delivery systems for morphine following subcutaneous administration in rats. Loading of the drug in NEs produced sustained release, and the *in vivo* analgesic duration of the drug by targeting the drug to the central nervous system could be prolonged from 1 to 3 h by the W/O NE (Wang et al., 2008a).

9.3.2 Ocular Application

The O/W NEs can also be used for ocular delivery to improve corneal penetration or sustain the pharmacological effect of drugs (Muchtar et al., 1992; Naveh et al., 1994). These emulsions could be advantageous because they are supposed to diminish vision-blurring effects (Schulz and Daniels, 2000).

These NEs can prolong the release of the drug and sustain the pharmacological effect of drugs in the eye following ocular application (Hagigit et al., 2008). Muchtar et al. (1992) and Navech et al. (1994) showed the application of NEs to prolong the response of antiglaucomatous drugs applied to rabbits. An NE of tetrahydrocannabinol, a highly liposoluble drug, reduced intraocular pressure in rabbit eye after a single ocular application, and an NE containing pilocarpine elicited higher and longer intraocular pressure reduction at the same dose. In contrast, when pilocarpine as an ion pair with monododecylphosphoric acid was used, the miotic effect observed in rabbits was the same for an aqueous solution or an NE, indicating that the increased partitioning of pilocarpine to the oily phase of the NE did not result in improved ocular bioavailability (Sznitowska et al., 2000).

Calvo et al. (1996) investigated the corneal penetration of a ^{14}C-indomethacin loaded NE prepared by the spontaneous emulsification technique. The mean size of the NE was found to be 216.74 nm; no physical problem was observed in long-term stability. The permeation coefficient of the drug was 3.65 times higher that obtained for the eye drop. Muchtar et al. (1997) had similar results. An NE prepared with Lipoid E-80 (Lipoid KG, Ludwigshafen, Germany), a lauroamphodiacetate, medium chain triglyceride, was found to be stable; the mean droplet size was 110 nm. The apparent corneal permeability coefficient of indomethacin incorporated in the NE was 3.8 times greater than that of the marketed aqueous solution.

Positively charged NEs were found to be suitable as topical ocular drug delivery systems (Hagigit et al., 2008; Klang et al., 1994, 1999). Cationic or positively charged NEs were well tolerated by the eye and were the most effective formulation in increasing the uptake of drugs in the ocular tissues following ocular application (Klang et al., 2000; Tamilvanan and Benita, 2004; Yang and Benita, 2000). Klang et al. (1994, 1999) prepared an NE using a combination of surfactants comprising poloxamer 188, Lipoid E-80 (Lipoid KG, Ludwigshafen, Germany), and stearylamine. The hydrophilic surfactant concentration was critical for prolonged system stability, and the incorporation of piroxicam in the system did not affect the physicochemical emulsion properties. The ocular tolerance study in rabbit eyes indicated that the NE was well tolerated without any toxic or inflammatory response in the rabbit eye. The positively charged NE of piroxicam had a pronounced effect on the ulceration rate and epithelial defects, and it was an effective formulation in lowering the ulcerative cornea score.

9.3.3 Oral Application

NE drug delivery systems are good vehicles for the oral delivery of poorly permeable or highly lipophilic drugs. Many new drugs or drug candidates show poor aqueous solubility (Kommuru et al., 2001; Strickley, 2004). NEs enhance the solubility of drugs, have permeation enhancing properties, and improve oral absorption (Rubinstein et al., 1991). An NE containing curcumin administered orally to mice had an enhanced antiinflammatory effect of the drug compared to curcumin in a surfactant solution according to the mouse ear inflammation model (Wang et al., 2008b). The activity was enhanced when the droplet size of the NE was reduced to below 100 nm.

NEs promote reduced inter- and intrasubject variability for some drugs, and they make the plasma concentrations and bioavailability of some drugs more reproducible (Constantinides, 1995; Kawakami et al., 2002a, 2002b; Kommuru et al., 2001; Lawrence and Rees, 2000).

Two studies found O/W NEs to be significantly effective for protecting the lipophilic molecule against enzymatic attack or hydrolysis (Crauste-Manciet et al., 1998; Nicolaos et al., 2003). To protect cefpodoxime-proxetil from enzymatic hydrolysis, the drug was included into the oil phase of an O/W NE. The NE had a significant increase of its area under the curve compared to the oral formulations (solution, suspension, and coarse emulsion) in rats. The mean absorption time and mean residence time of the NE containing cefpodoxime-proxetil was also significantly higher than those of oral formulations. Moreover, the area under the curve of the NE was not significantly different from that of the parenteral solution.

NEs can protect drugs' presystemic metabolism; they can be easily taken up by lipoprotein receptors in the liver, thus targeting the drug molecule toward the liver. An NE of primaquine was effective in antimalarial activity in mice at a 25% lower dose level compared to the conventional oral dose in the plain drug solution after oral administration. The system exhibited improved oral bioavailability, and higher drug levels were achieved in the liver. This targeting of the drug probably minimizes the systemic toxicity of the drug (Singh and Vingkar, 2008).

Some NEs tend to self-emulsify in aqueous media. This can be important for the absorption of drugs for oral formulation. Self-nanoemulsifying drug delivery systems (SNEDDSs), which are water-free systems that can be filled into soft or sealable hard capsules, form spontaneously in the aqueous fluid of the gastrointestinal tract after oral application (Date and Nagarsenker, 2007; Dixit and Nagarsenker, 2008; Nazzal and Khan, 2002). SNEDDSs have been found to solubilize lipophilic drugs and to increase the permeability of drugs (Brüsewitz et al., 2007; Dixit and Nagarsenker, 2008; Shafiq et al., 2007; Taha et al., 2004). Brüsewitz et al. (2007) discovered that self-nanoemulsifying systems had an active influence on intestinal permeation of transcellulary and paracellulary transported drugs. The apparent permeability of atenolol, danazol, and metoprolol by NEs was enhanced between 1.4- and 3.2-fold in the Caco-2 monolayer model of the small intestine. Shafiq et al. (2007) showed that a self-nanoemulsifying system of ramipril could be used as a delivery system for oral administration. *In vivo* studies revealed a significantly greater extent of absorption than the capsule formulation of ramipril in rats. The absorption of drugs from SNEDDSs resulted in a 2.94-fold increase in bioavailability compared to conventional capsules. Rao et al. (2008a, 2008c) also developed a self-nanoemulsifying system for protein drugs. All self-nanoemulsified systems of β-lactamase, a model protein, resulted in a higher transport rate than the free solution across the cell monolayer. The oral absorption of β-lactamase in rats was significantly increased by administration in the SNEDDS; the relative bioavailability, maximum drug concentration in plasma, and mean residence time were higher than that of the free solution.

Lipid-based W/O NEs have also been studied as delivery systems for hydrophilic and fragile molecules, such as peptides and proteins, to enhance oral bioavailability. Hydrophilic drugs have generally been successfully incorporated into the aqueous phase of W/O NEs for protection of drugs from harsh environmental conditions such as enzymatic activity or acidic pH (Çelebi et al., 2002; Çilek et al., 2005; Yetkin et al., 2004). Our previous studies investigated the effects of intragastric administration of nanosized lipid emulsions of epidermal growth factor (EGF) and transforming growth factor-α (TGF-α) on healing of acute gastric ulcers in rats. An NE containing EGF reduced basal gastric secretion after intragastrically administering the system in rats (Çelebi et al., 2002). The mean ulcer score was reduced in one week from 15.9 to 1.16 mm^2, and the gastric mucus secretion was increased significantly by intragastric treatment with the lipid-based W/O NE of EGF (Figure 9.4).

A similar result was found with an NE of TGF-α administered intragastrically that healed acute gastric ulcers induced by acetylsalicylic acid in rats (Yetkin et al., 2004). The basal gastric acid secretion after application of the NE containing TGF-α was lower than that of the TGF-α solution. The intragastric administration of acetylsalicylic acid (150 mg/kg) produced gastric ulcers in all rats with an average initial area of 89 mm^2. The gastric ulcers were significantly decreased in four days after ulcer induction (Figure 9.5).

In another study, we developed an NE formulation that improved the efficacy of recombinant human insulin administered orally in rats (Çilek et al., 2005). The NE formulations and insulin solution were administered orally and subcutaneously, respectively. The mean plasma insulin levels after administration to nondiabetic rats are shown in Figure 9.6. A significant difference was found

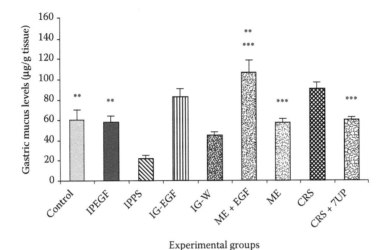

FIGURE 9.4 Gastric mucus levels in different experimental groups (mean ± standard error of the mean, $n = 8–10$ rats). ME + EGF group versus ME and CRS + 7UP groups (***$p < .001$), ME + EGF versus IPEGF and C groups (**$p < .01$); control—untreated rats; IPEGF—EGF solution administered intraperitoneally; IPPS—physiologic saline solution administered intraperitoneally; IG-EGF—EGF solution administered intragastrically; IG-W—water administered intragastrically; ME + EGF—NE containing EGF administered intragastrically; ME—plain NE; CRS—cold restraint stress; CRS + 7UO—untreated rats with cold restraint stress. (From Çelebi, N., et al. 2002. *Journal of Controlled Release* 83: 197–210. With permission.)

between the maximum drug concentration in the plasma values of the NE and the oral solution of insulin. Moreover, an NE containing aprotinin provided additional preservation effects on insulin levels.

Nevertheless, a phase inversion of the system may happen *in vivo* with a W/O NE: the degradation of molecules will occur and this leads to the loss of the peptides or proteins. Furthermore, oral

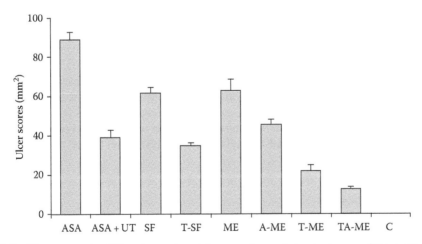

FIGURE 9.5 Ulcer scores in different experimental groups (mean ± standard error of the mean, $n = 5–10$ rats); $p < .01$ for T-ME/ASA-UT, T-ME/T-SF, and A-ME/ME; $p < .001$ for TA-ME/ASA, TA-ME/ASA-UT, TA-ME/ME, T-ME/ASA, T-ME/ME, T-SF/ASA, T-SF/SF, A-ME/ASA, A-ME/T-ME, and A-ME/TA-ME; ASA—acetylsalicylic acid, 150 mg/kg; ASA + UT—rats with ASA ulcer; SF—saline solution; T-SF—saline solution containing TGF-α; ME—plain NE; A-ME—NE containing aprotinin; T-ME—NE containing TGF-α; TA-ME—NE containing TGF-α and aprotinin. (From Yetkin, G., et al. 2004. *International Journal of Pharmaceutics* 277: 163–172. With permission.)

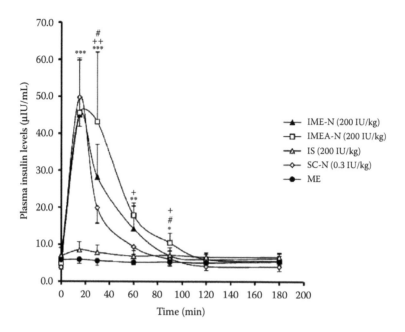

FIGURE 9.6 The plasma insulin levels for nondiabetic rats (mean ± confidence interval); ∗, +, and # represent IMEA-N versus IS, SC-N, and IME-N groups, respectively; ∗,+,#$p < .05$, ∗∗,++$p < .01$, ∗∗∗$p < .001$; IME-N—NE containing insulin; IMEA-N—NE containing insulin and aprotinin; IS—insulin solution administered orally; SC-N—insulin solution administered subcutaneously; ME—plain nanoemulsion. (From Çilek, A., et al. 2005. *International Journal of Pharmaceutics* 298: 176–185. With permission.)

administration of a W/O NE may cause problems with patient compliance because of the oily taste of the formulation (Rao et al., 2008b).

In several studies a model hydrophilic protein was loaded in the oil phase of a SNEDDS (Rao et al., 2008a, 2008b, 2008c) through the solid-dispersion technique. The SNEDDS increased the transport of protein molecules across the Madin–Darby canine kidney monolayer *in vitro* and significantly enhanced the oral bioavailability of the protein.

Paclitaxel is a very therapeutically effective drug that is currently administered by infusion of an ethanolic solution diluted in IV fluids. Because the application is difficult, researchers have been studying many alternative formulations for administration of paclitaxel. Khandavilli and Panchagnula (2007) prepared NEs for peroral bioavailability enhancement of paclitaxel. They observed that paclitaxel was rapidly absorbed; its absolute bioavailability was 70.62% when administered orally as an NE in rats. The plasma drug concentration was 3 μg/mL between 0.5 and 18 h, suggesting that the drug was absorbed throughout the gastrointestinal tract and the steady-state level persisted up to 18 h.

9.3.4 DERMAL APPLICATION

Many drugs exhibit low skin penetration that results in poor efficacy. The penetration and permeation of a poorly absorbed active ingredient can be improved however by the addition of specific enhancers to the formulations or by using colloidal delivery systems such as NEs (Benita, 1999).

The recent trends in lipid-based colloidal systems such as NEs are reported as one of the most promising formulations for enhancement of transdermal permeation and bioavailability of poorly soluble drugs. This system can enhance the penetration of active material and can lead it to a deeper layer of the skin because of its droplet size and low viscosity (Friedman et al., 1995). The large surface area of the emulsion system, the low surface tension of the whole system, and the low interfacial tension of the O/W droplets enhance the penetration of active agents (Bouchemal et al., 2004). The enhanced therapeutic effect with NEs is probably attributable to the enhanced permeation of the drug through skin; NEs themselves act as permeation enhancers. The low viscosity of NEs restrains their dermal application because of inconvenient use. Some hydrogel thickeners reportedly changed the viscosity of the system (Mou et al., 2008; Shakeel et al., 2008; Sonneville-Auburn et al., 2004; Yilmaz and Borchert, 2006).

For cosmetic products, NEs provide an elegant vehicle that can be used. The hydration power of NEs has been found to be much higher than the hydration power of MEs and conventional emulsions (Akkuş 2007; Guglielmini, 2008; Kim et al., 2004; Nohynek et al., 2007; Sonneville-Auburn et al., 2004; Zeevi et al., 1994).

As a topical carrier, NEs offer several significant advantages including low skin irritation, powerful permeation ability, and high drug-loading capacity for topical delivery when compared with other carriers such as MEs, liposomes, solid–lipid nanoparticles, classic emulsions, or gels (Hatziantoniou et al., 2007; Mou et al., 2008; Sonneville-Auburn et al., 2004; Tadros et al., 2004). Kotyla et al. (2008) compared the plasma level of an NE versus a coarse emulsion of δ-tocopherol after dermal application to hamsters. They demonstrated that the plasma concentration of δ-tocopherol increased 68-fold for hamsters after applying the NE, whereas there was an 11-fold increase for a coarse emulsion compared to baseline. This indicated that the NE system significantly increased the bioavailability of transdermally applied δ-tocopherol when compared to a micron-sized coarse emulsion of the drug. Shakeel et al. (2007) compared the *in vitro* skin permeation profile of an NE containing aceclofenac with that of a conventional gel. A significant increase in permeability parameters was observed with the NE formulation. Consequently, the antiinflammatory effect of NE significant increased after 24 h when compared with a conventional gel on paw edema in rats. The inhibition was high for NE at 82.2% compared with 41.8% for the conventional gel. In another study the use of NE improved the penetration of oxybenzone in skin when compared with classical emulsions (Fernandez et al., 2000). Unfortunately, the skin concentration and consequent systemic distribution of this sunscreen may occur at significant levels, so the authors pointed out that NEs are of little interest for sunscreen formulations.

Celecoxib is a highly lipophilic and poorly soluble drug with low oral bioavailability. Shakeel et al. (2008) investigated the bioavailability of celecoxib using a dermal NE. The absorption of celecoxib through a transdermally applied NE resulted in a 3.3-fold increase in bioavailability as compared to an oral capsule formulation in rats (Shakeel et al., 2008).

A dermal NE containing aspirin developed by Subramanian et al. (2008) increased the efficacy of the drug compared to an aspirin suspension; the dermal antiinflammatory effect was twofold greater than the suspension. They concluded that the NE of aspirin could reduce the adverse side effects associated with a high-dosage level of aspirin.

The W/O NEs have improved the dermal permeation of hydrophilic molecules included in the inner aqueous phase. Wu et al. (2001a, 2001b) prepared a W/O NE containing inulin or plasmid DNA. The rate and extent of inulin transport across hairy mouse skin was found to be highly dependent on the hydrophilic–lipophilic balance of the surfactant mixture in the NE. The delivery of plasmid DNA from the W/O NE into skin predominantly occurred via the follicular pathway. The levels of transgene expression observed using the NE were significantly greater than those observed using aqueous DNA at 24 h. This value that can be achieved in skin with normal follicular structure was also higher than that observed in the abnormal follicles present in the hairless mice.

NEs are suitable for efficient delivery of tocopherols through the skin. They significantly increase the bioavailability of transdermally applied δ-tocopherol and γ-tocopherol (Kotyla et al., 2008; Kuo et al., 2008). Kuo et al. (2008) investigated the antiinflammatory effect and bioavailability of NEs containing α-, δ-, and γ-tocopherol after dermal application in mice. Antioxidant NE formulations significantly reduced auricular thickness and the auricular concentration of cytokines compared to control and blank NEs. The NEs also increased bioavailability compared to suspensions of α-, δ-, and γ-tocopherol.

Friedman et al. (1995) observed improvement in the performance for transdermal delivery of steroidal and nonsteroidal antiinflammatory drugs when prepared in NEs rather than in topical cream formulations in rats. The improvement was more pronounced up to three- to fourfold. A decrease of the internal phase droplets caused the increase in activity for incorporated drugs.

The incorporation of a hydrophobic drug such as diazepam into an NE has been found to be an efficient way to significantly improve drug penetration through the stratum corneum (Schwarz et al., 1995). An NE of diazepam generated significant systemic activity compared with regular creams. Transdermal delivery of diazepam via an NE cream formulation was very effective; a single application of the formulation to mice skin provided pronounced dermal drug delivery and prolonged protective activity up to 6 h.

An NE of dacarbazine, which is highly lipid-soluble drug, was applied topically in a mouse xenograft model using a human melanoma cell line. The NE formulation caused a significant reduction in tumor size. In addition, during the cessation period (12 weeks) the NE of the drug showed fivefold greater efficacies in preventing tumor growth than the suspension preparations (Tagne et al., 2008a). Similar results were observed with an NE of another lipid-soluble drug, tamoxifen (Tagne et al., 2008b). The authors reported that the NE of the drug inhibited cell proliferation and increased cell apoptosis in the HTP-20 breast cancer cell line, and the results were statistically meaningful when compared to a suspension of tamoxifen.

Epithelial cells, including the skin, carry a negative charge on their surfaces, so that they are selective to positively charged solutes (Piemi et al., 1999; Rojanasakul et al., 1992; Yilmaz and Borchert, 2005). Positively charged NEs that will strongly interact with the cells result in better permeability of the drug and a prolonged pharmacological effect (Benita, 1999; Piemi et al., 1999). These systems can be effective vehicles to change the permeability of the skin. Piemi et al. (1999) reported that the effectiveness of a positively charged NE in promoting the *in vivo* efficacy of econazole and miconazole nitrate could be due to the binding affinity of the globules for the skin. Positively charged NE formulations were prepared using stearylamine, a cationic lipid that contributes the overall positive charge to the oil droplet faces (Piemi et al., 1999; Zeevi et al., 1994). A cell culture study carried out by Ezra et al. (1996) proved that positively charged α-tocopherol emulsions were effective in preventing oxidative damage, but the negatively charged emulsions were unable to provide such protection.

Ceramides have a major role in skin barrier homeostasis, and the reduction of the amount in the stratum corneum causes dry skin. Ceramides are insoluble compounds, and a suitable formulation is required to penetrate the stratum corneum. A positively charged NE was prepared to deliver ceramides into the skin, and this material could be successfully incorporated into the NE; good stability was observed that was due to the positive surface charge of the droplets (Yilmaz and Borchert, 2005, 2006).

In a different type of study, Primo et al. investigated the incorporation of Foscan, a classical and well-known photosensitizer for photodynamic therapy, in a magnetic NE (Primo et al., 2007). The magnetic nanoparticles used in the formulation were phosphate-coated maghemite stabilized at biological pH as an aqueous colloid. This system was added to a poloxamer 188 solution. Pig ear skin was used for the *in vitro* permeation study. The results showed that the presence of magnetic nanoparticles improved the penetration of the drug through the skin layer. The concentration of the drug in deep tissue layers was significantly higher than the NE that did not contain magnetic nanoparticles. The observed accumulation of Foscan with the magnetic NE in the epidermis and dermis layers was much higher than that of the NE in the absence of magnetic nanoparticles.

Some NEs with antimicrobial properties can be used as antibiofilm agents. Teixeira et al. (2007) and Hamouda et al. (1999) showed that an O/W-type NE containing soybean oil, tri-*n*-butyl phosphate, and Triton X-100 had antimicrobial properties. The rapid and nonspecific inactivation of vegetative bacteria and enveloped virus made this system a potential candidate for use as a topical biocidal agent (Hamouda et al., 2001).

9.3.5 OTHER APPLICATIONS

The targeting of drugs to the brain or cancer and tumor cells is also possible by administration of NEs (Desai et al., 2008; Kumar et al., 2008; Vyas et al., 2008). Moreover, NEs can also be prepared as vaccine delivery systems (Bivas-Benita et al., 2004).

NEs can enhance the brain disposition of drugs via oral application. Vyas et al. (2008) formulated saquinavir in different NEs made with oils rich in polyunsaturated fatty acids. The system was administered orally to mice to examine the oral bioavailability and distribution to the brain. NEs formulated with deoxycholic acid improved oral bioavailability and brain uptake of saquinavir. They pointed out that NEs can be extremely useful as a drug delivery system in patients infected with human immunodeficiency virus for specific delivery to viral reservoir sites that are hard to reach.

The intranasal administration of NEs can offer a practical and noninvasive route for delivery of drugs to the brain. Kumar et al. (2008) showed the delivery of risperidone to the brain via the nose with NE formulations in rats. The brain/blood uptake ratios of the nasal solution, nasal NE (plain), and nasal mucoadhesive NE were 0.617, 0.754, and 0.948, respectively, at 0.5 h. These data demonstrated the direct transport of drug molecules from the nose to the brain following nasal administration. A significant quantity of risperidone was quickly and effectively delivered to the brain by intranasal administration of the mucoadhesive NE compared to other formulations because of decreasing the mucociliary clearance of the drug.

Desai et al. (2008) investigated cytotoxicity and apoptosis in brain tumor cells (U-118 human glioblastoma cells) treated with paclitaxel and ceramide-C_6 in an O/W NE. The combination of drugs in the NE was significantly more effective than the aqueous solution. Paclitaxel and ceramide were delivered inside the U-118 cells and significantly decreased the cell viability to 33%.

Fang et al. (2008) developed an NE for camptothecin, which is a potent anticancer active agent. NEs were prepared using liquid perfluorocarbons and coconut oil as the inner oily phase. Camptothecin incorporated into NE was tested against melanoma cells and ovarian cancer cells. The drug-loaded NEs showed cytotoxicity against melanoma and ovary cancer cells *in vitro*. The NEs exhibited growth inhibition activity, and they entered cancer cells in a greater quantity.

NEs are also promising carriers for pulmonary mucosal DNA vaccines to the lungs. Bivas-Benita et al. (2004) evaluated a positively charged NE as a potential carrier for DNA vaccines. The NE formulation protected the DNA adsorbed to the cationic surface of the oil droplets from degradation in the presence of fetal calf serum in an *in vitro* stability experiment. Consequently, DNA was able to enter the nucleus and resulted in protein expression in the Calu-3 cell line (human bronchial epithelial cell line). Therefore, the DNA vaccine is expected to be stable upon exposure to the physiological environment and can be used for pulmonary immunization.

ABBREVIATIONS

DNA	Deoxyribonucleic acid
EGF	Epidermal growth factor
IV	Intravenous
ME	Microemulsion
NE	Nanoemulsion
O/W	Oil-in-water
SNEDDS	Self-nanoemulsifying drug delivery system
TGF-α	Transforming growth factor-α
W/O	Water-in-oil

REFERENCES

Akkuş, Ş. 2007. Development of nanoemulsion system containing kinetin for cosmetic usage. Master's dissertation, Gazi University.

Akkuş, Ş. and Tırnaksız, F. 2006. Formulations of o/w emulsions prepared using spontaneous emulsification technique. Poster presented at Skin and Formulation—2nd Symposium of Association de Pharmacie Galenique Industrielle, 9–10 October, Versailles, France.

Amani, A., York, P., Chrystyn, H., Clark, B. J., and Do, D. Q. 2008. Determination of factors controlling the particle size in nanoemulsions using artificial neural networks. *European Journal of Pharmaceutical Sciences* 35: 42–51.

Antonietti, M. and Landfester, K. 2002. Polyreactions in miniemulsions. *Progress in Polymer Science* 27: 689–757.

Araújo, S. C., Mattos, A. C. A., Teixeira, H. F., Coelho, P. M. Z., Nelson, D. L., and Oliveira, M. C. 2007. Improvement of *in vitro* efficacy of a novel schistosomicidal drug by incorporation into nanoemulsions. *International Journal of Pharmaceutics* 337: 307–315.

Benita, S. 1999. Prevention of topical and ocular oxidative stress by positively charged submicron emulsions. *Biomedicine and Pharmacotherapy* 53: 193–206.

Bivas-Benita, M., Oudshoorn, M., Romeijn, S., et al. 2004. Cationic submicron emulsions for pulmonary DNA immunization. *Journal of Controlled Release* 100: 145–155.

Bouchemal, K., Briançok, S., Perrier, E., and Fessi, H. 2004. Nano-emulsion formulation using spontaneous emulsification: Solvent, oil and surfactant optimization. *International Journal of Pharmaceutics* 280: 241–251.

Brüsewitz, C., Schendler, A., Funke, A., Wagner, T., and Lipp, R. 2007. Novel poloxamer-based nanoemulsions to enhance the intestinal absorption of active compounds. *International Journal of Pharmaceutics* 329: 173–181.

Calvo, P., Vila-Jato, J. L., and Alonso, M. J. 1996. Comparative *in vitro* evaluation of several colloidal systems, nanoparticles, nanocapsules, and nanoemulsions, as ocular drug carriers. *Journal of Pharmaceutical Sciences* 85: 530–536.

Çelebi, N., Türkyilmaz, A., Gönül, B., and Özogul, C. 2002. Effects of epidermal growth factor microemulsion formulation on the healing of stress-induced gastric ulcers in rats. *Journal of Controlled Release* 83: 197–210.

Chaix, C., Pacard, E., Elaïssari, A., Hilaire, J. F., and Pichot, C. 2003. Surface functionalization of oil-in-water nanoemulsion with a reactive copolymer: Colloidal characterization and peptide immobilization. *Colloids and Surfaces B: Biointerfaces* 29: 39–52.

Chiesa, M., Garg, J., Kang, Y. T., and Chen, G. 2008. Thermal conductivity and viscosity of water-in-oil nanoemulsions. *Colloids and Surfaces A: Physicochemical and Engineering Aspects* 326: 67–72.

Çilek, A., Çelebi, N., Tırnaksız, F., and Tay, A. 2005. A lecithin-based microemulsion of rh-insulin with aprotinin for oral administration: Investigation of hypoglycemic effects in non-diabetic and STZ-induced diabetic rats. *International Journal of Pharmaceutics* 298: 176–185.

Constantinides, P. P. 1995. Lipid microemulsions for improving drug dissolution and oral absorption and biopharmaceutical aspects. *Pharmaceutical Research* 12: 1561–1572.

Constantinides, P. P., Chaubal, M. V., and Shorr, R. 2008. Advances in lipid nanodispersions for parenteral drug delivery and targeting. *Advanced Drug Delivery Reviews* 60: 757–767.

Crauste-Manciet, S., Brossard, D., Decroix, M. O., Farinotti, R., and Chaumeil, J. C. 1998. Cefpodoximeproxetil protection from intestinal lumen hydrolysis by oil-in-water submicron emulsions. *International Journal of Pharmaceutics* 165: 97–106.

Date, A. A. and Nagarsenker, M. S. 2007. Design and evaluation of self-nanoemulsifying drug delivery systems (SNEDDS) for cefpodoxime proxetil. *International Journal of Pharmaceutics* 329: 166–172.

Desai, A., Vyas, T., and Amiji, M. 2008. Cytotoxicity and apoptosis enhancement in brain tumor cells upon coadministration of paclitaxel and ceramide in nanoemulsion formulations. *Pharmaceutical Nanotechnology* 97: 2945–2756.

Dixit, R. P. and Nagarsenker, M. S. 2008. Self-nanoemulsifying granules of ezetimibe: Design, optimization and evaluation. *European Journal of Pharmaceutical Sciences* 35: 183–192.

Ezra, R., Benita, S., Ginsburg, I., and Kohen, R. 1996. Prevention of oxidative damage in fibroblast cell cultures and rat skin by positively-charged submicron emulsion α-tocopherol. *European Journal of Pharmaceutics and Biopharmaceutics* 42: 291–298.

Fang, J.-Y., Hung, C.-F., Hua, S.-C., and Hwang, T.-L. 2008. Acoustically active perfluorocarbon nanoemulsions as drug delivery carriers for camptothecin: Drug release and cytotoxicity against cancer cells. *Ultrasonics* 49: 39–46.

Fang, J.-Y., Leu, Y. L., Chang, C. C., Lin, C. H., and Tsai, Y. H. 2004. Lipid nano/submicron emulsions as vehicles for topical flurbiprofen delivery. *Drug Delivery* 11: 97–105.

Fasolo, D., Schwingel, L., Holzschuh, M., Bassani, V., and Teixeira, H. 2007. Validation of an isocratic LC method for determination of quercetin and methylquercetin in topical nanoemulsions. *Journal of Pharmaceutical and Biomedical Analysis* 44: 1174–1177.

Fast, J. P., Perkins, M. G., Pearce, R. A., and Mecozzi, S. 2008. Fluoropolymer-based emulsions for the intravenous delivery of sevoflurane. *Anesthesiology* 109: 651–656.

Fernandez, C., Marti-Mestres, G., Ramos, J., and Maillols, H. 2000. LC analysis of bezophenone-3: II. Application to determination of "in vitro" and "in vivo" skin penetration from solvents, coarse and submicron emulsions. *Journal of Pharmaceutical and Biomedical Analysis* 24: 155–165.

Fernandez, P., André, V., Rieger, J., and Kühnle, A. 2004. Nano-emulsion formation by emulsion phase inversion. *Colloids and Surfaces A: Physicochemical Engineering Aspects* 251: 53–58.

Förster, Th. and Tesmann, H. 1991. Phase inversion emulsification. *Cosmetics and Toiletries* 106: 49–52.

Friedman, D. I., Schwarz, S., and Weisspapir, M. 1995. Submicron emulsion vehicle for enhanced transdermal delivery of steroidal and nonsteroidal antiinflammatory drugs. *Journal of Pharmaceutical Sciences* 84: 324–329.

Graves, S., Meleson, K., and Wilking, J. 2005. Structure of concentrated nanoemulsions. *Journal of Chemical Physics* 122: 1–6.

Guglielmini, G. 2008. Nanostructured novel carrier for topical application. *Clinics in Dermatology* 26: 341–346.

Hagigit, T., Nassar, T., Behar-Cohen, F., Lambert, G., and Benita, S. 2008. The influence of cationic lipid type on in-vitro release kinetic profiles of antisense oligonucleotide from cationic nanoemulsions. *European Journal of Pharmaceutics and Biopharmaceutics* 70: 248–259.

Hamouda, T., Hayes, M. M., Cao, Z., et al. 1999. A novel surfactant nanoemulsion with broad-spectrum sporicidal activity against Bacillus species. *Journal of Infectious Diseases* 180: 1939–1949.

Hamouda, T., Myc, A., Donovan, B., Shih, A. Y., Reuter, J. D., and Baker, J. R. 2001. A novel surfactant nanoemulsion with a unique non-irritant topical antimicrobial activity against bacteria, enveloped viruses and fungi. *Microbiological Research* 156: 1–7.

Hatziantoniou, S., Deli, G., Nikas, Y., Demetzos, C., and Papaioannou, G. Th. 2007. Scanning electron microscopy study on nanoemulsions and solid lipid nanoparticles containing high amounts of ceramides. *Micron* 38: 819–823.

Kawakami, K., Yoshikawa, T., Hayashi, T., Nishihara, Y., and Masuda, K. 2002a. Microemulsion formulation for enhanced absorption of poorly soluble drugs II. *In vivo* study. *Journal of Controlled Release* 81: 75–82.

Kawakami, K., Yoshikawa, T., Moroto, Y., et al. 2002b. Microemulsion formulation for enhanced absorption of poorly soluble drugs I. Prescription design. *Journal of Controlled Release* 81: 65–74.

Kelmann, R. G., Kuminek, G., Teixeira, H. F., and Koester, L. S. 2007. Carmazepine parenteral nanoemulsion prepared by spontaneous emulsification process. *International Journal of Pharmaceutics* 342: 231–239.

Khandavilli, S. and Panchagnula, R. 2007. Nanoemulsions as versatile formulations for paclitaxel delivery: Peroral and dermal delivery studies in rats. *Journal of Investigative Dermatology* 127: 154–162.

Kim, Y. D., Kim, J. S., Cho, I., and Kim, K. W. 2004. A novel nanoemulsification method of stirring at the Θ-point with the tocopherol-based block co-polymer nonionic emulsifier PPG-20 Tocophereth-50. *IFSCC Magazine* 7: 319–325.

Klang, S., Abdulrazik, M., and Benita, S. 2000. Influence on emulsion droplet surface charge on indomethacin ocular tissue distribution. *Pharmaceutical Development Technology* 5: 521–532.

Klang, S. and Benita, S. 1998. Design and evaluation of submicron emulsions as colloidal drug carriers for intravenous administration. In S. Benita (Ed.), *Submicron Emulsions in Drug Targeting and Delivery*, pp. 119–152. Amsterdam: Harwood Academic.

Klang, S. H., Frucht-Pery, J., Hoffman, A., and Benita, S. 1994. Physicochemical characterization and acute toxicity evaluation of a positively-charged submicron emulsion vehicle. *Journal of Pharmacy and Pharmacology* 46: 986–993.

Klang, S. H., Siganos, S., Benita, S., and Frucht-Pery, J. 1999. Evaluation of a positively charged submicron emulsion of piroxicam on the rabbit corneum healing process following alkali burn. *Journal of Controlled Release* 57: 19–27.

Kommuru, T. R., Gurley, B., Khan, M. A., and Reddy, I. K. 2001. Self-emulsifying drug delivery systems (SEDDS) of coenzyme Q_{10}: Formulation development and bioavailability assessment. *International Journal of Pharmaceutics* 212: 233–246.

Koo, O. M., Rubinstein, I., and Onyuksel, H. 2005. Role of nanotechnology in targeted drug delivery and imaging: A concise review. *Nanomedicine: Nanotechnology, Biology, and Medicine* 1: 193–212.

Kotyla, T., Kuo, F., Moolchandani, V., Wilson, T., and Nicolosi, R. 2008. Increased bioavailability of a transdermal application of a nano-sized emulsion preparation. *International Journal of Pharmaceutics* 347: 144–148.

Kumar, M., Misra, A., Babbar, A. K., Mishra, A. K., Mishra, P., and Pathak, K. 2008. Intranasal nanoemulsion based brain targeting drug delivery system. *International Journal of Pharmaceutics* 358: 285–291.

Kuo, F., Subramanian, B., Kotyla, T., Wilson, T. A., Yoganathan, S., and Nicolosi, R. J. 2008. Nanoemulsions of an anti-oxidant synergy formulation containing gamma tocopherol have enhanced bioavailability and anti-inflammatory properties. *International Journal of Pharmaceutics* 363: 206–213.

Lawrence, M. J. and Rees, G. D. 2000. Microemulsion-based media as novel drug delivery systems. *Advanced Drug Delivery Reviews* 45: 89–121.

Levy, M. Y. and Benita, S. 1989. Design and characterization of a submicronized o/w emulsion of diazepam for perenteral use. *International Journal of Pharmaceutics* 54: 103–112.

Levy, M. Y., Schutze, W., Fuhrer, C., and Benita, S. 1994. Characterization of diazepam submicron emulsion interface: Role of oleic acid. *Journal of Microencapsulation* 11: 79–92.

Mou, D., Chen, H., Du, D., et al. 2008. Hydrogel-thickened nanoemulsion system for topical delivery of lipophilic drugs. *International Journal of Pharmaceutics* 353: 270–276.

Muchtar, S., Abdulrazik, M., Frucht-Pery, J., and Benita, S. 1997. *Ex-vivo* permeation study of indomethacin from a submicron emulsion through albino rabbit cornea. *Journal of Controlled Release* 44: 55–64.

Muchtar, S., Almog, S., Torracca, M. T., Saettone, M. F., and Benita, S. 1992. A submicron emulsion as ocular vehicle for delta-8-tetrahydrocannabinol: Effect on intraocular pressure in rabbits. *Ophthalmic Research* 24: 142–149.

Naveh, N., Muchtar, S., and Benita, S. 1994. Pilocarpine incorporated into a submicron emulsion vehicle causes an unexpectedly prolonged ocular hypotensive effect in rabbits. *Journal of Ocular Pharmacology* 10: 509–520.

Nazzal, S. and Khan, M. A. 2002. Response surface methodology for the optimization of ubiquinone self-nanoemulsified drug delivery system. *AAPS PharmSciTech* 3: 23–31.

Nicolaos, G., Crauste-Manciet, S., Farinotti, R., and Brossard, D. 2003. Improvement of cefpodoxime proxetil oral absorption in rats by an oil-in-water submicron emulsion. *International Journal of Pharmaceutics* 262: 165–171.

Nohynek, G. J., Lademann, J., Ribaud, C., and Roberts, M. S. 2007. Grey goo on the skin? Nanotechnology, cosmetic and sunscreen safety. *Critical Reviews in Toxicology* 37: 251–277.

Piemi, M. P. Y., Korner, D., Benita, S., and Marty, J. P. 1999. Positively and negatively charged submicron emulsions for enhanced topical delivery of antifungal drugs. *Journal of Controlled Release* 58: 177–187.

Porras, M., Solans, C., Gonzáles, C., Martinez, A., Guinart, A., and Gutiérrez, J. M. 2004. Studies of formation of W/O nano-emulsions. *Colloids and Surfaces A: Physicochemical Engineering Aspects* 249: 115–118.

Primo, F. L., Ichieleto, L., Rodrigues, M. A. M., et al. 2007. Magnetic nanoemulsions as drug delivery system for Foscan®: Skin permeation and retention *in vitro* assays for topical application in photodynamic therapy (PDT) of skin cancer. *Journal of Magnetism and Magnetic Materials* 311: 354–357.

Rao, S. V. R., Agarwal, P. and Shao, J. 2008a. Self-nanoemulsifying drug delivery systems (SNEDDS) for oral delivery of protein drugs. II. In vitro transport study. *International Journal of Pharmaceutics* 362: 10–15.

Rao, S. V. R. and Shao, J. 2008b. Self-nanoemulsifying drug delivery systems (SNEDDS) for oral delivery of protein drugs. I. Formulation development. *International Journal of Pharmaceutics* 362: 2–9.

Rao, S. V. R., Yajurvedi, K., and Shao, J. 2008c. Self-nanoemulsifying drug delivery system (SNEDDS) for oral delivery of protein drugs. III. In vivo oral absorption study. *International Journal of Pharmaceutics* 362: 16–19.

Rojanasakul, Y., Wang, L. Y., Bhat, M., Glover, D. D., Malagna, C. J., and Ma, J. K. H. 1992. The transport barrier of epithelia: A comparative study on membrane permeability and charge selectivity in the rabbit. *Pharmaceutical Research* 9: 1029–1034.

Rubinstein, A., Pathak, Y. V., Kleinstern, J., Reches, A., and Benita, S. 1991. *In vitro* release and intestinal absorption of physostigmine salicylate from submicron emulsions. *Journal of Pharmaceutical Sciences* 80: 643–647.

Salager, J. L. and Marquez, L. 2003. Nanoemulsions: Where are they going to? *Colloidi Tpoint* 2: 12–14.

Santos-Magalhães, N. S., Pontes, A., Pereira, V. M.W., and Caetano, N. P. 2000. Colloidal carriers for benzathine penicillin G: Nanoemulsions and nanocapsules. *International Journal of Pharmaceutics* 208: 71–80.

Sarker, D. K. 2005. Engineering of nanoemulsions for drug delivery. *Current Drug Delivery* 2: 297–310.

Schulz, M. B. and Daniels, R. 2000. Hydroxypropylmethylcellulose (HPMC) as emulsifier for submicron emulsions: Influence of molecular weight and substitution type on the droplet size after high-pressure homogenization. *European Journal of Pharmaceutics and Biopharmaceutics* 49: 231–236.

Schwarz, J. S., Weisspapir, M. R., and Friedman, D. I. 1995. Enhanced transdermal delivery of diazepam by submicron emulsion (SE) creams. *Pharmaceutical Research* 12: 687–692.

Shafiq, S., Shakeel, F., Talegaonkar, S., Ahmad, F. J., Khar, R. K., and Ali, M. 2007. Development and bioavailability assessment of ramipril nanoemulsion formulation. *European Journal of Pharmaceutics and Biopharmaceutics* 66: 227–243.

Shafiq-un-Nabi, S., Shakeel, F., Talegaonkar, S., et al. 2007. Formulation development and optimization using nanoemulsion technique: A technical note. *AAPS PharmSciTech* 8: E1–E6.

Shakeel, F., Baboota, S., Ahuja, A., Ali, J., Aqil, M., and Shafiq, S. 2007. Nanoemulsions as vehicles for transdermal delivery of aceclofenac. *AAPS PharmSciTech* 8: 191–199.

Shakeel, F., Baboota, S., Ahuja A., Ali, J., and Shafiq, S. 2008. Skin permeation mechanism and bioavailability enhancement of celecoxib from transdermally applied nanoemulsion. *Journal of Nanobiotechnology* 6: 8.

Shinoda, K. and Saito, H. 1968. The effect of temperature on the phase equilibra and the types of dispersion of the ternary system composed of water, cyclohexane, and nonionic surfactant. *Journal of Colloid and Interface Science* 26: 70–74.

Sing, A. J.F., Graciaa, A., Lachaise, J., Brochette, P., and Salager, J. L. 1999. Interactions and coalescence of nanodroplets in translucent o/w emulsions. *Colloids and Surfaces A: Physicocchemical and Engineering Aspects* 152: 31–39.

Singh, K. S. and Vingkar, S. K. 2008. Formulation, antimalarial activity and biodistribution of oral lipid nanoemulsion of primaquine. *International Journal of Pharmaceutics* 347: 136–143.

Solans, C., Izquerdo, P., Nolla, J., Azemar, N., and Garcia-Celma, M. 2005. Nano-emulsions. *Current Opinion in Colloid and Interface Science* 10: 102–110.

Sonneville-Auburn, O., Simonnet, J. T., and L'Alloret, F. 2004. Nanoemulsions: A new vehicle for skincare products. *Advances in Colloid and Interface Science* 108–109: 145–149.

Strickley, R. G. 2004. Solubilizing excipients in oral and injectable formulations. *Pharmaceutical Research* 21: 201–230.

Subramanian, B., Kuo, F., Ada, E., et al. 2008. Enhancement of anti-inflammatory property of aspirin in mice by a nano-emulsion preparation. *International Immunopharmacology* 8: 1533–1539.

Sznitowska, M., Gajewska, M., Janicki, S., Radwanska, A., and Lukowski, G. 2001a. Bioavailability of diazepam from aqueous-organic solution, submicron emulsion and solid lipid nanoparticles after rectal administration in rabbits. *European Journal of Pharmaceutics and Biopharmaceutics* 52: 159–163.

Sznitowska, M., Janicki, S., Dabrowska, E., and Zurowska-Pryczkowska, K. 2001b. Submicron emulsions as drug carriers. Studies on destabilization potential of various drugs. *European Journal of Pharmaceutical Sciences* 12: 175–179.

Sznitowska, M., Zurowska-Pryczkowska, K., Dabrowska, E., and Janicki, S. 2000. Increased partitioning of pilocarpine to the oily phase of submicron emulsion does not result in improved ocular bioavailability. *International Journal of Pharmaceutics* 202: 161–164.

Tadros, T., Izquierdo, P., Esquena, J., and Solans, C. 2004. Formation and stability of nano-emulsions. *Advances in Colloid and Interface Science* 108–109: 303–318.

Tagne, J. B., Kakumanu, S., and Nicolosi, J. 2008a. Nanoemulsion preparations of the anticancer drug dacarbazine significantly increase its efficacy in a xenograft mouse melanoma model. *Molecular Pharmaceutics* 5: 1055–1063.

Tagne, J. B., Kakumanu, S., Ortiz, D., Shea, T., and Nicolosi, R. J. 2008b. A nanoemulsion formulation of tamoxifen increases its efficacy in a breast cancer cell line. *Molecular Pharmaceutics* 5: 280–286.

Taha, E. I., Al-Saidan, S., Samy, A. M., and Khan, M. A. 2004. Preparation and *in vitro* characterization of self-nanoemulsified drug delivery system (SNEDDS) of all-trans-retinol acetate. *International Journal of Pharmaceutics* 285: 109–119.

Tamilvanan, S. 2004. Oil-in-water lipid emulsions: Implications for parenteral and ocular delivering systems. *Progress in Lipid Research* 43: 489–533.

Tamilvanan, S. and Benita, S. 2004. The potential of lipid emulsion for ocular delivery of lipophilic drugs. *European Journal of Pharmaceutics and Biopharmaceutics* 58: 357–368.

Tamilvanan, S., Schmidt, S., Müller, R. H., and Benita, S. 2005. *In vitro* adsorption of plasma proteins onto the surface (charges) modified-submicron emulsion for intravenous administration. *European Journal of Pharmaceutics and Biopharmaceutics* 59: 1–7.

Teixeira, H., Dubernet, F., Puisieux, S., Benita, P., and Couvreur, P. 1999. Submicroncationic emulsions as a new delivery systems of oligonucleotides. *Pharmaceutical Research* 16: 30–36.

Teixeira, H., Dubernet, C., Rosilio, V., et al. 2001a. Factors influencing the oligonucleotides release from O-W submicron cationic emulsions. *Journal of Controlled Release* 70: 243–255.

Teixeira, H., Rosilio, V., Laigle, A., et al. 2001b. Characterization of oligonucleotide/lipid interactions in submicron cationic emulsions: Influence of the cationic lipid structure and the presence of PEG-lipids.
Biophysical Chemistry 92: 169–181.

Teixeira, P. C., Leite, G. M., Domingues, R. J., Silva, J., Gibbs, P. A., and Ferreira, J. P. 2007. Antimicrobial
effects of a microemulsion and a nanoemulsion on enteric and other pathogens and biofilms. *International
Journal of Food Microbiology* 118: 15–19.

Usón, N., Garcia, M. J., and Solans, C. 2004. Formation of water-in-oil (W/O) nano-emulsions in a water/
mixed non-ionic surfactant/oil systems prepared by a low-energy emulsification method. *Colloids and
Surfaces A: Physicochemical Engineering Aspects* 250: 415–421.

Vyas, T. K., Shahiwala, A., and Amiji, M. M. 2008. Improved oral bioavailability and brain transport of saquinavir upon administration in novel nanoemulsions formulations. *International Journal of Pharmaceutics*
347: 93–101.

Wang, J. J., Hung, C. F., Yeh, C. H., and Fang, J. Y. 2008a. The release and analgesic activities of morphine and
its ester prodrug, morphine propionate, formulated by water-in-oil nanoemulsions. *Journal of Drug
Targeting* 16: 294–301.

Wang, L., Li, X., Zhang, G., Dong, J., and Eastoe, J. 2007. Oil-in-water nanoemulsions for pesticide formulations. *Journal of Colloid and Interface Science* 314: 230–235.

Wang, X., Jiang, Y., Wang, Y. W., Huang, M. T., Ho, C. T., and Huang, Q. 2008b. Enhancing anti-inflammation
activity of curcumin through O/W nanoemulsions. *Food Chemistry* 108: 419–424.

Wu, H., Ramachandran, C., Bielinska, A. U., et al. 2001a. Topical transfection using plasmid DNA in a water-in
oil nanoemulsion. *International Journal of Pharmaceutics* 221: 23–34.

Wu, H., Ramachandran, C., Weiner, N. D., and Roessler, B. J. 2001b. Topical transport of hydrophilic compounds using water-in-oil nanoemulsions. *International Journal of Pharmaceutics* 220: 63–75.

Yang, S. C. and Benita, S. 2000. Enhanced absorption and drug targeting by positively charged submicron
emulsions. *Drug Development Research* 50: 476–486.

Yetkin, G., Çelebi, N., Özer, Ç., Gönül, B., and Özoğul, C. 2004. The healing effect of TGF-α on gastric ulcer
induced by acetylsalicylic acid in rats. *International Journal of Pharmaceutics* 277: 163–172.

Yilmaz, E. and Borchert, H. H. 2005. Design of a phytosphingosine-containing, positively-charged nanoemulsion as a colloidal carrier system for dermal application of ceramides. *European Journal of Pharmaceutics
and Biopharmaceutics* 60: 91–98.

Yilmaz, E. and Borchert, H. H. 2006. Effect of lipid-containing, positively charged nanoemulsions on skin
hydration, elasticity and erythema—An *in vivo* study. *International Journal of Pharmaceutics* 307:
232–238.

Yuan, Y., Gao, Y., Mao, L., and Zhao, J. 2008a. Optimization of conditions for the preparation of β-carotene
nanoemulsions using response surface methodology. *Food Chemistry* 107: 1300–1306.

Yuan, Y., Gao, Y., Zhao, J., and Mao, L. 2008b. Characterization and stability evaluation of β-carotene nanoemulsions prepared by high pressure homogenization under various emulsifying conditions. *Food
Research International* 41: 61–68.

Zeevi, A., Klang, S., Alard, V., Brossard, F., and Benita, S. 1994. The design and characterization of a positively-charged submicron emulsion containing a sunscreen agent. *International Journal of Pharmaceutics*
108: 57–68.

Zhao, D., Gong, T., Fu, Y., et al. 2008. Lyophilized Cheliensisin A submicron emulsion for intravenous injection: Characterization, *in vitro* and *in vivo* antitumor effect. *International Journal of Pharmaceutics* 357:
139–147.

10 Microemulsion Systems
Application in Delivery of Poorly Soluble Drugs

Ljiljana Djekic and Marija Primorac

CONTENTS

10.1 INTRODUCTION

The primary interest for pharmaceutical formulators is delivery of the active molecule to the target tissue at therapeutically relevant concentrations accompanied by negligible side effects. Drug delivery is often significantly influenced by the physicochemical properties of the active compound, particularly its solubility. The water solubility of drug molecules and gastrointestinal (GI) permeability are the basic criteria for the biopharmaceutical classification system of drugs comprising four classes (I–IV; Amidon et al., 1995). Class II and class IV cover drugs with low water solubility and high and low GI permeability, respectively. The primary limitation to the absorption of class II drugs is the slow dissolution rate in GI fluids, but poor absorption in class IV drugs is attributed to hindered permeability. Additionally, roughly 40% of the newly discovered chemical entities entering drug development programs are poorly water-soluble compounds. Thus, the solubility issue affects the bioavailability and efficacy of many existing and potential drugs. Various conventional formulation strategies for the improvement of the solubility and dissolution kinetics of drugs that have been developed (e.g., particle size reduction, synthesis of salts freely soluble in water, pH adjustment, addition of solubilizing agents) cannot provide satisfactory bioavailability enhancement in some cases. Therefore, researchers have made great efforts to develop new approaches for the successful delivery of poorly soluble drugs that will also be economically acceptable for the pharmaceutical industry. An important aspect of any new strategy for delivery of such drugs is to increase their solubility in biological milieu. Novel prospects to overcome problems regarding the limited solubility of drugs are based on their incorporation into an efficient formulation or a carrier system (Das and Das, 2006; Pouton, 2006). The concept of microemulsions represents a promising method that has been applied in the development of liquid vehicles and particulate carriers in order to improve the bioavailability of poorly soluble compounds (Datea and Patravale, 2004; Gasco and Trotta, 1986; Lawrence and Rees, 2000; Müller et al., 2000; Reis et al., 2006; Spernath and Aserin, 2006). Microemulsions are colloidal dispersions that usually consist of a water phase, an oil phase, and one or more amphiphiles; the physicochemical properties of the constituents and their concentrations are balanced to form thermodynamically stable liquids at a given temperature. Since these systems were introduced more than six decades ago by Hoar and Schulman (1943), their structure and processes on the molecular level have been the subject of intensive theoretical and experimental research. The results of these investigations disproved the early perception of microemulsions as *small emulsion-like structures* (Schulman et al., 1959) or *structured solutions* (Danielsson and Lindman, 1981) and pointed out the complexity of the phase behavior and diversity of the microstructure with a dynamic nature in microemulsion-forming systems (Ezrahi et al., 1999; Kahlweit, 1999; Sjoblom et al., 1996; Strey, 1994). Microemulsions are homogenous on a macroscopic level, but they are heterogenous on a microscopic level because the structure consists of diverse microdomains. There are three main types of microemulsion microstructures relevant to drug delivery: oil-in-water (O/W), water-in-oil (W/O), and bicontinuous structures. Droplet types of structures (O/W and W/O) contain spherical droplets of the water phase or oil phase (diameter = 10–100 nm) dispersed in a continuous oil or water phase, respectively, with the monomolecular film of amphiphiles at the water–oil interface. In bicontinuous systems, two immiscible liquids (oil and water) are interconnected and stabilized by a continuous flexible monomolecular film of amphiphiles at the interface (Ezrahi et al., 1999; Kahlweit, 1999; Sjoblom et al., 1996; Strey, 1994). The specific properties of microemulsions, such as thermodynamic stability, ultralow interfacial tension, small droplet size, low viscosity, and high interfacial area between the oil and water, provide tremendous benefits for their pharmaceutical applications including easy formation with little energy input (heat or mixing), long-term shelf life, filterability, sprayability, high solubilization capacity for drug molecules, and dose uniformity. Although microemulsions, particularly the O/W type, have gained increasing interest for use in improvement of the solubility and bioavailability of water-insoluble or sparingly water-soluble drugs (Bagwe et al., 2001; Kogan and Garti, 2006; Lawrence and Rees, 2000; Malmstein, 1999; Vandamme, 2002), this chapter highlights physicochemical and biopharmaceutical aspects of the microemulsions as colloidal systems currently of interest as templates for the development of improved formulations such as self-microemulsifying drug delivery systems (SMEDDSs), for synthesis of nanoparticulate drug carriers [solid lipid nanoparticles (SLNs) and polymeric nanoparticles], and for nanoengineering of poorly soluble drugs. In this scope, the most important features of microemulsion systems are increased drug solubilization capacity and the dynamic character of their microstructures. This provides the possibility for phase transformations from microemulsion preconcentrates (SMEDDSs) to O/W microemulsions when they are diluted with water phase or biological fluids, as well as the possibility for application of microemulsions as reactors for synthesis of nanoparticles. General observations obtained within an extensive research work have proved the potential of SMEDDSs as a promising strategy for promotion of the solubilization of drugs, manipulation of drug biodistribution, improvement of *in vivo* stability, and pharmacodynamics of poorly soluble drugs (Fatouros et al., 2007; Humberstone and Charman, 1997; Lawrence and Rees, 2000; Pouton, 2000, 2006; Pouton and Porter, 2008). SMEDDS formulations for oral delivery of cyclosporine (e.g., Neoral®, Novartis, Switzerland; Gengraf®, Abbot Laboratories, USA; and SangCya®, SangStat Medical Corporation, USA) have already entered the pharmaceutical market. There are still considerable issues with regard to the formulation, characterization, and *in vivo* performance of SMEDDS that need to be overcome before they may be considered as an important part of the drug market. Microemulsions have been successfully applied as current reaction media for the synthesis of SLNs and polymeric nanoparticles or utilized as a route to reduce the particle size of powdered drugs to less than 100 nm with very low polydispersity (Datea and

Patravale, 2004; Gasco and Trotta, 1986; Müller et al., 2000; Reis et al., 2006). SLNs produced by the microemulsion technique have been successfully applied in the pharmaceutical industry.

10.2 SELF-MICROEMULSIFYING DRUG DELIVERY SYSTEMS

SMEDDSs (microemulsion preconcentrates) are described as isotropic mixtures of oils, surfactants, and cosolvents with a solubilized drug substance, which form a microemulsion by diluting with an aqueous medium on mild agitation or by mixing with biological fluids under digestive motility after *in vivo* administration (Lawrence and Rees, 2000; Pouton, 2000, 2006; Pouton and Porter, 2008). The SMEDDS concept was introduced as a potential approach for improvement of the oral delivery of hydrophobic drugs (Shah et al., 1994). Although a certain number of reports in the scientific literature concerns parenteral (Lee et al., 2002; Park et al., 1999; von Corswant et al., 1998), dermal (Shukla et al., 2003), and transdermal (Fujii et al., 1996; Kemken et al., 1991, 1992) application of SMEDDS formulations, the majority of the work reported details the potential oral delivery of poorly soluble drugs by these systems (Borhade et al., 2008; Garti et al., 2006; Grove et al., 2006; Jing et al., 2006; Kang et al., 2004; Kommuru et al., 2001; Kovarik et al., 1994; Lu et al., 2008; Mandawgade et al., 2008; Mueller et al., 1994; Patel and Sawant, 2007; Patel and Vavia, 2007; Subramanian et al., 2004; Wei et al., 2005; Woo et al., 2008; Ying et al., 2008; Zhang et al., 2008). SMEDDSs for oral application usually have been reviewed as a type of so-called *lipid-based formulation* or *lipid-based drug delivery system* (Hauss, 2007; Jannin et al., 2008; Porter et al., 2008; Pouton, 2000, 2006; Pouton and Porter, 2008). However, the literature concerning pharmaceutical microemulsions has also often reviewed SMEDDS as systems related to microemulsions (Gupta and Moulik, 2008; Lawrence and Rees, 2000; Spernath and Aserin, 2006). Terminological duality is also evident and, although the term SMEDDS prevailed, self-nanoemulsifying drug delivery systems are used synonymously, considering the nanoscale diameter of the droplets (Jannin et al., 2008; Taha et al., 2004).

10.2.1 SELF-MICROEMULSIFYING DRUG DELIVERY SYSTEMS IN ORAL DELIVERY OF POORLY SOLUBLE DRUGS

SMEDDSs are classified as type III lipid formulations for the oral administration of drugs, according to the *lipid formulation classification system* proposed by Pouton (2000), which usually comprise up to 80% oil phase, 20–50% surfactants (hydrophilic–lipophilic balance >11), and 20–50% cosolvents. SMEDDS formulations are suitable for oral administration in the form of bulk solutions containing 1 µg/mL to 100 mg/mL of drug, hard or soft gelatin capsules, or hard hydroxypropyl methylcellulose capsules filled with liquid or semisolid (thermosoftening) formulations containing 0.25 µg to 500 mg of the drug per a unit-dose capsule product (Hauss, 2007). Aqueous dilution of SMEDDS formulations accompanied by gentle agitation results in the spontaneous formation of lipid droplets with diameters of usually less than 100 nm, depending on the excipient selection and relative composition of the formulation (Gursoy and Benita, 2004; Pouton, 2000).

The initial consideration in developing a SMEDDS formulation is the selection of excipients and their concentrations that are appropriate to solubilize the entire dose of the drug and provide the solubilized state for the drug during the absorption in the GI tract (GIT). The most important issues in the SMEDDS design are the pharmaceutical acceptability, biocompatibility, and physicochemical stability of the excipients; their mutual miscibility and compatibility; and their compatibility with the drug, solubility, and physicochemical stability of the drug substance; as well as the potential interactions of the components of the SMEDDS formulation with the GI environment.

For preparation of oral lipid-based formulations, the frequently chosen components of the oil phase are vegetable oils, composed predominantly of medium-chain triglycerides (e.g., coconut or palm oils) or long-chain triglycerides (e.g., corn, olive, peanut, rapeseed, sesame, or soybean oils).

Derivatives of natural oils are also used that are prepared by hydrogenation [hydrogenated vegetable oils: hydrogenated cottonseed oil (Lubritab™, Akofine™, or Sterotex™), hydrogenated palm oil (Dynasan™ P60, Softisan™ 154, or Suppocire®), hydrogenated castor oil (Cutina™ HR), or hydrogenated soybean oil (Sterotex HM, Hydrocote™, or Lipo™)] or by their separation into glyceride fractions [partial glycerides: glyceryl monocaprylocaprate (Capmul® MCM), glyceryl monostearate (Geleol™, Imwitor® 191, Cutina™ GMS, or Tegin™), glyceryl distearate (Precirol™ ATO 5), glyceryl monooleate (Peceol™), glyceryl monolinoleate (Maisine™ 35-1), or glyceryl dibehenate (Compritol® 888 ATO)], which have more suitable physicochemical and drug absorption-enhancing properties (Gibson, 2007).

The certain limitation ascribed to the SMEDDS formulations is a relatively high content of surfactants, frequently in combination with cosolvents, which are required for a successful microemulsification process as well as for complete solubilization of a given dose of the drug. Furthermore, the high concentrations of surfactants in the GIT lumen can inhibit digestion of the oil phase (Pouton, 2000). Surfactants often have potential toxic effects, particularly when they are used at high levels; thus, it is important to consider the toxicity and irritancy of these compounds. Therefore, the surfactants involved in the SMEDDS formulations are from the class of nonionic tensides that have been reported to have minimal toxicity. Pharmaceutically acceptable surfactants that are often employed in SMEDDS are those that are prepared usually by introduction of hydrophilic chemical entities in the molecules of medium-chain saturated fatty acids or glycerides derived from vegetable oils such as polyoxylglycerides (Labrasol®, Labrafil®-s, or Gelucire®-s), ethoxylated glycerides derived from castor oil (Cremophor® EL, RH40, or RH60), and esters of edible fatty acids and various alcohols [e.g., polyglyceryl oleate (Plurol™ Oleique CC497), propylene glycol monocaprylate (Capryol™ 90), propylene glycol monolaurate (Lauroglycol™ 90), poly(ethylene glycol) (PEG)-8 stearate and PEG-40 stearate (Mirj® 45 and Mirj 52), PEG-15 hydroxystearate (Solutol® HS15), sorbitan monooleate and sorbitan monolaurate (Span® 80 and Span 20), polyoxyethylene-20 sorbitan monooleate (polysorbate 80; Tween® 80), and polyoxyethylene-20 sorbitan monolaurate (polysorbate 20; Tween 20)]. The latter excipients are nonionic amphiphilic compounds with medium to high hydrophilic–lipophilic balance value, depending on the type of alcohols used and the degree of esterification; they may also act as efficient solubilizing agents, wetting agents, and surfactants (Constantinides and Scalart, 1997; Gibson, 2007; Jannin et al., 2008; Strickley, 2004). Among the polyoxyethylene-based surfactants, Labrasol and Cremophor EL are pharmaceutical excipients that have been well investigated as solubility and absorption enhancers in lipid-based formulations (Kommuru et al., 2001; Sha et al., 2005; Strickley, 2004).

SMEDDS formulations often contain, in addition to the oil and surfactant phase, hydrophilic cosolvents (e.g., low molecular weight PEGs, ethanol, propylene glycol, glycerin) (Gibson, 2007; Jannin et al., 2008). Although cosolvents are not amphiphilic molecules like the original cosurfactants, they may influence the solubility properties of both the water and oil phase and are used to facilitate the formation of microemulsions (Lawrence and Rees, 2000). Cosolvents also affect the capacity of SMEDDS formulations for solubilization of the drugs (Pouton, 2000). However, in some cases, cosolvents display significant chemical reactivity, bioincompatibility, or both. For example, PEGs are versatile, well-characterized solubilizers that are a widely applied class of pharmaceutical excipients (Rowe and Sheskey, 2003). However, when they are used in SMEDDS formulations, there is a risk of their irritating potential to the GI mucosa and relatively high chemical reactivity. PEGs contain peroxide impurities as well as secondary products formed by auto-oxidation, which can compromise the chemical stability of the oils and incorporated drug substance. Because of their hygroscopicity, PEGs may also affect the integrity of hard gelatin capsules (Hauss, 2007).

As already pointed out, the important task for formulators is to develop a SMEDDS formulation that has a sufficiently high solvent capacity for a given dose of a particular drug. Therefore, the generally accepted approach for screening of the excipients is the determination of the drug solubility in pure excipients (e.g., oils, surfactants) or preferably in the complete self-microemulsifying mixture (Jannin et al., 2008; Kang et al., 2004; Kommuru et al., 2001; Patel and Sawant, 2007; Ying et al.,

2008; Zhang et al., 2008). A recent comprehensive review by Rane and Anderson (2008) provides a summary of the fundamental factors that govern drug solubility in lipid mixtures; they also present models to estimate the solubility and describe the challenges involved in prediction of solubility. Pouton (2000) revealed earlier that triglycerides have a high solvent capacity for more hydrophobic drugs [log of the partition coefficient (P) > 4]. The solvent capacity for less hydrophobic drugs can be improved by blending triglycerides with mixed monoglycerides and diglycerides. Hydrophilic surfactants and cosolvents (propylene glycol, PEGs, ethanol, etc.) may also increase the solvent capacity of the SMEDDS formulation, particularly for the drugs with an intermediate log P value ($2 < \log P < 4$), which have limited solubility in both water and lipids (Pouton, 2000). Another important factor in the selection of the oil phase is the stability regarding oxidation. It is preferable to use hydrogenated vegetable oils, which may have better resistance to oxidative degradation. An additional aspect in the SMEDDS formulation is whether it should be liquid or semisolid, which will influence the options available for encapsulation. The melting temperature of semisolid (thermosoftening) materials (e.g., Gelucire 44/14, Gelucire 50/13, Gelucire 43/01, Gelucire 39/01, Labrafil M 2130CS) ranges from 26°C to 70°C, so they have a waxy consistency at room temperature and usually require melting at higher temperatures in lipid-based formulation manufacturing. Thermosoftening formulations are filled into capsules in the molten state and promptly solidify upon cooling to ambient temperature. In this case, issues arise regarding the temperature sensitivity of the drug or capsule shell. Thermosoftening formulations are frequently limited for hard capsules because of the lower melting point of the soft gelatin capsule shell, whereas liquid formulations are suitable for encapsulation in either hard capsules or soft gelatin capsules. Liquid formulations are more suited to thermolabile drugs compared to most thermosoftening formulations, the manufacturing of which is accompanied by relatively high temperatures during the filling process. However, liquid formulations may suffer from leakage of the capsule contents and thus require an additional capsule sealing operation (Hauss, 2007). Despite the improved physical stability profile of SMEDDSs in long-term storage compared to closely related ready to use microemulsions, these formulations generally suffer from susceptibility to oxidation. Some of the lipid-based products require low temperatures (2–8°C) for long-term storage, which is due to pronounced chemical or physical stability issues at room temperature (Hauss, 2007).

Optimum concentrations or the concentration ranges of the components of the SMEDDS formulation that are necessary to promote self-microemulsification are usually determined by phase behavior studies. The principles of microemulsion formation, preparation, and characterization of these systems, although outside the scope of this chapter, are elaborated in a numerous of comprehensive studies in the scientific literature (Evans and Wennerström, 1994; Ezrahi et al., 1999; Kahlweit, 1999; Sjoblom et al., 1996). The knowledge of phase transformations during the dilution of SMEDDS is extremely important. General observations from numerous studies pointed out two types of phase behavior during the dilution of an isotropic oil–surfactant mixture with water:

1. Transformation of W/O microemulsion into O/W microemulsion through a broad range of various polyphasic media (type S systems) (Clausse et al., 1987)
2. A continuous evolution from a W/O to an O/W microemulsion, which usually occurs through a bicontinuous state (described as type U systems) (Clausse et al., 1987)

SMEDDSs that generate type-S microemulsions undergo phase separation on dilution; that is, they form W/O and O/W microemulsions only within a specific and often narrow range of water concentrations. Conversely, Spernath and Aserin's (2006) comprehensive review documented that type-U microemulsions are preferred for SMEDDS formulations because of their capability for dilution with any given content of the water phase without disruption of the microemulsion state. Garti et al. (2003) patented type-U microemulsions as improved SMEDDS formulations that can be utilized for solubilization of both hydrosolubile and liposolubile drugs for promising oral and cutaneous delivery. Garti et al.'s (2006) recent study examined the potential of a U-type SMEDDS formulation, which consisted of R(+)-limonene, alcohol, propylene glycol, and Tween 60, in oral

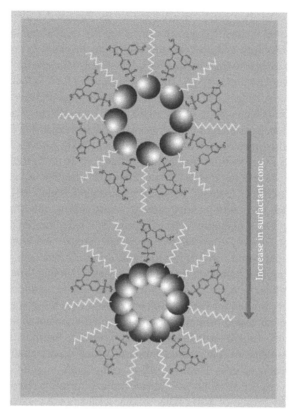

FIGURE 10.1 Schematic illustration (not to scale) of possible packing of celecoxib along dilution line 64 (60 wt % surfactant and 40 wt % oil phase) at three different dilution regions. Up: W/O microemulsion; middle: bicontinuous (not shown); down: O/W microemulsion. (From Garti, N., Avrahami, M., and Aserin A. *J. Colloid Interface Sci.* 2006; 299: 352–365. With permission.)

delivery of celecoxib. Celecoxib is a hydrophobic drug that has very variable absorption from the GIT. Particular attention was paid in this study to the characterization of the microemulsion structure and phase transitions along a suitable water dilution line, and a direct correlation between the microstructure and the drug solubilizing capacity was established. The transitions from the W/O to the bicontinuous phase and from this phase to the O/W microemulsion were identified at 30% (w/w) and 70% (w/w) of the aqueous phase, respectively. The solubilization capacity for celecoxib decreases as the water content increases. The self-diffusion study results showed that the drug is initially solubilized at the interface of the W/O droplets and then affects the bicontinuous structure; upon further dilution, the drug remains solubilized at the interface oriented with its hydrophilic part to the water, which strongly affected the inversion to O/W droplets (Figure 10.1).

Phase behavior studies should also assess the effect of drug loading on the efficiency of the self-emulsifying process and the diameters of the droplets (Shah et al., 1994). The construction of pseudoternary phase diagrams by Kang et al. (2004) in the presence of simvastatin was used to identify the self-emulsifying regions and to optimize the concentrations of oil (Capryol® 90), surfactant (Cremophor EL), and cosurfactant [diethyleneglycol monoethyl ether (Carbitol®)]. The authors found that the efficiency of the emulsification was good when the surfactant–cosurfactant mixture concentration was more than 40% (w/w) of the SMEDDS formulation and that the preferable ratio of surfactant/cosurfactant is 1:1. Similarly, Garti et al. (2006) investigated the interfacial behavior of celecoxib and estimated its effect on the phase behavior of an *R*-(+)-limonene/alcohol/propylene glycol/Tween 60 system. They found that the presence of celecoxib at a concentration of 4% in the oil phase did not affect the size of the isotropic region in the pseudo-ternary phase diagrams. Furthermore, the results of small angle x-ray scattering measurements indicate that the drug affects the structure of the microemulsion droplets, although the overall droplet sizes at high dilutions did not change very much. Djekic et al. (2008) recently presented a phase behavior study that may be useful in optimization of a cosurfactant and oil in potential SMEDDSs with maximized efficiency to form a microemulsion with a high content of the water phase. The results of this study revealed that microemulsion preconcentrates containing Labrasol (surfactant), isopropyl myristate (oil), and commercial mixtures of polyoxylethylene-type nonionic tensides [Solubilisant gamma® 2421 (Octoxynol-12 and Polysorbate 20) and Solubilisant gamma 2429 (Octoxynol-12, Polysorbate 20, and PEG-40 hydrogenated castor oil)] or PEG-40 hydrogenated castor oil (Cremophor RH 40) as cosurfactants were able to generate U-type microemulsions.

Numerous researchers have been involved in the formulation and characterization of different self-microemulsifying formulations as potential delivery systems for oral delivery of poorly soluble and poorly permeable drugs. It is evident from the literature that SMEDDS are a promising approach for promotion of solubilization and drug release at absorption sites, as well as the bioavailability of cyclosporine A (Kovarik et al., 1994; Mueller et al., 1994), simvastatin (Kang et al., 2004), coenzyme Q_{10} (CoQ$_{10}$; Kommuru et al., 2001), celecoxib (Garti et al., 2006; Subramanian et al., 2004), fenofibrate (Patel and Vavia, 2007), tacrolimus (Borhade et al., 2008), anethole trithione (Jing et al., 2006), acyclovir (Patel and Sawant, 2007), carvedilol (Wei et al., 2005), oridonin (Zhang et al., 2008), 9-nitrocamptothecin (Lu et al., 2008), vinpocetine (Ying et al., 2008), itraconazole (Woo et al., 2008), seocalcitol (Grove et al., 2006), β-artemether (Mandawgade et al., 2008), and exemestane (Singh et al., 2008). Furthermore, comparative biopharmaceutical studies demonstrated the significant potential for a more superior biopharmaceutical and pharmacokinetic profile of poorly soluble drugs with SMEDDS compared to both conventional oral solid dosage forms (tablets or capsules) and other lipid-based formulations such as lipid solutions and self-emulsifying drug delivery systems (SEDDSs; Fatouros et al., 2007; Jing et al., 2006; Lu et al., 2008; Mandawgade et al., 2008; Patel and Sawant, 2007; Patel and Vavia, 2007; Singh et al., 2008; Wei et al., 2005; Woo et al., 2008; Ying et al., 2008; Zhang et al., 2008).

There are two main aspects of the mechanism of action of oral SMEDDS formulations: the potential of the formulation to maintain the drug in the dissolved state in the GIT lumen during absorption and the enhancement of drug absorption (Pouton, 2000). The suggested critical factors of the *in vivo* performance of SMEDDSs are the solubility of the drug in SMEDDSs; the location of the drug in the microemulsion; the partition coefficient of the drug between the oil and water phase (i.e., GIT fluids); specific interactions of the drug with excipients; the size, polydispersity, and charge of the droplets; the site or path of absorption; the presence of components that can act as absorption enhancers; the influence of the variety of food materials; and the digestive fluids present in the GIT. The solubility of the drug in SMEDDS formulations is usually high, but the overall utility of each system will depend on its dilutability on use and whether the risk arises of the loss of solvent capacity on dilution followed by precipitation of the drug in the GIT (Narang et al., 2007; Pouton, 1997). The surfactants used in these formulations improve drug dissolution (Constantinides, 1985); however, their concentrations are occasionally insufficient to promote the self-emulsification process that will lead to drug precipitation upon dilution *in vivo* (Pouton, 2006). It is generally accepted that, during the dilution of SMEDDS, the hydrophilic surfactant will partition into the continuous (water) phase and percolation or phase separation may occur, which may affect the overall solubilization capacity for the drug. The extent of precipitation will mainly depend on the log P of the drug and to what extent the surfactant is involved in its solubilization within the formulation. In the cases where a water-soluble cosolvent (such as PEG or alcohol) is an important part of the formulation, dilution with water will usually result in partitioning of the cosolvent into the continuous phase, which may also lead to destruction of the microemulsion and to subsequent drug precipitation (Humberstone and Charman, 1997; Lawrence and Rees, 2000; Pouton, 1997, 2000). Thus, the development of methods for prediction of *in vivo* processes, particularly for assessment of the risk of drug precipitation, is an important task for researchers in the future (Narang et al., 2007). Most of the reported work investigating drug release from SMEDDS formulations actually deals with O/W microemulsions formed by dilution (Spernath and Aserin, 2006). The results of numerous studies demonstrates that the rate and extent of drug release from microemulsions depends on the physico-chemical properties of the drug as well as the composition and microstructure of the micro-emulsions (Bagwe et al., 2001; Kogan and Garti, 2006; Lawrence and Rees, 2000; Malmstein, 1999; Vandamme, 2002). Hydrophobic drugs are most likely dissolved in the oil phase of O/W microemulsions or are located at the interfacial film in the region of the hydrophobic tails of the amphiphilic molecules. Therefore, the diffusion of drug molecules is hindered and drug release is slow and predominantly depends on the partition coefficient of the drug. In some cases, the drug may interact with the other components of the microemulsion and significantly modify the phase behavior of the system as well as the dilution kinetics and drug release. This issue in particular may arise if the drug possesses amphiphilic properties (Djordjevic et al., 2004, 2005).

The significant improvement of the oral bioavailability of cyclosporine A when applied as a SMEDDS (Neoral, Novartis, Switzerland) compared to a SEDDS (Sandimmune®, Novartis, Switzerland) pointed out the general assumption that the main reason for this improvement is the small droplet size. Neoral is the first commercial SMEDDS formulation that was approved by the U.S. Food and Drug Administration in 1995 and entered successfully into the world market. Cyclosporine A is a hydrophobic cyclic undecapeptide that is often used as a potent immunosup-pressant in transplantation surgery (e.g., transplantation of liver or kidney), as well as in the treat-ment of some autoimmune disorders (psoriasis, rheumatoid arthritis). In contrast to most peptide drugs, cyclosporine A is insensitive to enzymatic degradation in the GIT. Thus, poor water solubility and high molecular weight are two main reasons for its poor oral bioavailability and high inter- and intraintraindividual variability in pharmacokinetics. From an historical perspective, the first regis-tered oral dosage form with cyclosporine A was Sandimmune (Novartis, Switzerland). Sandimmune is a solution of cyclosporine A in corn or olive oil, with the addition of a nonionic polyoxylglyceride type surfactant (Labrafil M 1944 CS) and ethanol (Klyashchitsky and Owen, 1998), in a dosage form of soft gelatin capsules (25 or 100 mg cyclosporine A/capsule) or as an oral liquid formula-tion containing 100 mg/mL of the drug. It has been shown that a Sandimmune emulsion precon-centrate improves the oral bioavailability of cyclosporine A, which is most likely attributable to lipolysis of the triglyceride to lower partial glycerides that may act as surfactants and thus enhance emulsification *in situ* as well as absorption. Furthermore, the variability of the pharmacokinetic properties is slightly reduced. Neoral is a self-microemulsifying isotropic mixture of cyclosporine A; mono-, di-, and triglycerides from corn oil; nonionic solubilizer polyoxyethylene-40 hydroge-nated castor oil (Cremophor RH40), propylene glycol, and ethanol (Klyashchitsky and Owen, 1998), which forms an O/W microemulsion in the GIT (Holt et al., 1994; Noble and Markham, 1995). Studies reported significant improvement of bioavailability and treatment efficacy with Neoral, and the superiority of Neoral over Sandimmune was reviewed (Fatouros et al., 2007; Gupta and Moulik, 2008; Lawrence and Rees, 2000). This improvement was connected with the microe-mulsion character of the administered drug (Kovarik et al., 1994). The delivery of cyclosporine A from Neoral was considered in detail and it was confirmed that the enhanced absorption of the drug is a function of nanometer-sized droplets (Kovarik et al., 1996; Ritschel, 1996; Vonderscher and Meinzer, 1994). Because the droplet surface area is inversely proportional to the diameter, smaller lipid droplets result in a large surface area from which the drug can partition, thus enhancing the rate of dissolution into a contacting aqueous phase. The small droplets also have a better chance to adhere to biological membranes and provide an increase in absorption and drug bioavailability (Humberstone and Charman, 1997). Therefore, it seems that the concept "smaller is better" is applicable for lipid-based formulations. However, in a recent study, Andrysek (2003) observed that there was no difference in the pharmacokinetic parameters between coarse dispersions (average particle size = 1–150 μm) and a Neoral microemulsion (particle size <0.15 μm) tested on healthy volunteers. Nevertheless, the absorption of cyclosporine A is more predictable and enhanced from

Neoral compared to Sandimmune and the pharmacokinetic variability is significantly reduced (Erkko et al., 1997; Noble and Markham, 1995; Ritschel, 1996). Neoral and Sandimmune are not bioequivalent, and it is necessary to adjust the dose and the regime of drug administration if there is a need to replace one product with other. Mueller et al. (1994) found that Neoral exhibited a linear dose/area under the curve relationship in the range of 200–800 mg in contrast to Sandimmune in fasted healthy volunteers and determined a dose-dependent relative bioavailability of Neoral versus Sandimmune of 174–239%, with the highest dose resulting in the largest difference in bio-availability. On the current market, there are several SMEDDS formulations of cyclosporine A that are bioequivalent to Neoral (Gengraf, Abbott Laboratories, USA), SangCya (SangStat Medical Corporation, USA), Consupren® (IVAX-CR, Czech Republic), and products under a generic name from different manufacturers (Eon Laboratories, USA; Apotex Corp., USA; Pliva Inc., Croatia; Bedford, USA). However, it has been established that the bioavailability of cyclosporine from these formulations compared to Neoral exhibited significant pharmacokinetic variations in differ-ent human subpopulations (Dunn et al., 2001; Pollard et al., 2003). It was recently revealed that the bioavailability of SangCya administered with apple juice is lower compared to Neoral, although these two formulations are bioequivalent when administered with chocolate milk; when mixed with apple juice, Neoral forms a microemulsion but SangCya forms a microdispersion (Chen, 2007). This case pointed out that the particular liquids that are recommended for SMEDDS dilu-tion prior to oral intake may affect the performance of the SMEDDS formulations, particularly the droplet size, as well as their potential for drug delivery. The influence of the droplet size on the oral bioavailability was also clearly observed for simvastatin, a cholesterol-lowering agent that is prac-tically insoluble in water and poorly absorbed from the GIT (Kang et al., 2004). The components of the optimal SMEDDS formulation with improved drug loading capabilities were Capryol 90 (oil), Cremophor EL (surfactant), and Carbitol (cosurfactant). The release of simvastatin from opti-mized SMEDDS formulations A (mean droplet size = 33 nm) and D (mean droplet size = 150 nm) was significantly higher than from the conventional tablet (Zocor®; 20 mg simvastatin) at pH 6.8. When administering the prefilled hard capsules to fasted beagle dogs, the relative bioavailability of simvastatin from SMEDDS formulations A and D compared to the conventional tablet was 159% and 143%, respectively. Furthermore, the release of simvastatin from the SMEDDS A formulation was faster than from the D formulation. Kang et al. (2004) suggest that simvastatin dissolved per-fectly in SMEDDS and the small droplet size enables a faster rate of drug release than the conven-tional tablet, leading to improved bioavailability. They also assumed that the drug release rate could be controlled by regulating the mean droplet size of the carrier. According to Hauss (2007), *in vitro* evaluation of the oil droplet size formed in a biorelevant aqueous test medium from lipid-based formulations that incorporate large amounts of surfactant could be very informative about drug release *in vivo*.

The mechanisms by which SMEDDSs influence drug delivery are complex and not yet fully eluci-dated. The mechanisms of enhanced drug absorption mediated by lipid-based formulations include (a) increased membrane fluidity facilitating transcellular absorption, (b) opening of tight junctions to allow paracellular transport, (c) inhibition of P-glycoprotein mediated drug efflux and/or metabolism by gut membrane-bound CYP450 enzymes, and (d) enhanced lymphatic drug transport occurring in conjunction with stimulation of lipoprotein and chylomicron production (O'Driscoll, 2002). Increased intestinal epithelial permeability, increased tight junction permeability, and decreased or inhibited P-glycoprotein drug efflux are frequently ascribed to the surfactants and cosolvents (Pouton, 2000, 2006; Pouton and Porter, 2008). Sha et al. (2005) prepared negatively or positively charged SMEDDS containing Labrasol as a surfactant to evaluate the effect of charge and dilution of formulations on Caco-2 cell monolayer permeability for mannitol (usually used as a marker of the permeability of hydrophilic molecules through the tight junction). They found that Labrasol at concentrations of 0.1% (w/w) and 1% (w/w) could increase the permeability of mannitol 4.6-fold and 33.8-fold, respectively, and concluded that a negatively or positively charged SMEDDS containing Labrasol enhanced the paracellular transport of mannitol at various dilutions. Moreover, the positively charged SMEDDS in higher dilutions opened the tight junctions reversibly, so the Caco-2 monolayer was not permanently damaged. Kommuru et al. (2001) investigated SMEDDS formulations for the delivery of CoQ_{10}, a potent high molecular weight antioxidant that is practically insoluble in water with poor absorption from the GIT. Four types of self-emulsifying formulations were prepared and characterized using two oils (Myvacet® 9-45, acetylated monoglycerides; and Captex®-200, mixed diesters of caprylic and capric acids of propylene glycol), two emulsifiers (Labrafac® CM-10 and Labrasol), and a cosurfactant (Lauroglycol®). In all formulations, the level of CoQ_{10} was fixed at 5.66% (w/w) of the vehicle. Efficient and better self-emulsification processes were observed for the systems containing Labrafac CM-10 than formulations containing Labrasol. Although upon dilution the optimized formulation consisting of Myvacet 9-45 (40%), Labrasol (50%), and Lauroglycol (10%) formed a coarse emulsion (25 μm), oral administration of the SMEDDS in dogs provided a twofold increase in bioavailability compared to a powder formulation. It appeared that the increased permeability of mannitol and improved bioavailability of CoQ_{10} is the result of the presence of the absorption enhancer Labrasol instead of the droplet size obtained upon dilution with water. Therefore, the presence of excipients that have the potential to enhance absorption is very important for the ability of the SMEDDS to improve the bioavailability of the drugs. Thus, the important challenge for formulators is to find effective absorption enhancers that can also form SMEDDSs over a wide range of water concentrations.

Certain lipidic excipients are associated with selective drug uptake into the lymphatic transport system, thereby reducing the effect of first-pass drug metabolism in the liver (Porter et al., 2007). A comparative study of the pharmacokinetics and bioavailability of silymarin (the drug primarily used for liver diseases) from a suspension, PEG 400 solution, and SMEDDS was reported (Wu et al., 2006). The SMEDDS consisted of silymarin, Tween 80, ethyl alcohol, and ethyl linoleate. Although *in vitro* release of silymarin from the SMEDDS was limited, incomplete, and typical of sustained characteristics, the relative bioavailability of silymarin evaluated in rabbits was dramatically enhanced by the SMEDDS on an average of 1.88- and 48.82-fold compared to the solution and suspension, respectively. The results pointed out that alternative mechanisms, such as an improved lymphatic transport pathway, other than improved release may contribute to the enhancement of the bioavailability of the drug. There were several estimations of the potential of SMEDDS formulations for improvement of the oral bioavailability of halofantrine, a highly hydrophobic antimalaria drug (Holm et al., 2003; Khoo et al., 1998, 2003). Khoo et al. (1998) achieved six- to eightfold higher bioavailability of halofantrine from SMEDDS formulations consisting of medium- or long-chain trglycerides, monoglyceride (Capmul MCM, mono- or diglycerides of caprylic acid), and ethanol compared to the tablet formulation. In a latter study in dogs, they showed that lymphatic transport plays a major role in the absorption of halofantrine from these SMEDDS based on long-chain triglycerides (Khoo et al., 2003). In a related study, Holm et al. (2003) demonstrated that the lymphatic transport and the absorption via the portal system were both affected after administration of halofantrine in two SMEDDS based on two different structural triglycerides, 1,3-dioctanoyl-2-linoleyl-*sn*-glycerol (MLM) and 1,3-dilinoyl-2-octanoyl-*sn*-glycerol, previously developed by Khoo et al. (1998), in a triple-cannulated canine model. The total bioavailability from the SMEDDS containing MLM (74.9%) was higher because of elevated portal transport of halofantrine when dosed in the MLM vehicle. This study suggests that the use of different structure triglycerides in a SMEDDS opens up the possibility to manipulate the relative contribution of the two absorption pathways within certain limits.

As holds in the case of cyclosporine A, absorption of bioactive molecules from SMEDDS occurs in a more controlled fashion. Studies with Sandimmune (300 mg cyclosporine A) and Neoral (180 mg cyclosporine A) in fasted healthy volunteers showed that the absorption is more rapid (time to reach maximum concentration of drug in the plasma) and peak concentrations of the drug are higher (higher maximum drug concentration in plasma and area under the curve) from the SMEDDS formulation (Holt et al., 1994; Kovarik et al., 1994; Mueller et al., 1994; Figure 10.2). Therefore, therapeutic levels of the drug could be achieved more rapidly and with lower doses after *de novo* administration of the SMEDDS (Amantea et al., 1997; Pouton, 2000).

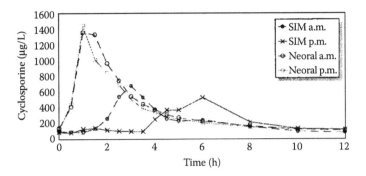

FIGURE 10.2 Cylcosporine blood concentration profiles from a renal transplant patient given Sandimmune (SIM) or Neoral without food (a.m.) or with food (p.m.). (From Holt, D. W., et al. *Transplant. Proc.* 1994; 26: 2935–2939. With permission.)

Subramanian et al. (2004) developed SMEDDS formulations with considerable potential to minimize the variability in the absorption after oral administration and to provide rapid onset of action of celecoxib, a specific cycloxygenase-2 inhibitor for the treatment of rheumatoid arthritis, osteoarthritis, and acute pain conditions. Celecoxib is a hydrophobic (log $P = 3.5$, aqueous solubility = 3–7 μg/mL) and highly permeable drug belonging to class II of the biopharmaceutical classification system. The SMEDDS formulation, which was optimized by a simplex lattice mixture design, consisted of PEG-8 caprylic/capric glycerides (49.5%), a 3:1 mixture of Tween 20 and propylene glycol monocaprylic ester (40.5%), and celecoxib (10%). The optimized formulation showed enhancement of the absorption and a subsequent increased bioavailability of 132% relative to the conventional solid dosage oral form (capsules). SMEDDS formulations are very useful in diminution of the effect of food on the absorption of poorly water-soluble drugs, as in the Neoral formulation of cyclosporine. Furthermore, unlike lipid solutions and SEDDSs, SMEDDS formulations are much less affected by digestion and the influence of bile salts in the GIT lumen (Goddeeris et al., 2007; Pouton, 1997; Tarr and Yalkowsky, 1989; Woo et al., 2008).

To summarize, SMEDDS formulations may be useful in oral delivery of poorly soluble drugs because of the significant potential for more superior biopharmaceutical and pharmacokinetic profiles compared to conventional oral solid dosage forms (tablets or capsules) and lipid-based formulations such as lipid solutions and SEDDS, including improved bioavailability and reduced variability of drug absorption. Moreover, from the industrial point of view, the manufacturing process of SMEDDS formulations requires readily available equipment and offers possibilities for reduction in, or elimination of, a number of development and processing steps (e.g., salt selection or identification of a stable crystalline form of the drug, coating, taste masking, dedusting). A review of the literature shows that SMEDDSs have attracted a lot of attention from researchers and companies as promising delivery systems, but a close examination of the formulations that are present in the market place reveals that the exploitation of this concept is still not equivalent to the documented drug delivery potential. A considerable gap still exists between the interest of the scientific community for the SMEDDS concept and a number of marketed products based on a self-microemulsification phenomenon. The major issues contributing to this discrepancy are the following:

1. The lack of clearly defined procedures for developing SMEDDS formulations, which involves the selection of biocompatible excipients with adequate functionality regarding the drug solubility, the physical and chemical stability of the solubilized drug and of the excipients, and the formulation performance

2. The complexity of the physicochemical and biopharmaceutical characterization of SMEDDS formulations using *in vitro* and animal studies, particularly including the assessment of the dilution kinetics and the fate of the formulation and the drug *in vivo* (Attama and Nkemnele, 2005; Fatouros et al., 2008; Goddeeris and Van den Mooter, 2008; Grove et al., 2007; Ljusberg-Wahren et al., 2005)

3. The lack of particular techniques for more precise characterization of SMEDDS performance during the development is a hurdle for numerous potential SMEDDS formulation to get onto the level of the clinical testing.

Although a number of *in vitro* tests to investigate the effects of dispersion, digestion, and gastric emptying that are useful in prediction of the fate of drug are being developed as methods for optimizing lipid-based formulations, it is necessary to establish standard test protocols, particularly in the case of the lipolytic digestion test, so that bioavailability data can be better understood and compared by investigators from various laboratories. Therefore, future research should be committed to the development of relevant *in vitro* tests that can predict the fate of the formulation and, most importantly, the drug and to conduct more studies on the mechanism of action in *in vivo* animal investigations as well as human bioavailability studies.

10.2.2 Self-Microemulsifying Drug Delivery Systems for Parenteral Administration

The development of formulations for parenteral administration of drugs with limited solubility, especially via the intravenous route, is a permanent challenge for formulators. Microemulsion preconcentrates indirectly exploit the benefits of microemulsions, such as high solubilization capacity, small droplet size, clarity, and low viscosity, thus representing an interesting strategy for the improvement of parenteral delivery for poorly soluble drugs. In addition, microemulsions open the possibility for sterilization by filtration and the avoidance of destabilization of the drug delivery system during autoclaving that occurs because of the presence of a phase inversion temperature in the systems stabilized by nonionic surfactants (Lawrence and Rees, 2000; Malmstein, 1999). The category of microemulsion preconcentrates for parenteral application actually includes both the formulations that are diluted with appropriate amounts of water phase forming a microemulsion prior to application and SMEDDS formulations for direct intravenous administration that will form an O/W microemulsion *in situ* on dilution by a biological aqueous phase. Although the literature contains details of many microemulsion systems, only a few of these are acceptable for parenteral administration because of the toxicity of the surfactant, cosurfactant, and oil phases. Therefore, the number of published studies concerning microemulsion preconcentrates for intravenous administration is rather small. Park et al. (1999) investigated the possibility for parenteral delivery of flurbiprofen using ethyl oleate, lecithin, distearoylphosphatidylethanolamine-*N*-PEG 2000 (DSPE-PEG), and ethanol as components of the preconcentrate. The phospholipid-based microemulsion preconcentrate solubilized more than 10 mg/mL of flurbiprofen at a vehicle/drug ratio of at least 10:1, and oil contents of 10% or 20%, were used. The drug concentrations in plasma and various organs after the intravenous administration of a flurbiprofen-loaded microemulsion were measured and compared with those after the intravenous administration of a commercial emulsion (Lipfen®, 10 mg/mL as flurbiprofen axetil) and flurbiprofen solution in rats. The half-life, area under the curve, and mean residence time of flurbiprofen loaded in the microemulsion (ethyl oleate/lecithin/ DSPE-PEG/flurbiprofen = 8:3:1:1.2) increased significantly, and the biodistribution of the drug was quite different from the emulsion and solution (Figure 10.3). Reticuloendothelial uptake of flurbiprofen loaded in the microemulsion was decreased compared with that in solution or Lipfen.

Similarly, a biocompatible microemulsion concentrate for intravenous administration composed of a medium-chain triglyceride, soybean phosphatidylcholine, PEG 660-12-hydroxystearate (12-HSA-EO15), PEG 400, and ethanol was developed by von Corswant et al. (1998). They showed that the microstructure of the concentrated system was bicontinuous, and on dilution it turned into

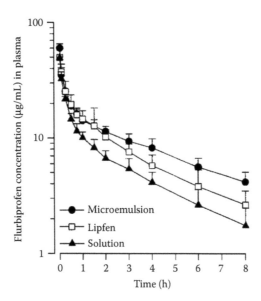

FIGURE 10.3 The plasma concentration versus time profiles of intravenous administered flurbiprofen-loaded microemulsion (ethyl oleate/lecithin/DSPE-PEG/flurbiprofen = 8:3:1:1.2), emulsion (Lipfen), and solution, equivalent to 2.5 mg/kg of flurbiprofen, in rats ($n = 5$). (From Park, K. M., et al. *Int. J. Pharm.* 1999; 183: 145–154. With permission.)

an O/W microemulsion with 60- to 200-nm droplets. Furthermore, the *in vivo* hemodynamic response after intravenous application to conscious rats showed that it was possible to administer up to 0.5 mL/kg of the formulation with an oil content of 50% without producing any significant effect on the acid–base balance, blood gases, plasma electrolytes, mean arterial blood pressure, heart rate, and time between depolarization of the atrium and chamber. This study evaluated the solubilization potential for felodipine, although *in vivo* investigations were conducted only for the drug-free vehicle.

Lee et al. (2002) incorporated clonixic acid into a premicroemulsion concentrate to overcome the poor solubility of the drug as well as to reduce the pain connected to intramuscular or intravenous application of a commercial dosage form. Although it was possible to incorporate 3.2 mg/mL of clonixic acid in the optimized formulation, which contained castor oil, Tween 20, and Tween 80 in a weight ratio of 5:12:18, respectively, the *in vivo* investigations indicated less pain on injection but the pharmacokinetic parameters were not significantly different.

In spite of the promising potential of SMEDDSs for improved solubilization of poorly water-soluble drugs for intravascular administration, which is documented in published studies, the lack of research in this field hinders the extraction of more general conclusions. The major obstacles that should be overcome for more extensive investigations of such formulations are a narrow choice of biocompatible surfactants and cosolvents that are acceptable for parenteral administration; the complicated physicochemical and biopharmaceutical characterization, including the performance in regard to the phase behavior and the droplet size; and the *in vivo* evaluation of the hemodynamic response after the administration of such formulations.

10.2.3 Self-Microemulsifying Drug Delivery Systems for Topical Delivery

There have been a few attempts to utilize the SMEDDS concept in cutaneous drug delivery of lidocaine/prilocaine (eutectic mixture) (Shukla et al., 2003), β-blockers (Kemken et al., 1991, 1992), and indomethacin (Fujii et al., 1996). The main idea in dermal administration of poorly soluble drugs using microemulsion preconcentrates was to achieve a supersaturation of the drug by hydration of the vehicle, resulting in an enhanced thermodynamic activity of the drug and improved pharmaco-dynamics (Kemken et al., 1992). In a recent study, Cirri et al. (2007) developed a liquid SMEDDS formulation intended for oral spray administration of xibornol, a lipophilic drug used for the local treatment of infection and inflammation of the throat. The selected formulations, which contained Labrafil M1944, Transcutol®, Labrafac PG, and a hydrophilic cosolvent (propylene glycol or PEG 200), allowed complete solubilization of the therapeutic concentration of xibornol (3%, w/v) and showed suitable viscosity and good physical stability.

In spite of promising results related to topical drug delivery, this strategy is rarely used in this field, mainly because of toxicity and irritability issues arising from high levels of surfactants as well as the obstacles for detailed investigations of the *in vivo* performance of the system.

10.3 MICROEMULSIONS AS MEDIA FOR SYNTHESIS OF NANOPARTICLES

Application of microemulsions as reaction media is currently of interest to the scientist as a prospec-tive relatively simple technique for synthesis of nanoparticulate drug carriers and as an approach for particle size reduction of poorly soluble drugs. Because of their unique combination of properties (thermodynamic stability, dynamic character of the microstructure, and high capacity for solubili-zation of both hydrophilic and lipophilic compounds), microemulsions have been used as reactors for the synthesis of nanoparticulate carriers of drugs such as SLNs or polymeric nanoparticles or as a method for the particle size reduction of solid drug substances (Datea and Patravale, 2004; Gasco and Trotta, 1986; Lopez-Quintela et al., 2004; Marengo et al., 2000; Reis et al., 2006). Nanoparticles obtained by the microemulsion technique usually have diameters of less than 100 nm. The direct correlation between the diameter of the particles and the droplets of the dispersed phase of micro-emulsions was established (Datea and Patravale, 2004; Gasco and Trotta, 1986; Lopez-Quintela et al., 2004; Marengo et al., 2000; Reis et al., 2006). Furthermore, scale up is relatively simple with appropriate optimization of the reaction medium. The main disadvantage of the microemulsion approach for nanoparticle synthesis is the high content of surfactants and cosurfactants. The use of biocompatible nonionic polyoxyethylene-type surfactants (e.g., poloxamers, polysorbates) and phos-phatidylcholine as well as more favorable solvents with respect to regulatory aspects (e.g., diethyl-eneglycol monoethyl ether, ethanol, isopropanol) were recommended. Another critical aspect is the required purification of the suspension of nanoparticles and their isolation from the reactor, usually by ultrafiltration, diultrafiltration, ultracentrifugation, or dialysis, that may significantly increase the costs of manufacturing process. Finally, the transfer of a nanoparticle dispersion into a product in the dry state is a necessary additional effort for elimination of huge amounts of water (e.g., by vac-uum evaporation or by liophilisation) (Datea and Patravale, 2004; Gasco and Trotta, 1986; Lopez-Quintela et al., 2004; Marengo et al., 2000; Reis et al., 2006). The next sections of this chapter provide a review of the achievements in this field.

10.3.1 SOLID LIPID NANOPARTICLES PRODUCED BY MICROEMULSION TECHNIQUE

Gasco (1993) developed a method for synthesis of SLNs from O/W microemulsions. The oil phase of the microemulsion reactors were fatty acids, glycerides, or both with melting temperature above 50°C. A relatively high concentration of the lipid phase (e.g., 30%) is preferred for technological reasons. The lipids and the water phase, which contained a mixture of lecithin (surfactant), bile salts, and butanol (cosurfactants), were heated separately at 60–70°C and then mixed until an O/W microemulsion formed. In the next step a hot microemulsion was dispersed into a huge amount of cold water (2–3°C) and the microemulsion broke up and lipid droplets recrystallized, forming lipid nanoparticles. The pharmaceutical industry (Vectorpharma, Trieste, Italy) expressed interest for production of SLNs by the microemulsion technique and recognized the process parameters that could be critical during the scaling up, such as the temperature of the microemulsion and water the temperature flows in the water medium, and the hydrodynamics of mixing (Marengo et al., 2000).

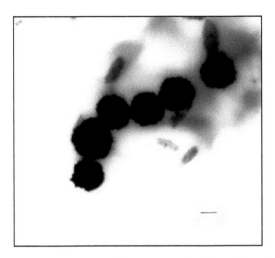

FIGURE 10.4 Photomicrography of SLNs containing cyclosporine A (scale bar = 100 nm). (From Ugazio, E., Cavalli, R., and Gasco, M. R. *Int. J. Pharm.* 2002; 241: 341–344. With permission.)

The microemulsion technique was applied for synthesis of SLNs loaded with the poorly soluble drugs cyclosporine A (Ugazio et al., 2002) and paclitaxel (Koziara et al., 2006). The diameter of the cyclosporine A SLN (Figure 10.4), which was synthesized in a matrix of stearic acid, taurocholate, and lecithin, was lower than 300 nm and the drug loading capacity was up to 13% (Ugazio et al., 2002). Although the *in vitro* release of cyclosporine A from the SLN was low, the authors proposed it for most administration routes, in particular, the duodenal route. Paclitaxel was entrapped in emulsifying wax-based nanoparticles by Koziara et al. (2006). This study demonstrated the efficacy of the obtained nanoparticles in an HCT-15 mouse xenograft model. They concluded that the inhibition in tumor growth of the nanoparticles over the commercial formulation (Taxol®) in the xenograft model was due to overcoming paclitaxel resistance and an antiangiogenic effect.

The microemulsion technique was similarly used for obtaining a biocompatible SLN consisting of emulsifying wax and polyoxyethylene alkyl ethers (Brij® 78) as nonionic surfactants (Lockman et al., 2003). This type of SLN was examined as a potential carrier for thiamine for specific brain drug delivery. The SLNs were coated with thiamine ligand (thiamine linked to DSPE via a PEG spacer). The mean diameter of the particles was 70 nm (Figure 10.5), and the polydispersity index was low. This study demonstrated the effectiveness of using microemulsions as precursors to engineering nanoparticles as well as the possible mechanism for drug delivery that is likely the promotion of the binding or association of the nanoparticles with blood–brain barrier thiamine transporters by the thiamine ligand.

The microemulsion technique was applied in the synthesis of positively charged SLNs. Cationic SLNs were investigated as potential carriers for deoxyribonucleic acid (DNA) molecules in gene therapy. The SLNs obtained in the system based on a nonionic surfactant (polysorbate 80) and butanol with the addition of stearylamine in the lipid phase (different triglycerides) ranged from 100 to 500 nm with a zeta potential of around +15 mV (Heydenreich et al., 2003). Zhengrong and Mumper (2002) developed a nanoengineered genetic vaccine for topical application by coating plasmid DNA (*p*DNA) on the surface of cationic nanoparticles obtained directly in the microemulsion system in which the lipid phase was emulsifying wax and the surfactant was cationic amphiphile cetyltrimethylammonium bromide. DNA molecules were attached on the surface of the particles by specific ligands. The humoral and proliferative immune responses were assessed in shaved *BALB/c* mice. All *p*DNA-coated nanoparticles resulted in significant enhancement in both the antigen-specific IgG titers (16-fold) and the splenocyte proliferation over *p*DNA alone.

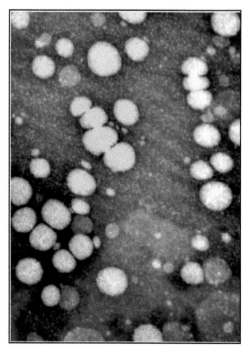

FIGURE 10.5 Photomicrography of thiamine-coated SLNs. (From Lockman, P. R., et al. *J. Control. Release* 2003; 93: 271–82. With permission.)

10.3.2 MICROEMULSIONS AS MEDIA FOR SYNTHESIS OF POLYMERIC NANOPARTICLES

Gasco and Trotta (1986) developed a simple method for synthesis of poly(alkyl cyanoacrylate) nanocapsules by using pharmaceutically acceptable W/O microemulsions (*microemulsion polymerization*). The polymerization process started in the microemulsion reaction medium when a cyanoacrylate monomer was activated by an initiator (ion or radical). The polymer formed *in situ* along the inner surface of the droplets of the W/O microemulsion (Figure 10.6). The nanocapsules were found to have a central cavity surrounded by a polymer wall. Furthermore, it was demonstrated that many pharmaceutically relevant physicochemical properties of poly(alkyl cyanoacrylate) nanocapsules prepared by polymerization of microemulsions can be easily adjusted by changing the pH of the aqueous phase, the water weight fraction of the microemulsion, or the mass of monomer used for polymerization (Watnasirichaikul et al., 2002).

The investigations of nanocapsules obtained by this technique revealed that if the drug is introduced in the reaction medium after the polymerization, it adsorbs on the surface of nanocapsules. In this case, the capacity for drug adsorption depends on the mutual affinity of the polymer and the drug. If the drug is introduced into the system before polymerization starts, it encapsulates inside the nanocapsules. Pitaksuteepong et al. (2002) established that the drug encapsulation efficacy depends on its molecular weight and charge. There was a high efficiency of encapsulation for hydrophilic peptide and protein drugs when applied this method (e.g., more than 80% for insulin) using microemulsions based on polyoxyethylene surfactants (Pitaksuteepong et al., 2002; Watnasirichaikul et al., 2000). In a related study Graf et al. (2008) showed that microemulsions containing sugar-based surfactants (decyl glucoside or capryl or caprylyl glucoside) are also suitable formulation templates for the formation of nanoparticles to deliver peptides. The obtained nanoparticles ranged from 145 to 660 nm with a unimodal size distribution depending on the type of monomer (ethyl-2- or butyl-2-cyanoacrylate, larger nanoparticles were formed by butyl-2-cyanoacrylate) and microemulsion template. Insulin, a model protein, did not alter the physicochemical properties of the microemulsions or the

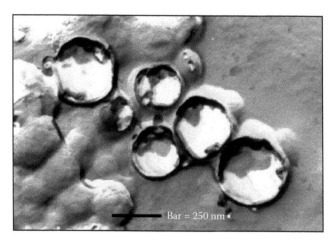

FIGURE 10.6 Photomicrography of poly(ethyl-2-cyanoacrylate) nanocapsules obtained by interfacial polymerization. (From Pitaksuteepong, T., et al. *Eur. J. Pharm. Biopharm.* 2002; 53: 335–342. With permission.)

morphology of the nanoparticles, although it induced a decrease in the nanoparticle size in the system containing capryl or caprylyl glucoside when using butyl-2-cyanoacrylate. In a recent study, Sahiner et al. (2007) applied the microemulsion polymerization method in the synthesis of poly(acrylonitrile-*co*-*N*-isopropylacrylamide) core–shell hydrogel nanoparticles. At the optimal reaction conditions, they prepared highly monodispersed nanoparticles with a size range of 50–150 nm. Furthermore, the hydrophobic core of the nanoparticles, which consists primarily of polyacrylonitrile, can be easily made highly hydrophilic by substituting the nitrile groups with amidoxime groups. They showed that the loading and release capacity of the nanoparticles for a model drug, propranolol, was increased almost twofold by the amidoximation of their core.

10.3.3 MICROEMULSIONS IN NANOENGINEERING OF POORLY SOLUBLE DRUGS

Microemulsions were successfully used for the preparation of nanoparticles of hydrophobic drugs via their direct precipitation in the water core of W/O microemulsions (*solvent diffusion technique*) (Debuigne et al., 2001; Trotta et al., 2003). The first step in this method is solubilization of the hydrophobic drug in the nonpolar or oil phase of the O/W microemulsion. In the second step, the microemulsion is diluted with the water phase. During the dilution, the dispersed nonpolar phase diffuses into a continuous phase. Therefore, the solubility of the drug decreases significantly leading to its precipitation and the formation of a suspension of nanodroplets. This technique was exploited successfully in the synthesis of nanoparticles of nimesulide (Debuigne et al., 2001) and griseofulvin (Trotta et al., 2003). Nanoparticles of nimesulide were obtained in W/O lecithin/isopropyl myristate/water/*n*-butanol or isopropanol microemulsion system. The size of the particles was 4–6 nm with a narrow distribution (Figure 10.7), which was independent of the formulation variables (Debuigne et al., 2001). In contrast, nanoparticles of griseofulvin with a diameter below 100 nm and low polydispersity were obtained with this technique after the optimization of a cosurfactant in a solvent-in-water microemulsion containing water, butyl lactate, lecithin, dipotassium glycyrrhizinate, and 1,2-propanediol or ethanol. Furthermore, the rate of dissolution of griseofulvin from the nanoparticles was increased threefold (Trotta et al., 2003).

Oyewumi et al. (2004) investigated the potential for targeted drug delivery in neutron capture therapy of tumors by gadolinium nanoparticles prepared from O/W microemulsion templates and coated with a folate ligand. Biodistribution and tumor retention studies were carried out at predetermined time intervals after injection of nanoparticles (10 mg/kg) into human nasopharyngeal carcinoma-bearing athymic mice. The gadolinium nanoparticles did not affect platelets or

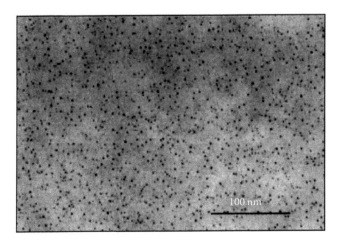

FIGURE 10.7 Photomicrography of nimesulide nanoparticles obtained by the solvent-diffusion method. (From Debuigne. F., et al. *J. Colloid Interface Sci.* 2001; 243: 90–101; Gasco, M. R. U.S. Patent 5250236, 1993. With permission.)

neutrophils. The retention of nanoparticles in the blood at 8, 16, and 24 h postinjection was 60%, 13%, and 11% of the injected dose, respectively. A maximum tumor localization of the drug (33 ± 7 μg/g) was achieved. The cell uptake and tumor retention of folate-coated nanoparticles was significantly enhanced over PEG-coated nanoparticles (as a control).

10.4   RECENT DEVELOPMENTS AND FUTURE DIRECTIONS

The development of strategies based on the microemulsification phenomenon for improvement of delivery of drugs with limited solubility such as SMEDDS and nanoparticles obtained in microemulsion reactors are closely connected to the development of the microemulsions themselves. The use of microemulsions in these strategies offers many potentially excellent advantages, but it is also an attractive field for research with many challenges to be overcome. The main objectives of the current investigations in this area are overcoming the obstacles related to the toxicological and regulatory aspects of potential excipients (surfactants, oils, and cosolvents) in order to improve the biocompatibility of the SMEDDS and microemulsions, development of methods for optimization of SMEDDS formulations or microemulsion media for nanoparticle synthesis (including the achievement of targeted delivery by coating of nanoparticles with ligands), and improvement of the performance of liquid and semisolid SMEDDSs by formulation of the solid dosage forms (e.g., tablets, capsules, granules, or pellets).

The persistent issue in the formulation of pharmaceutically acceptable microemulsions and SMEDDSs concerns the biological acceptability of the excipients, especially the surfactants and cosolvents, which may compromise their overall utility. The regulatory status of the different excipients will also depend on their intended use (e.g., oral, parenteral, and dermal). The development of biocompatible pharmaceutical microemulsions is becoming a very important technological challenge. Researchers are putting great efforts into the development of microemulsions based on excipients with *food grade*, *orally safe*, or *generally recognized as safe* status. Although ethoxylated alkyl ethers and sorbitan esters, which are synthetic nonionic surfactants that are often used in microemulsion systems, are generally less irritating and toxic than other classes of surfactants, microemulsions prepared from phospholipids and polyol-type nonionic surfactants, such as alkyl polyglucosides, sucrose esters, and polyglycerol esters of fatty acids (Chai et al., 2003; Fanun Al-Diyn, 2006; Garti et al., 1999; Glatter et al., 2001; Söderlind et al., 2003; Syamasri and Moulik, 2008), seem to be preferred from a toxicity and biodegradability viewpoint. These possible nontoxic alternative surfactants are prepared from natural resources that have hydrophilic and lipophilic properties that can be tuned by varying of the alkyl- or polyol-chain length. The phase behavior of systems involving these surfactants are much less influenced by temperature compared to systems based on polyoxyethylene surfactants (Chai et al., 2003; Garti et al., 1999). Therefore, there is a particular interest in the development of biocompatible microemulsions based on polyol surfactants. However, the preparation of microemulsions with these surfactants is not a simple task. The majority of polyol-type surfactants are not able to form microemulsions without the aid of cosurfactants such as short- or medium-chain length alcohols. Sugar-based surfactants cause severe hemolysis below or at the critical micelle concentration in contrast to the polyoxyethylene-based surfactants that are nonhemolytic in this concentration range (Söderlind et al., 2003). Further, the achievement of a uniform particle size in polyol-type surfactant-based microemulsions may requires complex size separation techniques. An additional factor affecting the use of SMEDDSs and microemulsions is the high content of the surfactants and cosolvents. These systems often require high surfactant concentrations in order to provide very low interfacial tension ($\leq 10^{-3}$ mN/m) as well as sufficient interfacial coverage to microemulsify the entire oil and water phases (Wennerström et al., 2006) and to provide a sufficiently high drug solubilizing capacity including the prevention of drug precipitation *in vivo*. However, high surfactant levels are not acceptable because of bioincompatibility, performance, or economic reasons. Furthermore, the purification of nanoparticles from microemulsion reaction media from surfactant and cosolvent contaminants is usually complex and expensive. Heydenreich et al. (2003) recently showed that the cell toxicity of SLNs prepared in Tween 80/butanol/stearylamine/triglycerides was dependent on both the SLN composition and the purification method. Therefore, there is a need for a suitable approach to maximize the efficiency of the surfactant, that is, to minimize the concentration required for stabilization of the microemulsion. Interest in using nonionic tensides as a surfactant and as a cosurfactant (so-called nonalcohol cosurfactants) is increasing because of the high stability, low toxicity, low irritancy, and biodegradability of many nonionic surfactants (Lawrence and Rees, 2000; Malmstein, 1999). This approach is also potentially useful for elimination of hydrophilic cosolvents from the systems and thus the improvement of their overall biocompatibility. Li et al. (2005) demonstrated the benefit of combining nonionic surfactants on the formation of U-type microemulsions generated from flurbiprofen-loaded preconcentrates containing Capmul PG8 (propylene glycol monocaprylate) as the oil phase and Tween 20 (polysorbate 20) and Cremophor EL (polyoxyl 35 castor oil) as surfactants. Systems stabilized by a mixture of both surfactants seem to exhibit a larger isotropic region and high capacity for drug loadings (up to 10%) with good stability on dilution compared to single surfactant-based preconcentrates (Figure 10.8).

Recent studies have investigated the potential of different nonionic polyoxyethylene and polyglycerol ester type surfactants to act as cosurfactants for Labrasol. The water solubilization capacity in microemulsion preconcentrates based on Labrasol, with isopropyl myristate as the oil phase, was enhanced with the addition of both groups of cosurfactants, which to an extent depended on the oil/(surfactant/cosurfactant) mass ratio (Djekic and Primorac, 2008; Djekic et al., 2008).

Another aspect in the area of microemulsion preconcentates and microemulsions is the assessment of the range of water–oil–surfactant–cosurfactant compositions, which can form microemulsions at a given temperature, and the effect of various formulation variables on the region of the existence of microemulsions, which is usually determined from phase behavior investigations and is represented in phase diagrams (Kahlweit, 1999). Although phase diagrams represent detailed compositional maps that are of great interest to the formulators, the construction of complete phase diagrams requires complex and time-consuming experimental work. Unfortunately, the majority of previous phase behavior investigations were related to surfactants and oils that do not have regulatory approval for pharmaceutical use. Certain data of the phase behavior of binary, ternary, quaternary, and more complex systems; their components and conditions of measurement; and complete bibliographic information may be extracted from the work of Koynova and Caffrey (2002). Although the main focus of their database is on lipids of membrane origin where water is the dispersing medium, it also

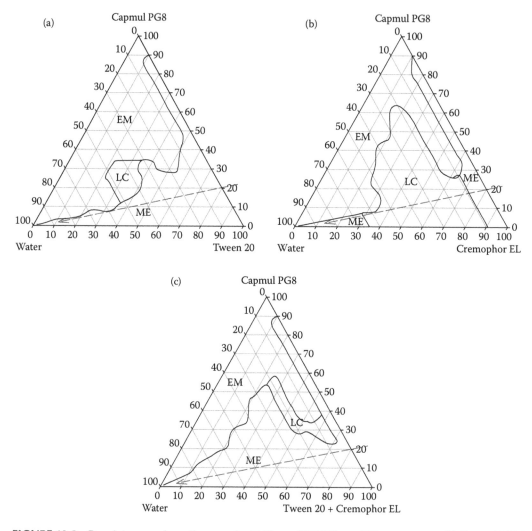

FIGURE 10.8 Pseudoternary phase diagrams for (a) Capmul PG8/Tween 20/water system, (b) Capmul PG8/Cremophor EL/water system, and (c) Capmul PG8/Tween 20/Cremophor EL/water system. (From Li, P., et al. *Int. J. Pharm.* 2005; 288: 27–34. With permission.)

includes information on acylglycerols, fatty acids, cationic lipids, and detergent-containing systems as well as the miscibility of synthetic and natural lipids with other lipids, water, drugs, organic solvents, and biomolecules (proteins, nucleic acids, carbohydrates, etc.). In contrast, there is growing interest in the development of artificial neural network models for computer simulation optimization of microemulsion systems using a limited number of experiments and inputs in order to minimize the experimental efforts as well as for successful prediction of *in vitro* and *in vivo* performance of the drug delivery systems (Agatonovic-Kustrin and Beresford, 2000; Agatonovic-Kustrin et al., 2003; Alany et al., 1999; Djekic et al., 2008; Mendyk and Jachowicz, 2007; Richardson et al., 1997).

The next big challenge, in addition to drug solubility and absorption enhancement, is the development of solid dosage forms from principally liquid or semisolid SMEDDS formulations. Semisolid formulations may be transformed into solid dosage forms using techniques such as melt granulation (the semisolid lipid excipient acts as a binder and solid granules are produced on cooling); solvents or supercritical fluids used with semisolid excipients, which are solubilized and then the solvent evaporated to produce a waxy powder; and spraying techniques. In many cases the SMEDDSs are liquid formulations; therefore, alternative approaches are required such as adsorption onto neutral carrier adsorbents (neutral silicate), encapsulation in solidified sodium alginate beads, and spray drying the liquid SMEDDS in a laboratory spray dryer using dextran as a solid carrier. These techniques enable the production of granules or powders that can then be compressed into tablet form or filled into capsules using conventional equipment. The results of the related studies demonstrated good flowability and storage stability. Furthermore, it was confirmed that the solid SMEDDS can preserve the improved bioavailability with releasing microemulsion lipid droplets from the formulation *in vivo* in the poorly soluble drugs piroxicam (Marchaud and Hughes, 2008), simvastatin (Kang et al., 2003), and nimodipine (Yi et al., 2008).

ABBREVIATIONS

CoQ_{10}	Coenzyme Q_{10}
DNA	Deoxyribonucleic acid
DSPE-PEG	Distearoylphosphatidylethanolamine-*N*-poly(ethylene glycol) 2000
GI	Gastrointestinal
GIT	Gastrointestinal tract
MLM	1,3-dioctanoyl-2-linoleyl-*sn*-glycerol
O/W	Oil-in-water microemulsions
*p*DNA	Plasmid deoxyribonucleic acid
PEG	Poly(ethylene glycol)
SEDDS	Self-emulsifying drug delivery system
SLN	Solid lipid nanoparticle
SMEDDS	Self-microemulsifying drug delivery system
W/O	Water-in-oil microemulsions

REFERENCES

Agatonovic-Kustrin, S. and Beresford, R. Basic concepts of artificial neural network (ANN) modeling and its application in pharmaceutical research. *J. Pharm. Biomed. Anal.* 2000; 22: 717–727.

Agatonovic-Kustrin, S., Glass, B. D., Wisch, M. H., and Alany, R. G. Prediction of stable microemulsion formulation for the oral delivery of a combination of antitubercular drugs using ANN methodology. *Pharm. Res* 2003; 20: 1760–1764.

Alany, R. G., Agatonovic-Kustrin, S., Rades, T., and Tucker, I. G. Use of artificial neural networks to predict quaternary phase systems from limited experimental data. *J. Pharm. Biomed. Anal.* 1999; 19: 443–452.

Amantea, J., Meier-Kriesche, H. U., Schoenberg, L., and Kahan, B. D. A pharmacokinetic comparison of the corn oil versus microemulsion gelcap formulation of cyclosporine used de novo after renal transplantation. *Transplant Int.* 1997; 10: 217–222.

Amidon, G. L., Lennernas, H., Shah, V. P., and Crison, J. R. A theoretical basis for a biopharmaceutic drug classification: The correlation of *in vitro* drug product dissolution and *in vivo* bioavailability. *Pharm. Res.* 1995; 12: 413–420.

Andrysek, T. Impact of physical properties of formulations on bioavailability of active substance: Current and novel drugs with cyclosporine. *Mol. Immunol.* 2003; 39: 1061–1065.

Attama, A. A. and Nkemnele, M. O. *In vitro* evaluation of drug release from self micro-emulsifying drug delivery systems using a biodegradable homolipid from *Capra hircus*. *Int. J. Pharm.* 2005; 304: 4–10.

Bagwe, R. P., Kanicky, J. R., Palla, B. J., Patanjali, P. K., and Shah, D. O. Improved drug delivery using microemulsions: Rationale, recent progress, and new horizons. *Crit. Rev. Ther. Drug Carrier Syst.* 2001; 18: 77–140.

Borhade, V., Nair, H., and Hegde, D. Design and evaluation of self-microemulsifying drug delivery system (SMEDDS) of tacrolimus. *AAPS PharmSciTech* 2008; 9: 13–21.

Chai, J.-L., Li, G.-Z., Zhang, G.-Y., Lu, J.-J., and Wang, Z.-N. Studies on the phase behavior of the system APG/alcohol/alkane/H2O with fishlike diagrams. *Colloids Surf. A: Physicochem. Eng. Aspects* 2003; 231: 173–180.

Chen, M.-L. Lipid excipients and delivery systems for pharmaceutical development—An FDA perspective. AAPS Workshop on the Effective Utilization of Lipid-Based Systems to Enhance the Delivery of Poorly Soluble Drugs, March 5–6, 2007, Bethesda, MD.

Cirri, M., Mura, P., and Mora, C. P. Liquid spray formulations of xibornol by using self-microemulsifying drug delivery systems. *Int. J. Pharm.* 2007; 340: 84–91.

Clausse, M., Nicolas Morgantini, L., Zradba, A., and Touraud, D. Water/ionic surfactant/alkanol/hydrocarbon system: Influence of certain constitution and composition parameters upon realms of existence and transport properties of microemulsion type media. In: HL Rosano and M Clausse (Eds.), *Microemulsion Systems*, pp. 15–63. New York: Marcel Dekker, 1987.

Constantinides, P. P. Lipid microemulsions for improving drug dissolution and oral absorption: Physical and biopharmaceutical aspects. *Pharm. Res.* 1985; 12: 161–172.

Constantinides, P. P and Scalart, J. P. Formulation and physical characterisation of water-in-oil microemulsions containing long versus medium chain length glycerides. *Int. J. Pharm.* 1997; 158: 57–68.

Danielsson, I. and Lindman, B. The definition of a microemulsion. *Colloids Surf.* 1981; 3: 391–392.

Das, N. G. and Das, S. K. Formulation of poorly soluble drugs. *Drug Deliv. Rep.* 2006; Spring/Summer: 52–55.

Datea, A. A. and Patravale, V. B. Current strategies for engineering drug nanoparticles. *Curr. Opin. Colloid Interface Sci.* 2004; 9: 222–235.

Debuigne, F., Cuisenaire, J., Jeunieau, L., Masereel, B., and Nagy, J. B. Synthesis of nimesulide nanoparticles in the microemulsion epikuron/isopropyl myristate/water/n-butanol or isopropanol. *J. Colloid Interface Sci.* 2001; 243: 90–101.

Djekic, L., Ibric, S., and Primorac, M. The application of artificial neural networks in the prediction of microemulsion phase boundaries in PEG-8 caprylic/capric glycerides based systems. *Int. J. Pharm.* 2008; 361: 41–46.

Djekic, L. and Primorac, M. The influence of cosurfactants and oils on the formation of pharmaceutical microemulsions based on PEG-8 caprylic/capric glycerides. *Int. J. Pharm.* 2008; 352: 231–239.

Djordjevic, L. J, Primorac, M., and Stupar, M. *In vitro* release of diclofenac diethylamine from caprylocaproyl macrogolglycerides based microemulsions. *Int. J. Pharm.* 2005; 296: 73–79.

Djordjevic, L. J., Primorac, M., Stupar, M., and Krajisnik, D. Characterization of caprylocaproyl macrogolglycerides based microemulsion drug delivery vehicles for an amphiphilic drug. *Int. J. Pharm.* 2004; 271: 11–19.

Dunn, C. J., Wagstaff, A. J., Perry, C. M., Plosker, G. L., and Goa, K. L. Cyclosporin: An updated review of the pharmacokinetic properties, clinical efficacy and tolerability of a microemulsion-based formulation (Neoral) in organ transplantation. *Drugs* 2001; 61: 1957–2016.

Erkko, P., Granlund, H., Nuutinen, M., and Reitamo, S. Comparison of cyclosporin A pharmacokinetics of a new microemulsion formulation and standard oral preparation in patients with psoriasis. *Brit. J. Dermatol.* 1997; 136: 82–88.

Evans, D. F. and Wennerström, H. *The Colloidal Domain: Where Chemistry, Physics, Biology and Technology Meet.* New York: VCH, 1994.

Ezrahi, S., Aserin, A. and Garti, N. Aggregation behaviour in one-phase (Winsor IV) microemulsion systems. In: P. Kumar and K .L. Mittal (Eds.), *Handbook of Microemulsions: Science and Technology*, pp. 185–246. New York: Marcel Dekker, 1999.

Fanun, M. and Al-Diyn, W. S. Electrical conductivity and self diffusion-NMR studies of the system: Water/sucrose laurate/ethoxylated mono-di-glyceride/isopropylmyristate. *Colloids Surf. A: Physicochem. Eng. Aspects* 2006; 277: 83–89.

Fatouros, D. G., Karpf, D. M., Nielsen, F. S., and Mullertz, A. Clinical studies with oral lipid based formulations of poorly soluble compounds. *Ther. Clin. Risk Manage.* 2007; 3: 591–604.

Fatouros, D. G., Nielsen, F. S., Douroumis, D., Hadjileontiadis, L. J., and Mullertz, A. *In vitro–in vivo* correlations of self-emulsifying drug delivery systems combining the dynamic lipolysis model and neuro-fuzzy networks. *Eur. J. Pharm. Biopharm.* 2008; 69: 887–898.

Fujii, M., Shiozawa, K., Henmi, T., Yamanouchi, S., Suzuki, H., Yamashita, N., and Matsumoto, M. Skin permeation of indomethacin from gel formed by fatty-acid ester and phospholipid. *Int. J. Pharm.* 1996; 137: 117–124.

Gasco, M. R. Method for producing solid lipid microspheres having a narrow size distribution. U.S. Patent 5250236, 1993.

Gasco, M. R. and Trotta, M. Nanoparticles from microemulsions. *Int. J. Pharm.* 1986; 29: 267–268.

Garti, N., Aserin, A., Spernath, A., and Amar, I. U.S. Patent Application 2003232095 A1 20031218, 2003.

Garti, N., Avrahami, M., and Aserin, A. Improved solubilization of celecoxib in U-type nonionic microemulsions and their structural transitions with progressive aqueous dilution. *J. Colloid Interface Sci.* 2006; 299: 352–365.

Garti, N., Clement, V., Leser, M. Aserin, A., and Fanun, M. Sucrose ester microemulsions. *J. Mol. Liq.* 1999; 80: 253–296.

Gibson, L. Lipid-based excipients for oral drug delivery. In: D. J. Hauss (Ed.), *Oral Lipid-Based Formulations: Enhancing the Bioavailability of Poorly Water-Soluble Drugs*, pp. 43–51. New York: Informa Healthcare, 2007.

Glatter, O., Orthaber, D., Stradner, A., Scherf, G., Fanun, M., Garti, N., Clément, V., and Leser, M. E. Sugar-ester nonionic microemulsion: Structural characterization. *J. Colloid Interface Sci.* 2001; 241: 215–225.

Goddeeris, C., Coacci, J., and Van den Mooter, G. Correlation between digestion of the lipid phase of SMEDDS and release of the anti-HIV drug UC 781 and the anti-mycotic drug enilconazole from SMEDDS. *Eur. J. Pharm. Biopharm.* 2007; 66: 173–181.

Goddeeris, C. and Van den Mooter, G. Free flowing solid dispersions of the anti-HIV drug UC 781 with poloxamer 407 and a maximum amount of TPGS 1000: Investigating the relationship between physicochemical characteristics and dissolution behaviour. *Eur. J. Pharm. Sci.* 2008; 35: 104–113.

Graf, A., Ablinger, E., Peters, S., Zimmer, A., Hook, S., and Rades, T. Microemulsions containing lecithin and sugar-based surfactants: Nanoparticle templates for delivery of proteins and peptides. *Int. J. Pharm.* 2008; 350: 351–360.

Grove, M., Müllertz, A., Nielsen, J. L., and Pedersen, G. P. Bioavailability of seocalcitol: II: Development and characterisation of self-microemulsifying drug delivery systems (SMEDDS) for oral administration containing medium and long chain triglycerides. *Eur. J. Pharm. Sci.* 2006; 28: 233–242.

Grove, M., Müllertz, A., Pedersen, G. P., and Nielsen, J. L. Bioavailability of seocalcitol: III. Administration of lipid-based formulations to minipigs in the fasted and fed state. *Eur. J. Pharm. Sci.* 2007; 31: 8–15.

Gupta, S. and Moulik, S. P. Biocompatible microemulsions and their prospective uses in drug delivery. *J. Pharm. Sci.* 2008; 97: 22–45.

Gursoy, R. N. and Benita, S. Self-emulsifying drug delivery systems (SEDDS) for improved oral delivery of lipophilic drugs. *Biomed. Pharmacother.* 2004; 58: 173–182.

Hauss, D. J. Oral lipid-based formulations. *Adv. Drug Deliv. Rev.* 2007; 59: 667–676.

Heydenreich, A. V., Westmeier, R., Pedersen, N., Poulsen, H. S., and Kristensen, H. G. Preparation and purification of cationic solid lipid nanospheres—Effects on particle size, physical stability and cell toxicity. *Int. J. Pharm.* 2003; 254: 83–87.

Hoar, T. P. and Schulman, J. H. Transparent water-in-oil dispersions: The oleopathic hydro-micelle. *Nature* 1943; 152: 102–103.

Holm, R., Porter, C. J. H., Edwards, G. A., Mullertz, A., Kristensen, H. G., and Charman, W. N. Examination of oral absorption and lymphatic transport of halofantrine in a triple-cannulated canine model after administration in self-microemulsifying drug delivery systems (SMEDDS) containing structured triglycerides. *Eur. J. Pharm. Sci.* 2003; 20: 91–97.

Holt, D. W., Mueller, E. A., Kovarik, J. M., van Bree, J. B., and Kutz, K. The pharmacokinetics of Sandimmune Neoral: A new oral formulation of cyclosporine. *Transplant. Proc.* 1994; 26: 2935–2939.

Humberstone, A. J. and Charman, W. N. Lipid-based vehicles for the oral delivery of poorly water soluble drugs. *Adv. Drug Deliv. Rev.* 1997; 25: 103–128.

Jannin, V., Musakhanian J., and Marchaud, D. Approaches for the development of solid and semi-solid lipid-based formulations. *Adv. Drug Deliv. Rev.* 2008; 60: 734–746.

Jing, Q., Shen, Y., Ren, F., Chen, J., Jiang, Z., Peng, B., Leng, Y., and Dong, J. HPLC determination of anethole trithione and its application to pharmacokinetics in rabbits. *J. Pharm. Biomed. Anal.* 2006; 42: 613–617.

Kahlweit, M. Microemulsions. *Annu. Rep. Prog. Chem. Sect. C* 1999; 95: 89–115.

Kang, B. K., Lee, J. S., Chon, S. K., Jeong, S. Y., Yuk, S. H., Khang, G., Lee, H. B., and Cho, S. H. Development of self-microemulsifying drug delivery systems (SMEDDS) for oral bioavailability enhancement of simvastatin in beagle dogs. *Int. J. Pharm.* 2004; 274: 65–73.

Kang, B. K., Yoon, B. Y., Seo, K. S., Jeung, S. Y., Kil, H. J., Khang, G., Lee, H. B., and Cho, S. H. Preparation of solid dosage form containing SMEDDS of simvastatin by microencapsulation. *J. Kor. Pharm. Sci.* 2003; 33: 121–127.

Kemken, J., Ziegler, A., and Müller, B. W. Investigations into the pharmacodynamic effects of dermally administered microemulsions containing β-blockers. *J. Pharm. Pharmacol.* 1991; 43: 679–684.

Kemken, J., Ziegler, A., and Müller, B. W. Influence of supersaturation on the pharmacodynamic effect of bupranol after dermal administration using microemulsions as a vehicle. *Pharm. Res.* 1992; 9: 554–558.

Khoo, S.-M., Humberstone, A. J., Porter, C. J. H., Edwards, G. A., and Charman, W. N. Formulation design and bioavailability assessment of lipidic self-emulsifying formulations of halofantrine. *Int. J. Pharm.* 1998; 167: 155–164.

Khoo, S.-M., Shackeleford, D., Porter, C. J. H., Edwards, G. A., and Charman, W. N. Intestinal lymphatic transport of halofantrine occurs after oral administration of a unit dose lipid-based formulation to fasted dogs. *Pharm. Res.* 2003; 20: 1460–1465.

Klyashchitsky, B. A. and Owen, A. J. Drug delivery systems for cyclosporine: Achievements and complications. *J. Drug Target.* 1998; 5: 443–458.

Kogan, A. and Garti, N. Microemulsions as transdermal drug delivery vehicles. *Adv. Colloid Interface Sci.* 2006; 123–126: 369–385.

Kommuru, T. R., Gurley, B., Khan, M. A., and Reddy, I. K. Self-emulsifying drug delivery systems (SEDDS) of coenzyme Q_{10}: Formulation development and bioavailability assessment. *Int. J. Pharm.* 2001; 212: 233–246.

Koynova, R. and Caffrey, M. An index of lipid phase diagrams. *Chem. Phys. Lipids* 2002; 115: 107–219.

Kovarik, J. M., Mueller, E. A., and Niese, D. Clinical development of a cyclosporin microemulsion in transplantation. *Ther. Drug Monit.* 1996; 18: 429–434.

Kovarik, J. M., Mueller, E. A., van Bree, J. B., Tetzloff, W., and Kutz, K. Reduced inter and intraindividual variability in cyclosporine pharmacokinetics from a microemulsion formulation. *J. Pharm. Sci.* 1994; 83: 444–446.

Koziara, J. M., Whisman, T. R., Tseng, M. T., and Mumper, R. J. *In-vivo* efficacy of novel paclitaxel nanoparticles in paclitaxel-resistant human colorectal tumors. *J. Control. Release* 2006; 112: 312–319.

Lawrence, M. J. and Rees, G. D. Microemulsion-based media as novel drug delivery systems. *Adv. Drug Deliv. Rev.* 2000; 45: 89–121.

Lee, J. M., Park, K. M., Lim, S. J., Lee, M. K., and Kim, C. K. Microemulsion formulation of clonixic acid: Solubility enhancement and pain reduction. *J. Pharm. Pharmacol.* 2002; 54: 43–49.

Li, P., Ghosh, A., Wagner, R. F., Krill, S., Joshi, Y. M., and Serajuddin, A. T. M. Effect of combined use of nonionic surfactant on formation of oil-in-water microemulsions. *Int. J. Pharm.* 2005; 288: 27–34.

Ljusberg-Wahren, H., Nielsen, F. S., Brogård, M., Troedsson, E., and Müllertz, A. Enzymatic characterization of lipid-based drug delivery systems. *Int. J. Pharm.* 2005; 298: 328–332.

Lockman, P. R., Oyewumi, M. O., Koziara, J. M., Roder, K. E., Mumper, R. J., and Allen, D. D. Brain uptake of thiamine-coated nanoparticles. *J. Control. Release* 2003; 93: 271–282.

Lopez-Quintela, M. A., Tojo, C., Blanco, M. C., Garcia Rio, L., and Leis, J. R. Microemulsion dynamics and reactions in microemulsions. *Curr. Opin. Colloid Interface Sci.* 2004; 9: 264–728.

Lu, J.-L., Wang, J.-C., Zhao, S.-X., Liu, X.-Y., Zhao, H., Zhang, X., Zhou, S.-F., and Zhang, Q. Self-microemulsifying drug delivery system (SMEDDS) improves anticancer effect of oral 9-nitrocamptothecin on human cancer xenografts in nude mice. *Eur. J. Pharm. Biopharm.* 2008; 69: 899–907.

Malmstein, M. Microemulsion in pharmaceuticals. In: P. Kumar and K.L. Mittal (Eds.), *Handbook of Microemulsion: Science and Technology*, pp. 755–772. New York: Marcel Dekker, 1999.

Mandawgade, S. D., Sharma, S., Pathak, S., and Patravale, V. B. Development of SMEDDS using natural lipophile: Application to β-artemether delivery. *Int. J. Pharm.* 2008; 362: 179–183.

Marchaud, D. and Hughes, S. Solid dosage forms from self-emulsifying lipidic formulations. Pharmaceutical Technology Europe, April 1, 2008.

Marengo, E., Cavalli, R., Caputo, O., Rodriguez, L., and Gasco, M. R. Scale-up of the preparation process of solid lipid nanospheres. Part II. *Int. J. Pharm.* 2000; 205: 3–13.

Mendyk, A. and Jachowicz, R. Unified methodology of neural analysis in decision support systems built for pharmaceutical technology. *Expert Syst. Appl.* 2007; 32: 1124–1131.

Mueller, E. A., Kovarik, J. M., van Bree, J. B., Tetzloff, W., Grevel, J., and Kutz, K. Improved dose linearity of cyclosporine pharmacokinetics from a microemulsion formulation. *Pharm. Res.* 1994; 11: 301–304.

Müller, R. H., Mäder, K., and Gohla, S. Solid lipid nanoparticles (SLN) for controlled drug delivery—A review of the state of the art. *Eur. J. Pharm. Biopharm.* 2000; 50: 161–177.

Narang, A. S., Delmarre, D., and Gao, D. Stable drug encapsulation in micelles and microemulsions. *Int. J. Pharm.* 2007; 345: 9–25.

Noble, S. and Markham, A. Cyclosporin: A review of the pharmacokinetic properties, clinical efficacy and tolerability of a microemulsion-based formulation (Neoral®). *Drugs* 1995; 50: 924–941.

O'Driscoll, C. M. Lipid-based formulations for intestinal lymphatic delivery. *Eur. J. Pharm. Sci.* 2002; 15: 405–415.

Oyewumi, M. O., Yokel, R. A., Jay, M., Coakley, T., and Mumper, R. J. Comparison of cell uptake, biodistribution and tumor retention of folate-coated and PEG-coated gadolinium nanoparticles in tumor-bearing mice. *J. Control. Release* 2004; 95: 613–626.

Park, K. M., Lee, M. K., Hwang, K. J., and Kim, C. K. Phospholipid-based microemulsions of flurbiprofen by the spontaneous emulsification process. *Int. J. Pharm.* 1999; 183: 145–154.

Patel, A. R. and Vavia, P. R. Preparation and *in vivo* evaluation of SMEDDS (self-microemulsifying drug delivery system) containing fenofibrate. *AAPS J.* 2007; 9(3): Article 41. http://www.aapsj.org

Patel, D. and Sawant, K. K. Oral bioavailability enhancement of acyclovir by self-microemulsifying drug delivery systems (SMEDDS). *Drug Dev. Ind. Pharm.* 2007; 33: 1318–1326.

Pitaksuteepong, T., Davies, N. M., Tucker, I. G., and Rades, T. Factors influencing the entrapment of hydrophilic compounds in nanocapsules prepared by interfacial polymerisation of water-in-oil microemulsions. *Eur. J. Pharm. Biopharm.* 2002; 53: 335–342.

Pollard, S., Nashan, B., Johnston, A., Hoyer, P., Belitsky, P., Keown, P., and Helderman, H. A pharmacokinetic and clinical review of the potential clinical impact of using different formulations of cyclosporin A. *Clin. Ther.* 2003; 25: 1654–1669.

Porter, C. J. H., Pouton, C. W., Cuine, J. F., and Charman, W. N. Enhancing intestinal drug solubilisation using lipid-based delivery systems. *Adv. Drug Deliv. Rev.* 2008; 60: 673–691.

Porter, C. J. H., Trevaskis, N. L., and Charman, W. N. Lipids and lipid-based formulations: Optimizing the oral delivery of lipophilic drugs. *Nat. Rev. Drug Discov.* 2007; 6: 231–248.

Pouton, C. W. Formulation of self-emulsifying drug delivery systems. *Adv. Drug Deliv. Rev.* 1997; 25: 47–58.

Pouton, C. W. Lipid formulations for oral administration of drugs: Non-emulsifying, self-emulsifying and "self-microemulsifying" drug delivery systems. *Eur. J. Pharm. Sci.* 2000; 11: S93–S98.

Pouton, C. W. Formulation of poorly water-soluble drugs for oral administration: Physicochemical and physiological issues and the lipid formulation classification system. *Eur. J. Pharm. Sci.* 2006; 29: 278–287.

Pouton, C. W. and Porter, C. J. H. Formulation of lipid-based delivery systems for oral administration: Materials, methods and strategies. *Adv. Drug Deliv. Rev.* 2008; 60: 625–637.

Rane, S. S. and Anderson, B. D. What determines drug solubility in lipid vehicles: Is it predictable? *Adv. Drug Deliv. Rev.* 2008; 60: 638–656.

Reis, C. P., Neufeld, R. J., Ribeiro, A. J., and Veiga, F. Nanoencapsulation I. Methods for preparation of drug-loaded polymeric nanoparticles. *Nanomed. Nanotechnol. Biol. Med.* 2006; 2: 8–21.

Richardson, C. J., Mbanefo, A., Aboofazeli, R., Lawrence, M. J., and Barlow, D. J. Prediction of phase behavior in microemulsion systems using artificial neural networks. *J. Colloid Interface Sci.* 1997; 187: 296–303.

Ritschel, W. A. Microemulsion technology in the reformulation of cyclosporine: The reason behind the pharmacokinetic properties of Neoral. *Clin. Transplant.* 1996; 10: 364–373.

Rowe, R. C., Sheskey, P. J., and Weller, P. J. (Eds.), *Handbook of Pharmaceutical Excipients*, 4th ed., pp. 454–459. London: Pharmaceutical Press/Washington, DC: American Pharmaceutical Association, 2003.

Sahiner, N., Alb, A. M., Graves, R., Mandal, T., McPherson, G. L., Reed, W. F., and John, V. T. Core–shell nanohydrogel structures as tunable delivery systems. *Polymer* 2007; 48: 704–711.

Schulman, J. H., Stoeckenius, W., and Prince, L. M. Mechanism of formation and structure of microemulsions by electron microscopy. *J. Phys. Chem.* 1959; 63: 1677–1680.

Sha, X., Yan, G., Wu, Y., Li, J., and Fang, X. Effect of self-microemulsifying drug delivery systems containing Labrasol on tight junctions in Cac o-2 cells. *Eur. J. Pharm. Sci.* 2005; 24: 477–486.

Shah, N. H., Carvajal, M. T., Patel, C. I., Infeld, M. H., and Malick, A. W. Self-emulsifying drug delivery systems with polyglycolyzed glycerides for improving *in vitro* dissolution and oral absorption of lipophilic drugs. *Int. J. Pharm.* 1994; 106: 15–23.

Shukla, A., Krause, A., and Neubert, R. Microemulsions as colloidal vehicle systems for dermal drug delivery. Part IV. Investigation of microemulsion systems based on a eutectic mixture of lidocaine and prilocaine as the colloidal phase by dynamic light scattering. *J. Pharm. Pharmacol.* 2003; 55: 741–748.

Singh, A. K., Chaurasiya, A., Singh, M., Upadhyay, S. C., Mukherjee, R., and Khar, R. K. Exemestane loaded self-microemulsifying drug delivery system (SMEDDS): Development and optimization. *AAPS PharmSciTech* 2008; 9: 628–634.

Sjoblom, J., Lindbergh, R., and Friberg, S. E. Microemulsions—Phase equilibria characterization, structures, applications and chemical reactions. *Adv. Colloid Interface Sci.* 1996; 95: 125–287.

Söderlind, E., Wollbratt, M., and von Corswant, C. The usefulness of sugar surfactants as solubilizing agents in parenteral formulations. *Int. J. Pharm.* 2003; 252: 61–71.

Spernath, A. and Aserin, A. Microemulsions as carriers for drugs and nutraceuticals. *Adv. Colloid Interface Sci.* 2006; 128–130: 47–64.

Strey, R. Microemulsion microstructure and interfacial curvature. *Colloid Polym. Sci.* 1994; 272: 1005–1019.

Strickley, R. G. Solubilizing excipients in oral and injectable formulations. *Pharm. Res.* 2004; 21: 201–230.

Subramanian, N., Ray, S., Ghosal, S. K., Bhadra, R., and Moulik, S. P. Formulation design of selfmicroemulsifying drug delivery systems for improved oral bioavailability of celecoxib. *Biol. Pharm. Bull.* 2004; 27: 1993–1999.

Syamasri, G. and Moulik, S. P. Biocompatible microemulsions and their prospective uses in drug delivery. *J. Pharm. Sci.* 2008; 97: 22–45.

Taha, E. I., Al-Saidan, S., Samy, A. M., and Khan, M. A. Preparation and *in vitro* characterization of self-nano-emulsified drug delivery system (SNEDDS) of all-transretinol acetate. *Int. J. Pharm.* 2004; 285: 109–119.

Tarr, B. D. and Yalkowsky, S. H. Enhanced intestinal absorption of cyclosporin in rats through the reduction of emulsion droplet size. *Pharm. Res.* 1989; 6: 40–43.

Trotta, M., Gallarate, M., Carlotti, M. E., and Morel, S. Preparation of griseofulvin nanoparticles from water-dilutable microemulsions. *Int. J. Pharm.* 2003; 254: 235–242.

Ugazio, E., Cavalli, R., and Gasco, M. R. Incorporation of cyclosporin A in solid lipid nanoparticles. *Int. J. Pharm.* 2002; 241: 341–344.

Vandamme, Th. F. Microemulsions as ocular drug delivery systems: Recent developments and future challenges. *Prog. Retin. Eye Res.* 2002; 21: 15–34.

von Corswant, C., Thoren, P., and Engstrom, S. Triglyceride-based microemulsion from intravenous administration of sparingly soluble substances. *J. Pharm. Sci.* 1998; 87: 200–208.

Vonderscher, J. and Meinzer, A. Rationale for the development of Sandimmune Neoral. *Transplant. Proc.* 1994; 26: 2925–2927.

Watnasirichaikul, S., Davies, N. M, Rades, T., and Tucker, I. G. Preparation of biodegradable insulin nanocapsules from biocompatible microemulsions. *Pharm. Res.* 2000; 17: 684–689.

Watnasirichaikul, S., Rades, T., Tucker, I. G., and Davies, N. M. Effects of formulation variables on characteristics of poly (ethylcyanoacrylate) nanocapsules prepared from w/o microemulsions. *Int. J. Pharm.* 2002; 235: 237–246.

Wei, L., Sun, P., Nie, S., and Pan, W. Preparation and evaluation of SEDDS and SMEDDS containing carvedilol. *Drug Dev. Ind. Pharm.* 2005; 31: 785–794.

Wennerström, H., Balogh, J., and Olsson, U. Interfacial tensions in microemulsions. *Colloids Surf. A: Physicochem. Eng. Aspects* 2006; 291: 69–77.

Woo, J. S., Song, Y.-K., Hong, J.-Y., Lim, S.-J., and Kim, C.-K. Reduced food-effect and enhanced bioavailability of a self-microemulsifying formulation of itraconazole in healthy volunteers. *Eur. J. Pharm. Sci.* 2008; 33: 159–165.

Wu, W., Wang, Y., and Que, L. Enhanced bioavailability of silymarin by self-microemulsifying drug delivery system. *Eur. J. Pharm. Biopharm.* 2006; 3: 288–294.

Yi, T., Wan, J., Xu, H., and Yang, X. A new solid self-microemulsifying formulation prepared by spray-drying to improve the oral bioavailability of poorly water soluble drugs. *Eur. J. Pharm. Biopharm.* 2008; 70: 439–444.

Ying, C., Gao, L., Xianggen, W., Zhiyu, C., Jiangeng, H., Bei, Q., Song, C., and Ruihua, W. Self-microemulsifying drug delivery system (SMEDDS) of vinpocetine: Formulation development and *in vivo* assessment. *Biol. Pharm. Bull.* 2008; 31: 118–125.

Zhang, P., Liu, Y., Feng, N., and Xu, J. Preparation and evaluation of self-microemulsifying drug delivery system of oridonin. *Int. J. Pharm.* 2008; 355: 269–276.

Zhengrong, C. and Mumper, R. J. Topical immunization using nanoengineered genetic vaccines. *J. Control. Release* 2002; 81: 173–84.

11 Diclofenac Solubilization in Mixed Nonionic Surfactants Microemulsions

Monzer Fanun

CONTENTS

11.1 INTRODUCTION

Diclofenac is a nonsteroidal compound with analgesic, antiinflammatory, and antipyretic properties. It is a weak acid (pK_a 4.0) used to relieve mild to moderate pain from injury, menstrual cramps, arthritis, and other musculoskeletal conditions. It is widely used because of its robust analgesic, antipyretic, and antiinflammatory effects (Kreilgaard, 2002; Lawrence and Rees, 2000; Steer et al., 2003), thus making it one of the most widely prescribed antiinflammatory drugs. The most common approaches to improving the solubility of drugs are the formation of salts (e.g., hydrochlorides, sulfates, nitrates, maleates, citrates, and tartarates) of the basic drugs possessing a net negative electrical charge and reduction of the particle size of the powdered drugs by new milling technologies or by applying new crystallization processes. Because of its poor solubility, short *in vitro* (shelf-life) and *in vivo* (half-life) stability, low bioavailability, and strong side effects, diclofenac delivery should be targeted. One approach to conquer these problems is to enfold the drug into a delivery service system. The integration of the drug into a delivery system can be envisaged to protect it against degradation *in vitro* as well as *in vivo*; the release can be controlled, and targeting can be achieved. Different colloidal systems were used as diclofenac delivery vehicles to improve its bioavailability (Beck et al., 2006; Cevc, 2004; Fanun, 2007a; Kantarci et al., 2005; Kreilgaard, 2002; Kweon et al.,

2004; Lawrence and Rees, 2000; Lindenstruth and Muller, 2004; Piao et al., 2006, 2007; Saravanan et al., 2004; Shakeel et al., 2007; Steer et al., 2003; Stevenson et al., 2005; Vucinic-Milankovic et al., 2007). These systems include nanoparticles (Beck et al., 2006), gelatin microspheres (Saravanan et al., 2004), solid-in-oil suspensions (Piao et al., 2006, 2007), emulsions (Vucinic-Milankovic et al., 2007), multiple emulsions (Lindenstruth and Muller, 2004), nanoemulsions (Shakeel et al., 2007), microemulsions (Fanun, 2007a; Kantarci et al., 2005; Kreilgaard, 2002; Kweon et al., 2004; Lawrence and Rees, 2000; Steer et al., 2003), liquid crystals (Stevenson et al., 2005), and vesicles (Cevc, 2004). The main requirements of the pharmaceutical market entering drug delivery formulations are ease of preparation, physical stability, excipients that are well tolerated and accepted by regulatory authorities, and the availability of large-scale production that is sanctioned by regulatory authorities (Muller and Keck, 2004). Microemulsions are clear, thermodynamically stable, isotropic mixtures of oil, water, and amphiphiles, frequently in combination with a cosurfactant. Introduced first by Hoar and Schulman (1943), microemulsions have been intensively studied during the last decade by many scientists and technologists because of their great potential in many applications (Fanun, 2008a; Kumar and Mital, 1999; Solans and Kunieda, 1997). The adsorption of drugs using microemulsion systems is influenced by the particle size, the partition coefficient of the drug between the two immiscible phases, the presence of the drug in the interface, the site or path of the absorption of the microemulsion components that can act as absorption enhancers, and the drug solubility in the microemulsion components. Microemulsions are (Bagwe et al., 2001; Lawrence and Rees, 2000; Tenjarla, 1999) effective vehicles for the solubilization of certain drugs because they provide all of the possible requirements of a liquid system including thermodynamic stability, ease of preparation, low viscosity, high surface area, and very small droplet size. Small droplets have a better chance to adhere to membranes and to transport bioactive molecules in a more controlled fashion. Microemulsions can be envisioned as media for the entrapped of drugs to protect them from degradation, hydrolysis, and oxidation. These systems can also provide prolonged release of the drug and prevent irritation, despite the toxicity of the drug. Microemulsions can be introduced into the body orally, topically on the skin, or nasally as an aerosol for direct entry into the lungs. They have been used for pulmonary (Courrier et al., 2003; Hiranita et al., 2003; Krafft and Goldmann, 2003; Patel et al., 2003), intravaginal, or intrarectal administration delivery vehicles for lipophilic drugs such as microcides, steroids, and hormones (D'Cruz et al., 1999, 2002a, 2002b; D'Cruz and Uckun, 2003); as well as intramuscular formulations of peptides or cell-targeting systems (Ho et al., 1996), among others. The increased absorption of solubilized drugs in microemulsions for topical applications is attributed to the enhancement of penetration through the skin by the carrier. Topical drug delivery has much compensation over the oral route of administration because it evades hepatic metabolism, the administration is easier and more convenient for the patient, and there is the possibility of immediate removal of the treatment if required. Microemulsions suffer from high surfactant contents and in most cases from high alcohol, solvent, and cosolvent contents. High levels of nonactive compounds are always a hazard. The characterization of microemulsions used as drug delivery systems is necessary to determine the locus of the drug in the loaded microemulsion. The properties of drug-loaded microemulsions can reveal the presence of molecular interactions between the loaded drug and the microemulsion. These properties include the electrical conductivity (Cametti et al., 1992; Kahlweit et al., 1993), viscosity (Berghenholtz et al., 1995; Matsumoto and Sherman, 1969; Ray et al., 1992), periodicity, correlation length (Choi et al., 1997; Glatter et al., 1996; Gradzielski et al. 1996; Mihailescu et al., 2002; Shukla et al., 2004a), diffusion (Fanun, 2007b; Fanun and Salah Al-Diyn, 2006, 2007; Olsson et al., 1986; Soderman and Nyden, 1999), droplet size (Goddeeris et al., 2006; Shukla and Neubert, 2005; Shukla et al., 2004b), and others (Kahlweit et al., 1987; Regev et al., 1996; Talmon, 1996). This chapter deals with the diclofenac solubilization capacity (SC) in three different microemulsion systems. The first is water/mixed nonionic surfactants/R-(+)-limonene (LIM) microemulsions. The second is a water/mixed nonionic surfactants/LIM + ethanol (EtOH) system. The third is microemulsion systems that are formed with isopropylmyristate (IPM) instead of LIM. The aims of this chapter are to outline our

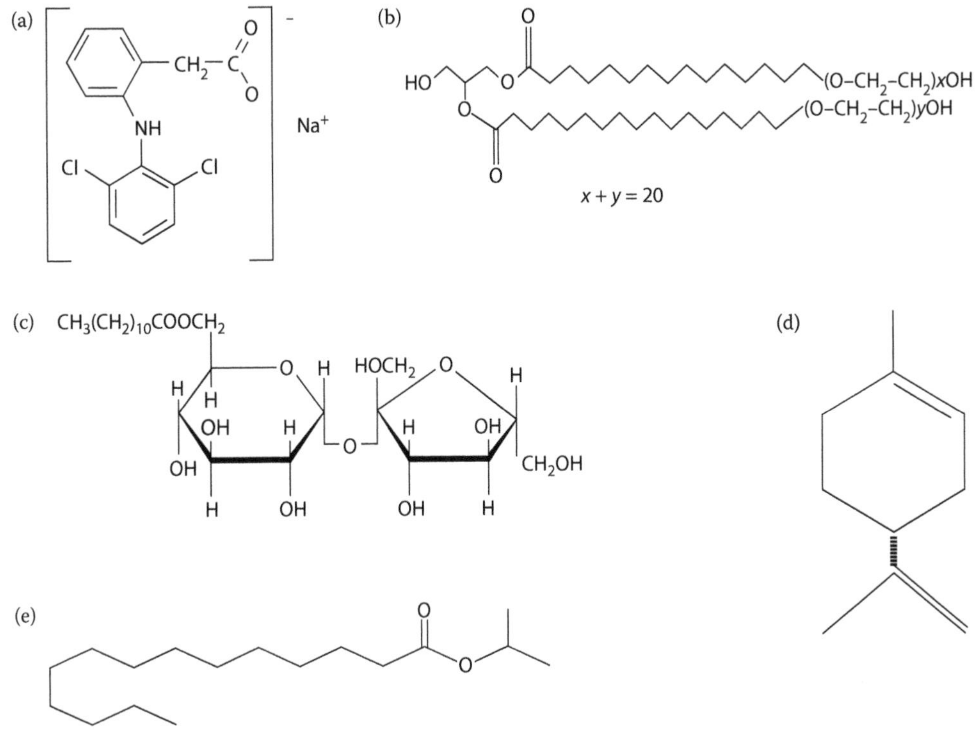

FIGURE 11.1 Chemical structures of (a) sodium diclofenac, (b) EMDG, (c) L1695, (d) LIM, and (e) IPM.

current knowledge on diclofenac solubilization in U-type microemulsions with cosmetically permitted nonionic surfactants and to discuss the influence of the microemulsion composition, components, and structure on drug solubilization and its delivery potential.

11.2 PHASE BEHAVIOR

To determine the SC of diclofenac (Figure 11.1a) in mixed nonionic surfactant microemulsions, we initially studied the phase behavior of water/mixed nonionic surfactants/oil and water/mixed nonionic surfactants/oil + EtOH. The mixed surfactants were ethoxylated mono- and diglyceride (EMDG; Figure 11.1b) and sucrose laurate (L1695; Figure 11.1c). The oils were LIM (Figure 11.1d) and IPM (Figure 11.1e). The mixing ratios (w/w) of L1695/EMDG and EtOH/oil equal unity. The phase diagrams shown in Figures 11.2 through 11.4 reveal the presence of an isotropic and low-viscosity area that is a microemulsion one-phase region (1ϕ); the remainder of the phase diagram represents a two-phase region designated as II ($W_m + O$), which signifies a water continuous micellar system with excess oil. A detailed discussion of the phase behavior was provided in our previous studies (Fanun, 2007a, 2008a, 2008b, 2008c).

11.3 SOLUBILIZATION CAPACITY

11.3.1 ALCOHOL-FREE SYSTEM

We evaluated the SC of diclofenac that was defined as the parts per million of solubilized diclofenac within the total formulation in the U-type microemulsions composed of water/L1695/EMDG/LIM (system A) at 25°C along the N60 dilution line. Figure 11.5 presents the SC as a function of the water volume fraction (ϕ), where we determined the maximum SC in each dilution point and calculate the average values from three independent preparations. The SC of sodium diclofenac in a micellar

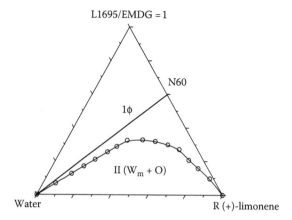

FIGURE 11.2 Pseudoternary phase behavior of the water/L1695/EMDG/LIM system at 25°C. The mixing ratio (w/w) of EMDG/L1695 equals unity. The one-phase region is designated by 1ϕ, and the two-phase region consisting of a water continuous micellar solution with excess oil is designated as (W_m + O). N60 is the water dilution line where the weight ratios of the L1695/EMDG/oleic phase equal 3/3/4.

solution containing mixed surfactants and LIM is 12,000 (12 wt%). Upon dilution with water the SC drops, indicating that the reverse micelles in the absence of water solubilize higher amounts of sodium diclofenac than in the presence of water once the water-in-oil droplets are formed. The reduction in the SC reflects both the dilution factor (DF) and the structural changes that occur upon dilution. The SC profile (Figure 11.5) reveals the presence of four different regions of solubilization within the phase diagram along the N60 dilution line. The first region (region I) is where the water volume fractions (ϕ) are lower than 0.21. In this region, the diclofenac SC drops dramatically from 12,000 to 11,100. For ϕ values higher than 0.21 and lower than 0.61, the diclofenac SC decreases gradually from 11,100 to 10,500 ppm (region II). A slight increase in the diclofenac SC is observed for water volume fractions between 0.61 and 0.72 (region III). In region IV (i.e., water volume fractions higher than 0.72), the diclofenac SC decreases dramatically from 10,500 to 8000 ppm.

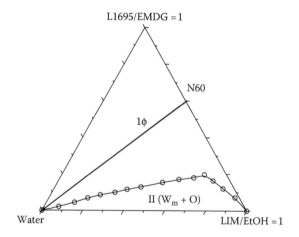

FIGURE 11.3 Pseudoternary phase behavior of the water/L1695/EMDG/LIM + EtOH system at 25°C. The mixing ratio (w/w) of LIM + EtOH and EMDG/L1695 equals unity. The one-phase region is designated by 1ϕ, and the two-phase region consisting of a water continuous micellar solution with excess oil is designated as (W_m + O). N60 is the water dilution line where the weight ratios of L1695/EMDG/oleic phase equal 3/3/4.

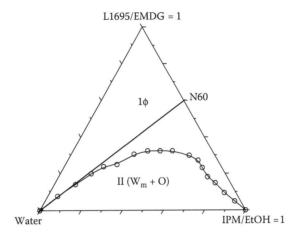

FIGURE 11.4 Pseudoternary phase behavior of the system water/L1695/EMDG/IPM + EtOH at 25°C. The mixing ratio (w/w) of IPM + EtOH and EMDG/L1695 equals unity. The one-phase region is designated by 1φ, and the two-phase region consisting of a water continuous micellar solution with excess oil is designated as $(W_m + O)$. N60 is the water dilution line where the weight ratios of L1695/EMDG/oleic phase equal 3/3/4.

11.3.2 Systems with Alcohol as Cosurfactant

The maximum SC of diclofenac found in Figures 11.6 and 11.7 was determined along dilution lines N60 in water/L1695/EMDG/oil + EtOH systems at 25°C. The oils were LIM (system B) and IPM (system C). The mixing ratios (w/w) of L1695/EMDG and EtOH/oil equal unity. Figures 11.6 and 11.7 show that dilution of microemulsions with water significantly decreases the SCs. The SCs of sodium diclofenac in a micellar solution containing mixed surfactants and an oil phase (see Figures 11.6 and 11.7) are 15,000 and 16,000 ppm (15 and 16 wt%) for LIM- and IPM-based systems, respectively. Upon dilution with water up to 0.2 φ, the SC drops dramatically from 15,000 to 11,800 in the LIM-based system and from 16,000 to 11,600 ppm in the IPM-based microemulsions. However, upon further dilution to 0.60–0.70 φ, the SC continues to decrease in the IPM system whereas in the LIM system it increases to a maximum and then decreases for φ above 0.8. The alcohol makes the formation

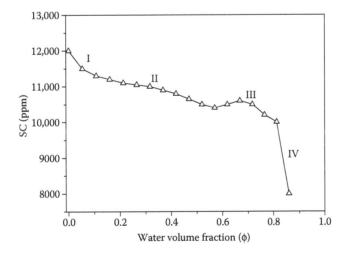

FIGURE 11.5 The SC of diclofenac sodium as a function of the water volume fraction along dilution line N60 at 25°C in the water/L1695/EMDG/LIM system. The mixing ratio (w/w) of EMDG/L1695 equals unity. The phase diagram is presented in Figure 11.2. The line is presented as a guide to the eye.

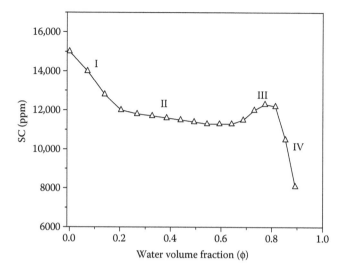

FIGURE 11.6 The SC of diclofenac sodium as a function of the water volume fraction along dilution line N60 at 25°C in the water/L1695/EMDG/LIM + EtOH system. The mixing ratios (w/w) of EMDG/L1695 and EtOH/oil equal unity. The phase diagram is presented in Figure 11.3. The line is presented as a guide to the eye.

of versatile transformable systems possible (Fletcher et al., 1984; Gradzielski and Hoffman, 1999; Saidi et al., 1990). In the water-poor region the alcohol migrates to the interface and competes with the drug on the free interfacial sites. However, at high levels of dilution, when oil-in-water droplets are formed, the EtOH departs from the interface and the micelles are tightly packed by the mixed surfactants alone. Comparing the SCs of solubilized sodium diclofenac in system A based on LIM and system B based on LIM + EtOH, this interesting alcohol effect is also reflected in the SCs (Figures 11.5 and 11.6). In the water-in-oil regions, the SC in alcohol-free systems was lower than in alcohol-based systems. This indicates that swollen reverse mixed micelles in the LIM-based system are more tightly packed and accommodate smaller amounts of solubilized diclofenac than those based on LIM + EtOH

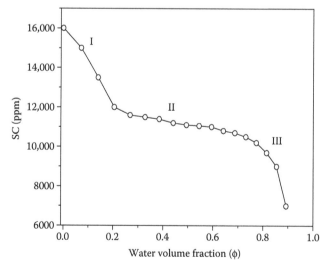

FIGURE 11.7 The SC of diclofenac sodium as a function of the water volume fraction along dilution line N60 at 25°C in the water/L1695/EMDG/IPM + EtOH system. The mixing ratios (w/w) of EMDG/L1695 and EtOH/oil equal unity. The phase diagram is presented in Figure 11.4. The line is presented as a guide to the eye.

that are less compact. Another explanation of this observation is that, at low water contents, the interface is concave toward the water and the oil is the continuous phase, so the SC of sodium diclofenac depends on the available space that then depends on the oil penetration at the interface. The penetration of LIM in system B is lower than in system A, causing higher amounts of sodium diclofenac to be loaded at the interface in system B that contains alcohol. However, as the dilution proceeds, the total solubilization decreases significantly because the alcohol content drops. It is important to note that the decrease in the total solubilization is stronger in the system that contains EtOH compared to the one free of alcohol.

11.4 SOLUBILIZATION EVALUATION

To better evaluate the role of the interface and the dilution by water on the SC along the N60 dilution line in these systems, we introduced three different parameters related to the drug maximum solubilization. Equation 11.1 was used to estimate the observed SC change factor (CF^{ϕ_n}) by dividing the measured SC at each water volume fraction (SC^{ϕ_n}) and by the measured SC at the previous water volume fraction ($SC^{\phi_{n-1}}$):

$$CF^{\phi_n} = \frac{SC^{\phi_n}}{SC^{\phi_{n-1}}}. \tag{11.1}$$

We also determined the "calculated DF" from the decrease in the SC as a function of dilution from one water volume fraction to the next using Equation 11.2:

$$DF^{\phi_n} = \frac{1 - \phi_n}{1 - \phi_{n-1}}. \tag{11.2}$$

We derived an "overall interfacial contribution factor" (IF) value that is strongly dependent on several structural and interfacial composition factors. The IF is calculated using Equation 11.3 by dividing the CF^{ϕ_n} by the DF^{ϕ_n} at each water volume fraction:

$$IF^{\phi_n} = \frac{CF^{\phi_n}}{DF^{\phi_n}}. \tag{11.3}$$

When IF < 1, the DF is dominant, it dictates the SC behavior, and the interface does not play a role in the SC, meaning that the drug is located mainly in the oil phase. At IF = 1, the interface and DFs contribute evenly, indicating that the drug is located partly at the interface and partly at the oil phase. When IF > 1, the interface plays a dominant role and the drug is mostly located at the interface. As the IF grows the interface plays a more significant role in the solubilization of the drug.

11.4.1 ALCOHOL-FREE SYSTEM

Diclofenac solubilization is affected by the addition of water and interfacial packing upon dilution. As the dilution with water advances, the calculated IF of the microemulsion increases. In the first region, the calculated IF value is higher than 1 and increases with dilution (Table 11.1 and Figure 11.8), confirming that, in spite of the dilution and swelling factors, there is a positive input in the SC because diclofenac contributes to the self-assembly of the water-in-oil microemulsions because its packing parameter is higher than 1; therefore, it is not "pressed away" or desorbed from the micellar assembly once water is added. The diclofenac along the bicontinuous phase (region II, 0.21–0.67 ϕ) behaves in a similar manner (Table 11.1 and Figure 11.8), where the SC continues to decrease but with a gentle slope. This is a clear indication that once the interface becomes flat (critical packing parameter ≈ 1) it pushes in fewer molecules of diclofenac (critical packing parameter > 1; Figure 11.1a). Suratkar and Mahapatra (2000) observed a similar change in the locus of solubilization of phenolic compounds in sodium dodecyl sulfate micelles. The calculated IF was greater than 1, meaning that the IFs are again dominant over the DF. However, it is clear

TABLE 11.1

Water Volume Fraction, Maximum Measured SC of Diclofenac, and the CF, DF, and IF along Dilution Line N60 in the Water/L1695/EMDG/LIM System at 25°C

ϕ	Max. Meas. SC (ppm)	CF	DF	IF
0	12,000	NA	NA	NA
0.05	11,500	0.96	0.95	1.01
0.11	11,300	0.98	0.94	1.04
0.16	11,200	0.99	0.94	1.05
0.21	11,100	0.99	0.94	1.06
0.27	11,050	1.00	0.93	1.07
0.32	11,000	1.00	0.93	1.07
0.37	10,900	0.99	0.92	1.07
0.42	10,800	0.99	0.92	1.08
0.47	10,650	0.99	0.91	1.08
0.52	10,500	0.99	0.91	1.09
0.57	10,400	0.99	0.90	1.10
0.62	10,500	1.01	0.89	1.14
0.67	10,600	1.01	0.87	1.16
0.72	10,500	0.99	0.85	1.16
0.76	10,200	0.97	0.83	1.17
0.81	10,000	0.98	0.80	1.23
0.86	8000	0.80	0.75	1.07

Note: The mixing ratio (w/w) of L1695/EMDG equals unity. NA—not available.

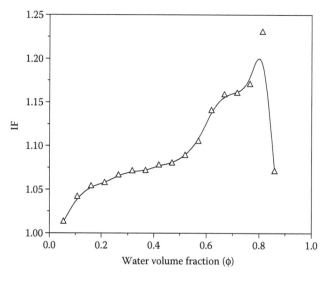

FIGURE 11.8 The IF of diclofenac sodium as a function of the water volume fraction along dilution line N60 in the water/L1695/EMDG/LIM system at 25°C. The mixing ratio (w/w) of EMDG/L1695 equals unity. The phase diagrams are presented in Figure 11.2. The line is presented as a guide to the eye.

TABLE 11.2
Diclofenac Sodium Solubility in the
Microemulsion Components at 25°C

Medium	Na Solubility (ppm)
LIM	<15
IPM	<10
EtOH	<600
Water	<200
EtOH/LIM = 1	<550
EtOH/IPM = 1	<500

(Figure 11.8) that, despite the reduction in the SC caused by the dilution, the IF increases slightly along the entire bicontinuous region (Table 11.1 and Figure 11.8), indicating that diclofenac molecules are still entrapped at the interface. Once the system inverts into oil-in-water droplets (region III, 0.67–0.86 ϕ; Table 11.1 and Figure 11.8), solubilization continues to decrease with a moderate slope. The oil is entrapped in the aqueous phase and its content (together with the mixed surfactants) progressively decreases. The oil-in-water interface requires surfactants with a critical packing parameter lower than one third, and such a hydrophilic interface is less capable of accommodating the lipophilic diclofenac molecules close to the head groups. The diclofenac has to be constricted between the mixed surfactants tails; the resulting total solubilization load decreases significantly. A closer look at the IF values reveals that, despite the structural limitations, the interfacial solubilization is a very dominant factor. The strong DF would have dropped the SC dramatically; with the very low oil content with such a dilution, we can expect minimal solubilization of the diclofenac. However, the solubilization is not zero (Figure 11.5). At 0.86 ϕ we found the SC to be 40-fold higher than the diclofenac solubility in water (Table 11.2). The IF is higher than 1 and contributes to the total solubilization of the drug. We should also not neglect additional factors that may affect the total SC that we could not estimate quantitatively.

11.4.2 SYSTEMS WITH ALCOHOL AS COSURFACTANT

In the LIM + EtOH based system (system B), we identified four major regions in the SC curve separated by SC deflection boundaries. In the IPM + EtOH based system, we identified three major regions in the SC curve. Diclofenac solubilization was influenced not only by the DF or the structural changes but also by the nature of the interface along the dilution line, which is shown in Tables 11.3 and 11.4 and Figures 11.9 and 11.10. At $\phi = 0$, the SC of diclofenac molecules is much higher than the diclofenac dissolution in the two oils (LIM and IPM, system C), which is attributable to its accommodation at the interface of many small reverse lipophilic mixed micelles. Diclofenac is embedded at the interface (and in the core of the mixed micelles) and contributes to the assembly of the reversed mixed micelles. Upon dilution the solubilization is affected by the addition of water and interfacial packing. As the dilution with water progresses, the SC of the microemulsion vehicles decreases. This decrease is affected by the dilution and swelling factors. The calculated IF value that equals 1 at very low water fractions (i.e., <0.07) decreases with dilution (Tables 11.3 and 11.4, Figures 11.9 and 11.10), confirming that the dilution has a negative contribution on the SC because of two factors: first, the diclofenac does not contribute to the self-assembly of the water-in-oil microemulsions, and it is "pushed away" or desorbed from the micellar assembly once water is added; second, the effect of the alcohol that is serving as a cosurfactant is to decrease the surface tension and break the lamellar structures, maintaining the droplet configuration where the surface area is

TABLE 11.3
Water Volume Fraction, Maximum Measured SC of Diclofenac, and the CF, DF, and IF along Dilution Line N60 in the Water/L1695/EMDG/LIM + EtOH System at 25°C

φ	Max. Meas. SC (ppm)	CF	DF	IF
0	15,000	NA	NA	NA
0.07	14,000	0.93	0.93	1.00
0.14	12,800	0.91	0.93	0.99
0.20	12,000	0.94	0.92	1.01
027	11,800	0.98	0.92	1.07
0.33	11,700	0.99	0.92	1.08
0.38	11,600	0.99	0.91	1.08
0.44	11,500	0.99	0.91	1.09
0.49	11,400	0.99	0.91	1.10
0.54	11,300	0.99	0.90	1.10
0.59	11,300	1.00	0.89	1.12
0.64	11,300	1.00	0.88	1.13
0.69	11,500	1.02	0.87	1.17
0.73	12,000	1.04	0.86	1.21
0.77	12,300	1.03	0.84	1.22
0.81	12,200	0.99	0.82	1.21
0.85	10,500	0.86	0.79	1.09
0.89	8100	0.77	0.74	1.05

Note: The mixing ratios (w/w) of L1695/EMDG and EtOH/LIM equal unity. NA—not available.

TABLE 11.4
Water Volume Fraction, Maximum Measured SC of Diclofenac, and the CF, DF, and IF along Dilution Line N60 in the Water/L1695/EMDG/IPM + EtOH System at 25°C

φ	Max. Meas. SC (ppm)	CF	DF	IF
0.00	16,000	NA	NA	NA
0.07	15,000	0.94	0.93	1.01
0.14	13,500	0.90	0.93	0.97
0.20	12,000	0.89	0.92	0.96
0.27	11,600	0.97	0.92	1.05
0.33	11,500	0.99	0.92	1.08
0.39	11,400	0.99	0.91	1.08
0.44	11,200	0.98	0.91	1.08
0.49	11,100	0.99	0.91	1.09
0.54	11,050	1.00	0.90	1.11
0.59	11,000	1.00	0.89	1.12
0.64	10,800	0.98	0.88	1.11
0.69	10,700	0.99	0.87	1.14
0.73	10,500	0.98	0.86	1.14
0.77	10,200	0.97	0.84	1.15
0.81	9700	0.95	0.82	1.16
0.85	9000	0.93	0.79	1.18
0.89	7000	0.78	0.74	1.05

Note: The mixing ratios (w/w) of L1695/EMDG and EtOH/IPM equal unity. NA—not available.

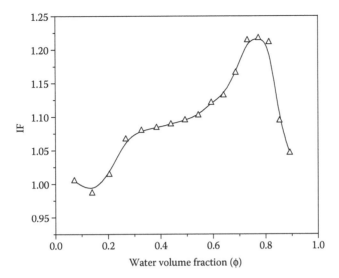

FIGURE 11.9 The IF of diclofenac sodium as a function of the water volume fraction along dilution line N60 in the water/L1695/EMDG/LIM + EtOH system at 25°C. The mixing ratios (w/w) of EMDG/L1695 and EtOH/ oil equal unity. The phase diagrams are presented in Figure 11.3. The line is presented as a guide to the eye.

kept high. We also observed that the minimum IF value is 0.99 at 0.14 ϕ in the LIM-based system, whereas the IF value reaches a minimum of 0.96 at 0.2 ϕ in the IPM-based system. This behavior is explained in terms of oil penetration at the interface. The diclofenac SC along the bicontinuous phase (region II, 0.20–0.61 ϕ) continues to decrease slowly. This is a clear indication that once the interface becomes flat it embeds less molecules of diclofenac. Suratkar and Mahapatra (2000) observed a similar change in the locus of solubilization of phenolic compounds in sodium dodecyl sulfate micelles. The calculated IF increases and reaches values higher than 1, meaning that the IFs are dominant over the DF. However, it is clear (Figures 11.9 and 11.10) that, despite the reduction in the SC caused by the dilution, the IF increases along the entire bicontinuous region, indicating that

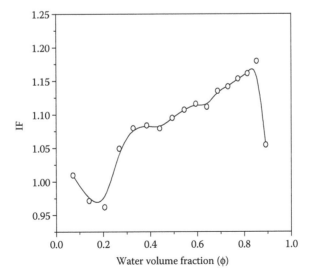

FIGURE 11.10 The IF of diclofenac sodium as a function of the water volume fraction along dilution line N60 in the water/L1695/EMDG/IPM + EtOH system at 25°C. The mixing ratios (w/w) of EMDG/L1695 and EtOH/ oil equal unity. The phase diagrams are presented in Figure 11.4. The line is presented as a guide to the eye.

diclofenac molecules are still entrapped at the interface (Tables 11.3 and 11.4, Figures 11.9 and 11.10). Once the system inverts into oil-in-water droplets (region III, 0.62–0.85 ϕ; Figure 11.6), the IF values increase dramatically, which indicates that the interfacial solubilization is a very dominant factor. In the oil-in-water microemulsions the oil is entrapped in the aqueous phase and its content (together with the surfactant) progressively decreases. The oil-in-water interface is a hydrophilic interface that is less capable of accommodating the lipophilic diclofenac molecules close to the head groups. The diclofenac has to be squeezed between the surfactant tails; the resulting total solubilization load decreases significantly. The IF values reveal that, regardless of the structural limitations, the interfacial solubilization is a very dominant factor. The strong DF would have dropped the SC very dramatically and with the very low oil content with such dilution, we can expect minimal solubilization of the diclofenac. However, the solubilization is not zero (Figure 11.3). At 0.85 ϕ we again found that the SC is much higher than the diclofenac solubility in water (see Table 11.4). The IF is higher than 1 and contributes to the total solubilization of the drug. We should also not neglect additional factors that may affect the total SC that we were not able to estimate quantitatively. Once sufficient water is added, the structure inverts into oil-in-water domains and some of the excess EtOH gradually migrates out of the interface and partitions between the oil and the water phases. The alcohol desorption leaves more space at the interface for the diclofenac and it is reflected in the IF value (IF = 1.05; Tables 11.1 and 11.3, Figures 11.8 and 11.9). The quantitative difference between the systems based on the two oils reflects the oil penetration contribution to the SC. The LIM penetration is smaller than that of the IPM and leaves more "solubilization room" for the diclofenac molecules. The overall effect is an increase in the SC and the IF. The IF value along all of dilution line N60 is higher in the LIM-based system compared to the IPM-based system. In the IPM-based system, the droplets swell and become very elongated—more elongated than in the LIM-based one (possibly wormlike micelles)—and are progressively transformed into bicontinuous domains. The domains' curvature is gradually reduced and results in fewer guest molecules per unit vehicle that are embedded at the interface, resulting in a significant decrease in the SC upon dilution. We note that the SC curves do not show any significant shift in the inversion boundaries in the two oils, but the IF calculation enables detection of the differences between the IFs in the systems based on the two oils. We conclude that the diclofenac solubilization is derived from its core solubilization along with its overall interfacial contribution entrapment and thus is strongly dependent on the number of mixed micelles or droplets, the interfacial curvature, and the composition of the interface, as well as the DF. The next section further explains the structural changes in the SC of diclofenac and determines the effect of the solubilized drug on the electrical conductivity and microstructure of the microemulsions.

11.5   ELECTRICAL CONDUCTIVITY OF DRUG-LOADED MICROEMULSIONS

To ascertain the effect of the solubilized drug on the electrical conductivity, which is a transport property of the microemulsions and the structural transitions occurring upon dilution with water, the electrical conductivities of the microemulsions were determined along water dilution line N60 in drug-loaded microemulsions and compared to the electrical conductivities of the drug-free systems. As reported previously (Fanun, 2008a, 2008b, 2008d), the electrical conductivities increase with the increase in the water volume fraction. Adding diclofenac to the microemulsions increases the electrical conductivity of the system compared to the drug-free system, which as shown in Figures 11.11 through 11.13. The effect of the drug on the electrical conductivity is small when the water is low (i.e., water is entrapped in the core of the reverse swollen micelles), but it becomes more pronounced once the continuous phase is water and the hydrophilic portion of the drug faces the water. A closer look at the conductivity profile reveals the existence of different solubilization regions manifested in different slopes. Tables 11.5 through 11.7 demonstrate these regions for the systems reported here. In the system based on LIM, the first region is from 0.01 to 0.22 ϕ and its slopes are $y = 54.23x$ and $75.92x$ for the free and drug-loaded microemulsions, respectively; in the

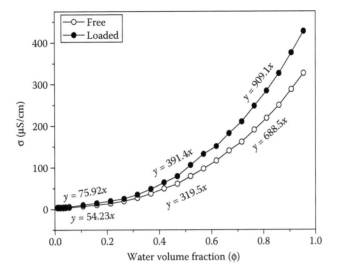

FIGURE 11.11   Electrical conductivity (σ) curves for samples with compositions lying along the N60 dilution line for the water/L1695/EMDG/LIM diclofenac-loaded and diclofenac-free microemulsions at 25°C. The mixing ratio (w/w) of EMDG/L1695 equals unity. The phase diagram is presented in Figure 11.2. The lines are presented as guides to the eye.

region of 0.26–0.67 ϕ the slopes are more gradual ($y = 319.5x$ and $391.4x$ for free and loaded systems, respectively), and in the region above 0.67 ϕ the slopes are steeper ($y = 688.5x$ and $909.1x$ for free and loaded systems, respectively). The LIM + EtOH based system loaded with diclofenac has three similar solubilization regions with slopes that are more pronounced compared to the previous system (see Figure 11.12 and Table 11.6). The IPM + EtOH based system loaded with diclofenac has two solubilization regions with slopes that are less pronounced compared to the LIM + EtOH based system (see Figure 11.13 and Table 11.7). The increases in the slopes in the drug-loaded systems indicate stronger electrical conductivity effects that are derived from the drug. The major difference

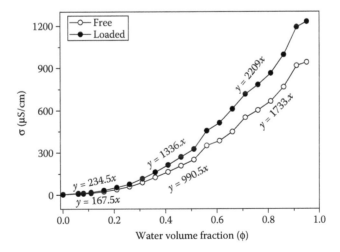

FIGURE 11.12   Electrical conductivity (σ) curves for samples with compositions lying along the N60 dilution line for the water/L1695/EMDG/LIM + EtOH diclofenac-loaded and diclofenac-free microemulsions at 25°C. The mixing ratios (w/w) of EMDG/L1695 and EtOH/ + LIM equal unity. The phase diagram is presented in Figure 11.3. The lines are presented as guides to the eye.

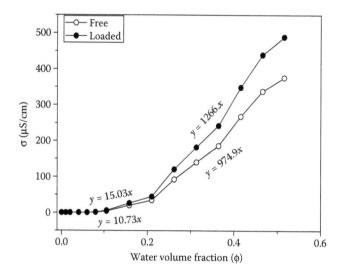

FIGURE 11.13 Electrical conductivity (σ) curves for samples with compositions lying along the N60 dilution line for the water/L1695/EMDG/IPM + EtOH diclofenac-loaded and diclofenac-free microemulsions at 25°C. The mixing ratios (w/w) of EMDG/L1695 and EtOH/IPM equal unity. The phase diagram is presented in Figure 11.4. The lines are presented as guides to the eye.

TABLE 11.5
Slopes of Electrical Conductivity Curves at Different Water Volume Fractions for the Water/L1695/EMDG/LIM Microemulsion System along Dilution Line N60 at 25°C

ϕ	Free	Loaded
0.01–0.22	54.23	75.92
0.26–0.67	319.5	391.4
0.72–0.95	688.5	909.1

Note: The mixing ratio (w/w) of EMDG/L1695 equals unity.

TABLE 11.6
Slopes of Electrical Conductivity Curves at Different Water Volume Fractions for the Water/L1695/EMDG/LIM + EtOH Microemulsion System along Dilution Line N60 at 25°C

ϕ	Free	Loaded
0.01–0.22	167.5	234.5
0.26–0.61	990.5	1336
0.66–0.95	1733	2209

Note: The mixing ratios (w/w) of EMDG/L1695 and EtOH/LIM equal unity.

TABLE 11.7
Slopes of Electrical Conductivity Curves at Different Water
Volume Fractions for the Water/L1695/EMDG/IPM + EtOH
Microemulsion System along Dilution Line N60 at 25°C

ϕ	Free	Loaded
0.01–0.15	10.73	15.03
0.15–0.52	974.9	1266

Note: The mixing ratios (w/w) of EMDG/L1695 and EtOH/IPM equal unity.

between the empty and loaded systems is the dilution point at which the slope changes and the transition occurs. We learned in previous studies (Fanun, 2008a, 2008b, 2008c) that in the U-type phase diagrams the reverse micelles that are formed in the oil phase slowly swell and deform upon the addition of water or an aqueous phase, and at a certain point the water migrates out of the inner phase and spongelike domains are formed, which are termed the bicontinuous phase. Full inversion occurs upon further dilution and the water entrapped within the oil becomes a continuous phase. The structural transition from the water-in-oil to the bicontinuous phase occurs in both the empty and loaded systems at about 0.22–0.26 ϕ. The first structural transition is seemingly not significantly affected by the drug. The water reservoirs are equally deformed and they no longer exhibit spherical or disklike shapes upon the gradual addition of more water, instead turning into "wormlike" water domains that are dispersed into the oil continuous domains. At a certain concentration, the water phase starts to migrate to the outer phase, the curvatures change, and the conductivity increases. The bicontinuous region is quite large and exists within 0.22–0.62 and 0.26–0.27 ϕ in the LIM and LIM + EtOH based systems, respectively, and within 0.2–0.52 ϕ in the IPM + EtOH based system. The water and oil phases are interwoven within this region. The second transition in the LIM-based systems takes place at about 0.62–0.67 wt% water and is difficult to detect because the process is quite gradual. This slope deflection reflects the full transformation of the bicontinuous domain into oil-in-water droplets, and the water becomes the continuous phase. The transition in systems loaded with the drug seems to occur at somewhat higher water content and seems to be completed at about 0.7 ϕ. The drug appears to affect the curvature of the bicontinuous domains, retarding their transition into oil-in-water droplets.

11.6 DYNAMIC VISCOSITY OF DRUG-LOADED MICROEMULSIONS

The dynamic viscosity of microemulsions is structure dependent, and its measurement can provide significant information about the structural transitions in microemulsions (Fanun et al., 2001; Yaghmur et al., 2003). The variation in the dynamic viscosity as a function of the water volume fraction in the diclofenac-loaded and diclofenac-free microemulsions reported in the previous sections is illustrated in Figures 11.14 through 11.16. The investigated samples show a Newtonian flow behavior. The viscosity behavior of free and loaded microemulsions appears to be parallel with higher values given by the diclofenac-loaded microemulsions compared to the diclofenac-free ones. Once the droplets swell, the droplet–droplet interactions increase and the viscosity ascends roughly (Gradzielski and Hoffman, 1999; Saidi et al., 1990). Fletcher et al. (1984) studied an *n*-heptane/ Aerosol® OT/glycerol system and observed strong attractive interactions between the droplets, which led to the formation of clusters. Mathew et al. (1991) support this interpretation. With the increase in water volume fraction, bicontinuous structures are subsequently formed. In this region the interconnected water and oil "channels" progressively increase the structural interactions and therefore the viscosity. The interaction between the mixed surfactant tails is maximal and a strong impediment is expected, leading to an increase in viscosity. The viscosity is maximal once the

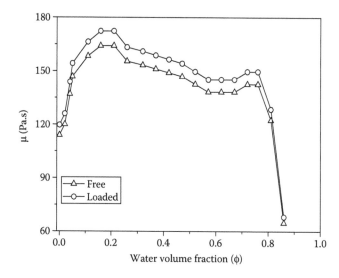

FIGURE 11.14 Dynamic viscosity (μ) curves for samples with compositions lying along the N60 dilution line for the water/L1695/EMDG/LIM diclofenac-loaded and diclofenac-free microemulsions at 25°C. The mixing ratio (w/w) of EMDG/L1695 equals unity. The phase diagram is presented in Figure 11.2. The lines are presented as guides to the eye.

system is totally inverted into the bicontinuous domains. However, with further dilution, the bicontinuous structure gradually disintegrates and a transition into an oil-in-water microemulsion occurs, which is reflected in a sharp decrease in the viscosity. In this region, the head groups' hydration is strongly increased and the packing parameter drops to values of less than one third. The viscosity continues to decrease as the nanodroplets' number and size decrease and become increasingly diluted with the aqueous phase. The progressive dilution decreases the interdroplet interactions. When diclofenac is loaded into the reverse mixed micelles in region I, the viscosity of the system is only slightly affected because the diclofenac molecules are located mostly in the oil continuous

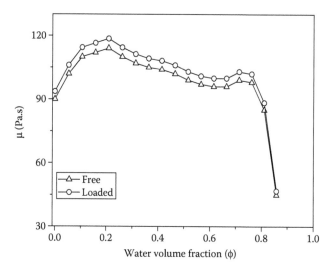

FIGURE 11.15 Dynamic viscosity (μ) curves for samples with compositions lying along the N60 dilution line for the water/L1695/EMDG/LIM + EtOH diclofenac-loaded and diclofenac-free microemulsions at 25°C. The mixing ratios (w/w) of EMDG/L1695 and EtOH/LIM equal unity. The phase diagram is presented in Figure 11.3. The lines are presented as guides to the eye.

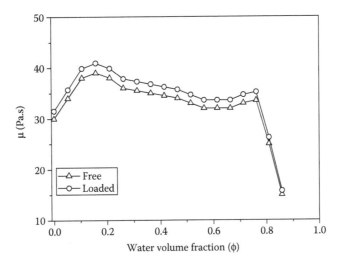

FIGURE 11.16 Dynamic viscosity (μ) curves for samples with compositions lying along the N60 dilution line for the water/L1695/EMDG/IPM + EtOH diclofenac-loaded and diclofenac-free microemulsions at 25°C. The mixing ratios (w/w) of EMDG/L1695 and EtOH/IPM equal unity. The phase diagram is presented in Figure 11.4. The lines are presented as guides to the eye.

phase and the interfacially absorbed tails are not interacting. The viscosities of the diclofenac-loaded systems are relatively larger than the viscosity of the diclofenac-free ones (Figures 11.14 through 11.16). The explanation for this difference in behavior is that, once diclofenac is solubilized (in greater excess), it pushes away the mixed surfactant tails that tend to entrap and increase the viscosity. Consequently, the interdomain interactions are stronger than before, which causes an increase in the viscosity in the diclofenac-loaded systems over the diclofenac-free ones; and an inversion to the bicontinuous structure, which is the addition of diclofenac, precedes the full structure transformation (the transition occurs at ~50 and 40 wt% of the aqueous phase in the empty and loaded systems, respectively). It should be noted therefore that, although the electrical conductivity reflects structural transitions, the viscosity reflects mostly structural interaction variations and is indicative of the sharp collapse of the oil-in-water bicontinuous network. If we examine the behavior of the three systems (i.e., microemulsions based on LIM, LIM + EtOH, and IPM + EtOH) in the presence of solubilized diclofenac, it is clear that the presence of EtOH and the type of oils strongly affects the interdroplet interaction and the viscosity values are different. These behaviors can again be explained in terms of EtOH and oil penetration effects on the interdroplet interactions.

11.7   MICROSTRUCTURE OF DRUG-LOADED MICROEMULSIONS

We used small-angle x-ray scattering (SAXS) measurements to learn about the microstructure parameters (periodicity and correlation length) of the sodium diclofenac loaded microemulsions. Figure 11.17 presents the characteristic profiles as an example of what happens in the diclofenac-loaded water/L1695/EMDG/LIM microemulsion system as a function of the water volume fraction along dilution line N60. The mixing ratio of EMDG/L1695 also equals unity. According to the Teubner–Strey (1987) equation (Equation 11.4), we were able to derive the scattering intensity [$I(q)$] from the values of the periodicity (d), correlation length (ξ), and amphiphilicity factor (f_a) as described in the experimental section (Equations 11.4 through 11.8):

$$I(q) = \left(\frac{1}{a_2} + c_1 q^2 + c_2 q^4 \right) + b,$$

(11.4)

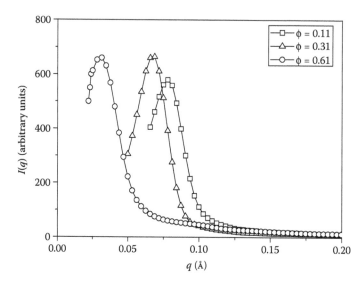

FIGURE 11.17 SAXS curves for samples with compositions lying along the N60 dilution line for the water/L1695/EMDG/LIM diclofenac-loaded microemulsions at 25°C. The mixing ratio (w/w) of EMDG/L1695 equals unity. The phase diagram is presented in Figure 11.2. The lines are presented as guides to the eye.

where q is the scattering vector and the constants a_2, c_1, c_2, and b were obtained by using the Levenburg–Marquardt procedure (Teukolsky et al., 1992). Such a functional form is simple and convenient for the fitting of spectra. Equation 11.5 corresponds to a real space correlation function [$\gamma(r)$] of the form

$$\gamma(r) = \left(\sin \frac{\lambda r}{\lambda r} \right) \exp^{-r/\xi}. \tag{11.5}$$

The correlation function describes a structure with $d = 2\pi/\lambda$ damped as a function of ξ. This formalism also predicts the surface/volume ratio. However, because this ratio is inversely related to the correlation length and therefore must go to zero for a perfectly ordered system, the calculated values are frequently found to be too low (Billman and Kaler, 1991). The values of d and ξ are related to the constants in Equation 11.4 by Teubner and Strey (1987):

$$d = 2\pi \left[\left(\frac{1}{2} \right) \left(\frac{a_2}{c_2} \right)^{1/2} - \left(\frac{c_1}{4c_2} \right) \right]^{-1/2}, \tag{11.6}$$

$$x = \left[\left(\frac{1}{2} \right) \left(\frac{a_2}{c_2} \right)^{1/2} + \left(\frac{c_1}{4c_2} \right) \right]^{-1/2}. \tag{11.7}$$

A third parameter that can also be defined is the amphiphilicity factor (f_a; Gradzielski et al., 1996; Schubert and Strey, 1991; Schubert et al., 1994; Teubner and Strey, 1987), which relates to the behavior of the correlation function and reflects the ability of the surfactant to impose order on the microemulsion:

$$f_a = \frac{c_1}{(4a_2c_2)^{1/2}}. \tag{11.8}$$

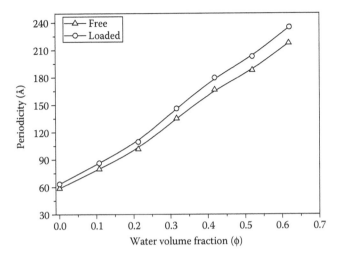

FIGURE 11.18 Drug-loaded and drug-free microemulsion periodicities as a function of the volume fraction of water for samples with compositions lying along the N60 dilution line for the water/L1695/EMDG/LIM systems at 25°C. The mixing ratio (w/w) of EMDG/L1695 equals unity. The phase diagram is presented in Figure 11.2. The lines are presented as guides to the eye.

The periodicity calculated from Teubner and Strey's model (1987) was plotted against the water volume fraction as shown in Figure 11.18. The periodicity values increase as the water content increases. A comparison of the periodicity values of the drug-loaded microemulsions with those of drug-free microemulsions for the systems presented in our previous work (Fanun, 2008b) allows us to conclude that the drug affects the periodicity values (size of the domains). The droplets with the drug swell, and the average size increases by 8–10% (depending on the dilution). Figure 11.19 provides the correlation lengths of the microemulsion systems loaded with drug. The different regions observed in these curves correspond to the presence of structural transitions along the dilution line from water-in-oil to bicontinuous to oil-in-water microemulsions. The correlation length reflects the

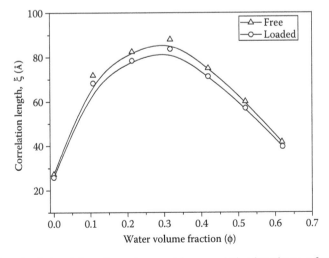

FIGURE 11.19 Drug-loaded and drug-free microemulsion correlation lengths as a function of the volume fraction of water for samples with compositions lying along the N60 dilution line for the water/L1695/EMDG/LIM systems at 25°C. The mixing ratio (w/w) of EMDG/L1695 equals unity. The phase diagram is presented in Figure 11.2. The lines are presented as guides to the eye.

TABLE 11.8
Values of the Amphiphilicity Factor for the Diclofenac-Loaded Water/L1695/EMDG/LIM Microemulsion System as a Function of the Water Volume Fraction along Dilution Line N60

ϕ	f_a
0	−0.75
0.11	−0.90
0.21	−0.89
0.32	−0.80
0.42	−0.68
0.52	−0.67
0.62	−0.67

Note: The mixing ratio (w/w) of EMDG/L1695 equals unity. The values of f_a were calculated from Equation 11.8 from data obtained at 25°C.

degree of order in the microemulsion. The systems have a typical bell-shaped curve for the correlation length. As observed by the other analytical tools, the maximum correlation length is obtained at 0.35 ϕ. The measurements indicate that the degree of order is lower in the presence of the drug compared to the systems free of drug (Fanun, 2008b). The drug causes significant disorder in the microemulsion structures. The amphiphilicity factor values (Table 11.8) calculated using Equation 11.8 in the drug-loaded systems are less negative than those observed in the drug-free systems as reported previously (Fanun, 2008b). This behavior again indicated that the drug-loaded systems are less ordered than the drug-free ones. Figure 11.20 provides the characteristic profiles for the drug-loaded water/L1695/EMDG/LIM + EtOH microemulsions. The dependence of the periodicity and correlation length as a function of the progressive dilution in the loaded systems is plotted in Figures 11.21

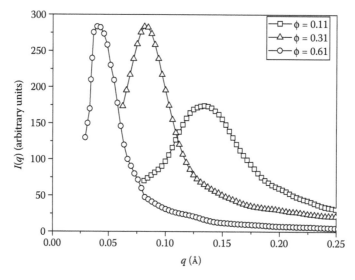

FIGURE 11.20 SAXS curves for samples with compositions lying along the N60 dilution line for the water/L1695/EMDG/LIM + EtOH diclofenac-loaded microemulsions at 25°C. The mixing ratios (w/w) of EMDG/L1695 and EtOH/oil equal unity. The phase diagram is presented in Figure 11.3. The lines are presented as guides to the eye.

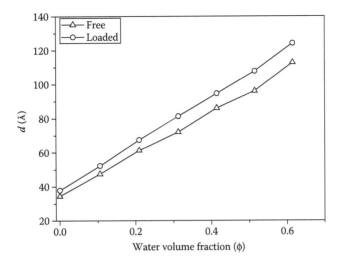

FIGURE 11.21 Drug-loaded and drug-free microemulsion periodicities as a function of the volume fraction of water for samples with compositions lying along the N60 dilution line for the water/L1695/EMDG/LIM + EtOH system at 25°C. The mixing ratios (w/w) of EMDG/L1695 and EtOH/oil equal unity. The phase diagram is presented in Figure 11.3. The lines are presented as guides to the eye.

and 11.22. The periodicity value increases monotonically over the whole range of water dilutions along the N60 dilution line. The periodicity values of the drug-loaded system are about 10–13 times higher than those observed in the drug-free system (Fanun, 2008a, 2008b, 2009). Figure 11.23 shows the characteristic profiles for the drug-loaded water/L1695/EMDG/IPM + EtOH microemulsions. The dependence of the periodicity and correlation length as a function of the progressive dilution in the loaded systems is plotted in Figures 11.24 and 11.25. The periodicity values of the drug-loaded system are about 10–15 times higher than those observed in the drug-free system (Fanun, 2008a,

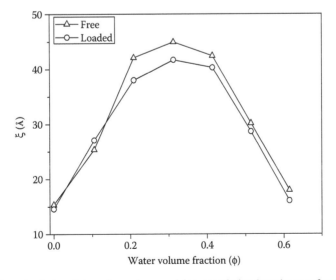

FIGURE 11.22 Drug-loaded and drug-free microemulsion correlation lengths as a function of the volume fraction of water for samples with compositions lying along the N60 dilution line for the water/L1695/EMDG/LIM + EtOH system at 25°C. The mixing ratio (w/w) of EMDG/L1695 equals unity. The phase diagram is presented in Figure 11.3. The lines are presented as guides to the eye.

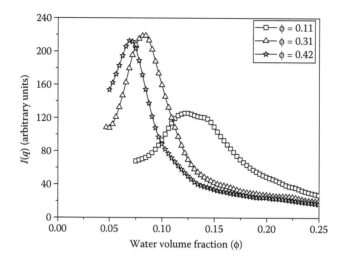

FIGURE 11.23   SAXS curves for samples with compositions lying along the N60 dilution line for the water/L1695/EMDG/IPM + EtOH diclofenac-loaded microemulsions at 25°C. The mixing ratios (w/w) of EMDG/L1695 and EtOH/oil equal unity. The phase diagram is presented in Figure 11.3. The lines are presented as guides to the eye.

2008b, 2009). The trend in the values of the amphiphilicity factor calculated using Equation 11.8 is the same as that observed in the LIM-based system (Tables 11.9 and 11.10). For all systems the values of the loaded microemulsions differ slightly (up to 15%) from the reported values of the free systems (Fanun, 2008a, 2008b, 2009); in other words, the drug does not have a dramatic effect on the periodicity (size of the domains). This behavior of the microemulsions can be correlated with the structural transitions and can be explained in the following way: when water is the internal phase, increasing the aqueous phase content swells the surfactant aggregates and results in an increasingly ordered system (ξ increases); when water is the continuous phase, an added aqueous phase dilutes the system and the order decreases (ξ decreases). On this basis, we suggest that the inversion for this type of

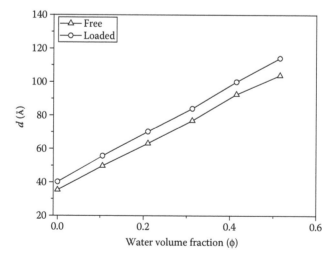

FIGURE 11.24   Drug-loaded and drug-free microemulsion periodicities as a function of the volume fraction of water for samples with compositions lying along the N60 dilution line for the water/L1695/EMDG/IPM + EtOH system at 25°C. The mixing ratios (w/w) of EMDG/L1695 and EtOH/oil equal unity. The phase diagram is presented in Figure 11.4. The lines are presented as guides to the eye.

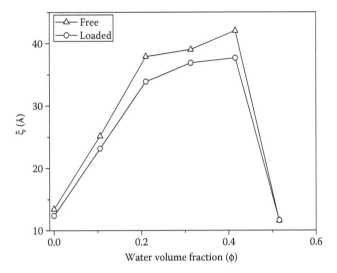

FIGURE 11.25 Drug-loaded and drug-free microemulsion correlation lengths as a function of the volume fraction of water for samples with compositions lying along the N60 dilution line for the water/L1695/EMDG/ IPM + EtOH system at 25°C. The mixing ratio (w/w) of EMDG/L1695 equals unity. The phase diagram is presented in Figure 11.4. The lines are presented as guides to the eye.

microemulsion occurs where the correlation length achieves its maximum value (Fanun, 2008a, 2008b, 2009). In the LIM + EtOH based microemulsions the values of correlation length along the dilution line in the loaded microemulsion are only slightly different from the values obtained from the drug-free system (Fanun, 2008a, 2008b, 2009). The inversion occurs when the maximum value of the correlation length is reached at 0.32 ϕ. Because the drug participates at the interface, it increases the interdroplet interactions and precedes the inversion from the bicontinuous domains to the water-in-oil droplets. In the IPM-based system, the correlation length reaches its maximum at 0.41 ϕ. The correlation length values of the loaded IPM-based microemulsions are smaller than the values of the LIM-based microemulsion. This means that the increase of the oil penetration decreases the order of the system. A similar effect was achieved at the water-in-oil and bicontinuous domains

TABLE 11.9
Values of the Amphiphilicity Factor for the Diclofenac-Loaded Water/L1695/EMDG/LIM + EtOH Microemulsion System as a Function of the Water Volume Fraction along Dilution Line N60

ϕ	f_a
0	−0.78
0.11	−0.82
0.21	−0.88
0.31	−0.86
0.41	−0.79
0.51	−0.70
0.61	−0.65

Note: The mixing ratios (w/w) of EtOH/oil and L1695/EMDG equal unity. The values of f_a were calculated from Equation 11.8 from data obtained at 25°C.

TABLE 11.10
Values of the Aphiphilicity Factor for the Diclofenac-Loaded Water/
L1695/EMDG/IPM + EtOH Microemulsion System as a Function of
the Water Volume Fraction along Dilution Line N60

ϕ	f_a
0	−0.75
0.11	−0.80
0.21	−0.86
0.31	−0.83
0.41	−0.76
0.52	−0.60

Note: The mixing ratios (w/w) of EtOH/oil and L1695/EMDG equal unity. The values of f_a
were calculated from Equation 11.8 from data obtained at 25°C.

by the addition of diclofenac to the microemulsion system. The measurements indicate that the degree of order in all of the loaded systems with any dilution composition that was tested is lower in the presence of the drug, which means that the drug causes significant disorder. Figure 11.26 presents a schematic illustration (not to scale) of the possible packing of diclofenac within water-in-oil, bicontinuous, and oil-in-water microemulsions along dilution line N60. This schematic illustrates that in the water-in-oil microstructure the diclofenac molecules can simply penetrate between the mixed surfactants' hydrophobic chains. Upon water dilution and formation of bicontinuous microstructures, the ability of diclofenac to be set in between the surfactant hydrophobic chains decreases. Note that two different bicontinuous microstructures could be formed as the water dilution advanced. Initially, elongated wormlike reverse micelles are dispersed in continuous oil phase; upon further dilution the microstructure is converted to elongated wormlike direct micelles dispersed in the water continuous phase. The diclofenac solubilization seems to be higher in the water-in-oil bicontinuous microstructure than in the oil-in-water bicontinuous one, given that in the water-in-oil bicontinuous

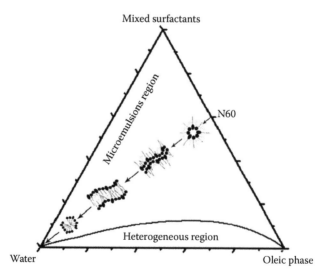

FIGURE 11.26 Schematic presentation (not to scale) of possible packings of diclofenac along dilution line N60 at the different dilution regions: water-in-oil, bicontinuous, and oil-in-water microemulsions.

microstructure the mixed surfactants tails are hanging in the oil continuous phase and the diclofenac molecules can penetrate into the mixed surfactants' film more easily. Conversely, in the oil-in-water bicontinuous microstructure the mixed surfactants' tails are confined to the inner oil phase, which has less space to contain the diclofenac molecules. Upon the inversion into oil-in-water-like droplets, the mixed surfactants are closely packed around the oil droplets, leaving less free space for diclofenac. The packing parameters of the mixed surfactants seem lower than that of diclofenac and, because the packing of diclofenac into the interface is not favorable, its SC is thus the lowest.

11.8 CONCLUSIONS

We demonstrated the utilization of U-type mixed nonionic surfactant based microemulsions as vehicles for solubilizing diclofenac sodium. We concluded that the diclofenac solubility in microemulsions is much higher compared to its solubility in both oil and water. Concentrates (water-free) loaded with very high levels of the drug can be prepared easily. The diclofenac solubility in microemulsions does not affect the extent of the one-phase microemulsion region. This is an important advantage of such systems that is reflected in the ability of the end user to progressively dilute the formulation with an aqueous phase to any required dilution without phase separation. The SC of diclofenac is highly dependent on the composition of the microemulsion system. The presence of alcohol as a cosurfactant and the oil type and its penetration at the interface both greatly affect the degree of diclofenac solubilization. The SC of diclofenac is dependent on the microstructure, which means that the microemulsion type strongly influences the extent of diclofenac solubilization. Solubilized diclofenac influences both the microstructure and diffusion parameters of the loaded microemulsions compared to the drug-free microemulsions. The drug interacts at the interface with the microemulsion components at any specific microstructure of the investigated vehicles and affects the water contents at which the transitions from water-in-oil to bicontinuous to oil-in-water microemulsions, so the drug release kinetics from these microemulsions should be affected. The drug remains solubilized at the interface upon dilution with water and is oriented with its hydrophilic part facing the water. In many of the formulations the drug is soluble in the concentrated capsule, but it precipitates at once if diluted with water. Our formulations are based on nonionic surfactants and therefore are more resistant to low pH and can survive the stomach dilution and acidity, especially the LIM-based system.

ABBREVIATIONS

EMDG   Ethoxylated mono- and diglyceride
EtOH   Ethanol
IPM   Isopropylmyristate
L1695   Sucrose laurate
LIM R-(+)-limonene
SAXS   Small angle x-ray scattering
SC   Solubilization capacity

SYMBOLS

d Periodicity or characteristic length
ξ Correlation length
ϕ Water volume fraction
f_a Amphiphilicity factor
λ X-ray wavelength
$I(q)$ Scattering intensity
μ Dynamic viscosity

1φ   One-phase microemulsion region
q Scattering vector = $(4\pi/\lambda)\sin\theta$
σ   Electrical conductivity
2θ   Scattering angle
$W_m + O$ Water continuous micellar system with excess oil

REFERENCES

Bagwe, R. P., Kanicky, J. R., Palla, B. J., Patanjali, P. K., and Shah, D. O. 2001. Improved drug delivery using microemulsions: Rationale, recent progress, and new horizons. *Crit. Rev. Ther. Drug Carr. Syst.* 18: 77–144.

Beck, R., Haas, S., Guterres, S., Ré, M., Benvenutti, E., and Pohlmann, A. R. 2006. Nanoparticle-coated organic–inorganic microparticles: Experimental design and gastrointestinal tolerance evaluation. *Quim. Nova* 29: 990–996.

Berghenholtz, J., Romagnoli, A., and Wagner, N. 1995. Viscosity, microstructure, and interparticle potential of AOT/H$_2$O/n-decane inverse microemulsions. *Langmuir* 11: 1559–1570.

Billman, J. F. and Kaler E. W 1991. Structure and phase behavior in four-component nonionic microemulsions. *Langmuir* 7: 1609–1617.

Cametti, C., Codastefano, P., Tartaglia, P., Chen, S. H., and Rouch, J. 1992. Electrical conductivity and percolation phenomena in water-in-oil microemulsions. *J. Phys. Rev. A* 45: R5358–R5361 and references therein.

Cevc, G. 2004. Lipid vesicles and other colloids as drug carriers on the skin. *Adv. Drug Deliv. Rev.* 56: 675–711.

Choi, S. M., Chen, S. H., Sottmann, T., and Strey, R. 1997. Measurement of interfacial curvatures in microemulsions using small-angle neutron scattering. *Phys. B: Condens. Matter* 241–243: 976–978.

Courrier, H. M., Krafft, M. P., Butz, N., Porté, C., Frossard, N., Rémy-Kristensen, A., Mély, A. Y., Pons, F., and Vandamme, Th.F. 2003. Evaluation of cytotoxicity of new semi-fluorinated amphiphiles derived from dimorpholinophosphate. *Biomaterials* 24: 689.

D'Cruz, O. J., Erbeck, D., Waurzyniak, B., and Uckun, F. M. 2002b. Two-year toxicity and carcinogenicity studies in B$_6$C$_3$F$_1$ mice with 5-bromo-6-methoxy-5,6-dihydro-3′-azidothymidine-5′-(p-bromophenyl) methoxyalaninyl phosphate (WHI-07), a novel anti-HIV and contraceptive agent. *Toxicology* 179: 61–77.

D'Cruz, O. J. and Uckun, F. M. 2003. Contraceptive activity of a spermicidal aryl phosphate derivative of bromo-methoxy-zidovudine (compound WHI-07) in rabbits. *Fertil. Steril.* 79: 864–872.

D'Cruz, O. J., Waurzyniak, B., and Uckun, F. M. 2002a. Subchronic (13-week) toxicity studies of intravaginal administration of spermicidal vanadocene acetylacetonato monotriflate in mice. *Toxicology* 170: 31–43.

D'Cruz, O. J, Zhu, Z., Yiv, S. H., Chen, C. L., Waurzyniak, B., and Uckun, F. M. 1999. WHI-05, a novel bromo-methoxy substituted phenyl phosphate derivative of zidovudine, is a dual-action spermicide with potent anti-HIV activity. *Contraception* 59: 319–331.

Fanun, M. 2007a. Conductivity, viscosity, NMR and diclofenac solubilization capacity studies of mixed nonionic surfactants microemulsions. *J. Mol. Liquids* 135: 5–13.

Fanun, M. 2007b. Structure probing of water/mixed nonionic surfactants/caprylic capric triglyceride. *J. Mol. Liquids* 133: 22–27.

Fanun M. (Ed.). 2008a. *Microemulsions: Properties and Applications. Surfactants Science Series*, Vol. 144. Boca Raton, FL: Taylor & Francis/CRC Press.

Fanun, M. 2008b. A study of the properties of mixed nonionic surfactants microemulsions by NMR, SAXS, viscosity and conductivity. *J. Mol. Liquids* 142: 103–110.

Fanun, M. 2008c. Water solubilization in mixed nonionic surfactants microemulsions. *J. Disp. Sci. Technol.* 29: 1043–1052.

Fanun, M. 2008d. Conductive flow parameters of mixed nonionic surfactants microemulsions. *J. Disp. Sci. Technol.* 29: 1426–1434.

Fanun, M. 2009. Microstructure of mixed nonionic surfactants microemulsions studied by SAXS and DLS. *J. Disp. Sci. Technol.* 30: 115–123.

Fanun, M. and Salah Al-Diyn, W. 2006. Electrical conductivity and self diffusion NMR studies of the system: Water/sucrose laurate/ethoxylated mono-di-glyceride/isopropylmyristate systems. *Colloids Surf. A: Physicochem. Eng. Aspects* 277: 83–89.

Fanun, M. and Salah Al-Diyn, W. 2007. Structural transitions in the system water/mixed nonionic surfactants/R-(+)-limonene studied by electrical conductivity and self diffusion-NMR. *J. Disp. Sci. Technol.* 28: 165–174.

Fanun, M., Wachtel, E., Antalek, B., Aserin, A., and Garti, N. 2001. A study of the microstructure of four-component sucrose ester microemulsions by SAXS and NMR. *Colloids Surf. A: Physicochem. Eng. Aspects* 180: 173–186.

Fletcher, P. D. I., Galal, M. F., and Robinson, B. H. 1984. Structural study of aerosol-OT-stabilised microemulsions of glycerol dispersed in *n*-heptane. *J. Chem. Soc. Faraday Trans. I* 80: 3307–3314.

Glatter, O., Strey, R., Schubert, K.-V., and Kaler, E. W. 1996. Small angle scattering applied to microemulsions. *Ber. Bunsenges. Phys. Chem.* 100: 323–335.

Goddeeris, C., Cuppo, F., Reynaers, H., Bouwman, W. G., and Van Den Mooter, G. 2006. Light scattering measurements on microemulsions: Estimation of droplet sizes. *Int. J. Pharm.* 312: 187–195.

Gradzielski, M. and Hoffman, H. 1999. Rheological properties of microemulsions. In: P. Kumar and K. L. Mittal (Eds.), *Handbook of Microemulsion Science and Technology*, pp. 357–386. New York: Marcel Dekker.

Gradzielski, M., Langevin, D., Sottmann, T., and Strey, R. 1996. Small angle neutron scattering near the wetting transition: Discrimination of microemulsions from weakly structured mixtures. *J. Chem. Phys.* 104: 3782–3787.

Hiranita, T., Nakamura, S., Kawachi, M., Courrier, H. M., Vandamme, Th. F., Krafft, M. P., and Shibata, O. 2003. Miscibility behavior of dipalmitoylphosphatidylcholine with a single-chain partially fluorinated amphiphile in Langmuir monolayers. *J. Colloid Interface Sci.* 265: 83–92.

Ho, H., Hsiao, C. C., and Sheu, M. T. 1996. Preparation of microemulsions using polyglycerin fatty acid esters as surfactant for the delivery of protein drugs. *J. Pharm. Sci.* 85: 138–143.

Hoar, T. P. and Schulman, J. H. 1943. Transparent water-in-oil dispersions: The oleopathic hydro-micelle. *Nature* 152: 102–103.

Kahlweit, M., Busse, G., and Winkler, J. 1993. Electric conductivity in microemulsions. *J. Chem. Phys.* 99: 5605–5614.

Kahlweit, M., Strey, R., Haase, D., Kunieda, H., Schmeling, T., Faulhaber, B., Borkovec, M., et al. 1987. How to study microemulsions. *J. Colloid Interface Sci.* 118: 436–453.

Kantarci, G., Ozguney, I., Karasulu, H. Y., Guneri, T., and Basdemir, G. 2005. *In vitro* permeation of diclofenac sodium from novel microemulsion formulations through rabbit skin. *Drug Dev. Res.* 65: 17–25.

Krafft, M. P. and Goldmann, M. 2003. Monolayers made from fluorinated amphiphiles *Curr. Opin. Colloid Interface Sci.* 8: 243–250.

Kreilgaard, M. 2002. Influence of microemulsions on cutaneous drug delivery. *Adv. Drug Deliv. Rev.* 54: S77–S98.

Kumar, P. and Mital, K. L. (Eds.). 1999. *Handbook of Microemulsion Science and Technology*. New York: Marcel Dekker .

Kweon, J.-H., Chi, S.-C., and Park, E.-S. 2004. Transdermal delivery of diclofenac using microemulsions. *Arch. Pharm. Res.* 27: 351–356.

Lawrence, M. J. and Rees, G. D. 2000. Microemulsion-based media as novel drug delivery systems. *Adv. Drug Deliv. Rev.* 45: 89–121.

Lindenstruth, K. and Muller, B. W. 2004. W/O/W multiple emulsions with diclofenac sodium. *Eur. J. Pharm. Biopharm.* 58: 621–627.

Mathew, C., Saidi, Z., Peyrelasse, J., and Boned, C. 1991. Viscosity, conductivity, and dielectric relaxation of waterless glycerol–sodium bis(2-ethylhexyl) sulfosuccinate–isooctane microemulsions: The percolation effect. *Phys. Rev. A* 43: 873–882.

Matsumoto, S. P. and Sherman, P. 1969. The viscosity of microemulsions. *J. Colloid Interface Sci.* 30: 525–536.

Mihailescu, M., Monkenbusch, M., Allgaier, J., Frielinghaus, H., Richter, D., Jakobs, B., and Sottmann, T. 2002. Neutron scattering study on the structure and dynamics of oriented lamellar phase microemulsions. *Phys. Rev. E* 66: 041504/1–041504/13. DOI:10.1103/PhysRevE.66.041504.

Muller, R. H. and Keck, C. M. 2004. Challenges and solutions for the delivery of biotech drugs—A review of drug nanocrystal technology and lipid nanoparticles. *J. Biotechnol.* 113: 151–170.

Olsson, U., Shinoda, K., and Lindman, B. 1986. Change of the structure of microemulsions with the hydrophile–lipophile balance of nonionic surfactant as revealed by NMR self-diffusion studies. *J. Phys. Chem.* 90: 4083–4088.

Patel, N., Marlow, M., and Lawrence, M. J. 2003. Formation of fluorinated nonionic surfactant microemulsions in hydrofluorocarbon 134a (HFC 134a). *J. Colloid Interface Sci.* 258: 345–353.

Piao, H., Hirata, A., Yokoyama, H., Fujii, T., Shimizu, I., Ito, S., Kamiya, N., and Goto, M. 2007. Reduction of gastric ulcerogenicity during multiple administration of diclofenac sodium by a novel solid-in-oil suspension. *Pharm. Dev. Technol.* 12: 321–325.

Piao, H., Kamiya, N., Watanabe, J., Yokoyama, H., Hirata, A., Fujii, T., Shimizu, I., and Goto, M. 2006. Oral delivery of diclofenac sodium using a novel solid-in-oil suspension. *Int. J. Pharm.* 313: 159–162.

Ray, S., Bisal, S., and Moulik, S. 1992. Studies on structure and dynamics of microemulsions II: Viscosity behavior of water-in-oil microemulsions. *J. Surf. Sci. Technol.* 8: 191–208.

Regev, O., Ezrahi, S., Aserin, A., Garti, N., Wachtel, E., Kaler, E., Khan, A., and Talmon, Y. 1996. A study of the microstructure of a four-component nonionic microemulsion by cryo-TEM, NMR, SAXS, and SANS. *Langmuir* 12: 668–674.

Saidi, Z., Mathew, C., Peyrelasse, J., and Boned, C. 1990. Percolation and critical exponents for the viscosity of microemulsions. *Phys. Rev. A* 42: 872–876.

Saravanan, M., Bhaskar, K., Maharajan, G., and Pillai, K. S. 2004. Ultrasonically controlled release and targeted delivery of diclofenac sodium via gelatin magnetic microspheres. *Int. J. Pharm.* 283: 71–82.

Schubert, K.-V. and Strey, R. 1991. Small-angle neutron scattering from microemulsions near the disorder line in water/formamide-octane-C_iE_j systems. *J. Chem. Phys.* 95: 8432–8445.

Schubert, K.-V., Strey, R., Kline, S. R., and Kaler, E. W. 1994. Small angle neutron scattering near Lifshitz lines: Transition from weakly structured mixtures to microemulsions. *J. Chem. Phys.* 101: 5343–5355.

Shakeel, F., Baboota, S., Ahuja, A., Ali, J., Aqil, M., and Shafiq, S. 2007. Nanoemulsions as vehicles for transdermal delivery of aceclofenac. *AAPS PharmSciTech* 8: 191–199.

Shukla, A., Graener, H., and Neubert, R. H. H. 2004b. Observation of two diffusive relaxation modes in microemulsions by dynamic light scattering. *Langmuir* 20: 8526–8530.

Shukla, A., Kiselev, M. A., Hoell, A., and Neubert, R. H.H. 2004a. Characterization of nanoparticles of lidocaine in w/o microemulsions using small-angle neutron scattering and dynamic light scattering. *Pramana J. Phys.* 63: 291–295.

Shukla, A. and Neubert, R. H. H. 2005. Investigation of W/O microemulsion droplets by contrast variation light scattering. *Pramana J. Phys.* 65: 1097–1108.

Soderman, O. and Nyden, M. 1999. NMR in microemulsions. NMR translational diffusion studies of a model microemulsion. *Colloids Surf. A: Physicochem. Eng. Aspects* 158: 273–280.

Solans, C. and Kunieda, H. (Eds.). 1997. *Industrial Applications of Microemulsions. Surfactant Science Series*, Vol. 66. New York: Marcel Dekker.

Steer, P., Millgard, J., Basu, S., Lithell, H., Vessby, B., Berne, C., and Lind L. 2003. Vitamin C, diclophenac, and L-arginine protect endothelium-dependent vasodilation against elevated circulating fatty acid levels in humans. *Atherosclerosis* 168: 65–72.

Stevenson, C. L., Bennett, D. B., and Lechuga-Ballesteros, D. 2005. Pharmaceutical liquid crystals: The relevance of partially ordered systems. *J. Pharm. Sci.* 94: 1861–1880.

Suratkar, V. and Mahapatra, S. 2000. Solubilization site of organic perfume molecules in sodium dodecyl sulfate micelles: New insights from proton NMR studies. *J. Colloid Interface Sci.* 225: 32–38.

Talmon, Y. 1996. Transmission electron microscopy of complex fluids: The state of art. *Ber. Bunsenges. Phys. Chem.* 100: 364–372.

Tenjarla, S. 1999. Microemulsions: An overview and pharmaceutical applications. *Crit. Rev. Ther. Drug Carr. Syst.* 16: 461–525.

Teubner, M. and Strey, R. 1987. Origin of the scattering peak in microemulsions. *J. Chem. Phys.* 87: 3195–3200.

Teukolsky, S. A., Vetterling, W. T., and Flannery, B. P. 1992. *Numerical Recipes in C: The Art of Scientific Computing*, 2nd ed, pp. 683–688. New York: Cambridge University Press.

Vucinic-Milankovic, N., Savić, S., Vuleta, G., and Vucinić, S. 2007. The physicochemical characterization and *in vitro/in vivo* evaluation of natural surfactants-based emulsions as vehicles for diclofenac diethylamine. *Drug Dev. Ind. Pharm.* 33: 221–234.

Yaghmur, A., Aserin, A., Antalek, B., and Garti, N. 2003. Microstructure considerations of new five-component Winsor IV food-grade microemulsions studied by pulsed gradient spin-echo NMR, conductivity, and viscosity. *Langmuir* 19: 1063–1068.

12 Self-Emulsifying Drug Delivery Systems

D. P. Maurya, Yasmin Sultana, and M. Abul Kalam

CONTENTS

12.1   INTRODUCTION

Oral delivery of various drugs is altered because of their low water solubility (Lipinski, 2002; Palmer, 2003), causing poor bioavailability, high subject (intra- or inter-) variability, and lack of dose proportionality (Hauss, 2007). Therefore, during the production of a suitable formulation it is very important to improve the solubility and bioavailability of these drugs. One of the most popular and commercially suitable approaches to solve these problems is to formulate self-emulsifying drug delivery systems (SEDDSs). SEDDSs improve the oral bioavailability of less water-soluble and hydrophobic drugs (Gursoy and Benita, 2004). They are isotropic mixtures of oil and a nonionic emulsifier. One feature of these mixtures is their ability to form oil-in-water (O/W) emulsions with only gentle agitation when they are exposed to aqueous media. Because of this property, SEDDSs are good candidates for the oral delivery of hydrophobic drugs with adequate oil solubility. After oral administration, the soft gelatin capsules encasing SEDDSs readily disperse in the stomach to form a fine emulsion. The digestive motility of the stomach and the intestine provide necessary agitation for self-emulsification (Groves and De Galindez, 1976; McClintic, 1976). At a given temperature, self-emulsification occurs when the entropy change that favors dispersion is greater than the energy required to increase the surface area of the dispersion (Reiss, 1975). The performance of SEDDSs is dependent upon two main factors: the ability of the SE mixture to form an emulsion of fine particles and the polarity of the resulting oil droplets to promote a fast release rate of the drug into the aqueous phase. The efficiency of emulsifiers in SEDDSs is commonly related to their ability to form a fine droplet size in the emulsion on exposure to water, and they have a polarity favoring a faster rate of the drug release (Charman et al., 1992; Groves and De Galindez, 1976; Pouton, 1985a, 1985b). For drugs subject to dissolution rate limited absorption, SEDDSs may improve the rate and extent of absorption, as well as the reproducibility of the blood level time profile (Pouton, 1985a, 1985b; Charman et al., 1992). SEDDSs containing medium-chain monoglycerides (Capmul MCM90), and poly(ethylene glycol)-25 (PEG-25) trioleate has been reported to form small (submicron) droplet O/W emulsions (Bachynsky et al., 1989; Charman et al., 1992; Shah et al., 1990).

12.2   SELF-EMULSIFYING DRUG DELIVERY SYSTEMS

12.2.1   EXCIPIENTS

The process of self-emulsification is specific to the nature of the surfactant–oil pair, their ratio, the concentration of surfactant, and the temperature at which it occurs (Pouton, 1985a, 1985b; Wakerly et al., 1986, 1987). It has been reported that only very specific pharmaceutical excipients can result in an efficient SE system (SES; Chanana and Sheth, 1995; Hauss et al., 1998; Karim et al., 1994; Kimura et al., 1994; Shah et al., 1994). Examples of drug-loaded SEDDSs are depicted in Table 12.1.

12.2.1.1   Oils

Oil is one of the most important excipients in the SEDDSs formulation because of its solubilizing and self-emulsification facility nature for marked amounts of the lipophilic drugs. It also increases the amount of hydrophobic drugs that are transported through the intestinal lymphatic system, resulting in an increase in the absorption of drugs from the gastrointestinal (GI) tract according to the molecular nature of the triglyceride (Charman and Stella, 1991; Gershanik and Benita, 2000; Holm et al., 2002; Lindmark et al., 1995). Oils of long- and medium-chain triglycerides with different degrees of saturation have been used for the formulation of SEDDSs. Edible oils are not frequently selected because of their poor ability to dissolve large amounts of lipophilic drugs. Hydrogenated vegetable oils have been widely used because of their better drug solubility properties and ability to form good emulsion systems with a huge number of approved surfactants for oral administration (Constantinides, 1995; Kimura et al., 1994; Hauss et al., 1998). Novel semisynthetic, amphiphilic compounds (medium-chain derivatives) with surfactant properties replace the

TABLE 12.1
Examples of SEDDSs Prepared for Oral Administration of Poorly Water-Soluble Drugs

Active Ingredients	Drug Content (%)	Solvents	Oil	Surfactants	Reference
CsA	10	Ethanol	Olive oil	Polyglycolyzed glycerides	Grevel et al., 1986, Meinzer et al., 1995
Progesterone	2.5	Ethanol	Ethyl oleate	Tween 80	Charman et al., 1992
Saquinavir	16	Ethanol	D,L-α Tocopherol	Medium-chain mono- and diglycerides	NA
Coenzyme Q₁₀	5.66	Ethanol	Captex 200	Labrafac CM10, lauroglycol	Dabros et al., 1999
Ritonavir	8	NA	Oleic acid	Polyoxyl 35 castor oil	NA
CsA	10	NA	Ethyl oleate	Tween 80	Malcolmson et al., 1993
A naphthalene derivative	5	NA	Fatty acids, peanut oil	Medium-chain mono- and diglycerides, Tween 80, PEG-25 glyceryl trioleate, polyglycolyzed glycerides	Shah et al., 1994
Ontazolast	7.5	NA	A mixture of mono- and diglycerides of oleic acid	Solid, polyglycolyzed mono-, di-, and triglycerides; Tween 80	Hauss et al., 1998

CsA, Cyclosporin A; NA—not available.

regular medium-chain triglyceride oils for the formulation of SEDDSs (Constantinides, 1995; Karim et al., 1994).

12.2.1.2 Cosolvents

Relatively high concentrations of surfactants are required to produce optimized SEDDSs (Table 12.1). Organic solvents like ethanol, propylene glycol, and PEG are used to produce oral SEDDSs, because of their ability to dissolve a large amount of drug and hydrophilic surfactant in the lipid base. They also act as cosurfactants for microemulsions.

12.2.1.3 Surface Active Agents

The surfactants that are commonly used to formulate SEDDSs are nonionic surfactants with a relatively high hydrophilic–lipophilic balance (HLB; Table 12.1). They also have hydrophilicity so that they may immediately form O/W emulsion droplets and the formulation rapidly spreads in the aqueous media of the GI tract. For proper absorption of drugs in the GI tract, the drug should be in solubilized form for a prolonged period of time at the site of absorption (Serajuddin et al., 1988; Shah et al., 1994). Because of their amphiphilic nature, surfactants can dissolve or solubilize high amounts of hydrophobic drugs. The commonly used emulsifiers are ethoxylated polyglycolyzed glycerides and polyoxyethylene 20 oleate (Tween 80). Natural origin emulsifiers are mostly preferred because they are safer than synthetic surfactants (Constantinides, 1995; Georgakopoulos et al., 1992; Hauss et al., 1998; Yuasa et al., 1994). Nonionic surfactants are less toxic than ionic surfactants, but they may change the permeability of the intestinal lumen (Swenson et al., 1994). The surfactant concentration usually used to form stable SEDDSs ranges between 30% (w/w) and

60% (w/w). The concentration of surfactant should be properly determined because large amounts of surfactants may cause GI tract irritation and may affect the size of the droplets. Few reports showed that the mean droplet size may be reduced by increasing the concentration of the surfactant and vice versa. The explanation of the first phenomenon is that the oil droplets are stabilized, causing localization of the surfactant molecules at the oil–water interface (Craig et al., 1995; Kommuru et al., 2001; Levy and Benita, 1990; Pouton, 1992). This is a result of the interfacial disruption caused by enhanced water penetration into the oil droplets that is attributable to the increased concentration of the surfactant and causing the ejection of oil droplets into the aqueous phase (Pouton, 1997).

12.3 THEORETICAL ASPECTS OF SELF-EMULSIFYING DRUG DELIVERY SYSTEMS

Two factors that affect the performance of SEDDSs are the production of a uniform fine particle size of oil droplets during the exposure to aqueous media and the polarity of the produced oil droplets. Both factors affect the release rate of drugs from the oil droplets to the aqueous phase. SEDDSs form O/W emulsions on exposure to the aqueous phase. The O/W emulsions are produced spontaneously due to the thermodynamic stability of SEDDSs compared to the regular emulsions, which are thermodynamically unstable. The stability of the O/W emulsion is favored by two factors: the narrow range of the size distribution of oil droplets and the small volume of the dispersed oil phase. Emulsions with small and uniform sizes of oil droplets take more time to break. The larger droplets are less stable compared to smaller droplets because of their larger area/volume ratio (Shaw, 1980). The smaller oil droplets have a larger interfacial surface area per unit volume. The diffusion path for a drug will be decreased with the reduction of the radius of the oil droplets. The polarity of the oil droplets is the other important factor for the SEDDSs to work properly. The HLB, chain length, degree of unsaturation of the fatty acid, molecular weight of the hydrophilic portion and concentration of the emulsifier are the main factors that back the polarity of the oil droplets. The combination of droplets together with the appropriate polarity of the small oil droplets permits a reliable drug release rate. The oil–water partition coefficient of the lipophilic drug is also another method to determine the polarity of the oil droplets.

12.4 MECHANISM OF SELF-EMULSIFICATION

The mechanism of self-emulsification is not well established. It is believed that self-emulsification occurs when the entropy change of dispersion is more than the energy needed to increase the dispersion surface area (Reiss, 1975). A conventional emulsion formulation requires free energy to create a new surface between the oil and water phases. The phases of the emulsion are separated with time to reduce the interfacial area as well as the free energy of the systems. Emulsifying agents stabilize emulsions by forming a monolayer around the emulsion droplets, resulting in a reduction in the interfacial energy and a reduction in the tendency to coalescence the emulsion droplets. Emulsification in SEDDS occurs spontaneously because much less emulsification energy is required (Dabros et al., 1999). Emulsification has been related to the easy penetration of water into the various phases on the droplet surface (Groves et al., 1974; Rang and Miller, 1999). When a mixture of oil and a nonionic surfactant is added to water, an interface forms between the oil and water phases. As a result of aqueous penetration through the interface, solubilization of water within the oil phase occurs until the solubilization is close to the interphase. Interface disruption and droplet formation occur on gentle agitation of the SES in water because of penetration into the aqueous phase. The SEDDS becomes very stable to coalescence because of interface formation around the oil droplets (Pouton et al., 1987). Low frequency dielectric spectroscopy and particle size analysis were used to determine the self-emulsification properties of a mixture of Tween 80 and mono- and diglycerides of capric and caprylic acid systems. The drug compound may also affect the emulsion characteristics by interacting with the interfacial phase (Gursoy et al., 2003).

12.5 CHARACTERIZATION OF DELIVERY SYSTEM

The ability for self-emulsification of SEDDS can be estimated by determining the oil droplet size distribution and rate of emulsification. The charge on the oil droplets of SEDDS also needs to be determined (Gursoy and Benita, 2004). The melting properties and polymorphism of the lipid or drug in SEDDS can be determined by x-ray diffraction and differential scanning calorimetry.

12.6 DRUG ENTRAPMENT INTO A SELF-EMULSIFYING DRUG DELIVERY SYSTEM

The problem of poor aqueous solubility of hydrophobic drugs can be solved by dissolving them in novel synthetic hydrophilic oils and surfactants. The addition of solvents like ethanol and PEG may also improve the solubility of the drug in the lipid vehicle. The efficiency of drug entrapment into a SEDDS is generally dependent on the physicochemical compatibility of the drug and excipient system. The efficiency of a SEDDS may be hindered by altering the charge movement in the system by complexation of the drug compound with some of the components in the mixture. A change in the droplet size distribution may be caused by the interference of the drug compound with the self-emulsification process that can be a function of the drug concentration. Emulsions with smaller oil droplets in more complex formulations are more prone to changes.

12.7 SOLID SELF-EMULSIFYING DRUG DELIVERY SYSTEM

A SEDDS may exist in either a liquid or solid state. Because most of the excipients that are used in the formulation of SEDDS are liquid at room temperature, the SEDDS is usually limited to liquid dosage forms. Solid SEDDSs (S-SEDDSs) have been extensively exploited in recent years because of the well-known advantage of being a solid dosage form.

In S-SEDDS, incorporation of liquid or semisolid SE ingredients into nanoparticles or powders occurs by different solidification techniques that are usually further processed into other solid SE dosage forms or filled into capsules (Attama and Mpamaugo, 2006). In this way, S-SEDDSs are combinations of SEDDSs and solid dosage forms. SE pellets can be characterized by the assessment of self-emulsification, friability, and surface roughness. Initially, S-SEDDS existed in the form of SE capsules, solid dispersions, and dry emulsions; in recent years, pellets and tablets, microspheres and nanoparticles, and suppositories and implants also emerged.

12.8 SOLIDIFICATION OF LIQUID OR SEMISOLID SELF-EMULSIFYING DRUG DELIVERY SYSTEM TO SOLID SELF-EMULSIFYING DRUG DELIVERY SYSTEM

12.8.1 SPRAY DRYING

In this technique the formulation was prepared by mixing lipids, surfactants, drug, and solid carriers followed by solubilization. Then the solubilized liquid formulation is spray dried. The volatile phase evaporates in the drying chamber, leaving the dry particles under controlled temperature and airflow conditions. Such dry particles can be further converted into capsules or tablets. Variables like the selection of atomizer, temperature, suitable airflow pattern, and design of the drying chamber are considered according to the specifications of the product.

12.8.2 ENCAPSULATION OF LIQUID OR SEMISOLID SELF-EMULSIFYING DRUG DELIVERY SYSTEM

Encapsulation of liquid or semisolid SEDDS in the capsules is the most commonly employed procedure to convert them into S-SEDDS. Capsule filling of semisolid formulations involves a four-phase process: heating the semisolid excipient at least above its melting point, loading the active ingredients and stirring it, filling the molten mixture into capsules, and cooling the system at room temperature. For liquid formulations, it is a two-phase process of filling the capsules with liquid preparation and then sealing the body and cap of the capsule by using either a microspray or banding process (Jannin et al., 2008). With the advancement in capsule technology, Alza Corporation has designed a controlled delivery system for peptides or insoluble drug substances known as liquid-Oros technology. It possess an osmotic layer, which expands when it comes in contact with water and expels the drug formulation through an orifice from the soft or hard capsule (Dong et al., 2000, 2001). The important factor that has to be considered during capsule filling is the compatibility of the formulation excipients with the capsule shell. Semisolid or liquid lipophilic vehicles that are compatible with hard gelatin capsules are described by Cole et al. (2008).

12.8.3 Adsorption to Solid Carriers

Free flowing powders with significant content uniformity may be obtained by adsorption of a liquid formulation to the solid carriers. Then the free flowing powders may be filled directly into capsules or mixed with suitable excipients before converting into tablets (Ito et al., 2005).

Solid carriers that can be used to adsorb the liquid formulations may include high surface area colloidal inorganic adsorbent substances, microporous inorganic substances, and cross-linked polymers or nanoparticle adsorbents, for example, silicates, silica, magnesium hydroxide, magnesium trisilicate, crospovidone, talc, cross-linked poly(methyl methacrylate), and cross-linked sodium carboxymethyl cellulose (Fabio and Elisabetta, 2003; Venkatesan et al., 2005). Cross-linked polymers may sustain the drug reprecipitation as well as the drug dissolution rate (Boltri et al., 1997).

12.8.4 Extrusion Spheronization

This is a solvent-free process resulting in content uniformity as well as efficiency for high drug incorporation. The extrusion spheronization process converts a raw material with plastic properties into a product with uniform density and shape by passing it through a die under controlled product flow, pressure, and temperature conditions (Verreck and Brewster, 2004). The resulting spheroid size depends on the size of the extruder aperture. This process is commonly used to produce uniform size spheroids. The process requires the following six steps: preparation of a homogeneous powder by dry mixing of the active ingredients and excipients, wetting by adding binder, extrusion into an extrudate, spheronization from the extrudate to uniform size spheroids, drying of spheroids, and size separation. In the wet masses, the relative quantities of water and SES had a significant effect on the size spread, extrusion force, surface roughness of the pellets, and disintegration time. Studies suggested that the maximum 42% (w/w) of the dry pellet weight of wet masses can be solidified by extrusion spheronization (Newton et al., 2001). Generally, the high water level is indicative of a sustained disintegration time (Newton et al., 2005).

12.8.5 Melt Granulation

In this process powder agglomerates are prepared by the addition of a binder that melts or softens at very low temperatures. It is one-step process, and it is more advantageous than conventional wet granulation because of liquid addition and omission of the drying steps. The impeller speed, mixing time, binder particle size, and viscosity of the binder are the main parameters that control the granulation process. A wide range of semisolid and solid lipids can be used as meltable binders to produce granules. Some lipid-based excipients such as lecithin, Gelucire (Seo et al., 2003), polysorbates, or partial glycerides are also evaluated for melt granulation to create a solid SES. The melt granulation process is usually used to adsorb the SES onto solid neutral carriers (Gupta et al., 2001, 2002).

12.9 DOSAGE FORMS OF SOLID SELF-EMULSIFYING DRUG DELIVERY SYSTEM

12.9.1 SUSTAINED RELEASE TABLETS

To reduce the amount of solidifying excipients necessary for converting SEDDS into solid dosage forms, a gelled SEDDS was developed by Patil et al. (2004) by using colloidal silicon dioxide as a gelling agent for the oil-based systems, which reduces the amount of required solidifying excipients as well as sustaining the drug release. The resultant SE sustained release tablets consistently maintained a higher active ingredient concentration in blood plasma over a prolonged period of time compared to a nonemulsifying tablet. The latest development in this field is the SE osmotic pump tablet that has outstanding stable plasma concentrations and a controllable drug release rate.

12.9.2 CAPSULES

To solve the problem of irreversible phase separation of conventional liquid SE formulation containing capsules, sodium dodecyl sulfate was added into the SE formulation (Itoh et al., 2002). Another approach was the formulation of supersaturatable SEDDS by using a small quantity of hydroxypropyl methyl cellulose and a small amount of surfactant to prevent precipitation of the drug (Gao and Morozowich, 2006; Gao et al., 2003). Like liquid filling, liquid SE ingredients can also be filled into capsules in a solid or semisolid state obtained by adding solid carriers like solid PEG; these solid carriers will neither interfere with the process of self-microemulsification nor with the solubility of the drug upon mixing with water (Li et al., 2007). Oral administration of SE capsules enhanced patient compliance compared to the previously used parenteral route of administration (Ito et al., 2006).

12.9.3 DRY EMULSIONS

These are powdered dosage forms that spontaneously emulsify *in vivo* and/or on exposure to aqueous medium. Dry emulsions can also be converted into tablets and capsules. These formulations are prepared from O/W emulsions consisting of a solid carrier in the aqueous phase by freeze-drying (Bamba et al., 1995), rotary evaporation (Myers and Shively, 1992), or spray drying (Christensen et al., 2001; Hansen et al., 2004; Jang et al., 2006). Myers and Shively (1992) obtained solid-state dry "foam" glass emulsions of heavy mineral oil and sucrose by rotary evaporation. The O/W emulsion was formulated by the spray-drying method. One of the major developments in this field is the formulation of an enteric-coated dry emulsion by using a surfactant, a vegetable oil, and a pH-responsive polymer for the oral delivery of peptide and protein drugs (Toorisaka et al., 2005). Recently, Cui et al. (2007) prepared dry emulsions by drying liquid O/W emulsions on a glass plate and triturating it into powders.

12.9.4 SUSTAINED RELEASE PELLETS

Serratoni et al. (2007) prepared SE controlled-release pellets that showed enhanced drug release by the extrusion or spheronization method incorporating two water-insoluble drugs (methyl paraben and propyl paraben) into an SES containing mono- and diglycerides and polysorbate 80. The pellets were then coated with a water-insoluble polymer that reduced the rate of drug release.

12.9.5 BEADS

Patil and Paradkar (2006) prepared microchannels of porous polystyrene beads with the solvent evaporation technique by using copolymerizing styrene and divinyl benzene. These excipients are inert, stable over a wide pH range, and stable to extreme conditions of humidity and temperature. This research concluded that porous polystyrene beads were potential carriers for solidification of an SES (Patil and Paradkar, 2006).

12.9.6 Solid Dispersions

Serajuddin (1999) and Vasanthavada and Serajuddin (2007) pointed out that the stability problems of solid dispersions could be overcome by the use of SE excipients. These excipients have the potential to increase the absorption of poorly water-soluble drugs and may also be filled directly into hard gelatin capsules in the molten state, thus bypassing the milling and blending steps before filling. The most widely used SE excipients are Gelucire 50/02, Gelucire 44/14, Transcutol, Tocopheryl PEG 1000 succinate, and Labrasol (Khoo et al., 2000).

12.9.7 Sustained-Release Microspheres

You et al. (2006) prepared solid SE sustained-release microspheres of zedoary turmeric oil by using the quasiemulsion–solvent-diffusion method of the spherical crystallization technique. The release behavior of zedoary turmeric oil could be controlled by the ratio of hydroxypropyl methylcellulose acetate succinate/Aerosil 200 in the formulation. The plasma drug concentration that was achieved was better than conventional products after oral administration of such microspheres to rabbits (You et al., 2006).

12.9.8 Nanoparticles

Nanoparticle techniques are also useful in the production of SESs. One of the preparation techniques of nanoparticles is solvent injection. In this process, excipients and drugs are melted together and injected dropwise into a nonsolvent stirred at constant speed. The resulting SE nanoparticles are centrifuged and freeze-dried (Attama and Nkemnele, 2005). A second technique, sonication emulsion–diffusion–evaporation, was used to coload the active ingredients in biodegradable nanoparticles (Hu et al., 2005). Trickler et al. (2008) recently developed novel chitosan and glyceryl monooleate nanoparticles of paclitaxel with nearly 100% entrapment efficiencies.

12.9.9 Implants

The S-SEDDS also showed its utility to formulate SE implants. For example, the effectiveness of carmustine, a chemotherapeutic agent used to treat malignant brain tumors, was decreased because of its short half-life. To increase its stability, SES was formulated with tributyrin, Cremophor RH 40, and Labrafil 1944. Thus, the SES increased the *in vitro* half-life of 1,3-bis(2-chloroethyl)-1-nitrosourea (BCNU) up to 130 min compared to 45 min of BCNU only. *In vitro* release of BCNU from SE poly(lactic-*co*-glycolic acid) implants were sustained up to one week with higher *in vitro* antitumor activity and less susceptibility to hydrolysis (Chae et al., 2005).

12.9.10 Suppositories

A S-SEDDS can also increase rectal and vaginal absorption (Kim and Ku, 2000). For example, Glycyrrhizin can produce satisfactory therapeutic activity for chronic hepatic diseases by either vaginal or rectal SE suppositories.

12.10 CONCLUSION

Numerous studies have confirmed that SEDDS substantially improved the solubility, absorption, and bioavailability of drug compounds with poor aqueous solubility. In most instances, the composition of the SEDDS formulation should be determined very carefully because the efficiency of the SEDDS formulation is mostly case specific. The toxicity of the surfactant must be taken into account because a high concentration of surfactants is generally used to formulate the SEDDS. Two other important factors that affect the GI efficiency of SEDDS are the size and charge of the oil droplet in the emulsion. Several modified formulations have also been developed as an alternative to conventional SEDDS. All of these novel modified formulations will produce micelle dispersions or fine oil droplets when they are diluted with aqueous media. Because nearly 40% of the new drug compounds are poorly water soluble, it is necessary that more drug products be formulated as SEDDS. Compared to conventional liquid SEDDSs, S-SEDDSs are better in terms of lower production cost, simple manufacturing process, and improved patient compliance as well as stability. The S-SEDDSs are easily developed solid dosage forms for parenteral as well as oral administration. Major research is needed before more solid SE dosage forms appear on the market.

ABBREVIATIONS

BCNU	1,3-bis(2-chloroethyl)-1-nitrosourea
GI	Gastrointestinal
HLB	Hydrophilic–lipophilic balance
O/W	Oil-in-water
PEG	Poly(ethylene glycol)
SE	Self-emulsifying
SES	Self-emulsifying system
SEDDS	Self-emulsifying drug delivery system
S-SEDDS	Solid self-emulsifying drug delivery system

GLOSSARY

SEDDS: SEDDS can be described as an isotropic solution of oil and surfactant that form O/W microemulsions on mild agitation in the presence of water.

HLB: Griffin developed a scale based on the balance between hydrophobic and lipophilic solution tendencies of surface active agents. This so-called HLB scale extends from 1 to 50. The more hydrophilic surfactants have high HLB numbers whereas lipophilic surfactants have HLB numbers from 1 to 10. Surfactants with a proper balance in their hydrophilic and lipophilic affinities are effective emulsifying agents because they concentrate at the oil–water interface.

S-SEDDS: S-SEDDS represents the solid dosage form with self-emulsification properties. They can be prepared by incorporation of liquid or semisolid SE ingredients into powders or nanoparticles by different solidification techniques. Such powders or nanoparticles are usually further processed into other solid SE dosage forms or filled into capsules.

REFERENCES

Attama, A. A. and Mpamaugo, V. E. 2006. Pharmacodynamics of piroxicam from self-emulsifying lipospheres formulated with homolipids extracted from *Capra hircus. Drug. Deliv.* 13, 133–137.

Attama, A. A. and Nkemnele, M. O. 2005. *In vitro* evaluation of drug release from self micro-emulsifying drug delivery systems using a biodegradable homolipid from *Capra hircus. Int. J. Pharm.* 304, 4–10.

Bachynsky, M., Shah, N. H., Infeld, M. H., Margolis, R. J., and Malick, A. W. 1989. Oral delivery of lipophilic drugs in self emulsifying liquid formulations. Presented at the AAPS Annual Meeting, Atlanta, GA, October 1989.

Bamba, J. et al., 1995. Cryoprotection of emulsions in freeze-drying: Freezing process analysis. *Drug. Dev. Ind. Pharm.* 21, 1749–1760.

Boltri, L., Coceani, N., De Curto, D., Dobetti, L., and Esposito, P. 1997. Enhancement and modification of etoposide release from crospovidone particles loaded with oil–surfactant blends. *Pharm. Dev. Technol.* 2, 373–381.

Chae, G. S. et al., 2005. Enhancement of the stability of BCNU using self-emulsifying drug delivery systems (SEDDS) and *in vitro* antitumor activity of self-emulsified BCNU-loaded PLGA wafer. *Int. J. Pharm.* 301, 6–14.

Chanana, G. D. and Sheth, B. B. 1995. Particle size reduction of emulsions by formulation design. II: Effect of oil and surfactant concentration. *PDA J. Pharm. Sci. Technol.* 49, 71–76.

Charman, S. A., Charman, W. N., Rogge, M. C., Wilson, T. D., Dutko, F. J., and Pouton, C. W. 1992. Self-emulsifying drug delivery systems: Formulation and biopharmaceutic evaluation of an investigational lipophilic compound. *Pharm. Res.* 9, 87–93.

Charman, W. N. and Stella, V. J. 1991. Transport of lipophilic molecules by the intestinal lymphatic system. *Adv. Drug Deliv. Rev.* 7, 1–14.

Christensen, K. L. et al., 2001. Technical optimization of redispersible dry emulsions. *Int. J. Pharm.* 212, 195–202.

Cole, E. T. et al., 2008. Challenges and opportunities in the encapsulation of liquid and semi-solid formulations into capsules for oral administration. *Adv. Drug Deliv. Rev.* 60, 747–756.

Constantinides, P. P. 1995. Lipid microemulsions for improving drug dissolution and oral absorption: Physical and biopharmaceutical aspects. *Pharm. Res.* 12, 1561–1572.

Craig, D. Q.M., Barker, S. A., Banning, D., and Booth, S. W. 1995. An investigation into the mechanisms of self-emulsification using particle size analysis and low frequency dielectric spectroscopy. *Int. J. Pharm.* 114, 103–110.

Cui, F. D. et al., 2007. Preparation of redispersible dry emulsion using Eudragit E100 as both solid carrier and unique emulsifier. *Colloids Surf. A: Physicochem. Eng. Aspects* 307, 137–141.

Dabros T., Yeung, A., Masliyah, J., and Czarnecki, J. 1999. Emulsification through area contraction. *J. Colloid Interface Sci.* 210, 222–224.

Dong, L. et al., 2000. A novel osmotic delivery system: L-OROS SOFTCAP. Proceedings of the International Symposium on Controlled Release of Bioactive Materials, Paris, July 2000 (CD ROM).

Dong, L. et al., 2001. L-OROS HARDCAP: A new osmotic delivery system for controlled release of liquid formulation. Proceedings of the International Symposium on Controlled Release of Bioactive Materials, San Diego, June 2001 (CD-ROM).

Fabio, C. and Elisabetta, C. 2003. Pharmaceutical composition comprising a water/oil/water double micro-emulsion incorporated in a solid support. Patent WO2003/013421.

Gao, P. and Morozowich, W. 2006. Development of supersaturatable selfemulsifying drug delivery system formulations for improving the oral absorption of poorly soluble drugs. *Expert Opin. Drug. Discov.* 3, 97–110.

Gao, P. et al., 2003. Development of a supersaturable SEDDS (S-SEDDS) formulation of paclitaxel with improved oral bioavailability. *J. Pharm. Sci.* 92, 2386–2398.

Georgakopoulos, E., Farah, N., and Vergnault, G. 1992. Oral anhydrous non-ionic microemulsions administered in softgel capsules. *Bull. Tech. Gattefosse* 85, 11–20.

Gershanik, T. and Benita, S. 2000. Self-dispersing lipid formulations for improving oral absorption of lipophilic drugs. *Eur. J. Pharm. Biopharm.* 50, 179–188.

Grevel, J., Nuesch, E., Abisch, K., Kutz, K. 1986. Pharmacokinetics of oral cyclopsorinA (Sandimmun) in healthy subjects. *Eur. J. Clin. Pharmacol.* 31, 211–216.

Groves, M. J. and De Galindez, D. A. 1976. The selfemulsifying action of mixed surfactants in oil. *Acta Pharm. Suec.* 13, 361–372.

Groves, M. J., Mustafa, R. M.A., and Carless, J. E. 1974. Phase studies of mixed phosphated surfactants, *n*-hexane and water. *J. Pharm. Pharmacol.* 26, 616–623.

Gupta, M. K. et al., 2001. Enhanced drug dissolution and bulk properties of solid dispersions granulated with a surface adsorbent. *Pharm. Dev. Technol.* 6, 563–572.

Gupta, M. K. et al., 2002. Hydrogen bonding with adsorbent during storage governs drug dissolution from solid-dispersion granules. *Pharm. Res.* 19, 1663–1672.

Gursoy, N., Garrigue, J. S., Razafindratsita, A., Lambert, G., and Benita, S. 2003. Excipient effects on *in vitro* cytotoxicity of a novel paclitaxel self emulsifying drug delivery system. *J. Pharm. Sci.* 92, 2420–2427.

Gursoy, R. N. and Benita, S. 2004. Self-emulsifying drug delivery systems (SEDDS) for improved oral delivery of lipophilic drugs. *Biomed. Pharmacother.* 58, 173–182.

Hansen, T. et al., 2004. Process characteristics and compaction of spray-dried emulsions containing a drug dissolved in lipid. *Int. J. Pharm.* 287, 55–66.

Hauss, D. J. (Ed.), 2007. *Oral lipid-based formulations: enhancing the bioavailability of poorly water soluble drugs*, pp. 1–339. New York: Informa Healthcare, Inc.

Hauss, D. J., Fogal, S. E., Ficorilli, J. V., Price, C. A., Roy, T., Jayaraj A. A., and Keirns, J. J. 1998. Lipid-based delivery systems for improving the bioavailability and lymphatic transport of a poorly water-soluble LTB4 inhibitor. *J. Pharm. Sci.* 87, 164–169.

Holm, R., Porter, C. J.H., Müllertz, A., Kristensen, H. G., and Charman, W. N. 2002. Structured triglyceride vehicles for oral delivery of halofantrine: Examination of intestinal lymphatic transport and bioavailability in conscious rats. *Pharm. Res.* 19, 1354–1361.

Hu, Y. X. et al., 2005. Preparation and evaluation of 5-FU/PLGA/gene nanoparticles. *Key Eng. Mater.* 288–289, 147–150.

Ito, Y. et al., 2005. Oral solid gentamicin preparation using emulsifier and adsorbent. *J. Control. Release* 105, 23–31.

Ito, Y. et al., 2006. Preparation and evaluation of oral solid heparin using emulsifier and adsorbent for *in vitro* and *in vivo* studies. *Int. J. Pharm.* 317, 114–119.

Itoh, K. et al., 2002. Improvement of physicochemical properties of N-4472. Part I: Formulation design by using self-microemulsifying system. *Int. J. Pharm.* 238, 153–160.

Jang, D. J. et al., 2006. Improvement of bioavailability and photostability of amlodipine using redispersible dry emulsion. *Eur. J. Pharm. Sci.* 28, 405–411.

Jannin, V. et al., 2008. Approaches for the development of solid and semi-solid lipid-based formulations. *Adv. Drug Deliv. Rev.* 60, 734–746.

Karim, A., Gokhale, R., Cole, M., Sherman, J., Yeramian, P., Bryant, M., and Franke, H. 1994. HIV protease inhibitor SC-52151: A novel method of optimizing bioavailability profile via a microemulsion drug delivery system. *Pharm. Res.* 11, S368.

Khoo, S. M. et al., 2000. The formulation of halofantrine as either non-solubilising PEG 6000 or solubilising lipid based solid dispersions: Physical stability and absolute bioavailability assessment. *Int. J. Pharm.* 205, 65–78.

Kim, J. Y. and Ku, Y. S. 2000. Enhanced absorption of indomethacin after oral or rectal administration of a self-emulsifying system containing indomethacin to rats. *Int. J. Pharm.* 194, 81–89.

Kimura, M., Shizuki, M., Miyoshi, K., Sakai, T., Hidaka, H., Takamura, H., and Matoba, T. 1994. Relationship between the molecular structures and emulsification properties of edible oils. *Biosci. Biotechnol. Biochem.* 58, 1258–1261.

Kommuru, T. R., Gurley, B., Khan, M. A., and Reddy, I. K. 2001. Self-emulsifying drug delivery systems (SEDDS) of coenzyme Q10: Formulation development and bioavailability assessment. *Int. J. Pharm.* 212, 233–246.

Levy, M. Y. and Benita, S. 1990. Drug release from submicronized o/w emulsion: A new *in vitro* kinetic evaluation model. *Int. J. Pharm.* 66, 29–37.

Li, P. et al., 2007. Development and characterization of a solid microemulsion preconcentrate system for oral delivery of poorly water soluble drugs. Controlled Release Society Annual Meeting, Long Beach, CA, June 2007.

Lindmark, T., Nikkila, T., and Artursson, P. 1995. Mechanisms of absorption enhancement by medium chain fatty acids in intestinal epithelial Caco-2 monolayers. *J. Pharmacol. Exp. Ther.* 275, 958–964.

Lipinski, C. 2002. Poor aqueous solubility—An industry wide problem in drug discovery. *Am. Pharm. Rev.* 5, 82–85.

Malcolmson, C. and Lawrence, M. J. 1993. A comparison of the incorporation of model steroids into non-ionic micellar and microemulsion systems. *J Pharm Pharmacol* 45, 141–143.

McClintic, J. R. 1976. *Physiology of the Human Body*, 2nd edition, p. 189. New York: Wiley.

Meinzer, A., Mueller, E., and Vonderscher, J. 1995. Microemulsion—A suitable galenical approach for the absorption enhancement of low soluble compounds? *Bull. Tech. Gattefosse* 88, 21–26.

Myers, S. L. and Shively, M. L. 1992. Preparation and characterization of emulsifiable glasses: Oil-in-water and water-in-oil-in-water emulsion. *J. Colloid Interface Sci.* 149, 271–278.

Newton, J. M. et al., 2005. Formulation variables on pellets containing self emulsifying systems. *Pharm. Tech. Eur.* 17, 29–33.

Newton, M. et al., 2001. The influence of formulation variables on the properties of pellets containing a self-emulsifying mixture. *J. Pharm. Sci.* 90, 987–995.

Palmer, A. M. 2003. New horizons in drug metabolism, pharmacokinetics and drug discovery. *Drug News Perspect.* 16, 57–62.

Patil, P. and Paradkar, A. 2006. Porous polystyrene beads as carriers for self emulsifying system containing loratadine. *AAPS PharmSciTech* doi:10.1208/pt070128 http://www.aapspharmscitech.org/articles/pt0701/pt070128/pt070128.pdf

Patil, P. et al.,, 2004. Effect of formulation variables on preparation and evaluation of gelled self-emulsifying drug delivery system (SEDDS) of ketoprofen. *AAPS PharmSciTech* doi:10.1208/pt050342 http://www.aapspharmscitech.org/articles/pt0503/

Pouton, C. W. 1985a. Effects of the inclusion of a model drug on the performance of self-emulsifying formulations. *J. Pharm. Pharmacol.* 37, 1–11.

Pouton, C. W. 1985b. Self-emulsifying drug delivery systems: Assessment of the efficiency of emulsification. *Int. J. Pharm.* 27, 335–348.

Pouton, C. W. 1997. Formulation of self-emulsifying drug delivery systems. *Adv. Drug Deliv. Rev.* 25, 47–58.

Pouton, C. W., Charman, S. A., Charman, W. N., Rogge, M. C., Wilson, T. D., Dutko, F. J. 1992. Self-emulsifying drug delivery systems: formulation and biopharmaceutic evaluation of an investigational lipophilic compound. *Pharm Res* 9, 87–93.

Pouton, C. W., Wakerly, M. G., and Meakin, B. J. 1987. Self-emulsifying systems for oral delivery of drugs. *Proc. Int. Symp. Control. Release Bioact. Mater.* 14, 113–114.

Rang, M. J. and Miller, C. A. 1999. Spontaneous emulsification of oils containing hydrocarbon, non-ionic surfactant, and oleyl alcohol. *J. Colloid Interface Sci.* 209, 179–192.

Reiss, H. 1975. Entropy-induced dispersion of bulk liquids. *J. Colloid Interface Sci.* 53, 61–70.

Seo, A. et al., 2003. The preparation of agglomerates containing solid dispersions of diazepam by melt agglomeration in a high shear mixer. *Int. J. Pharm.* 259, 161–171.

Serajuddin, A. T.M. 1999. Solid dispersion of poorly water-soluble drugs: Early promises, subsequent problems, and recent breakthroughs. *J. Pharm. Sci.* 88, 1058–1066.

Serajuddin, A. T.M., Shee, P. C., Mufson, D., Bernstein, D. F., and Augustine, M. A. 1988. Effect of vehicle amphiphilicity on the dissolution and bioavailability of a poorly water-soluble drug from solid dispersion. *J. Pharm. Sci.* 77, 414–417.

Serratoni, M. et al., 2007. Controlled drug release from pellets containing water insoluble drugs dissolved in a self-emulsifying system. *Eur. J. Pharm. Biopharm.* 65, 94–98.

Shah, N. H., Bachynsky, M., Lazzara, F., Patel, C. L., Infeld, M. H., and Malick, A. W. 1990. Factors affecting the *in vitro* efficiency of self-emulsifying delivery systems. AAPS Annual Meeting, Las Vegas, NV, November 1990.

Shah, N. H., Carvajal, M. T., Patel, C. I., Infeld, M. H., and Malick, A. W. 1994. Self emulsifying drug delivery systems (SEDDS) with polyglycolized glycerides for improving *in vitro* dissolution and oral absorption of lipophilic drugs. *Int. J. Pharm.* 106, 15–23.

Shaw, D. J. 1980. *Introduction to Colloid and Surface Chemistry*, 3rd edition. London: Butterworths.

Swenson, E. S., Milisen, W. B., and Curatolo, W. 1994. Intestinal permeability enhancement: Efficacy, acute local toxicity and reversibility. *Pharm. Res.* 11, 1132–1142.

Toorisaka, E. et al., 2005. An enteric-coated dry emulsion formulation for oral insulin delivery. *J. Control. Release* 107, 91–96.

Trickler, W. J. et al., 2008. A novel nanoparticle formulation for sustained paclitaxel delivery. *AAPS PharmSciTech* doi:10.1208/s12249-008-9063-7.

Vasanthavada, M. and Serajuddin, A. T. M. 2007. Lipid-based self-emulsifying solid dispersions. In: D. J. Hauss, (Ed.) *Oral Lipid-Based Formulations: Enhancing Bioavailability of Poorly Water-Soluble Drugs*, pp. 149–184. Informa Healthcare.

Venkatesan, N. et al., 2005. Liquid filled nanoparticles as a drug delivery tool for protein therapeutics. *Biomaterials* 26, 7154–7163.

Verreck, G. and Brewster, M. E. 2004. Melt extrusion-based dosage forms: Excipients and processing conditions for pharmaceutical formulations. *Bull. Tech. Gattefosse* 97, 85–95.

Wakerly, M. G., Pouton, C. W., and Meakin, B. J. 1987. Evaluation of the self-emulsifying performance of a non-ionic surfactant–vegetable oil mixture. *J. Pharm. Pharmacol.* 39, 6.

Wakerly, M. G., Pouton, C. W., Meakin, B. J., and Morton, F. S. 1986. Self-emulsification of vegetable oil-nonionic surfactant mixtures. *ACS Symp. Ser.* 311, 242–255.

You, J. et al., 2006. Study of the preparation of sustained-release microspheres containing zedoary turmeric oil by the emulsion–solvent-diffusion method and evaluation of the self-emulsification and bioavailability of the oil. *Colloid. Surf. B* 48, 35–40.

Yuasa, H., Sekiya, M., Ozeki, S., and Watanabe, J. 1994. Evaluation of milk fat globule membrane (MFGM) emulsion for oral administration: Absorption of α-linolenic acid in rats and the effect of emulsion droplet size. *Biol. Pharm. Bull.* 17, 756–758.

13 Liquid Crystals and Their Application in the Field of Drug Delivery

Rakesh Patel and Tanmay N. Patel

CONTENTS

13.1 INTRODUCTION

The discovery of liquid crystals (LCs) is thought to have occurred nearly 150 years ago, although its significance was not fully realized until more than 100 years later. Around the middle of the last century Virchow, Mettenheimer, and Valentin found that the nerve fiber they were studying formed a fluid substance when left in water that exhibited strange behavior when viewed under polarized light. They did not realize that this was a different phase, but they are attributed with the first observation of LCs.

In 1877 Otto Lehmann used a polarizing microscope with a heated stage to investigate the phase transitions of various substances (Lehmann, 1889). He found that one substance changed from a clear liquid to a cloudy liquid before crystallizing, but he thought that this was simply an imperfect phase transition from liquid to crystalline. In 1888 Reinitzer conducted similar experiments and was the first to suggest that this cloudy fluid was a new phase of matter (Reinitzer, 1888). He has consequently been given the credit for the discovery of the liquid crystalline phase. Up until 1890 all of the liquid crystalline substances that had been investigated were naturally occurring, and it was then that the first synthetic LC, *p*-azoxyanisole, was produced by Gatterman and Ritschke. More LCs were subsequently synthesized, and it is now possible to produce LCs with specific predetermined material properties.

George Freidel conducted many experiments on LCs in the beginning of the 20th century, and he was the first to explain the orienting effect of electric fields and the presence of defects in LCs. In 1922 he proposed a classification of LCs based upon the different molecular orderings of each substance. Between 1922 and World War II Oseen and Zöcher developed a mathematical basis for the study of LCs.

After the start of the World War II, many scientists believed that the important features of LCs had now been discovered. It was not until the 1950s that work by Brown in America, Chistiakoff in the Soviet Union, and Gray and Frank in England led to a revival of interest in LCs. In 2006, Jákli and Saupe (Jákli and Saupe, 2006) formulated a microscopic theory of LCs. The interest in LCs has grown ever since, which is partly due to the great variety of phenomena exhibited by LCs.

In the 1960s, the French theoretical physicist Pierre-Gilles de Gennes, who had been working with magnetism and superconductivity, turned his interest to LCs and soon found fascinating analogies between LCs and superconductors as well as magnetic materials. His work was rewarded with the Nobel Prize in Physics in 1991. The modern development of LC science has since been deeply influenced by his work (Gennes and Prost, 1993).

Today, thanks to Reinitzer, Lehmann, and their followers, we know that literally thousands of substances have a diversity of other states. Some of them have been found to be very usable in several technical innovations, among which LC substances and LC thermometers may be the best known.

13.2 INTRODUCTION TO LIQUID CRYSTALLINE SYSTEMS

LCs are intermediate states of matter or mesophases, which are halfway between an isotropic liquid and a solid crystal. In nature, some substances, or even mixtures of substances, present these mesomorphic states. Of the many liquid crystalline structures self-assembled from aqueous surfactant systems, bicontinuous cubic phases possess a special status. As natural exhibitions of differential geometry, cubic phases are composed of contorted bilayers that partition hydrophobic and hydrophilic regions into continuous but nonintersecting spaces. Their discovery and subsequent structural characterization probably sparked the current broad exploration of the complex relationship between surfactant structures from the molecular to nanometer to millimeter scales. Of their many unique properties, the ability of cubic phases to exist as discrete dispersed colloidal particles, or cubosomes, is perhaps the most intriguing. Whereas most concentrated surfactants that form cubic LCs lose these phases to micelle formation at high dilutions, a few surfactants have optimal water insolubility. Their cubic phases exist in equilibrium with excess water and can be dispersed to form cubosomes.

Just as small-angle x-ray scattering was crucial to the discovery and structural characterization of bulk cubic phases, cryo-transmission electron microscopy (cryo-TEM) has been central to studies of cubosome dispersions. The field evolved rapidly, new intermediate and equilibrium structures are still being discovered, and direct cryo-TEM observation of new morphologies has no substitute. In addition to providing direct measures of cubosome size and shape, cryo-TEM also serves as a quantitative probe of the liquid crystalline structural periodicity.

Applications often determine the rate and volume of research in a given area, and the study of cubosomes has been similarly driven. Despite the breadth of work already performed, no commercial product is known to incorporate cubosomes, although patent art exists. In that light, this review chapter selectively examines recent research to provide a subjective discussion of areas where cryo-TEM and other microscopy techniques can provide insight. Three intertwined application areas are the focus: active ingredient delivery, cubic phase–tissue interfaces, and material synthesis. Recent cubosome reviews exist, and this chapter builds on these to examine more recent literature and subjectively identify new or promising directions of cubosome research.

13.2.1 Definition

The liquid crystalline state combines the properties of both liquid and solid states. The liquid state is associated with the ability to flow, whereas the solid state is characterized by an ordered, crystalline structure.

LCs are intermediate states of matter or mesophases, which are halfway between an isotropic liquid and a solid crystal. Some substances in nature, or even mixtures of substances, present these mesomorphic states. This picture leads to the concept of ordering (Figure 13.1).

- In a solid crystal, the basic units display translational long-range order, with the center of mass of atoms or molecules located on a crystal lattice; in some cases, the basic units also display orientational order.
- In an isotropic liquid, the basic units do not present either positional or orientational long-range order.
- In plastic crystals, the basic units are located on a lattice but without any orientational order.
- In *LCs*, the basic units display orientational order and even positional order along some directions. These materials flow like an isotropic fluid and have the characteristic optical properties of solid crystals.

Accordingly, liquid crystalline phases represent intermediate states that are also called mesophases. Molecules that can form mesophases are called mesogens. Depending on the molecular

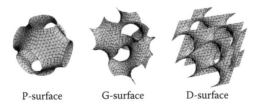

P-surface   G-surface   D-surface

FIGURE 13.1 Basic structure of LCs.

shape, rodlike mesogens form calamitic mesophases and disclike mesogens form discotic meso-phases. Rod-shaped molecules are often excipients of drugs (e.g., surfactants). Even drug compounds themselves (e.g., the salts of organic acids or bases with anisometric molecular shape) fulfill the requirements for the formation of calamitic mesophases.

13.2.2 An Introductory Example

It is interesting to point out that there is a family of complex isotropic fluids, which we can call micro-emulsions, whose characteristics overlap with those of lyotropics in some respects. Microemulsions are mixtures of oil, water, and amphiphilic molecules, which behave as an optically isotropic and thermodynamically stable liquid solution. These systems differ from the emulsions, which are kineti-cally stable. In microemulsions, the typical size of the basic units (self-assembled molecular aggre-gates) is about 10 nm, which makes the mixture transparent to visible light. In contrast, emulsions diffuse visible light, displaying a milky or cloudy aspect, which indicates that their basic units are larger, typically with micron dimensions.

The conceptual boundaries between lyotropics, particularly the isotropic phases, and micro-emulsions are not sharp; sometimes the isotropic phases of the same mixture, with oil as one of the components, are included in different sides of this border. To differentiate them, we point out that microemulsions are two-phase systems and lyotropics are one-phase systems. In this chapter we always refer to lyotropics and use their nomenclature to describe the isotropic micellar and bicon-tinuous phases, even if oil is present in the mixture.

13.3 CLASSIFICATION OF LIQUID CRYSTALLINE SYSTEM

LCs are mainly classified as *thermotropics* and *lyotropics*, depending on the physicochemical parameters responsible for the phase transitions. With their formation starting with the crystalline state, the mesophase is reached either by increasing the temperature or by adding a solvent, which corresponds to the differentiation between thermotropic and lyotropic LCs, respectively. As with thermotropic LCs, a variation in temperature can also cause a phase transformation between differ-ent mesophases with lyotropic LCs.

13.3.1 Thermotropic Liquid Crystals

The phase transition of thermotropic LCs depends on temperature. There are mainly two types of thermotropic LCs: *nematic* and *smectic*.

- As temperature increases, the first LC phase is *smectic A*, where there is layerlike arrange-ment as well as translational and rotational motion of the molecules.
- A further increase in temperature leads to the *nematic phase*, where the molecules rapidly diffuse out of the initial lattice structure and from the layerlike arrangement as well.
- At the highest temperatures, the material becomes an *isotropic* liquid where the motion of the molecules changes yet again.

13.3.1.1 Nematic Phase
- The simplest form is a nematic LC, that is, long-range orientational order but no positional order.
- The preferred direction is known as the director.
- Despite the high degree of orientational order, the nematic phase as a whole is in disorder, that is, no macroscopic order (orientation within a group is similar but not from one group to another).

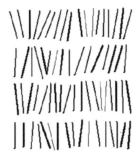

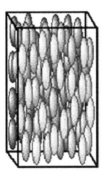

A schematic representation of the nematic phase (left) and a photo of a nematic liquid crystal (above).

FIGURE 13.2 Nematic LC. (Photograph courtesy of Dr. Mary Heuben, Liquid Crystal Institute, Kent State University.)

- The structure of the nematic phase can be altered in a number of ways, for example, an electric or magnetic field or treatment of surfaces of the sample container.
- Thus, it is possible to have microscopic order and macroscopic order (Figure 13.2).

13.3.1.2 Cholesteric Liquid Crystals

1. The first LC that was observed through a polarizing microscope is cholesteryl benzoate. Thus, it is known as a cholesteric LC or a chiral nematic LC (Kelker and Hatz, 1980). For example, cholesteryl benzoate is an LC at 147°C and isotropic at 186°C.
2. Cholesteric LCs have great potential uses for the following:
   - Drug delivery
   - Sensors
   - Thermometer
   - Fashion fabrics that change color with temperature
   - Display devices
3. In the cholesteric phase, there is orientational order and no positional order, but the director is in helical order.
4. The structure of the cholesteric LC depends on the pitch (Kelker and Hatz, 1980), which is the distance over which the director makes one complete turn (Figure 13.3). One pitch comprises several hundred nanometers.
5. Pitch is affected by the following:
   - Temperature
   - Pressure
   - Electric and magnetic fields

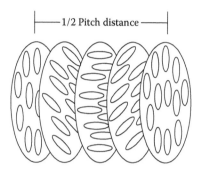

FIGURE 13.3 Cholesteric LC.

13.3.1.2.1 Intrinsic Cholesterics

These are obtained by adding a chiral amphiphilic molecule to a nematic lyotropic LC. These molecules take part in the micellar structure as a nonchiral amphiphile. Examples of chiral molecules used for this purpose are *l-n*-lauroyl potassium alaninate, 2-sodium decylsulfate, and L,D-octanol.

13.3.1.2.2 Extrinsic Cholesterics

These are obtained by adding a chiral nonamphiphilic molecule to a lyotropic nematic. Depending on the electrostatic characteristics of the molecule (polar or nonpolar), it can be accommodated either in the inner or the outer part of the micelle. Examples of chiral molecules used for this purpose are brucine sulfate heptahydrate, tartaric acid, cholesterol, L-sorbose, diacetone sorbose, and diacetone-2-ceto potassium gulonate.

13.3.1.3 Smectic Phase

The smectic phase (Kelker and Hatz, 1980) occurs at a temperature below the nematic or cholesteric phase. The molecules align themselves approximately parallel and tend to arrange in layers (Figure 13.4). Not all positional order is destroyed when a crystal melts to form a smectic LC. Chiral smectic C LCs are useful in drug delivery (LCDs).

13.3.2 EXAMPLES OF PHASE CHANGES

1. Cholesteryl myristate:

Solid $\xleftrightarrow{74°C}$ Smectic C $\xleftrightarrow{94°C}$ Nematic $\xleftrightarrow{124°C}$ Isotropic.

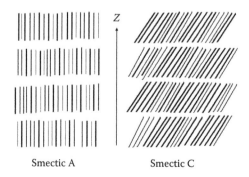

Smectic A   Smectic C

FIGURE 13.4 Smectic phase.

2. 4, 4'-Di-heptyloxyazoxybenzene:

$$\text{Solid} \xleftrightarrow{74°C} \text{Smectic C} \xleftrightarrow{94°C} \text{Nematic} \xleftrightarrow{124°C} \text{Isotropic.}$$

Different smectics may be distinguished on the basis of a variety of arrangements with varying temperatures. Phase transitions occur with increasing temperature, for example, crystalline to smectic C to smectic A to nematic to isotropic, or crystalline to nematic to isotropic. These examples demonstrate that all possible transitions do not necessarily occur. Depending on the number of mesophases occurring, thermotropic mono-, di-, tri-, or tetramorphism may be distinguished.

Discotic LCs arise from disc-shaped molecules as nematic or cholesteric mesophases. Their structural characteristics (Kelker and Hatz, 1980) are similar to that of their respective calamitic mesophases, that is, the normals of the discs are oriented parallel. Instead of the smectic mesophases, discotic columnar LCs arise from stapling the discs one on the other. The columns of the discotic columnar mesophase form a two-dimensional (2D) lattice that is in either a hexagonal or a rectangular modification.

In thermotropic LCs the basic units are molecules, and phase transitions depend on temperature and pressure. A pronounced shape anisotropy (in other words, the anisometry) is the main feature of the molecules that gives rise to a thermotropic mesophase. Rod, disc, and banana shapes are examples of molecular geometries associated with thermotropic LCs. In addition to pure substances, mixtures of molecules can also present thermotropic mesomorphic properties.

13.3.3 LYOTROPIC LIQUID CRYSTALS

For lyotropic LCs, the phase transition depends on the temperature and concentration. Lyotropic LCs, shortly called lyotropics or lyomesophases (Jákli and Saupe, 2006), are mixtures of amphiphilic molecules and solvents at given temperatures and relative concentrations. The mesomorphic properties change with the temperature, pressure, and relative concentrations of the different components of the mixture. An important feature of lyotropics is the self-assembly of the amphiphilic molecules as supermolecular structures, which are the basic units of these mesophases. The physicochemical properties of the lyotropics have an interesting interface with biology; the understanding of these properties has been relevant for improving some technological aspects of cosmetics, soaps, food, crude oil recovery, and detergent production.

Lyotropic LCs differ from thermotropic LCs. They are formed by mesogens, which are not the molecules themselves but their hydrates or solvates, and by associates of hydrated or solvated molecules. In the presence of water or a mixture of water and an organic solvent as the most important solvents for drug molecules, the degree of hydration or salvation depends on the amphiphilic properties of a drug molecule. Hydration and salvation of the mostly rod-shaped molecules result in different geometries such as cones and cylinders.

Cylinders arrange in layers; this results in a lamellar phase with alternating polar and nonpolar layers. Water and aqueous solutions can be included in the polar layers, resulting in an increase of the layer thickness. Analogously, affinic molecules (Jákli and Saupe, 2006) can be included in the nonpolar layers. In addition to the increased layer thickness of the lamellar phase, lateral inclusion between molecules is also possible with an increase in the solvent concentration, which transforms the rod shape of the solvated molecules to a cone shape, thereby leading to a phase change. Depending on the polar or nonpolar character of the solvating agent and the molecule itself, the transition results in a hexagonal or an inverse hexagonal phase.

The hexagonal phase is named after the hexagonally packed rod micelles of solvated molecules, whereby their polar functional groups point either to the outside or to the inside of the structure (inverse hexagonal phase). In the hexagonal phase, the additional amount of water or unpolar solvent that can be included is limited. As the molecular geometry (Gennes and Prost, 1993) changes further during solvation, another phase transformation to a cubic form (type I) or inverse cubic form (type IV) takes place, consisting of spherical or ellipsoidal micelles, inverse micelles, or both.

In addition to the cubic and inverse cubic forms described previously, further transitional forms exist between the lamellar phase and the hexagonal mesophase (cubic, type II) or inverse hexagonal mesophase (cubic, type III). In contrast to the discontinuous phases of types I and IV, cubic mesophases of types II and III belong to the bicontinuous phases. A range of lyotropic mesophases are possible, depending on the mesogen concentration and the lipophilic or hydrophilic characteristics of the solvent and the molecule itself.

13.4 AMPHIPHILIC MOLECULES

Amphiphilic molecules are always present in the composition of lyotropic LCs. They may be synthesized for different purposes, ranging from interests in basic science to technological applications in various branches of industry.

The name amphiphilic comes from the Greek prefix *amphi*, which means both or double, and the word *phile*, which means like or love. This word is applied to a compound that displays a double "preference," "loving both," from the electrostatic point of view. It is used to name a molecule with a polar water-soluble group attached to a water-insoluble hydrocarbon chain. An example of these types of molecule is sodium decylsulfate (Figueiredo Neto and Salinas, 2005; Figure 13.5). These molecules are surfactants (from surface-active agents) because they can modify the properties of surfaces and interfaces between different media, such as solid–liquid or liquid–gas interfaces (Mlodozeniec, 1978).

There are different types of naturally and chemically synthesized amphiphilic molecules: anionic amphiphiles (soaps of fatty acids, e.g., potassium laurate), detergents (e.g., sodium decylsulfate); cationic amphiphiles (e.g., hexadecyl trimethylammonium bromide), nonionic amphiphiles (e.g., pentaethyleneglycol dodecyl ether), and zwitterionic amphiphiles (which develop an electric dipole in the presence of water, e.g., lysolecithin; Figure 13.6).

Another type of surfactant molecules that gives rise to lyotropic mesophases are the anelydes. These molecules are able to selectively complex some metallic ions, which are then incorporated in their structure (Figueiredo Neto and Salinas, 2005).

In addition to these so-called classical amphiphiles, there are molecules with a more complex topology, with more than one polar group, which also yield lyotropic mesophases, for example, gemini surfactants, rigid spiro-tensiles, and phospholipids, which have molecules of the hydrophilic group grafted in a position lateral to a rodlike rigid core.

Facial amphiphiles are block molecules in which two alkyl chains are placed in both sides of a calamitic core and the polar group is attached to the core, perpendicular to the sticklike molecule. In bolaamphiphiles, there are two polar heads in both sides of the sticklike molecule and the alkyl chain is perpendicularly attached to the core (Figueiredo Neto and Salinas, 2005; Figure 13.7). In the presence of polar and nonpolar solvents, these molecules form lyotropic mesophases with nanosegregation properties.

As a final remark, it is important to note that a polar group is not always required to be hydrophilic (nor is a nonpolar group always hydrophobic). The topology of the molecule and its insertion into the water network are also important to characterize the solubility in water.

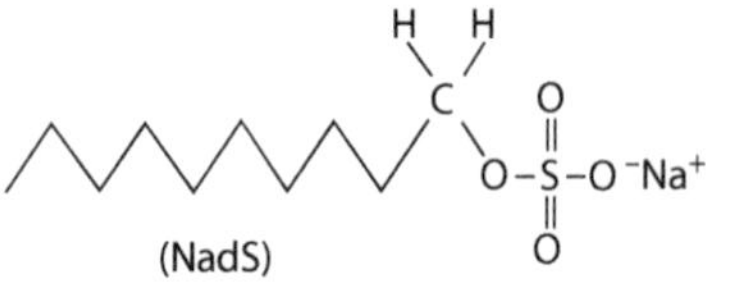

FIGURE 13.5 Amphiphilic molecule of sodium decylsulfate. (Figueiredo Neto, A. M. and Salinas, S. R. A., *The Physics of Lyotropic Liquid Crystals*, p. 17, 18. New York: Oxford University Press, 2005.)

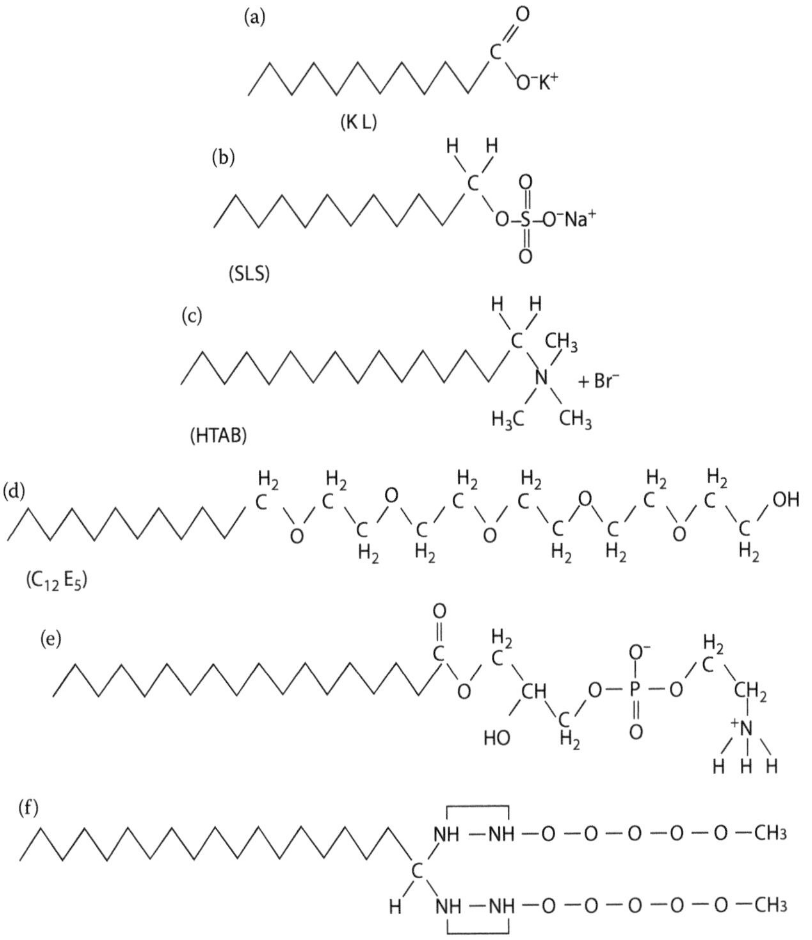

FIGURE 13.6 Examples of different amphiphiles: (a) anionic, potassium laurate (KL); (b) detergent, sodium lauryl sulfate (SLS); (c) cationic, cetyl trimethylammonium bromide (CTAB) or hexadecyl trimethylammonium bromide (HTAB); (d) nonionic, pentaoxyethylene dodecyl ether; (e) zwitterionic; and (f) anelydes. (Figueiredo Neto, A. M. and Salinas, S. R. A., *The Physics of Lyotropic Liquid Crystals*, p. 17, 18. New York: Oxford University Press, 2005.)

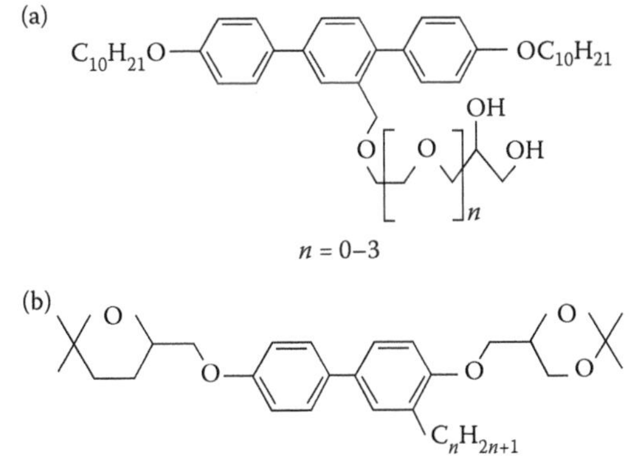

FIGURE 13.7 (a) Facial amphiphile and (b) bolaamphiphile.

13.5 LYOTROPIC MIXTURE

Under suitable conditions of temperature and relative concentrations, mixtures of amphiphilic molecules and solvents can give rise to a lyotropic mesophase. In this type of system, amphiphilic molecules form self-assembled superstructures with several shape anisotropies and sizes.

Lyotropics can be classified into three large families as follows:

1. Micellar systems with molecular aggregates called micelles, which have small shape anisotropy, as sketched in Figure 13.8a (Figueiredo Neto and Salinas, 2005). These micelles are aggregates of amphiphilic molecules with typical dimensions of about 10 nm and shape anisotropy on the order of 1:2 in linear dimensions.
2. Systems with aggregates of large shape anisotropy, of a typical order of 1:100 in terms of linear dimensions. These aggregates are sometimes called infinite, but we do not use this nomenclature. Figure 13.8b shows a long cylindrical aggregate (Figueiredo Neto and Salinas, 2005).
3. Bicontinuous systems, in which the amphiphilic molecules self-assemble as a three-dimensional (3D) continuous structure at large scales (>103 nm). Figure 13.8c is a sketch of a bicontinuous molecular aggregate with cubic symmetry (Figueiredo Neto and Salinas, 2005).

13.6 ACTIVE INGREDIENT DELIVERY THROUGH LIQUID CRYSTALS

Drugs and other active ingredients require a range of delivery techniques and vehicles because of their variable solubility, availability, and stability. The field is an extremely active one encompassing numerous devices and methods. Starting from their discovery in studies of fat digestion, the potential of cubosomes as encapsulation and delivery vehicles was obvious. Cubic phases with a bicontinuous structure (Jákli et al., 2001) have a high solidlike viscosity, so cubosomes reflect the bulk phase's native symmetry in their cubic faceted shapes. The bicontinuous structure of the cubic phases enables solubilization of diverse molecules ranging from proteins to small-molecule drugs, and the bicontinuous structure's tortuosity leads to diffusion-controlled release of the solubilizates. It is not surprising that the first and still the broadest application pursued for cubosomes has been as controlled-release vehicles, despite some inherent limitations.

It is common to hear cubosomes described as excellent controlled-release vehicles (Mueller-Goymann and Hamann, 1993) for delivery of active materials, especially drugs. Although the diffusivity of solubilized molecules is reduced by about 33% in bulk cubic phases, the much

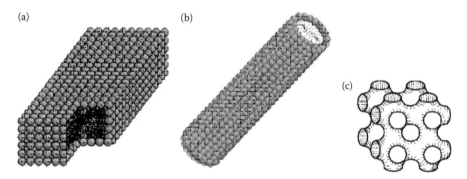

FIGURE 13.8 (a) Sketch of an orthorhombic micelle, where the cut in the bottom right side shows the paraffinic chains in its inner part; (b) a large anisotropic cylindrical aggregate; and (c) a sketch of a bicontinuous molecular aggregate with cubic symmetry. (Figueiredo Neto, A. M. and Salinas, S. R. A., *The Physics of Lyotropic Liquid Crystals*, p. 17, 18. New York: Oxford University Press, 2005.)

smaller length scale of cubosomes makes it difficult to use them directly for controlled release. Boyd examines a range of hydrophilic and hydrophobic drugs and finds only burst release from cubosomes in aqueous media. Apparently, earlier claims of controlled release from cubosomes did not accurately characterize the transport behavior in dispersions. Still, altering the cubosome charge, viscosity, and structure can improve the release kinetics, but controlling the responsiveness of cubosomes *in vivo* might do more. Beyond controlled release is targeted delivery of actives, for example, to cancerous tumor vasculature (Mueller-Goymann and Hamann, 1993). Here DNA–lipid complexes in the form of liposomes have shown promise during local injections by clustering around tumor sites and exhibiting efficient transfection. No clinical work on DNA–lipid complexes in cubosome form is known, but the extension is an obvious one and recent DNA–lipid cubic phase discoveries offer intriguing promise. Some interactions with biological systems can be undesirable, as when lipase enzyme is present. When injected into plasma, cubosomes are apparently hydrolyzed, but the solubilized actives remain encapsulated to some extent (Jákli et al., 2001). Through what phases do these particles pass as they degrade and how do solubilized proteins and nucleic acids affect the progression? The actual interactions in blood plasma should be quantified with cryo-TEM and compared to the work of Borné et al. on lipase degradation of cubosomes.

Beyond the effects of the immediate environment on cubosomes, microscopy can also convey the sometimes profound structural influences that solubilized molecules can have on the cubic phase. Such effects are exemplified in the work of Angelova et al. who created a new class of particles called "proteocubosomes," which are cubosomes fragmented from the bulk cubic phase by incorporated proteins. Proteocubosomes possess a tertiary structure hierarchically constructed of ordered and disordered subunits called "nanocubosomes." Although the thermodynamic stability of such hybrid structures is an open question, their fascinating formation from cubosome building blocks is one of the first and most successful attempts to alter and customize the structure of cubosomes. The relationship between proteocubosomes and the intermediates observed by Borné is not known, if any exists. A large factor in the success and performance of cubosomes as delivery vehicles is their interactions with biological surfaces.

13.7 INTERFACE WITH BIOLOGY

Amphiphilic and water molecules are important components of the human body. The structures formed by amphiphilic molecules in the human body and the processes involving the connections with water, other polar, and nonpolar molecules are essential ingredients for the existence of life. Because some of these processes are essential for the stability of lyotropic mixtures, there is a large interface between research in lyotropics and in biological sciences. We now present some examples of this interface.

A lipid is an important structural component of living tissues in plants, animals, and microorganisms. Lipid molecules may be neutral, phosphatides, sphingolipids, glycolipds, and terpenoids. Phospholipids have polar and nonpolar parts; under dispersion in an aqueous medium, they self-organize in micelles and more complex structures, depending on the relative concentrations and temperature. In the presence of water, porphyrins, which are essential for life, also form liquid crystalline structures. In particular, phospholipids in the presence of increasing amounts of water were shown to reduce their marked endothermic transition (Jákli et al., 2001), which is usually observed in differential thermal analysis experiments, corresponding to the melt of the hydrocarbon chains. Mixtures of natural phospholipids, with highly unsaturated chains, and water display a transition between crystalline and LC phases at temperatures lower than in the biological environment. At room temperature, the chains of these molecules have high mobility, which improves the permeability of the membranes that are formed. One of the most expressive similarities between living systems and lyotropics, especially in water-based mixtures, is the dominant presence of water that, in some mixtures, reaches concentrations larger than 96 mol%.

If water is mixed with suitable solvents, biological compounds or even molecules commonly found in the human body may present liquid crystalline phases. An interesting example of liquid crystalline polymorphism with biological lipids is a mixture of the mitochondrial lipids and water; there were observations of three lamellar and one hexagonal phases, depending on the temperature and relative concentrations of lipids and water. Polypeptides are systems containing two or more amino acids and one or more peptide groups, such as poly-γ-benzyl-L-glutamate; if they are mixed with different solvents, polypeptides may present liquid crystalline phases (Jákli et al., 2001). Cholesterol and fatty acids form esters, which also have liquid crystalline properties. In particular, cholesterol is used to induce cholesteric mesophases in originally lyotropic nematic structures.

Micellar solutions of amphiphiles also play an important role in biological processes: the digestion of fats by the human body requires solubilization by bile salt micelles, which act as the detergent of the body through the digestive tract; and micellar catalysis, which exploits the large surface available in micellar systems (e.g., an ester hydrolysis by an acid). Particularly in the case of the fat solubilization problem, *in vitro* micellar solutions of lyotropics are used to study the temperature dependence of the solubilization rate of fat acids in micelles. In actual lyotropic mixtures this rate can increase by a factor of 4 at temperatures above the penetration temperature (at which the liquid crystalline phases appear).

Phospholipids and proteins are the basic structural blocks of the cell membrane. About 50% of the mass of most animal cell plasma membranes are made of lipidic membranes. Lipids form a bimolecular layer (bilayer) that controls the traffic of substances to and from the inner part of the cell. Despite the differences between the liquid crystalline structures and the living cell, lyotropics can be used as models to study some aspects of cell membranes. Biological membranes inside cells serve different purposes, such as dividing compartments with different functions and as regulators of the flux of substances with different electric properties. These membranes usually have encrusted proteins. Investigations of the permeability of macromolecules through a lyotropic model membrane, in a lamellar-type phase, bring important insights into problems regarding the form (structure) versus functionality in actual membranes.

Amphiphilic molecules of biological membranes usually have two hydrocarbon chains from glycerol derivatives. The membrane prevents transport from one compartment to the other without a biological command. From this point of view, the equilibrium in terms of the concentration gradients of different substances is not achieved passively. In plants and bacteria, the dominant lipid membranes are glycolipids. In animals, the most common membranes are phospholipids.

In the human body, molecular species with multiple positive charges are rare and biological membranes have a net negative surface charge. The diffusion coefficients of molecules across the membrane and on its surface can be very different. Because this behavior resembles the behavior of lamellar LCs, it is one of the reasons why lyotropics can be used as a model system to study diffusion processes in biological membranes. For example, the permeability to both strongly polar (water) and nonpolar (hydrocarbons) molecules by the stratum corneum (the outer part of the human skin) can be studied by using lyotropic lamellar model systems.

One of the most interesting and important problems in living systems is the investigation of the interaction between lipids and proteins. According to a current view on this problem, proteins and cholesterol are incorporated into the bilayer structure and consequently disrupt it. Cholesterol molecules are interdigitated between lipid molecules of the bilayer, and the hydroxyl groups are in the external water layer. The bilayer is conceived as a dispersion medium for proteins and cholesterol. The stability of the structure and the permeability properties if the bilayer is in contact with hosts, such as proteins and cholesterol, can also be investigated in lyotropic model membranes and in cholesteric phases (in the particular case of chiral molecules incorporated into micelles or even in bicontinuous structures). The presence of cholesterol in lipid bilayers was shown to lower the transition temperature between the gel and liquid crystalline phases. It disturbs the crystalline arrangement of the hydrocarbon chains in the gel phase structure. This process can be studied *in vitro* by using a lipid/cholesterol/water mixture.

Another aspect of the similarity between lyotropics and living systems is related to the process of aging of the cell. As the living system becomes older, the cell membrane becomes stiffer and drifts toward crystallinity. In a typical binary lyotropic water-based mixture at a fixed temperature and for increasing the concentration of amphiphiles, there is a phase sequence from micellar isotropic to hexagonal, then lamellar, and finally to a crystalline phase. The aging of a cellular membrane has some similarities with the concentration-driven phase transition sequences of lyotropic systems. In lyotropics, the loss of water is the way to crystallinity. In human cell membranes, the aging process is certainly more complex than a simple loss of water, so that many other aspects have to be considered. However, the role of the water molecules and their interactions with the membrane structure can certainly be one of the main aspects of this problem. For example, the investigation of the role of the hydration layers around different amphiphilic molecular aggregates in lyotropics, from micelles to the lamellae, can provide relevant insights on the time dependence of the characteristics of membranes of living systems.

Shape transformations in biological lyotropic-like systems play a key role in many processes in the human body, for example, in the propagation of nerve impulses over the synaptic clef separating nerve cells by vesicles. Vesicles, which are a blister-type bilayered structure ranging in diameter from microns to millimeters, undergo a transformation and participate in this propagation under the action of an electric excitation. This process is reversible and, due to the hydrolysis by an appropriate enzyme, the system is ready to receive and deliver new signals. Amphiphilic biocompatible molecular aggregates are used as drug vectors (Jákli et al., 2001).

Medicines can be encapsulated inside micelles or vesicles and injected in the living system. These closed structures are used, for example, as simple containers to transport drugs within the blood system.

13.8 CUBIC PHASE–TISSUE INTERFACES

Some of the earliest observations of cubosomes and related structures were in biological systems, including plant membranes and human digestion models. Building on liposome usage as model "cells," the thought came to incorporate proteins into cubic phases. Buchheim and Larsson allow the protein casein to diffuse into a cubic phase and then carry out a beautiful work of freeze-fracture electron microscopy by showing the boundary between two different cubic phases at the limit of casein infiltration. This pioneering work established the idea of probing the dynamics of the interface between a biological system and a liquid crystalline phase. A step beyond such work is the now well-established success in meso crystallization of membrane proteins in cubic phases for structural determination.

Cryo-TEM imaging of crystal nucleation and growth in cubosomes could provide verification of the proposed need for a local lamellar phase inclusion within the cubic phase to feed monomers to the growing crystals. Other progress in cubic–biological interfaces highlights the unique bioadhesive nature of cubosomes. Although the mechanism is not yet understood, imaging studies of the adhesion, deformation, and failure of tissue–cubosome bonds could provide a dynamic window into the adhesion phenomenon and its roots. Recent work on adhesion and failure of vesicle–surface bonds offers an excellent basis to understand the wetting and adhesion physics of such interactions in the context of tissue–cubic liquid crystalline interfaces.

For example, what sort of deformation occurs as bioadhesive cubic phases are removed from biological surfaces as a function of the interfacial energy? The work of Carlos Rodriguez (2006) offers a tantalizing means of blurring the boundary between cubic self-assembly structures in surfactant and biological systems in his theory of a bicontinuous cubic structure of the stratum corneum in human epidermis. Rodriguez's use of cryoelectron microscopy of vitreous sections allows direct imaging of the epidermal spaces and finds a distinct similarity between cubosomes and the cubic structures of the intermediate filaments in skin. Recent reviews build on these new insights to examine the broad applicability of cubic phases as substrates for bioelectrodes.

Hoath et al. have a vision of using cubosomes as an adapter that allows continuity between electrical and other medical sensors and the interior of the human body without invasive punctures or other modifications. They show that the cubic phase on human skin is superior to existing skin treatments because it simultaneously acts as a protective barrier, a moisture sink, and perhaps most importantly as a permeable mass. With applied cubosomes the skin is still able to transpire but is protected and moisturized. Despite the broad interest in cubosomes and their "soft" nature, commercial applications have also driven the pursuit of these phases in a more permanent form.

13.9 MATERIAL FOR PREPARATION OF LIQUID CRYSTALS

The materials for synthesis of LCs have tremendously increased nowadays. The bicontinuous cubic phase structure was recognized early as a fascinating template for more rigid, high surface area nanometer-scale structures. The pioneering work at Mobil of Kresge et al. formed a new aluminosilicate zeolite matrix by initiating a sol–gel ceramic formation from precursors that were solubilized and thus shaped by the self-assembly forces at work on the cetyl trimethylammonium bromide surfactants themselves. The new material, MCM-48, is the monolithic substrate for catalytic reactions used in petroleum cracking operations. Templating of bulk bicontinuous cubic and other self-assembled phases has broadened into a self-sustaining field of its own, and a recent review exists.

A logical extension of bulk cubic phase templating is templating of cubosomes. Jákli et al. (2001) exploit liquid crystalline phase transition kinetics in evaporation-induced self-assembly processes for the aerosol synthesis of nanostructured particles. More recently, Yang et al. polymerized cubosomes and preserved the local microstructure. It is also possible to extend beyond the well-known cubosome symmetries using both kinetic and equilibrium structures on the colloidal and larger length scales. Lynch et al. followed the formation of unique pyramidal droplets during temperature-induced shifts along the cubic phase boundary in the C12E2–water system. They also observed coupling of the structures formed and their orientation with the surface of the capillary containing the phase, perhaps early evidence that such structures can be tuned by surface forces as suggested above.

Staggeringly complex geometries have been formed in lyotropic and thermotropic cubic phase particles by Imperor-Clerc and coworkers, again using careful temperature schedules to form all of the crystalline facets. As new triply continuous cubic phases are explored, the range of geometric templates will expand further. As mentioned in the above, proteocubosomes also expand the available structural templates and suggest the use of proteins not as delivery passengers but as structural adjuncts that broaden the symmetries and defects contributing to available cubosome structures and shapes. Here the use of 2D cryo-TEM images in conjunction with newly developed models to extract the 3D structure could rapidly accelerate the exploration of previously unknown kinetic and thermodynamically stable geometries.

Experimental microscopy could truly blend with theoretical models. Cubic phases possess broad potential as templates at both the molecular and colloidal length scales. Recent breakthroughs in cubosome yield are tantalizing in the possibilities that they offer to apply cubosomes in the way that monodisperse polymer colloids are now. For example, photonics researchers produce ordered structures by first equilibrating polymer colloids in colloidal crystal phases in a metal oxide solution, then solidifying the structure and burning out the polymer particles. However, some symmetries have a favorable band gap for photonics applications but are difficult to access using colloidal crystals based on spherical particles. It would be interesting to assemble monodisperse cubosomes, explore their colloidal self-assembly structures, and template them by infiltration of silica solutions. Then the cubosomes could simply be dissolved away with organic solvents, leaving behind rigid structures based on cubic building blocks rather than spherical ones. A crude proof of concept shows a 100-μm cube formed by coating a large cubosome with silica sol, evaporating the water to consolidate the silica, and dissolving the cubosome. Larger-scale structures would require ordering the cubosomes first and then templating; but this could be done by forming the cubosomes from vesicle or emulsion precursors, ordering these, and then slowly crystallizing them into cubosomes using the techniques.

The importance of templating to material science is growing. As researchers become increasingly adept, a shift in emphasis from equilibrium phases to kinetic structures is logical. An example is the use of kinetic myelinic structures as templates during lamellar phase formation. Recent work has shown that certain kinetic pathways leading to cubosomes through lamellar intermediates produce structures clearly influenced by myelinic structure formation. What has not been done is to probe the internal structure of such kinetically intermediate forms using cryo-TEM. Even simple systems exhibit the potential for much structural complexity, and future microstructural probes, for example, of poorly resolved regions of the ternary ethanol–monoolein–water phase diagram, may yield additional kinetic and possibly equilibrium structures of interest.

Cubosome research currently progresses through incremental advances and breakthroughs. Much knowledge exists in proprietary circles given the patent and literature activity, but without a well-known commercialized example it is difficult to coalesce the field around some fundamental and central challenges. Nevertheless, the areas discussed above are just a few of the potential applications of cubosomes where microscopy could directly address unknowns. Other areas of potential include a reexamination of the key conclusions of the last 20 years using the new freeze-fracture direct imaging techniques that avoid some of the artifacts associated with cryo-TEM and freeze-fracture TEM. Finally, the rheological and flow behavior of cubic phases is at the core of their uniqueness; but few fundamental studies exist that link the flow, deformation, and hydrodynamic behavior of cubosome dispersions with their structure and performance. Examples from the liquid mixing community offer the most hope for physically based interpretations.

13.10 METHODS FOR CHARACTERIZATION OF LIQUID CRYSTALLINE SYSTEMS

Methods appropriate for the investigation and characterization of lyotropic LCs are frequently used in drug development and may thus be employed in pharmaceutical laboratories. These methods are both macroscopic and microscopic.

13.10.1 Polarized Light Microscopy

Lyotropic LCs except for cubic mesophases show birefringence just like real crystals do. Birefringence can be observed in a polarization microscope. Two polarizers in cross-position are mounted below and above the birefringent object being examined. The cross-position of the polarizers provides plane polarized waves perpendicular to each other. Therefore, the light passing the polarizer below an isotropic object cannot pass the polarizer in cross-position above the object. In an anisotropic material, some parts of the light are able to pass the second polarizer because the plane polarized beam has been rotated by an angle relative to the plane of the incoming beam.

Each LC shows typical black and white textures. The addition of an l-plate with strong birefringent properties enables the observation of color effects of the textures in yellow, turquoise, and pink. Color effects arise because rotation of the plane polarized light depends on the wavelength. The thickness of the l-plate is suited for a 550-nm wavelength. After leaving the plate, this wavelength swings in the same plane as the incoming polarized white light does. Therefore, it is totally absorbed by the second polarizer in the cross-position. All of the other wavelengths of the white light except for the 550-nm one are more or less rotated with respect to the polarization plane. Hence, they pass the polarizer with various intensities. White light minus the 550-nm wavelength (green yellow) gives the impression of a pink color. With an additional birefringent liquid crystalline material in the microscope, small deviations of the wavelengths being absorbed occur; thereby, turquoise and yellow textures can be observed. The smectic mesophases of the thermotropic LCs show a variety of textures but resemble the fan-shaped texture of the lyotropic hexagonal mesophase.

13.10.2 Transmission Electron Microscopy

Because of the high magnification power of the electron microscope, the microstructure of LCs can be visualized. However, aqueous samples do not survive the high vacuum of an electron microscope without loss of water, thus changing their microstructure. Therefore, special techniques of sample preparation are necessary prior to electron microscopy. The freeze-fracture technique has proven to be successful in this regard. For this purpose, a replicum of the sample is produced and viewed in the electron microscope. To preserve the original microstructure of the sample during the replication, the first step is to shock freeze the sample. For high freezing rates up to 105–106 K/s, the sample is sandwiched as a thin layer between two gold plates and then shock frozen with either nitrogen-cooled liquid propane at −196°C or slush nitrogen at −210°C. If the temperature of the cooling medium is far below its boiling temperature, an efficient freezing rate can be obtained (Burylov and Raikher, 1995).

The frozen sample within the sample holder is transduced into the recipient of a freeze-fracture apparatus, in which the fracture is performed at a temperature of −100°C and with a vacuum between 10–6 and 5–7 bar. Within a homogeneous material, the fracture occurs randomly because all structural elements have equal probabilities for fracturization. However, even a homogeneous material often consists of more or less polar areas. Within polar areas, stronger interactions via hydrogen bonds prevent the fracture; thus, fracture within polar areas is less probable than fracture within apolar areas. Therefore, the sample profile obtained after fracturization represents the microstructure of the sample just qualitatively and not quantitatively.

Following this, the sample surface is shadowed with 2 nm thick platinum under an angle of 45°. Additional vertical shadowing with a 10 times thicker carbon layer of 20-nm platinum provides high mechanical stability of the replicum, which means easier handling in regard to removing, cleaning, drying, and finally observing in the transmission electron microscope (Burylov and Raikher, 1995).

The shadowing with platinum under an angle of 45° provides differences in contrast because platinum precipitation takes place preferably at sample positions that face the platinum source in luff, whereas sample positions in lee are less or not shadowed. In TEM, these different thicknesses of platinum absorb the electron beam to different extents, thus forming shadows. This phenomenon results in the formation of a plastic impression of the transmission electron micrographs of the replicum. It represents transmission electron micrographs of different lyotropic LCs after freeze-fracture without etching. The layer structure of the lamellar mesophase (Burylov and Raikher, 1995), including confocal domains, hexagonal arrangement of the rodlike micelles within the hexagonal mesophase, and closely packed spherical micelles within the cubic LC, was clearly seen.

13.10.3 X-Ray Scattering

With x-ray scattering experiments, characteristic interferences are generated from an ordered microstructure. A typical interference pattern arises that is attributable to specific repeat distances of the associated interlayer spacing d. According to Bragg's equation, d can be calculated as $d = \frac{1}{4}$ $n(l/2)\sin W$, where l is the wavelength of the x-ray (e.g., 0.145 nm by using a copper anode or 0.229 nm by using a chromium anode), n is an integer that denotes the order of the interference, and W is the angle under which interference occurs (i.e., reflection conditions are fulfilled).

Bragg's equation points at the inverse proportionality between d and W. Large terms for d in the region of long-range order are registered by the small-angle x-ray diffraction (SAXD) technique, whereas small terms for d in the region of short-range order are registered by the wide-angle x-ray diffraction technique. SAXD is important for the exact determination of the distances of d of liquid crystalline systems. With wide-angle x-ray diffraction, the loss of short-range order of liquid crystalline systems can be recognized in terms of the absence of interferences, which are characteristic of the crystalline state (Fontell and Gray, 1974).

Interferences can be detected in two ways: film detection and registration of x-ray counts with scintillation counters or position-sensitive detectors.

However, SAXD not only detects interferences from which the interlayer spacings can be calculated, but it also enables us to decide the type of LC from the sequence of the interferences.

13.10.4 DIFFERENTIAL SCANNING CALORIMETRY

Phase transitions correspond with changes in the energy content of the respective system. This phenomenon is caused by changing either the enthalpy or the entropy. Enthalpy changes cause endothermic or exothermic signals, depending on whether the transition is due to consumption of energy (e.g., melting of a solid) or release of energy (e.g., recrystallization of an isotropic melt).

Note that the transition from crystalline to amorphous requires much energy; but the transition from crystalline to liquid crystalline, from liquid crystalline to amorphous (Burylov and Raikher, 1995), and particularly the transition between different LCs consume low amounts of energy. Therefore, care has to be taken about the appropriate sensitivity of the measuring device as well as a sufficiently low detection limit.

Entropically caused phase transitions may be recognized by a change in baseline slope according to a change in the specific heat capacity. In particular, the phase transitions of liquid crystalline polymers result from entropic reasons, thus being considered transitions of the second order. These are usually called glass transitions. They can be overlaid from an enthalpic effect, so their detection can be complicated.

13.10.5 RHEOLOGY

Different types of LCs exhibit different rheological properties. With an increase in the microstructural organization of the LC, its consistency increases and the flow behavior becomes more viscous. The coefficient of dynamic viscosity Z, although a criterion for the viscosity of ideal viscous flow behavior (Newtonian systems), is rather high for cubic and hexagonal LCs but fairly low for lamellar ones; however, the flow characteristics are not Newtonian but plastic for cubic and hexagonal crystals or pseudoplastic for lamellar ones (Roux and Nallet, 1993).

For thermotropic LCs, the viscosity increases in the following sequence:

$$\text{Nematic} < \text{Smectic A} < \text{Smectic C.}$$

The low flowability of lyotropic LCs such as cubic and hexagonal mesophases is due to their 3D and 2D order (Jákli and Saupe, 2006), respectively. Lamellar mesophases with one-dimensional long-range order have a fairly high flowability (Roux and Nallet, 1993). Because of their gel character, cubic and hexagonal mesophases even exhibit a yield stress until flow occurs. Unlike the corresponding inverse LCs, the gel character is much more pronounced because of the interactions between polar functional groups located at the surface of the associates. Via polar interactions, for example, hydrogen bonds, the associates may form strong networks with each other. In contrast, the surface of the associates of inverse mesophases consists of apolar groups of the associated molecules. Thus, the resulting interactions are less strong and the gel can become deformed more easily.

A mechanical oscillation measurement is the method of choice for determining the elasticity of liquid crystalline gels. Without applying a superposition of shear strain, the viscoelastic properties of LCs may be studied without a change in network microstructure, which usually occurs in terms of mechanical deformation with rheological investigations (Roux and Nallet, 1993). With the oscillation experiments, the viscoelastic character of cubic and hexagonal mesophases as well as that of lamellar mesophases and highly concentrated dispersions of vesicles (which also show viscoelastic behavior) can be quantified. A vesicle dispersion of low content of the inner phase, however, exhibits an ideal viscous flow property. According to the Einstein equation, Z is larger than Z_0 of the continuous phase, which is usually pure water or solvent, by the multifactor 2.5 volume ratio of the dispersed phase f.

13.10.6 DETERMINATION OF VESICLE SIZE BY LASER LIGHT SCATTERING

Vesicle size is an important parameter in not only in-process control but also quality assurance in particular because the physical stability of the vesicle dispersion depends on the particle size and particle size distribution. An appropriate and particularly quick method is laser light scattering (for particle size) or diffraction (for particle size distribution). Laser light diffraction can be applied for >1-mm particles; according to the diffraction theory of Fraunhofer, this refers to the proportionality between the intensity of the diffraction and the square of the particle diameter.

Rayleigh's theory holds for <200-nm particles, which considers scattering intensity to be proportional to the sixth potency of the particle diameter. Both Fraunhofer's and Rayleigh's theories are only approximations of Mie's theory, which claims that scattering intensity depends on the scattering angle, absorption, and size of the particles as well as on the refractive indices of both the particles and the dispersion medium. Unfortunately, the latter parameters are difficult to determine. Furthermore, most vesicle dispersions consist of a dispersed mesophase with particle sizes of <200 nm up to 1 mm. Therefore, photon correlation spectroscopy (PCS) that is based on laser light scattering provides an appropriate method of investigation.

Dynamically raised processes in the dispersion, such as Brownian molecular motion, cause variations in the intensities of the scattered light with time, which is measured by PCS. The smaller the particle is, the higher the fluctuations by Brownian motion (Roux and Nallet, 1993). Thus, a correlation between the different intensities that are measured is only possible for short time intervals. In a monodisperse system following first-order kinetics, the autocorrelation function decreases rather fast. In a half-logarithmic plot of the autocorrelation function, the slope of the graph enables the calculation of the hydrodynamic radius by the Stokes–Einstein equation. Commercial PCS devices determine the z-average, which corresponds to the hydrodynamic radius.

In a polydisperse system, the calculation of the particle size distribution is possible by using special transformation algorithms. Certain requirements need to be fulfilled for this: a spherical particle shape, sufficient dilution, and a large difference between the refractive indices of the inner and outer phase. Because not all requirements can usually be fulfilled, the z-average as a directly accessible parameter is preferred to the distribution function, depending on the models.

13.11 APPLICATIONS OF LIQUID CRYSTALS IN DRUG DELIVERY

Some drug substances are able to form mesophases with either a solvent or alone. An increase in temperature causes the transition from the solid state to the liquid crystalline one, which is called thermotropic mesomorphism. Lyotropic mesomorphism occurs in combination with a solvent, usually water. Furthermore, a change in temperature may cause additional transitions. Thermotropic and lyotropic liquid crystalline mesophases of drug substances may interact with mesomorphous vehicles as well as with liquid crystalline structures in humans.

Arsphenamin was the first drug substance with thermotropic mesomorphism to be therapeutically used as Salvarsan during the first half of the 19th century. The drug is effective against microorganisms and offered the first efficient therapy for venereal diseases such as syphilis. Today it has been replaced by antibiotics with less serious side effects.

The molecular structure of arsphenamin is typical of a thermotropic mesogen. With its symmetrical arrangement of atoms, the same holds for disodium cromoglicinate (DNCG), which forms thermotropic LCs and lyotropic mesophases with water. If micronized DNCG powder is applied to the mucosa of the nose or the bronchi, the powder absorbs water from the high relative humidity of the respiratory tract, first transforming into a lyotropic mesophase and then into a solution.

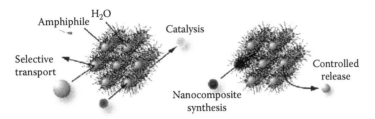

FIGURE 13.9 Functional LC assembly (Douglas, L. G., et al., 2007).

DNCG serves as a mast cell stabilizer. Mast cells are located on the mucosa of the respiratory tract and act by releasing the mediator substance histamine on contact with an allergen, provided the patient was sensitized previously. Because of its mast cell stabilizing effect, DNCG acts as a prophylactic against allergic reactions associated with asthma and hay fever. In addition to this prophylactic effect, DNCG exhibits a second mode of action in causal therapy of asthma, which has not yet been fully clarified. According to recent findings, DNCG has a positive effect on the inflammation of the mucosa of bronchi.

For therapeutic purposes, a similar frequently used group of drug compounds are nonsteroidal antiinflammatory drugs (NSAIDs). The best-known representatives of the aryl acetic acid derivatives and aryl propionic acid derivatives are diclofenac and ibuprofen, respectively. Both have acidic properties, so they dissociate while being dissolved and may form salts with amphiphilic properties. Together with appropriate counterions, these amphiphilic organic acids may form lyotropic mesophases with water at even room or body temperature, for example, diclofenac diethylamine or ibuprofen lysinate. Furthermore, some anhydrates of NSAIDs, for example, fenoprofen calcium, exhibit thermotropic mesomorphism after thermal dehydration of the crystalline salt (Figure 13.9).

13.11.1 LIQUID CRYSTALLINE FORMULATIONS FOR TOPICAL APPLICATION

As long as drug molecules with an amphiphilic character form lyotropic mesophases, amphiphilic excipients in drug formulations form lyotropic LCs. Especially surfactants, which are commonly used as emulsifiers in dermal formulations, associate to micelles after dissolution in a solvent. With increasing concentration of these micelles, the probability of interaction between these micelles increases, formatting LCs.

13.11.2 SURFACTANT GEL

Monophasic systems of lyotropic LCs are relatively seldom used and are limited to gels. A variety of polar surfactants (e.g., ethoxylated fatty alcohols) are hydrated in the presence of water and form spherical or ellipsoidal micelles (Mueller-Goymann, 1984). At high surfactant concentrations, these associates are densely packed and are thus identified as cubic LCs. Such gels are optically transparent. If agitated mechanically, their elastic properties become evident. Because of resonance effects in the audible range, they are also called ringing gels. The lipophilic components are solubilized together with the active ingredients in hydrated associates of the surfactants. However, the solubilization capacity for lipophilic components is generally limited. By exceeding this capacity, the excess of the lipophilic component will be dispersed dropwise in the liquid crystalline phase. Such systems exhibit a white appearance according to the change in the refractive index at the interface between the continuous liquid crystalline and dispersed oil phases. In addition, the dispersed drops are mechanically stabilized because the liquid crystalline phase of either a hexagonal or cubic character has a high yield stress.

Ringing gels with cubic liquid crystalline microstructure are used as commercial drug formulations, especially for topical NSAID formulations. Examples in the German market include

Contrheuma Gel Forte N, Trauma-Dolgit Gel, and Dolgit Mikrogel. Dolgit Mikrogel was introduced in 1996 (Mueller-Goymann, 1984) and contains ibuprofen as an active ingredient. The high surfactant concentration of such gels is necessary to verify the liquid crystalline microstructure, but this concentration influences the microstructure of the stratum corneum lipids via an increase in permeability. This effect is also achieved by alcohol, which is also solubilized in the formulation. The amount of ibuprofen permeating per unit time and surface area is much higher for Dolgit Mikrogel than for an aqueous mixed micellar solution of the drug. Although relatively high permeation rates are possible for the liquid preparation, the commercial formulation is significantly more effective because the high surfactant content and the alcohol favor high permeability.

A ringing surfactant gel of liquid crystalline microstructure containing the antimycotic bifonazole (Bifomyk gel) was introduced in 1995 into the German market. Similar to surfactant gels (Mueller-Goymann, 1984) containing NSAIDs, an improved penetration of the active ingredient is desired in antifungal therapy of the dermis as well. However, because the LC structure only forms with a relatively high surfactant concentration, the positive effect of improved penetration must be considered together with the potential of irritation. The objective is to achieve improved penetration with minimum irritation via a change in the skin structure. Because hyphae fungi (mycelium) can penetrate deep into the epidermal layers by sliding past corneocytes of the horny layer, improvement of antimycotic therapy is of particular importance. The same holds true for penetration of NSAIDs through several epidermal layers because they have to arrive at the deeply located muscle and joint tissues.

13.11.3 Ointments and Creams

Usually the surfactant concentration in ointments and creams is significantly lower than in surfactant gels. Ointments are nonaqueous preparations, whereas creams are ointments (Wahlgren et al., 1984) with water added. The microstructure of both ointments and creams may consist of LCs, as long as a liquid crystalline network or matrix is formed by amphiphilic molecules. In a liquid crystalline matrix, it is easier to deform the system by shear; such formulations show plastic and thixotropic flow behavior on shear (Roux and Nallet, 1993). In comparison to systems with a crystalline matrix that are usually destroyed irreversibly by shear, those with a liquid crystalline matrix exhibit a short regeneration time of the sheared matrix. To obtain a liquid crystalline matrix, amphiphilic surfactants that form lyotropic LCs at room temperature have to be selected. Preferably, lamellar LCs should be formed that are able to solubilize high amounts of further ingredients and spread through the whole formulation as a network forming a cross-linked matrix. In contrast, ointments that contain long-chain fatty alcohols such as cetyl or stearyl alcohol have a crystalline structure at room temperature (Wahlgren et al., 1984).

Although the so-called a-phase of the fatty alcohols (a thermotropic-type smectic B LC with hexagonal arrangement of molecules within the double layers) is initially formed from the melt during the manufacturing process, it normally transforms into a crystalline modification as it cools. However, the crystallization of the gel matrix can be avoided if the a-phase can be kept stable as it cools to room temperature. This can be achieved by combining appropriate surfactants such as myristyl or lauryl alcohol and cholesterol, a mixture of which forms a lamellar LC at room temperature. Because of depression of the melting point, the phase transition temperature of crystalline to liquid crystalline as well as liquid crystalline to isotropic decreases. Therefore, a liquid crystalline microstructure is obtained at room temperature.

The polar character of a surfactant molecule enables the addition of water to form creams (Tiddy and Blackmore, 1980). Depending on whether the surfactant or the surfactant mixture has a strong or weak polar character, oil-in-water or water-in-oil type creams are formed. Water-in-oil type creams are produced from systems that are stabilized solely with weakly polar surfactants such as fatty alcohols, cholesterol, glycerol monostearate, or sorbitan fatty acid esters (Mueller-Goymann, 1984). The surfactant or surfactant mixtures are adsorbed at the interface of the dispersed aqueous phase and the continuous lipophilic phase. Even multiple layers of the surfactant will be adsorbed if the concentration of mesogenic molecules is high enough to form their own liquid crystalline phase. Apart from the reduction of surface tension or surface energy, the liquid crystalline interface also has a mechanically stabilizing effect on the emulsion drops (Mueller-Goymann, 1984).

Surfactants such as sulfated fatty alcohols may be hydrated to a higher extent than fatty alcohols alone, thus stabilizing oil-in-water emulsions. The combination of an anionic and a nonionic surfactant has proven to be particularly effective, because the electrostatic repulsion forces among the ionic surfactant molecules at the interface are reduced by the incorporation of nonionic molecules, thereby improving the emulsion stability. The combination of cetyl/stearyl sulfate (Lanette E) and cetyl/stearyl alcohol (Lanette O) to yield an emulsifying cetyl/stearyl alcohol (Lanette N) is an example of this approach. The polar properties of this surfactant mixture are dominant, so oil-in-water creams are formed. In contrast to water-in-oil systems, the stabilizing effect of the surfactant mixture is not mainly attributable to adsorption at the interface. Instead, the mixed surfactants are highly hydrated and form a lamellar network, which is dispersed throughout the continuous aqueous phase, whereas the dispersed lipophilic components are immobilized within the gel network (Mueller-Goymann, 1984). However, this hydrated gel matrix is not crystalline at room temperature as are the corresponding water-in-oil creams with cetyl/stearyl alcohol but is instead in its *a*-phase, which belongs to the thermotropic smectic LCs and exhibits a strong similarity to lyotropic lamellar LCs.

Analogous gel matrices of liquid crystalline lamellar phases can also be formed with nonionic mesogens, for example, with the combination of cetyl/stearyl alcohol and ethoxylated fatty alcohol, provided the hydrophilic and lipophilic properties of the surfactant molecules are more or less balanced to favor the formation of lamellar structures.

13.11.4 LIPOSOME DISPERSION

Although liposomes have been extensively studied since 1970, only a few commercial drug formulations contain liposomes as drug carriers. The first commercial drug formulation with liposomes for topical administration was registered in Italy. The antimycotic econazol was encapsulated in liposomes being dispersed in a hydrogel (Ecosom Liposomengel, formerly Pevaryl Lipogel). Because of the formation of a highly hydrated gel network of the hydrophilic polymers, liposomes are immobilized within the gel network and thus mechanically stabilized (Larsson, 1972). This stabilization via gelation of the continuous aqueous phase can also be applied to other dispersion systems (e.g., suspensions or emulsions). An example of such an emulsion/hydrogel combination that contains heparin sodium as an active ingredient and liposomes as the additional dispersed phase (the latter only since 1995) is Hepaplus Liposom. A formulation with an analogous emulsion/hydrogel combination but without additional liposomes is Voltaren Emulgel. A transmission electron micrograph reveals adsorption of lamellar LCs at the interface of dispersed oil drops and the aqueous continuous phase. The aqueous continuous phase is again a hydrogel based on polyacrylate in which the lipophilic phase is immobilized. The interface consists of multilamellar layers comprising both surfactant and drug molecules. Thus, the hydrogel is not only stabilized by the hydrogel network itself but also by the liquid crystalline interface (Larsson, 1972). The active ingredient diclofenac diethylamine diffuses slowly from the dispersed phase via the multilamellar interface into the continuous phase, from where it penetrates into the epidermis.

Similar to Voltaren Emulgel, oily droplets of a eutectic mixture of lidocaine and prilocaine are dispersed in a hydrogel to provide local anesthesia of the skin for injections and surgical treatment (Emla cream). A further possibility is the dermal administration of a liposome dispersion as a spray. After administration, water and isopropyl alcohol partially evaporate and result in an increase of concentration and thereby in a transition from the initial liposome dispersion to a lamellar LC. The therapeutic effect thus appears to be influenced favorably by the presence of lecithins alone, rather than by the degree of dispersion of liposomes.

Liposome dispersions for parenteral administration depend on their size and surface charge; parenterally administered liposomes interact with the reticuloendothelial system and provoke an immunological response (Larsson, 1972). After being marked by the adsorption of certain serum proteins, called opsonins, they are identified as an invader and destroyed by specific immune cells mainly in the liver, spleen, and bone marrow.

This passive drug targeting enables an efficient therapy of diseases of these organs or their affected cells. Clinical tests in the therapy of parasitic diseases of the liver and spleen have proved most efficient by having encapsulated the drug substance in liposomes. Apart from passive drug targeting, drug encapsulation within liposomes offers a modification of the therapeutic effect in terms of intensity and duration, together with a minimization of undesired side effects. For this purpose, liposomes have to circulate for as long as possible in the vascular system and remain unrecognized by phagocytic cells.

The antimycotic amphotericine is encapsulated in liposomes and marketed as AmBisome to treat severe systemic mycosis. The liposomal encapsulation reduces the toxicity of amphotericine while increasing the half-life of the drug and plasma level peaks. For stability reasons, the parenteral formulation is a lyophilized powder, which has to be reconstituted by adding the solvent just before administration.

The cytostatic daunorubine, which is administered in the later stage of Kaposi's sarcoma of acquired immunodeficiency syndrome patients, is encapsulated in liposomes that are about 45 nm in size. The liposome dispersion is marketed as a sterile, pyrogen-free concentrate (DaunoXome) and has to be diluted with a 5% glucose solution just before being administered as an infusion. Although daunorubicine by itself is cardiotoxic, the liposomal formulation attacks cardiac tissue only insignificantly but strongly affects the tumor cells by being taken up preferentially. It is postulated that small unilamellar vesicles may pass through endothelial gaps in recently formed capillaries of the tumor, thereby entering the tumor tissue. At this site, the drug is released from the liposomal carrier and inhibits the proliferation of the tumor cells.

13.11.5 Liposome Dispersions for an Installation into the Lung

A liposomal formulation consisting of a surfactant, which usually coats the mucosa of the bronchi and prevents a collapse of the alveolar vesicles of the lung, has been developed for patients who suffer either from infant respiratory distress syndrome (IRDS) or adult/acquired respiratory distress syndrome (ARDS). IRDS often affects premature babies who have not yet developed a functional lung surfactant and therefore develop a failure in pulmonary gas exchange. ARDS is also a life-threatening failure or loss of lung function, which is usually acquired by illness or accident. Clinical trials with liposomal surfactants have proven to be efficient in prophylactic treatment of IRDS and ARDS.

The surfactant is obtained by extraction from the lungs of cattle, which is washed and centrifuged several times. This raw extract is treated with appropriate organic solvents, sterilized by filtration, dried by solvent evaporation under aseptic conditions, resuspended in water, and finally homogenized in a French press under cooling. Care has to be taken to maintain sterility of the extract during all procedures. Special attention has to be paid to transmissible spongiform encephalopathies. The whole manufacturing process has been validated in terms of a decrease of infectious material by a factor of 1021, although a factor of 108 is sufficient. The result is a formulation (Alvefact) that is considered safe in the context of transmissible spongiform encephalopathies and viruses, and it also contains all relevant components of a lung surfactant in terms of pulmonary exchange of gas.

13.11.6 Transdermal Patch

To obtain a systemic effect via percutaneous penetration of a drug compound, high permeability through the stratum corneum and the living tissue beneath it is required as well as high potency of the drug for a low dose to be administered. For an additional short biological half-life, the development of a controlled release transdermal system is a good choice.

Transdermal patches are high-tech devices, which contain the drug substance in a reservoir from which the drug is released in a controlled manner (i.e., zero-order kinetics). The control element is either a membrane or a matrix. Membrane-controlled patches were the first to be marketed. A major disadvantage of these is the so-called dose dumping that occurs in membrane damage during handling. To ensure the desired drug control, even liquid crystalline polymers have been examined with regard to their usefulness in membrane-controlled transdermal patches (Tiemessen, 1989). The matrix-controlled transdermal patch consists of only one functional element, a porous polymer matrix, which not only controls drug release but also simultaneously acts as a drug reservoir and adhesive element.

Transdermal patches are marketed worldwide with the drug substances glycerole trinitrate, estradiol, testosterone, clonidine, scopolamine, fentanyl, and nicotine. The patch has to remain for up to one week at the appropriate body site. In this case the drug amount in the reservoir is rather high. Because liquid crystalline vehicles with lamellar microstructure have high solubilization capacities, they are recommended as reservoirs for transdermal patches (Tiemessen, 1989), although the high surfactant concentration of the lamellar LC might have an irritating effect on the skin. Especially in terms of the membrane-controlled patch, the liquid crystalline vehicle is not in direct contact with the skin and thus will not exhibit an irritating effect on the skin.

13.11.7 Sustained Release from Solid, Semisolid, and Liquid Formulations

The therapy of a chronic disease requires repeated dosing of a drug. Drugs with a short biological half-life have to be administered up to several times daily within short intervals. Sustained formulations have been developed to reduce the application frequency. Liquid crystalline excipients are appropriate candidates for this because in a liquid crystalline vehicle the drug diffusion is reduced by a factor of 10–1000 in comparison with a liquid vehicle such as a solution. The factor depends on the kind of LC being employed.

13.11.7.1 Solid Formulations

Solid formulations for sustained drug release may contain mesogenic polymers as excipients. The mesogenic polymers form a matrix, which is usually compressed into tablets. Some of the most frequently used excipients for sustained release matrices include cellulose derivatives, which behave like lyotropic LCs when they are gradually dissolved in aqueous media. Cellulose derivatives such as hydroxypropyl cellulose or hydroxypropylmethyl cellulose form gel-like lyotropic mesophases in contact with water, through which diffusion takes place relatively slowly. Increasing the dilution of the mesophase with water transforms the mesophase to a highly viscous slime and then to a colloidal polymer solution.

13.11.7.2 Semisolid Formulations

The solubilization of a drug substance in monophasic liquid crystalline vehicles results in semisolid formulations, which are preferably used for topical application (Mueller-Goymann, 1984; see Sections 13.11.2 and 13.11.6).

13.11.7.3 Liquid Formulations

Sustained release from disperse systems such as emulsions and suspensions can be achieved by the adsorption of appropriate mesogenic molecules at the interface. The drug substance that either forms the inner phase or is included in the dispersed phase cannot pass the LCs at the interface easily and thus diffuses slowly into the continuous phase and from there further into the organism via the site of application. Such a sustained drug release is especially pronounced in the case of multi-lamellar LCs at the interface.

A further possibility is the formation of LCs on contact with body fluids at the site of application. The applied drug solution interacts with body fluids such as plasma, tears, fluids, or skin lipids and undergoes a phase transition to a mono- or multiphasic system of LCs. For example, oily solutions of reverse micellar solutions of phospholipids, which solubilize any additional drug, transform into liquid crystalline lamellar phases by the absorption of water when applied to the mucosa. Drug release is controlled by the LCs because diffusion within the liquid crystalline phase is slowest and thus rate controlling. This principle can be used for ophthalmological administration as well as for nasal, buccal, rectal, vaginal, or even parenteral subcutaneous application. However, the peroral administration of such reverse micellar solutions either directly or encapsulated within soft gelatin capsules is not recommended because the sustained release effect is limited by interindividual variations in digestion, such as the amount and composition of the gastric fluid as well as its ability to emulsify and solubilize in terms of enteral absorption.

The chemotherapeutic metronidazole has proven to be effective for the treatment of paradontitis of infected gum pockets. The crystalline prodrug metronidazole benzoate, which has to provide the active metronidazole through dissolution and hydrolysis, is suspended in an oleogel (Elyzol Dentalgel). The oleogel consists of glycerol monooleate and sesame oil, which are immobilized within the matrix structure of the surfactant. The base melts at body temperature and spreads evenly over the inner surface of the gum pockets. The molten system absorbs water and transforms into a reverse hexagonal phase. This liquid crystalline structure has high viscosity. The resulting system adheres well to the surface of the mucosa and releases the active ingredient slowly.

13.11.8 LIQUID CRYSTALS IN COSMETICS

LCs are mainly used for decorative purposes in cosmetics. Cholesteric LCs are particularly suitable because of their iridescent color effects and find applications as coloring for nail varnishes, eye shadows, and lipsticks. The structure of these thermotropic LCs changes with body temperature, resulting in the required color effect. In recent times, such thermotropic cholesteric LCs have also been included in body care cosmetics, where they are dispersed in a hydrogel. Depending on whether this dispersion in the hydrogel involves stirring or a special spraying process, the iridescent liquid crystalline particles are distributed throughout the gel or concentrated locally to give the formulation the required appearance. Tests of the cosmetic efficiency of the liquid crystalline constituents have not yet been published.

13.12 CONCLUDING REMARKS

LCs offer tremendous opportunities in drug delivery systems. Because of their highly uniform, porous nanoscale structures, LC phases and LC-based materials have been proposed for use in a number of materials applications. However, LC materials with functional properties and demonstrated applications of LC systems have been realized only during the last two decades. This work provides an overview of functional LC materials and the areas of application where they have made an impact. As new functional properties and capabilities are realized in LC materials, it is almost certain that they will play more prevalent roles in nanoscience and nanotechnology in the near future.

ABBREVIATIONS

ARDS   Adult/acquired respiratory distress syndrome
DNCG   Disodium cromoglicinate
IRDS   Infant respiratory distress syndrome
LC   Liquid crystal
NSAID   Nonsteroidal antiinflammatory drug

PCS   Photon correlation spectroscopy
SAXD   Small-angle x-ray diffraction
TEM   Transmission electron microscopy

SYMBOLS

d Interlayer spacing
W Angle under which interference occurs
Z Coefficient of dynamic viscosity

REFERENCES

Angelova, A., Angelov, B., Papahadjopoulos-Sternberg, B., Ollivon, M., Bourgaux, C., Proteocubosomes: nanoporous vehicles with tertiary organized fluid interfaces. *Langmuir*, 21, 38–43, 2005.

Blumstein, A. (Ed.), *Liquid Crystalline Order in Polymers*. New York: Academic Press, 1978.

Brown, G. H. and Wolker, J. J., *Liquid Crystals and Biological Structures*, pp. 78, 194–201. New York: Academic Press, 1979.

Buchheim, W., Larsson, K., Cubic lipid-protein-water phases. *Journal of Colloid and Interface Science*, 117, 582–586, 1987.

Burylov, S. V. and Raikher, Y. L., Orientation of a solid particle embedded in a monodomain nematic liquid crystal. *Physical Review*, E-50, 358–367, 1994.

Borné, J., Effect of lipase on monoolein-based cubic phase dispersion (cubosomes) and vesicles. *Journal of Physical Chemistry B*, 106, 492–500, 2002.

Borné, J., Nylander, T., Khan, A., Effect of lipase on different lipid liquid crystalline phases formed by oleic acid based acylglycerols in aqueous systems. *Langmuir*, 18, 72–81, 2002.

Boyd, B. J., Characterisation of drug release from cubosomes using the pressure ultrafiltration method. *International Journal of Pharmaceutics*, 260, 239–247, 2003.

Douglas, L. G., Pecinovsky, C. S., Bara, J. E., and Kerr, R. L., Functional lyotropic liquid crystal materials. *Structure and Bonding*, 128, 181–222, 2007.

Figueiredo Neto, A. M., Levelut, A. M., Liébert, L., and Galerne, Y., Molecular crystallinity. 119, 191–120, 1985.

Figueiredo Neto, A. M. and Salinas, S. R. A., *The Physics of Lyotropic Liquid Crystals*, p. 17, 18. New York: Oxford University Press, 2005.

Fontell, K., Gray, G. W. and Winsor, P. A. X-ray diffraction by liquid crystals–amphiphilic systems. *Liquid Crystals and Plastic Crystals- Physicochemical Properties and Methods of Investigation*, 2, 81–109, 1974.

Gennes, P. G. de and Prost, J. *The Physics of Liquid Crystals*, pp. 25–26, 175–179. Oxford, U.K.: Clarendon Press, 1993.

Hoath, S. B., Norlen, L., Cubic phases and human skin: theory and practice, *Journal of surfactant science series*, 127, 41–58, 2005.

Imperor-Clerc, M., Veber, M., Levelut, A. M., Phase transition between single crystals of two thermotropic cubic. *Chem Phys Chem*, 2, 533–535, 2001.

Jákli, A., Cao, W., Huang, Y., Lee, C. K., and Chien, L. C., Liquid crystal behavior. 28, 289–301, 2001.

Jákli, A. and Saupe, A. *One- and Two-Dimensional Fluids: Physical Properties of Smectic, Lamellar and Columnar Liquid Crystals*, pp. 256–270. Boca Raton, FL: CRC Press/Taylor & Francis, 2006.

Kelker, H. and Hatz, R. *Handbook of Liquid Crystals*, pp. 121–134, 325. Weinheim, Germany: Chemie Verlag, 1980.

Khoo, I. C., *Liquid Crystals: Physical Properties and Nonlinear Optical Phenomena*. New York: Wiley–Interscience, 1987.

Kresge, C. T., Leonowicz, M. E., Roth, W. J., Vartuli, J. C., Beck, J. S., Ordered mesoporous molecular sieves synthesized by a liquid-crystal template mechanism. *Nature*, 359, 710–712, 1992.

Larsson, K., Structure of isotropic phases in lipid–water systems. *Chemistry and Physics of Lipids*, 53, 237–243, 1972.

Lehmann, O. U., Über fließende kristalle. *Zeitschrift für Physikalische Chemie*, 4, 462–471, 1889.

Lynch, M. L., Kochvar, K. A., Burns, J. L., Laughlin, R. G., Aqueous-phase behavior and cubic phase-containing emulsions in the C12E2-water system. *Langmuir*, 16, 537-542, 2000.

Marcondes Helene, M. E. and Figueiredo Neto, A. M, Topology of a Surface of the Phase Diagram of the Cholesteric Lyotropic Mesophase: Potassium Laurate/1-Decanol/Water/I-N-Lauroyl Potassium Alaninate, *Molecular crystal and liquid crystal*, 162, 127–138, 1988.

Mlodozeniec, A. R., Thermodynamics and physical properties of a lyotropic mesophase (liquid crystal) and micellar solution of an ionic amphiphile. *Journal of the American Chemical Society*, 29, 659–683, 1978.

Mueller-Goymann, C. C., Liquid crystals in emulsions, creams and gels, containing ethoxylated sterols as surfactant. *Pharmaceutical Research,* 1(4), 154–158, 1984.

Mueller-Goymann, C. C. and Hamann, H. J., Sustained release from reverse micellar solutions by phase transformation into lamellar liquid crystals. *Journal of Controlled Release*, 23, 165–174, 1993.

Papell, S. S., U.S. Patent 3215572, October 9, 1963.

Priestley, E. B., Wojtowicz, P. I., and Sheng, P., *Introduction to Liquid Crystals*. New York: Plenum Press, 1979.

Reinitzer, F., Beiträge zur kenntnis des cholesterins. *Monatshefte für Chemie*, 9, 412–441, 1888.

Rieker, T. and Marcondes Helene, M. E., Liquid crystal. *Journal of Physical Chemistry*, 61, 143–144, 1987.

Rodriguez, C., Formation and properties of reverse micellar cubic liquid crystals and derived emulsion, *Langmuir*, 23, 11007–11014, 2007.

Roux, D. and Nallet, F., Rheology of lyotropic lamellar phases. *Europhysics Letters*, 15, 53–58, 1993.

Tiddy, G. J. T. and Blackmore, E. S., Surfactant–water liquid crystal phases. *Physics Reports*, 96, 389, 1980.

Tiemessen, H. L. G. M., Non-ionic surfactant systems for transdermal drug delivery. PhD thesis. Leiden University, 1989.

Usoltseva, N., Praefcke, K., Smirnova, A., and Blunk, D., Fundamentals of liquid crystals. 26, 223, 1999.

Wahlgren, S., Lindstrom, A. L., and Friberg, S., Liquid crystals as a potential ointment vehicle. *Journal of Pharmaceutical Science*, 73(10), 1484–1486, 1984.

Yang, D., O'Brien, D. F., Marder, S. R., Polymerized bicontinuous cubic nanoparticles (cubosomes) from a reactive monoacylglycerol. *Journal of American Chemical Society*, 124, 388–389, 2002.

14 Liquid Crystalline Nanoparticles as Drug Nanocarriers

Anan Yaghmur and Michael Rappolt

CONTENTS

14.1 INTRODUCTION

In living cells, the lamellar–nonlamellar transitions are fundamentally important in modulating the physicochemical properties of biomembranes and consequently in regulating different biological processes (Landh, 1995; Luzzati, 1997; Patton and Carey, 1979; Simidjiev et al., 2000). In this respect, it is important to understand how lipid composition and protein content modulate the membrane structure and its vital function. Stimulated by biomembranes' excellent example of the possible formation of fascinating lipidic domains with internal self-assembled structures as a result of membrane fusion (Ortiz et al., 1999; Szule et al., 2003), fat digestion (Patton and Carey, 1979), or under different pathophysiological conditions (Almsherqi et al., 2006; Deng et al., 2002), much effort has been devoted in the last two decades to forming various nanostructured aqueous dispersions of model surfactant-like lipids that mimick biological systems (Larsson, 1989, 2000; Yaghmur and Glatter, 2008; Yang et al., 2004). For instance, we recently investigated the aqueous dispersion of a model monoelaidin (ME)/water system that is a good example for lamellar–nonlamellar transitions analogous to certain steps in membrane fusion (Yaghmur et al., 2008a). Our experiments demonstrated the direct transition from vesicles to particles with an internal bicontinuous cubic phase (cubosomes) by heating these aqueous dispersions.

As a good example of self-assembled nanostructures formed in biomembranes, Almsherqi et al.'s (2006) recent review pointed out that cubic assemblies form in various cells as a result of cell stress, starvation, or lipid and protein alterations. It is also intriguing that these cubic biomembranes can be utilized for loading significantly short deoxyribonucleic acid segments and thus may offer new systems for gene transfection (Almsherqi et al., 2008).

Nanostructured aqueous dispersions of model lipid systems with internal hierarchical self-assemblies are receiving much attention because of their potential applications in various areas, including the formulation of functional food and drug nanocarriers (Drummond and Fong, 1999; Larsson, 2008; Spicer, 2005; Yaghmur and Glatter, 2008). Cubosomes (aqueous dispersions of inverted-type bicontinuous cubic phases, V_2; Barauskas et al., 2005; de Campo et al., 2004; Gustafsson et al., 1997; Larsson, 2000; Spicer et al., 2001), hexosomes (aqueous dispersions of inverted-type hexagonal phase, H_2; Dong et al., 2006; Yaghmur et al., 2005), micellar cubosomes (aqueous dispersions of inverted-type discontinuous cubic phase, I_2; Nakano et al., 2002; Yaghmur et al., 2006b), emulsified microemulsions (EMEs; Yaghmur et al., 2005), and dispersed sponge phases (L_3; Barauskas et al., 2005, 2006) are good examples of such dispersions in which the dispersed particles with a size of a few hundred nanometers embed distinctive internal well-ordered nanostructures. Under specific conditions, the fully hydrated inverted-type bulk phases of liquid crystalline systems and microemulsions formed through the self-assembly of lipids are internally confined in the kinetically dispersed particles as high-energy input is applied in the presence of a suitable stabilizer (Gustafsson et al., 1997; Larsson, 1989). In general, a hydrophilic polymeric stabilizer is used to efficiently cover the outer surface of the dispersed particles and to retain the structures of these self-assembled systems.

Since the pioneering studies of Larsson and coworkers (Gustafsson et al., 1996, 1997; Larsson, 1989) on the formation and characterization of monoolein (MO)-based cubosomes and hexosomes, various nanostructured particles based on other surfactant-like lipids were described (Abraham et al., 2005; Boyd et al., 2006; Dong et al., 2006; Johnsson et al., 2005a; Yaghmur and Glatter, 2008). In addition, different investigations were carried out on characterizing various nanostructured aqueous dispersions as attractive media for solubilizing proteins (Angelova et al., 2005a, 2005b, 2008) and as drug nanocarriers (Boyd et al., 2006; Johnsson et al., 2006; Lopes et al., 2006; Malmsten, 2006; Yaghmur and Glatter, 2008).

Figure 14.1 gives an overview of different lipid drug carrier systems. These *soft* lipidic drug nanoparticles have attracted broad research interest for the formulation of safe and efficient drug delivery (Boyd, 2008; Couvreur and Vauthier, 2006; Malmsten, 2006). In particular, it is an ambitious goal for many research groups to enhance the solubilization of poorly water-soluble drugs such

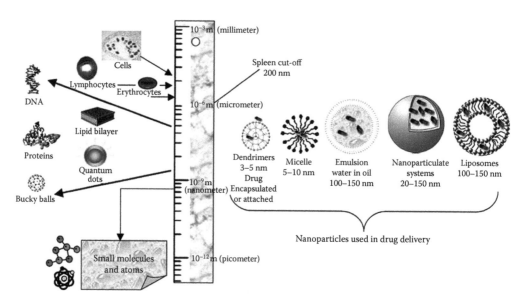

FIGURE 14.1 Lipid-based drug nanocarriers and their nanoscale dimensional comparison with a biological cell, deoxyribonucleic acid, lipid bilayers, and atoms. (From Arruebo M., et al., 2007. *Nano Today* 2: 22–32. With permission.)

as anticancer and antifungal bioactive materials (Couvreur and Vauthier, 2006). This is also stimulated by the urgent needs for curing serious life-threatening diseases without side effects and thus saving the lives of many patients. In this respect, lipidic nanoparticulate drug carriers based on cubosomes, hexosomes, and micellar cubosomes could offer a promising strategy for forming such efficient delivery systems (Boyd et al., 2006; Chung et al., 2002; Johnsson et al., 2006; Malmsten, 2006; Yaghmur and Glatter, 2008).

This chapter presents an overview of the role of the packing frustration and the curvature elasticity in the formation of self-assembled lamellar and nonlamellar phases. Recent studies are described that deal with the formation of different nanoparticulate carriers based on cubosomes, hexosomes, and micellar cubosomes for enhancing the solubilization capacity of active guest molecules. We summarize investigations carried out on the structural transitions within this colloidal family of dispersions after loading them with guest biomolecules. In particular, we emphasize the attractiveness of these nanocarriers for enhancing the solubilization of poorly water-soluble drugs in an attempt to increase their bioavailability and to prevent unwanted side effects.

14.2 NONLAMELLAR LIQUID CRYSTALLINE PHASES

Rational drug design based on nonlamellar phases is still a relatively young research field. Although great progress has been made in the formulation of various nanoparticulate carriers, which we will outline in the following sections, some fundamental obstacles have to be considered: *biocompatibility* and *stability* are in many cases far from being solved, and *drug targeting* and *release on-demand* are often not more than wishful thinking (Boyd, 2003, 2004). Hence, various model membranes are still the main workhorses in this field. It is clear that learning from nature, especially in drug delivery research, is a constant source of inspiration, and further prominent examples will underline the importance of bionic approaches. The biological significance of planar fluid bilayers in nature is unquestioned, and therefore finding the right model of biological membranes has been a considerable target for scientists since the end of the 19th century. It took almost 100 years of research from the first idea that cells and cell compartments are enclosed by a membrane (Pfeffer, 1877), to the first claim of a bilayer structure (Gorter and Grendel, 1925), followed by the sandwich model predicting the location of proteins at the outer and inner interface of the lipid bilayer (Danielli and Davson, 1935), to the formulation of the now legendary fluid mosaic model of Singer and Nicholson (1972). Today the research on membrane microdomains, so-called lipid rafts, has regained scientific interest (Simons and Toomre, 2000; Simons and van Meer, 1988). In addition, there is growing awareness that nonlamellar biomembranes play an important role in cell life. Although, broadly speaking, the planar plasma membrane helps to the keep the status quo of the cell intact, nonlamellar membrane formations such as in bilayer nanotubules (Kralj-Iglic et al., 2002), narrow necks of phospholipid bilayers connecting buds to the mother membrane (Kralj-Iglic et al., 2006), and fusion intermediates such as pores (Hui et al., 1981) definitely play an important role in dynamic processes like vesicle fusion (e.g., in endo- and exocytosis) and cell communication in a broader sense. Fortunately, improved imaging techniques like two-photon laser scanning fluorescence microscopy (Denk et al., 1990) have made it possible to visualize living matter in even greater detail. It was recently demonstrated how material is transported from one cell to another via microvesicles that travel like gondolas along tubular connections (Veranic et al., 2008; Figure 14.2a through c). However, locally curved membranes have their function in cell life, and larger domains of stable nonlamellar phases have been identified. For instance, domains of H_2 phases have been found in paracrystalline inclusions of the retina (Corless and Costello, 1981) and the existence of saddle-like curved membrane phases have been proved by a lot of examples (for a review, see Almsherqi et al., 2006). Remarkably, later curved membrane morphologies have been published in numerous transmission electron microscopy (TEM) studies without understanding the three-dimensional (3D) structures. For instance, they have been described as "paracrystalline membranes," "lattice organelles," and "undulating membranes," to mention a

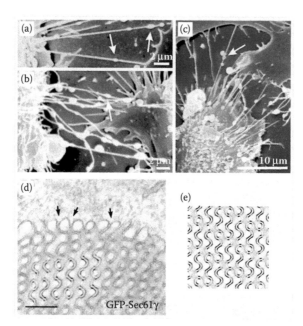

FIGURE 14.2 Examples of nonlamellar membranes in nature. (a–c) Membrane nanotubes with gondolas (white arrows) observed between cells in the human urothelial cell lines (a) RT4 and (b and c) T24 by SEM under physiological conditions. Note that the gondolas are an integral part of the tubes. (From Veranic, P., et al., 2008. *Biophysical Journal* 95: 4416–4425. With permission.) (d and e) Green flurorescent proteins induce the formation of distinct forms of smooth endoplasmic reticulum. (d) A bicontinuous cubic phase in contact with the cytoplasm. Note that the 8-nm electron-dense space between membranes is continuous with the cytoplasm (arrows). Scale bar = 160 nm. (From Snapp, E. L., et al., 2003. *Journal of Cell Biology* 163: 257–269. With permission.) (e) The original TEM micrograph (d) matches perfectly to the theoretical double diamond projection. (From Almsherqi, Z. A., S. D. Kohlwein, and Y. Deng, 2006. *Journal of Cell Biology* 173: 839–844. With permission.)

few. However, since the ground-breaking analysis of electron micrographs by Landh (1995), these phases have been unequivocally identified as nonlamellar phases with bicontinuous cubic symmetry (Figure 14.2d and e). Today it is well accepted that the curved bilayers of the V_2 lipid/water phases are draped around periodic minimal surfaces (Figure 14.3, second from top), creating two distinct continuous systems of water channels (Hyde et al., 1997).

In nature, continuous membranes sometimes divide the space into more than two physically distinct subspaces. The lattice dimensions of cubic biomembranes are usually greater than 1000 Å, whereas synthetic V_2 phases have lattice constants of about 100–200 Å (e.g., see Mariani et al., 1988). Although the biological function of cubic phases is mostly still unknown, cubic biomembranes surely add a new dimension to cell life through their extraordinary efficiency as subcellular space organizers (Hyde et al., 1997).

Next to directly investigate biological membranes, it is indispensable for the understanding of the formation and structural stability of curved membranes to also study the phase behavior of model binary surfactant-like lipid–water systems, which have the tendency to form inverted-type nonlamellar phases. It is important to mention that the experimental phase diagrams of these systems may already seem complicated, because the phase behavior depends on a complex interplay of several factors such as temperature, pressure, water concentration, and lipid chain length (Seddon and Templer, 1995). In this respect it is instructive to take a closer look at the membrane–water interface of the different model membranes and especially see how different parameters affect the mean interfacial curvature (H). Note that membrane curvatures are described using the geometrical concepts of principal curvatures c_1 and c_2. The mean curvature is defined as the average of the principal

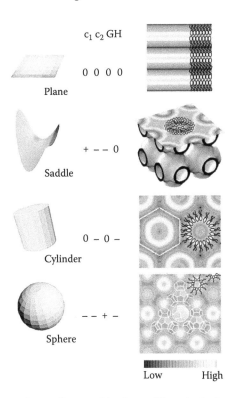

$$c_1 \; c_2 \; G \; H$$

Plane   0   0   0   0

Saddle   +   −   −   0

Cylinder   0   −   0   −

Sphere   −   −   +   −

Low   High

FIGURE 14.3 Basic membrane shapes. Two combinations of the principal curvatures are commonly used to describe membranes: the mean curvature H and the Gaussian curvature G. For biologically relevant membranes, there are four basic membrane shapes: planar ($c_1 = c_2 = 0$: isotropic surface), saddle-like ($c_1 = -c_2 > 0$: anisotropic surface), cylindrical ($c_1 = 0$ and $c_2 < 0$: anisotropic surface), and spherical shapes ($c_1 = c_2 < 0$: isotropic surface). (Right) Examples of electron density maps from different model membranes are shown. High density areas are given in white and lower density regions in darker shades. For ease of interpretation, the lipid molecule models with white head groups and black hydrocarbon tails are superimposed on the electron density maps. (Top to bottom) A fluid L_α, a primitive V_2, an H_2 (with Wigner–Seitz cell, white), and an I_2 phase. Two subsequent layers of micelles are visualized as polyhedrons (Fd3m packing). All structural examples including amplitude and phase listings can be found in Rappolt (2006). (Data from Rappolt, M. 2006. In *Planar Lipid Bilayers and Liposomes*, A. Leitmannova (Ed.), Vol. 5, pp. 253–283. Amsterdam: Elsevier.)

curvatures: $H = (c_1 + c_2)/2$. Another useful parameter is given by the Gaussian curvature G, which is defined as the product of the principal curvatures (see Hyde et al., 1997). A few examples will illustrate how surface curvature is influenced by external parameters. Increasing the temperature acts mainly on the chain region of the lipids and results in greater chain disorder, thus favoring a bending of the lipid head-group surface toward the water. The opposite tendency accounts for an increase of the hydrostatic pressure: the lipid chain packing condenses with augmenting pressure and thus an induction of convex interfaces is favored. The existence of a natural sequence for the appearance of lipid mesophases should be clear, for example, for a hypothetical lipid–water phase diagram in which the transitions are driven by temperature. From the planar fluid lamellar phase (L_α), the V_2 phase may form, followed by the columnar H_2 phase, which finally may convert into an I_2 phase (Figure 14.3, from top to bottom). All of these phases have increasingly negative mean interfacial curvatures, which can be estimated easily (Seddon and Templer, 1993). Obviously $H = 0$ for planar lamellar phases. Definitely more complicated is the geometry of bicontinuous cubic phases, but it can be estimated for this mesophase that $H \sim lc_1c_2$, where c_1 and c_2 are the principal curvatures of the bilayer midplane and l is the monolayer thickness (lipid length). The averaged interface curvature of the inverted hexagonal and micellar phases depends solely on the water core radius (R_W) and is easily found as $-1/(2R_W)$ and $-1/R_W$, respectively. Realistic values of H for phospholipid systems in the bicontinuous cubic, H_2, and inverse micellar phases are -0.015 Å^{-1} (Harper and Gruner, 2000), -0.03 Å^{-1} (Rappolt et al., 2003), and -0.07 Å^{-1} (Luzzati et al., 1992), respectively.

The interfacial mean curvature is tightly connected to the molecular shape of the lipids (Hyde, 2001). Alternatively, the membrane curvature can also be rationalized using the critical packing parameter concept of Isrealachvili (1991). This parameter is also known as the shape parameter of molecules (s) and is defined as the hydrophobic chain volume divided by the lipid chain length and its optimal head-group area. For instance, cylindrical lipid molecules ($s = 1$) will lead to planar membranes and cone-shaped ($s < 1$) and wedge-shaped molecules ($s > 1$) will induce convex and concave interfaces, respectively. In this respect, the typical values of s for the V_2 and H_2 phases are 1.3 and 1.7, respectively (Larsson, 1989). However, s does not supply information on the radial symmetry of the molecules. In contrast, molecular deformations do not necessarily preserve rotational symmetry (Templer, 1998). Note for instance that the principal curvatures of the monolayers in the V_2 as well as in H_2 phases are different (Figure 14.3, second and third row of illustrations from top); that is, their surface curvatures are anisotropic. This has as a consequence that the involved lipid molecules also exhibit anisotropic shapes. Recent theories also consider anisotropic molecular shapes, and especially the stability of strong bent membranes as in necks or tubular protrusions can be simulated satisfactorily (Kralj-Iglic et al., 2002, 2006). Because phase diagrams on nonlamellar phases can be predicted in a better manner and refined molecular shapes result from these simulations (Mares et al., 2008), hopefully future rational design of new cubosomes and hexosomes will profit from this newest bottom-up approach.

14.3 LIPIDIC NANOSTRUCTURED AQUEOUS DISPERSIONS

Liposomes are used extensively as a model phospholipid dispersion that mimicks biomembranes (Ortiz et al., 1999; Siegel and Epand, 1997; Yaghmur et al., 2008a, 2008b) and for drug delivery formulations (Couvreur and Vauthier, 2006). They are formed when the fully hydrated one-dimensional lamellar phase is dispersed in a continuous aqueous medium. In a similar manner, the nanostructured aqueous dispersions confining internal two-dimensional (2D) or 3D inverted-type liquid crystalline phases are formed. Under full hydration conditions, the emulsification of these self-assembled nanostructures is accomplished in the presence of suitable stabilizers (Gustafsson et al., 1996, 1997; Larsson, 2000; Yaghmur and Glatter, 2008). Two prominent examples on these aqueous dispersions are cubosomes and hexosomes. In general, small-angle x-ray scattering (SAXS) and cryo-TEM techniques are excellent tools to characterize the morphology of the dispersed particles and their internal nanostructures (Gustafsson et al., 1996, 1997; Larsson, 2000; Yaghmur and Glatter, 2008). Figure 14.4 provides two typical examples for cryo-TEM images that were taken for cubosomes and hexosomes (de Campo et al., 2004).

In recent years, there has been increasing interest in fully understanding the dynamic behavior of these dispersions under varying temperature conditions (Yaghmur et al., 2005, 2006b, 2008a; Yaghmur and Glatter, 2008) or the addition of different solutes (amphiphilic molecules, hydrophilic or hydrophobic additives, salts, etc.; Abraham et al., 2005; Barauskas 2005; Dong et al., 2006; Yaghmur et al., 2005, 2006a; Yaghmur and Glatter, 2008). The reported nanostructured aqueous dispersions are based on amphiphilic lipids with a tendency to form nonlamellar phases. Among these lipids, the most reported lipid classes in the literature are monoglycerides (de Campo et al., 2004; Dong et al., 2006; Gustafsson et al., 1996, 1997), glycolipids (Abraham et al., 2005), phosphatidylethanolamines (Johnsson et al., 2005a), and urea-based amphiphiles (Gong et al., 2008). Nanostructured dispersions based on mixtures of MO with copolymers bearing blocks of lipid-mimetic anchors (Rangelov and Almgren, 2005) or mixed binary catanionic surfactants (Rosa et al., 2006) were also recently described. In addition to cubosomes and hexosomes, different studies reported on the formation of micellar cubosomes (Nakano

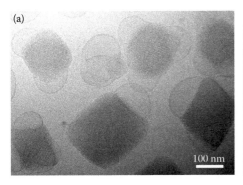 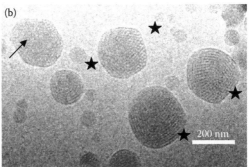

FIGURE 14.4 Cryo-TEM micrographs of (a) cubosomes at 25°C and (b) hexosomes at 55°C. Some of the nanostructured hexosome particles exhibit an internal hexagonal symmetry (arrows); others exhibit both hexagonal symmetry and curved striations (stars). (From de Campo, L., et al., 2004. *Langmuir* 20: 5254–5261. With permission.)

et al., 2002; Yaghmur et al., 2006b), sponge particles (emulsified L_3 phase; Barauskas et al., 2005, 2006), emulsified inverted types of micelles (L_2; de Campo et al., 2004), and swollen micelles (EMEs; Yaghmur et al., 2005; Yaghmur and Glatter, 2008).

The effect of lipid composition (Barauskas et al., 2005; Johnsson et al., 2005b; Kamo et al., 2003; Yaghmur et al., 2006a) and solubilizing guest molecules (Dong et al., 2006; Johnsson et al., 2006; Nakano et al., 2002; Yaghmur et al., 2005; Yaghmur and Glatter, 2008) on the confined nanostructures of monoglycerides-based cubosomes has been investigated intensively in recent years. For example, different studies were carried out on the effect of solubilizing oil on the structure of monolinolein (MLO) dispersed and nondispersed systems (Yaghmur et al., 2005). For the oil-free aqueous dispersions of the MLO–water system (de Campo et al., 2004), by varying the temperature the dispersed particles undergo a reversible transition from cubosomes (internal cubic Pn3m phase) via hexosomes (internal H_2 phase) to an emulsified L_2 phase (internal water/MLO inverse micellar solution) at relatively high temperatures (Figure 14.5a). In these dispersions, the polymeric stabilizer Pluronic polymer poly(ethylene oxide)$_{99}$–poly(propylene oxide)$_{67}$–poly(ethylene oxide)$_{99}$ (F127) was an effective stabilizer covering the outer surface of the dispersed particles without a significant impact on the confined nanostructure. In analogy with temperature, loading these dispersions at ambient temperatures with oil induces (Figure 14.5b) internal structural transitions in the order V_2 (Pn3m) (cubosomes) $\rightarrow$ H_2 (hexosomes) $\rightarrow$ I_2 [discontinuous micellar cubic phase (Fd3m), micellar cubosomes] $\rightarrow$ water-in-oil microemulsion (EMEs; Yaghmur et al., 2005, 2006b). These structural transitions are reversible and identical to those observed in the corresponding nondispersed phases coexisting with an excess of water. This means that there is a reversible exchange of water inside and outside the internally confined nanostructures as the temperature is switched up and down during the heating and cooling cycles. The internal nanostructures of these oil-free and oil-loaded particles can also be modulated by varying the lipid composition (Yaghmur et al., 2006a).

The common tendency of water-deswelling behavior in fully hydrated monoglyceride-based systems and their nanostructured aqueous dispersions was observed in most studies on varying temperatures (de Campo et al., 2004; Misquitta and Caffrey, 2001; Qiu and Caffrey, 2000; Yaghmur et al., 2005, 2007). This is due to dehydration (a reduction in the hydrophilic head-group area) and simultaneous enhancement of the hydrophobic chain volume. The most investigated monoglycerides containing a cis-monounsaturated acyl chain such as MO, MLO, and phytantriol expel water and induce the formation of more negative spontaneous curvatures with increasing temperature (de Campo et al., 2004; Qiu and Caffrey, 2000; Yaghmur et al., 2005). However, we recently observed

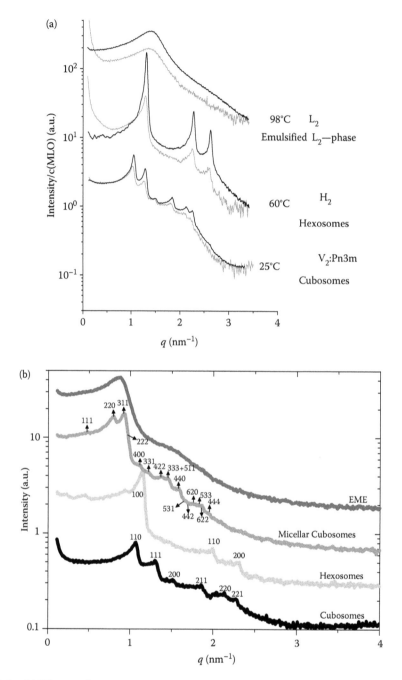

FIGURE 14.5 SAXS scattering curves of different MLO/water systems. (a) SAXS scattering curves for a MLO-based dispersion (gray lines) in comparison with those from a nondispersed bulk phase (black lines) at three different temperatures, and (b) the effect of the amount of solubilized oil on MLO-based emulsified systems at 25°C. In these dispersions, the α ratio [$100 \times$ (mass of oil)/(mass of MLO)] varied from 0 (cubosomes) to 28 (hexosomes), 40 (micellar cubosomes), and 60 (EMEs). (From Yaghmur, A., et al. 2006b. *Langmuir* 22: 517–521. With permission.)

significantly different water behavior in ME (a rodlike monoglyceride containing a trans-monounsaturated acyl chain) dispersed and nondispersed samples (Yaghmur et al., 2008a). It is interesting that the heating-induced structural transition of a fluid lamellar (L_α) to bicontinuous cubic phase of the symmetry Im3m involves a significant enlargement of the nanostructure (Im3m phase) as a result of water uptake (Figure 14.6).

Important recent achievements on cubosomes and hexosomes concern their 3D complex morphology characterization (Boyd et al., 2007; Rizwan et al., 2007). Cryo field emission scanning electron microscopy (cryo-FESEM) was recently used to directly provide further information on the 3D morphology of cubosome (Rizwan et al., 2007) and hexosome particles (Boyd et al., 2007). In the cubosome particles, the cryo-FESEM images reveal "ball-like" cubosome particles enclosing aqueous water channels in good agreement with the proposed mathematical models using a nodal surface representation (Figure 14.7a; Rizwan et al., 2007). The observed 3D morphology of hexosome particles resembles "spinning-top-like" structures (Figure 14.7b; Boyd et al., 2007).

Of particular interest is the possible utilization of cubosomes for confining protein molecules and retaining their native structure and function (Larsson, 2008). Protein-loaded cubosome particles (proteocubosomes; Angelova et al., 2005a, 2005b) were characterized as hierarchical well-ordered nanostructures modifying the native protein-free cubosome particles. They offer promising systems for the formation of protein delivery carriers.

The formation of dispersed nanostructure particles requires the application of high-energy input (ultrasonication, microfluidization, and homogenization; Gustafsson et al., 1997; Larsson, 2000; Yaghmur and Glatter, 2008) or the addition of a suitable hydrotrope (the dilution process as described

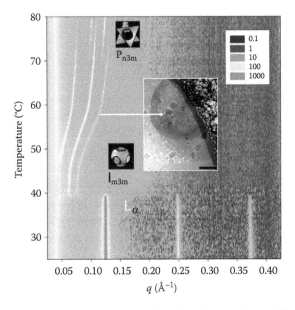

FIGURE 14.6 Contour x-ray diffraction plots covering the phase regime of the L_α, Im3m, and Pn3m phases. In a temperature scan from 25°C to 80°C at a rate of 1°C/min, heating the ME aqueous dispersion stabilized by F127 induced a direct transformation of liposomes to cubosomes: an internal L_α via Im3m to Pn3m structural transition. The inset shows a stabilized ME-based particle with the inner cubic structure highlighted in dark cyan. The inner nanotubular network is clearly visible. Judging from the x-ray data, the inner lattice should be Im3m symmetry. The overall shape however is not well defined compared to other cubosomes known from the literature (de Campo et al., 2004; Gustafsson et al., 1997; Larsson, 2000). The color scale of the diffraction intensity is also mentioned. (From Yaghmur, A., et al., 2008a. *PLoS ONE* 3: e3747. With permission.)

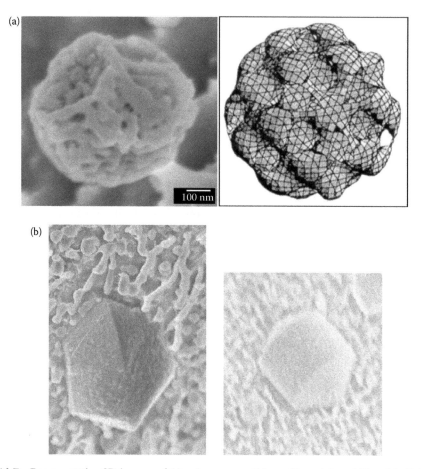

FIGURE 14.7 Representative 3D images of (a) cubosome particles with an internal V_2 cubic Pn3m phase (Rizwan et al., 2007) and (b) oil-loaded hexosome particles (Boyd et al., 2007). The imaging was obtained using cryo-FESEM. (Reproduced from Rizwan, S. B., et al., 2007. *Micron* 38: 478–485; Boyd, B. J., et al., 2007. *Langmuir* 23: 12461–12464. With permission.)

by Spicer et al., 2001) to induce the spontaneous formation of the dispersions. The latter technique avoids dispersing at high temperatures, which is extremely important for temperature-sensitive materials. Salentinig et al.'s (2008) recent work introduced a new and fast emulsification technique for the formation of relatively highly concentrated aqueous dispersions with a controllable size of dispersed particles. These concentrated dispersions were formed by using an approach similar to that proposed by Bibette's method, which is well known for the preparation of monodispersed emulsions. The ability to form nanostructured aqueous dispersions with a uniform dispersed particle size is a desirable goal for many pharmaceutical applications.

14.4 ADSORPTION OF LIQUID CRYSTALLINE NANOPARTICLES ON SURFACES

The application of nanostructured aqueous dispersions for various purposes in the pharmaceutical industry is undeniably linked to the adsorption mechanism of the dispersed particles to the biological surfaces (Svensson et al., 2008; Vandoolaeghe et al., 2006, 2008). For instance, the utilization of these nanostructured aqueous dispersions as mucosal drug nanocarriers requires the understanding of their mucoadhesive properties. Svensson et al. (2008) recently studied the interaction of MO cubosomes stabilized in the presence of the polymeric stabilizer F127 with mucin. They found that cubosome particles behave similarly to materials bearing poly(ethylene oxide) chains. They are weakly mucoadhesive and the adsorption of mucin is pH dependent. As observed for poly(ethylene oxide)-coated polymeric particles, these results suggest the possible enhancement of the transport of active molecules loaded to cubosomes through the mucous gel layer.

In general, the interaction between dispersed cubosome particles and hydrophilic or hydrophobic surfaces can be systematically affected by varying the surface properties or parameters that also influence the stability of the dispersions such as changing the pH value or ionic strength (Vandoolaeghe et al., 2006).

Vandoolaeghe et al. (2008) recently carried out *in situ* experiments by using different techniques to monitor the interfacial behavior of cubosome particles that interact strongly with supported phospholipid bilayer membranes. Important aspects are outlined concerning lipid exchange, the ability to trigger the release, and further interesting features of the dynamic process that occur when cubosomes adsorb to the lipid bilayer. In brief, it was clearly shown that the strong adsorption of the cubosome particles onto the lipid bilayer involves the interfacial exchange of materials between the dispersed particles and the supported lipid bilayer: the cubosome particles expel MO to the bilayer, whereas the investigated phospholipid (dioleoylphosphatidlycholine) moves from the bilayer to the particles. In addition, the obtained results allow the interpretation that the trigger release of the cubosome particles depends on the concentration of these particles in the dispersion. The release of these particles can be triggered by adjusting this concentration to a certain threshold value.

14.5 NANOSTRUCTURED AQUEOUS DISPERSIONS AS DRUG NANOCARRIERS

The feasibility of using nanostructured aqueous dispersions as drug nanocarriers for different treatment purposes has been investigated in recent years by different research groups (Boyd et al., 2006; Chung et al., 2002; Esposito et al., 2005; Kuntsche et al., 2008; Lopes et al., 2006; Swarnakar et al., 2007; Yaghmur and Glatter, 2008). The main motivation for such studies arises from the potential of these aqueous dispersions to solubilize drugs that are poorly soluble in both aqueous and organic media and to form efficient delivery systems. Therefore, it is assumed that this colloidal family could possess superior efficacy in delivering solubilized bioactive materials than conventional emulsions and liposomal formulations. They are attractive for different administration routes commonly used in the pharmaceutical industry such as oral, injectable, and topical applications. As proposed in different studies, the advantages of utilizing dispersions such as cubosomes, hexosomes, or micellar cubosomes based on lipids are the following: solubilization of poorly water-soluble drugs could improve bioavailability, formation of more efficient delivery systems, decrease of unwanted side effects, improvement of intracellular penetration, protection against degradation, and possible control of drug release (Boyd et al., 2006; Johnsson et al., 2006; Lopes et al., 2006; Malmsten, 2006; Yaghmur and Glatter, 2008). To date, the number of studies on these drug-loaded nanoparticulate dispersions is still very limited despite their potential in drug related applications. There is need for further investigations to fully understand their behavior under different conditions, to study the interaction of the lipids with new bioactive materials such as various designer peptides, and to address different basic challenges such as stability and drug targeting. The following are a few representative examples of these nanoparticulate systems for drug delivery.

1. *Oral insulin formulation based on cubosomes.*   Chung et al. (2002) recently proposed a new oral lipid cubosomal formulation for the delivery of insulin. By carrying out *in vitro* and *in vivo* studies, this formulation induced a significant reduction in the blood glucose concentration that was maintained at a low level for at least 6 h. Thus, nanocarriers offer a promising approach for the delivery of insulin.

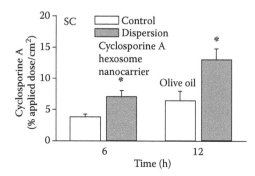

FIGURE 14.8 *In vitro* penetration of the peptide cyclosporin A in the stratum corneum at 6 and 12 h following the topical application of both cyclosporin A loaded hexosomes and control sample (cyclosporin A dissolved in olive oil; Lopes et al., 2006). (Reproduced from Lopes, L. B., et al., 2006. *Pharmaceutical Research* 23: 1332–1342. With permission.)

   2. *Cubosomes and hexosomes as topical delivery system.*   Cubosomes (Esposito et al., 2005) and hexosomes (Lopes et al., 2006) were proposed as alternative approaches for the topical delivery of drugs. Indomethacin and cyclosporin A were chosen as model drugs in two different studies. Indomethacin is an antiinflammatory drug with potential side effects when administrated orally, whereas cyclosporin A is a peptide with poor skin penetration. The indomethacin-loaded cubosomes were suitable nanocarriers for efficiently prolonging antiinflammatory activity and for controlling drug release (Esposito et al., 2005). Cyclosporin A loaded hexosomes demonstrated that this drug nanocarrier significantly enhances the penetration of the peptide through the skin, as illustrated in Figure 14.8. Hexosomes seem to be a promising media for improving the topical delivery of the peptide without causing skin irritation.
   3. *Hexosomes and micellar cubosomes as injectable nanocarriers.*   Hexosomal and micellar cubosomal formulations were also investigated as injectable nanocarriers for the delivery of irinotecan, an anticancer drug (Boyd et al., 2006), and propofol, an anesthetic agent (Johnsson et al., 2006), respectively. When irinotecan is accommodated into the hexosome particles (Boyd et al., 2006), its retention in lactone form at neutral pH is a promising alternative approach for forming injectable nanocarriers when compared with the marketed formulation Camptosar® that is formulated at acidic pH. The propofol-bearing micellar cubosomes (Johnsson et al., 2006) are a good example of the different advantages that can be obtained when the nanostructured aqueous dispersions are compared with conventional dispersions of liposomal formulations and emulsions. The studied micellar cubosomes enhanced the drug solubilization capacity and constituted the formation of highly effective delivery systems (Johnsson et al., 2006).

14.6   SUMMARY AND OUTLOOK

Lipidic nanoparticulate systems have recently attracted much attention for the delivery of peptides, drugs, and functional foods. Among these systems, dispersions confining well-ordered nanostructures such as cubosomes, hexosomes, and micellar cubosomes are the subject of various studies aiming to evaluate their efficacy as drug or food nanocarriers. The nanostructured aqueous dispersions in these studies are a promising strategy for loading drugs and offer an alternative method for overcoming the problems of the administration of drugs that are poorly soluble in aqueous medium. They are potential candidates for enhancing the bioavailability of loaded drugs and avoiding serious side effects.
   Nevertheless, there is need for further future investigations to address the challenges of enhancing the stability of the dispersed particles after administration and modulating their nanostructure in order to optimize their interaction with different biological surfaces. For instance, we

recently found that the addition of short charged designer peptide surfactants mimicking biological phospholipids can be used to *functionalize* nonlamellar mesophases and to enhance the loading capacity of charged active molecules (Yaghmur et al., 2007). The term *functionalization* means to control the solubilization capacity of the liquid crystalline phases by the inclusion of specific anchors such as charged or long-chain amphiphilic molecules. It will also be appealing to study the potential of these dispersions in loading new bioactive materials, to investigate the impact of such molecules on the internal nanostructures, and to study their interaction with different biological surfaces.

ABBREVIATIONS

Cryo-FESEM	Cryo-field emission scanning electron microscopy
Cryo-TEM	Cryo-transmission electron microscopy
EME	Emulsified microemulsion
F127	Pluronic polymer poly(ethylene oxide)$_{99}$–poly(propylene oxide)$_{67}$–poly(ethylene oxide)$_{99}$
Fd3m	Discontinuous micellar cubic phase, also called I_2
Im3m	The primitive type bicontinuous cubic phase
ME	Monoelaidin
MLO	Monolinolein
MO	Monoolein
Pn3m	The Schwarz diamond type bicontinuous cubic phase
SAXS	Small-angle X-ray scattering
TEM	Transmission electron microscopy

SYMBOLS

c_1, c_2	Principal curvatures
G	Gaussian curvature
H	Mean interfacial curvature
H_2	Inverted-type hexagonal liquid crystalline phase
I_2	Inverted-type discontinuous micellar cubic phase
l	Monlayer thickness
L_α	Lamellar phase
L_2	Reverse micellar system (water-in-oil micelles)
L_3	Dispersed sponge phases
R_W	Water core radius in the inverted-type self-assembled systems
s	Shape parameter of molecules
V_2	Inverted-type bicontinuous cubic phase

REFERENCES

Abraham, T., M. Hato, and M. Hirai, 2005. Polymer-dispersed bicontinuous cubic glycolipid nanoparticles. *Biotechnology Progress* 21: 255–262.

Almsherqi, Z., S. Hyde, M. Ramachandran, and Y. Deng, 2008. Cubic membranes: A structure-based design for DNA uptake. *Journal of the Royal Society Interface* 5: 1023–1029.

Almsherqi, Z. A., S. D. Kohlwein, and Y. Deng, 2006. Cubic membranes: A legend beyond the flatland of cell membrane organization. *Journal of Cell Biology* 173: 839–844.

Angelova, A., B. Angelov, S. Lesieur, R. Mutafchieva, M. Ollivon, C. Bourgaux, R. Willumeit, and P. Couvreur, 2008. Dynamic control of nanofluidic channels in protein drug delivery vehicles. *Journal of Drug Delivery Science and Technology* 18: 41–45.

Angelova, A., B. Angelov, B. Papahadjopoulos-Sternberg, C. Bourgaux, and P. Couvreur, 2005a. Protein driven patterning of self-assembled cubosomic nanostructures: Long oriented nanoridges. *Journal of Physical Chemistry B* 109: 3089–3093.

Angelova, A., B. Angelov, B. Papahadjopoulos-Sternberg, M. Ollivon, and C. Bourgaux, 2005b. Proteocubosomes: Nanoporous vehicles with tertiary organized fluid interfaces. *Langmuir* 21: 4138–4143.

Arruebo, M., R. Fernández-Pacheco, M. R. Ibarra, and J. Santamaría, 2007. Magnetic nanoparticles for drug delivery. *Nano Today* 2: 22–32.

Barauskas, J., M. Johnsson, and F. Tiberg, 2005. Self-assembled lipid superstructures: Beyond vesicles and liposomes. *Nano Letters* 5: 1615–1619.

Barauskas, J., A. Misiunas, T. Gunnarsson, F. Tiberg, and M. Johnsson, 2006. "Sponge" nanoparticle dispersions in aqueous mixtures of diglycerol monooleate, glycerol dioleate, and polysorbate 80. *Langmuir* 22: 6328–6334.

Boyd, B. J., 2003. Characterisation of drug release from cubosomes using the pressure ultrafiltration method. *International Journal of Pharmaceutics* 260: 239–247.

Boyd, B. J., 2004. Controlled release from cubic liquid crystalline particles (cubosomes). In *Bicontinuous Structured Liquid Crystals*, P. T. Spicer and M. L. Lynch (Eds.), pp. 285–304. New York: Marcel Dekker.

Boyd, B. J., 2008. Past and future evolution in colloidal drug delivery systems. *Expert Opinion on Drug Delivery* 5: 69–85.

Boyd, B. J., S. B. Rizwan, Y. D. Dong, S. Hook, and T. Rades, 2007. Self-assembled geometric liquid-crystalline nanoparticles imaged in three dimensions: Hexosomes are not necessarily flat hexagonal prisms. *Langmuir* 23: 12461–12464.

Boyd, B. J., D. V. Whittaker, S. M. Khoo, and G. Davey, 2006. Hexosomes formed from glycerate surfactants—Formulation as a colloidal carrier for irinotecan. *International Journal of Pharmaceutics* 318: 154–162.

Chung, H., J. Kim, J. Y. Um, I. C. Kwon, and S. Y. Jeong, 2002. Self-assembled "nanocubicle" as a carrier for peroral insulin delivery. *Diabetologia* 45: 448–451.

Corless, J. M. and M. J. Costello, 1981. Paracrystalline inclusions associated with the disk membranes of frog retinal rod outer segments. *Experimental Eye Research* 32: 217–228.

Couvreur, P. and C. Vauthier, 2006. Nanotechnology: Intelligent design to treat complex disease. *Pharmaceutical Research* 23: 1417–1450.

Danielli, J. F. and H. Davson, 1935. A contribution to the theory of permeability of thin films. *Journal of Cellular and Comparative Physiology* 5: 495–508.

de Campo, L., A. Yaghmur, L. Sagalowicz, M. E. Leser, H. Watzke, and O. Glatter, 2004. Reversible phase transitions in emulsified nanostructured lipid systems. *Langmuir* 20: 5254–5261.

Deng, Y. R., S. D. Kohlwein, and C. A. Mannella, 2002. Fasting induces cyanide-resistant respiration and oxidative stress in the amoeba *Chaos carolinensis*: Implications for the cubic structural transition in mitochondrial membranes. *Protoplasma* 219: 160–167.

Denk, W., J. H. Strickler, and W. W. Webb, 1990. Two-photon laser scanning fluorescence microscopy. *Science* 248: 73–76.

Dong, Y. D., I. Larson, T. Hanley, and B. J. Boyd, 2006. Bulk and dispersed aqueous phase behavior of phytantriol: Effect of vitamin E acetate and F127 polymer on liquid crystal nanostructure. *Langmuir* 22: 9512–9518.

Drummond, C. J. and C. Fong, 1999. Surfactant self-assembly objects as novel drug delivery vehicles. *Current Opinion in Colloid and Interface Science* 4: 449–456.

Esposito, E., R. Cortesi, M. Drechsler, L. Paccamiccio, P. Mariani, C. Contado, E. Stellin, E. Menegatti, F. Bonina, and C. Puglia, 2005. Cubosome dispersions as delivery systems for percutaneous administration of indomethacin. *Pharmaceutical Research* 22: 2163–2173.

Gong, X., S. Sagnella, and C. J. Drummond, 2008. Nanostructured self-assembly materials formed by nonionic urea amphiphiles. *International Journal of Nanotechnology* 5: 370–392.

Gorter, E. and F. Grendel, 1925. On bimolecular layers of lipoids on the chromocytes of the blood. *The Journal of Experimental Medicine* 41: 439–443.

Gustafsson, J., H. Ljusberg-Wahren, M. Almgren, and K. Larsson, 1996. Cubic lipid–water phase dispersed into submicron particles. *Langmuir* 12: 4611–4613.

Gustafsson, J., H. Ljusberg-Wahren, M. Almgren, and K. Larsson, 1997. Submicron particles of reversed lipid phases in water stabilized by a nonionic amphiphilic polymer. *Langmuir* 13: 6964–6971.

Harper, P. E. and S. M. Gruner, 2000. Electron density modeling and reconstruction of infinite periodic minimal surfaces (IPMS) based phases in lipid–water systems. I. Modeling IPMS-based phases. *European Physical Journal E* 2: 217–228.

Hui, S. W., T. P. Stewart, L. T. Boni, and P. L. Yeagle, 1981. Membrane fusion through point defects in bilayers. *Science* 212: 921–923.

Hyde, S., S. Andersson, K. Larsson, Z. Blum, T. Landh, and B. W. Ninham, 1997. *The Language of Shape. The Role of Curvature in Condensed Matter: Physics, Chemistry, Biology.* Amsterdam: Elsevier.

Hyde, S. T., 2001. Identification of lyotropic liquid crystalline mesophases. In *Handbook of Applied Surface and Colloid Chemistry*, K. Holmberg (Ed.), pp. 299–332. New York: Wiley.

Isrealachvili, J. N., 1991. *Intermolecular and Surface Forces* (2nd ed.). London: Academic Press.

Johnsson, M., J. Barauskas, A. Norlin, and F. Tiberg, 2006. Physicochemical and drug delivery aspects of lipid-based liquid crystalline nanoparticles: A case study of intravenously administered propofol. *Journal of Nanoscience and Nanotechnology* 6: 3017–3024.

Johnsson, M., J. Barauskas, and F. Tiberg, 2005a. Cubic phases and cubic phase dispersions in a phospholipid-based system. *Journal of the American Chemical Society* 127: 1076–1077.

Johnsson, M., Y. Lam, J. Barauskas, and F. Tiberg, 2005b. Aqueous phase behavior and dispersed nanoparticles of diglycerol monooleate/glycerol dioleate mixtures. *Langmuir* 21: 5159–5165.

Kamo, T., M. Nakano, W. Leesajakul, A. Sugita, H. Matsuoka, and T. Handa, 2003. Nonlamellar liquid crystalline phases and their particle formation in the egg yolk phosphatidylcholine/diolein system. *Langmuir* 19: 9191–9195.

Kralj-Iglic, V., B. Babnik, D. R. Gauger, S. May, and A. Iglic, 2006. Quadrupolar ordering of phospholipid molecules in narrow necks of phospholipid vesicles. *Journal of Statistical Physics* 125: 727–752.

Kralj-Iglic, V., A. Iglic, G. Gomiscek, F. Sevsek, V. Arrigler, and H. Hägerstrand, 2002. Microtubes and nanotubes of a phospholipid bilayer membrane. *Journal of Physics A: Mathematical and General* 35: 1533–1549.

Kuntsche, J., H. Bunjes, A. Fahr, S. Pappinen, S. Ronkko, M. Suhonen, and A. Urtti, 2008. Interaction of lipid nanoparticles with human epidermis and an organotypic cell culture model. *International Journal of Pharmaceutics* 354: 180–195.

Landh, T., 1995. From entangled membranes to eclectic morphologies: Cubic membranes as subcellular space organizers. *FEBS Letters* 369: 13–17.

Larsson K., 1989. Cubic lipid–water phases: Structures and biomembrane aspects. *Journal of Physical Chemistry* 93: 7304–7314.

Larsson, K., 2000. Aqueous dispersions of cubic lipid–water phases. *Current Opinion in Colloid and Interface Science* 5: 64–69.

Larsson, K., 2008. Lyotropic liquid crystals and their dispersions relevant in foods. *Current Opinion in Colloid and Interface Science* 14: 16–20.

Lopes, L. B., D. A. Ferreira, D. de Paula, M. T. J. Garcia, J. A. Thomazini, M. C. A. Fantini, and M. V. L. B. Bentley, 2006. Reverse hexagonal phase nanodispersion of monoolein and oleic acid for topical delivery of peptides: *In vitro* and *in vivo* skin penetration of cyclosporin A. *Pharmaceutical Research* 23: 1332–1342.

Luzzati, V., 1997. Biological significance of lipid polymorphism: The cubic phases. *Current Opinion in Structural Biology* 7: 661–668.

Luzzati, V., R. Vargas, A. Gulik, P. Mariani, J. M. Seddon, and E. Rivas, 1992. Lipid polymorphism: A correction. The structure of the cubic phase of extinction symbol Fd– consists of two types of disjointed reverse micelles embedded in a three-dimensional hydrocarbon matrix. *Biochemistry* 31: 279–285.

Malmsten, M., 2006. Soft drug delivery systems. *Soft Matter* 2: 760–769.

Mares, T., M. Daniel, S. Perutkova, A. Perne, G. Dolina, A. Iglic, M. Rappolt, and V. Kralj-Iglic, 2008. Role of phospholipid asymmetry in the stability of inverted hexagonal mesoscopic phases. *Journal of Physical Chemistry B* 112: 16575–16584.

Mariani, P., V. Luzzati, and H. Delacroix, 1988. Cubic phases of lipid-containing systems. Structure analysis and biological implications. *Journal of Molecular Biology* 204: 165–189.

Misquitta, Y. and M. Caffrey, 2001. Rational design of lipid molecular structure: A case study involving the C19:1c10 monoacylglycerol. *Biophysical Journal* 81: 1047–1058.

Nakano, M., T. Teshigawara, A. Sugita, W. Leesajakul, A. Taniguchi, T. Kamo, H. Matsuoka, and T. Handa, 2002. Dispersions of liquid crystalline phases of the monoolein/oleic acid/Pluronic F127 system. *Langmuir* 18: 9283–9288.

Ortiz, A., J. A. Killian, A. J. Verkleij, and J. Wilschut, 1999. Membrane fusion and the lamellar-to-inverted-hexagonal phase transition in cardiolipin vesicle systems induced by divalent cations. *Biophysical Journal* 77: 2003–2014.

Patton, J. S. and M. C. Carey, 1979. Watching fat digestion. *Science* 204: 145–148.

Pfeffer, W., 1877. Osmotische Untersuchungen. *Studien zur Zellmechanik* (1st ed.). Leipzig, Germany: Engelmann.

Qiu, H. and M. Caffrey, 2000. The phase diagram of the monoolein/water system: Metastability and equilibrium aspects. *Biomaterials* 21: 223–234.

Rangelov, S. and M. Almgren, 2005. Particulate and bulk bicontinuous cubic phases obtained from mixtures of glyceryl monooleate and copolymers bearing blocks of lipid-mimetic anchors in water. *Journal of Physical Chemistry B* 109: 3921–3929.

Rappolt, M. 2006. The biologically relevant lipid mesophases as "seen" by X-rays. In *Planar Lipid Bilayers and Liposomes*, A. Leitmannova (Ed.), Vol. 5, pp. 253–283. Amsterdam: Elsevier.

Rappolt, M., A. Hickel, F. Bringezu, and K. Lohner, 2003. Mechanism of the lamellar/inverse hexagonal phase transition examined by high resolution X-ray diffraction. *Biophysical Journal* 84: 3111–3122.

Rizwan, S. B., Y. D. Dong, B. J. Boyd, T. Rades, and S. Hook, 2007. Characterisation of bicontinuous cubic liquid crystalline systems of phytantriol and water using cryo field emission scanning electron microscopy (cryo FESEM). *Micron* 38: 478–485.

Rosa, M., M. R. Infante, M. D. Miguel, and B. Lindman, 2006. Spontaneous formation of vesicles and dispersed cubic and hexagonal particles in amino acid-based catanionic surfactant systems. *Langmuir* 22: 5588–5596.

Salentinig, S., A. Yaghmur, S. Guillot, and O. Glatter, 2008. Preparation of highly concentrated nanostructured dispersions of controlled size. *Journal of Colloid and Interface Science* 326: 211–220.

Seddon, J. M. and R. H. Templer, 1993. Cubic phases of self-assembled amphiphilic aggregates. *Philosophical Transactions of the Royal Society A: Mathematical Physical and Engineering Sciences* 344: 377–401.

Seddon, J. M. and R. H. Templer, 1995. Polymorphism of lipid water systems. In *Structure and Dynamics of Membranes*, R. Lipowsky and E. Sackmann (Eds.), pp. 97–160. Amsterdam: North-Holland.

Siegel, D. P. and R. M. Epand, 1997. The mechanism of lamellar-to-inverted hexagonal phase transitions in phosphatidylethanolamine: Implications for membrane fusion mechanisms. *Biophysical Journal* 73: 3089–3111.

Simidjiev, I., S. Stoylova, H. Amenitsch, T. Javorfi, L. Mustardy, P. Laggner, A. Holzenburg, and G. Garab, 2000. Self-assembly of large, ordered lamellae from non-bilayer lipids and integral membrane proteins *in vitro*. *Proceedings of the National Academy of Sciences of the United States of America* 97: 1473–1476.

Simons, K. and D. Toomre, 2000. Lipid rafts and signal transduction. *Nature Reviews Molecular Cell Biology* 1: 31–39.

Simons, K. and G. van Meer, 1988. Lipid sorting in epithelial cells. *Biochemistry* 27: 6197–6202.

Singer, S. J. and G. L. Nicolson, 1972. The fluid mosaic model of the structure of cell membranes. *Science* 175: 720–731.

Snapp, E. L., R. S. Hegde, M. Francolini, F. Lombardo, S. Colombo, E. Pedrazzini, N. Borgese, and J. Lippincott-Schwartz, 2003. Formation of stacked ER cisternae by low affinity protein interactions. *Journal of Cell Biology* 163: 257–269.

Spicer, P. T. 2005. Progress in liquid crystalline dispersions: Cubosomes. *Current Opinion in Colloid and Interface Science* 10: 274–279.

Spicer, P. T., K. L. Hayden, M. L. Lynch, A. Ofori-Boateng, and J. L. Burns, 2001. Novel process for producing cubic liquid crystalline nanoparticles (cubosomes). *Langmuir* 17: 5748–5756.

Svensson, O., K. Thuresson, and T. Arnebrant, 2008. Interactions between drug delivery particles and mucin in solution and at interfaces. *Langmuir* 24: 2573–2579.

Swarnakar, N. K., V. Jain, V. Dubey, D. Mishra, and N. K. Jain, 2007. Enhanced oromucosal delivery of progesterone via hexosomes. *Pharmaceutical Research* 24: 2223–2230.

Szule, J. A., S. E. Jarvis, J. E. Hibbert, J. D. Spafford, J. E. A. Braun, G. W. Zamponni, G. M. Wessel, and J. R. Coorssen, 2003. Calcium-triggered membrane fusion proceeds independently of specific presynaptic proteins. *Journal of Biological Chemistry* 278: 24251–24254.

Templer, R. H., 1998. Thermodynamic and theoretical aspects of cubic mesophases in nature and biological amphiphiles. *Current Opinion in Colloid and Interface Science* 3: 255–263.

Vandoolaeghe, P., A. R. Rennie, R. A. Campbell, R. K. Thomas, F. Höök, G. Fragneto, F. Tiberg, and T. Nylander, 2008. Adsorption of cubic liquid crystalline nanoparticles on model membranes. *Soft Matter* 4: 2267–2277.

Vandoolaeghe, P., F. Tiberg, and T. Nylander, 2006. Interfacial behavior of cubic liquid crystalline nanoparticles at hydrophilic and hydrophobic surfaces. *Langmuir* 22: 9169–9174.

Veranic, P., M. Lokar, G. J. Schutz, J. Weghuber, S. Wieser, H. Hägerstrand, V. Kralj-Iglic, and A. Iglic, 2008. Different types of cell-to-cell connections mediated by nanotubular structures. *Biophysical Journal* 95: 4416–4425.

Yaghmur, A., L. de Campo, L. Sagalowicz, M. E. Leser, and O. Glatter, 2005. Emulsified microemulsions and oil-containing liquid crystalline phases. *Langmuir* 21: 569–577.

Yaghmur, A., L. de Campo, L. Sagalowicz, M. E. Leser, and O. Glatter, 2006a. Control of the internal structure of MLO-based isasomes by the addition of diglycerol monooleate and soybean phosphatidylcholine. *Langmuir* 22: 9919–9927.

Yaghmur, A., L. de Campo, S. Salentinig, L. Sagalowicz, M. E. Leser, and O. Glatter. 2006b. Oil-loaded mono-linolein-based particles with confined inverse discontinuous cubic structure (Fd3m). *Langmuir* 22: 517–521.

Yaghmur, A. and O. Glatter, 2008. Characterization and potential applications of nanostructured aqueous dispersions. *Advances in Colloid and Interface Science* 147–148: 333–342.

Yaghmur, A., P. Laggner, M. Almgen, and M. Rappolt, 2008a. Self-assembly in monoelaidin aqueous dispersions: Direct vesicles to cubosomes transition. *PLoS ONE* 3: e3747.

Yaghmur, A., P. Laggner, B. Sartori, and M. Rappolt, 2008b. Calcium triggered L alpha-H_2 phase transition monitored by combined rapid mixing and time-resolved synchrotron SAXS. *PLoS ONE* 3: e2072.

Yaghmur, A., P. Laggner, S. Zhang, and M. Rappolt, 2007. Tuning curvature and stability of monoolein bilayers by designer lipid-like peptide surfactants. *PLoS ONE* 2: e479.

Yang, D., B. Armitage, and S. R. Marder, 2004. Cubic liquid-crystalline nanoparticles. *Angewandte Chemie— International Edition* 43: 4402–4409.

15 Niosomal Delivery System for Macromolecular Drugs

Yongzhuo Huang, Faquan Yu, and Wenquan Liang

CONTENTS

15.1 INTRODUCTION

Niosomes, which are the self-assembly of nonionic amphiphiles, are closed bilayer vesicles ranging in size from microns to nanometers. Niosomes are attractive because of their structural similarity to liposomes and their alternative application potential in the cosmetic and pharmaceutical industries. Analogous to liposomes, niosomes show great advantages in stability, cost, and safety. The mechanism of nonionic surfactant vesicle formation in aqueous media involves their amphiphilic nature and helper additives like cholesterols, which were well reviewed by Uchegbu and Vyas (1998).

Like liposomes, niosomes represent a highly flexible platform that is due to their unique bilayer structure. First, the bilayer structure is composed of two layers of nonionic amphiphiles; their hydrophobic tails face one another and form the lipophilic core region, and their hydrophilic heads are exposed to the interfacial aqueous environment. Therefore, such a unique structure makes niosomes a promising drug carrier because the interior is able to encapsulate aqueous solutes, and hydrophobic molecules can be incorporated within the lipophilic region of the membrane. Second, unstable drugs (e.g., peptides, proteins, and genetic materials) can be isolated from the adverse environment by being encapsulated into the interior vehicles. Third, the bilayer provides great potential for surface modification to achieve special functions. Almost all of the methods used in liposomal surface modification can be easily transferred to the application of niosomal system. For instance, covalent attachment of poly(ethylene glycol) (PEGylation) has been widely used in liposomal drug delivery systems for the past two decades, because it decreases the interaction between liposomes and serum proteins, with a consequent long-circulation effect. Similarly, PEGylated niosomes have also been developed for specific delivery purposes.

15.2 BRIEF INFORMATION OF NIOSOME PREPARATION

Phospholipids can form bilayer vesicles (liposomes) via self-assembly (Figure 15.1), although the input of physical energy is often involved during the process to promote this formation and reduce the size of the particles. Liposome preparation includes two major steps: enclosing vesicles and sizing some specific particle diameters. The general principles of liposome formation are basically similar to that of niosomes. Therefore, the methods for liposome preparation are usually applied to niosome preparation, which include thin-film hydration, ether injection, reverse phase evaporation, and organic solvent free preparation for formation of vesicles and sonication, high-pressure homogenization, and extrusion for reducing the particle size. These techniques are well reviewed in the niosome monograph edited by Uchegbu (2000) and in many liposome books, so a detailed discussion is not included here. However, it is worthwhile to note that nonionic surfactants display superior physicochemical stability over phospholipids, thereby gaining greater advantages in preparations involving hot aqueous solvents or vigorous heat-producing processes (e.g., sonication and homogenization).

15.3 SURFACE MODIFICATION

Surface modification plays an important role in niosomal drug delivery systems, which can be functioned for a variety of purposes, and a brief summary follows. We hope that it may provide a guide to rational design of a niosomal system for specific delivery.

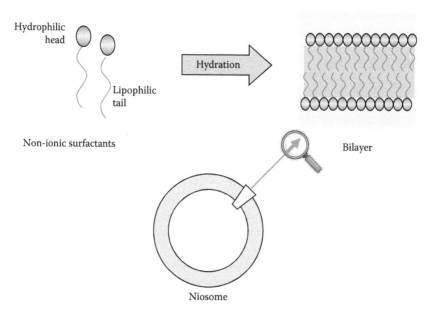

FIGURE 15.1   Niosome formation.

15.3.1 Modification of Surface Charge

Surface charge is a key factor influencing the physical stability and biofate of nanocarriers. In pharmaceutical sciences, the zeta potential is commonly used as an indicator of the surface charge of drug particulate carriers and is widely applied to predict the physical stability of particulate systems. In general, a high zeta potential is accompanied by great stability because of the electrostatic repulsion against the natural tendency to aggregate. A rough rule is often used to predict the storage stability: a zeta potential of >30 mV indicates full electrostatic stability, 5–15 mV indicates limited flocculation, and 3–5 mV indicates maximum flocculation (Heurtault et al., 2003).

The charge property of particulates has also been put to good use for loading drugs with opposite charges to the carriers. A typical application is in gene delivery. Cationic lipids are necessarily involved in colloidal gene delivery systems, which consequently bear positive charges. Because of their opposite charge, cationic carriers can form positively charged complexes with negatively charged genes, which are consequently compacted.

In general, the surface charge of niosomes can be modified by incorporating charged amphiphiles into the bilayer membrane. Based on this method, we developed cationic niosomes for gene delivery. Cholesterol is the most common additive in niosomal systems, and it was thus included in a 1:1 molar ratio to build less leaky bilayer membranes (Uchegbu and Vyas, 1998). We used 3β-[N-(N′,N′-dimethylaminoethane)-carbamoyl]-cholesterol (DC-Chol) as a substitute for cholesterol to prepare cationic niosomes in our study. With a 1:1 molar ratio to different nonionic surfactants (Span 80, Span 60, Span 40, and Span 20), stable cationic niosomal systems were prepared with zeta potentials all above +35 mV (Huang et al., 2005a). A cationic surfactant was also used as an additive to modify the surface charge of the niosomes. For gene delivery purposes, we incorporated cetyltrimethylammonium bromide (CTAB), which has a high efficiency for deoxyribonucleic acid (DNA) binding, in a 1:2:1 molar ratio (Span 40/cholesterol/CTAB) into niosomes. Cationic niosomes with a zeta potential of about +45 mV were obtained.

In addition, modification of surface charge also has a significant impact on the disposition of particles for their *in vivo* application. There have been numerous studies on liposomal systems but few investigations of niosomes.

15.3.2 Modifying the Hydrophilic and Hydrophobic Properties

Hydrophilic–lipophilic balance (HLB) values are often used as a prediction parameter of the niosome forming ability of any surfactant (Uchegbu and Vyas, 1998). The HLB is also an indicator of the hydrophilic and hydrophobic properties of vesicles. Hydrophobicity plays an important role in cell adhesion. For particulate drug carriers, the first step in phagocytosis of a cell is the adhesion of particles to the cell. It has been widely proved that the process of cell adhesion and phagocytosis for bacteria is dependent on hydrophobicity (Liu et al., 2004). Furthermore, compared to Tween 20 modified lipsomes, the higher fusion interaction observed among Tween 80 modified lipsomes was due to the increase in hydrophobic interactions by the longer alkyl chain of Tween 80 (Tasi et al., 2003). We prepared a series of Tween niosomes for antisense oligodeoxynucleotide cellular delivery and found out that their cellular uptake efficiencies were well in accordance with the hydrophilic and hydrophobic properties of vesicles; from high to low, the order is Tween 85 niosomes (HLB = 11.0), Tween 80 niosomes (HLB = 14.9), Tween 40 niosomes (HLB = 15.6), and Tween 20 niosomes (HLB = 16.7; Huang et al., 2005b). This correlation revealed that optimized surface hydrophobicity of vesicles could promote the cellular uptake efficiency through a possible mechanism of increasing the interactions between vesicles and cells.

15.3.3 Sterically Stabilized Effect

Nanovesicles are naturally prone to adhere and fuse to form larger ones, essentially because of their nanosize effect. During adhesion and fusion, vesicles experience "open-and-fuse" processes that result in inclusion leakage (Zhang and Granick, 2006). Many attempts have been made to develop a stabilization method for nanovesicles. Among them, surface modification with hydrophilic polymers has become the most common method, especially since its great success in liposomal products with the landmark stealth liposome being marketed. A variety of hydrophilic polymers, such as pullulan (Sihorkar and Vyas, 2000), mannan (Jain et al., 2005), and PEG (Beugin et al., 1998), have been investigated for their potential to improve liposomal physical stability. PEG is the most widely used hydrophilic polymer for the steric stabilization of nanoparticle drug delivery systems (Torchilin and Trubetskoy, 1995). PEG displays excellent water solubility in all proportions over a wide range of molecular weights. The meander structure of its linear backbone provides comprehensive surface coverage, thus forming a hydrophilic brushlike configuration around the vesicle. However, there is a concentration restriction for PEG—a high PEG content results in vesicle disruption and disk formation. Such a concentration restriction is dependent on the PEG molecular weight. In a diglycerol hexadecyl ether niosomal system, nonaggregated, stable, unilamellar vesicles were obtained at low polymer levels with optimal shape and size homogeneity at cholesterol conjugate/lipid ratios of 10 mol% M-PEG1000-Chol or 5 mol% M-PEG2000-Chol, corresponding to the theoretically predicted brush conformational state of the PEG chains (Beugin et al., 1998).

Span/DC-Chol cationic niosomes were developed for oligonucleotide (OND) delivery; but these carriers also have a stability problem common to cationic drug carriers, which are prone to bind with serum proteins in the physiological environment and thus result in aggregation. Therefore, PEGylated cationic niosomes, composed of DC-Chol, PEG2000–1,2-distearoyl-*sn*-glycero-3-phosphoethanolamine, and nonionic surfactant (Spans) were developed for increasing the stability of niosomes in the physiological environment (Huang et al., 2008). These PEGylated cationic niosomes remained relatively stable in the presence of serum for up to 20 h, whereas the controlled cationic niosomes precipitated during the same time frame. PEGylated cationic niosomes could also provide comprehensive protection for a payload (DNA) against nuclease degradation due to the hydrophilic and sterically stabilized structure of PEG that prevents the near approach of enzymes, thereby protecting DNA from degradation.

15.3.4 Specific Targeting

PEGylated liposomes have been named "stealth" liposomes because of their ability to avoid opsonin binding and subsequently escape the capture of the reticuloendothelial system, thus leading to long circulation, which in turn could favor specific tissue accumulations (Ishida et al., 2002). PEGylation was also applied to niosomal systems for the same purpose. Alain Luciani et al. (2004) developed a tumor-oriented magnetic resonance contrast agent based on PEGylated niosomes bearing N-palmitoyl glucosamine ligands targeted to the glucose receptors that are overexpressed in tumor cells. PEGylated niosomes endured an increased plasma half-life when compared with nonconjugated niosomes. Furthermore, the tumor-targeting ligands on the niosome surface significantly increased the drug accumulation in the tumor. Brain accumulation was also observed. However, there was relatively weak enhancement of brain uptake for PEGylated niosomes because continuous capillaries of the blood–brain barrier do not favor passive targeting. The effect of the chain length of PEG on poly[methoxy-poly(ethylene glycol) cyanoacrylate-*co-n*-hexadecyl cyanoacrylate] (PHDCA) niosomes was investigated; the results demonstrated the optimal coating was PEG5000 in aspects of *in vitro* drug release, phagocytic uptake, pharmacokinetics, and antitumor activity (Shi et al., 2006). The PEG5000-PHDCA niosomes with particle sizes of 92.5 nm had the least phagocytic uptake, the longest half-life of 11.46 h, and the best tumor inhibition rate of 97.1%. The tumor-targeting effect of PEG-PHDCA niosomes may be attributable to the enhanced permeability and retention effect.

Polysaccharide mannan was also used to modify the noisome surface to not only achieve stratum corneum penetration enhancement but also target the Langerhans cells, the major antigen-presenting cells found in abundance beneath the stratum corneum (Jain et al., 2005).

15.3.5 Modifying the Membrane Permeability

In bilayer-structural vesicles-like liposomes and niosomes, cholesterol is the most commonly used membrane stabilizer, preventing leakiness and retarding permeation of solutes enclosed in the aqueous core of these vesicles. Other materials (e.g., fatty alcohols) with good bilayer-stabilizing properties were also identified (Devaraj et al., 2002). Niosomes incorporating fatty alcohols showed a release pattern similar to that of cholesterol. However, their release rates were slower than that of cholesterol. Generally, the smaller the chain length was, the higher the release rate. Based on the chain length, release rates could be ranked as stearyl (C18) < myristyl (C14) < cetyl (C16) < lauryl (C12) < cetostearyl (C16 + C18). The relatively higher release rate observed with cetostearyl alcohol and lauryl alcohol could be due to the perturbations in the bilayer structure brought about by these alcohols. Another report revealed that nisomes modified with fatty alcohols could provide controlled release of the encapsulated drugs (Bandyopadhyay and Johnson, 2007). The fatty alcohol based system exhibited controlled release phenomena as it released ~20% of encapsulated carboxyfluorescein (CF) after 6 h compared to the release of ~35% encapsulated CF from sorbitol and ~65% encapsulated CF from the fatty acid based system over the same time period. This difference was attributable to fatty alcohols participating in the bilayer structure, thus stabilizing the system and lowering the release rate, whereas fatty acids were destabilizing the vesicular system.

Many lipid-soluble compounds penetrate the bilayer and perturb the organization of the lipid acyl chains; these include fatty acids, alkanols, detergents, organic solvents, inorganic ions, and fluorescent probes (Devaraj et al., 2002). However, a detailed investigation for niosomes has yet to be revealed.

15.3.6 Mucoadhesion

Bioadhesive materials have been widely used for surface modification of colloidal drug delivery systems in order to increase mucosal absorption. Chitosan and carbopol were applied to modify niosomal timolol maleate (0.25%) formulations (Aggarwal and Kaur, 2005). The *in vitro* release phase of timolol (91% release in 2 h) was extended significantly by its incorporation into niosomes and further by the polymer coating (40–43% release up to 10 h). The developed formulations were applied to albino rabbits, and chitosan-coated formulation showed a more sustained effect of up to 8 h (vs. 6 h for carbopol-coated niosomes). More efficient control of intraocular pressure was observed in polymer-coated niosomes.

15.4 NIOSOMES FOR DELIVERY OF GENETIC DRUGS

15.4.1 Niosomes for Antisense Oligonucleotide Drugs

We first reported using Span cationic niosomes as antisense OND carriers (Huang et al., 2005b). The cationic lipid DC-Chol, a cholesterol derivative with low toxicity, was the first cationic lipid approved for use in clinical trials (Caplen et al., 1995). We used it as a substitute for cholesterol, which was the most common membrane additive in niosomal systems. The niosomes of Spans showed high efficiency for mediating the cellular uptake of OND. Compared with cationic liposomes composed of soybean phospholipids and DC-Chol of equal moles, niosomes of Span 40 and Span 60 showed a significant enhancement in this study ($p < 0.05$) whereas those of Span 20 and Span 80 also had some positive enhancement but without significant differences. In addition, these cationic niosomes showed low cellular toxicity that the cell viability was above 90% when the lipid concentration was less than 40 μM. However, positively charged vesicles usually display unstable properties in the physiological environment. To improve the stability of this niosomal system, PEGylation modification was applied to the Span cationic nisome; then these novel PEGylated cationic niosomes were used for antisense OND delivery (Huang et al., 2008). Complexes of PEGylated cationic niosomes and OND showed a neutral zeta potential with a particle size of around 300 nm. PEG modification significantly decreased the binding of serum proteins and prevented particle aggregation in the serum. Compared with cationic niosomes, the PEGylated niosomes showed a higher efficiency of OND cellular uptake in the serum. Two methods were applied to prepare OND/ PEGylated niosome complexes: *precoating* and *postcoating*. PEGylated niosomes from the postcoating method showed a greater capability of antisense oligodeoxynucleotide delivery than those from the precoating method. Therefore, in terms of their stable physiochemical properties in storage and the physiological environment, as well as low cost and widely available materials, PEGylated cationic niosomes are promising drug delivery systems for improving OND potency *in vivo*.

15.4.2 Niosomes for Plasmid Deoxyribonucleic Acid Carrying Reporter Genes

The extremely unstable properties of plasmid DNA impose a major restriction for their application as therapeutic agents. Therefore, DNA delivery systems are required to not only carry the gene to the specific sites but also protect its stability. PEGylated cationic niosomes were also tested for their DNA delivery ability in the presence of serum (Huang et al., 2008). Serum proteins in blood play a key role in quenching the cellular penetration ability of cationic carriers because of their charge neutralization. The stability of DNA/PEGylated cationic niosome complexes in the presence of serum was significantly increased compared to non-PEGylated complexes. However, the cellular transfection efficiency, which was evaluated by green fluorescent protein expression, was low. This compromised result was partially due to the PEG coating essentially becoming a barrier for the DNA complexes to access cells. When incorporating CTAB into PEGylated niosomal formulations at a concentration of 10 mol%, the transfection efficiency increased significantly. Therefore, further investigation of a helper additive will be beneficial.

Luciferase plasmid (pLuc) is another commonly used reporter gene in gene delivery studies. The entrapment efficiency and storage stability of pLuc entrapped in cationic bilayer vesicles prepared from various molar ratios of amphiphiles [dipalmitoyl phosphaditylcholine (DPPC), Tween 61, or Span 60], cholesterol, and a cationic charge lipid dimethyl dioctadecyl ammonium bromide (DDAB) was investigated (Manosroi et al., 2008). All of the tested cationic bilayer vesicles gave the pLuc an entrapment efficiency of 100%. However, the stability of pLuc can be further enhanced by entrapping it in cationic liposomes rather than in niosomes.

15.4.3 Niosomes for Plasmid Deoxyribonucleic Acid Vaccines

Vyas' group reported on topical DNA delivery using niosomal systems for hepatitis B immunization (Vyas et al., 2005). DNA encoding hepatitis B surface antigen (HBsAg) was encapsulated in Span 85 niosomes with an entrapment efficiency of 45.4%. These DNA encapsulated niosomes were capable of inducing comparable serum anti-HBsAg titers and cytokines levels in BALB/c mice after topical application compared to intramuscular recombinant HBsAg and topical liposomes. However, the niosomal vaccination shows the additional advantages of a noninvasive vaccination and low cost as well as high stability. Another application of niosomal DNA vaccine delivery for oral immunization was reported by the same group (Jain et al., 2005). DNA-loaded Span 60 niosomes were coated with polysaccharide O-palmitoyl mannan in order to protect them from dissolution and enzymatic degradation in the gastrointestinal tract caused by bile salts and to enhance their affinity toward the antigen-presenting cells of Peyer's patches. O-Palmitoyl mannan coated niosomes produced strong humoral (both systemic and mucosal) and cellular immune responses upon oral administration, whereas intramuscular naked DNA and recombinant HBsAg did not elicit a secretory immunoglobulin A titer in mucosal secretions and pure HBsAg also failed to elicit a cellular response (cytokines level).

Perrie et al. (2004) reported on an influenza DNA niosomal vaccine delivery system. The plasmid pI.18Sfi/NP containing the nucleoprotein gene of A/Sichuan/2/87 (H3N2) influenza virus in the pI.18 expression vector was incorporated by the dehydration–rehydration method into various surfactant vesicle formulations and yielded high DNA vaccine incorporation values (85–97% of the

DNA used). Using these systems for subcutaneous delivery of DNA vaccines enhanced both humoral and cell-mediated immune responses to the encoded antigens.

15.5 NIOSOMES FOR PEPTIDE AND PROTEIN DELIVERY

15.5.1 NIOSOMES FOR PEPTIDES

Niosomal insulin delivery systems have been widely investigated in the past decade. Khaksa et al. developed injectable niosomal insulin formulations for diabetic therapy (Khaksa et al., 2000). Niosomes significantly reduced the blood glucose level in diabetic rats. These blood glucose levels were almost 92% of the initial value. The half-life of insulin was prolonged by 4–5 h in niosomal form in contrast to 2 h for the free drug. Niosomes maintained the plasma insulin level for up to 12 h, but the free drug was cleared quickly.

Niosomal insulin formulations with the potential of sustained-release oral delivery were also reported (Pardakhty et al., 2007). Entrapment of insulin in niosomes of polyoxyethylene alkyl ethers (Brij) protected it against the proteolytic activity of α-chymotrypsin, trypsin, and pepsin *in vitro*. The maximum protection activity was seen in Brij 92/cholesterol (7:3 molar ratios) in which only $26.3 \pm 3.98\%$ of entrapped insulin was released during 24 h in simulated intestinal fluid. Span surfactant series niosomes were also investigated as insulin oral delivery carriers (Varshosaz et al., 2003). Span 60 niosomes showed the highest protection of insulin against different proteolytic enzymes and good physical stability.

Ning et al. (2005) explored the possible application of mucosal insulin delivery. Span niosomal insulin formulations were applied in vaginal delivery and showed Span 60 and Span 40 niosomes were both higher than blank Span 40 and Span 60 vesicles, as well as free insulin physical mixture groups ($p < 0.05$). Compared with subcutaneous administration of the insulin solution, the relative pharmacological bioavailability and relative bioavailability of the insulin–Span 60 vesicles group were 8.43% and 9.61%, respectively, and the insulin–Span 40 niosomes were 9.11% and 10.03% ($p > 0.05$), respectively, indicating niosomes had an enhancing effect on vaginal delivery of insulin.

15.5.2 NIOSOMES FOR PROTEINS

Streptokinase is utilized as a therapeutic agent because of its fibrinolytic activity. Niosomes and other lipid vesicles were used as drug carriers for streptokinase to achieve slow release of entrapped proteins in circulation to increase their half-life, to mask immunogenic properties, and to protect against the loss of enzymatic activity (Erdogan et al., 2006). Streptokinase niosomes showed an average size of 190 nm and entrapment efficiency of 13%. The highest concentration of niosomes was in the spleen at levels above 60% per gram at 1 and 4 h after administration. With entrapment of streptokinase in the niosomes, thrombus uptake and imaging quality were improved; at 4 h after administration, a higher thrombus/vein ratio was obtained when compared with free streptokinase ($p < 0.05$).

Niosomal immunization has gradually attracted scientists' interests since the late 1990s. A variety of niosomes have been shown to be versatile in their ability for the incorporation of a diverse range of antigens. Some model antigens, for example, bovine serum albumin (BSA; Brewer and Alexander, 1992; Murdan et al., 1999), ovalbumin (Rentel et al., 1999), and hemagglutinin (Murdan et al., 1999), were encapsulated within niosomes; these novel vaccine delivery systems revealed promising immunization outcomes. Furthermore, viral influenza antigen (Chattaraj and Das, 2003), malarial antigens (Vangala et al., 2006), and HBsAg (Vangala et al., 2007) were encapsulated in niosomal carriers for immunization purposes. Vangala et al. reported that dimethyldioctadecylammonium-based niosomes showed the ability to promote both cell-mediated and humoral immune responses to the HBsAg and malarial antigens in mice.

Noninvasive routes have been explored for niosomal vaccine delivery. Niosomes show great potential for transdermal delivery, and some reviews have surfaced in this field (Choi and Maibach, 2005). In recent years, niosomes have been found to have great potential in transcutaneous immunization (Jain and Vyas, 2005). Mannosylated niosomes composed of sorbitan monostearate/ sorbitan trioleate (Span 60/Span 85), cholesterol, and stearylamine have been developed as a topical vaccine delivery carrier and adjuvant with a model antigen of BSA. The encapsulation efficiency of BSA was $44.3 \pm 3.8\%$ and $41.2 \pm 3.2\%$ for Span 60 and Span 85 based plain niosomes, respectively. Niosomal formulations elicited a significantly higher serum immunoglobulin G (IgG) titer upon topical application compared to topically applied alum adsorbed BSA ($p < 0.05$). The serum IgG levels were significantly higher for the mannosylated niosomes compared to plain uncoated niosomes ($p < 0.05$). Combined serum IgG2a/IgG1 responses revealed that the formulations were capable of eliciting both humoral and cellular responses.

The adjuvant effect of a particular system has been well recognized in vaccine delivery systems. There are mainly three mechanisms involved: slow release of antigens at the injection site (a depot effect); targeting of antigens to the relevant antigen-presenting cells of the immune system, that is, macrophages; and direct activation of cells in the immune system (e.g., bacterial adjuvants and cytokines; Sinyakov et al., 2006). Niosomes have also been widely examined for their adjuvant function. Brewer and Alexander reported that niosomes were generally better stimulators of IgG2a than Freund's complete adjuvant but poorer stimulators of IgG1, which means niosomes are effective stimulators of the Th1 lymphocyte subset, and by inference, potent stimulators of cellular immunity (Brewer and Alexander, 1992). However, the adjuvant activity of niosomes was dependent on factors such as administration routes and the entrapment of the antigen within niosomes. Intraperitoneal inoculation of niosomes did not generate strong immunostimulation; but subcutaneous inoculation, oral immunization (Jain et al., 2005), and topical immunization (Jain and Vyas, 2005) had niosomal adjuvant activity.

15.6 CONCLUSION

Biomacromolecules have opened an era for therapy with the great advance of biotechnology. However, challenges have been raised in the development of carriers for these biomacromolecular drugs. Niosomes are produced from entirely synthetic and well-defined surfactants. Numerous studies have demonstrated that niosomes have advantages over liposomes in such aspects as low cost, storage, and industrial-scale production. Niosomal delivery has seen advances and new applications in biomacromolecular drug delivery. We believe that surface modification is one of the most important techniques applied in niosomal drug delivery systems, because the niosomal systems can thereby be tailored for some specific purposes as in physical forms (e.g., stability, drug release) or biological performance (e.g., biodistribution, half-life, and tissue or cellular penetration). Therefore, multifunctional niosomal delivery systems for biomacromolecules can be developed using suitable modification techniques.

ABBREVIATIONS

Brij	Polyoxyethylene alkyl ethers
BSA	Bovine serum albumin
CF	Carboxyfluorescein
CTAB	Cetyltrimethylammonium bromide
DC-Chol	3β-[N-(N',N'-dimethylaminoethan)-carbamoyl]-cholesterol
DNA	Deoxyribonucleic acid
HBsAg	Hepatitis B surface antigen
HLB	Hydrophilic–lipophilic balance
IgG	Immunoglobulin G
OND	Antisense oligonucleotide
PEG	Poly(ethylene glycol)
PHDCA	Poly[methoxy-poly(ethylene glycol) cyanoacrylate-co-n-hexadecyl cyanoacrylate]
pLuc	Luciferase plasmid

REFERENCES

Aggarwal, D. and Kaur, I. P. 2005. Improved pharmacodynamics of timolol maleate from a mucoadhesive niosomal ophthalmic drug delivery system. *Int. J. Pharm.*, 290: 155–159.

Bandyopadhyay, P. and Johnson, M. 2007. Fatty alcohols or fatty acids as niosomal hybrid carrier: Effect on vesicle size, encapsulation efficiency and *in vitro* dye release. *Colloids Surf. B Biointerfaces*, 58: 68–71.

Beugin, S., Edwards, K., Karlsson, G., Ollivon, M., and Lesieur, S. 1998. New sterically stabilized vesicles based on nonionic surfactant, cholesterol, and poly(ethylene glycol)-cholesterol conjugates. *Biophys. J.*, 74: 3198–3210.

Brewer, J. M. and Alexander, J. 1992. The adjuvant activity of non-ionic surfactant vesicles (niosomes) on the BALB/c humoral response to bovine serum albumin. *Immunology*, 75: 570–575.

Caplen, N. J., Alton, E. W., Middleton, P. G., Dorin, J. R., Stevenson, B. J., Gao, X., Durham, S. R., et al. 1995. Liposome-mediated CFTR gene transfer to the nasal epithelium of patients with cystic fibrosis. *Nat. Med.*, 1: 39–46.

Chattaraj, S. C. and Das, S. K. 2003. Physicochemical characterization of influenza viral vaccine loaded surfactant vesicles. *Drug Deliv.*, 10: 73–77.

Choi, M. J. and Maibach, H. I. 2005. Liposomes and niosomes as topical drug delivery systems. *Skin Pharmacol. Physiol.*, 18: 209–219.

Devaraj, G. N., Parakh, S. R., Devraj, R., Apte, S. S., Rao, B. R., and Rambhau, D. 2002. Release studies on niosomes containing fatty alcohols as bilayer stabilizers instead of cholesterol. *J. Colloid Interface Sci.*, 251: 360–365.

Erdogan, S., Ozer, A. Y., Volkan, B., Caner, B., and Bilgili, H. 2006. Thrombus localization by using streptokinase containing vesicular systems. *Drug Deliv.*, 13: 303–309.

Heurtault, B., Saulnier, P., Pech, B., Proust, J. E., and Benoit, J. P. 2003. Physico-chemical stability of colloidal lipid particles. *Biomaterials*, 24: 4283–4300.

Huang, Y., Chen, J., Chen, X., Gao, J., and Liang, W. 2008. PEGylated synthetic surfactant vesicles (niosomes): Novel carriers for oligonucleotides. *J. Mater. Sci. Mater. Med.*, 19: 607–614.

Huang, Y. Z., Han, G., Wang, H., and Liang, W. Q. 2005a. Cationic niosomes as gene carriers: Preparation and cellular uptake *in vitro*. *Pharmazie*, 60: 473–474.

Huang, Y. Z., Liang, W. Q., and Yang, V. C. 2005b. Tween non-ionic surfactant vesicles as gene carriers: Preparation and cellular study *in vitro*. Paper presented at the Symposium of the National Doctoral Academic Forum for Chemical Engineering and Technology, Tianjin University, Tianjin, China.

Ishida, T., Harashima, H., and Kiwada, H. 2002. Liposome clearance. *Biosci. Rep.*, 22: 197–224.

Jain, S., Singh, P., Mishra, V., and Vyas, S. P. 2005. Mannosylated niosomes as adjuvant-carrier system for oral genetic immunization against hepatitis B. *Immunol. Lett.*, 101: 41–49.

Jain, S. and Vyas, S. P. 2005. Mannosylated niosomes as carrier adjuvant system for topical immunization. *J. Pharm. Pharmacol.*, 57: 1177–1184.

Khaksa, G., D'Souza, R., Lewis, S., and Udupa, N. 2000. Pharmacokinetic study of niosome encapsulated insulin. *Indian J. Exp. Biol.*, 38: 901–905.

Liu, Y., Yang, S. F., Li, Y., Xu, H., Qin, L., and Tay, J. H. 2004. The influence of cell and substratum surface hydrophobicities on microbial attachment. *J. Biotechnol.*, 110: 251–256.

Luciani, A., Olivier, J. C., Clement, O., Siauve, N., Brillet, P. Y., Bessoud, B., Gazeau, F., Uchegbu, I. F., Kahn, E., Frija, G., and Cuenod, C. A. 2004. Glucose-receptor MR imaging of tumors: Study in mice with PEGylated paramagnetic niosomes. *Radiology*, 231: 135–142.

Manosroi, A., Thathang, K., Werner, R. G., Schubert, R., and Manosroi, J. 2008. Stability of luciferase plasmid entrapped in cationic bilayer vesicles. *Int. J. Pharm.*, 356: 291–299.

Murdan, S., Gregoriadis, G., and Florence, A. T. 1999. Sorbitan monostearate/polysorbate 20 organogels containing niosomes: A delivery vehicle for antigens? *Eur. J. Pharm. Sci.*, 8: 177–186.

Ning, M., Guo, Y., Pan, H., Yu, H., and Gu, Z. 2005. Niosomes with sorbitan monoester as a carrier for vaginal delivery of insulin: Studies in rats. *Drug Deliv.*, 12: 399–407.

Pardakhty, A., Varshosaz, J., and Rouholamini, A. 2007. *In vitro* study of polyoxyethylene alkyl ether niosomes for delivery of insulin. *Int. J. Pharm.*, 328: 130–141.

Perrie, Y., Barralet, J. E., McNeil, S., and Vangala, A. 2004. Surfactant vesicle-mediated delivery of DNA vaccines via the subcutaneous route. *Int. J. Pharm.*, 284: 31–41.

Rentel, C. O., Bouwstra, J. A., Naisbett, B., and Junginger, H. E. 1999. Niosomes as a novel peroral vaccine delivery system. *Int. J. Pharm.*, 186: 161–167.

Shi, B., Fang, C. and Pei, Y. 2006. Stealth PEG-PHDCA niosomes: Effects of chain length of PEG and particle size on niosomes surface properties, *in vitro* drug release, phagocytic uptake, *in vivo* pharmacokinetics and antitumor activity. *J. Pharm. Sci.*, 95: 1873–1887.

Sihorkar, V. and Vyas, S. P. 2000. Polysaccharide coated niosomes for oral drug delivery: Formulation and *in vitro* stability studies. *Pharmazie*, 55: 107–113.

Sinyakov, M. S., Dror, M., Lublin-Tennenbaum, T., Salzberg, S., Margel, S., and Avtalion, R. R. 2006. Nano- and microparticles as adjuvants in vaccine design: Success and failure is related to host natural antibodies. *Vaccine*, 24: 6534–6541.

Tasi, L.-M., Liu, D.-Z., and Chen, W.-Y. 2003. Microcalorimetric investigation of the interaction of polysorbate surfactants with unilamellar phosphatidylcholines liposomes. *Colloids Surf. A: Physicochem. Eng. Aspects*, 213: 7–14.

Torchilin, V. P. and Trubetskoy, V. S. 1995. Which polymers can make nanoparticulate drug carriers long-circulating? *Adv. Drug Deliv. Rev.*, 16: 141–155.

Uchegbu, I. F. 2000. *Synthetic Surfactant Vesicles: Niosomes and Other Non-Phospholipid Vesicular Systems.* Amsterdam: Harwood Academic.

Uchegbu, I. F. and Vyas, S. P. 1998. Non-ionic surfactant based vesicles (niosomes) in drug delivery. *Int. J. Pharm.*, 172: 33–70.

Vangala, A., Bramwell, V. W., McNeil, S., Christensen, D., Agger, E. M., and Perrie, Y. 2007. Comparison of vesicle based antigen delivery systems for delivery of hepatitis B surface antigen. *J. Control. Release*, 119: 102–110.

Vangala, A., Kirby, D., Rosenkrands, I., Agger, E. M., Andersen, P., and Perrie, Y. 2006. A comparative study of cationic liposome and niosome-based adjuvant systems for protein subunit vaccines: Characterisation, environmental scanning electron microscopy and immunisation studies in mice. *J. Pharm. Pharmacol.*, 58: 787–799.

Varshosaz, J., Pardakhty, A., Hajhashemi, V. I., and Najafabadi, A. R. 2003. Development and physical characterization of sorbitan monoester niosomes for insulin oral delivery. *Drug Deliv.*, 10: 251–262.

Vyas, S. P., Singh, R. P., Jain, S., Mishra, V., Mahor, S., Singh, P., Gupta, P. N., Rawat, A., and Dubey, P. 2005. Non-ionic surfactant based vesicles (niosomes) for non-invasive topical genetic immunization against hepatitis B. *Int. J. Pharm.*, 296: 80–86.

Zhang, L. and Granick, S. 2006. How to stabilize phospholipid liposomes (using nanoparticles). *Nano Lett.*, 6: 694–698.

16 A New Class of Mesoscopic Aggregates as a Novel Drug Delivery System

Federico Bordi, Cesare Cametti, and Simona Sennato

CONTENTS

16.1 INTRODUCTION

With the progress of molecular biology and genetics, particularly with the identification, sequencing, and characterization of an increasing number of pathogenic genes, genetic therapy has appeared as a therapeutic modality with a tremendous potential impact on the quality of human life. In general, genetic therapy consists of delivering nucleic acids to the interior of specific target cells in the organism (Rayburn et al., 2005). In the more traditional approach of gene therapy, the gene of interest is inserted into the genetic code to restore or correct some functions in the cell. In a different approach, ribonucleic acid (RNA) silencing molecules are delivered to the cell cytoplasm and produce, on the basis of different mechanisms not yet completely understood, the inhibition of the expression of pathogenic genes. In both cases, however, a proper *vector* transporting the therapeutic nucleic acids through the cell membrane and delivering the genetic material to the cytoplasm must be employed.

To reach their target, all drugs have to be transported through a complex "aqueous environment" system, such as a living body, and have to cross a series of different "barriers." In genetic materials, the main barriers are the cell membranes. In some cases, many effective drugs are hydrophobic or amphiphilic and the barrier is represented by the difficulty of transporting these insoluble substances through the blood stream or through extracellular fluid. In other cases the drug, although easily dissolved in an aqueous solvent, is immediately degraded by enzymes present in the body fluids. Moreover, even if the drug is able to cross the barriers without being degraded or without producing undesired side effects, it is still possible that its local bioavailability is not sufficient for effective pharmacokinetics or even for effecting its function. In all of these cases, the active substance has to be encapsulated in a proper vector in order to favor the effective transport through the different barriers and to reduce its toxicity.

For these reasons, there is a general tendency to employ specific vectors for targeting active substances to the designed organs or to the corporeal districts in the attempt to modify the pharmaceutical index of the drugs, minimizing unwanted side effects. As an example, in the case of doxorubicin, an effective anticancer drug but with important side effects against the cardiac muscle (Arancia et al., 1995), its formulation with poly(ethylene glycol) (PEG) liposomes (PEGylated liposomes) proved very effective in reducing these undesirable effects, even in critical situations as in the case of human immunodeficiency virus (HIV)-infected patients (Stevenson, 2003).

Finally, there is the rapidly developing technique of genetic vaccination. Over the last few years, advances in gene-based delivery technology contributed to the revitalization of the field of vaccine development. Genetic vaccinations, encoding antigens from bacteria, virus, and cancer, have been shown to be very promising in enhancing protective immunity. However, again the lack of effective intracellular vectors has reduced the value of the deoxyribonucleic acid (DNA) vaccine approach. Genetic immunization by the use of plasmid DNA has often led to protective humoral and cell-mediated immunity, but "naked DNA" vaccines can be easily degraded by nucleases *in situ*. Moreover, naked DNA is obviously unable to target antigen presenting cells. To optimize antigen delivery efficiency, as well as vaccine efficacy, there is a strong need for effective vectors to be employed as vaccine carriers (El-Aneed, 2004).

The systems currently in use can be divided into two broad categories: viral vectors and nonviral vectors. Viruses have naturally evolved to efficiently infect eukaryotic cells, transferring their genetic materials into the host cell. Different viruses, both RNA and DNA, have been evaluated as possible carriers for gene therapy, but they have proved of limited use. Their main limitations are the reduced size of the genetic material that can be transported and, from a practical point of view, the difficulties of large-scale production and purification. Moreover, they present significant risks of toxicity and immunogenicity and their acceptance by the patients is generally low (Wasungu and Hoekstra, 2006; Zhang et al., 2004).

Synthetic nonviral vectors, such as cationic lipids and polymers, have several potential advantages compared with viral systems. They include lower toxicity and immunogenicity, simpler quality control and regulatory requirements, and no limitation in the size of the genetic material to be transported (from oligonucleotides to artificial chromosomes; Wasungu and Hoekstra, 2006; Zhang et al., 2004).

Among the different systems that have been proposed, cationic lipids appear to be the most probable alternative to viral delivery systems, and their use in transfection protocols *in vitro* and *in vivo* shows an increasing diffusion. However, the transport efficiency of these vectors still remains unsatisfactory (Safinya, 2001), and much research effort has been directed at improving this aspect (Lin et al., 2003; Hoekstra et al., 2007) as well as reducing their toxicity (Rayburn et al., 2005).

The accomplishment of this task is made more difficult by our incomplete understanding of the mechanisms that govern the transfection. For example, although endocytosis is considered the major pathway of access to the cytoplasm, the relative contribution of distinct endocytic pathways (including clathrin- and caveolae-mediated endocytosis or macropino cytosis) is poorly defined. For this reason, there is much interest in developing new vector systems. Lipids and lipid encapsulation technologies could furnish an effective solution to this problem.

In his interesting book *Life—As a Matter of Fat*, Ole G. Mouritsen (2005) writes:

First of all, lipids are amphiphiles designed to mediate hydrophobic and amphiphile environments, which makes them perfect emulsifiers. Secondly, many lipids are biocompatible and biodegradable and hence harmless to biological systems. Thirdly, lipids are a rich class of molecules allowing for a tremendous range of possibilities. Finally and possibly most important, lipids are the stuff out of which the barriers that limit drugs transport and delivery are themselves made. Therefore by using lipids for transport and delivery of the drugs, one can exploit nature's own tricks to interact with cells, cell membranes, and receptors for drugs.

A further advantage of the structures formed by lipid molecules for their use as vectors for drug delivery is that these structures form spontaneously or self-assemble into vesicular monolayer or multilayer aggregates that encompass an aqueous core (liposomes).

This chapter focuses on a new class of colloids built up by the aggregation of cationic liposomes stuck together by oppositely charged linear polyions. This aggregation process gives rise to an equilibrium cluster phase controlled by an interparticle potential with a short-range attraction and a screening repulsion, both of electrostatic origin. The balance between electrostatic repulsion and attraction favors the formation of relatively large, stable, equilibrium aggregates where each liposome maintains its integrity and consequently the ability to transfer both hydrophilic (within the aqueous core) and hydrophobic (within the lipidic bilayer) molecules.

These self-assembling, supermolecular complexes represent a new class of colloids, which have intriguing properties that are not yet completely investigated and are far from being thoroughly understood. These structures should have wide potential technological applications in the area of nanoparticles and nanostructured materials and represent a very promising research field for biotechnological and biomedical applications. The assembled liposome structures may have the advantage to act as a multicomponent drug delivery system, among others. In contrast to what happens in *vesosomes*, it is possible to build up vesicles (liposomes) containing different therapeutic agents and stuck together by means of an electrostatic glue, without the need of an outer lipidic membrane that encompasses the whole aggregate.

Biocompatibility, stability, and, above all, the ability to deliver a broad range of bioactive molecules make these colloidal aggregates a versatile drug delivery system with the possibility of efficient targeting to different organs.

We will review some recent results concerning the hydrodynamic, electrical, and structural properties of these aggregates in different environmental conditions, with a special attention to our work. We expect that this chapter will provide a comprehensive overview of electrostatic stabilized polyion-induced liposome aggregates, encouraging the proposed strategy into clinical reality.

16.2 POLYION–LIPOSOME COMPLEXES

Polyelectrolyte and oppositely charged lipid complexes show a rich and interesting phenomenology that has recently attracted much interest, because of a variety of implications in the most disparate fields of technology, as well as more fundamental research, from membrane biophysics and DNA condensation to technical issues of interest in waste treatment or oil extraction. In particular, when lipids are structurally organized in vesicles, their interactions with linear polyelectrolytes (polyions) show peculiar aspects, suggesting that these systems belong to the class of colloids characterized by long-range electrostatic repulsions and short-range attraction interactions (Bordi et al., 2005a; Sanchez and Bartlett, 2005; Sciortino et al., 2005), although with a very distinctive characteristics, because repulsion and attraction both share the same electrostatic nature (Bordi et al., 2005a).

This latter aspect is particularly intriguing because, thanks to these particular electrostatic interactions, liposomes maintain their individuality within the clusters, which appear as "multicompartment" aggregates of water-filled vesicles (Bordi et al., 2006a). Recent years have witnessed a growing interest in the role of interparticle potential in controlling the structure and dynamics of colloidal dispersions, both for its fundamental implications in soft matter physics and to develop new routes toward advanced materials.

When linear polyelectrolytes are added to a suspension of oppositely charged mesoscopic particles, the polyion chains rapidly adsorb onto the particle surface (Bordi et al., 2004a; Ciani et al., 2004; Gonçalves et al., 2004; Piedade et al., 2004; Sennato et al., 2005a) and repel each other, reconfiguring themselves in more or less orderly patterns (Dobrynin et al., 2000; Yu et al., 2002) to gain some energy.

This *lateral* correlation of the adsorbed polyelectrolyte produces two effects: a phenomenon of *overcharging* and the appearance of a short-range attractive potential between different *polyelectrolyte-decorated* particles (pd-liposomes) that produces the effect of *reentrant condensation*. Overcharging is obtained when, with the increase of the polyelectrolyte concentration, the adsorption of the polyion chains at the particle surface progressively neutralizes the particle charges. However, depending on the relative size and the charge density of the particles and the polyelectrolytes, polyion chains might continue adsorbing even beyond the neutralization point, so that the sign of the net charge of the whole assembly is reverted (charge inversion). In other words, more polyelectrolyte adsorbs at the particle surface than is needed for neutralizing its original charge (giant charge inversion). Concomitant to this effect and as a direct consequence of it, the average size of the aggregate increases with the increase of the polyion concentration, reaches a maximum in correspondence to the charge neutralization condition, and then decreases toward values similar to the ones of isolated liposomes (reentrant condensation).

Briefly, the actual origin of the attraction could be traced back to the nonhomogeneous distribution of adsorbed polyions that repel each other and form "structures" at the particle surface, resembling a Wigner crystal. The precise conformation of these structures is sensitively dependent on the relative size of both dispersed particles and polyions, their valence, and charge density. However, because they are "similar," that is, showing, on average, an identical periodicity, these nonuniformly charged regions form interlocking patterns on different particle surfaces, so that a short-range attraction can arise when a "counterion domain" on one particle corresponds to a "counterion-free domain" on the other particle.

More generally, from a nonuniform distribution of charges at a microscopic level, an attraction can arise ("charge patch" attraction; Miklavic et al., 1994; Khachatourian and Wistrom, 1998) and this mechanism has been invoked to explain the effect of different polyelectrolytes in inducing aggregation of colloidal particles (Leong, 2001; Walker and Grant, 1996).

We recently demonstrated (Bordi et al., 2004a, 2005a, 2006a; Sennato et al., 2005a) in complexes formed by charged liposomes with oppositely charged polyelectrolytes that in a broad polyelectrolyte concentration range around the point of charge inversion (isoelectric point) there is a cluster phase that appears to be formed by *equilibrium* aggregates (Bordi et al., 2005a). The observed reversible aggregation was due to the balance of the long-range electrostatic repulsion and the short-range attraction (Bordi et al., 2005a; Sennato et al., 2005a) arising from the correlated distribution of electrostatic charges at the pd-liposome surface, where the polyelectrolyte domains "alternate" with an oppositely charged polyelectrolyte-free domain (Khachatourian and Wistrom, 1998; Leong, 2001; Miklavic et al., 1994; Walker and Grant, 1996). In these systems, close to the neutralization point, a reentrant condensation is observed (i.e., at a fixed liposome concentration) by increasing the polyelectrolyte content; and the size of the clusters formed by the pd-liposomes increases, reaches a maximum at the neutralization point, and decreases monotonously beyond this point as the degree of overcharging increases.

The aim of this chapter is to summarize the main phenomena occurring in colloidal suspensions of charged mesoscopic particles in the presence of oppositely charged linear polyions. We will discuss in detail overcharging and reentrant condensation, which are both governed by electrostatic interactions, on the basis of experimental evidence derived from the measurement of the relevant parameters, such as the average size and size distribution of the aggregates [from dynamic light scattering (DLS) experiments], cluster morphology [from transmission electron microscopy (TEM)], and electrical parameters (from electrical conductivity measurements and electrophoretic mobility measurements).

Section 16.3 briefly discusses the structural and electrical characteristics of polyelectrolytes and liposomes as separate objects in aqueous suspensions. Section 16.4 is devoted to polyelectrolyte–liposome interactions. The dynamics of the formation of aggregates that result from this interaction is discussed in Sections 16.4 and 16.5, whereas the morphology of the aggregates is described in Section 16.7. Because of its importance for possible biotechnological applications and its novelty for these systems, the multicompartment structure of the aggregates will be described in detail and experimental evidence of this structure will be presented and commented upon at some length.

16.3 POLYELECTROLYTES AND CHARGED LIPOSOMES

16.3.1 Polyelectrolytes

Polyelectrolytes are macromolecules bearing numerous ionizable groups along their backbone. When dissolved in a polar solvent such as water, these groups dissociate and counterions, which diffuse into the bulk solution, are left behind oppositely charged groups on the polyion backbone (Dobrynin and Rubinstein, 2005). Because of the fine interplay between the electrostatic attraction of counterions to a polyion and the loss of translational entropy that is attributable to their localization in the vicinity of the polyion chain, these solutions display peculiar behaviors that differ from neutral polymer solutions and from simple electrolyte solutions.

These systems experience a relevant phenomenon known as "counterion condensation." In a very dilute polyelectrolyte solution, the entropic penalty for the counterions being "trapped" close to a polyion is very high and virtually all counterions leave the polymer chains, "freely" diffusing in the solution. As the polymer concentration increases, the entropic penalty for counterion localization decreases, resulting in a gradual increase in the number of counterions that "condense" in a volume close to each polyion. This phenomenon is known as the Manning–Oosawa counterion condensation (see Bordi et al., 2004b and references cited therein; Dobrynin and Rubinstein, 2005). According to this theory, the counterions accumulate in a "condensed layer" along the polyion chain exactly to the point that the parameter $\xi = b/l_B$ is reduced to $\xi = 1$ (for monovalent counterions), that is, the "effective separation" between charges along the polyion is increased from b to l_B, where b is the distance between two neighboring charges along the chain and l_B is the Bjerrum length, which is the distance at which electrostatic interactions between two particles bearing an elementary charge e and suspended in a medium of permittivity ε reduces to the thermal energy K_BT. In other words, if the charge spacing is too small, the electric field becomes so strong that the system reduces its free energy by "condensing" some counterions on the polyion chain. In these conditions, each polyion bears an effective charge $Q_p = eNf$, while the remaining fraction $f = b/l_B$ of counterions is "free" in the solution.

If a simple electrolyte is added to the suspension, the increased ionic strength screens the electrostatic interactions, influencing the configuration of the polyion chains and the properties of the solution as a whole. For sufficiently small fixed charge density, this screening, which is described by the linearized form of the Poisson–Boltzmann equation, is quantified by a screening length k_D^{-1} (the Debye length) defined as

$$k_D^{-1} = (4^1 l_B \sum z_i^2 c_i)^{-1}, \tag{16.1}$$

where c_i is the number density of ions of valence z_i.

The conformation assumed by a polymer in solution is usually described in terms of a persistence length L_p, which is a measure of the chain stiffness. In polyelectrolytes, the persistence length, according to the Odijk–Skolnick–Fixman theory (Odijk, 1977), can be decomposed into structural (L_0) and electrostatic (L_e) contributions:

$$L_p = L_0 + L_e = L_0 + \frac{l_B}{(2bk_D)^2}. \tag{16.2}$$

In a regime of sufficiently low added salt, intra- and interchain electrostatic interactions strongly influence both the chain conformation and the properties of the solution. In order to describe the conformation of polyions in solution, a scaling model for polymer solutions was originally proposed by de Gennes in 1976 (Pincus et al., 1976). This approach is based on the existence of different length scales and on the concept of an "electrostatic blob" as the elementary unit of the chain conformation.

On very small scales (on the order of a few monomers), because of the insufficient charge repulsion necessary to modify its conformation, the chain forms little coils or "blobs," and inside these blobs its conformation is unperturbed by electrostatic interactions. If the solvent is a "good solvent" for the chain, that is, if the solvent swells the uncharged chain as it happens for highly charged polyions, the electrostatic blob size is determined by a balance between the electrostatic energy inside the blob and the thermal energy $K_B T$:

$$\frac{(efg_e)^2}{\varepsilon D_e} \approx K_B T, \tag{16.3}$$

where e is the elementary charge, f is the fraction of monomers bearing an effective charge, ε is the permittivity of the solvent, and D_e is the size of the blob containing g_e monomers. In most cases, the solvent is a "poor solvent" for the uncharged chain, that is, the uncharged polymer would not dissolve in the solvent. For example, water is a poor solvent for most polymers. In this case, the size of the electrostatic blobs D_e is given by a balance between the electrostatic energy inside the blob, which favors its swelling, and the excess free energy γD_e^2 due to the unfavorable interaction with the solvent, which tends to collapse the blob (Dobrynin et al., 1995; Grosberg and Khokhlov, 1994):

$$\frac{(efg_e)^2}{\gamma D_e} \approx \gamma D_e^2. \tag{16.4}$$

This excess free energy can be expressed in the form of interfacial tension $\gamma = (\tau/b)^2 K_B T$. Here, $\tau = (\theta - T)/\theta$ is the reduced temperature, where θ is the temperature at which the net interaction between uncharged polymer and the solvent is zero. In any polar solvent and in a dilute solution, that is, in the condition largely encountered in liposome interactions, a flexible polyelectrolyte with no added salt adopts a highly extended conformation with a length L determined by the strong electrostatic repulsion between the $N_D = N/g_e$ electrostatic blobs in the chain $L \approx N_D D_e$.

With this "blob picture" in mind for polyions dissolved in a polar solvent, we now consider the case of polyion chains adsorbed at an interface. The adsorption of polyelectrolytes at charged and neutral surfaces from aqueous solutions has been the subject of extensive theoretical and experimental investigations (Dobrynin and Rubinstein, 2005; Netz and Andelman, 2003).

In aqueous solutions, most polyelectrolytes behave as a "surface-active" (surfactant) substance, that is, they adsorb at the free water surface (Yim et al., 2002). This behavior is essentially due to the presence of hydrophobic moieties along the chain (de Meijere et al., 1999; Yim et al., 2002). Usually, at very low concentrations, polyions adsorb at the air–water surface only for very low molecular weights (Caminati and Gabrielli, 1993). The profiles of the adsorbed layer also depend on the bulk concentration. At a low polymer concentration, the profiles at the air surface comprise a thin layer of high concentration, where segments of the chains lie flat (trains), and a second layer of much lower segment concentration that extends into the liquid (loops and tails; Yim et al., 2002).

Polyelectrolytes spontaneously adsorb to a surface of opposite charge. The adsorbed layers can be thin with the chains lying flat on the surface, or they can be more fluffy with the chains forming loops and dangling ends between "adsorption trains" in a "pseudobrush" configuration (Théodoly et al., 2001). Which conformation is favored mostly depends on the linear charge density of the polyion and on the charge density of the surface. In other words, this conformation depends on a balance between the strength of the attraction (electrostatic and nonelectrostatic) between the chain and surface and the increase of free energy of the adsorbed chains that is attributable to the loss of conformational entropy. Depending on the stiffness of the polyelectrolyte, the layer can be flat and compressed or coiled and extended.

Several theories have been proposed to describe the coupling of polyelectrolytes to a charged interface bathed by the polyion solution (Borukhov et al., 1998; Dobrynin et al., 2000; Dobrynin and Rubinstein, 2005; Netz and Andelman, 2003; Netz and Joanny, 1999). The main questions that have been addressed concern the conformation of the molecules in the adsorbed layer and the origin and amplitude of the *charge overcompensation* or overcharging effect, that is, the possibility that more polyions adsorb than are needed to neutralize the interface (overcompensation), so that the overall net charge of the surface changes its sign (charge inversion). Theoretical studies predicted that the thickness of an adsorption layer of a strong polyelectrolyte in a solution of low ionic strength is proportional to the inverse square root of the charged polymer fraction (Borukhov et al., 1998). This prediction has been confirmed experimentally (de Meijere et al., 1999).

More recently, the dependence of the conformation of the chain on both the polyion linear charge density and the surface charge density has been thoroughly studied on the basis of the scaling model for flexible, highly charged polyelectrolytes (Dobrynin and Rubinstein, 2005; Netz and Andelman, 2001).

In the low salt regime (k_D^{-1} larger than the thickness D of the film) and at sufficiently low surface charge density σ, the polyelectrolytes, which are strongly attracted to the surface and strongly repel each other, lie flat at the charged surface. As a result of the balance between the electrostatic attraction of the chains to the surface and their confinement entropy, the layer thickness D decreases with increasing σ (chains lie flatter and flatter) as $D \approx (f\sigma l_B/b^2)^{-1/3}$ (in a θ solvent). In this regime, at the lowest order in σ, the charge of adsorbed polyelectrolytes, $f\Sigma$ (where Σ is the surface density of adsorbed monomers) exactly compensates the surface charge, that is, $\sigma = f\Sigma$. However, at higher order in k_D, adsorbed polyions *overcompensate* the surface charge and $f\Sigma = (\sigma + d\sigma)$. The excess charge $d\sigma$ is due to the presence of loops, as a result of the conformational entropy of the chains, and depends on the σ (Dobrynin and Rubinstein, 2005; Netz and Andelman, 2001) as

$$\frac{\delta\sigma}{\sigma} = k_D D \left(\frac{D}{D_e} \right)^2. \tag{16.5}$$

At increasing charge density, a value of $\sigma = \sigma_e = f/b^2$ is reached when the adsorbed polymer chains come into close contact. For σ f σ_e, the polyions cannot lie flat at the surface any longer and they form a self-similar carpet. In this regime, the electrostatic attraction between the polyelectrolytes and charged surface is balanced by the short-range repulsion between monomers. The D increases now with σ as $D \approx D_e(\sigma/\sigma_e)^{1/3}$. In this "carpet regime" Dobrynin et al. (2000) and Dobrynin and Rubinstein (2005) predict an overcharging

$$\delta\sigma \approx \sigma_e k_D D_e \left[1 - k_D D_e \left(\frac{\sigma}{\sigma_e} \right)^{4/3} \right] \tag{16.6}$$

that increases with σ. As an example, for sodium polyacrylate (NaPAA), which is characterized by a monomer size $b = 1.8$ Å (Schmitz and Yu, 1988), the charge fraction f on the polyion calculated from the Manning theory is $f = 0.25$, corresponding to a crossover surface charge density σ_e equal to one elementary charge e per 13 Å^2. This surface charge density is higher, for example, than the one reached when a 1,2-dioleoyl-3-trimethylammonium-propane (DOTAP) lipid film is at it maximum compression before collapse (about 1/60 Å^2; Bordi et al., 2003a).

In the adsorption process, part of the counterions condensed on the polyions is probably released. However, as recently discussed by Grosberg et al. (2002) for counterion release, the key role is played by the correlation between adsorbed macroions at the charged surface, and the counterion release is a consequence rather than the driving force for adsorption (Wagner et al., 2000).

16.3.2 CATIONIC LIPOSOMES

Liposomes are vesicular structures formed by a closed lipid bilayer, encompassing an aqueous core. In an appropriate environment, these structures spontaneously self-assemble, because of the amphiphilic character of their component molecules. Amphiphilic molecules, which are composed of different hydrophobic and hydrophilic parts, in an aqueous solution dissolve as isolated monomers only at very low concentrations. Above the critical micelle concentration (CMC), these molecules aggregate, forming large structures. The CMC is characteristic of the different substances. However, it is invariably very low and on the order of a few micromoles per liter for short-chain lecithins and even lower when the chain length and hence the volume of the hydrophobic part increases (Jones and Chapman, 1995).

Among the morphologically different structures that, under different thermodynamic conditions and for different molecules, result from this spontaneous aggregation, unilamellar or multilamellar vesicles (liposomes) are of peculiar interest in several aspects. Liposomes represent a good model of biomembranes, particularly for studying the interactions between membrane proteins and different lipid environments, because they offer the unique advantage that the composition of their bilayer can be varied in a well-defined and controlled way. For example, liposomes can be doped with receptor moieties to mimic cell adhesion and used as targets for membrane-active peptides.

Adsorption and adhesion of liposomes play essential roles in many biological processes such as exo- and endocytosis and membrane trafficking. Moreover, liposomes are frequently used as vehicles for drug delivery or as reaction compartments on a nanoscale level. As noted earlier, because the lipid bilayer allows the entrapment of hydrophobic material within the hydrocarbon chain phase and hydrophilic material within the aqueous core, liposomes represent a versatile drug delivery system (Gregoriadis, 1993; Lasic, 1998).

The lipid composition, together with the characteristics of the aqueous phase, defines the physicochemical properties of these structures, such as their stability, the surface charge density (at the lipid–aqueous phase interface), surface hydrophilicity/hydrophobicity ratio, and bilayer rigidity, and their properties as colloidal particles, such as size, electrophoretic mobility, interparticle interactions, and interactions with ions and other molecules in the solution.

The possibility of fine-tuning these properties suggested a number of applications in various fields. There has been considerable development of liposome applications in the last few years, especially in the clinical and biotechnological fields (Gregoriadis, 1993; Lasic, 1998).

Liposomes are usually built up by phospholipids (which are also the main components of biological cell membranes) with variable amounts of cholesterol and other surfactants. Phospholipids are derivatives of a trivalent alcohol: glycerol. Two of the alcoholic moieties are esterified with the fatty acids, which form the hydrophobic part (usually referred to as a "hydrophobic tail") and can have variable length and degrees of unsaturation. The hydrophilic "head," which is made up of a phosphate group and a variable polar residue, is bonded to the third alcoholic group.

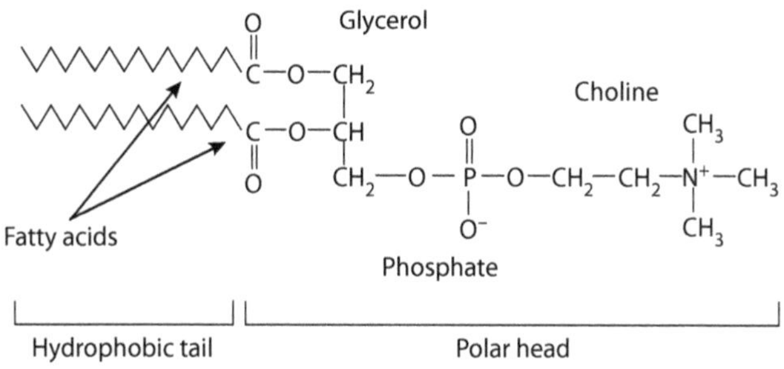

FIGURE 16.1 Structure of phospholipid, di-palmitoyl-phosphatidyl-choline. A trivalent alcohol (glycerol) links the two fatty acids that form the hydrophobic part. The acyl chains can be saturated or unsaturated, and in naturally occurring phospholipids generally have chain lengths from C_{10} to C_{28}. The hydrophilic "head," made up of a phosphate group and a variable polar residue (choline, in this example), is bonded to the third alcoholic group.

The name of this class of substances is derived from the simultaneous presence within the molecule of a phosphate (phospho-) group and a lipidic (fatty) part. Figure 16.1 shows the structure of di-palmitoyl-phosphatidyl-choline, which is derived from palmitic acid (*n*-hexadecanoic acid), a saturated fatty acid with 16 carbon atoms. The last part, choline, is the name of the group that constitutes the characteristic part of the hydrophilic head.

Natural phospholipids are mostly unsaturated; that is, they contain one or more double carbon–carbon bonds in one or in both of the hydrophobic chains. Molecules obtained from different natural sources show characteristic "patterns" in terms of chain length and degree of unsaturation. The higher the degree of unsaturation is, the lower the main transition temperature T_m. Below T_m, the hydrophobic tails within a double layer arrange themselves in a more ordered state (gel state); above T_m, the tails are more disordered (liquid state) and the fluidity of the bilayer increases.

Concerning the polar head, natural phospholipids either bear a net negative charge (an anionic phospholipid, e.g., phosphatidic acid, or phosphatidylinositol) or are zwitterionic, such as phosphatidylcholine. In zwitterionic molecules, although they are neutral as a whole, a unit charge appears displaced, so that the molecule acquires a strong dipole moment. In phospholipids, the phosphate usually loses a proton, which is acquired by the polar head group (e.g., choline). This effect is strongly dependent on the pH of the solutions, so a zwitterionic phospholipid can assume a negative net charge in basic environments and a positive one in acidic environments at lower pH values.

Cationic lipids used in gene therapy are synthetic molecules that generally maintain a close similarity to the overall structure of their natural homologues. Figure 16.2 shows the structure of two different cationic lipids widely employed in gene therapy: DOTAP (Figure 16.2a) with two long unsaturated chains and a trimethylammonium cationic group as the polar head and 3β-[N-(N′,N′-dimethylaminoethane)-carbamoyl]-cholesterol (Figure 16.2b), which is the cationic homologue of neutral cholesterol.

Although the basic "driving force" for the self-assembling of amphiphilic lipids is represented by the entropic gain that occurs when water is not "forced" to assume a structure compatible with a nonpolar surface (Finkelstein and Ptitsyn, 2002), electrostatic interactions play a fundamental role in the aggregation process and in determining the properties of the resulting particles as a whole. For this reason, the presence of an external electric field can have a strong influence on the organization of the lipids within the bilayer. Correspondingly, the coupling of an external electric field to the system furnishes a valuable physical probe to investigate the structure and the dynamical properties of the liposome–water interface (Bordi et al., 2006b).

Liposomes can be prepared by different methods to control their dimension and the number of the lamellae. Various experimental procedures have been developed to obtain homogeneous and reproducible liposome suspensions of appropriate size range (Barenholtz and Crommelin, 1994;

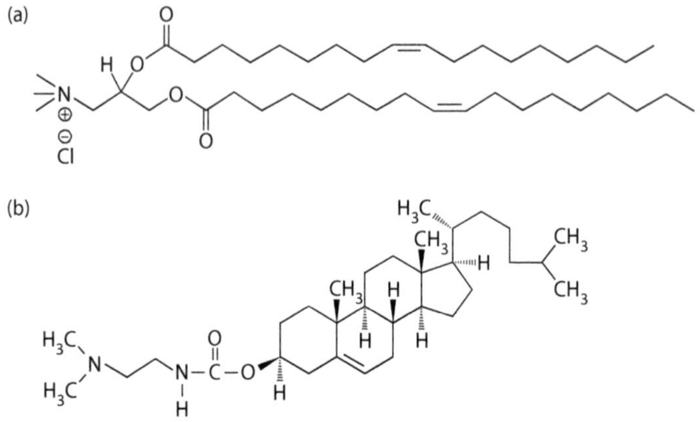

FIGURE 16.2 Structure of typical cationic lipids widely employed in gene therapy: (a) DOTAP and (b) 3β-[N-(N',N'-dimethylaminoethane)-carbamoyl]-cholesterol.

Gregoriadis, 1993; Hope et al., 1983; Lichtenberg and Barenholtz, 1998; New, 1990; Storm et al., 1988; Szoka and Papahadjopolous, 1981).

16.3.2.1 Effect of Surface Charge Density on Spontaneous Vesicle Formation

Relatively large vesicles, such as those required in biomedical applications, usually form under nonequilibrium conditions and the deformation of the flat lipid double layer into a spherical one is obtained at the cost of some bending energy (Gradzielski, 2003). For smaller vesicles, the free energy increase due to bending can be compensated if the outer layer contains significantly more molecules than the inner layer. Symmetrical bilayers consisting of two identical monolayers have zero spontaneous curvature and their bending elasticity can be written (Helfrich, 1973) as

$$dE = \left[\frac{1}{2}\rho(c_1 + c_2)^2 + \bar{\rho}c_1c_2 \right]dA, \tag{16.7}$$

where dE is the free energy cost for bending the unit area dA with principal local curvatures c_1 and c_2 and ρ and $\bar{\rho}$ represent the mean and the Gaussian curvature moduli, respectively. Because pure phospholipid bilayers have bending moduli on the order of 5–25 $K_B T$ or more (Brannigan et al., 2004; Kumaran, 1993, 2001a, 2001b) at room temperature, vesicles ≥100 nm should rarely form spontaneously in these conditions. However, when the bilayer consists of a mixture of different amphiphiles, characterized by different packing parameter values (Israelachvili, 1985), equilibrium vesicles form spontaneously (Safran et al., 1991).

In charged bilayers, the bending moduli ρ and $\bar{\rho}$ should be written as the sum of two contributions: a specific electrostatic contribution, ρ_{elec} and $\overline{\rho_{elec}}$, should be added to the curvature moduli representing the "intrinsic" bending elasticity arising from packing constraints, ρ_p and $\bar{\rho}_p$. The comprehensive mean and Gaussian bending moduli are thus rewritten as

$$\rho = \rho_p + \rho_{elec},$$
$$\bar{\rho} = \bar{\rho}_p + \overline{\rho_{elec}}. \tag{16.8}$$

This makes it evident that the charge density asymmetries on the two layers contribute to the vesicle stabilization (Kumaran, 2000, 2001a, 2001b; Lau and Pincus, 1998). When the charge density on the outer surface of the vesicle is larger than the one on the inner surface, the energy expenditure for bending the bilayer can be compensated by a reduction in electrostatic repulsion in the outer monolayer, which is due to its deformation. This is also true when the charges are allowed to flip from one side to the other. In addition, the entropic disadvantage attributable to the unequal charge distribution cannot be sufficient to compensate the reduction in electrostatic energy (Kumaran, 1993).

Even in the absence of charge distribution asymmetries, there could be spontaneous formation of vesicles when the surface charge density exceeds a limiting value (Winterhalter and Helfrich, 1988). Analyzing the effect of charge density on membrane curvature moduli shows that, although ρ_{elec} is always positive, $\overline{\rho_{elec}}$ may be negative. At a sufficiently high surface charge density and low ionic strength, the negative contribution of $\overline{\rho_{elec}}$ becomes comparable to the positive ρ_{elec}. In these conditions, polydisperse, entropically stabilized vesicles could form spontaneously. At higher charge densities, $\overline{\rho_{elec}}$ dominates and the flat bilayer becomes unstable against spontaneous bending, leading to the formation of small, thermodynamically stable vesicles. The effect of the surface charge in inducing double-layer bending and in stabilizing liposomes has been observed in different systems (Brasher et al., 1995; Claessens et al., 2004; Hao et al., 2001; Hoffmann et al., 1994; Kim and Sung, 2002; Oberdisse, 1998; Oberdisse et al., 1996; Oberdisse and Porte, 1997).

16.3.2.2 Counterion Condensation and Effective Charge of Colloidal Particles

Because simple ions in electrolyte solutions accumulate around highly charged colloidal particles, the relevant parameter to compute particle–particle interactions is not their bare charge, but an effective (or renormalized) quantity, whose value is sensitive to the geometry of the colloidal particle, the temperature, and the presence of added salt. This nonlinear screening effect is a central feature in the field of colloidal suspensions.

A quantitative analysis of the charge renormalization in colloidal suspensions is comparatively more recent and was initiated by the pioneering work of Alexander et al. (1984). As a result, an "effective charge" is associated with the "decorated" object made of the charged particle plus its condensed counterions.

The extent of counterion condensation clearly depends on the intrinsic charge density at the particle surface. However, when an *isolated* charged object is surrounded by an *unbounded* domain, its geometry plays an important role as well. It is easy to show that at infinite dilution, counterion condensation takes place only at the surface of an infinitely extended plane or a cylinder but not around a spherical particle. For a charged plate, the electric potential felt by the single counterion grows linearly with the size of the plate. In this case, an unbounded domain means an infinitely extended plate that, with its infinite surface potential, can "condense" any counterion. In cylinders, the surface potential depends logarithmically on the cylinder length. Because the entropy associated with the counterion is proportional to $K_B T \ln V$, where V is the available volume, at infinite dilution, the logarithmic divergence of entropy is balanced by the divergence of the electrostatic gain, as the length of the cylinder goes to infinity. For this reason, Manning (1969, 1978) and Oosawa (1971) indicated that the line charge density of a polyelectrolyte is limited to a maximum value determined by this balance. All other counterions are "condensed" close to the polyelectrolyte. However, in an isolated charged sphere, the potential energy of a counterion on its surface remains finite also at infinite dilution, so that, in this limit, a charged sphere is unable to bind counterions at any finite temperature. Nevertheless, although counterions do not condensate around an isolated charged sphere, a finite counterion concentration exists, because the counterion entropy in this condition is also finite in all practical salt-free colloidal systems and counterion condensation occurs for sufficiently high charge densities.

In practice, in colloidal suspensions made of spherical charged particles the notion of *effective charge* is widely used in the literature at finite concentration, both for the equilibrium and dynamical properties. Because of all of these considerations, in spherical geometry, charge renormalization should only be considered at intermediate salt concentrations (Netz and Andelman, 2003). In the absence of a "limiting condition" similar to the Manning condition, the problem remains of defining how much of the counterions are effectively condensed and which is the effective charge of a colloidal particle.

As a further difficulty, for liposomes, Taheri-Araghi and Ha (2005) calculated that the dielectric discontinuity at the surface (the dielectric permittivity within the double layer is much lower than the permittivity of water) and the charge correlations among lipids and condensed counterions influence the effective charge of the surface. In particular, for monovalent counterions, dielectric discontinuities can enhance counterion condensation. Moreover, the effects of dielectric discontinuities and surface-charge distributions seem to be correlated. Dielectric discontinuity diminishes the condensation if the surface charge is uniformly smeared out, whereas counterions are localized in space. They can, however, enhance condensation when a discrete surface charge is considered.

From an experimental point of view, it is noteworthy that, for charged liposomes, the effective charge obtained from electrophoretic mobility measurements is in good agreement with the values that can be calculated from the measured (by light scattering) static structure factor of a liposome suspension, assuming a Debye–Huckel type pair interaction potential, with the Ornstein–Zernike equation and the hypernetted chain approximation as a closure relation (Haro-Pérez et al., 2003). In contrast, the effective charge estimated from the measured shear modulus in deionized aqueous suspensions of highly charged spherical latex colloids was systematically smaller than the effectively transported charge by about 30% (Wette et al., 2002). These findings again point out that the value of the effective charge determined by different techniques and on different systems can differ significantly, so this concept should be used with caution.

16.4 POLYELECTROLYTE–LIPOSOME INTERACTIONS

When polyelectrolytes and charged surfactants are added to the same aqueous solution, they strongly interact, affecting both the bulk and the surface properties of their environment. For example, because of a strong electrostatic interaction, polyelectrolytes and surfactants of opposite charge can form hydrophobic complexes at surfactant concentrations lower than the CMC or their coadsorption at the air–water interface can lead to interfacial gels (Monteux et al., 2004).

In the presence of an oppositely charged surface in an aqueous solution, polymers readily adsorb on the surface itself, although usually with a very slow kinetics (Noskov et al., 2004; Théodoly et al., 2001). The adsorbed layers can be thin with the chains lying flat on the surface or can be more thick and fluffy with chains forming loops and dangling ends between the adsorption trains, adhering to the surface in a pseudobrush configuration (Millet et al., 2002; Yim et al., 2002). The favored conformation depends mostly on the charge fraction of the polymer and the surface charge density of the substrate (Dobrynin and Rubinstein, 2005; Klebanau et al., 2005).

When polyelectrolytes are added to a suspension of oppositely charged liposomes, the adsorption is an almost instantaneous phenomenon (Bordi et al., 2003b, 2004a) in striking contrast to the slow kinetics observed for the adsorption on a monolayer surface. In fact, in liposomes, the adsorption process is strongly enhanced by the huge surface area available (in 1 mL of a liposome suspension 1% in volume with an average particle size of 100 nm, the total area of the external surface is about 3×10^3 cm^2) and the adsorption is favored by the fine dispersion in the whole suspension volume (in the same example the average distance between two liposomes = ~700 nm).

In some cases, colloidal particles are stabilized by polymers grafted or adsorbed on their surface (Russel et al., 1989). If two polymer-coated particles approach each other, the respective surface layers overlap and this usually leads to a repulsive interaction between the particles ("steric interaction" or "steric stabilization"). Conversely, in polyelectrolytes adsorbed onto oppositely charged liposomes, the interaction phenomenology is quite different and the pd-liposomes rapidly aggregate (Bouyer et al., 2001; Gregory, 1973, 1976; Leong, 1999, 2001; Leong et al., 1995; Walker and Grant, 1996).

The interaction mechanism appears to be related to the correlated adsorption of the polyelectrolytes at the particle surface. Because adsorbed polyions are not positioned at random, but correlations occur, because they reconfigure themselves in an orderly way to gain some energy, polymers form charged "patches" of one sign (negative in anionic polyelectrolytes) on otherwise oppositely charged particles (e.g., cationic liposomes). In other words, avoiding each other and residing as far away as possible to minimize their reciprocal electrostatic repulsion, adsorbed polyions leave the particle surface "partially uncovered." Hence, the particle surface appears to be decorated by a more or less ordered "patchwork-like" pattern, with excess negative charge domains (polyelectrolyte domains) and excess positive charge domains (polyelectrolyte free). When two such particles approach closely enough, the oppositely charged patches on the different particles can attract each other, even though the overall net charge on the two particles has the same sign. This picture was invoked by Gregory (1973, 1976) in the early 1970s to rationalize his experimental results on the flocculating effect of polyelectrolytes on colloidal suspensions.

The possibility that a short-range attractive potential can arise between two nonuniformly charged surfaces bathed by an electrolyte ("charge patch" attraction) recently gained a solid theoretical basis (Khachatourian and Wistrom, 1998; Miklavic et al., 1994; Velegol and Thwar, 2001). In its original formulation, the charge patch model was concerned with intrinsic nonhomogeneities in the distribution of charged groups on the surfaces, but it can be easily generalized to the case where the nonhomogeneities are due to adsorbed polyions. Velegol and Thwar (2001) showed that the interaction between two nonuniformly charged colloidal surfaces is always attractive and that the length scale of this interaction is on the order of the Debye length.

In recent years, increasing evidence from experiments (Bloomfeld, 1996; Olvera de la Cruz et al., 1995; Strey et al., 1998) and numerical simulations (Linse, 2002; Messina et al., 2000; Najia and Netz, 2004) showed that like-charged particles can attract each other via effective forces of electrostatic origin. Numerous theoretical studies sought to clarify the possible underlying mechanisms of these counterintuitive interactions. Within the framework of a mean-field approach, that is, within the classical Poisson–Boltzmann theory (Neu, 1999; Trizac, 2000), only repulsive forces between like-charged objects can be expected. Different models have been proposed that, incorporating the effect of the correlations, justify most of the features of this like-charge attraction on the basis of a rigorous statistical mechanics theory (see, e.g., Boroudjedi et al., 2006; Linse, 2002).

Recent Monte Carlo simulations of complexes formed by polyelectrolytes and oppositely charged colloids pointed out the possibility of a different mechanism that results in an effective attraction between the complexes (Dzubiella et al., 2003). Simulations demonstrated that, for sufficiently short polyion chains, the existence of a permanent dipole moment of the complexes (again due to the correlated adsorption of polyelectrolytes on the colloidal particles) leads to a van der Waals type short-range attraction. This attractive interaction vanishes for very long chains (larger than the particle size), where the permanent dipole moment is negligible. At short distances, the complexes interact with a deep short-range attraction, which is due to an "energetic bridging" (Podgornik et al., 1995; i.e., a polyelectrolyte chain is "shared" by two adjacent particles, partially neutralizing both colloids) for short chains and to an entropic bridging (part of a chain is adsorbed on one particle and the remaining on another particle) for long chains.

From the above discussion it appears that, although theoretical approaches differ on several aspects, correlations among the charged particles, which are neglected in the mean-field approximation, play a fundamental role. Whatever the details of this interaction may be, the simple picture of charge patch attraction appears to be able to intuitively capture the basis of the attraction mechanism.

Returning to the observed phenomenology, adding increasing amounts of polyions in a suspension of oppositely charged liposomes, the polyion chains keep adsorbing, in a correlated way, on the liposome surface. Close enough to the isoelectric point, where the charge of the adsorbed polyelectrolyte compensates the original liposome charge, the charge patch attraction prevails on the relatively small electrostatic repulsion and the polyion-decorated liposomes begin to aggregate. Surprisingly enough, aggregation stops when the clusters reach an equilibrium average size that depends on the polyion/liposome charge ratio and the liposome concentration (Bordi et al., 2004a, 2004b), which is discussed in the next section.

16.5   AGGREGATION DYNAMICS: REENTRANT CONDENSATION AND CHARGE INVERSION

Now that we have discussed the basis of the polyion-induced charged particle aggregation, we are able to present the phenomenology of the charge inversion and reentrant condensation of a particular system consisting of clusters of DOTAP liposomes stuck together by an oppositely charged polyion. Because of this system's aggregate size (and size distribution) and average electrical charge properties, it is particularly suitable to be employed as a drug delivery system.

Within the framework of the classical Derjaguin–Landau–Verwey–Overbeek (DLVO) theory of colloid stability, interactions between colloidal particles are described as the superposition of long-range electrostatic repulsions and short-range attractions. When attractive interactions have sufficient strength (some $K_B T$ per particle, where $K_B T$ is the thermal energy) and repulsions are sufficiently weak, an irreversible aggregation process occurs (Russel et al., 1989). An example of this process is the destabilizing effect when a simple salt is added to a colloidal suspension. In this case, because of the increase of the ionic strength of the suspension, electrostatic repulsions between the particles are increasingly screened. Above a certain amount of added salt, repulsions are no longer effective in avoiding collisions among the particles and an irreversible aggregation process begins. The process eventually yields colloid flocculation (the formation of large aggregates that are visible to the naked eye as flakes or flocks and ultimately separate from the suspending medium).

Polyelectrolytes can effectively induce the irreversible aggregation of oppositely charged liposomes (Bordi et al., 2004c), but only when a large excess of polyions is added to the suspension. In this case, the aggregation proceeds until the clusters flocculate and precipitate and a further increase of the polyion concentration above an "irreversible aggregation threshold" does not change the final outcome (flocculation), instead only influencing the kinetics of the process.

However, the aggregation of pd-liposomes observed at moderate polyelectrolyte/liposome charge ratios, where the total charge on the polymer is comparable to the total charge on the liposomes (i.e., close to the isoelectric point), is a completely different phenomenon (Bordi et al., 2004a, 2004c, 2005a, 2005b, 2005c, 2006a; Sennato et al., 2004, 2005a, 2005b). Figure 16.3

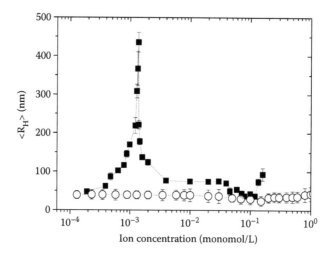

FIGURE 16.3   Different effects of adding a simple electrolyte (NaCl) or a polyelectrolyte (NaPA) to a suspension of cationic liposomes (DOTAP). (Open symbols) At increasing NaCl concentration, the average hydrodynamic radius $\langle R_H \rangle$ of the suspended particles obtained from DLS is approximately constant. (Filled symbols) Completely different behavior is observed when the polyelectrolyte NaPA is added at the same monomolar concentration to a DOTAP liposome suspension. In this case, a striking "reentrant" condensation is observed, with the formation of large stable clusters, the radius of which does not evolve with time and depends, in a rather reproducible manner, on the polyelectrolyte/liposome concentration ratio.

compares these two different behaviors (simple salt and polyion-induced aggregation). By adding increasing amounts of NaCl to a suspension of cationic DOTAP liposomes, the average hydrodynamic radius $<R_H>$ of the suspended particles (obtained from DLS measurements) remains constant in a wide concentration range. At an NaCl concentration higher than 1 mol/L, liposomes begin to aggregate irreversibly (data not shown); from this point onward, the $<R_H>$ values are no longer equilibrium values.

Different behavior is observed when a polyelectrolyte is added to the liposome suspension, and a new phenomenon appears. Adding monotonously increasing amounts of polyelectrolyte to equal volumes of the liposome suspensions causes stable clusters to form rapidly, the radius of which does not evolve further with time; but its value depends reproducibly on the polyion/liposome concentration ratio ξ (Figure 16.3). The behavior undergone by the average hydrodynamic radius is typical of the reentrant condensation— increasing with the increase of the polyion concentration, reaching a maximum close to the isoelectric point, and then decreasing again toward its initial value—when the polyion concentration continues to increase. The parameter ξ is defined as the ratio between the total number of the negative charges N^- on the polyions and the total number of the positive charges N^+ on DOTAP molecules in the whole suspension according to

$$\xi = \frac{N^-}{N^+} = \frac{C_P}{M_{WP}} \frac{M_{WL}}{C_L}, \tag{16.9}$$

where C_P and C_L are the (weight) concentrations of the polyelectrolyte and the lipid into the sample solution, respectively, and M_{WP} and M_{WL} are the molecular weights of the repeating unit of the polyelectrolyte and the lipid, respectively.

This "reentrant" condensation is different from the irreversible aggregation predicted by the DLVO theory, which was observed at higher polyelectrolyte concentrations. In addition to the evident reentrant behavior, there are at least two differences. The first difference is that the clusters form very rapidly, reach a given size, and remain stable for weeks or longer periods of time. In the DLVO model the aggregation is an irreversible process that eventually ends with flocculation or with "kinetically stabilized" large aggregates, which stop to aggregate only because they become too large to diffuse and meet each other on the timescale of the experiment. In contrast, in the case shown in Figure 16.3, liposome aggregates are stable; once an appropriate size is reached, its value is maintained constant over a long period of time. The second important difference is that the "equilibrium" size reached by the complexes depends reproducibly on the polyion/liposome ratio and the overall liposome concentration (Sennato et al., 2004). Finally, for each charge ratio, the aggregates are relatively monodisperse whereas irreversible aggregation (both diffusion-limited and reaction-limited cluster aggregation; Russel et al., 1989) is usually accompanied by high polydispersity of the aggregates, with monomers that continue to coexist with the larger aggregates, even in the later stages of the process.

All of these features can be accounted for by means of a semiquantitative model that satisfactorily describes the observed behavior of the system, although to a first approximation (Bordi et al., 2005a, 2005c). In this model, assuming that the clusters are equilibrium aggregates, their average size R_H is calculated as the result of a balance between the screened electrostatic repulsion, attributable to the residual overall charge of the pd-liposomes, and a short-range attraction (due to the charge patch).

The balance is attained when the last pd-liposome of radius R_0 sticks to a cluster already containing $(N-1)$ pd-liposomes. From there onward the repulsion due to the charge built up on the whole cluster exceeds the short-range attraction and no other pd-liposome can stick. Details of the model can be found in the works of Sennato et al. (2004) and Bordi et al. (2004c, 2005c). With the increase of the polyion/liposome ratio, the pd-liposomes go through the "charge inversion" sequence, and the charge of the adsorbed polyelectrolyte increasingly compensates and then exceeds the original liposome charge. For this reason, the neat charge on each primary particle composing the cluster decreases in absolute value, goes through zero, and inverts its sign. Hence, because of the charge inversion, the system shows a reentrant condensation with the size of the aggregates that goes through a maximum when the charge on the individual primary particles (pd-liposomes) is approximately neutral.

Concomitant to the reentrant condensation, the system undergoes a relevant charge inversion effect. An example of this complex behavior is provided in Figure 16.4, where in correspondence to the isoelectric condition, characterized by the maximum aggregate size, the zeta potential inverts its sign and passes from positive to negative values. The zeta potential is an electrokinetic parameter that depends on the electrical charge distribution at the particle surface. The inversion of its value, which is close to the isoelectric point, means that the single aggregate during the cluster formation changes its charge and at the end of the process it bears an excess of polyions, causing an excess of negative charges (see Figure 16.4). These two peculiar features of this system (charge inversion and reentrant condensation) are of relevant importance when these aggregates are considered as drug carriers in drug delivery. It is well known that the aggregate size is the major determinant of *in vitro* (and *in vivo*) transfection efficiency. However, the appropriate choice of this parameter is not enough to ensure the maximum of transfection; different studies suggest that an optimization of size and charge needs to be individualized to significantly enhance the activity of these lipidic vectors. Optimization will remain largely a result of trial and error until the importance of the correlation between size and charge is recognized. These systems offer the possibility to go into this direction because they allow the simultaneous modulation of charge and size to the desired value (at least to

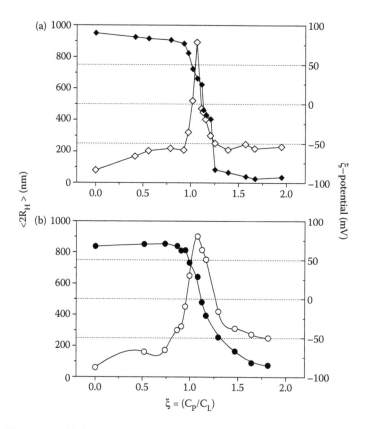

FIGURE 16.4 (Open symbols) Average hydrodynamic diameter $<2R_H>$ and (filled symbols) zeta potential of NaPA-DOTAP complexes as a function of the polyion/DOTAP charge ratio parameter ξ, for two different molar concentrations of a simple electrolyte (CsCl) employed in liposome preparation: (a) 0.1 M CsCl and (b) 0.01 M CsCl. Lines are to guide the eye only.

a large extent) in a relatively simple and reproducible way. This important aspect has never been studied seriously.

16.6 EQUILIBRIUM VERSUS KINETICALLY STABILIZED AGGREGATES

Whether the observed clusters are "true" equilibrium or kinetically stabilized aggregates still remains a unsettled matter (Bordi et al., 2005a; Volodkin et al., 2006). Within the framework of the model described in the previous section (Bordi et al., 2005a), the clusters would result from a force balance. The whole aggregation process should thus be reversible and the aggregates should resize as a consequence of any change in the equilibrium conditions. Such an unbalance can be realized, for example, by varying the net charge on the primary particles. However, the simple addition of a proper amount of the polyelectrolyte to already formed aggregates did not have the expected result, and the size of the aggregates did not change according to the new polyion/liposome ratio.

However, resizing of the clusters can be obtained by following an appropriate experimental procedure. The equilibrium can be shifted to change, for example, the value of the screening length κ_D^{-1}, by adding a simple electrolyte to the suspension. Thus, resizing of the clusters can be induced without any modification of the polyion adsorption, but only by changing the range of the repulsive interaction. By adding a simple electrolyte at different concentrations to the liposome suspension (keeping the charge ratio ξ constant), clusters of different sizes are obtained. Figure 16.5 shows the equilibrium size distribution of the aggregates formed at the same charge ratio ($\xi = 0.5$) but with NaCl concentrations of 0.004 mol/L (Figure 16.5a) and 0.35 mol/L (Figure 16.5b). The larger aggregates (Figure 16.5b) are obtained for the larger electrolyte concentration, where the range of repulsive interactions is smaller.

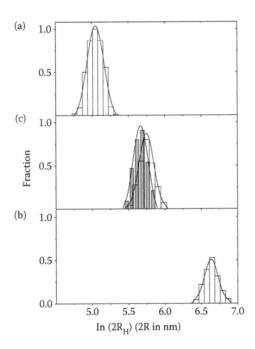

FIGURE 16.5 Particle size distribution of pd-liposome clusters at different NaCl electrolyte concentrations: (a) 0.004 mol/L; (b) 0.35 mol/L; and (c) (open bars) 1:1 (v/v) mixture of (a) and (b), resulting in a final concentration of 0.17 mol/L, and (filled bars) the sample obtained by directly adding NaCl to the liposome suspension to a final concentration of 0.17 mol/L. The polyion/lipid ratio is 0.5. The autocorrelation functions of the intensity of the scattered light were analyzed by using a nonnegatively linear sampling algorithm Lawson and Morrison, 1974. The distributions are normalized to the intensity of the scattered light measured for sample (a).

By mixing these two suspensions, after some time (on the order of minutes) the expected intermediate size clusters are yielded, corresponding to the final intermediate NaCl concentration of 0.17 mol/L (Figure 16.5c).

Although the adsorption of polyions onto the liposome surface cannot be considered an equilibrium process (at least at these charge ratio values), this experiment clearly shows that the aggregation of the pd-liposomes appears as an equilibrium, reversible process.

This aspect further reinforces the possibility of utilizing these systems in drug delivery techniques because the size (and consequently the charge) can be conveniently tuned by varying the environmental parameters, such as the ionic strength of the aqueous dispersing medium.

16.7 MORPHOLOGY AND STRUCTURE OF THE AGGREGATES

The morphology of the pd-liposome clusters presents intriguing aspects. The structures of the aggregates formed by cationic lipids and DNA have been thoroughly investigated in recent years (Harries et al., 1998; Koltover et al., 1999; May et al., 2000; Rädler et al., 1997, 1998; Safinya, 2001; 199) because of the increasing acceptance of these complexes as preferential DNA delivery vehicles in gene therapy (Ferber, 2001; Pedroso De Lima et al., 2001; Woodle and Scaria, 2001). Although it has been shown that under appropriate conditions liposome restructuring occurs during the formation of liposome–DNA complexes (lipoplexes; Harries et al., 1998; Koltover et al., 1999; May et al., 2000; Rädler et al., 1997, 1998; Safinya, 2001), there is mounting evidence that, at least in a low concentration range, aggregates form whereas liposomes maintain their individuality (Bordi et al., 2006a, 2004a; Volodkin et al., 2006; Yaroslavov et al., 1998).

Direct and unequivocal evidence of the formation of such stable clusters of vesicles was recently produced by using a sophisticated electron microscopy technique and employing different concentrations of a high atomic number electrolyte, such as CsCl, contained within the aqueous core of the liposomes as a contrast medium (Bordi et al., 2006a). In this experiment, DOTAP liposomes were prepared in aqueous suspensions at two different CsCl electrolyte concentrations (0.1 and 0.01 M) following the standard procedure. Equal amounts of the two liposome suspensions were mixed together immediately before the addition of the NaPA polyions inducing the aggregation to have heavily and lightly Cs-loaded liposomes within the same cluster. The evolution of the hydrodynamic radius for the complexes of Cs-loaded liposomes with the polyelectrolyte at increasing polymer concentrations shows the usual reentrant condensation and charge inversion effects.

Within this procedure, electron spectroscopy imaging (ESI) was utilized to obtain a direct image of the liposomes in a typical cluster without any need to stain the sample (Figure 16.6). Moreover, the ESI technique furnishes a "map" of the presence of Cs within the sample, allowing the unambiguous identification of the structures that contain this element. ESI evolved from electron energy loss spectroscopy, which measures the energy loss suffered by the high-energy incident electrons when transmitted across the sample. In this technique, only the electrons that show the energy loss characteristic of their interaction with a specified element contribute to the image formation and a topographic map of that particular element within the sample can be obtained. As a further advantage, energy filtering allows the suppression of the contribution of inelastic scattering that typically occurs when the sample is mainly made by light elements, as in the case of biological samples (Diociaiuti, 2005). Figure 16.6 shows an ESI-TEM image of a typical aggregate induced by adding the polyelectrolyte to a mixed suspension of heavily and lightly Cs-loaded liposomes. The aggregate clearly appears as a cluster of small globular particles (~30 -nm diameter). Darker and lighter zones are clearly visible within the aggregate. The sample is not stained, so the strong contrast is only attributable to differences in the Cs concentration (in these conditions non-Cs-loaded liposomes would be invisible to the electron microscope).

The clear-cut edges of the different regions reveal the contour of intact liposomes filled with the two different Cs concentrations that were employed. This high-contrast appearance of the

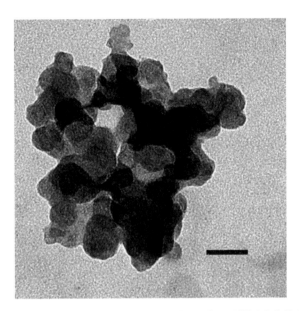

FIGURE 16.6 ESI-TEM image of a typical aggregate of "heavy" and "lightly" Cs-loaded liposomes. The aggregate appears to be built up by globular particles: darker and lighter individual globules are clearly recognized. In the absence of any staining, the observed contrast is due to differences in the elastic scattering of electrons caused by the different Cs concentrations. Scale bar = 100 nm.

individual liposomes allows us to exclude a rearrangement of the bilayers within the aggregate and particularly the fusion of the vesicles, a process that would be accompanied by mixing the content of the different compartments, with a blurring of the contours.

The overall aspect and the shape of the aggregates can be better appreciated at a smaller magnification (Figure 16.7). Smaller aggregates appear to have a more compact and approximately spheroidal shape, but larger aggregates rapidly become elongated.

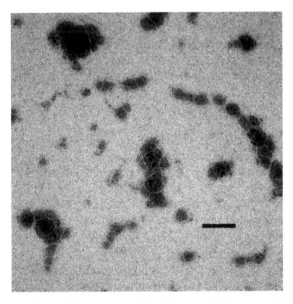

FIGURE 16.7 TEM image of typical aggregates of polyion-coated liposomes. Smaller aggregates have a more compact and approximately spheroidal shape, but larger aggregates are more elongated. Darker and lighter liposomes (high Cs and low Cs concentrations, respectively) are clearly distinguishable. Scale bar = 200 nm.

16.8   INTACT LIPOSOMES: CONDUCTOMETRIC EVIDENCE

Further evidence for the existence of a cluster phase, where intact pd-liposomes are the primary particles, can be obtained from low-frequency electrical conductivity measurements. We recenty investigated (Bordi et al., 2006c) the conductivity behavior of suspensions of single liposomes built up in a high-conductivity salty aqueous solution (0.3 M NaCl) as a function of the difference in the salt concentration between the internal aqueous compartment of the liposomes and the (external) dispersing medium.

Because low-frequency electrical conductivity measurements primarily probe the concentration of small ions in bulk solution, a change in the overall electrical conductivity effectively points out a leakage from the inner core when the salt concentration difference is sufficiently high between the inner compartment and the external medium. Measuring the electrical conductivity of the different suspensions before and after an ultrasound stimulus provides clear evidence of the release of the ionic content induced by the mechanical stress in the individual liposomes, both isolated and within the clusters. Conversely, the aggregation process induces negligible effects on the ion permeability of the liposome membranes, thus excluding fusion or complete restructuring of the liposomes, which should have been accompanied by an almost complete mixing of the internal solution with the external medium. This finding gives further evidence on the existence of a cluster phase of intact liposomes as opposed to the formation of aggregates where the liposomes undergo a restructuring process, resulting in multilayered heaps. Moreover, it opens new and interesting perspectives for practical applications of these clusters as, for example, multifunctional vectors in drug delivery (Bordi et al., 2005b; Volodkin et al., 2006).

A summary of the results of this approach is illustrated in Figure 16.8, where the electrical conductivity σ of three liposome suspensions measured in three different aqueous environments is shown: liposomes built up in a 0.3 M NaCl electrolyte solution and dispersed in the same solution; liposomes with a 0.3 M NaCl aqueous solution in their core, but dispersed in deionized water; and this latter liposome suspension after a sonication cycle where, because of the mechanical (transient) rupture of the membranes, the core content is expected to have mixed with the dispersing solution (Figure 16.8, right panel).

Unilamellar liposomes that are 100–200 nm in size are capable of withstanding significant osmotic stress (they are "osmolitically inert"; Bangham et al., 1967; Fettiplace, 1978; Price and Thompson, 1969; Yaroslavov et al., 1998), and significant transient membrane lysis only occurs at considerable osmotic pressures, depending on their size and composition (Logisz and Hovis, 2005). For these reasons, the sudden increase of about 40% in the low-frequency electrical conductivity (from 0.0173 mho/m measured after the dialysis process in Figure 16.8a, right panel, to 0.0243 mho/m in Figure 16.8b, right panel) was ascribed to the transient lysis of the membrane bilayer with the consequent release of the liposome core content that was shared with the external medium.

This simple experiment shows that single liposomes are able to maintain the imposed differential of osmolality without any appreciable leakage and, conversely, that leakages can be easily induced by an appropriate ultrasound stress. By adding appropriate amounts of NaPA polyions to the dialyzed liposome suspension, there is the usual reentrant aggregation.

When the polyelectrolyte is added to the liposome suspension, there is an increase of the measured conductivity, which is mainly due to the contribution of the counterions of both the polyelectrolyte and the liposomes that are partially released in the bulk solution and partially remain condensed at the pd-liposome surface. However, upon sonication, the σ is further increased, because now the electrolyte content contained in the internal core of the liposomes within the aggregates has been released to the external medium. In the example shown in Figure 16.8 (left panel), the σ increases from a value of 0.0691 mho/m, measured for the aggregate suspension (Figure 16.8c) to a value of 0.0776 mho/m (Figure 16.8d) after the sonication procedure. On the assumption that all of the liposomes release their ionic content upon sonication, the expected σ calculated on the basis of

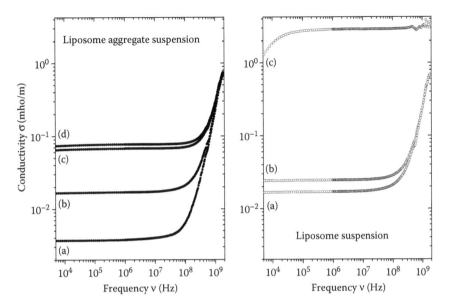

FIGURE 16.8   The electrical conductivity (σ) of liposome aqueous suspensions is a function of the frequency at 25.0 ± 0.2°C. The lipid concentration is 0.68 mg/mL. (Right) (a) A liposome suspension after exhaustive dialysis against deionized water, (b) a liposome suspension after sonication (the electrolyte content in the liposome core has been released in the external medium, with a consequent increase of the conductivity), and (c) a liposome suspension in 0.3 M NaCl electrolyte solution. Although only the low-frequency limit of the conductivity is employed in the characterization of the structure of the vesicle aggregates, the whole measured conductivity spectrum is shown. (Left) (a) Liposome suspension dispersed in deionized water solution; (b) polyion solution; (c) liposome aggregate suspension, where the aggregation was induced by NaPA; and (d) liposome aggregate suspension after sonication. The electrolyte content inside each liposome is shared with the external medium.

heterogeneous system mixture equations (Asami, 2002) is 0.078 mho/m, which fits surprisingly well with what was found experimentally.

Moreover, the size distributions of pd-liposome aggregates compared before and after sonication are very similar; both of them are monomodal and there is only a little shift toward a smaller average size after sonication. In these conditions, sonication does not produce, other than the transient opening of the membranes and the release of the ionic content of the liposomes, a significant restructuring of the clusters.

These results give further evidence to the existence of an equilibrium cluster phase where intact and separate liposomes are "glued together" by polyelectrolytes. The possibility of controlling this by means of thermodynamic parameters such as the ionic strength, size, and electrical charge of clusters composed of liposomes that maintain their integrity offers new and interesting applications in different fields of chemistry, soft-matter physics, and life sciences.

16.9 LIPOSOME CLUSTERS AS A MULTIDRUG DELIVERY SYSTEM: A PERSPECTIVE

For successful delivery and an efficient expression of its action, a nonviral vector must be able to overcome many different barriers in order to protect its payload and to deliver it to target cells as specifically as possible. Different strategies to obtain these ideal performances are currently under investigation, but most of them are based on the peculiar characteristic of different nanostructured or mesostructured materials. Among them, the strategy based on the supramolecular structures formed between charged polymers and oppositely charged particles, particularly liposomes, shows promising potential. The main advantage of these supramolecular assemblies resides in their intrinsic modularity and flexibility. In fact, because they are based on physical mechanisms, their self-assembling ability is maintained regardless of the specific chemical nature of the components, the only requisites for the formation of the polyion-decorated clusters being the presence of charged colloidal particles and oppositely charged linear polyelectrolytes. Moreover, the average size (and size distribution) of the resulting aggregates, in addition to their electrical charge, can be varied and controlled by means of different environmental parameters. In particular, the system we described allows us to obtain both positively and negatively charged aggregates of the desired size (and size distribution) by simply changing the basic liposomal vesicles (positively or negatively charged) and by using appropriate oppositely charged polyions. This means that the chemical composition can be varied to a large extent and adjusted to meet the requirements of different applications.

In addition to charged liposomes, other colloidal particles can be employed, such as polymeric nanocapsules or different nanostructured polymers, and some of them are already employed as drug carriers (e.g., chitosan nanoparticles; Masotti et al., 2008). In the case of liposomes, the huge potential of these versatile lipid assemblies can be fully exploited as drug carriers (Huwyler and Krähenbuhl, 2008). For example, by using hybrid niosomes as primary particles to form the clusters (Sennato et al., 2008), the characteristic of "stealth carriers" of these vesicles are partly transferred to their aggregates. Hybrid niosomes are built up by a mixture of PEGylated lipids (e.g., Tween 20) and cholesterol (the nonionosomes) to which an ionic lipid (e.g., phosphatidic acid or dicetylphosphate) is added to confer the desired net electrical charge. The presence of the PEG moieties at the particle surface is very effective in reducing the undesired interactions with the extracellular environment, with the result of a prolonged circulation lifetime (Lasic and Needham, 1995).

"Conventional" liposomes have been modified in different ways to improve the efficient transfer of their payloads to the cell cytoplasm. For example, decorating the liposome surface with the oligopeptide octaarginine (Khalil et al., 2008) greatly enhanced their cellular uptake and, by optimizing the density of the peptide, liposomes could be internalized via clathrin-independent pathways, a mechanism that improves intracellular trafficking, avoiding lysosomal degradation. Many of the interesting features of the variously modified liposomes could in principle be transferred to their polyion-induced aggregates that, as an extra bonus, have the advantage of being multicompartment structures. In their different compartments, each one formed by an individual liposome, aggregates can transport different active substances. Because of its multicompartment structure, a single aggregate is able to separately but simultaneously transport and deliver different active substances to a single cell, including possibly diagnostic probes for a direct "real-time" visualization of the effective drug transfer and downstream processes. Moreover, the procedure for assembling these multivectors is primarily straightforward. Ionic liposomes differing in composition or payload can be prepared separately and then mixed in proper proportion to obtain the desired stoichiometry of the various components in the suspension. At that point, the aggregation process is initiated by adding the oppositely charged polyions that act as a sort of electrostatic glue and induce the formation of multiliposome clusters. For sufficiently large clusters, the defined stoichiometry of the different components previously established in the initial suspension is reproduced within each aggregate. In this way, by controlling the stoichiometry at a nanoscale level, it can be reasonably assumed that all of the different substances to be delivered reach each cell. This peculiar aspect is particularly appealing in developing new strategies in antiretroviral therapies. For example, several biological and nonbiological carrier systems have been developed for anti-HIV therapy in the last few years. Among these, liposomes showed excellent potential and were tested with various drugs, antisense oligonucleotides, ribozymes, and therapeutic genes (Lichterfeld et al., 2005).

Generally, nanoparticles are cell-specific transporters of drugs against macrophage-specific infections, such as HIV. In particular, the relatively large polyion-decorated liposome aggregates are rapidly captured by the macrophages, an elective reservoir for the virus, that are difficult to reach with the traditional pharmacological approaches. In fact, large lipid–peptide structures, such as erythrocyte ghosts and bacterial ghosts, are a promising delivery system for therapeutic peptides and HIV vaccines. The pd-liposome clusters could also be effectively employed in this case, being similar in size and possibly in composition, because various lipidic and peptidic components of these cells could be easily incorporated in the liposome structure.

The HIV virus is characterized by an extreme genetic variability and a high mutation rate that provokes a rapid escape from adaptive immune responses (Walker and Burton, 2008). For this reason, cocktails of different drugs are usually employed that are potentially active against the different strains of the virus. In this case, the practical advantage is plainly apparent for a multiple vector that is able to deliver all of the components of the cocktail to a single macrophage with a definite and controlled stoichiometry at the level of the single vector particle.

A major problem for the use of pd-liposome clusters as vectors is the high efficiency of the macrophages in capturing these relatively large particles, even when they are stealth (PEGylated). However, even though their systemic administration could appear problematic, other routes appear to be very promising. One of the most noninvasive approaches is via inhalation. The delivery of genes via aerosol holds promise for the treatment of a broad spectrum of pulmonary disorders and offers numerous advantages over other more invasive delivery modes. After the cloning of the cystic fibrosis gene, there was great interest in the delivery of genes directly to lung surfaces via inhalation and many efforts focused on the use of nonviral vectors, particularly cationic lipids (Densmore, 2006). Because one of the viral reservoirs of HIV that is less accessible to traditional pharmacological approaches is represented by pulmonary macrophages, the use of pd-liposome aggregates is interesting. However, in addition to the ability to deliver different substances separately but simultaneously, the multicompartment structures formed by polyion-decorated liposomes show other potential advantages. Supramolecular hydrogels based on the self-assembly of complexes between various biomacromolecules have been actively investigated for their use as implantable drug delivery systems for controlled release of macromolecular drugs (Li and Loh, 2008). In general, scaffolds serve a central role in many technological strategies by providing the means to control the local environment. Moreover, in this case, scaffolds also provide the means of prolonging the release of the active macromolecules by simply trapping them within a matrix that delays their delivery to the surrounding environment.

The multicompartment clusters formed by pd-liposomes are, in a sense, nanoscaffolds that, in principle, offer the possibility of extending the duration of the release period of the different substances enclosed in their compartments. For example, water-soluble active substances entrapped within the aqueous core of the liposomal compartments will be released to the surrounding aqueous medium only progressively, as the degradation and the rupture of the single vesicles proceeds from the more external compartments toward the interior of the cluster. The pd-liposome cluster could serve simultaneously as a multiload cargo, transporting several different drugs or diagnostic probes, shielding the active substances from the aggression of the degrading enzymes of the extracellular fluids, and finally maintaining the proper "nanoenvironment" for drug storage and controlled release.

Prolonged delivery is attractive in all cases where repeated applications are not feasible or impractical. For instance, the eye is an attractive target for gene therapy strategies because of its accessibility and immune-privileged characteristics. This latter property is ideal for local gene therapy because it is expected that the inflammatory and immune reactions toward the gene vector or transgenic protein will not be a major drawback. Because of its relatively small size, effective treatment of the ocular tissues will require a smaller drug concentration, wheras the diffusion from the eye into the circulation and vice versa is limited (Bloquel et al., 2005). Topical instillation of active compounds is the easiest method of delivery for ocular therapy. However, the eye has very strong defenses against the entry of foreign compounds. Thus, the intraocular availability of instilled compounds is generally very poor. Intracameral and intracorneal injections are more effective but invasive. Subconjunctival injection is a more promising strategy because this procedure is less invasive, allows for larger injection volumes, and can be repeated more frequently. However, in this case and especially in intracameral and intracorneal injections, the possibility of employing vectors that allow prolonged release appears to be an interesting perspective.

16.10   CONCLUSIONS

This chapter compiled relevant results that were recently observed in aqueous suspensions of charged liposomes and oppositely charged polyelectrolytes. Although we mainly utilized the experimental results from our group, we tried to draw a comprehensive picture of the state of art of electrostatically interacting colloidal mixtures that are of great importance in technological and biotechnological applications.

We demonstrated that electrical conductivity measurements give further independent support to the hypothesis that the complexes that form in the suspensions as a result of the interaction of the oppositely charged macroions are clusters of liposomes. Within the clusters, the liposomes, which are simply glued together by the polyions, are able to maintain their structural integrity. Moreover, they keep the content of their inner aqueous core separate from the surrounding medium at all stages of the aggregation process without any significant restructuring of the double layer.

The morphological study of these aggregates by electron microscopy techniques had already furnished strong evidence for a multicompartment organization of these aggregates. However, electrical conductivity measurements allowed a "functional characterization" of these structures, demonstrating that, after the formation of the clusters, the liposomes can be still induced to release and share their inner content with the suspending medium by means of appropriate external stimuli.

This result opens interesting avenues for employing these structures as systems for drug delivery that could allow the simultaneous transport of different active substances within the different compartments (multidrug vectors).

The process that results in the formation of these multicompartment aggregates was discussed in the light of recent theories on the adsorption of polyelectrolytes at oppositely charged interfaces and the more advanced models for aggregation in colloidal systems, where the interparticle potential is characterized by the simultaneous presence of short-range repulsion and long-range attraction. This viewpoint enabled us to effectively describe the interweaved phenomena of the overcharging of the aggregates and their reentrant condensation, which have both been observed for a long time, in suspensions of cationic liposomes and DNA that has eluded a convincing explanation in these systems for quite some time.

The charge inversion that we observed has a simple and straightforward explanation in view of the general theory of polyion *correlated* adsorption onto an oppositely charged surface: when linear polyions are added to a solution bathing a charged surface, they adsorb on it; but, because they repel each other, to gain some energy they reconfigure themselves at the surface in more or less orderly patterns. By the same token, correlated adsorption of the polyions at the liposome surface is also the origin of the reentrant condensation. From lateral correlations, or in other words from local nonhomogeneities of the net charge distribution at the polyion-decorated surface, short-range attractive interactions can arise; and the observed reentrant behavior can be described in terms of a balance of these short-range attractions and the screened electrostatic repulsions attributable to the residual net charge of the complexes.

In this context, the electrical behavior of these systems during the polyion-induced liposome aggregation allows the addition of relevant tesserae to the whole mosaic. In particular, the conductometric behavior seemed to be consistent with the description of both the single liposomes and the pd-liposome aggregates as different entities able to encompass an aqueous core from the surrounding aqueous solution. This finding points out the nonnegligible role that electrostatic interactions among charged surfaces and small counterions play in determining the whole dynamics of the system.

ABBREVIATIONS

CMC   Critical micelle concentration
DLS   Dynamic light scattering
DLVO   Derjaguin–Landau–Verwey–Overbeek model of colloidal stability

DNA	Deoxyribonucleic acid
DOTAP	1,2-Dioleoyl-3-trimethylammonium-propane
ESI	Electron spectroscopy imaging
HIV	Human immunodeficiency virus
NaPA	Polyacrylate sodium salt
pd	Polyelectrolyte-decorated particles
PEG	Poly(ethylene glycol)
RNA	Ribonucleic acid
TEM	Transmission electron microscopy

SYMBOLS

b	Effective distance between adjacent charges in the polyion chain
c_1, c_2	Principal local curvatures
c_i	Concentration of the ith species of ions
C_p, C_L	Concentrations (weight) of polyelectrolytes and lipids, respectively
D	Thickness of the adsorbed polyion layer at aqueous interfaces
D_e	Size of an electrostatic blob
ε	Permittivity of the aqueous phase
f	Fraction of free counterions
γ	Interfacial tension
g_e	Number of monomers inside an electrostatic blob
κ_D^{-1}	Debye screening length
$K_B T$	Thermal energy
l_B	Bjerrum length
L_0	Structural persistence length
L_e	Electrostatic contribution to the persistence length
L_p	Polyion persistence length
M_{WP}, M_{WL}	Molecular weights of the repeating unit of polyions and lipids, respectively
$\rho, \bar{\rho}$	Mean and Gaussian curvature moduli
R_H	Hydrodynamic radius of the aggregates
σ	Electrical conductivity
τ	Reduced temperature $[\tau = (\theta - T)/T]$
ξ	Negative to positive charge ratio
z_i	Valence of ith ion

REFERENCES

Alexander, S., P. M. Chaikin, P. Grant, G. J. Morales, P. Pincus, and D. Hone, Charge renormalization, osmotic pressure, and bulk modulus of colloidal crystals: Theory. *J. Chem. Phys.*, 80: 5776–5786, 1984.

Arancia, G., F. Bordi, A. Calcabrini, M. Diociaiuti, and A. Molinari, Ultrastructural and spectroscopic methods in the study of anthtracycline–membrane interaction. *Pharmacol. Res.*, 32: 255–272, 1995.

Asami, K., Characterization of heterogeneous systems by dielectric spectroscopy. *Prog. Polym. Sci.*, 27: 1617–1659, 2002.

Bangham, A., J. De Gier, and D. Greville, Osmotic properties and water permeability of phospholipid liquid crystals. *Chem. Phys. Lipids*, 1: 225–246, 1967.

Barenholtz, Y. and D. J. A. Crommelin, Chapter: Liposomes as pharmaceutical dosage forms. In A. Swarbrick and J. C. Boylan (Eds.), *Encyclopedia of Pharmaceutical Technology*, pp. 1–39. Marcel Dekker, New York, 1994.

Bloomfeld, V. A., DNA condensation. *Curr. Opin. Struct. Biol.*, 6: 334–341, 1996.

Bloquel, C., J. Bourges, E. Touchard, M. Berdugo, D. Ben Ezra and F. Behar-Cohen, Non viral ocular therapeutic avenues. *Adv. Drug Deliv. Rev.*, 15: 1224–1242, 2005.

Bordi, F., C. Cametti, and R. H. Colby, Dielectric spectroscopy and conductivity of polyelectrolyte solutions. *J. Phys.: Condens. Matter*, 16: R1423–R1449, 2004b.

Bordi, F., C. Cametti, F. De Luca, T. Gili, D. Gaudino, and S. Sennato, Charged lipid monolayers at the air–water interfaces: Coupling to polyelectrolytes. *Colloid Surf. B*, 29: 149–157, 2003a.

Bordi, F., C. Cametti, M. Diociaiuti, D. Gaudino, T. Gili, and S. Sennato, Complexation of anionic polyelectrolytes with cationic liposomes: Evidence of reentrant condensation and lipoplex formation. *Langmuir*, 20: 5214–5227, 2004a.

Bordi, F., C. Cametti, M. Diociaiuti, and S. Sennato, Large equilibrium clusters in low-density aqueous suspensions of polyelectrolyte–liposome complexes: A phenomenological model. *Phys. Rev. E*, 71: 050401(R), 2005a.

Bordi, F., C. Cametti, T. Gili, D. Gaudino, and S. Sennato, Time evolution of the formation of different size cationic liposome–polyelectrolyte complexes. *Bioelectrochemistry*, 59: 99–107, 2003b.

Bordi, F., C. Cametti, C. Marianecci, and S. Sennato, Equilibrium particle aggregates in attractive colloidal suspensions. *J. Phys.: Condens. Matter*, 17: S3423–S3432, 2005c.

Bordi, F., C. Cametti, and S. Sennato, Polyions act as an electrostatic glue for mesoscopic particle aggregates. *Chem. Phys. Lett.*, 409: 134–138, 2005b.

Bordi, F., C. Cametti, and S. Sennato, Electrical properties of aqueous liposome suspensions. In A. L. Liu (Ed.), *Advances in Planar Lipid Bilayers and Liposomes*, Vol. 4, pp. 281–320. Academic Press, New York, 2006b.

Bordi, F., C. Cametti, S. Sennato, and M. Diociaiuti, Direct evidence of multicompartment aggregates in polyelectrolyte-charged liposome complexes. *Biophys. J.*, 91: 1513–1520, 2006a.

Bordi, F., C. Cametti, S. Sennato, and D. Viscomi, Conductometric evidence for intact polyion-induced liposome clusters. *J. Colloid Interface Sci.*, 304: 512–517, 2006c.

Bordi, F., S. Sennato, and C. Cametti, Correlated adsorption of polyelectrolytes in the "charge inversion" of colloidal particles. *Europhys. Lett.*, 68: 296–302, 2004c.

Boroudjedi, H., Y.-W. Kim, A. Naji, R. R. Netz, X. Schlagberger, and A. Serr, Statics and dynamics of strongly charged soft matter. *Phys. Rep.*, 416(3–4): 129–199, 2006.

Borukhov, I., D. Andelman, and H. Orland, Scaling laws of polyelectrolyte adsorption. *Macromolecules*, 31: 1665–1671, 1998.

Bouyer, F., A. Robben, W. L. Yu, and M. Borkovec, Aggregation of colloidal particles in the presence of oppositely charged polyelectrolytes: Effect of surface charge heterogeneities. *Langmuir*, 17: 5225–5231, 2001.

Brannigan, G., A. C. Tamboli, and F. L. H. Brown, The role of molecular shape in bilayer elasticity and phase behaviour. *J. Chem. Phys.*, 121: 3259–3271, 2004.

Brasher, L. L., K. L. Herrington, and E. W. Kaler, Electrostatic effects on the phase behaviour of aqueous cetylmethylammonium bromide and sodium octyl sulfate mixtures with added sodium bromide. *Langmuir*, 11: 4267–4277, 1995.

Caminati, G. and G. Gabrielli, Polystyrene sulfonate adsorption at water–graphon and water–air interfaces. *Colloids Surf. A*, 70: 1–14, 1993.

Ciani, L., S. Ristori, A. Salvati, L. Calamai, and G. Martini, DOTAP/DOPE and DCchol/DOPE lipoplexes for gene delivery: Zeta potential measurements and electron spin resonance spectra. *Biochim. Biophys. Acta*, 1664: 70–79, 2004.

Claessens, M. M. A. E., B. F. van Oort, F. A. M. Leermakers, F. A. Hoekstra, and M. A. Cohen Stuart, Charged lipid vesicles: Effects of salts on bending rigidity, stability and size. *Biophys. J.*, 87: 3882–3893, 2004.

de Meijere, K., G. Brezesinski, T. Pfohl, and H. Möhwald, Influence of the polymer charge density on lipid–polyelectrolyte complexes at the air/water interface. *J. Phys. Chem. B*, 103: 8888–8893, 1999.

Densmore, C. L., Advances in noninvasive pulmonary gene therapy. *Curr. Drug Deliv.*, 3: 55–63, 2006.

Diociaiuti, M., Electron energy loss spectroscopy microanalysis and imaging in the transmission electron microscope: Example of biological applications. *J. Electron Spectrosc.*, 143: 189–203, 2005.

Dobrynin, A., R. H. Colby, and M. Rubinstein, Scaling theory of polyelectrolyte solutions. *Macromolecules*, 28: 1859–1869, 1995.

Dobrynin, A. V., A. Deshkovski, and M. Rubinstein, Adsorption of polyelectrolytes at an oppositely charged surface. *Phys. Rev. Lett.*, 84: 3101–3104, 2000.

Dobrynin, A. V. and M. Rubinstein, Theory of polyelectrolytes in solutions and at surfaces. *Prog. Polym. Sci.*, 30: 1049–1118, 2005.

Dzubiella, J., A. G. Moreira, and P. A. Pincus, Polyelectrolyte–colloid complexes: Polarizability and effective interaction. *Macromolecules*, 36: 1741–1752, 2003.

El-Aneed, A., An overview of current delivery systems in cancer gene therapy. *J. Control. Release*, 94: 1–14, 2004.

Ferber, D., Gene therapy: Safer and virus-free? *Science*, 294: 1638–1642, 2001.

Fettiplace, R., The influence of the lipid on the water permeability of artificial membranes. *Biochim. Biophys. Acta*, 513: 1–10, 1978.

Finkelstein, A. and O. B. Ptitsyn, *Protein Physics*. Academic Press, San Diego, CA, 2002.

Gonçalves, E., R. J. Debs, and T. D. Heath, The effect of liposome size on the lipid/DNA ratio of cationic lipoplexes. *Biophys. J.*, 86: 1554–1563, 2004.

Gradzielski, M., Vesicles and vesicle gels—Structure and dynamics of formation. *J. Phys.: Condens. Matter*, 15: R655–R697, 2003.

Gregoriadis, G. (Ed.), *Liposome Technology. Volume 1: Liposome Preparation and Related Techniques*, 2nd ed. CRC Press, Boca Raton, FL, 1993.

Gregory, J., Rates of flocculation of latex particles by cationic polymers. *J. Colloid Interface Sci.*, 42: 448–456, 1973.

Gregory, J., The effect of cationic polymers on the colloidal stability of latex particles. *J. Colloid Interface Sci.*, 55: 35–44, 1976.

Grosberg, A. Yu. and A. R. Khokhlov, *Statistical Physics of Macromolecules*. AIP Press, New York, 1994.

Grosberg, A. Yu., T. T. Nguyen, and B. I. Shklovskii, Colloquium: The physics of charge inversion in chemical and biological systems. *Rev. Mod. Phys.*, 74: 329–345, 2002.

Hao, J., H. Hoffmann, and K. Horbaschek, A novel cationic/anionic surfactant system from a zwitterionic alkyldimethylamine oxide and dihydroperfluorooctanoic acid. *Langmuir*, 17: 4151–4160, 2001.

Haro-Pérez, C., M. Quesada-Pérez, J. Callejas-Fernández, E. Casals, J. Estelrich, and R. Hidalgo-Álvarez, Liquidlike structures in dilute suspensions of charged liposomes. *J. Chem. Phys.*, 118: 5167–5173, 2003.

Harries, D., S. May, W. M. Gelbart, and A. Ben-Shaul, Structure, stability, and thermodynamics of lamellar DNA–lipid complexes. *Biophys. J.*, 75: 159–168, 1998.

Helfrich, W., Elastic properties of lipid bilayers. Theory and possible experiments. *Z. Naturforsch. C*, 28: 693–703, 1973.

Hoekstra, D., J. Rejman, L. Wasungu, F. Shi, and I. Zuhorn, Gene delivery by cationic lipids: In and out of an endosome. *Biochem. Soc. Trans.*, 35: 68–71, 2007.

Hoffmann, H., C. Thunig, P. Schmiedel, and U. Munkert, Surfactant system with charged multilamellar vesicles and their rheological properties. *Langmuir*, 10: 3972–3981, 1994.

Hope, M. J., M. B. Bally, G. Webb, and P. R. Cullis, Production of large unilamellar vesicles by a rapid extrusion procedure. Characterization of size distribution, trapped volume and ability to maintain a membrane potential. *Biochim. Biophys. Acta*, 812: 55–65, 1983.

Huwyler, J. J. D. and S. Krähenbuhl, Tumor targeting using liposomal antineoplastic drugs. *Int. J. Nanomed.*, 1: 21–29, 2008.

Israelachvili, J., *Intermolecular and Surface Forces*. Academic Press, London, 1985.

Jones, M. N. and D. Chapman, *Micelles, Monolayers and Biomembranes*. Wiley–Liss, New York, 1995.

Khachatourian, A. V. M. and A. O. Wistrom, Electrostatic interaction force between planar surfaces due to 3-D distribution of sources of potential (charge). *J. Phys. Chem. B*, 102: 2483–2493, 1998.

Khalil, I., K. Kogure, S. Futaki, and H. Harashima, Octaarginine modified liposomes: Enhanced cellular uptake and controlled intracellular trafficking. *Int. J. Pharm.*, 354: 39–48, 2008.

Kim, Y. W. and W. Sung, Effects of charge and its fluctuation on membrane undulation and stability. *Europhys. Lett.*, 58: 147–152, 2002.

Klebanau, A., N. Kliabanova, F. Ortega, F. Monroy, R. G. Rubio, and V. Starov, Equilibrium behaviour and dilational rheology of polyelectrolyte/insoluble surfactant adsorption films: Didodecyldimethylammonium bromide and sodium poly(styrenesulfonate). *J. Phys. Chem. B*, 109: 18316–18323, 2005.

Koltover, I., T. Salditt, and C. R. Safinya, Phase diagram, stability, and overcharging of lamellar cationic lipid–DNA self-assembled complexes. *Biophys. J.*, 77: 915–924, 1999.

Kumaran, V., Spontaneous formation of vesicles by weakly charged membranes. *J. Chem. Phys.*, 99: 5490–5499, 1993.

Kumaran, V., Instabilities due to charge–density–curvature coupling in charged membranes. *Phys. Rev. Lett.*, 85: 4996–4999, 2000.

Kumaran, V., Effect of surface charge on the curvature moduli of a membrane. *Phys. Rev. E*, 64: 51922–51932, 2001a.

Kumaran, V., Electrohydrodynamic instability of a charged membrane. *Phys. Rev. E*, 64: 11911–11923, 2001b.

Lasic, D. and D. Needham, The "stealth" liposome: A prototypical biomaterial. *Chem. Rev.*, 95: 2601–2628, 1995.

Lasic, D., Novel applications of liposomes. *Trends Biotechnol.*, 16: 307–321, 1998.

Lau, A. W. C. and P. Pincus, Charge-fluctuation-induced non-analytic bending rigidity. *Phys. Rev. Lett.*, 81: 1338–1341, 1998.

Lawson, C. L. and I. D. Morrison, *Solving Least Squares Problems. A FORTRAN Program and Subroutines Called NNLS*. Prentice–Hall, Englewood Cliffs, NJ, 1974.

Leong, Y. K., Interparticle forces arising from an adsorbed strong polyelectrolyte in colloidal dispersions: Charged patch attraction. *Colloid Polym. Sci.*, 277: 299–305, 1999.

Leong, Y. K., Charged patch attraction in dispersion: Effect of polystyrene molecular weight on patch size. *Colloid Polym. Sci.*, 279: 82–87, 2001.

Leong, Y. K., P. J. Scales, T. W. Healy, and D. V. Boger, Interparticle forces arising from adsorbed polyelectrolytes in colloidal suspensions. *Colloids Surf. A*, 95: 43–52, 1995.

Li, J. and X. J. Loh, Cyclodextrin-based supramolecular architectures: Synthesis, structure and applications for drug and gene delivery. *Adv. Drug Deliv. Rev.*, 60: 1000–1017, 2008.

Lichtenberg, D. and Y. Barenholtz, Liposomes: Preparation, characterization and preservation. In D. Glick (Ed.), *Methods of Biological Analysis*, Vol. 33. Wiley, New York, 1988.

Lichterfeld, M., N. Qurishi, C. Hoffmann, B. Hochdorfer, N. H. Brockmeyer, K. Arasteh, S. Mauss, and J. K. Rockstroh, Treatment of HIV-1-associated Kaposi's sarcoma with PEGylated liposomal doxorubicin and haart simultaneously induces effective tumour remission and CD4 + T cell recovery. *Infection*, 33: 140–147, 2005.

Lin, A. L., N. L. Slack, A. Ahmad, C. X. George, C. E. Samuel, and C. R. Safinya, Three-dimensional imaging of lipid gene-carriers: Membrane charge density controls universal transfection behaviour in lamellar cationic–DNA complexes. *Biophys. J.*, 84: 3307–3316, 2003.

Linse, P., Mean force between like-charged macroions at high electrostatic coupling. *J. Phys.: Condens. Matter*, 14: 13449–13457, 2002.

Logisz, C. C. and J. S. Hovis, Effect of salt concentration on membrane lysis pressure. *Biochim. Biophys. Acta*, 1717: 104–112, 2005.

Manning, G. S., Limiting laws and counterion condensation in polyelectrolyte solutions: 1. Colligative properties. *J. Chem. Phys.*, 51: 924–933, 1969.

Manning, G. S., The molecular theory of polyelectrolyte solutions with applications to the electrostatic properties of polynucleotides. *Q. Rev. Biophys.*, 11: 179–246, 1978.

Masotti, A., F. Bordi, G. Ortaggi, F. Marino, and C. Palocci, A novel method to obtain chitosan/DNA nanospheres and a study of their release properties. *Nanotechnology*, 19: 55302–55308, 2008.

May, S., D. Harries, and A. Ben-Shaul, The phase behaviour of cationic lipid DNA complexes. *Biophys. J.*, 78: 1681–1694, 2000.

Messina, R., C. Holm, and K. Kremer, Strong attraction between charged spheres due to metastable ionized states. *Phys. Rev. Lett.*, 85: 872–876, 2000.

Miklavic, S. J., D. Y. C. Chan, L. R. White, and T. W. Healy, Double layer forces between heterogeneous charged surfaces. *J. Phys. Chem.*, 98: 9022–9032, 1994.

Millet, F., P. Perrin, M. Merlange, and J. J. Benattar, Logarithmic adsorption of charged polymeric surfactants at the air–water interface. *Langmuir*, 18: 8824–8828, 2002.

Monteux, C., C. E. Williams, J. Meunier, O. Anthony, and V. Bergeron, Adsorption of oppositely charged polyelectrolyte/surfactant complexes at the air/water interface: Formation of interfacial gels. *Langmuir*, 20: 57–63, 2004.

Mouritsen, O. G., *Life—As a Matter of Fat*. Springer, Berlin, 2005.

Najia, A. and R. R. Netz, Attraction of like-charged macroions in the strong-coupling limit. *Eur. Phys. J. E*, 13: 43–59, 2004.

Netz, R. and D. Andelman, Polyelectrolytes in solution and at surfaces. In A. J. Bard, M. Stratmann, E. Gileadi, and M. Urbakh (Eds.), *Encyclopedia of Electrochemistry*, Vol. 1, Wiley–VCH, Weinheim, 2002.

Netz, R. R. and D. Andelman, Neutral and charged polymers at interfaces. *Phys. Rep.*, 380: 1–95, 2003.

Netz, R. R. and J.-F. Joanny, Complexation between a semi-flexible polyelectrolyte and an oppositely charged sphere. *Macromolecules*, 32: 9026–9040, 1999.

Neu, J. C., Wall-mediated forces between like-charged bodies in an electrolyte. *Phys. Rev. Lett.*, 82: 1072–1076, 1999.

New, R. R. C., *Liposomes: A Practical Approach*. IRL Press at Oxford University Press, Oxford, UK, 1990.

Noskov, B. A., G. Loglio, and R. Miller, Dilational viscoelasticity of polyelectrolyte/surfactant adsorption films at the air/water interface: Dodecyltrimethyl ammonium bromide and sodium poly(styrenesulfonate). *J. Phys. Chem. B*, 108: 18615–18622, 2004.

Oberdisse, J., Transition from small to big charged unilamellar vesicles. *Eur. Phys. J. B*, 3: 463–469, 1998.

Oberdisse, J., C. Couve, J. Appell, J. F. Berret, C. Ligoure, and G. Porte, Vesicles and onions from charged surfactant bilayers: A neutron scattering study. *Langmuir*, 12: 1212–1218, 1996.

Oberdisse, J. and G. Porte, Size of micro-vesicles from charged surfactant bilayers: Neutron scattering data compared to an electrostatic model. *Phys. Rev. E*, 56: 1965–1975, 1997.

Odijk, J., Polyelectrolytes near the rod limit. *J. Polym. Sci.: Polym. Phys.*, 15: 477–483, 1977.

Olvera de la Cruz, M., L. Belloni, M. Delsanti, J. P. Dalbiez, O. Spalla, and M. Driford, Precipitation of highly charged polyelectrolyte solutions in the presence of multivalent salts. *J. Chem. Phys.*, 103: 5781–5791, 1995.

Oosawa, F., *Polyelectrolytes*. Marcel Dekker, New York, 1971.

Pedroso De Lima, M. C., S. Simoes, P. Pires, H. Faneca, and N. Duzgunes, Cationic lipid–DNA complexes in gene delivery: From biophysics to biophysical applications. *Adv. Drug Deliv. Rev.*, 47: 277–294, 2001.

Piedade, J. A. P., M. Mano, M. C. Pedroso De Lima, T. S. Oretskaya, and A. M. Oliveira-Brett, Electrochemical sensing of the behaviour of oligonucleotide lipoplexes at charged interfaces. *Biosens. Bioelectron.*, 20: 975–985, 2004.

Pincus, P., P. G. de Gennes, R. M. Velasco, and F. Brochard, Remarks on polyelectrolyte conformation. *J. Phys. (France)*, 37: 1461–1473, 1976.

Podgornik, R., T. Akesson, and B. Jönsson, Colloidal interactions mediated via polyelectrolytes. *J. Chem. Phys.*, 102: 9423–9434, 1995.

Price, H. and T. Thompson, Properties of liquid bilayer membranes separating two aqueous phases: Temperature dependence of water permeability. *J. Mol. Biol.*, 41: 443–457, 1969.

Rädler, J. O., I. Koltover, A. Jamieson, T. Salditt, and C. R. Safinya, Structure and interfacial aspects of self-assembled cationic lipid–DNA gene carrier complexes. *Langmuir*, 14: 4272–4283, 1998.

Rädler, J. O., I. Koltover, T. Salditt, and C. R. Safinya, Structure of DNA–cationic liposome complexes: DNA intercalation in multi-lamellar membranes in distinct interhelical packing regimes. *Science*, 275: 810–814, 1997.

Rayburn, E., H. Wang, J. He, and R. Zhang, RNA silencing technologies in drug discovery and target validation. *Lett. Drug Des. Discov.*, 2: 1–18, 2005.

Russel, W. B., D. A. Saville, and W. R. Schowalter, *Colloidal Dispersions*. Cambridge University Press, Cambridge, UK, 1989.

Safinya, C. R., Structures of lipid–DNA complexes: Supra-molecular assembly and gene delivery. *Curr. Opin. Struct. Biol.*, 11: 440–452, 2001.

Safran, S. A., P. A. Pincus, D. Andelman, and F. C. MacKintosh, Stability and phase behaviour of mixed surfactant vesicles. *Phys. Rev. A*, 43: 1071–1078, 1991.

Sanchez, R. and P. Bartlett, Equilibrium cluster formation and gelation. *J. Phys.: Condens. Matter*, 17: S3551–S3556, 2005.

Schmitz, K. S. and J.-W. Yu, On the electrostatic contribution to the persistence length of flexible polyelectrolytes. *Macromolecules*, 21: 484–493, 1988.

Sciortino, F., P. Tartaglia, and E. Zaccarelli, One-dimensional cluster growth and branching gels in colloidal systems with short-range depletion attraction and screened electrostatic repulsion. *J. Phys. Chem. B*, 109: 21942–21953, 2005.

Sennato, S., F. Bordi, and C. Cametti, On the phase diagram of reentrant condensation in polyelectrolyte–liposome complexation. *J. Chem. Phys.*, 121: 4936–4952, 2004.

Sennato, S., F. Bordi, C. Cametti, A. Di Biasio, and M. Diociaiuti, Polyelectrolyte–liposome complexes: An equilibrium cluster phase close to the isoelectric condition. *Colloids Surf. A*, 270: 138–147, 2005b.

Sennato, S., F. Bordi, C. Cametti, M. Diociaiuti, and M. Malaspina, Charge patch attraction and reentrant condensation in DNA–liposome complexes. *Biochim. Biophys. Acta*, 1714: 11–22, 2005a.

Sennato, S., F. Bordi, C. Cametti, C. Marianecci, M. Carafa, and M. Cametti, Hybrid niosome complexation in the presence of oppositely charged polyions. *J. Phys. Chem. B*, 112: 3720–3727, 2008.

Stevenson, M., HIV-1 pathogenesis. *Nat. Med.*, 9: 853–860, 2003.

Storm, G., P. A. Peeters, U. K. Nassander, and D. J. A. Crommelin, *Liposomes. Biodegradable Polymers as Drug Delivery Systems*. Marcel Dekker, New York, 1988.

Strey, H. H., R. Podgornik, D. C. Rau, and V. A. Parsegian, DNA–DNA interactions. *Curr. Opin. Struct. Biol.*, 8: 309–313, 1998.

Szoka, F. and D. Papahadjopolous, *Liposomes: Preparation and Characterization*, pp 51–82. Elsevier North-Holland, Amsterdam, 1981.

Taheri-Araghi, S. and B.-Y. Ha, Charge renormalization and inversion of a highly charged lipid bilayer: Effects of dielectric discontinuities and charge correlations. *Phys. Rev. E*, 72: 21508–21514, 2005.

Théodoly, O., R. Ober, and C. E. Williams, Adsorption of hydrophobic polyelectrolytes at the air/water interface: Conformational effect and history dependence. *Eur. Phys. J. E*, 5: 51–61, 2001.

Trizac, E., Effective interactions between like-charged macromolecules. *Phys. Rev. E*, 62: R1465–R1469, 2000.

Velegol, D. and P. K. Thwar, Analytical model for the effect of surface charge non-uniformity on colloidal interactions. *Langmuir*, 17: 7687–7693, 2001.

Volodkin, D., V. Ball, P. Schaaf, J.-C. Voegel, and H. Möhwald, Complexation of phosphocholine liposomes with polylysine: Stabilization by surface coverage versus aggregation. *Biochim. Biophys. Acta*, 1423: 143–154, 2006.

Wagner, K., D. Harries, S. May, V. Kahl, J. O. Rädler, and A. Ben-Shaul, Direct evidence for counterion release upon cationic lipid–DNA condensation. *Langmuir*, 16: 303–306, 2000.

Walker, B. and D. R. Burton, Towards an AIDS vaccine. *Science*, 320: 760–764, 2008.

Walker, W. H. and S. B. Grant, Factors influencing the flocculation of colloidal particles by a model anionic polyelectrolyte. *Colloids Surf. A*, 119: 229–239, 1996.

Wasung, L. and D. Hoekstra, Cationic lipids, lipoplexes and intracellular delivery of genes. *J. Control. Release*, 116: 255–264, 2006.

Wette, P., H. J. Schöpe, and T. Palberg, Comparison of colloidal effective charges from different experiments. *J. Chem. Phys.*, 24: 10981–10988, 2002.

Winterhalter, M. and W. Helfrich, Stability and phase behaviour of mixed surfactant vesicles. *J. Phys. Chem.*, 92: 6865–6867, 1988.

Woodle, M. C. and P. Scaria, Cationic liposomes and nucleic acids. *Curr. Opin. Colloid Interface Sci.*, 6: 77–86, 2001.

Yaroslavov, A., E. Kiseliova, O. Udalykh, and V. Kabanov, Integrity of mixed liposomes contacting a polycation depends on the negatively charged lipid content. *Langmuir*, 14: 5160–5163, 1998.

Yim, H., M. S. Kent, A. Matheson, M. J. Stevens, R. Ivkov, S. Satija, J. Majewski, and G. S. Smith, Adsorption of sodium poly(styrenesulfonate) to the air surface of water by neutron and x-ray reflectivity and surface tension measurements: Polymer concentration dependence. *Macromolecules*, 35: 9737–9747, 2002.

Zhang, S. B., Y. M. Xu, B. Wang, W. H. Qiao, D. L. Liu, and Z. S. Li, Cationic compounds used in lipoplexes and polyplexes for gene delivery. *J. Control. Release*, 100: 165–180, 2004.

17 Liposomes and Biomacromolecules
Effects of Preparation Protocol on In Vitro *Activity*

Paola Luciani, Debora Berti, and Piero Baglioni

CONTENTS

17.1 INTRODUCTION

Liposomal solutions are colloidal dispersions in which self-closed bilayered membranes composed of lipid molecules are suspended in an aqueous solution (Lasic and Templeton, 1996). The interior of the lipid vesicles is an aqueous core, the chemical composition of which corresponds in a first approximation to the chemical composition of the aqueous solution where the vesicles are prepared.

Liposomes are typically made of natural, biodegradable, nontoxic, and nonimmunogenic lipid molecules and can encapsulate or bind a variety of drug molecules within or onto their membranes. Usually they are composed of natural and synthetic lipids (phospho- and sphingolipids) and may also contain other bilayer constituents such as cholesterol (CH) and hydrophilic polymer-conjugated lipids. The net physicochemical properties of the lipids composing the liposomes, such as membrane fluidity, charge density, steric hindrance, and permeability, determine the interactions of the liposomes with blood components and other tissues after systemic administration (Sharma and Sharma, 1997). All of these properties make them attractive candidates for drug delivery vehicles (Lasic, 1998). Prolific fields of investigation and applications originated from the pioneering study by Bangham et al. (1967).

The drugs which are to be encapsulated may be included in the aqueous hydration buffer for hydrophilic drugs or in the lipid film for lipophilic drugs. Liposomes of different sizes and characteristics usually require different methods of preparation. The most simple and widely used method for the preparation of liposomes is the thin-film hydration procedure (Bangham et al., 1965). A thin film of lipids is hydrated with an aqueous buffer at a temperature above the transition temperature of the lipids. The thin-film hydration method produces a heterogeneous population of multilamellar vesicles (MLVs; 1–5 µm diameter), which can be sonicated or extruded through polycarbonate filters to produce a smaller (up to 0.025 µm) and more uniformly sized population of small unilamellar vesicles (SUVs). Several methods have been developed for the preparation of large unilamellar vesicles (LUVs), including solvent (ether or ethanol) injection (Hope et al., 1986), detergent dialysis (Bosworth et al., 1982; Matsumoto et al., 1977), calcium-induced fusion (Papahadjopoulos et al., 1977), and reverse-phase evaporation (Szoka and Papahadjopoulos, 1978).

Since their discovery as possible delivery systems, biophysical and biological investigations have focused on the interaction of liposomes with biomacromolecules. In this review chapter, we utilize the term *biomacromolecules* to refer to macromolecules that present biological activity: primarily nucleic acids and proteins and, more broadly, pharmaceutical compounds. Preservation of biomacromolecule functionality is one of the essential parameters that have to be monitored in drug delivery, gene therapy, and protein reconstitution. Thus, in addition to novel design of liposomal components (surfactants and lipids; Bombelli et al., 2009; Karmali et al., 2006; Saily et al., 2006; Sen and Chaudhuri, 2005), structural investigations of complexes (Bombelli et al., 2005b; Caracciolo et al., 2007; Ewert et al., 2004; Ma et al., 2007; Safinya, 2001), and biological tests of novel and various formulations (Bombelli et al., 2005c; Lonezet al., 2008; Pastorino et al., 2007), a correlation between the physicochemical characterizations and *in vitro* activities of the tested systems must be established to achieve consistent advances in liposome technology.

17.2   LIPOSOMES AND NUCLEIC ACIDS

Compaction of deoxyribonucleic acid (DNA) offers protection from nuclease degradation, and it represents a first crucial step in evaluating the efficiency of a gene delivery system. The condensation of the nucleic acids seems to be essential, which is also demonstrated by biological systems. DNA is condensed by positively charged proteins in the nucleus of eukaryotic cells or in the viral core, where part of the capsid functions as a shell to protect the viral genome from nucleases (Gelderblom, 1996; Yang and Huang, 1998). Mimicking nature, it is possible to achieve such an important structural and functional change upon interaction of the nucleic acid with specific agents. The interactions with cationic species reduce the electrostatic repulsion on the DNA chain causing a structural transition. It is believed that one of the major effects of positively charged systems is their ability to effectively reduce DNA size in terms of the radius of gyration, which facilitates cell internalization. However, this interaction must be reversible, because DNA has to be decondensed for transfection and gene expression.

Cationic lipids are widely used to complex DNA, because they provide a positively charged molecular portion that binds DNA, as well as a hydrophobic moiety that participates in the formation of highly ordered, compact lipid–DNA particles (lipoplexes). The formation of the lipoplex is a spontaneous process; the positive charge of the polar head group of the cationic lipid binds through charge–charge interactions with the negative charge of the DNA strand and thus condensates DNA. This process is time dependent and generally occurs within a timescale ranging from seconds to minutes (Eastman et al., 1997). The kinetics and thermodynamics of this complexation, which depend on the relative concentrations of the lipids and DNA, rate and order of mixing, temperature, salt concentration, and so forth, are still under debate. The exhaustive reviews of Pedroso de Lima et al. (2001) and Bally et al. (1999) provide additional information on this topic.

A fundamental challenge for the effective targeted delivery of DNA is to control the surface and colloidal properties of plasmids in a biological environment, because these properties influence their biological distribution, cellular uptake, intracellular trafficking, and nuclear translocation (Rolland, 1998). The flexibility in the design of a cationic lipid structure and liposome composition and the diversity of methods for their preparation, combined with the *in vivo* efficiency, have promoted the notion that cationic lipids can be efficiently used in human gene transfer (Tomlinson and Rolland, 1996).

The transfection ability of lipoplexes *in vitro* depends on many parameters, such as their physicochemical characteristics, type of cells, and incubation conditions (Felgner et al., 1996; Lasic, 1997; Lasic and Templeton, 1996). From the vast amount of studies performed on lipofection over the last few decades, it is clear that even *in vitro* (and more so *in vivo*) the lipofection process is multifactorial. Some of the factors involved are external and related to the type of cells and lipofection medium. Other factors are intrinsic and directly related to the physicochemical properties and mode of lipoplex preparation (Zuidam et al., 1999).

The challenges facing the development of liposomal gene delivery systems are similar to those for liposomal drug delivery. With gene-based drugs, however, delivery into appropriate cells represents only part of the problem; a number of intracellular barriers exist that can inhibit the biological activity of gene-based drugs (Chonn and Cullis, 1998).

17.2.1 Liposomes and Deoxyribonucleic Acid

A reproducible production of a well-characterized lipoplex dispersion is critical for the understanding of the molecular, biophysical, and biological mechanisms of lipoplex formation and action (Zelphati et al., 1998). Control over the biophysical and molecular parameters influencing lipoplex formation represents a central issue to consistently obtain well-defined, stable, and monodisperse formulations with reproducible biological activity. The concentration of lipid and DNA at the time of mixing, cationic lipid/DNA charge ratio, order of addition, mixing rate, vesicle size, and ionic strength of the hydration buffer represent a minimal set of parameters that affects lipoplex characteristics (Zuhorn and Hoekstra, 2002). However, when the same reagents and the same concentrations but different formulation procedures are used, the resulting thermodynamic and biological stability of lipoplexes may again differ greatly (Ferrari et al., 2001; Thierry et al., 1997; Zelphati et al., 1998).

Rakhmanova et al.'s (2004) definitive work pointed out how formulation procedures might affect the transfection efficiency even when the changes in the preparation protocol may appear modest. They explored how different methods of lipid dispersion and different modes of mixing DNA and lipids (together with size determination) might influence the transfection efficiency. Their investigations were referred to 1,2-dioleoyl-*sn*-glycero-3-ethylphosphocholine as phospholipids and baby hamster kidney cells. By leaving the molar lipid concentration and lipid/DNA ratio constant, they prepared lipoplexes using eight different procedures: the addition order was varied (vortexed or extruded liposomes added to DNA and vice versa) rapidly or dropwise at a constant rate. According to their reports, the formulation procedures have large effects on the lipoplex size, and we should recall here that the size of the complex is key in order to have an effective transfection. For the examined complexes and cells, a lower threshold in lipoplex size that was fixed at around 650 nm seemed to exist. Smaller and larger populations were quite ineffective; DNA added to vortexed liposomes (MLV) by rapid mixing appeared to be the best formulation procedure as far as efficient transfection was concerned. However, the authors specified that the size of most lipoplex formulations varies as a natural consequence of the physical and geometric properties of DNA, lipids, and liposomes.

Thierry et al. (1997) varied both the lipid/DNA ratio and the methods of lipoplex preparation and obtained an improved delivery system suitable for systemic administration. Two different methods of preparation were used throughout all the *in vitro* studies: (1) complex formation between preformed cationic dioctadecylamidoglycylspermine–dioleoylphosphatidylethanolamine vesicles or (2) hydration of a dioctadecylamidoglycylspermine–dioleoylphosphatidylethanolamine dry lipid film with an aqueous DNA solution. Preliminary morphological and size characterizations were done by means of transmission electron microscopy. *In vitro* transfection, cytoplasmic and nuclear transport, and reporter gene expression were compared to *in vivo* evaluation of DNA protection, pharmacokinetics, and transgene expression in mouse tissues. Reporter gene expression was detected in mouse tissues after intravenous administration of complexed DNA. Its level varied with the DNA/lipid ratio of the injected plasmid, the procedure used in preparing complexed DNA, and the tissue being tested. The lipoplex physicochemical properties affected the efficiency of *in vivo* and *in vitro* transgene expression differently. Consequently, the transfection efficiency of lipoplex formulations determined *in vitro* could not be extrapolated in an *in vivo* setting. The order of mixing reportedly influences other systems (Zelphati et al., 1998) tested on different cell cultures, which were prepared with the aim of developing a physically stable lipoplex formulation. Without the addition of any other components, monodisperse and stable lipoplexes with high transfection activities could be prepared only by combining two very simple procedures that are able to avoid aggregation problems: liposome extrusion and controlled mixing. Lack of attention to these preparation variables can lead to extensive aggregation. When negative complexes are formed, smaller aggregates can be obtained if liposomes are added to excess DNA; in positive complexes, the addition order has to be reversed. Large aggregation results in lower transfection activity. The physical properties and transfection activities of lipoplexes prepared from heterogeneous MLVs were compared with lipoplexes prepared from homogeneous extruded liposomes with well-defined sizes. Because the order of the addition of DNA and cationic liposomes was shown to be critical for controlling the size of the lipoplexes, Zelphati et al. (1998) developed a method to mix both components simultaneously at a fixed ratio and a controlled speed. The production of homogeneous and monodisperse lipoplexes was dependent on the concentration of DNA. According to tests of the *in vitro* transfection efficiency of lipoplexes formed from unextruded or extruded liposomes using the controlled mixing method, lipoplexes prepared with larger liposomes were more active than those prepared from smaller liposomes. However, for *in vivo* applications, the benefits of using small versus large lipoplexes may depend on the route of administration and on the intended cell or tissue targeted.

Ferrari et al. (2001) underscored the importance of the formulation procedure with respect to lipoplex structure by carrying out a systematic investigation of the effects of the components and the preparation technique on DNA-binding properties and on *in vitro* transfection of lipoplexes. The authors used a group of structurally related cationic lipids and varied the injection media. Even if phosphate was found to be a unique ion in terms of its effect on lipoplexes, different chemical substitution in the lipids reacted differently to the phosphate ion concentration in the injection medium. Among the examined formulation variables, the mixing method had the greatest effect on transgene expression *in vitro*. However, no direct correlation was found between the physicochemical trend and the biological activity for the examined systems (Ferrari et al., 2001).

Zuidam et al. (1999) used only one type of cell (mouse embryonic fibroblast NIH-3T3 cells), one type of medium, and one cationic lipid {N-[1-(2,3-dioleoyloxy)propyl]-N,N,N-trimethylammonium (DOTAP)]} to change the mode of lipoplex preparation. They used these few variables together with the DOTAP/DNA molar ratio, the presence of a helper lipid, and the lamellarity of liposomes to try to understand the effects on the *in vitro* efficacy of lipoplexes. The authors concluded that lipoplexes should be physicochemically characterized at two different levels of structural properties: the macrolevel that relates to size and size instability and the microlevel that relates to the intimate interaction between the plasmid DNA (pDNA) and the lipids. At the microlevel, all parameters are reversible, independent of history, and determined by the DNA/DOTAP mole ratio. In contrast, the macrolevel (which is the most important for transfection efficiency) is dependent on history and not reversible. Lipofection followed the macrostructure and not the microstructure. Even if LUVs are undoubtedly considered superior to MLVs as pharmaceuticals, Zuidam et al. (1999) reported that MLV transfection efficiency is less dependent on DNA/DOTAP concentration than that of LUV. It can be assumed that lipofection efficiency is related to the fraction of a specific population of lipoplexes (DNA/DOTAP ratio = ~0.5, excess cationic lipid), as in the previously shown case of

Rakhmanova et al. (2004). According to Zuidam et al., at this ratio the lipoplexes reach a high level of instability that is due to lipid packing defects, which results from a lateral phase separation between regions of bilayers that were condensed by DNA and those that were not (Hirsch-Lerner and Barenholz, 1998). The authors propose that defects in lipid packing, which lead to inherent instability, are the common denominator for optimizing transfection.

When the formulation of lipoplexes has to be varied, as in poly(ethylene glycol) (PEG)ylated lipoplexes, the importance of the order of mixing must not be undervalued. As recent and distinct papers attest (Bombelli et al., 2007; Peeters et al., 2007), different physicochemical features and, most importantly, different biological activity can be obtained by adding a PEG derivative at different stages of the complexation. Bombelli et al. (2005c, 2007) utilized complexes made of DNA and mixed vesicles containing cationic gemini surfactants to investigate the connection between the lipid molecular architectures and the transfection efficiency displayed by lipoplexes. Their papers demonstrate that the biological properties of lipoplexes can be correlated in some cases to the chiro-optical features of DNA in cationic complexes and to the preparation protocol of the complexes. The secondary structure of DNA was previously demonstrated (Bombelli et al., 2005a) to be dependent on the stereochemistry of the gemini surfactant spacer added in mixed liposomes, and the strong compaction of DNA typical of the ψ-phase can be related to the transfection efficiency (Bombelli et al., 2005c). A modification of the liposome surface by inserting alkylated PEG monolaurate (PEG-ML) polymers was introduced to reduce opsonization for future *in vivo* applications. Samples were prepared according to three different strategies (PEG-ML added to preformed lipoplexes, PEG-ML added to the hydration buffer, and PEG-ML codissolved in the dry lipid film). The addition of PEG-ML to the preformed lipoplexes provided a better insertion of the PEG on the surface of the lipoplexes. When the PEG-ML inclusion is achieved by means of the other two tested protocols, part of the poly(ethylenglycol), needing a longer time for reorganization, is entrapped in the internal core of the vesicles and it is, very likely, no longer able to interact with DNA. The highest transfection efficiency was observed in correspondence with the addition of PEG after the formation of lipoplexes. Peeters et al. (2007) studied the effect of pre- and post-PEGylation of lipoplexes on cell adhesion and on retinal pigment epithelium (RPE) cell transfection. It turned out that pre-PEGylated lipoplexes did not succeed in transfection, likely because, among other possible reasons, pre-PEGylated lipoplexes were unable to escape from the RPE endosomes and a high amount of PEG at the surface of the lipoplexes also reduced their cellular internalization. According to Peeters et al., post-PEGylated lipoplexes obtained by preparing the lipoplexes and subsequently coating them with PEG chains, using PEG ceramides (short acyl-chain ceramides) that spontaneously adsorb to the lipoplexes, are expected to lose the PEG ceramides (and thus the PEG coating) upon contact with cellular membranes. Post-PEGylated lipoplexes (only 100–200 nm in size) do not aggregate in the vitreous body, a crucial requirement when the lipoplexes are applied intravitreally for targeting inner retinal cells or even RPE cells. Their transport through the neural retina will hopefully occur much more efficiently than that of the non-PEGylated lipoplexes that form micron-sized aggregates in the vitreous body and will be strongly blocked by the neural retina.

Freeze–thaw PEGylated liposomes (Ueno and Sriwongsitanont, 2005) may have different consequences on fusion and fission of phospholipid vesicles, and this has to be taken into account when using PEGylated liposomes for complexing DNA.

In addition to insertion of polymers like PEG to achieve a prolonged circulation lifetime, other ternary systems were studied such as the one described by Penacho et al. (2008). Their system is composed of cationic liposomes/pDNA and transferrin (Tf): association of this blood protein, responsible for the transport of iron into cells, to lipoplexes, results in a significant enhancement of transfection in a large variety of cell types (da Cruz et al., 2005; Neves et al., 2006). Different ionic strengths of the medium and different modes of preparation were tested to understand how the different modes of preparation of ternary complexes composed of a cationic lipid, pDNA, and a ligand (e.g., Tf) affect their physicochemical properties and consequently influence their biological activity. The order of addition seemed to be extremely important in terms of facilitating DNA release from the complex into the cell cytoplasm: when Tf was added to liposomes prior to pDNA, the complexes were able to mediate transfection with better efficacy.

17.2.2 LIPOSOMES AND OLIGONUCLEOTIDES

Oligonucleotides are appealing and promising DNA substitutes. They are widely studied for bio-technological and pharmaceutical applications, mostly in their antisense form. Aliño et al. (1999) investigated the effect of the lipid composition and preparation method on the pharmacokinetics of oligonucleotide–liposomes complexes. The authors tested two methods of preparation: preparation of MLVs and preparation of liposomes by the dehydration–rehydration method. Liposomes obtained by the dehydration–rehydration method were characterized by higher encapsulation efficiency and yielded lower oligodeoxynucleotide plasma clearances than liposomes prepared with the MLV method, regardless of the chemical composition of the liposomes and the type of oligonucleotide involved. More recently, Ciani et al. (2007) investigated the effect of the preparation procedure on the structural properties of oligonucleotide–cationic liposome complexes. They did not reveal any influence of the preparation procedures on the physicochemical features of the lipoplexes, but unfortunately no *in vitro* or *in vivo* test was performed.

The successful delivery of nucleic acids to cells requires a synthetic carrier that is able to complex DNA, the formation of a supramolecular DNA–carrier assembly, and the uptake of this complex in the cell. Cationic-based synthetic nonviral systems are routinely used *in vitro* as convenient biological tools for transporting nucleic acids to cells; however, their proven toxicity at high doses (Anderson and Borlak, 2006) drove several scientists to explore alternative routes by designing and synthesizing neutral or anionic lipids. One of the most challenging issues is designing robust neutral supramolecular complexes that are more easily targetable to specific cells, able to deliver other biologically active compounds, and able to overcome the drawback of an aspecific binding to cell surfaces. As extensively reviewed by Barthélémy and coworkers (Gissot et al., 2008), nucleolipids represent a fascinating system that may modulate the interactions between DNA and a synthetic carrier. In nucleolipids, functionalization with pyrimidine or purine bases of the amphiphile brings about new favorable H-bonding and π-stacking interactions (Banchelli et al., 2007; Berti, 2006; Berti et al., 1998; Moreau et al., 2005). Berti and coworkers (Milani et al., 2007) recently demonstrated that a novel lamellar phase is obtained from the spontaneous ordering of polyuridylic acid between the lamellae formed by the anionic 1-palmitoyl-2-oleoylphosphatidyl-adenosine nucleolipid without any mediation from divalent cations. This nucleolipid is biocompatible and can be enzymatically cleaved by phospholipases once inside living organisms to release the polynucleotide. Their study opens new perspectives in the fabrication of lipoplexes, which were up to now obtained by favorable coloumbic interactions or the presence of divalent cations. The proof of the principle that molecular recognition can be the driving force for bioconjugate constructs establishes a new paradigm with potentially important impact in the biomedical field.

17.2.3 LIPOSOMES AND RIBONUCLEIC ACID

Ribonucleic acid interference (RNAi) is a mechanism for RNA-guided regulation of gene expression in which double-stranded RNA inhibits the expression of genes with complementary nucleotide sequences. There is a widespread mistaken belief that all nucleic acids are similar and a general lack of notion that different delivery systems may deliver RNA and DNA to different intracellular pathways, with consequent possible different transcription efficacies (Tagami et al., 2007). Tagami et al. (2007) demonstrated that simultaneous transfection with the target gene pDNA and the gene-silencing small interference RNA (siRNA) can be achieved by means of cationic liposomes. The cotransfection was accomplished following two different approaches: method I in which pDNA and siRNA were in the same carrier complex, and method II in which a mixture of pDNA–carrier complex and siRNA–carrier complex was used for transfection. Thanks to the significant chemical and structural differences between pDNA and siRNA, the physicochemical features of the three complexes (pDNA–siRNA–liposome, siRNA–liposome, and pDNA–liposome) are likely to be significantly different: they have anionic phosphodiester backbones with identical negative charge/nucleotide ratios ready to interact with cationic liposomes to form lipoplexes, but their molecular topography and molecular weight are very different. According to Tagami et al., this is likely to lead to essentially distinct interactions of the complexes with cells, their cellular uptake, and the intracellular distribution and ultimate fate of siRNA and pDNA, which will largely determine the effectiveness of the gene-silencing effect. The siRNA lipoplexes produced by method I showed a much stronger reporter-gene-silencing effect than those produced by method II. In terms of physicochemical characterization, no significant differences between complexes prepared by the two different methods were pointed out as far as particle diameter and zeta potential were concerned. The major cause of the different gene-silencing effects of siRNA observed in this study did not derive from a difference in physicochemical properties. This interpretation was supported by the constant values of the control exogenous gene expression levels that were observed and by the similar amounts of siRNA associated with the cells, irrespective of the preparation method.

17.3 LIPOSOMES AND PROTEINS

17.3.1 PROTEIN ENCAPSULATION

One of the most systematic studies that investigated the effect of the preparation technique on protein encapsulation in liposomes was carried out by Colletier et al. (2002). Proteins can be extremely sensitive to chemical and physical treatments: freeze–thaw cycles, extrusion, and thermal vortexing may affect native conformation of proteins and enzymes that may be needed to encapsulate or reconstitute in lipid bilayers. One of the most popular methods is the lipid film's hydration: a phospholipid solution in chloroform is dried under vacuum to obtain a lipid film, and the lipid film is subsequently hydrated in a solution containing the protein (Anselem et al., 1993; Kirby and Gregoriadis, 1984). This method allows the encapsulation of the protein in its functional form, but the efficiency of encapsulation is generally weak. The authors' purpose was to improve the film hydration method for efficient protein (acetylcholinesterase) upload in liposomes while preserving its functionality. The effects of the lipid concentration and composition, number of freeze–thaw cycles, extrusion, composition, and coencapsulation of stabilizers on acetylcholinesterase activity were systematically investigated. Interesting conclusions were drawn: encapsulation is proportional to the number of lipids and thus proportional to the surface; an increase of encapsulation efficiency occurs according to the number of freeze–thaw cycles without significant denaturation of the enzyme; retention of some vesicles or lipids may occur on extrusion filters with specific compositions; the observed decrease of the encapsulation efficiency while increasing salt concentration suggests that encapsulation might be related to a specific interaction between the phospholipids and protein; and the encapsulation efficiency does not depend on the hydrophobic component of the phospholipids, whereas it appears to be dependent on the electrostatic interactions between the enzyme peripheral surface and the polar head group of the phospholipids.

To improve the encapsulation efficiency of proteins in a size-regulated phospholipids vesicle by using an extrusion method, Sou et al. (2003) conducted a very interesting investigation on the effect of the frozen–thawed vesicles on water-soluble protein hemoglobin encapsulation, which unfortunately lacked biological activity tests.

Among the possible interesting proteins for pharmaceutical applications, synthetic peptides must be mentioned: synthetic peptides of antigenic epitopes are receiving considerable interest as the basis for "next generation" highly specific and safe vaccines (Arnon and Ben-Yedidia, 2003). Conjugation of peptides to lipoamino acids confers increased lipophilicity to the peptides and therefore increases their membrane permeability (Toth et al., 1994) in addition to protecting it from enzymatic digestion. The presence of lipidic moieties appears to overcome the need for additional adjuvants. However, as a result of their increased lipophilicity, the aqueous solubility of the constructs can be dramatically reduced, frequently requiring administration as a suspension. To overcome this formulation issue and to further enhance their bioactivity, lipid-conjugated peptides have been encapsulated within liposomes (Babu et al., 1995; Haro et al., 2003). Liang et al. (2005) investigated the effect of the chemical nature of the lipid on the efficiency of entrapment of a lipopeptide within a liposome, and the effect of the mode of preparation was also considered. Preparing liposomes by means of hydration of dried lipid films results in differing rates of precipitation upon solvent evaporation and the formation of a lipid film in which the lipopeptide is not homogeneously dispersed because of the differential solubility of lipopeptides and phospholipids. Lipopeptide entrapment within liposomes prepared by this method is incomplete and dependent on the structure of the lipopeptide and loading. Conversely, when liposomes are prepared by hydration of freeze-dried monophase systems, entrapment is high and less affected by the lipopeptide structure and higher loadings can be achieved. Unfortunately, no biological activity of these complexes was monitored.

An extensive study about enzyme encapsulation in lipid bilayers was reported by Walde and Ichigawa (2001). The issues addressed in the paper are strictly limited to vesicle encapsulation of water-soluble enzymatic active proteins (enzymes), although the authors signaled a number of other papers dealing with the entrapment of other nonenzymatically active, water-soluble proteins (Adrian and Huang, 1979; Meyer et al., 1994; Zheng et al., 1994). These extensive and detailed reviews provide further insights on these topics.

17.3.2 PROTEIN ANCHORING

The key role of G proteins in complex intracellular signaling explains their importance as pharmaceutical targets. The correct physiological functionality of G proteins requires anchoring to a lipid bilayer: mutations that prevent myristoylation or palmitoylation may severely interfere with protein functionality. Luciani et al. (2009) recently carried out a study on $G\alpha$ subunit functionality, including fatty acid modifications and taking into proper account the importance of a hydrophobic environment by means of protein reconstitution in a lipid bilayer. They sought to improve the protocols of $G\alpha$-mir reconstitution in lipid vesicles with preservation of protein functionality, so the effects of freeze–thaw cycles, lamellarity, extrusion procedures, and lipid chain length on $G\alpha i$ activity were evaluated. The reconstitution of $G\alpha$-mir protein in a liposomal structure represents a more physiological system that improves the protein basal activity. To the best of the authors' knowledge, this was the first complete characterization of a $G\alpha$ subunit reconstitution in model membranes, in terms of protein activity as a function of the reconstitution protocol. The importance of correlating the variations in preparation methods to the functionality maintenance was underscored.

17.4 LIPOSOMES AND DRUGS

Liposomal drug delivery is one of the principal routes that is currently being explored to achieve increased therapeutic indices for existing drugs and for some new classes of drugs whose clinical application has been prevented by rapid breakdown or a lack of solubility. Liposomes might improve the bioavailability of pharmacologically active molecules by solubilizing those with reduced aqueous solubility, by entrapping rapidly degradable drugs, by reducing the accumulation of drugs in sensitive tissues thanks to the alteration of the biodistribution of their associated drugs, and by opportunely targeting the lipid carriers to specific cells and tissues (Allen and Moase, 1996). It is possible to improve the degree of interaction and capture of liposomes by phagocytic cells by using a suitable liposome design.

Different methods of liposome preparation may markedly affect the relative stability and percentage of encapsulation of markers and drugs. This consequently affects *in vitro* activity, as in the case investigated by Vitas et al. (1996) in which the therapeutic capacity of gentamicin was evaluated by encapsulation in different types of liposomes for monocytes infected with *Brucella abortus*. Standard MLVs and standard stable plurilamellar vesicles (SPLVs) were prepared together with freeze–thaw MLVs and modified SPLVs. Their stability in the presence of high-density lipoprotein with and without charged molecules was evaluated: comparative studies of the stabilities of vesicles prepared by different techniques indicate that, in addition to the composition of the liposomes, the method of preparation might also affect the physical and chemical characteristics of the vesicles, thus altering their stability and encapsulating efficiency. The best dispersion for antibiotic transport was modified SPLVs. The preparation method might affect liposome capture.

Another study (Zeisig et al., 1998) focused on the influence of the composition of liposomes as drug carriers, especially with regard to CH content, charge, size, and sterical stabilization, on the physical properties and the cytotoxic effect *in vitro*, as well as the therapeutic activity *in vivo*. Two different preparation procedures were also used in this case, and their effect on biological activity was investigated. MLVs and LUVs containing with a novel cancerostatic alkylphospholipid [octa-decyl-(1,1-dimethyl-piperidino-4-yl)-phosphate] were prepared, physically characterized, and tested on four different human breast cancer cell lines. The effect of CH on the stability of the different dispersions and on the capacity of increasing release was evaluated; the effects of the different preparations against tumor cells *in vitro* were investigated. The reduction of liposomal CH correlated with an increase of the cytotoxic effect *in vitro*. This could be explained by the increase in membrane fluidity caused by CH reduction (Nagayasu et al., 1996), which probably facilitates (specific) interactions between the components of the liposomal membrane and those of the cellular membrane. In terms of cytotoxicity, size dependency is involved: large MLVs are less toxic, most likely by the inhibition of the transbilayer transport of the MLVs into the cells. Octadecyl-(1,1-dimethyl-piperidino-4-yl)-phosphate–LUVs showed high stability in serum, but no further correlation between the preparation procedures and *in vitro* and *in vivo* tests was established in this study.

Maestrelli et al. (2006) evaluated complexes of cyclodextrin–ketoprofen–liposomes to improve drug solubility. The aim of their work was to find the most effective operative conditions to improve the effectiveness of the ketoprofen–cyclodextrin–liposome system in terms of drug encapsulation efficiency and permeation properties. Toward this purpose, liposomes of constant composition [i.e., 60/40 (w/w) phosphatidylcholine/CH] were prepared by using different preparation methods (MLVs, frozen–thawed MLVs, SUVs, and LUVs). The percentage of encapsulation efficiency depended on both the liposome preparation method and the complex concentration in the aqueous phase used for liposome preparation and was in the order MLVs > LUVs > SUVs. A different drug permeation rate was observed from the different liposomal formulations (i.e., solutions > SUVs > frozen–thawed MLVs = MLVs > LUVs). MLVs exhibited the highest value of encapsulation efficiency, so they represent the best formulation to optimize the drug entrapment efficiency and permeation rate when a prolonged drug release is desirable. Unfortunately, *in vitro* or *in vivo* tests were not performed.

The physicochemical properties of bioactive components of natural traditional herbs may also be affected by different methods of inclusion of the compound in lipid bilayer. Fan et al. (2007) studied salidroside encapsulation in lipid vesicles by preparing them according to five different methods: thin-film evaporation, sonication, reverse-phase evaporation, melting, and freezing–thawing. A physicochemical characterization of the complexes was performed: a relationship between the encapsulating efficiency of salidroside liposomes and different preparation methods and loading capacities was established. Moreover, morphological and dimensional studies were performed together with leakage and temperature-stability characterization. Salidroside liposomes prepared by the melting method showed better encapsulating efficiency and physicochemical stability. The loading and leakage characteristics of salidroside liposomes showed that salidroside loading efficiency highly influenced the encapsulating efficiency, and the preparation method had a great effect on the leakage of salidroside from liposomes. Unfortunately, no biological evaluation was done.

17.5 CONCLUSIONS

This chapter summarizes recent systematic approaches in rationalizing parameters during preparation of liposomes as drug and gene delivery systems. Biological response may be affected by several variables, but undoubtedly biomacromolecule functionality must be preserved. Thus, understanding how different preparation techniques may affect physicochemical features and how changed characteristics of the biomacromolecule–liposome complexes may impact cell and tissue interaction, as well as pharmaceutical and biological applications in general, should be mandatory. Among the open challenges in this multidisciplinary field, the understanding of the role played by molecular interactions in complex systems as supramolecular aggregates, biomacromolecules, and cells represents the most exciting milestone to be overcome.

ACKNOWLEDGMENT

Consorzio Interuniversitario per lo Sviluppo dei Sistemi a Grande Interfase (CSGI) is acknowledged for financial support.

ABBREVIATIONS

CH	Cholesterol
DNA	Deoxyribonucleic acid
DOTAP	N-[1-(2,3-Dioleoyloxy)propyl]-N,N,N-trimethylammonium
LUV	Large unilamellar vesicle
MLV	Multilamellar vesicle
pDNA	Plasmid deoxyribonucleic acid
PEG	Poly(ethylene glycol)
PEG-ML	Poly(ethylene glycol) monolaurate
RNA	Ribonucleic acid
RNAi	Interference ribonucleic acid
RPE cells	Retinal pigment epithelium cells
siRNA	Small interference ribonucleic acid
SPLV	Stable plurilamellar vesicle
SUV	Small unilamellar vesicle
Tf	Transferrin

REFERENCES

Adrian, G., and Huang, L. 1979. Entrapment of proteins in phosphatidylcholine vesicles. *Biochemistry* 18: 5610–5614.

Aliño, S. F., Crespo, J., Tarrason, G., Blaya, C., Adan, J., Escrig, E., Benet, M., Crespo, A., Peris, J. E., and Piulats, J. 1999. Pharmacokinetics of oligodeoxynucleotides encapsulated in liposomes: Effect of lipid composition and preparation method. *Xenobiotica* 29: 1283–1291.

Allen, T. M., and Moase, E. H. 1996. Therapeutic opportunities for targeted liposomal drug delivery. *Advanced Drug Delivery Reviews* 21: 117–133.

Anderson, N., and Borlak, J. 2006. Drug-induced phospholipidosis. *FEBS Letters* 580: 5533–5540.

Anselem, S., Gabizon, A., Barenholz, Y., and Gregoriadis, G. 1993. *Liposome Technology*. Boca Raton, FL: CRC Press.

Arnon, R., and Ben-Yedidia, T. 2003. Old and new vaccine approaches. *International Immunopharmacology* 3: 1195–1204.

Babu, J. S., Nair, S., Kanda, P., and Rouse, B. T. 1995. Priming for virus-specific CD8+ but not CD4+ cytotoxic T lymphocytes with synthetic lipopeptide is influenced by acylation units and liposome encapsulation. *Vaccine* 13: 1669–1676.

Bally, M. B., Harvie, P., Wong, F. M. P., Kong, S., Wasan, E. K., and Reimer, D. L. 1999. Biological barriers to cellular delivery of lipid-based DNA carriers. *Advanced Drug Delivery Reviews* 38: 291–315.

Banchelli, M., Berti, D., and Baglioni, P. 2007. Molecular recognition drives oligonucleotide binding to nucleo-lipid self-assemblies. *Angewandte Chemie International Edition* 46: 3070–3073.

Bangham, A. D., Hill, M. W., and Miller, N. G. A. 1967. Preparation and use of liposomes as models of biological membranes. In E. D. Korn (Ed.), *Methods in Membrane Biology*, Vol. 1. New York: Plenum Press.

Bangham, A. D., Standish, M. M., and Watkins, J. C. 1965. Diffusion of univalent ions across the lamellae of swollen phospholipids. *Journal of Molecular Biology* 13: 238.

Berti, D. 2006. Self assembly of biologically inspired amphiphiles. *Current Opinion in Colloid and Interface Sciences* 11: 74.

Berti, D., Baglioni, P., Bonaccio, S., Barsacchi-Bo, G., and Luisi, P. L. 1998. Base complementarity and nucleoside recognition in phosphatidylnucleoside vesicles. *Journal of Physical Chemistry B* 102: 303–308.

Bombelli, C., Borocci, S., Diociaiuti, M., Faggioli, F., Galantini, L., Luciani, P., Mancini, G., and Sacco, M. G. 2005a. Role of the spacer of cationic gemini amphiphiles in the condensation of DNA. *Langmuir* 21: 10271–10274.

Bombelli, C., Caracciolo, G., DiProfio, P., Diociaiuti, M., Luciani, P., Mancini, G., Mazzuca, C. et al. 2005b. Inclusion of a photosensitizer in liposomes formed by DMPC/gemini surfactant: Correlation between physicochemical and biological features of the complexes. *Journal of Medicinal Chemistry* 48: 4882–4891.

Bombelli, C., Faggioli, F., Luciani, P., Mancini, G., and Sacco, M. G. 2005c. Efficient transfection of DNA by liposomes formulated with cationic gemini amphiphiles. *Journal of Medicinal Chemistry* 48: 5378–5382.

Bombelli, C., Faggioli, F., Luciani, P., Mancini, G., and Sacco, M. G. 2007. PEGylated lipoplexes: Preparation protocols affecting DNA condensation and cell transfection efficiency. *Journal of Medicinal Chemistry* 50: 6274–6278.

Bombelli, C., Giansanti, L., Luciani, P., and Mancini, G. 2009. Gemini surfactant based carriers in gene and drug delivery. *Current Medicinal Chemistry* 16, 171–183.

Bosworth, M. E., Hunt, C. A., and Pratt, D. 1982. Liposome dialysis for improved size distributions. *Journal of Pharmaceutical Sciences* 71: 806–812.

Caracciolo, G., Pozzi, D., Caminiti, R., Mancini, G., Luciani, P., and Amenitsch, H. 2007. Observation of a rectangular DNA superlattice in the liquid-crystalline phase of cationic lipid/DNA complexes. *Journal of the American Chemical Society* 129: 10092–10093.

Chonn, A., and Cullis, P. R. 1998. Recent advances in liposome technologies and their applications for systemic gene delivery. *Advanced Drug Delivery Reviews* 30: 73–83.

Ciani, L., Ristori, S., Bonechi, C., Rossi, C., and Martini, G. 2007. Effect of the preparation procedure on the structural properties of oligonucleotide/cationic liposome complexes (lipoplexes) studied by electron spin resonance and zeta potential. *Biophysical Chemistry* 131: 80–87.

Colletier, J.-P., Chaize, B., Winterhalter, M., and Fournier, D. 2002. Protein encapsulation in liposomes: Efficiency depends on interactions between protein and phospholipid bilayer. *BMC Biotechnology* 2: 9.

da Cruz, M. T., Cardoso, A. L. C., de Almeida, L. P., Simoes, S., and Pedroso de Lima, M. C. 2005. Tf-lipoplex-mediated NGF gene transfer to the CNS: Neuronal protection and recovery in an excitotoxic model of brain injury. *Biochimica et Biophysica Acta* 12: 1242–1252.

Eastman, S. J., Siegel, C., Tousignant, J., Smith, A. E., Cheng, S. H., and Scheule, R. K. 1997. Biophysical characterization of cationic lipid: DNA complexes. *Biochimica et Biophysica Acta* 1325: 41–62.

Ewert, K., Slack, N. L., Ahmad, A., Evans, H. M., Lin, A. J., Samuel, C. E., and Safinya, C. R. 2004. Cationic lipid–DNA complexes for gene therapy: Understanding the relationship between complex structure and gene delivery pathways at the molecular level. *Current Medicinal Chemistry* 11: 133–149.

Fan, M., Xu, S., Xia, S., and Zhang, X. 2007. Effect of different preparation methods on physicochemical properties of salidroside liposomes. *Journal of Agricultural and Food Chemistry* 55: 3089–3095.

Felgner, P. L., Tsai, Y. J., and Felgner, J. H. 1996. Advances in the design and application of cytofectin formulations. In D. D. Lasic and Y. Barenholz (Eds.), *Handbook of Nonmedical Applications of Liposomes*. Boca Raton, FL: CRC Press.

Ferrari, M. E., Rusalov, D., Enas, J., and Wheeler, C. J. 2001. Trends in lipoplex physical properties dependent on cationic lipid structure, vehicle and complexation procedure do not correlate with biological activity. *Nucleic Acids Research* 29: 1539–1548.

Gelderblom, H. R. 1996. Structure and classification of viruses. In S. Baron (Ed.), *Medical Microbiology*. Galveston, TX: University of Texas Medical Branch at Galveston.

Gissot, A., Camplo, M., Grinstaff, M. W., and Barthélémy, P. 2008. Nucleoside, nucleotide and oligonucleotide based amphiphiles: A successful marriage of nucleic acids with lipids. *Organic and Biomolecular Chemistry* 6: 1324–1333.

Haro, I., Pèrez, S., Garcìa, M., Chan, W. C., and Ercilla, G. 2003. Liposome entrapment and immunogenic studies of a synthetic lipophilic multiple antigenic peptide bearing VP1 and VP3 domains of the hepatitis A virus: A robust method for vaccine design. *FEBS Letters* 540(1–3): 133–140.

Hirsch-Lerner, D., and Barenholz, Y. 1998. Probing DNA–cationic lipid interactions with the fluorophore trimethylammonium diphenyl-hexatriene (TMADPH). *Biochimica et Biophysica Acta* 1370: 17–30.

Hope, M. J., Bally, M. B., Mayer, L. D., Janoff, A. S., and Cullis, P. R. 1986. Generation of multilamellar and unilamellar phospholipid vesicles. *Chemistry and Physics of Lipids* 40: 89–107.

Karmali, P. P., Majeti, B. K., Sreedhar, B., and Chaudhuri, A. 2006. *In vitro* gene transfer efficacies and serum compatibility profiles of novel mono-, di-, and tri-histidinylated cationic transfection lipids: A structure–activity investigation. *Bioconjugate Chemistry* 17: 159–171.

Kirby, C. J., and Gregoriadis, G. 1984. Preparation of liposomes containing factor VIII for oral treatment of haemophilia. *Journal of Microencapsulation* 1: 33–45.

Lasic, D. D. 1997. *Liposomes in Gene Delivery*. Boca Raton, FL: CRC Press.

Lasic, D. D. 1998. Novel applications of liposomes. *Trends in Biotechnology* 16: 307–321.

Lasic, D. D., and Templeton, N. S. 1996. Liposomes in gene therapy. *Advanced Drug Delivery Reviews* 20: 221–266.

Liang, M. T., Davies, N. M., and Toth, I. 2005. Encapsulation of lipopeptides within liposomes: Effect of number of lipid chains, chain length and method of liposome preparation. *International Journal of Pharmaceutics* 301: 247–254.

Lonez, C., Vandenbranden, M., and Ruysschaert, J.-M. 2008. Cationic liposomal lipids: From gene carriers to cell signaling. *Progress in Lipid Research* 47: 340–347.

Luciani, P., Berti, D., Fortini, M., Baglioni, P., Ghelardini, C., Pacini, A., Manetti, D., Gualtieri F., Bartolini, A., and Di Cesare Mannelli, L. 2009. Receptor-independent modulation of reconstituted Gα_i protein mediated by liposomes. *Molecular BioSystems* 5: 356–367.

Ma, B., Zhang, S., Jiang, H., Zhao, B., and Lv, H. 2007. Lipoplex morphologies and their influences on transfection efficiency in gene delivery. *Journal of Controlled Release* 123: 184–194.

Maestrelli, F., Gonzalez-Rodriguez, M. L., Rabasco, A. M., and Mura, P. 2006. Effect of preparation technique on the properties of liposomes encapsulating ketoprofen–cyclodextrin complexes aimed for transdermal delivery. *International Journal of Pharmaceutics* 312: 53–60.

Matsumoto, S., Kohda, M., and Murata, S.-I. 1977. Preparation of lipid vesicles on the basis of a technique for providing W/O/W emulsions. *Journal of Colloid and Interface Science* 62: 149–157.

Meyer, J., Whitcomb, L., and Collins, D. 1994. Efficient encapsulation of proteins within liposomes for slow release *in vivo*. *Biochemical and Biophysical Research Communications* 199: 433–438.

Milani, S., Bombelli, F. B., Berti, D., and Baglioni, P. 2007. Nucleolipoplexes: A new paradigm for phospholipid bilayer–nucleic acid interactions. *Journal of the American Chemical Society* 129: 11664–11665.

Moreau, L., Barthélémy, P., Li, Y., Luo, D., Prata, C. A. H., and Grinstaff, M. W. 2005. Nucleoside phosphocholine amphiphile for *in vitro* DNA transfection. *Molecular Biosystems* 1: 260–264.

Nagayasu, A., Uchiyama, K., Nishida, T., Yamagiwa, Y., Kawai, Y., and Kiwada, H. 1996. Is control of distribution of liposomes between tumors and bone marrow possible? *Biochimica et Biophysica Acta* 1278: 29–34.

Neves, S. S., Sarmento-Ribeiro, A. B., Simoes, S. P., and Pedroso de Lima, M. C. 2006. Transfection of oral cancer cells mediated by transferrin-associated lipoplexes: Mechanisms of cell death induced by herpes simplex virus thymidine kinase/ganciclovir therapy. *Biochimica et Biophysica Acta* 1758: 1703–1712.

Papahadjopoulos, D., Vail, W. J., Newton, C., Nir, S., Jacobson, K., Poste, G., and Lazo, R. 1977. Studies on membrane fusion. III. The role of calcium-induced phase changes. *Biochimica et Biophysica Acta* 465: 579–598.

Pastorino, F., Marimpietri, D., Brignole, C., Paolo, D. D., Pagnan, G., Daga, A., Piccardi, F., Cilli, M., Allen, T. M., and Ponzoni, M. 2007. Ligand-targeted liposomal therapies of neuroblastoma. *Current Medicinal Chemistry* 14: 3070–3078.

Pedroso de Lima, M. C., Simoes, S., Pires, P., Faneca, H., and Duzgunes, N. 2001. Cationic lipid–DNA complexes in gene delivery: From biophysics to biological applications. *Advanced Drug Delivery Reviews* 47: 277–294.

Peeters, L., Sanders, N. N., Jones, A., Demeester, J., and De Smedt, S. C. 2007. Post-PEGylated lipoplexes are promising vehicles for gene delivery in RPE cells. *Journal of Controlled Release* 121: 208–217.

Penacho, N., Filipe, A., Simões, S., and Pedroso de Lima, M. C. 2008. Transferrin-associated lipoplexes as gene delivery systems: Relevance of mode of preparation and biophysical properties. *Journal of Membrane Biology* 221: 141–152.

Rakhmanova, V. A., Pozharski, E. V., and MacDonald, R. C. 2004. Mechanisms of lipoplex formation: Dependence of the biological properties of transfection complexes on formulation procedures. *Journal of Membrane Biology* 200: 35–45.

Rolland, A. P. 1998. From genes to gene medicines: Recent advances in nonviral gene delivery. *Critical Reviews in Therapeutic Drug Carrier Systems* 15: 143–198.

Safinya, C. R. 2001. Structures of lipid–DNA complexes: Supramolecular assembly and gene delivery. *Current Opinion in Structural Biology* 11: 440–448.

Saily, V. M. J., Ryhanen, S. J., Lankinen, H., Luciani, P., Mancini, G., Parry, M. J., and Kinnunen, P. K. J. 2006. Impact of reductive cleavage of an intramolecular disulfide bond containing cationic gemini surfactant in monolayers and bilayers. *Langmuir* 22: 956–962.

Sen, J., and Chaudhuri, A. 2005. Design, syntheses, and transfection biology of novel non-cholesterol-based guanidinylated cationic lipids. *Journal of Medicinal Chemistry* 48: 812–820.

Sharma, A., and Sharma, U. S. 1997. Liposomes in drug delivery: Progress and limitations. *International Journal of Pharmaceutics* 154: 123–140.

Sou, K., Naito, Y., Endo, T., Takeoka, S., and Tsuchida, E. 2003. Effective encapsulation of proteins into size-controlled phospholipid vesicles using freeze–thawing and extrusion. *Biotechnology Progress* 19: 1547–1552.

Szoka, Jr., F., and Papahadjopoulos, D. 1978. Procedure for preparation of liposomes with large internal aqueous space and high capture by reverse-phase evaporation. *Proceeding of the National Academy of Science* 75: 4194–4198.

Tagami, T., Barichello, J. M., Kikuchi, H., Ishida, T., and Kiwada, H. 2007. The gene-silencing effect of siRNA in cationic lipoplexes is enhanced by incorporating pDNA in the complex. *International Journal of Pharmaceutics* 333: 62–69.

Thierry, A. R., Rabinovich, P., Peng, B., Mahan, L. C., Bryant, J. L., and Gallo, R. C. 1997. Characterization of liposome-mediated gene delivery: Expression, stability and pharmacokinetics of plasmid DNA. *Gene Therapy* 4: 226–237.

Tomlinson, E., and Rolland, A. P. 1996. Controllable gene therapy pharmaceutics of non-viral gene delivery systems. *Journal of Controlled Release* 39: 357–372.

Toth, I., Flinn, N., Hillery, A., Gibbons, W. A., and Artursson, P. 1994. Lipidic conjugates of luteinizing hormone releasing hormone (LHRH)+ and thyrotropin releasing hormone (TRH)+ that release and protect the native hormones in homogenates of human intestinal epithelial (Caco-2) cells. *International Journal of Pharmaceutics* 105: 241–247.

Ueno, M., and Sriwongsitanont, S. 2005. Effect of PEG lipid on fusion and fission of phospholipid vesicles in the process of freeze–thawing. *Polymer* 46: 1257–1267.

Vitas, A. I., Diaz, R., and Gamazo, C. 1996. Effect of composition and method of preparation of liposomes on their stability and interaction with murine monocytes infected with *Brucella abortus*. *Antimicrobial Agents and Chemotherapy* 40: 146–151.

Walde, P., and Ichikawa, S. 2001. Enzymes inside lipid vesicles: Preparation, reactivity and applications. *Biomolecular Engineering* 18: 143–177.

Yang, J.-P., and Huang, L. 1998. Novel supramolecular assemblies for gene delivery. In A. K. Kabanov, P. L. Felgner, and L. W. Seymour (Eds.), *Self-Assembling Complexes for Gene Delivery—From Laboratory to Clinical Trial*. Chichester, UK: Wiley.

Zeisig, R., Arndt, D., Stahn, R., and Fichtner, I. 1998. Physical properties and pharmacological activity *in vitro* and *in vivo* of optimised liposomes prepared from a new cancerostatic alkylphospholipid. *Biochimica et Biophysica Acta* 1414: 238–248.

Zelphati, O., Nguyen, C., Ferrari, M., Felgner, J., Tsai, Y. J., and Felgner, P. L. 1998. Stable and monodisperse lipoplex formulations for gene delivery. *Gene Therapy* 5: 1272–1282.

Zheng, S., Zheng, Y., Beissinger, R. L., and Fresco, R. 1994. Microencapsulation of hemoglobin in liposomes using a double emulsion, film dehydration/rehydration approach. *Biochimica et Biophysica Acta* 1196: 123–130.

Zuhorn, I. S., and Hoekstra, D. 2002. On the mechanism of cationic amphiphile-mediated transfection. To fuse or not to fuse: Is that the question? *Journal of Membrane Biology* 189: 167–179.

Zuidam, N. J., Hirsch-Lerner, D., Margulies, S., and Barenholz, Y. 1999. Lamellarity of cationic liposomes and mode of preparation of lipoplexes affect transfection efficiency. *Biochimica et Biophysica Acta* 1419: 207–220.

18 Colloidal Nanocarrier Systems as a Tool for Improving Antimycobacterial and Antitumor Activities and Reducing the Toxicity of Usnic Acid

N. S. Santos-Magalhães, N. P. S. Santos, M. C. B. Lira,
M. S. Ferraz, E. C. Pereira, and N. H. Silva

CONTENTS

18.1 INTRODUCTION

Of the hundreds of known secondary lichen metabolites, usnic acid is certainly the most extensively studied because it exhibits antimicrobial activity (Correché et al., 1998; Francolini et al., 2004; Lauterwein et al., 1995; Perry et al., 1999; Ribeiro et al., 2006), particularly an antimycobacterial effect (Ingólfsdóttir et al., 1998), and antitumor activity (Pereira et al., 1994; Santos et al., 2005), among other properties. However, the therapeutic use of usnic acid is limited by its very poor water solubility (Kristmundsdóttir, 2005; Kristmundsdóttir et al., 2002) and hepatotoxicity (Favreau et al., 2002; Han et al., 2004; Pramyothin et al., 2004; Ribeiro-Costa et al., 2004; Santos et al., 2006). The use of usnic acid in a safe and efficient approach therefore requires appropriate nano- or microparticulated systems to increase the therapeutic index of this drug by improving the efficacy and reducing the toxicity. In this challenging scenario, liposomal and nano- and microparticulated formulations containing usnic acid or usnic acid–β-cyclodextrin inclusion complex were developed and characterized for the *in vitro* kinetic release, biological activity, and toxicity.

This review chapter was designed to show the overall results of the *in vitro* release profile, cytotoxicity, and antitumor and antimycobacterial activity of usnic acid encapsulated into liposomes, nanocapsules, and microspheres. The cellular uptake of usnic acid loaded liposomes by J774 macrophages was also investigated, exploiting the intrinsic fluorescence of the drug. Bearing in mind the issues of tuberculosis chemotherapy, studies were carried out to examine the antimycobacterial activity of usnic acid encapsulated in liposomes against the *Mycobacterium tuberculosis* H37Rv strain. Finally, the progress and issues involved in the encapsulation of usnic acid as an antimycobacterial agent, as well as further prospects in this field, are presented and discussed.

18.2 CHEMICAL AND PHARMACOLOGICAL PROPERTIES OF USNIC ACID

18.2.1 CHEMICAL PROPERTIES OF USNIC ACID

All of the earlier reports on the properties of usnic acid and its derivatives pointed out only the pharmacological activities, but the chemical properties are highlighted here because of the benefits of understanding the pharmacological activities and mechanism of action of usnic acid.

Usnic acid [(9bR)-2,6-diacetyl-3,7,9-trihydroxy-8,9b-dimethyl-dibenzofuran-1-one] is a phenolic compound formed by two cycles of six carbon atoms fused with a furan ring with the chemical structure $C_{18}H_{16}O_7$ (Figure 18.1). It exists in two enantiomeric forms, (+) and (−) usnic acid, which differ in orientation from the methyl group located in the chiral carbon at the 9b position (Asahina and Shibata, 1954; Huneck and Yoshimura, 1996).

Usnic acid is widely distributed in lichen species such as *Cladonia*, *Usnea*, *Lecanora*, *Ramalina*, *Evernia*, and *Parmotrema* (Ingólfsdóttir, 2002) as a yellowish water-insoluble pigment produced in the secondary metabolism of lichens through the biosynthetic route of acetate-polymalonate. Monocyclic phenol structures are initially formed from carboxylic acids derived from acetil-coenzyme A and from malonil-coenzyme A units catalyzed by aromatic synthetases. Usnic acid exists in tautomeric β-diketone form, sometimes as the 1-OH, 3-oxo-tautomer; the Chemical Abstracts Service name refers to the 1,3-dioxo tautomer (Rashid et al., 1999). The presence of different hydroxyl groups, a carbonyl group, and an enol function in the molecule of usnic acid leads to keto-enol tautomerism (Roach et al., 2006). Usnic acid is a dimer of methylfluoracetophenone and its chirality is produced during the dimerization process after two units of methylflouroacetophenone are stereospecifically coupled, producing hydrated usnic acid. Finally, dehydration occurs, leading to an ether bound formation (Nash, 1996).

The crystals of usnic acid are characterized by yellow needles (13.5 g), a melting point of 203–204°C, and $[\alpha]_{20}^{D} + 495°$ in chloroform (Culberson, 1963; Huneck and Yoshimura, 1996). The molecular weight of usnic acid is 344.31, and its melting point is around 203°C (Huneck and Yoshimura, 1996). The theoretical partition coefficient (log *P*) of usnic acid is 2.679 ± 0.631, and the log *P* of (−)-usnic acid in the octanol–water system (25°C) was found to be 2.88 by direct measurement (Takai et al., 1979). Apparently, the intramolecular hydrogen bonds in the usnic acid molecule not only contribute to the poor solubility in hydrophilic solvents (water <10 mg/100 mL, 25°C; ethanol 20 mg/100 mL, 25°C; Stark et al., 1950) but also lead to a log *P* value that is well into the lipophilic range. The hydrophobic character of usnic acid is due to the presence of the β-triketonic groups and the furan ring linking both cyclic rings (Asahina and Shibata, 1954; Huneck and Yoshimura, 1996). The intramolecular hydrogen bonds contribute to its lipophilic nature (Bertilsson and Wachtmeister, 1968), and the acidity is justified by the presence of the unstable enol ring (Shibata, 2000). An electron paramagnetic resonance spectrum of usnic acid confirmed that the dissociation constant

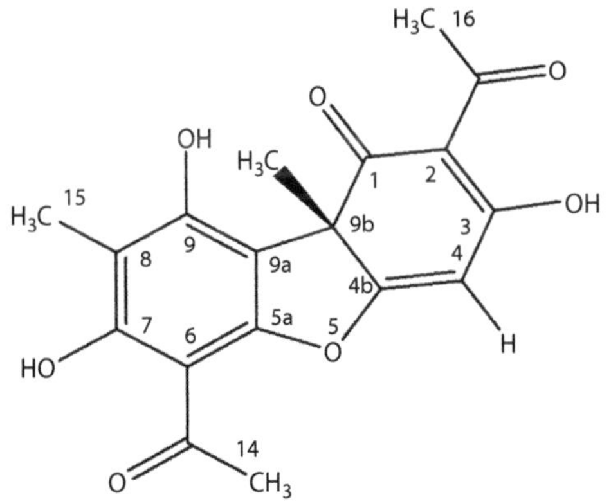

FIGURE 18.1 The chemical structure of usnic acid and its C atom assignments.

(pK_a) for the enol hydroxyl group located at position 3 is 4.4 and the pK_a value for the phenolic hydroxyl groups located at positions 9 and 7 are 8.8 and 10.7, respectively. It was found that the usnic acid is entirely unprotonated at pH 11.5 (Hauck and Jürgens, 2008; Sharma and Jannke, 1966).

18.2.1.1 Characterization of Usnic Acid

The chemical structure of usnic acid was characterized and confirmed by vibrational spectroscopic analyses, including infrared (IR) and Raman, nuclear magnetic resonance (NMR), mass spectroscopy, and x-ray diffraction (Campanella et al., 2002; Rashid et al., 1999; Takani et al., 2006). The IR spectrum of usnic acid presents a characteristic band of the conjugated ketone cyclic group at 1694 cm^{-1}. The conjugation, electron donation of the constituent rings, and possible hydrogen bonds contributed to the small wavelength of the aromatic methyl ketone at 1627 cm^{-1}. It is possible to assign the conjugated cyclic ketone group to the 1694 cm^{-1} band. Weak bands at 1716 and 1676 cm^{-1} in the IR spectrum are assigned to the ν(C=O) nonconjugated cyclic ketone and the nonaromatic methyl ketone, respectively. Conjugation, the electron donating ring substituent, and possible intramolecular hydrogen bonds all contribute to the lower wavenumber position of the aromatic methyl ketone at 1627 cm^{-1}. In addition, the IR spectrum of usnic acid included a band of hydroxyl phenolic groups at 3150 cm^{-1} and bands of unsaturated α-β-carbonyl groups and aromatic rings between 1800 and 1550 cm^{-1}. It is also possible to assign the antisymmetric and symmetric stretching ν(COC) aryl alkyl ether bands at approximately 1288 and 1070 cm^{-1}, respectively, with the aid of the IR spectrum (Edwards et al., 2003). The spin–lattice relaxation proton NMR (^{1}H NMR) spectrum of usnic acid showed that hydroxyl groups are found in regions of high chemical shifting and are highly influenced by the solvent polarity (Campanella et al., 2002; Rashid et al., 1999).

The mass spectrum of usnic acid was particularly informative because it showed the molecular ion at a mass/charge ratio (m/z) of 344, in agreement with the molecular formula $C_{18}H_{16}O_7$ (calculated molecular weight = 344.31) and a base peak at m/z 217 (Huneck and Yoshimura, 1996). Moreover, more recently mass spectroscopic analysis of usnic acid revealed other ion fragmentations m/z 329, 260, and 233 (Huneck and Schmidt, 1980; Kutney et al., 1974; Letcher, 1968; Shukla et al., 2004; Talvitie, 1983). A summary of the overall spectroscopic assigned data of usnic acid is described as follows: IR (KBr) typical vibrations centered on 3150, 1695, 1630, 1550, 1465, 1300, 1200, 1075, 1040, 965, and 900 cm^{-1}; ^{1}H NMR [dimethyl sulfoxide (DMSO), chemical shifts (δ)]: 6.32 (1H, s, ϕH), 2.62 (6H, s, 2COCH$_3$), 1.98 (3H, s, -ϕ-CH$_3$), and 1.69 (3H, s, =C–CH$_3$); mass spectrometry parental ion peak at 344 and fragmentation ion peaks at m/z 233(100), 260(60), 344(55), 232(31), 217(29), 234(20), 215(13), 261(11), and 345(10).

The optical properties of usnic acid isolated from the lichens *Ramalina reticulata*, *Parmelia molliuscula*, and *Usnea californica* Herre have been determined. The crystals obtained from chloroform solution are orthorhombic with indices α = 1.611, β = 1.710, and γ = 1.772. X-ray diffraction photographs of these crystals give a = 19.10 Å parallel to β, b = 20.39 Å parallel to γ, and c = 8.09 Å parallel to α. The space group is probably D_4^2-P2$_1$2$_1$2$_1$ but may be D_2^3-P2$_1$2$_1$2, and the density is 1.46 g/mL (Jones and Palmer, 1950).

18.2.1.2 Solubility Properties of Usnic Acid

Usnic acid is poorly soluble in water at 25°C as shown by the different reported values approximating 0.01% (100 µg/mL; see Merck Index, Budavari, 1989) and 5 µg/mL (Kristmundsdóttir et al., 2002). In addition, usnic acid is partially soluble in ethanol and considerably soluble in organic solvents such as ether, acetone, benzene, chloroform, and DMSO. The water solubility of usnic acid (5 µg/mL) was increased 70-fold in the presence of 10% hydroxypropyl-β-cyclodextrin (350 µg/mL; Kristmundsdóttir et al., 2002).

18.2.1.3 Chemical Derivatives of Usnic Acid

A huge limitation to the practical use of usnic acid is low solubility in organic solvents and in water. Accordingly, interest has arisen in the preparation of derivatives maintaining the antibiotic activity, but with more favorable solubility. For the purpose of increasing the solubility, salts of usnic acid were obtained in the form of the drugs Usno (Grasso et al., 1989) and Binan (Najdenova et al., 2001), preserving the antimicrobial activity. In addition, the preparation of new esters was investigated starting from R-(+)-usnic acid with an esterification reaction dependent on the kind of acylating agent. The use of aliphatic acyl chlorides afforded the expected diesters at positions 7 and 9, involving only the phenolic hydroxyl groups. However, when aromatic acyl chlorides are used, the reactivity is shifted in favor of the enolic hydroxyl groups; that is, the reaction occurred at position 3 and on the enolized form of the acetyl group at position 2 (Erba et al., 1998). Usimine derivatives of usnic acid were also synthesized recently to improve the solubility and pharmacological properties (Erba et al., 1998; Luzina et al., 2007; Seo et al., 2008; Takai et al., 1979). The synthesis of polyamine usnic acid derivatives (diamines and triamines), including a dimeric product of usnic acid, was carried out and the derivatives were evaluated against *Staphylococcus aureus*. The addition of an amine chain to the usnic acid molecule improved its antimicrobial activity, which was possibly due to an increase in its hydrosolubility (Tomasi et al., 2006). Further, the greatest activity was observed for the triamine derivatives, indicating the importance of the aminopropyl chain.

18.2.1.4   Chemical Methods for Quantifying Usnic Acid

The methods proposed for determining usnic acid content in lichens or pharmaceutical dosage forms were different conventional techniques such as spectroscopy (Siqueira-Moura et al., 2008; Takani et al., 2002) gas chromatography (Rashid et al., 1999), and high performance liquid chromatography (HPLC; Ribeiro-Costa et al., 2004; Roach et al., 2006). Note that HPLC is the current method for quantifying usnic acid at 233 nm with a retention time of approximately 13.3 min (Cansaran et al., 2007; Erba et al., 1998; Ji and Khan, 2005; Luzina et al., 2007; Takai et al., 1979; Tomasi et al., 2006) or at 235 nm with a retention time of 7.1 min (Huneck and Yoshimura, 1996; Kristmundsdóttir et al., 2005). Moreover, HPLC was proposed for determining usnic acid concentration in human plasma (Venkataramana and Krishna, 1992) and rabbit plasma (Venkataramana and Krishna, 1993). Furthermore, other unconventional methods such as capillary electrophoresis were applied in the analysis of usnic acid using reverse polarity capillary zone electrophoresis (Kreft and Štrukelj, 2001). More recently, usnic acid was determined using a micellar electrokinetic chromatography method (Falk et al., 2008).

18.2.1.5   Interactions of Usnic Acid and Phospholipids

The binding interactions of usnic acid and phospholipids were first investigated in lipidic vesicles mixed in an usnic acid solution by measuring spin–lattice NMR relaxation rates, which provide an evaluation of motional and environmental features of bound usnic acid (Campanella et al., 2002). The presence of bilayer membranes induces enhancement of the R value of the methyl and aromatic protons of usnic acid. The values are consistent with the slowing down of the reorientational molecular tumbling rate that in turn results from the binding interaction of usnic acid with bilayers based on the nonpolar interactions in the form of hydrophobic forces.

The interfacial properties and thermodynamic parameters of usnic acid in phospholipid monolayers were investigated, bearing in mind the use of liposomes as nanocarrier systems for this drug (Andrade et al., 2006). The results showed a better interaction between usnic acid and dipalmitoylphosphatidylcholine than with dioleylphosphatidylcholine, which might be attributable to the higher fluidity of the dioleylphosphatidylcholine monolayer in the presence of an unsaturated phospholipid chain. The surface pressure ($\Delta\Pi$) and surface potential (ΔV) compression isotherms as a function of the molecular area revealed in all mixed monolayers that there is clear evidence of interactions between the molecules of usnic acid and phospholipids. The variations found in the isotherms are presumed to result from the differences in the longer phospholipid hydrocarbon chains. The assumption that the usnic acid molecules interact with the hydrocarbon chains of the phospholipids agrees with reported results (Campanella et al., 2002).

18.2.2 Pharmacological Properties of Usnic Acid

A renewed interest in usnic acid has emerged, which is corroborated by an increasing number of publications on its biological properties and pharmacological activities such as the reviews at the beginning of this decade (Cocchietto et al., 2002; Ingólfsdóttir, 2002; Müller, 2001). Usnic acid has been extensively studied because it exhibits a number of pharmacological activities such as antimicrobial (Francolini et al., 2004; Ingólfsdóttir et al., 1998), antiprotozoal (De Carvalho et al., 2005), antiproliferative (Cardarelli et al., 1997; Correcché et al., 2004; Kumar and Müller, 1999; Mayer et al., 2005; Ögmundsdóttir et al., 1998; Santos et al., 2005), antitumor (Fujita and Nagao, 1977; Kupchan and Kopperman, 1975; Ribeiro-Costa et al., 2004; Santos et al., 2006; Takai et al., 1979), antiinflammatory (Vijayakumar et al., 2000), and gastroprotective and antioxidant (Odabasoglu et al., 2007). Moreover, usnic acid is proposed as a noteworthy candidate for clinical trials on patients with vancomycin-resistant enterococci and methicillin-resistant *Staphylococcus aureus* (Elo et al., 2007). The antiinflammatory activity of usnic acid was confirmed in a study with patients with human papilloma virus genital infection. Usnic acid was administered in association with zinc sulfate by the intravaginal route as an adjuvant therapy combined with radiosurgery. The administration of usnic acid and zinc sulfate improved reepithelization time 1 month after radiosurgery, and the differences between treated and untreated control patients remained significant even 6 months after radiosurgery (Scirpa et al., 1999).

Moreover, ever since 1949, a number of studies have investigated the antimicrobial activity of lichen compounds, especially their antimycobacterial activity (Ingólfsdóttir et al., 1998; Marshak et al., 1949; Vartia, 1950, 1973). Usnic acid is a strong inhibitor of Gram-positive bacteria (Vartia, 1973; Yamamoto et al., 1993, 1985) and aerobic and anaerobic microorganisms (Lauterwein et al., 1995), because it is a selective agent against species of *Streptococcus mutans* (Grasso et al., 1989) and against clinical isolates of vancomycin-resistant enterococci and methicillin-resistant *Staphylococcus aureus* (Elo et al., 2007). As previously mentioned, (−)-usnic acid exhibited inhibitory activity against tubercle bacilli strains at a titer of 1:60,000 (16.6 µg/mL; Vartia, 1973) and (+)-usnic acid from *Cladonia arbuscula* exhibited the highest activity with a minimum inhibitory concentration (MIC) of 32 µg/mL against *Mycobacterium aurum*, a nonpathogenic organism with a similar sensitivity profile to *M. tuberculosis* (Ingólfsdóttir et al., 1998).

Although the toxicity of usnic acid is well established and its mechanism of action remains somewhat unclear, several topical formulations containing this drug are available either as such or in salt form (Cocchietto et al., 2002; Kilpio, 1956; Lodetti et al., 2000). This drug is also used in antifeedant products (Durazo et al., 2004), mouth rinses, and toothpaste (Grasso et al., 1989), as well as cosmetics (Najdenova et al., 2001). It should be emphasized that usnic acid is the antimicrobial agent in products for personal care and cosmetics. Products such as deodorants, creams, toothpaste, mouthwash, and sunscreens containing usnic acid as a preservative have been developed and are widely used, especially in Europe (Ingólfsdóttir, 2002).

18.2.3 Pharmacokinetics of Usnic Acid

The only report on the pharmacokinetics of usnic acid, which was evaluated in rabbits after intravenous (5 mg/kg) and oral (20 mg/kg) administrations, showed mean terminal half-life constants ($t_{1/2}$) of 10.7 ± 4.6 and 11.4 ± 2.8 h, respectively (Venkataramana and Krishna, 1993). The dose-normalized absolute oral bioavailability of usnic acid was 77.8% with a plasma maximal concentration (C_{max}) of 32.45 ± 6.84 µg/mL and a time corresponding to C_{max} (t_{max}) of 16.6 ± 3.5 h. A plasma concentration of approximately 30 µg/mL was maintained for 24 h, followed by a decrease to around 10 µg/mL at 48 h.

18.2.4 Mechanism of Action of Usnic Acid

Although a number of studies have reported the *in vitro* mechanism of action of usnic acid, its specific mechanism of action remains to be more accurately elucidated for different biological properties and pharmacological activities. Further studies are required to clarify the mode of action of usnic acid on a molecular and therapeutic basis. Such studies are mandatory to guarantee its therapeutic use as an antibiotic or antitumor drug.

As previously reported, the mechanism of the antibiotic activity of (+)-usnic acid against Gram-positive bacteria was attributed to its protonophoric properties as an uncoupler of oxidative phosphorylation (Abo-Khatwa et al., 1996). This effect was remarkably more pronounced than that produced by dinitrophenol, a well-known uncoupler agent that increases the permeability of mitochondria to protons, reducing the electrochemical potential and thus inhibiting adenosine triphosphate synthesis. Two research groups found that usnic acid played an active role in the transport of protons through the membranes of isolated mouse liver mitochondria, uncoupling the electron flow through respiratory electron transport chains from the generation of an acidic gradient and thereby inhibiting the synthesis of adenosine triphosphate (Abo-Khatwa et al., 1996; Bouaid and Vicente, 1998). The ability of usnic acid to shuttle protons through membranes was confirmed by studies with artificial phospholipid membranes (Backor et al., 1998). A model for the function of usnic acid in the control of the intracellular pH of lichens was recently proposed (Hauck and Jürgens, 2008) that involves two hypotheses: (1) at the optimum pH near the pK_{a1} value of usnic acid (4.4), buffering is assumed to be compensated by the usnic acid mediated proton transport into the cell; (2) at low pH (<3.5), the equilibrium between usnic acid and usneate shifts toward the usnic acid and protons are increasingly shuttled into the cells, considering that usnic acid dissociates to usneate at cytosolic pH 7.4 (Raven, 1986). This implies that more protonated molecules would be able to cross the membrane and release protons into the cell. As a result, the intracellular pH would decrease and lead to the death of the cells (Hauck and Jürgens, 2008).

With reference to the antiviral activity of usnic acid on the proliferation of mouse polyomavirus in 3T6 cells, it was shown that Py deoxyribonucleic acid replication is severely inhibited at a concentration of 10 µg/mL. Usnic acid acts as a generic repressor of ribonucleic acid transcription (Campanella et al., 2002).

The mechanism of antiproliferative activity of usnic acid was investigated in regard to apoptosis via protein 53 (p53), on wild-type p53 (MCF7), as well as the nonfunctional p53 (MDA-MB-231) breast cancer cell lines and the lung cancer cell line H1299, which is null for p53 (Mayer et al., 2005). Cells exposed to usnic acid showed an accumulation of p53, but there was no increase in p53 transcriptional activity. The results demonstrated that there was no phosphorylation of p53 at Ser15 after treatment of MCF7 cells with 29 mM usnic acid, indicating that deoxyribonucleic acid damage is not involved in the oxidative stress and disruption of the normal metabolic processes of cells triggered by usnic acid. Usnic acid is thus a nongenotoxic antiproliferative drug that acts in a p53-independent manner.

18.2.5 Toxicology of Usnic Acid

The toxicological data of usnic acid, expressed as the lethal dose in 50% of subjects, were determined in different animals after intravenous administration. These lethal dose values were 25, 30, and 40 mg/kg in mice and rats, rabbits, and dogs, respectively (Virtanen and Kärki, 1956).

The allergenic side effect of (−)-usnic acid has been known since the 1960s (Mitchell, 1965), but not until the 1980s was it demonstrated that both (+)- and (−)-isomers of usnic acid are allergenic and individuals may react to one or both enantiomers (Salo et al., 1981). More recently, allergic contact dermatitis caused by a deodorant containing usnic acid was reported in patients who manifested bilateral axillary or ear contact dermatitis after using a natural odorless deodorant (Sheu et al., 2006).

In addition to allergenic reactions, usnic acid induces hepatotoxicity. This was confirmed in experimental *in vivo* studies (Pramyothin et al., 2004) as well as in patients after chronic use or those undergoing high-dose treatments (Durazo et al., 2004; Han et al., 2004; Neff et al., 2004). With regard to the mechanism of hepatotoxicity, *in vitro* studies showed that usnic acid can uncouple mitochondrial oxidative phosphorylation and generate oxidative stress in a dose-dependent manner

(Abo-Khatwa et al., 1997; Han et al., 2004). This mechanism was recently investigated *in vivo* in rats (200 mg/kg), as well as *in vitro* in isolated rat hepatocytes and isolated rat liver mitochondria (Pramyothin et al., 2004). They found that the hepatotoxic effect of (+)-usnic acid might be mediated by itself or by its reactive metabolite(s), causing loss of membrane integrity and destruction of mitochondrial functions.

Regarding *in vivo* toxicity of usnic acid, terpenoid-containing dietary supplements are implicated as the cause of severe and sometimes fatal hepatotoxicity. Dietetic supplement products containing usnic acid for weight loss caused hepatotoxicity, and a number of clinical cases of severe liver failure events were described (Arneborn et al., 2005; Chitturi and Farrell, 2008; Foti et al., 2008; Lazerow et al., 2005; Neff et al., 2004; Sanchez et al., 2006). Twelve clinical cases of hepatotoxicity associated with weight loss supplements containing ma huang or usnic acid were reported (Neff et al., 2004). Further, a case of liver failure in a 28-year-old woman taking only usnic acid for weight loss was described (Durazo et al., 2004). The authors noted that usnic acid is a component of many weight loss supplements, but hepatotoxicity from these supplements was historically attributed to other components. However, in this instance, the patient took only usnic acid, which resulted in fulminate liver failure requiring a liver transplant, establishing the first case report of liver failure from usnic acid alone. In addition to these cases, Sanchez and colleagues (2006) described two patients who developed severe hepatotoxicity while using a dietary supplement containing usnic acid, one of whom experienced fulminant liver failure requiring emergency transplantation. There is a full report on the toxicity of usnic acid in a recent review on the hepatotoxicity of usnic acid and other herbal hepatotoxins (Chitturi and Farrell, 2008).

18.3 NANOTECHNOLOGY AS A TOOL FOR IMPROVING ANTIMYCOBACTERIAL AND ANTITUMOR ACTIVITIES OF USNIC ACID

Usnic acid is one of the most extensively studied lichen derivatives because it exhibits, among other properties, antimycobacterial activity (Ingólfsdóttir et al., 1998; Lira et al., 2009). However, the potential therapeutic benefits of usnic acid are limited by its unfavorable solubility (Kristmundsdóttir et al., 2002) and hepatotoxicity (Han et al., 2004; Pramyothin et al., 2004; Ribeiro-Costa et al., 2004; Santos et al., 2006). To overcome these drawbacks, one possible strategy is to find a suitable solubilizer agent or to develop nanocarrier systems for improving the therapeutic index, thereby ensuring the safe and efficient clinical use of usnic acid.

As part of our continuing research, we have developed different approaches on liposomes and polymeric nanocarrier systems in order to introduce usnic acid in the treatment of tuberculosis by exploiting the main biological effects of usnic acid's antimycobacterial and antitumor activities. Before the presentation of our results, we provide a brief discussion of the issues related to the treatment of tuberculosis.

Tuberculosis, a widespread intracellular infectious disease, can be treated with rifampicin, isoniazid, streptomycin, and pyrazinamide. Nonetheless, it requires a high-dose treatment over a minimum period of 6–12 months, which is usually associated with undesirable side effects. The failure of chemotherapy in tuberculosis is related to poor patient compliance, degradation of the drugs before reaching their site of action, and the drugs' low solubility. In addition, the incorrect use of antibiotics has contributed particularly to the development of *M. tuberculosis* multiresistance (Nguyen and Thompson, 2006). From our point of view, there are two promising strategies for achieving real progress in the chemotherapy of tuberculosis: the search for new natural or synthetic compounds with antimycobacterial activity and the design of novel dosage forms of well-established drugs based on nanotechnological approaches. In this challenging scenario of tuberculosis treatment, different nanocarrier systems have been proposed because of their ability to enhance drug concentration in infected cells by improving their uptake by the mononuclear phagocytic system. Colloidal nanocarriers such as liposomes (Chono et al., 2008; Düzgünes et al., 1996; Salem and Düzgünes, 2003; Vyas et al.,

2004; Wijagkanalan et al., 2008) and nanoparticles (Fawaz et al., 1998) containing antimicobacterial agents have been studied *in vitro* and *in vivo* for the treatment of tuberculosis.

The literature results are promising and have encouraged us to develop colloidal nanosystems to provide safe and efficient dosage forms of usnic acid for tuberculosis therapy. Therefore, the results from the encapsulation of usnic acid in liposomes, nanocapsules, and microspheres will be presented and discussed in the next sections.

18.3.1 Encapsulation of Usnic Acid in Liposomes

18.3.1.1 Usnic Acid Loaded Liposomes

Because of the requirement to use nanocarrier systems to decrease hepatotoxicity and possibly enhance the efficacy of usnic acid, usnic cid was encapsulated into liposomes. Usnic acid loaded liposomes were prepared using the dried-lipid film hydration method followed by sonication (Lira et al., 2009). A typical stable liposomal formulation was prepared with soybean phosphatidylcholine, cholesterol, and stearylamine in a 7:2:1 molar ratio (42 μM of lipids for each 10 μL of encapsulated water) and 1.2 mg/mL of usnic acid (1:12 drug/lipid molar ratio). To improve the stability, liposomes were lyophilized using 1% trehalose as a cryoprotectant.

18.3.1.2 Characterization of Liposomes Containing Usnic Acid

Usnic acid loaded liposomes with a mean vesicle diameter of 90 ± 20 nm and a surface charge expressed by a zeta potential of −9.37 ± 3.6 mV were utilized by Lira et al. (2009). The content of usnic acid in the liposomes was 92.5 ± 0.3% and the drug encapsulation ratio was 99.6 ± 0.2%. It should be emphasized that the stability of usnic acid loaded liposomes was assured by the presence of stearylamine, a positively charged lipid, in the bilayer of phospholipids. Stearylamine was essential for maintaining the molecules of usnic acid imbibed in the phospholipid bilayer of liposomes. An enhancement of the forces of attraction between negatively charged usnic acid and positively charged stearylamine molecules might well be produced. Moreover, it is well known that charged lipids induce an increase in the electrostatic repulsion forces between vesicles and prevent their aggregation. Moreover, taking into account the fundamentals of physicochemical interfacial studies of usnic acid in lipidic vesicles (Campanella et al., 2002) and phospholipid monolayers (Andrade et al., 2006), it is expected that at high concentrations the usnic acid molecules are squeezed out of the bilayer of liposomes and precipitate as aggregates. Based on these findings, we hypothesize that the stability of usnic acid loaded liposomes is dependent not only on the presence of positive charges at the surface of liposomes but also on the type of phospholipids, with a more favorable structure in unsaturated phospholipids.

In terms of long-term stability, positively charged usnic acid loaded liposomes maintained their stability for over 24 months with an usnic acid content of 90 ± 2.5% and the drug encapsulation ratio remained about 100% after resuspension of the lyophilized liposomal form. The results confirmed that the liposomal encapsulation of usnic acid is especially encouraging and demonstrated the stability of liposomes and their ability to be used as nanocarriers of usnic acid. They also clearly showed the influence of the constituent concentrations and charge of phospholipids on the stability of usnic acid loaded liposomes.

18.3.2 Encapsulation of Usnic Acid in Poly(Lactic-co-Glycolic Acid) Nanocapsules

18.3.2.1 Usnic Acid Loaded Nanocapsules

Given that usnic acid is a lipophilic molecule, its encapsulation into the oily cavity of polymeric nanocapsules seems to be a good strategy for enhancing its solubility, cellular uptake, and antimicrobial efficacy. Usnic acid extracted from *Cladonia substellata* (Pereira et al., 1994) was thus encapsulated into nanocapsules prepared with poly(lactic-*co*-glycolic acid) (PLGA 50:50), soybean oil as the oily core, and soybean phosphatidylcholine and poloxamer F68 as surfactants. The method of interfacial deposition of a preformed polymer on the surface of an oil-in-water nanoemulsion was used for the manufacture of nanocapsules (Fessi et al., 1989). Stable PLGA nanocapsules containing usnic acid (1 mg/mL) were obtained with a 1:15 drug/polymer weight ratio and a 1:10 drug/oil weight ratio (Santos et al., 2005). The usnic acid content ($101.7 \pm 1.7\%$) and encapsulation efficiency ($99.4 \pm 0.2\%$) were determined by HPLC. However, the drug content of usnic acid loaded nanocapsules in suspension form decreased to 64.3% during storage at 4°C for 2 months. To overcome the instability of the colloidal nanocapsule suspension, usnic acid loaded nanocapsules were lyophilized using 1% trehalose as a cryoprotectant. In this case, the drug content of the lyophilized form of usnic acid loaded nanocapsules was $77 \pm 3.6\%$ after 36 months of storage at 4°C.

18.3.2.2 Characterization of Usnic Acid Loaded Nanocapsules

Usnic acid loaded nanocapsules in a colloidal suspension dosage form presented a surface charge of -28.4 ± 8 mV, which was determined by measuring the zeta potential using the electrophoresis technique. The analysis of the particle size of PLGA nanocapsules containing usnic acid was performed using laser light scattering, and a mean particle size of 214 ± 75 nm with a narrow polydispersity index (PDI = 0.26) was determined (Santos et al., 2006). The encapsulation of usnic acid did not affect the mean size diameter of particles, and unloaded PLGA nanocapsules exhibited a mean diameter of 167 ± 55 nm (PDI = 0.19). Moreover, the particle size evolution of the usnic acid loaded nanocapsule suspension was negligible for 120 days (285 ± 116 nm, PDI = 0.43) when stored at 4°C.

18.3.3 ENCAPSULATION OF USNIC ACID IN POLY(LACTIC-CO-GLYCOLIC ACID)–POLY(ETHYLENE GLYCOL) BLEND MICROSPHERES

18.3.3.1 Usnic Acid Loaded Microspheres

To improve the encapsulation efficiency and to provide an oral controlled release dosage form of usnic acid, microspheres containing usnic acid from *C. substellata* were prepared with a physical polymeric blend of PLGA and poly(ethylene glycol) (PEG 4000) using poly(vinyl alcohol) as a stabilizer (Ribeiro-Costa et al., 2004). The emulsification and solvent evaporation technique was used for manufacturing usnic acid loaded microspheres. Usnic acid loaded microspheres were obtained with a 1:45 drug/polymer ratio (10 mg of usnic acid and a physical blend of 450 mg PLGA and 400 mg PEG). A drug encapsulation efficiency of $99 \pm 0.46\%$ was achieved.

18.3.3.2 Physicochemical Characterization of Microspheres Containing Usnic Acid

A morphological characterization of lyophilized usnic acid loaded microspheres prepared with or without PEG was performed through scanning electron microscopy (SEM). As shown in Figure 18.2, PLGA-PEG microspheres were spherical in shape and a relative homogeneity in size distribution of particles was noted (Ribeiro-Costa et al., 2004). Further, the role of PEG in the stability of PLGA microspheres containing usnic acid was clearly established, as demonstrated by SEM images (Figure 18.2a and b). The presence of PEG was essential for preventing the formation of microspheres with a porous surface and avoiding their aggregation. A mean diameter of 7.02 ± 2.74 μm was estimated from the SEM analysis of the microspheres. The content of usnic acid in the microspheres was $105 \pm 5\%$ after manufacture. However, a slight decrease in drug content to 90% was determined in lyophilized microspheres after 7 months of storage at 4°C.

18.4 *IN VITRO* KINETICS OF USNIC ACID FROM COLLOIDAL NANO- AND MICROCARRIER SYSTEMS

18.4.1 KINETICS OF USNIC ACID LOADED LIPOSOMES

The *in vitro* release profile of usnic acid from liposomes was assessed using a dialysis technique in which 2 mL of usnic acid loaded liposomes was placed inside a dialysis sac (cellulose membrane,

FIGURE 18.2 SEM photographs of PLGA microspheres containing usnic acid prepared (a) with a PLGA-PEG blend or (b) without PEG. Original magnification × 100. [(a) Reproduced from Ribeiro-Costa, R. M., et al. 2004. *Journal of Microencapsulation* 21: 371–384. With permission. (b) Reproduced from Lira, M. C. B., et al. 2009. *Journal of Liposome Research* 19: 49–58. With permission.]

cutoff = 12,400 Da, Sartorius, Germany), sealed, and immersed in a vessel containing 200 mL of release medium (pH 7.4 phosphate buffer solution; Lira et al., 2009). The kinetic pattern of usnic acid from liposomes presented bimodal behavior (Figure 18.3). As can be seen, an initial burst release of 20.6 ± 0.4% occurred during the first 8 h, followed by a first-order kinetic with a constant rate of 32.92 ± 0.75 μg/h (correlation coefficient r^2 = 0.9753) for the interval from 8 to 72 h. Almost all of the payload drug was released (98.5 ± 2.7%) after 96 h of the kinetic process. An interesting controlled release profile of usnic acid from liposomes was found, which could be explained by both the arrangement and interactions of the drug molecules with the phospholipid bilayer of the liposomes. As expected, the usual burst effect occurred that was due to the release of the drug molecules located at the outer monolayer of the phospholipid bilayer of liposomes. Conversely, a remarkable controlled release of usnic acid took place between 8 and 72 h, which can be attributed to the release of drug molecules located on the inner phospholipid monolayer of the liposomal bilayer. In addition, a strong interaction of usnic acid molecules with phospholipids may control the release. Andrade et al. (2006) showed that the presence of usnic acid in phospholipid monolayers enhanced their interfacial properties and contributed to their stability.

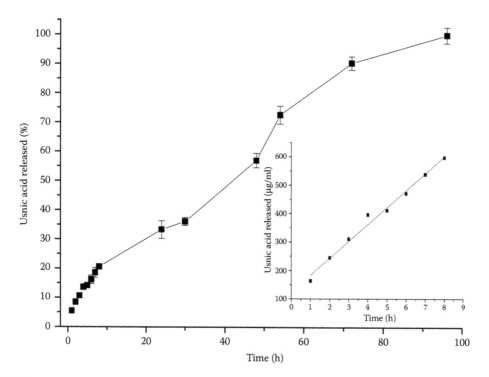

FIGURE 18.3 *In vitro* release profile of usnic acid encapsulated into positively charged liposomes (1.5 mg/mL) in pH 7.4 phosphate buffer solution at 37°C. The results are the mean values of three assays. The inset shows the release profiles over the first 8 h ($p < 0.05$). (From Santos, N. P., et al. 2005. *Journal of Drug Delivery Science and Technology* 15: 355–361 and Ribeiro-Costa, R. M., et al. 2004. *Journal of Microencapsulation* 21: 371–384. With permission.)

18.4.2 KINETICS OF USNIC ACID LOADED NANOCAPSULES

The *in vitro* release profile of usnic acid from PLGA nanocapsules was evaluated using the dialysis technique (Santos et al., 2005). An initial burst effect of $15.1 \pm 0.1\%$ occurred 1 h after the start of the kinetic process. This effect can be attributed to the prompt release of drug molecules adsorbed on the polymeric surface of the nanocapsules or located on the polymeric wall of the nanocapsules. The second stage of the kinetics occurred with slow and gradual releases attaining $78.30 \pm 0.08\%$ of the drug payload in the nanocapsules. The kinetic release data of usnic acid from PLGA nanocapsules (Santos et al., 2005) were fitted using an exponential model according to

$$M_t = M_\infty (1 - k_1 e^{-k_2 t}),\qquad(18.1)$$

where M_t and M_∞ are the mass of the drug released at a determined time (t) and at an infinite time (t_∞) of the kinetic process, respectively; k_1 is a fitting constant; and k_2 is the kinetic rate constant. The kinetic parameters were thus determined, yielding the following values: $M_\infty = 78.84 \pm 1.41$ µg, $k_1 = 0.928 \pm 0.02$, and $k_2 = 0.134 \pm 0.01$ h^{-1} with $r^2 = 0.9979$ (Figure 18.4). The plot $dM_t/dt \times t$ of the kinetics of usnic acid from PLGA nanocapsules showed an exponential decrease in the release rate starting with at 9.8 µg/h (Figure 18.4, inset). The respective constant of time ($1/k_2$) at 7.5 h was estimated and consequently an elapsed time of 30 h, assessed by computing the elapsed time to achieve a kinetic process ($4/k_2$), was required to achieve the kinetic process of usnic acid from PLGA nanocapsules. It should be emphasized that about 20% of the drug remained in the device and this profile is characteristic of reservoir systems such as nanocapsules, which are formed by an oily core containing the dissolved drug that is enclosed by a thin polymeric wall. We can postulate a diffusion

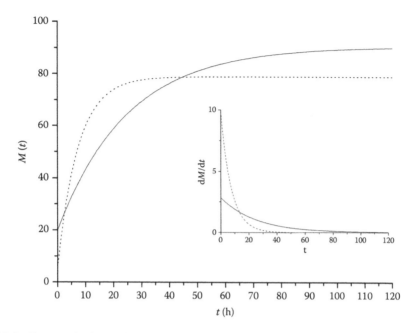

FIGURE 18.4 Exponential fitting of the *in vitro* kinetic release data of usnic acid from (—) PLGA nanocapsules and (• • •) PLGA-PEG blend microspheres according to Equation 18.1: $M_t /M_\infty = (1 - k_1 e^{-k_2 t})$ ($n = 3$). The inset shows the release rate pattern ($dM_t/dt \times t$) of usnic acid from (—) PLGA nanocapsules and (• • •) PLGA–PEG blend microspheres. The experimental kinetic data were taken from Santos et al. (2005) and Ribeiro-Costa et al. (2004). (Reproduced from Lira, M. C. B., 2009. *Journal of Liposome Research* 19: 49–58. With permission.)

mechanism of drug molecules throughout the polymeric thin shell of nanocapsules during the two first stages of the kinetic process. However, a saturation process throughout the polymeric barrier must have occurred at the end of the kinetic process, corresponding to the device exhaustion, preventing the release of the remainder of the drug.

18.4.3 KINETICS OF USNIC ACID LOADED MICROSPHERES

The *in vitro* release of usnic acid from PLGA-PEG microspheres was performed using the dissolution technique (Ribeiro-Costa et al., 2004). The release pattern of usnic acid was characterized by a large burst effect of $35 \pm 0.13\%$ occurring at 1 h, which may have been caused by a variety of factors. Among these, the presence of drug molecules on or near the surface of microspheres, the small size of the microspheres (7 μm), and the presence of porous particles in the PLGA matrix might be considered. The burst effect was followed by the controlled release of usnic acid that reached $92 \pm 0.4\%$ within 5 days. After this period, a slight decrease in the usnic acid in the dissolution medium was found that may be ascribed to drug photodegradation and chemical instability. Similar behavior was observed in the kinetic study of usnic acid release from nanocapsules (Santos et al., 2005).

The kinetic release data of usnic acid from PLGA-PEG blend microspheres were also fitted according to Equation 18.1, yielding $M_\infty = 90.54 \pm 6.39$ μg, $k_1 = 0.781 \pm 0.073$, and $k_2 = 0.04 \pm 0.02$ h^{-1} with $r^2 = 0.948$, as shown in Figure 18.4. The plot $dM_t/dt \times t$ of the kinetic of usnic acid from microspheres showed an exponential decrease with an initial release rate of 2.8 μg/h (Figure 18.4, inset). The respective $1/k_2$ at 25 h was estimated; as a result, an elapsed time of 100 h, which was assessed by computing $4/k_2$, was required to achieve the kinetic process of usnic acid from microspheres. The mechanism of usnic acid release from microspheres might be governed by the diffusion of drug molecules throughout the polymeric matrix, followed by an erosion of microspheres beginning on the surface and spreading toward the matrix core.

The *in vitro* kinetic studies of usnic acid, which used the dialysis method for liposomes (theoretical $C_\infty \cong 12\text{--}13$ µg/mL) and nanocapsules (theoretical $C_\infty \cong 10$ µg/mL) and the dissolution technique for microspheres (theoretical $C_\infty \cong 10$ µg/mL), revealed different patterns of usnic acid release that were dependent on the type of colloidal nano- and microcarrier systems. Considering the reservoir systems as lipidic vesicles (liposomes) or polymeric capsules (nanocapules) and a polymeric matrix (microspheres), the kinetic profiles were different. The encapsulation of usnic acid into liposomes provides controlled the release, and exponential release profiles were obtained for nanocapsules and microspheres. Moreover, the release of usnic acid from the microspheres was very slow in comparison with the release from nanocapsules. All of these findings should be taken into account in further kinetic studies of usnic acid from colloidal dosage forms. Furthermore, the development of different-sized microspheres prepared with different types of polymeric matrices might change the *in vitro* kinetic profile of usnic acid, providing controlled drug delivery systems.

18.5 CYTOTOXICITY OF USNIC ACID ENCAPSULATED IN COLLOIDAL NANO- AND MICROCARRIER SYSTEMS

18.5.1 CYTOTOXICITY OF USNIC ACID LOADED LIPOSOMES

In relation to cytotoxicity, the encapsulation of usnic acid into conventional liposomes improved its cytotoxic effect on J774 macrophages evaluated by the 3-(4,5-dimethylthiazol-2-yl)-2,5-diphenyl-tetrazolium bromide (MTT) assay (Lira et al., 2009). As shown in Figure 18.5, the cytotoxicity of liposomal usnic acid on macrophages, expressed as the concentration of the drug needed to inhibit 50% of cell proliferation (IC_{50}), was dose dependent and twice as high (12.5 ± 0.26 µg/mL) as that of pure usnic acid (22.5 ± 0.60 µg/mL). As expected, unloaded liposomes did not have any effect on the proliferation macrophages. These results suggest that the encapsulation of usnic acid into liposomes enhanced its antiproliferative effect because of the increase in the uptake of liposomes by macrophages. The results for the cytotoxicity of usnic acid from *C. substellata* are in agreement with those of a previous report (Mayer et al., 2005) in which the IC_{50} values of commercial usnic

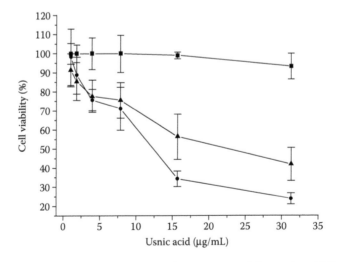

FIGURE 18.5 (•) Cytotoxicity of positively charged liposomes containing usnic acid (1.5 mg/mL) on J774 macrophages. (▲) Usnic acid in 0.2% DMSO and (■) unloaded liposomes were used as controls. Cell viability was determined by the MTT method at 37°C in a 5% CO_2 atmosphere. The vehicle control (0.2% DMSO) did not show any cell toxicity (data not shown). The values are the mean ± SD of three sets of experiments ($n = 18$). (From Santos, N. P., et al. 2005. *Journal of Drug Delivery Science and Technology* 15: 355–361. With permission.)

acid (Sigma–Aldrich) on MCF7 and MDA-MB-231 breast cancer cell lines and the lung cancer cell line H1299 were around of 25 μM (10 μg/mL).

18.5.2 CYTOTOXICITY OF USNIC ACID LOADED NANOCAPSULES

The cytotoxicity of usnic acid from *C. substellata* on NCI-H292 or HEp-2 cells, which were evaluated using the standard MTT method, was dose dependent (Santos et al., 2005). Usnic acid loaded nanocapsules and free usnic acid both presented IC_{50} values of the same magnitude of 10 and 13.5 μg/mL, respectively (Figure 18.6). Note that unloaded nanocapsules have no cytotoxicity effect on NCI-H292 cells. The morphological aspects of NCI-H292 cells were evaluated after treatment with free and encapsulated usnic acid. Cell abnormalities such as irregular cytoplasm, vacuolization, and inhibition of proliferation were found after treatment with usnic acid at <2.5 μg/mL. At a high usnic acid concentration (5 μg/mL) a drastically antiproliferative effect was verified in comparison with the untreated control cells. The cells exhibited substantial alterations and cell adherence was strongly affected by the formation of cell clusters. Other cells presented irregular cytoplasm, vacuolization with the presence of basophilic material, and naked nuclei. Conversely, the treatment of cells with encapsulated usnic acid (5 μg/mL) induced practically no morphological alterations in the cells.

18.5.3 CYTOTOXICITY OF USNIC ACID LOADED MICROSPHERES

The cytotoxicity of usnic acid from *C. substellata* pure and encapsulated into PLGA microspheres on HEp-2 cells was also evaluated using the MTT method (Ribeiro-Costa et al., 2004). IC_{50} values of 12.6 and 14.4 μg/mL were found for usnic acid and usnic acid loaded microspheres, respectively. No cytotoxicity was found for unloaded microspheres, and practically 100% of cell viability was observed even for a large amount of microspheres (15 μg/mL of usnic acid) placed in contact with the cells.

18.5.4 UPTAKE OF USNIC ACID LOADED LIPOSOMES BY MACROPHAGES

The uptake of usnic acid loaded liposomes by J774 macrophages was evaluated through fluorescence spectroscopy to exploit the intrinsic fluorescence of usnic acid (Lira et al., 2009). A time-dependent increase in the fluorescence intensity of usnic acid was observed in cells treated with usnic acid loaded liposomes or pure usnic acid. An enhancement in the intracellular uptake

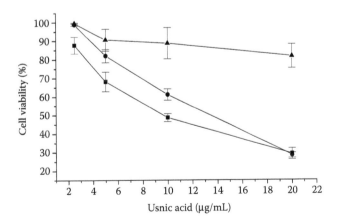

FIGURE 18.6 Cytotoxicity of usnic acid (1.5 mg/mL) on NCI-H292 cells: (■) free usnic acid in 0.4% DMSO, (•) usnic acid loaded nanocapsules, and (▲) unloaded nanocapsules. The cell viability was verified by the MTT method at 37°C in a 5% CO_2 atmosphere for 72 h. The values are the mean ± SD of three sets of experiments (*n* = 12). (From Santos, N. P., et al. 2005. *Journal of Drug Delivery Science and Technology* 15: 355–361. With permission.)

of usnic acid loaded liposomes was evidenced by the highest detected fluorescence emission $(21.6 \times 10^4 \pm 28.3 \times 10^2$ counts/s) at 10 h, representing about 100% uptake of usnic acid loaded liposomes in comparison with the pure drug. The uptake of pure and encapsulated usnic acid by J774 macrophages was statistically different $(p < 0.05)$ throughout the kinetic process (Figure 18.7). Moreover, the liposomal formulation insured that the usnic acid remained inside the macrophages longer (92% at 30 h, corresponding to $19.8 \times 10^4 \pm 10.2 \times 10^2$ counts/s; $p < 0.05$). An increase in the interaction of cells incubated with free usnic acid was detected, attaining a maximum of 46% $(9.5 \times 10^4 \pm 11.4 \times 10^2$ counts/s) at 8 h. However, after 10 h of incubation there was a decrease in the uptake of usnic acid molecules by macrophages, and the uptake reached only 24% at 30 h. Furthermore, even after 48 h of incubation the uptake of usnic acid loaded liposomes was 2.4 times greater $(12.6 \times 10^4 \pm 27.5 \times 10^2$ counts/s) than that detected for pure usnic acid. The *in vitro* uptake studies thus revealed that usnic acid was poorly taken up by J774 macrophages. In contrast, J774 cells substantially took up usnic acid encapsulated into liposomes.

There is clear evidence of the advantage of encapsulating usnic acid into liposomes and nanocapsules in order to improve its cytotoxicity as a result of the greater cellular uptake of colloidal nanocarriers, which is attributable to their smaller size (<200 nm). It is interesting that the encapsulation of usnic acid into microspheres $(7.02 \pm 2.72 \mu m)$ was also efficient for improving its cytotoxicity. Taken together, the earlier reports on the antiproliferative effect of usnic acid give rise to considerable expectations regarding the use of usnic acid as an anticancer drug. Usnic acid in 10% aqueous solutions of 2-hydroxypropyl-β-cyclodextrin presented an IC_{50} value of 4.7 µg/mL on K-562 cells (Kristmundsdóttir et al., 2002). More recently, Kristmundsdóttir et al. (2005) found IC_{50} values of 2.9, 4.3, and 8.2 µg/mL for usnic acid in T-47D breast cancer cells, Panc-1 pancreas cancer cells, and PC-3 prostate cancer cells, respectively. It was suggested that the antiproliferative activity of usnic acid is to a certain extent more cytostatic than cytotoxic, because no damage to plasmatic membrane of cells was confirmed (Ingólfsdóttir, 2002). Although a number of studies on the cytotoxicity of usnic acid have already been undertaken on different cell lines, its exact mechanism of action still has to be elucidated. The acidic character and high hydrophobicity of usnic acid could account for its lower incorporation into cells and lower cytotoxicity compared with standard anticancer drugs.

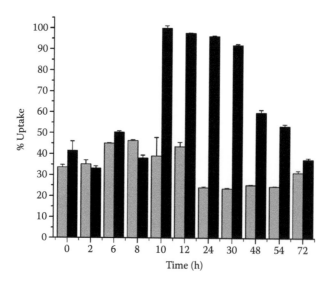

FIGURE 18.7 Uptake kinetics of usnic acid loaded liposomes (black) and pure usnic acid (grey) by J774 macrophages derived from fluorescence intensity measurements. The macrophages were incubated in a medium containing 15 µg/mL of pure usnic acid (0.2% DMSO) or a liposomal dosage form (1.5 mg/mL). Cellular uptakes are given as mean ± SD of three sets of experiments ($n = 18$). (Modified from Lira, M. C. B., 2009. *Journal of Liposome Research* 19: 49–58. With permission.)

18.6 ANTIMYCOBACTERIAL AND ANTITUMOR ACTIVITIES OF NANOENCAPSULATED USNIC ACID

18.6.1 ANTIMYCOBACTERIAL ACTIVITY OF USNIC ACID LOADED LIPOSOMES

The antimycobacterial activity of usnic acid was evaluated on *M. tuberculosis* $H_{37}Rv$, a commonly used virulent laboratory strain, through determination of the MIC and the minimal bactericidal concentration. MIC values of 6.5 and 5.8 µg/mL for pure and encapsulated usnic acid, respectively, were found against *M. tuberculosis* $H_{37}Rv$ (Lira et al., 2009). The bactericidal activity of usnic acid loaded liposomes was 16 µg/mL, whereas the minimal bactericidal concentration of pure usnic acid was twice as high (32 µg/mL). The results clearly showed that liposomal encapsulation of usnic acid provided an enhancement of its antimycobacterial activity on *M. tuberculosis* $H_{37}Rv$. The MIC of encapsulated usnic acid on *M. tuberculosis* was about 5 times lower than that obtained for pure usnic acid (32 µg/mL) on *M. aurum*, a nonpathogenic species of mycobacterium (Ingólfsdóttir et al., 1998).

18.6.2 ANTITUMOR ACTIVITY OF USNIC ACID LOADED NANOCAPSULES

The antitumor activity of usnic acid loaded nanocapsules was assessed in male Swiss mice bearing Sarcoma 180 (Santos et al., 2006). The animals were treated with an usnic acid suspension (0.5% Tween 80) or usnic acid loaded nanocapsules with daily intraperitoneal (i.p.) doses of 15 mg/kg for 7 days. All of the animals survived to the end of the treatment, and no clinical or behavioral abnormalities were observed for either treatment. After treatment of the animals with usnic acid loaded nanocapsules, a marked decrease in tumor mass was observed (0.281 ± 0.013 g) compared to the drug suspension (0.995 ± 0.050 g). Conversely, untreated animals presented a progressive increase in tumor growth, achieving a weight of 1.07 ± 0.15 g on average at 7 days after inoculation with tumor cells. The treatment with usnic acid loaded nanocapsules induced a $69.7 \pm 6\%$ tumor inhibition compared with the control group. This tumor inhibition is clear confirmation that nanoencapsulation improved the activity of usnic acid by 26.4% in contrast with its suspension form, which presented an antitumor activity of $43.3 \pm 4\%$ compared to the control group. Moreover, a delay in the development of the tumor and a reduction in its size, as well as increased infiltration by lymphoid cells, granulation tissue, and fibrosis surrounding the tumor, were detected with the usnic acid treatment.

The histopathological analyses of the tumor and liver of the animals treated with free and encapsulated usnic acid showed the presence of both typical and atypical cells in constant mitosis after treatment with usnic acid (Santos et al., 2006). However, more extensive necrotic areas with uncharacterized pyknotic nuclear cells were observed on the tumor tissue of animals treated with usnic acid loaded nanocapsules. There were a few residual neoplastic cells inserted into a large central area of the necrosis in the tumor. Tumor histopathological studies thus confirmed a great reduction in tumor size after treatment with usnic acid loaded nanocapsules in comparison with the pure drug. Moreover, histological observations indicated an increase in the number of lymphocytes surrounding the tumor tissue. These findings suggest that the immune system was stimulated, providing an increased host response to the tumor. It should be emphasized that these *in vivo* results corroborate the mechanism of the antitumor activity of usnic acid via necrosis without apoptosis involvement.

18.6.3 ANTITUMOR ACTIVITY OF USNIC ACID LOADED MICROSPHERES

The antitumor activity of usnic acid loaded microspheres was evaluated in male Swiss mice against a Sarcoma 180 tumor (Ribeiro-Costa et al., 2004). Chemotherapy was started 24 h after the inoculation of tumor cells, and animals received daily i.p. injections of 15 mg/kg usnic acid suspension (0.5% Tween 80) or usnic acid loaded microspheres for 7 days. The treatment with usnic acid loaded microspheres caused a 63% tumor inhibition, wheras the tumor inhibition after treatment with usnic acid suspension was only 42%. An improvement of 21% in the antitumor activity of usnic acid was achieved by its encapsulation into PLGA microspheres.

A comparison of the antitumor effects of the encapsulation of usnic acid into PLGA microspheres and into PLGA nanocapsules (Santos et al., 2006), taking into account that the same experimental protocols were adopted for both studies, indicates that nanoencapsulation seems to be more effective than microencapsulation for improving the antitumor activity of usnic acid. These findings could be explained by the faster uptake of nanocapsules (<200 nm) than microspheres (7 μm) by the tumor cells, which can accumulate in the peritoneal cavity, which makes their uptake by the tumor cells more difficult. These results clearly reveal the effect of the particle size of nanocarrier systems on the bioavailability of usnic acid.

18.7 TOXICITY OF USNIC ACID ENCAPSULATED IN NANO- AND MICROCARRIERS

In addition to the evaluation of antitumor activity, histopathological, hematological, and biochemical analyses were carried out to ascertain the effect of nano- and microencapsulation on the toxicity of usnic acid. A histopathological evaluation showed morphological alterations in the liver of the animals undergoing either pure or encapsulated usnic acid treatments with daily i.p. doses of 15 mg/kg/day for 7 days (Santos et al., 2006). Vacuolization of hepatocytes and an intensive lymphocyte infiltration in portal spaces was found in the livers of the animals treated with usnic acid suspension in comparison with the controls. Conversely, the liver of animals treated with usnic acid loaded nanocapsules presented only morphologically uncharacterized hepatocytes and a mild lymphocyte infiltration in the portal space. Moreover, a process of hepatocytic necrosis was also present. However, this hepatotoxicity was substantially reduced when the animals were treated with usnic acid loaded nanocapsules. The liver histopathological analysis thus showed that the nanoencapsulation of usnic acid was able to reduce the hepatotoxicity compared to the usnic acid suspension. Concerning renal toxicity, no histological changes were noticed in the kidneys of any of the animals treated with either free or encapsulated usnic acid.

Similar results were obtained in the study of usnic acid loaded microspheres. A histopathological analysis of a tumor after treatment with free and microencapsulated usnic acid revealed extensive necrotic areas on the tumor tissue after treatment with the usnic acid suspension. The liver histopathological analysis also showed extensive areas of necrosis after treatment with usnic acid, whereas only morphological uncharacterized hepatocytes were found after treatment with usnic acid loaded microspheres. This effect was considerably reduced after treatment with the microencapsulated dosage form of usnic acid. These results corroborated the advantages of using microencapsulation for reducing the toxic effect of usnic acid (Ribeiro-Costa et al., 2004).

The hematology profile of treated and untreated Sarcoma 180 bearing animals showed no significant alterations in red blood cell levels (from 5.42 to 6.57×10^{-6} cells/mL) in comparison with the negative control group ($6.6 \pm 0.7 \times 10^{-6}$ cells/mL). However, the hematological findings did not show any treatment-related effects and were not significantly different from those of the positive control group. No significant reduction in white blood cells levels occurred in the animals treated with either an usnic acid suspension or usnic acid loaded nanocapsules in comparison with the positive controls (Santos et al., 2006). As expected, lymphocyte levels were increased in tumor-bearing animals compared with healthy animals. Moreover, no significant differences in lymphocytes and neutrophils were found between the treated groups and the untreated positive control group. These results suggested that usnic acid in suspension or nanoencapsulated caused no hematological toxicity in the treated animals. Furthermore, free or encapsulated usnic acid has no immunological effects, because the same quantity of leukocytes was found in the untreated tumor-bearing animals. However, further investigations should be carried out to substantiate the argument that usnic acid has no effect on the immune response of the host.

A subchronic toxicity study was performed in healthy male Swiss mice receiving daily doses of 15 mg/kg/day for 15 days (Santos et al., 2006). The results showed that there were no statistically significant differences in body weight between the control and treated animal groups. No significant clinical signs were observed in the animals, and no deaths occurred during treatment. The serum levels of blood urea nitrogen (BUN), creatinine (CRT), and transaminases (alanine aminotransferase and aspartate aminotransferase) of the control group were 140 mg/dL, 1.07 mg/dL, 289.4 IU/L, and 241.5 IU/L, respectively. The serum levels of BUN, CRT, alanine aminotransferase, and aspartate aminotransferase of the animals treated with usnic acid suspension were 156 mg/dL, 0.94 mg/dL, 552 IU/L, and 403.1 IU/L, respectively; but the encapsulated usnic acid produced values of 138 mg/dL, 0.82 mg/dl, 406 IU/L, and 322.5 IU/L, respectively. No alterations in the serum levels of BUN and CRT were observed with the usnic acid treatment, strongly suggesting that the renal function of the animals was preserved with this treatment. Usnic acid therefore does not appear to cause any renal toxicity when administered as a long-term treatment. The serum transaminase activity in the treated groups was significantly higher than in the control group. Serum transaminase activities were increased with the usnic acid treatment, indicating liver cell injury. Moreover, histological observation of the liver of the animals treated with the usnic acid suspension and usnic acid loaded nanocapsules for 15 days revealed extensive necrotic areas on the liver tissue after treatment with the usnic acid suspension (data not shown), although this abnormality was substantially reduced with the treatment using usnic acid loaded nanocapsules.

The hepatotoxicity of usnic acid was thus confirmed in the subchronic toxicity study. High serum levels of transaminases suggest a chronic hepatic dysfunction caused by usnic acid. Nevertheless, the encapsulation of usnic acid was able to reduce its hepatotoxicity.

18.8 CONCLUSION

This review brings together the results obtained from research using the nanotechnology approach to reduce hepatotoxicity and ameliorate the antitumor and antimycobacterial activities of usnic acid. The results demonstrate the feasibility of usnic acid loaded liposomes with high encapsulation efficiency. A controlled release profile of usnic acid from liposomes was found, which might be explained by both the arrangement and interactions of the drug molecules with the phospholipid bilayer of liposomes. An enhancement of the intracellular uptake of liposomal usnic acid was evidenced, and its encapsulation insured that it remained inside the macrophages for longer than was the case with pure usnic acid. Furthermore, the results indicated a strong interaction between liposomes and J774 macrophages, thereby facilitating usnic acid penetration into cells and considerably improving its activity against *M. tuberculosis*. Moreover, the encapsulation of usnic acid into PLGA nanocapsules led to an increase in the antitumor activity of the drug and improved its antibacterial activity (30%) against *M. tuberculosis*. These *in vitro* findings are remarkable, and further *in vivo* studies should be carried out to confirm the antitubercular activity of nanoencapsulated usnic acid. Finally, the encapsulation of usnic acid into microspheres may offer an effective strategy for the oral administration of usnic acid in the treatment of tuberculosis. In the light of these findings and considering the advantages and the particular features of nanocarrier systems, we hypothesize that usnic acid encapsulated into liposomes or nanocapsules is feasible and has a potential application in the treatment of human pulmonary tuberculosis. However, further *in vivo* studies of the activity, bioavailability, and toxicity must be carried out using long-circulating and targeted nanocarriers containing usnic acid.

ACKNOWLEDGMENTS

The first author is grateful to the Brazilian Council for Scientific and Technological Development (CNPq/MCT) for Grant 301771/2006-5. All of the authors thank the Brazilian Network on

Nanobiotechnology—Nanobiotec (CNPq/MCT) and the Science Foundation of the State of Pernambuco (FACEPE) for its financial support. A grant provided by the French-Brazilian Research Program CAPES-COFECUB 535/06 of the Brazilian Education Ministry was much appreciated.

ABBREVIATIONS

BUN	Blood urea nitrogen
CRT	Creatinine
DMSO	Dimethyl sulfoxide
HPLC	High performance liquid chromatography
^{1}H NMR	Proton nuclear magnetic resonance
IC_{50}	Concentration of a drug needed to inhibit 50% of cell proliferation
i.p.	Intraperitoneal
IR	Infrared
MIC	Minimal inhibitory concentration
MTT	3-(4,5-Dimethylthiazol-2-yl)-2,5-diphenyltetrazolium bromide
NMR	Nuclear magnetic resonance
p53	Protein 53
PDI	Polydispersity index
PEG	Poly(ethylene glycol)
PLGA	Poly(lactic-co-glycolic acid)
SEM	Scanning electron microscopy

SYMBOLS

A	Molecular area
C_∞	Concentration of a drug released at an infinite time (t_∞)
C_{max}	Plasma maximal concentration of a drug
δ	Chemical shifts
k_1	Kinetic fitting constant
k_2	Kinetic rate constant
$1/k_2$	Constant of time
$4/k_2$	Elapsed time required to achieve a kinetic process
$\log P$	Partition coefficient
M_t	Mass of the drug released at a determined time (t)
M_∞	Mass of a drug released at an infinite time (t_∞)
m/z	Mass/charge ratio
ν	Antisymmetric and symmetric stretching
$\Delta\Pi$	Surface pressure
pK_a	Dissociation constant
r^2	Correlation coefficient
$t_{1/2}$	Half-life
t_{max}	Time corresponding to C_{max}
ΔV	Surface potential

REFERENCES

Abo-Khatwa, A. N., Al-Robai, A. A., and Al-Jawhari, D. A. 1996. Lichen acids as uncouplers of oxidative phosphorylation of mouse-liver mitochondria. *Natural Toxins* 4: 96–102.

Abo-Khatwa, A. N., Al-Robai, A. A., and Al-Jawhari, D. A. 1997. Isolation and identification of usnic acid and atranorin from some Saudi-Arabian lichens. *Arab Gulf Journal of Scientific Research* 15: 15–28.

Andrade, C. A. S., Santos-Magalhães, N. S., and de Melo, C. P. 2006. Thermodynamic characterization of the prevailing molecular interactions in mixed floating monolayers of phospholipids and usnic acid. *Journal of Colloid and Interface Science* 298: 145–153.

Arneborn, P., Jansson, A., and Bottiger, Y. 2005. Acute hepatitis in a woman after intake of slimming pills bought via Internet. *Lakartidningen* 102: 2071–2072.

Asahina, Y. and Shibata, S. 1954. *Chemistry of Lichen Substances.* Tokyo: Japan Society for the Promotion of Science.

Backor, M., Hudak, J., Repcak, M., Ziegler, W., and Backorova, M. 1998. The influence of pH and lichen metabolites (vulpinic acid and (+) usnic acid) on the growth of the lichen photobiont *Trebouxia irregularis. Lichenologist* 30: 577–582.

Bertilsson, L. and Wachtmeister, C. A. 1968. Methylation and racemisation studies on usnic acid. *Acta Chemica Scandinavica* 22: 1791–1800.

Bouaid, K. and Vicente, C. 1998. Chlorophyll degradation effected by lichen substances. *Annales Botanici Fennici* 35: 71–74.

Budavari, S. (Ed.). 1989. *Merck Index* (11th ed.). Whitehouse Station, NJ: S. Merck & Co.

Campanella, L., Delfini, M., Ercole, P., Iacoangeli, A., and Risuleo, G. 2002. Molecular characterization and action of usnic acid: A drug that inhibits proliferation of mouse polyomavirus *in vitro* and whose main target is RNA transcription. *Biochimie* 84: 329–334.

Cansaran, D., Atakol, O., Halici, M. G., and Aksoy, A. 2007. HPLC analysis of usnic acid in some Ramalina species from Anatolia and investigation of their antimicrobial activities. *Pharmaceutical Biology* 45: 77–81.

Cardarelli, M., Serino, G., Campanella, L., Ercole, P., De Cicco Nardone, F., Alesiani, O., and Rossiello, F. 1997. Antimitotic effects of usnic acid on different biological systems. *Cellular and Molecular Life Sciences* 53: 667–672.

Chitturi, S. and Farrell, G. C. 2008. Hepatotoxic slimming aids and other herbal hepatotoxins. *Journal of Gastroenterology and Hepatology* 23: 366–373.

Chono, S., Tanino, T., Seki, T., and Morimoto, K. 2008. Efficient drug targeting to rat alveolar macrophages by pulmonary administration of ciprofloxacin incorporated into mannosylated liposomes for treatment of respiratory intracellular parasitic infections. *Journal of Controlled Release* 127: 50–58.

Cocchietto, M., Skert, N., Nimis, P., and Sava, G. 2002. A review on usnic acid, an interesting natural compound. *Naturwissenschaften* 89: 137–146.

Correcché, E. R., Enriz, R. D., Piovano, M., Garbarino, J., and Gómez-Lechón, M. J. 2004. Cytotoxic and apoptotic effects on hepatocytes of secondary metabolites obtained from lichens. *Alternatives to Laboratory Animals* 32: 605–615.

Correché, E. R., Carrasco, M., Escudero, M. E., Velazquez, L., De Guzman, A. M. S., Giannini, F., Enriz, R. D., Jauregui, E. A., Cenal, J. P., and Giordano, O. S. 1998. Study of the cytotoxic and antimicrobial activities of usnic acid and derivatives. *Fitoterapia* 69: 493–501.

Culberson, C. F. 1963. The lichen substances of the genus Evernia. *Phytochemistry* 2: 335–340.

De Carvalho, E. A. B., Andrade, P. P., Silva, N. H., Pereira, E. C., and Figueiredo, R. C. B. Q. 2005. Effect of usnic acid from the lichen *Cladonia substellata* on *Trypanosoma cruzi* in vitro: An ultrastructural study. *Micron* 36: 155–161.

Durazo, F. A., Lassman, C., Han, S. H. B., Saab, S., Lee, N. P., Kawano, M., Saggi, B., et al. 2004. Fulminant liver failure due to usnic acid for weight loss. *American Journal of Gastroenterology* 99: 950–952.

Düzgünes, N., Flasher, D., Reddy, M. V., Luna-Herrera, J., and Gangadharam, P. R. J. 1996. Treatment of intracellular *Mycobacterium avium* complex infection by free and liposome-encapsulated sparfloxacin. *Antimicrobial Agents and Chemotherapy* 40: 2618–2621.

Edwards, H. G. M., Newton, E. M., and Wynn-Williams, D. D. 2003. Molecular structural studies of lichen substances II: Atranorin, gyrophoric acid, fumarprotocetraric acid, rhizocarpic acid, calycin, pulvinic dilactone and usnic acid. *Journal of Molecular Structure* 651–653: 27–37.

Elo, H., Matikainen, J., and Pelttari, E. 2007. Potent activity of the lichen antibiotic (+)-usnic acid against clinical isolates of vancomycin-resistant enterococci and methicillin-resistant *Staphylococcus aureus. Naturwissenschaften* 94: 465–468.

Erba, E., Pocar, D., and Rossi, L. M. 1998. New esters of *R*-(+)-usnic acid. *Farmaco* 53: 718–720.

Falk, A., Green, T. K., and Barboza, P. 2008. Quantitative determination of secondary metabolites in *Cladina stellaris* and other lichens by micellar electrokinetic chromatography. *Journal of Chromatography A* 1182: 141–144.

Favreau, J. T., Ryu, M. L., Braunstein, G., Orshansky, G., Park, S. S., Coody, G. L., Love, L. A., and Fong, T. L. 2002. Severe hepatotoxicity associated with the dietary supplement LipoKinetix. *Annals of Internal Medicine* 136: 590–595.

Fawaz, F., Bonini, F., Maugein, J., and Lagueny, A. M. 1998. Ciprofloxacin-loaded polyisobutylcyanoacrylate nanoparticles: Pharmacokinetics and *in vitro* antimicrobial activity. *International Journal of Pharmaceutics* 168: 255–259.

Fessi, H., Piusieux, F., Devissaguet, J. P., Ammoury, N., and Benita, S. 1989. Nanocapsule formation by interfacial polymer deposition following solvent displacement. *International Journal of Pharmaceutics* 55: R1–R4.

Foti, R. S., Dickmann, L. J., Davis, J. A., Greene, R. J., Hill, J. J., Howard, M. L., Pearson, J. T., et al. 2008. Metabolism and related human risk factors for hepatic damage by usnic acid containing nutritional supplements. *Xenobiotica* 38: 264–280.

Francolini, I., Norris, P., Piozzi, A., Donelli, G., and Stoodley, P. 2004. Usnic acid, a natural antimicrobial agent able to inhibit bacterial biofilm formation on polymer surfaces. *Antimicrobial Agents and Chemotherapy* 48: 4360–4365.

Fujita, E. and Nagao, Y. 1977. Tumor inhibitors having potential for interaction with mercapto enzymes and/or coenzymes. A review. *Bioorganic Chemistry* 6: 287–309.

Grasso, L., Ghirardi, P. E., and Ghione, M. 1989. Usnic acid, a selective antimicrobial agent against *Streptococcus mutans*: A pilot clinical study. *Current Therapeutic Research: Clinical and Experimental* 45: 1067–1070.

Han, D., Matsumaru, K., Rettori, D., and Kaplowitz, N. 2004. Usnic acid-induced necrosis of cultured mouse hepatocytes: Inhibition of mitochondrial function and oxidative stress. *Biochemical Pharmacology* 67: 439–451.

Hauck, M. and Jürgens, S. R. 2008. Usnic acid controls the acidity tolerance of lichens. *Environmental Pollution* 156: 115–122.

Huneck, S. and Schmidt, J. 1980. Lichen substances—126 mass spectroscopy of natural products—10. Comparative positive and negative ion mass spectroscopy of usnic acid and related compounds. *Biomedical Mass Spectrometry* 7: 301–308.

Huneck, S. and Yoshimura, Y. 1996. *Identification of Lichen Substances*. New York: Springer.

Ingólfsdóttir, K. 2002. Usnic acid. *Phytochemistry* 61: 729–736.

Ingólfsdóttir, K., Chung, G. A. C., Skúlason, V. G., Gissurarson, S. R., and Vilhelmsdóttir, M. 1998. Antimycobacterial activity of lichen metabolites in vitro. *European Journal of Pharmaceutical Sciences* 6: 141–144.

Ji, X. and Khan, I. A. 2005. Quantitative determination of usnic acid in usnea lichen and its products by reversed-phase liquid chromatography with photodiode array detector. *Journal of AOAC International* 88: 1265–1268.

Jones, F. T. and Palmer, K. J. 1950. Optical, crystallographic and X-ray diffraction data for usnic acid. *Journal of the American Chemical Society* 72: 1820–1822.

Kilpio, O. 1956. Antibacterial effects of an usnic acid derivative, usno, and its clinical uses with cases of pyoderma. *Nordisk Hygienisk Tidskrift* 37(11–12): 289–294.

Kreft, S. and Štrukelj, B. 2001. Reversed-polarity capillary zone electrophoretic analysis of usnic acid. *Electrophoresis* 22: 2755–2757.

Kristmundsdóttir, T., Aradóttir, H. A., Ingólfsdóttir, K., and Ögmundsdóttir, H. M. 2002. Solubilization of the lichen metabolite (+)-usnic acid for testing in tissue culture. *Journal of Pharmacy and Pharmacology* 54: 1447–1452.

Kristmundsdóttir, T., Jónsdóttir, E., Ögmundsdóttir, H. M., and Ingólfsdóttir, K. 2005. Solubilization of poorly soluble lichen metabolites for biological testing on cell lines. *European Journal of Pharmaceutical Sciences* 24: 539–543.

Kumar, K. C. S. and Müller, K. 1999. Lichen metabolites. 2. Antiproliferative and cytotoxic activity of gyrophoric, usnic, and diffractaic acid on human keratinocyte growth. *Journal of Natural Products* 62: 821–823.

Kupchan, S. M. and Kopperman, H. L. 1975. *l*-Usnic acid: Tumor inhibitor isolated from lichens. *Experientia* 31: 625.

Kutney, J. P., Sanchez, I. H., and Yee, T. H. 1974. Mass-spectral fragmentation studies in usnic acid and related compounds. *Organic Mass Spectrometry* 8: 129–146.

Lauterwein, M., Oethinger, M., Belsner, K., Peters, T., and Marre, R. 1995. *In vitro* activities of the lichen secondary metabolites vulpinic acid, (+)-usnic acid, and (−)-usnic acid against aerobic and anaerobic microorganisms. *Antimicrobial Agents and Chemotherapy* 39: 2541–2543.

Lazerow, S. K., Abdi, M. S., and Lewis, J. H. 2005. Drug-induced liver disease 2004. *Current Opinion in Gastroenterology* 21: 283–292.

Letcher, R. M. 1968. Chemistry of lichen constituents. 6. Mass spectra of usnic acid lichexanthone and their derivatives. *Organic Mass Spectrometry* 1: 551–561.

Lira, M. C. B., Siqueira-Moura, M. P., Rolim-Santos, H. M. L., Galetti, F. C. S., Simioni, A. R., Santos, N.P., Egito, E. S. T., Silva, C. L., Tedesco, A. C., and Santos-Magalhães, N. S. 2009. *In vitro* uptake and anti-mycobacterial activity of liposomal usnic acid formulation. *Journal of Liposome Research* 19: 49–58.

Lodetti, G., Gigola, P., Bertasi, B., D'Ambrosa, F., Ponchio, G., Fishbach, M., and Losio, M. N. 2000. Valutazione *in vitro* della citotossicità indotta da prodotti per l'igiene orale contenenti acido usnico, agente antimicrobico naturale estratto dai licheni. *Gimmoc* 4: 67–72.

Luzina, O. A., Polovinka, M. P., Salakhutdinov, N. F., and Tolstikov, G. A. 2007. Chemical modification of usnic acid 2. Reactions of (+)-usnic acid with amino acids. *Russian Chemical Bulletin* 56: 1249–1251.

Marshak, A., Schaefer, W. B., and Rajagopalan, S. 1949. Antibacterial activity of *d*-usnic acid and related compounds on M-Tuberculosis. *Proceedings of the Society for Experimental Biology and Medicine* 70: 565–568.

Mayer, M., O'Neill, M. A., Murray, K. E., Santos-Magalhães, N. S., Carneiro-Leão, A. M. A., Thompson, A. M., and Appleyard, V. C. L. 2005. Usnic acid: A non-genotoxic compound with anti-cancer properties. *Anti-Cancer Drugs* 16: 805–809.

Mitchell, J. C. 1965. Allergy to lichens. Allergic contact dermatitis from usnic acid produced by lichenized fungi. *Archives of Dermatology* 92: 142–146.

Müller, K. 2001. Pharmaceutically relevant metabolites from lichens. *Applied Microbiology and Biotechnology* 56: 9–16.

Najdenova, V., Lisickov, K., and Zoltan, D. 2001. Antimicrobial activity and stability of usnic acid and its derivatives in some cosmetic products. *Olaj, Szappan, Kozmetika* 50: 158–160.

Nash, T. H. 1996. *Lichen Biology* (2nd ed.). New York: Cambridge University Press.

Neff, G. W., Rajender Reddy, K., Durazo, F. A., Meyer, D., Marrero, R., and Kaplowitz, N. 2004. Severe hepatotoxicity associated with the use of weight loss diet supplements containing ma huang or usnic acid. *Journal of Hepatology* 41: 1062–1064.

Nguyen, L. and Thompson, C. J. 2006. Foundations of antibiotic resistance in bacterial physiology: The mycobacterial paradigm. *Trends in Microbiology* 14: 304–312.

Odabasoglu, F., Aygun, H., Yildirim, O. S., Halici, Z., Aslan, A., Cakir, A., Halici, M., and Cadirci, E. 2007. Effect of usnic acid on tissue caspase activity and glutathione level in titanium-implanted subjects. *FEBS Journal* 274(Suppl. 1): 269.

Ögmundsdóttir, H. M., Zoega, G. M., Gissurarson, S. R., and Ingolfsdóttir, K. 1998. Anti-proliferative effects of lichen-derived inhibitors of 5-lipoxygenase on malignant cell-lines and mitogen-stimulated lymphocytes. *Journal of Pharmacy and Pharmacology* 50: 107–115.

Pereira, E. C., Nascimento, S. C., Lima, R. C., Silva, N. H., Oliveira, A. F. M., Bandeira, E., Boitard, M., Beriel, H., Vicente, C., and Legaz, M. E. 1994. Analysis of *Usnea fasciata* crude extracts with antineoplasic activity. *Tokai Journal of Experimental and Clinical Medicine* 19: 47–52.

Perry, N. B., Benn, M. H., Brennan, N. J., Burgess, E. J., Ellis, G., Galloway, D. J., Lorimer, S. D., and Tangney, R. S. 1999. Antimicrobial, antiviral and cytotoxic activity of New Zealand lichens. *Lichenologist* 31: 627–636.

Pramyothin, P., Janthasoot, W., Pongnimitprasert, N., Phrukudom, S., and Ruangrungsi, N. 2004. Hepatotoxic effect of (+) usnic acid from *Usnea siamensis* Wainio in rats, isolated rat hepatocytes and isolated rat liver mitochondria. *Journal of Ethnopharmacology* 90: 381–387.

Rashid, M. A., Majid, M. A., and Quader, M. A. 1999. Complete NMR assignments of (+)-usnic acid. *Fitoterapia* 70: 113–115.

Raven, J. A. 1986. Biochemical disposal of excess H^+ in growing plants? *New Phytologist* 104: 175–206.

Ribeiro, S. M. A., Pereira, E. C. G., Gusmão, N. B., Falcão, E. P. S., and Silva, N. H. 2006. Produção de metabólitos bioativos pelo líquen *Cladonia substellata* Vainio. *Acta Botanica Brasilica* 20: 265–272.

Ribeiro-Costa, R. M., Alves, A. J., Santos, N. P., Nascimento, S. C., Gonçalves, E. C. P., Silva, N. H., Honda, N. K., and Santos-Magalhães, N. S. 2004. *In vitro* and *in vivo* properties of usnic acid encapsulated into PLGA-microspheres. *Journal of Microencapsulation* 21: 371–384.

Roach, J. A. G., Musser, S. M., Morehouse, K., and Woo, J. Y. J. 2006. Determination of usnic acid in lichen toxic to elk by liquid chromatography with ultraviolet and tandem mass spectrometry detection. *Journal of Agricultural and Food Chemistry* 54: 2484–2490.

Salem, I. I. and Düzgünes, N. 2003. Efficacies of cyclodextrin-complexed and liposome-encapsulated clarithromycin against *Mycobacterium avium* complex infection in human macrophages. *International Journal of Pharmaceutics* 250: 403–414.

Salo, H., Hannuksela, M., and Hausen, B. 1981. Lichen picker's dermatitis (*Cladonia alpestris* (L.) Rab.). *Contact Dermatitis* 7: 9–13.

Sanchez, W., Maple, J. T., Burgart, L. J., and Kamath, P. S. 2006. Severe hepatotoxicity associated with use of a dietary supplement containing usnic acid. *Mayo Clinic Proceedings* 81: 541–544.

Santos, N. P., Nascimento, S. C., Silva, J. F., Pereira, E. C. G., Silva, N. H., Honda, N. K., and Santos-Magalhães, N. S. 2005. Usnic acid-loaded nanocapsules: An evaluation of cytotoxicity. *Journal of Drug Delivery Science and Technology* 15: 355–361.

Santos, N. P. D., Nascimento, S. C., Wanderley, M. S. O., Pontes, N. T., da Silva, J. F., de Castro, C. M. M. B., Pereira, E. C., da Silva, N. H., Honda, N. K., and Santos-Magalhães, N. S. 2006. Nanoencapsulation of usnic acid: An attempt to improve antitumour activity and reduce hepatotoxicity. *European Journal of Pharmaceutics and Biopharmaceutics* 64: 154–160.

Scirpa, P., Scambia, G., Masciullo, V., Battaglia, F., Foti, E., Lopez, R., Villa, P., Malecore, M., and Mancuso, S. 1999. A zinc sulfate and usnic acid preparation used as post-surgical adjuvant therapy in genital lesions by human papilloma virus. *Minerva Ginecologica* 51: 255–260.

Seo, C., Jae, H. S., Seong, M. P., Joung, H. Y., Hong, K. L., and Oh, H. 2008. Usimines A–C, bioactive usnic acid derivatives from the Antarctic lichen *Stereocaulon alpinum*. *Journal of Natural Products* 71: 710–712.

Sharma, R. K. and Jannke, P. J. 1966. Acidity of usnic acid. *Indian Journal of Chemistry* 4: 16.

Sheu, M., Simpson, E. L., Law, S. V., and Storrs, F. J. 2006. Allergic contact dermatitis from a natural deodorant: A report of 4 cases associated with lichen acid mix allergy. *Journal of the American Academy of Dermatology* 55: 332–337.

Shibata, S. 2000. Yasuhiko Asahina (1880–1975) and his studies on lichenology and chemistry of lichen metabolites. *Bryologist* 103: 710–719.

Shukla, V., Negi, S., Rawat, M. S. M., Pant, G., and Nagatsu, A. 2004. Chemical study of *Ramalina africana* (Ramalinaceae) from the Garhwal Himalayas. *Biochemical Systematics and Ecology* 32: 449–453.

Siqueira-Moura, M. P., Lira, M. C. B., and Santos-Magalhães, N. S. 2008. Validação de método analítico espectrofotométrico UV para determinação de ácido úsnico em lipossomas. *Brazilian Journal of Pharmaceutical Sciences* 44: 622–628.

Stark, J. B., Walter, E. D., and Owens, H. S. 1950. Method of isolation of usnic acid from *Ramalina reticulata*. *Journal of the American Chemical Society* 72: 1819–1820.

Takai, M., Uehara, Y., and Beisler, J. A. 1979. Usnic acid derivatives as potential antineoplastic agents. *Journal of Medicinal Chemistry* 22: 1380–1384.

Takani, M., Takeda, T., Yajima, T., and Yamauchi, O. 2006. Indole rings in palladium(II) complexes. Dual mode of metal binding and aromatic ring stacking causing syn-anti isomerism. *Inorganic Chemistry* 45: 5938–5946.

Takani, M., Yajima, T., Masuda, H., and Yamauchi, O. 2002. Spectroscopic and structural characterization of copper(II) and palladium(II) complexes of a lichen substance usnic acid and its derivatives. Possible forms of environmental metals retained in lichens. *Journal of Inorganic Biochemistry* 91: 139–150.

Talvitie, A. 1983. Reassignment of the H-1-NMR signals of the methyl-groups of usnic acid. *Finnish Chemical Letters* 10: 8–9.

Tomasi, S., Picard, S., Laine, C., Babonneau, V., Goujeon, A., Boustie, J., and Uriac, P. 2006. Solid-phase synthesis of polyfunctionalized natural products: Application to usnic acid, a bioactive lichen compound. *Journal of Combinatorial Chemistry* 8: 11–14.

Vartia, K. O. 1950. Antibiotics in lichens. II. *Annals of Medical Experimental Biology Fenniae* 28: 7–19.

Vartia, K. O. 1973. Antibiotics in lichens. In V. Ahmadjian and M. E. Hale (Eds.), *The Lichens* (pp. 547–561). New York: Academic Press.

Venkataramana, D. and Krishna, D. R. 1992. High-performance liquid chromatographic determination of usnic acid in plasma. *Journal of Chromatography: Biomedical Applications* 575: 167–170.

Venkataramana, D. and Krishna, D. R. 1993. Pharmacokinetics of *D*-(+)-usnic acid in rabbits after intravenous administration. *European Journal of Drug Metabolism and Pharmacokinetics* 18: 161–163.

Vijayakumar, C. S., Viswanathan, S., Kannappa Reddy, M., Parvathavarthini, S., Kundu, A. B., and Sukumar, E. 2000. Anti-inflammatory activity of (+)-usnic acid. *Fitoterapia* 71: 564–566.

Virtanen, O. E. and Kärki, N. 1956. On the toxicity of an usnic acid preparation with the trade name USNO. *Suom Kemistilehti* 29B: 225–226.

Vyas, S. P., Kannan, M. E., Jain, S., Mishra, V., and Singh, P. 2004. Design of liposomal aerosols for improved delivery of rifampicin to alveolar macrophages. *International Journal of Pharmaceutics* 269: 37–49.

Wijagkanalan, W., Kawakami, S., Takenaga, M., Igarashi, R., Yamashita, F., and Hashida, M. 2008. Efficient targeting to alveolar macrophages by intratracheal administration of mannosylated liposomes in rats. *Journal of Controlled Release* 125: 121–130.

Yamamoto, Y., Miura, Y., Higuchi, M., Kinoshita, Y., and Yoshimura, I. 1993. Using lichen tissue-cultures in modern biology. *Bryologist* 96: 384–393.

Yamamoto, Y., Mizuguchi, R., and Yamada, Y. 1985. Tissue cultures of *Usnea rubescens* and *Ramalina yasudae* and production of usnic acid in their cultures. *Agricultural and Biological Chemistry* 49: 3347–3348.

19 Dendrimers in Drug Delivery

Hu Yang

CONTENTS

19.1 INTRODUCTION

The success of the synthesis of dendritic polymers was first reported in the early 1980s (Newkome et al., 1985; Tomalia et al., 1985). Dendritic polymers are now commonly referred to as dendrimers, which have a highly branched, three-dimensional, nanoscale architecture with very low polydispersity and high functionality, comprising a central core, internal branches, and a number of reactive surface groups (Figure 19.1). Further, the number of branches and surface groups exponentially increases along with the generation, thus allowing for high drug payload and multimodality. The emergence of dendrimers has greatly expanded the pool of carriers for drug delivery and led to the development of more efficient drug delivery systems. Because of their unique structure and properties, dendrimer drug delivery is undergoing rapid development. This chapter not only traces the evolution of the field of dendrimers in drug delivery but also reflects the journey through the subject from inception to contemporary progress.

19.2 INCEPTION OF DENDRIMER DRUG DELIVERY

The inception of dendrimer drug delivery began with the demonstration of the encapsulation of guest molecules within the internal cavity of dendrimers. Although the use of dendrimers for drug encapsulation was not suggested in the work of Naylor et al. (1989), they found the void space of dendrimers to be capable of encapsulating smaller guest molecules such as 2,4-dichlorophenoxy-acetic acid and acetylsalicylic acid. They studied the morphological changes of dendrimers having an NH_3 initiator core and β-alanine branch units in the presence of guest molecules. Their simulation and nuclear magnetic resonance results revealed that the structures of low generation dendrimers (1–3) are open and domed shapes. In contrast, higher generations (4–7) have more dense spheroid-like topologies. In 1993 Frechet's group reported that the water solubility of pyrene, a

435

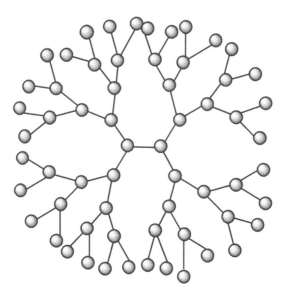

FIGURE 19.1 A schematic of a dendritic structure.

hardly water-soluble compound, was significantly increased through encapsulation by dendritic poly-ether macromolecules composed of 3.5-dihydroxybenzyl alcohol building blocks and carboxylate surface groups (Hawker et al., 1993). This amphiphilic dendrimer was named a unimolecular micelle because one single dendrimer molecule behaves like an assembly of amphiphilic molecules in an aqueous environment to encapsulate hydrophobic compounds (Figure 19.2). Unimolecular micelles are independent of concentration and temperature. Therefore, they overcome the thermodynamic instability present in polymeric and conventional micelles. Unimolecular micelles are able to solubi-lize hydrophobic molecules even at very low concentrations. For example, the concentration of a satu-rated pyrene solution in the presence of this dendrimer at a concentration of 2.13×10^{-4} mol/mL increased 120-fold to 9.47×10^{-5} mol/mL when compared to pure water. However, neither of the above systems allows for controlled drug delivery because both systems are dynamic processes and depend on the equilibrium conditions, thus making drug encapsulation and release uncontrollable.

It was not until the emergence of the elegant design of the dendritic box that the potential of dendrimers as vehicles for delivery of other molecules was truly demonstrated (Jansen et al., 1994). In this design, poly(propylene imine) (PPI) dendrimers are first synthesized following the divergent method and utilized as a flexible core. A subsequent coupling reaction is performed to introduce *tert*-butyloxycarbonyl-protected L-phenylalanine amino acid derivative to the dendrimer surface in the presence of guest molecules. The *tert*-butyloxycarbonyl-protected L-phenylalanine residues immobilized on the dendrimer surface are bulky and sterically crowed, thus forming a rigid shell at the outer layer. The authors named this core–shell structure a dendritic box because it has a rigid shell and a flexible core with internal cavities available for guest molecules (Figure 19.3). They used an electron paramagnetic resonance probe, 3-carboxy-proxyl, for the demonstration of the principle and found that 3-carboxy-proxyl could be encapsulated into a 64-branched dendritic box rather than in an 8-branched dendritic box. Further, they confirmed that extensive washing caused the complete removal of the electron paramagnetic resonance probe from the 64-branched dendritic box, indicat-ing that the probe was locked into the dendritic box physically instead of through covalent bonds. Jansen et al. (1994) also showed that a variety of dye molecules could be encapsulated into the den-dritic box, including Bengal Rose. Further, stable encapsulation was confirmed because the release of any of the encapsulated guest molecules from the box was unmeasurably slow, even when the system was subjected to different solvents, heating, and prolonged dialysis. A later report from Jansen et al. (1995) elucidated that the number of guest molecules is determined by the shapes of the

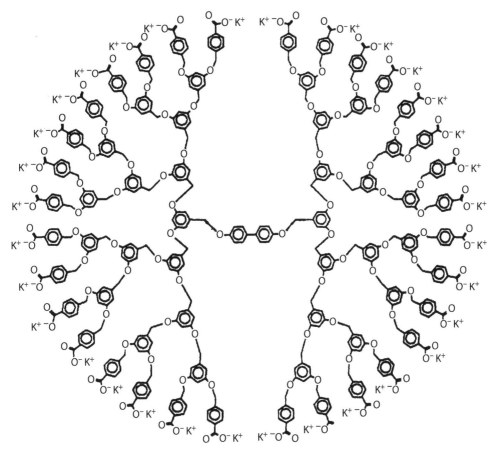

FIGURE 19.2 A water-soluble unimolecular dendritic polyether micelle. (From Hawker, C. J., Wooley, K. L., and Frechet, J. M. J. *Journal of the Chemical Society, Perkin Transactions 1* 1993, 12, 1287–1297. With permission.)

guest and the cavity. They demonstrated that a shape-selective liberation can be achieved by removing the shell in two steps: partial perforation of the shell to liberate small guest molecules followed by complete removal of the shell to liberate larger guest molecules. However, the release of drug molecules from the dendritic box in a physiological environment seems impossible because hydrolysis of the outer shell requires a strong acid (e.g., 12*N* HCl) under reflux for hours. Therefore, dendritic boxes that can be opened enzymatically, photochemically, or with a milder condition are highly desirable. Although there have been few studies using dendritic boxes for biomedical applications, the unique optical properties of dye-encapsulated dendritic boxes may be used for imaging. Further surface modifications of dendritic boxes with bioactive entities are necessary in order to tailor dendritic boxes for specific biomedical applications. The rapid development of dendrimers in drug delivery has been witnessed over the past decade.

19.3 MULTIMODALITY OF DENDRIMERS IN DRUG DELIVERY

Dendrimers provide a unique platform for developing nanostructured drug delivery systems. Because the size, geometry, and nature of the internal cavity of dendrimers as well as the functionality and quantity of their surface groups are well controlled, dendrimers have great structural adaptability and can be tailored to develop a variety of drug delivery systems to improve therapeutic

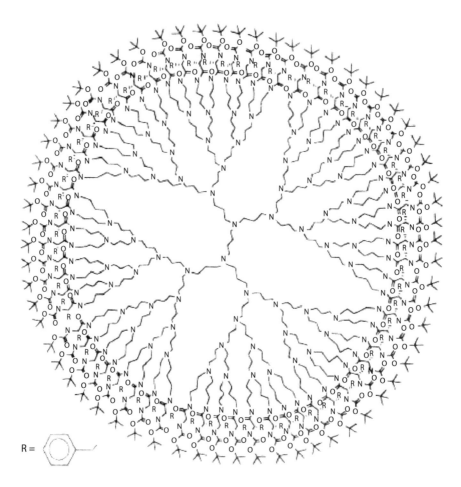

R = ⬡—/

FIGURE 19.3 A two-dimensional schematic of a dendritic box. (From Jansen, J. F. G. A., Meijer, E. W., and de Brabander van den Berg, E. M. M. *Journal of the American Chemical Society* 1995, 117, 4417–4418. With permission.)

efficacy or to meet specific pharmaceutical and biomedical needs, which are otherwise rarely achieved with other means.

19.3.1 DRUG DELIVERY THROUGH NONCOVALENT INTERACTIONS

Nearly 40% of newly developed drugs cannot be marketed because of low solubility (Svenson and Chauhan, 2008). In addition to polymeric micelles and liposomes that are applied to enhance drug solubility, dendrimers are able to enhance the solubility of poorly water-soluble drugs through encapsulation within the void spaces of the dendrimer, such as dendritic boxes, and noncovalent interactions, such as hydrogen bonding, van der Waals interactions, or electrostatic attractions.

Polyamidoamine (PAMAM) dendrimers with poly(ethylene glycol) (PEG) immobilized on the surface (i.e., PEGylated PAMAM dendrimers) can be used to enhance the water solubility of anti-cancer drugs, such as adriamycin and methotrexate (MTX). PEGylated dendrimers achieve enhanced drug-loading capacity, make the drug release rate slow, and decrease the hemolytic toxicity of dendrimers (Bhadra et al., 2003). By comparing unmodified dendrimers, PEGylated dendrimers increased 5-fluorouracil entrapment 12-fold (Figure 19.4; Bhadra et al., 2003). The encapsulation ability of PEGylated PAMAM dendrimers is determined by the PEG chain length

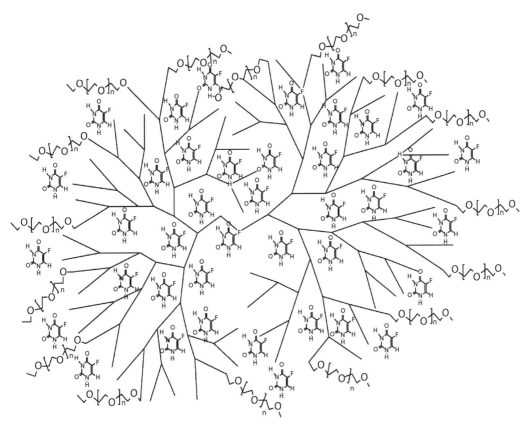

FIGURE 19.4 Encapsulation of 5-fluorouracil by PEGylated dendrimers. (From Bhadra, D., et al. *International Journal of Pharmaceutics* 2003, 257, 111–124. With permission.)

and dendrimer generation. For instance, up to 6.5 adriamycin molecules and up to 26 MTX molecules could be encapsulated by 1 PEGylated dendrimer composed of a generation 4.0 (G4.0) PAMAM core and a layer of PEG with a molecular weight of 2000 Da (Kojima et al., 2000). Yang et al. (2004) found that a longer PEG length (molecular weight = 5000 Da) resulted in deceased encapsulation efficiency that was presumably because longer chains may prevent drug uptake. Sarkar and Yang (2008) also demonstrated that the water solubility of anastrozole, which is used to treat postmenopausal women who are estrogen-receptor positive and need hormone-sensitive breast cancer treatment, could be enhanced by PEGylated dendrimers. They achieved an extended release of anastrozole for up to two days.

A poly(glycerol-succinic acid) dendrimer was recently synthesized, in which building blocks, such as glycerol and succinic acid, are natural metabolites (Morgan et al., 2006). This dendrimer was capable of encapsulating camptothecin (CPT), which is an anticancer drug with low water solubility (Figure 19.5). The aqueous solubility of CPT was increased by approximately one order of magnitude. Further, *in vitro* studies revealed that the potency of CPT could be increased because the dendrimer enhances both the uptake and retention of these compounds within cancer cells (Morgan et al., 2006). Furosemide is used orally to treat hypertension and edematous states associated with cardiac, renal, and hepatic failure. The effectiveness of this drug is limited because of its poor water solubility. Devarakonda et al. (2007) demonstrated that low generation PAMAM dendrimers (<G4) could encapsulate furosemide in the dendrimer cavity. They discovered that the increase in solubility of the drug was attributed to the pH of the medium, dendrimer generation, and type and number of internal amine groups. Because furosemide is a weak acid, they found that ionic charge–charge interactions are a driving force because careful selection of the pH value ionizes both the dendrimer

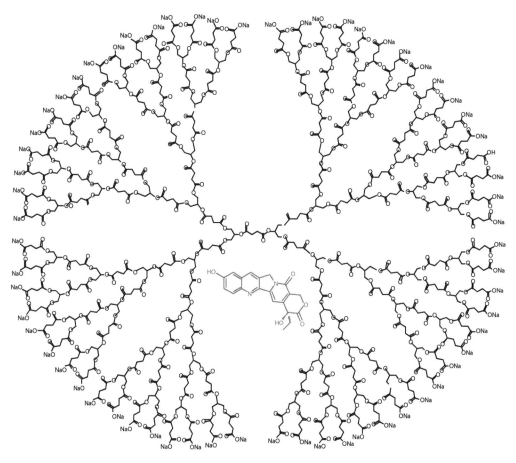

FIGURE 19.5 Encapsulation of camptothecin by poly(glycerol-succinic acid) dendrimer. (From Morgan, M. T., et al. *Cancer Research* 2006, 66, 11913–11921. With permission.)

and the drug. For example, neutral pH makes tertiary amines in the dendrimer cavity positively charged and carboxylate group of furosemide negatively charged. They further proved that the most efficient drug encapsulation was achieved in slightly acidic conditions (pH 4.0–6.0), where partially ionized amide groups in the dendrimer cavities were able to attract positively charged furosemide [pH > dissociation constant (pK_a)].

Cheng et al. (2008) compared the solubility enhancement of phenobarbital and primidone by PAMAM dendrimers. These two drug molecules share a very similar chemical structure, but primidone is more hydrophobic than phenobarbital (Figure 19.6). Further, primidone has a pK_a of >13, whereas phenobarbital has a pK_a between 7.2 and 7.4. Although primidone is more hydrophobic than phenobarbital and both have a similar size, they found that more phenobarbital molecules were entrapped compared to primidone by dendrimers in a basic solution (pH ~10). They reasoned that phenobarbital and primidone molecules display different forms in basic solutions: noncharged forms and negatively charged forms. In particular, primidone molecules exist in noncharged form. In contrast, phenobarbital molecules display negatively charged forms such that they can be attracted to the positively charged surface of the full generation PAMAM dendrimer. By excluding the negligible influence of molecular size, they believed that it was strong electrostatic interactions that caused the significant solubility enhancement of phenobarbital. They further validated the relationship of the electrostatic interaction and encapsulation between PAMAM dendrimers and negatively charged drug molecules by comparing the solubility enhancement of sulfamethoxazole and

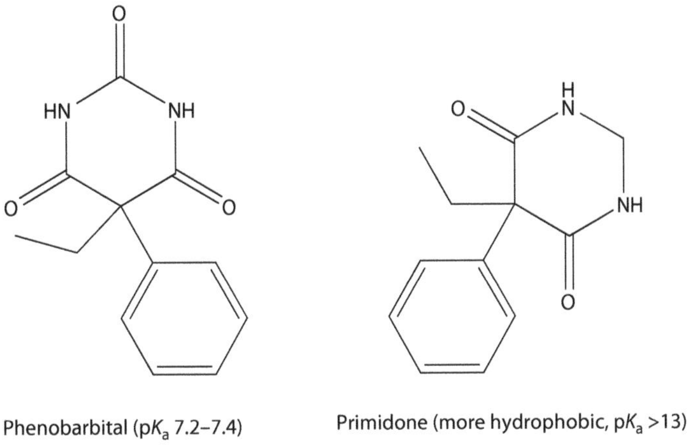

Phenobarbital (pK_a 7.2–7.4)   Primidone (more hydrophobic, pK_a >13)

FIGURE 19.6 Structures of phenobarbital and primidone.

trimethoprim in the presence of PAMAM dendrimers. Sulfamethoxazole and trimethoprim are both folate antagonists with extremely low water solubility. Cheng et al. (2008) observed a significant increase in the water solubility of sulfamethoxazole in the presence of PAMAM dendrimers, which was attributed to the ionization of the acidic sulfamoyl group ($-SO_2NH-$) in the sulfamethoxazole molecule (Figure 19.7). Nevertheless, the solubility of trimethoprim was not increased by PAMAM dendrimers because trimethoprim can only present a hydrophobic form.

Klajnert et al. (2008) built up a rigid maltose shell on the surface of PPI dendrimers and used 1-anilinonaphthalene-8-sulfonic acid (ANS) as a fluorescence probe to investigate the encapsulation performance of maltose-modified PPI dendrimers. Maltose is a neutral molecule. ANS is negative charged such that it can bind to the primary amine groups of the unmodified PPI dendrimers. After

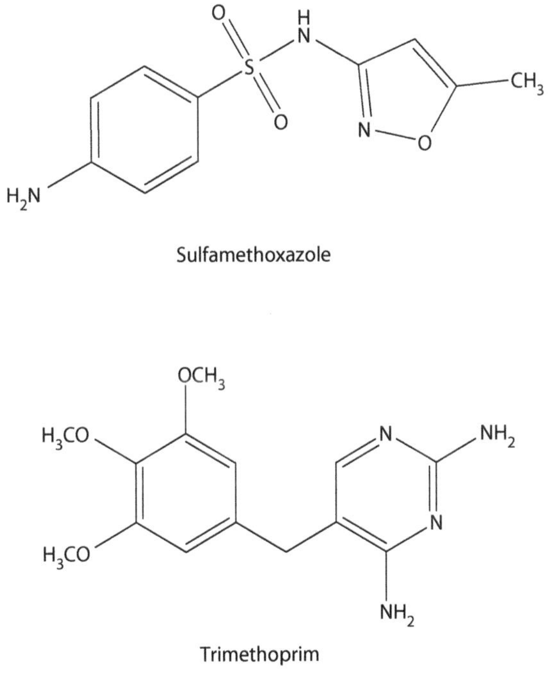

Sulfamethoxazole

Trimethoprim

FIGURE 19.7 Chemical structures of sulfamethoxazole and trimethoprim.

modification, the surface charges decreased because of the presence of maltose molecules on the surface as confirmed by zeta potential and polyelectrolyte titration measurements. Although the surface of the PPI dendrimer has diminished ionic charges, ANS achieved approximately twofold higher fluorescence intensity with maltose-modified dendrimers. The enhanced encapsulation of ANS was attributed to the deeper incorporation of ANS into a more hydrophobic microenvironment. Higher generation PPIs encapsulated more ANS drugs because of their extended dendritic structure.

Although most dendrimers have been synthesized to have a hydrophobic internal void space to encapsulate hydrophobic drugs, dendritic structures containing a hydrophilic interior have also been reported lately for encapsulation of hydrophilic drugs. Dhanikula and Hildgen (2006) synthesized polyester-*co*-polyether dendrimers consisting of a hydrophilic interior. The core was synthesized from butanetetracarboxylic acid and aspartic acid and the dendrons were synthesized from poly(ethylene oxide), dihydroxybenzoic acid or gallic acid, and PEG monomethacrylate (Figure 19.8). They used rhodamine as a hydrophilic model compound and β-carotene as a hydrophobic model compound to evaluate the encapsulation efficiency of this dendrimer. Dhanikula and Hildgen demonstrated that incorporation of poly(ethylene oxide) in the core allows for delivery of either

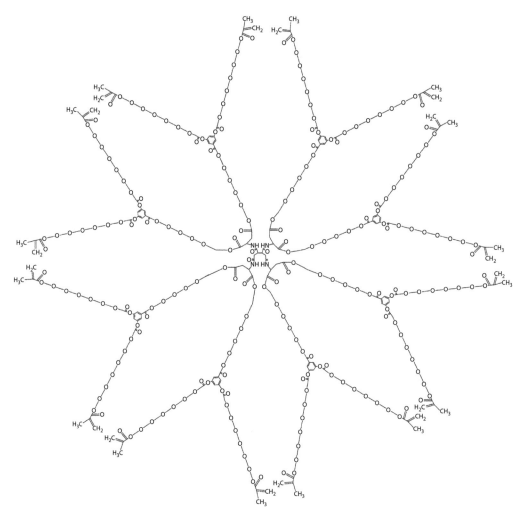

FIGURE 19.8 The structure of a polyester-*co*-polyether dendrimer molecule consisting of a hydrophilic interior. (From Dhanikula, R. S. and Hildgen, P. *Bioconjugate Chemistry* 2006, 17, 29–41. With permission.)

water-soluble or poorly water-soluble compounds. For instance, when the drug/dendrimer feed ratio was 1:4.5, a loading of 10.76% (w/w) was obtained for rhodamine and 5.86% (w/w) for β-carotene. They also found that more compounds could be taken up by the dendrimer with an increase in the drug/dendrimer feed ratio before a maximal drug/dendrimer feed ratio was reached. When the drug/dendrimer feed ratio was increased further, no increase in drug loading was observed for either of the two compounds. Neither rhodamine nor β-carotene has ionizable groups. Therefore, they believed that physical entrapment as well as hydrogen bonding caused more rhodamine molecules to be encapsulated by this dendrimer than β-carotene. Encapsulation of β-carotene simply relied on physical entrapment.

Dendrimers were also found to be capable of encapsulating silver ions in the core and through surface electrostatic attraction to formulate novel antimicrobial agents. Reduction of silver ions to silver particles in the presence of dendrimers leads to the formation of silver–dendrimer nanocomposites. Silver–dendrimer complexes and nanocomposites display stronger antimicrobial activity than silver alone, which is highly desirable for wound healing (Balogh et al., 2001; Ghosh and Banthia, 2004). The reason for the enhanced antimicrobial activity is that dendrimers provide extremely high local concentrations. Further, the movement of silver is retarded and the silver adsorption at a wound site is minimized because of strong interactions with dendrimers such that the likelihood of silver toxicity is reduced.

Kolhatkar et al. (2008) prepared 7-ethyl-10-hydroxy-camptothecin (SN-38)–dendrimer complexes for possible oral delivery. The formation of the complexes was presumably based on the ionic interaction between the phenolic OH of the drug and the surface amine groups on the PAMAM dendrimer (Figure 19.9). Neutralization of phenolic OH at acidic pH deceased the ionic interactions between the deprotonated phenolic OH and dendrimer, thus releasing complexed SN-38 from the dendrimer. Polycationic dendrimers have also been extensively studied for delivery of genes because they aid deoxyribonucleic acid in being efficiently internalized following endocytosis and membrane destabilization (Bielinska et al., 2000; Eichman et al., 2000; Haensler and Szoka, 1993; Liu and Frechet, 1999; Schatzlein et al., 2005). Genes are complexed with cationic dendrimers through electrostatic interactions. Further, secondary and tertiary amines within the inner core of dendrimers act as a "proton sponge" to facilitate the escape of gene–dendrimer polyplexes from endosomes and lysosomes (Sonawane et al., 2003; Takahashi et al., 2005). Because of the unique characteristics of dendrimers in gene delivery, their use for gene delivery has been extensively studied and reviewed (Bielinska et al., 2000; Braun et al., 2005; Brazeau et al., 1998; Dufes et al., 2005; Eichman et al., 2000; Huang et al., 2007; Kim et al., 2004, 2007; Liu and Frechet, 1999; Paleos et al., 2007; Schatzlein et al., 2005).

19.3.2 Drug Delivery via Covalent Linkage

Assembling therapeutic and bioactive molecules on the dendrimer surface through proper covalent linkage has become a common way to develop dendrimer-based drug delivery systems. A major advantage of drug delivery based on dendrimers is high drug payload that is enabled by dendrimers. A high drug payload will enhance therapeutic effectiveness, reduce dosing frequency, and improve patient compliance. The number of surface groups exponentially increases along with the increase in the dendrimer generation. For instance, ethylenediamine (EDA) core PAMAM dendrimer G0 has four terminal groups. By G10 there are an impressive 4096 terminal groups. However, the diameter of the EDA dendrimer increases only by 10-fold to 13.5 nm for G10 from 1.5 nm for G0. Low generation EDA core PAMAM dendrimers are commonly used as carriers to construct drug delivery systems as they provide a large quantity of surface groups for drug loading and display that have more favorable biological properties than high generation dendrimers. Since dendrimers containing various functional surface groups have been synthesized, the pool of drug molecules that can be covalently conjugated to the dendrimer has been greatly expanded. In addition, linker or spacer molecules can be utilized to covalently couple drug molecules to the dendrimer. The linkage

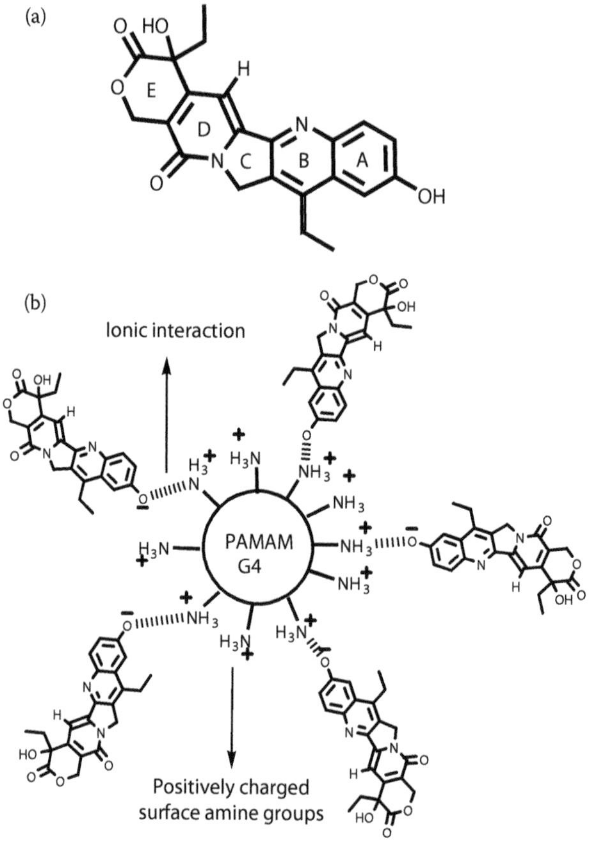

FIGURE 19.9 (a) The structure of SN-38 and (b) a complex of SN-38–G4.0 PAMAM dendrimer. (From Kolhatkar, R. B., Swaan, P., and Ghandehari, H. *Pharmaceutical Research* 2008, 25, 1723–1729. With permission.)

between the drug and the dendrimer carrier varies vastly, including amide bonds, hydrolytic ester bonds, acid-labile hydrazone, and enzyme-cleavable disulfide bonds.

A wide range of small molecular weight drugs can be directly coupled to the surface sites of dendrimers, thus creating a high drug payload on the dendrimer surface. Various functional groups of drugs such as hydroxyl (OH), carboxyl (COOH), primary amine (NH2), and thiol (SH) can be utilized to couple with dendrimer surface groups (Yang and Kao, 2006). Venlafaxine, a third generation antidepressant, can be directly conjugated to half-generation PAMAM dendrimers (i.e., G2.5) through dicyclohexyl-carbodiimide (DCC)/4-dimethylaminopyridine coupling chemistry (Yang and Lopina, 2005). In this reaction, the hydroxyl group of venlafaxine reacted with one carboxylate group of G2.5 PAMAM dendrimer to generate a hydrolytic ester bond. Nearly 100% of the dendrimer surface sites were conjugated with venlafaxine molecules. The hydrolysis study showed that venlafaxine conjugated with dendrimer was released over an extended period of time: only 50% of conjugated drug molecules were released within 18 h. Similarly, ibuprofen can be loaded onto the surface of hydroxyl-terminated PAMAM dendrimers (i.e., G4-OH) through an ester linkage with the aid of DCC. A high drug payload (i.e., 59 ibuprofen molecules/dendrimer) was identified by Kolhe et al. (2006). Carboxylate-containing drugs can be directly conjugated to amine-terminated dendrimers. For example, *N*-succinyl-Ala-Ala-Pro-Phe-*p*-nitroanilide can be conjugated to the G3 PAMAM dendrimer using the DCC/hydroxybenzotriazole method (Yang and Lopina, 2006). Note that if the conjugated drug molecule has a nonpolar bulky group, the high drug payload may lead to a decrease in the water solubility of dendrimer–drug conjugates. Amine-carrying compounds can be directly conjugated to amine-terminated or carboxylate-terminated dendrimers, depending on the coupling chemistry. For example, carboxylate surface groups of half-generation PAMAM dendrimers can be converted into active esters by first using the N-(3-dimethylaminopropyl)-N′-ethylcarbodiimide/N-hydroxysuccinimide (NHS) method. Then the N-terminus of Arg-Gly-Asp (RGD) can react with the NHS ester directly because this NHS-activated ester has high reactivity toward amine groups (Scheme 19.1a; Yang and Kao, 2007). Scheme 19.1b shows that the amine surface groups activated by differential scanning calorimetry/triethylamine can also display high reactivity toward the N-terminus of RGD (Scheme 19.1b; Yang and Kao, 2007). The methods described by Yang and Kao are efficient in creating high drug payload on the dendrimer surface and can be applied to couple dendrimers with other drug molecules containing same functional groups. MTX can be directly conjugated to amine- or hydroxyl-terminated dendrimers through amide or ester linkages following the activation by 1-[3-(dimethylamino) propyl]-3-ethylcarbodiimide hydrochloride (Quintana et al., 2002). The reaction conditions including the pH and reaction medium also affect the drug-loading efficiency. For example, 97% incorporation of doxorubicin to G4.0 PAMAM dendrimer was achieved when the molar ratio of doxorubicin to dendrimer was 3:1 in N-tris(hydroxymethyl) methyl-2-aminoethanesulfonic acid (pH 7.5; Papagiannaros et al., 2005). Doxorubicin incorporation decreased when the coupling reaction proceeded at a lower pH (i.e., 4.5) and higher molar ratio of doxorubicin to dendrimer (i.e., 6:1).

Drug molecules can also be conjugated to the dendrimer through various spacer molecules. Quinidine is an orally administered antiarrhythmic drug for the correction of abnormal heart rhythms. It contains a hydroxyl functional group, which is sterically hindered by adjacent quinoline and peperidine groups. The embedded hydroxyl group makes quinidine hardly react with a traditional polymer molecule because of its limited functional groups, thus leading to poor drug-loading efficiency. To overcome steric hindrance and improve the loading efficiency, a short glycine spacer was added to extend the quinidine's hidden hydroxyl group (Yang and Lopina, 2007). Paclitaxel (Taxol), a novel anticancer drug, was attached to a G5.0 PAMAM dendrimer using succinic anhydride as a spacer (Majoros et al., 2006). The linkage between Taxol and the spacer is an ester linkage such that the drug can be released through enzymatic hydrolysis. Majoros et al. used glycidol as a spacer to link MTX to the dendrimer via an ester linkage and to block the remaining amine groups to avoid unwanted nonspecific cellular uptake (Majoros et al., 2005). PEG serves as an important class of spacer. Its major function as a spacer is to link the targeting ligand to the dendrimer and give the conjugated ligand sufficient spatial freedom to maintain its targeting ability. For example, folate, a widely used ligand to target many cancer cell lines, can be attached to the dendrimer via a PEG spacer (Singh et al., 2008). Its role as a spacer and other functions are now discussed.

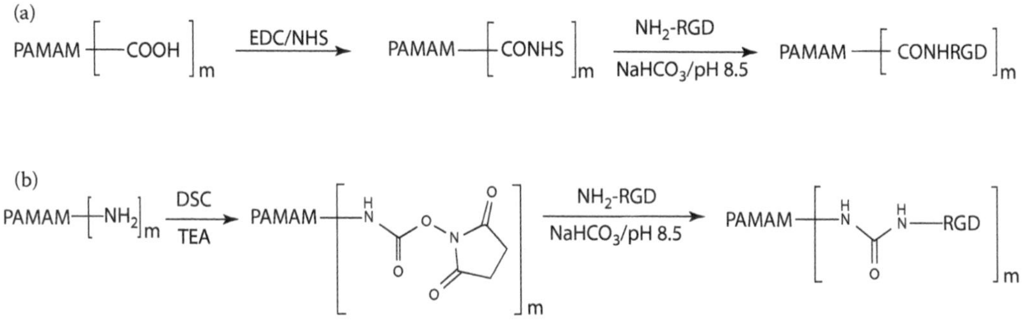

SCHEME 19.1 Synthesis of dendrimer-RGD conjugates based on (a) carboxylate-terminated dendrimers and (b) amine-terminated dendrimers. (From Yang, H. and Kao, W. J. *International Journal of Nanomedicine* 2007, 2, 89–99. With permission.)

19.3.3 PEGYLATED DENDRIMERS

PEG is a U.S. Food and Drug Administration approved biocompatible polymer that has been widely used in pharmaceutical and biomedical applications. In addition to enhancing the drug-loading capacity of dendrimers, coupling PEG to the dendrimer modifies the properties of dendrimers and provides many other functions:

1. It serves as a spacer to provide spatial flexibility for coupling the targeting ligand to the dendrimer and retains the binding ability of the conjugated targeting ligand, which otherwise is diminished because of steric hindrance
2. It overcomes the shortcomings commonly associated with nanostructured materials including reticuloendothelial system uptake, drug leakage, immunogenicity, hemolytic toxicity, and cytotoxicity
3. It provides the delivery system with improved water solubility
4. It protects conjugated drugs from degradation
5. It generates favorable pharmacokinetic (e.g., half-lives) and tissue distribution and eliminates the undesirable potential accumulation of side effects of dendrimers

Transferrin is a brain-targeting ligand. Huang et al. (2007) used PEG as a spacer for conjugation of transferrin to a dendrimer. The constructed delivery system showed efficacy in delivering therapeutic genes to the brain (Huang et al., 2007). Guillaudeu et al. (2008) synthesized a family of PEGylated ester dendrimers, composed of a pentaerythritol core and a bis-2,2-hydroxymethylpropanoic acid building block. After PEG was conjugated to the surface of this dendrimer, its blood circulation half-life was found to be 23.8 ± 2.5 h. This long circulation time is believed to lead to higher accumulation of drug-bearing dendrimers that is attributable to the enhanced permeability and retention effect (Guillaudeu et al., 2008).

Yang and Lopina (2003) designed PAMAM dendrimer drug delivery systems in which PEG was used as spacer. Because half- and full-generation PAMAM dendrimers have carboxylic and primary amine groups on their surfaces, respectively, they explored various conjugation methods to link the drug to the dendrimer using carboxylate-containing penicillin V as a model drug (Yang and Lopina, 2003). The linkage between penicillin V and PEG is the amide bond in Scheme 19.2, and the linkage between the drug and PEG is the ester bond in Scheme 19.3. The carboxylic acid group is essential for the antibacterial activity of penicillin V. The hydrolytic release of penicillin V from the PEGylated dendrimer was evaluated, and Yang and Lopina found that the modification of penicillin through esterification (Figure 19.10) did not alter the penicillin activity after hydrolysis. They also pointed out that penicillin V bound through an amide bond was not expected to be released from the carrier because the amide bond is stable *in vivo*. The synthesis strategy presented in Scheme 19.2 would provide a guide for preparing amide linkage between a carboxylic group containing drug and PEGylated dendrimers. This synthesis route would be useful when stable linkage between a drug and carrier is required and biological activity is less dependent or not dependent on the carboxylic group.

PEG conjugated on the dendrimer surface can provide a shielding layer to protect peptides bound to the dendrimer surface from enzymatic degradation. Yang and Lopina (2006) used *N*-succinyl-Ala-Ala-Pro-Phe-*p*-nitroanilide as a model peptide drug and synthesized three types of dendritic peptides: peptide–PAMAM, in which peptide was directly conjugated the dendrimer without PEGylation; peptide–PAMAM–PEG, in which both peptide and PEG were conjugated to the dendrimer surface; and peptide–PEG–PAMAM, in which PEG was a spacer to link the peptide to the dendrimer. When the peptide was linked to the dendrimer via PEG, the enzymatic hydrolysis of the peptide was increased. However, when PEG and peptide were both conjugated to dendrimer surface sites, the enzymatic hydrolysis of the peptide was decreased. According to this finding, Yang and Lopina suspected that it was the PEG long arms that influenced the enzyme–substrate complex. PEG could protect the conjugated peptides from attack by the enzyme. Therefore, the enzymatic

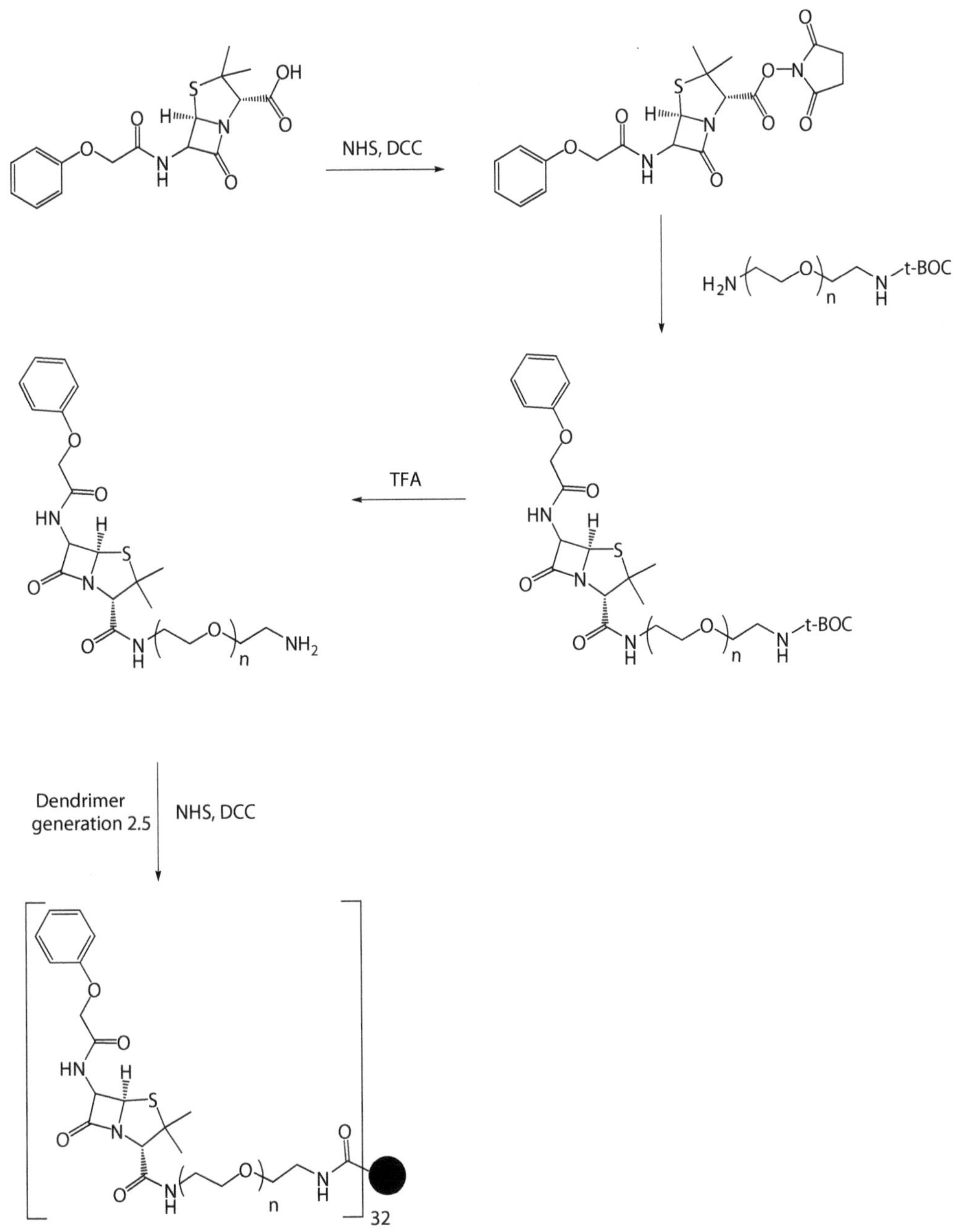

SCHEME 19.2 Synthesis of penicillin V–PEG–PAMAM (G2.5) conjugates (TFA—trifluoroacetic acid). (From Yang, H. and Lopina, S. T. *Journal of Biomaterials Science: Polymer Edition* 2003, 14, 1043–1056. With permission.)

stability of peptide–PAMAM–PEG was higher than that of any other types. In general, the enzymatic stability of dendritic peptides from high to low was peptide–PAMAM–PEG, peptide–PAMAM, free peptide, and peptide–PEG–PAMAM, respectively. Selectively increasing or decreasing the enzymatic stability of dendritic peptides is possible by varying the coupling sequence of peptide, PEG, and dendrimer. The stability of the dendritic peptides was evaluated only within a few minutes following the addition of enzyme to the solution, so the long-term stability (up to hours)

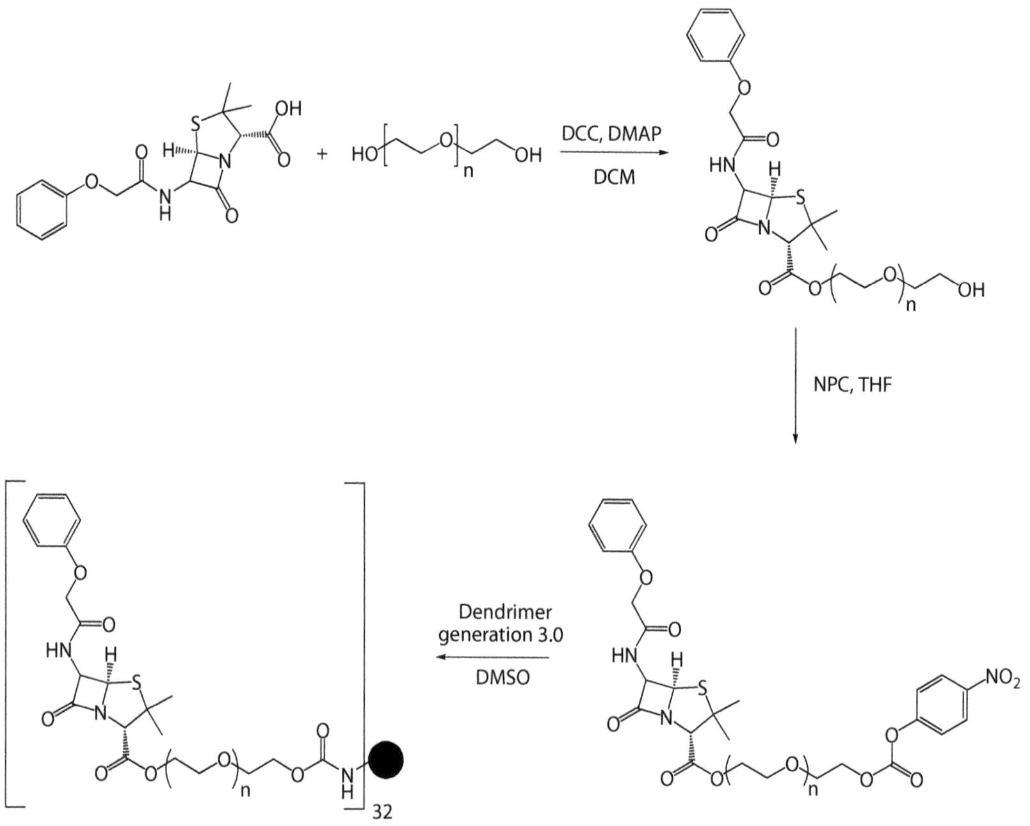

SCHEME 19.3 Synthesis of penicillin V–PEG–PAMAM (G3.0) conjugates (DCM—dichloromethane, DMSO—dimethyl sulfoxide). (From Yang, H. and Lopina, S. T. *Journal of Biomaterials Science: Polymer Edition* 2003, 14, 1043–1056. With permission.)

and how much peptides remained intact were not known. Because the enzyme concentration used in Yang and Lopina's study (2006) was much higher than that found in a physiological environment, a high degree of peptide protection may have resulted. Further, if peptide–PAMAM–PEG is used for delivering a peptide drug safely to a target cell or inside a target cell, the protection by PEG chains will also suppress the effectiveness of the peptide drug. Such issues need to be addressed before it is applied for drug delivery.

Kono et al. (2008) developed PEGylated PAMAM dendrimers for the delivery of the anticancer drug adriamycin. They modified every chain end of the dendrimer with glutamic acid (Glu). Both PEG and the drug were conjugated to each Glu residue on the dendrimer surface. The linkage between the drug and Glu was controlled to be either an amide bond or a hydrazone bond. They found that adriamycin was released slightly at pH 7.4 from the dendrimer–adriamycin conjugates that had either amide bonds or hydrazone bonds. In contrast, a significant release of the drug from the hydrazone linkage based dendrimer drug delivery system was identified at pH 5.5, but drug release from the dendrimer–adriamycin conjugates with amide linkage was still minimal. Kono et al. further demonstrated that these conjugates with hydrazone bonds exhibited higher cytotoxic effects than those with amide bonds because pH-labile hydrazone bonds allow the drug to be released quicker to take effect.

Incorporation of PEG into dendrimer drug delivery systems has proven to be a necessary procedure to make dendrimer nanoparticles more biocompatible. As the main function of dendrimer carriers to deliver therapeutic or bioactive molecules, the degree of PEGylation needs to be reduced to a minimal level such that more surface sites can be utilized for the creation of a high drug

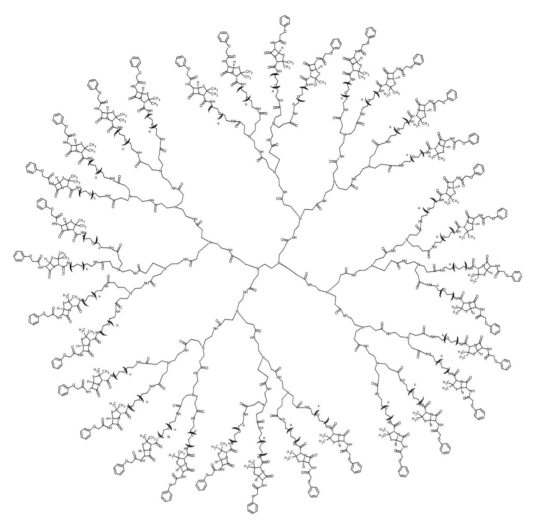

FIGURE 19.10 The chemical structure of a penicillin V bearing PEGylated dendrimer molecule. (From Yang, H. and Lopina, S. T. *Journal of Biomaterials Science: Polymer Edition* 2003, 14, 1043–1056. With permission.)

payload or conjugation with many other molecules. To efficiently utilize PEGylated dendrimers for drug delivery, Yang et al. (2008) quantitatively assessed the PEGylation efficacy in improving the cytocompatibility of low generation polycationic PAMAM dendrimers (i.e., G3.0 and G4.0). Both 3PEG–G3.0 (i.e., the degree of PEGylation was 9%) and 8PEG–G3.0 (i.e., the degree of PEGylation was 25%) significantly improved the cytocompatibility of G3.0 PAMAM dendrimers. Further, they found that the degree of enhanced cytocompatibility of G3.0 by 3PEGs–G3.0 was comparable to that of G3.0 by 8PEGs–G3.0. The cytocompatibility of G4.0 was also significantly increased through PEGylation. Similarly, a low degree of PEGylation (16%, i.e., 10 PEGs per G4.0) could sufficiently maintain the G4.0 cytocompatibility. Yang et al.'s study demonstrated that PEGylation could be correlated with cytocompatibility and drug payload to maximize the drug payload and keep the cytocompatibility of the drug carrier. Correlating the PEGylation, cytocompatibility, and drug payload integratively provides a means to understand the structure–property relationships of PEGylated dendrimers and allows for quantitative optimization of PEGylated dendrimers for drug delivery. In addition, PEG helps to improve the gene transfection efficiency of dendrimers, for example, PAMAM dendrimers, as demonstrated by Luo et al. (2002). They found that the increase in transfection efficiency relies on the PEGylation as well as the size of the dendrimer and the ratio of PEGylated PAMAM to deoxyribonucleic acid.

19.3.4 TARGETED DELIVERY

A variety of ligands have been conjugated to the dendrimer to create targeted dendritic drug delivery systems. In most cases, ligands are conjugated to the dendrimer through a spacer of appropriate length as discussed in Section 19.3.2. The methods employed for coupling the ligand to the dendrimer should not alter the binding ability of the ligand.

Folate receptors are overexpressed in several types of human cancer cells. Dendrimers conjugated with folic acid can specifically target tumor cells through folate receptors and significantly enhance drug uptake by cells. Quintana et al. (2002) designed a targeted dendrimer drug delivery system carrying the anticancer drug MTX and targeting ligand folic acid. They found that targeted delivery improved the cytotoxic response of the cells to MTX 100-fold over the free drug. Turk et al. (2002) recently found that the folate receptor was also overexpressed on the surface of activated macrophages in inflammatory tissue due to rheumatoid arthritis. Chandrasekar et al. (2007) developed a folate-mediated dendrimer drug delivery system to treat rheumatoid arthritis by coupling folate to a G3.5 PAMAM dendrimer via a PEG spacer and encapsulating indomethacin inside the dendrimer core. They found that the number of drug molecules that were encapsulated increased by 10- to 20-fold for this new delivery system as compared to the unmodified PAMAM dendrimer. The folate-coupled dendrimer delivery system had an increased half-life and residence time. As a result, indomethacin molecules in significant amounts were delivered to the target site in their animal study. Prostate-specific membrane antigen (PSMA) is highly expressed by prostate cancers. Antibodies against PSMA such as J591 can bind the extracellular PSMA domain, leading to efficient and selective internalization of dendrimers modified with J591 (Figure 19.11; Patri et al., 2004). A galactosyl residue was utilized to target gene-carrying dendrimer to hepatocytes through binding of galactosyl to asialo-glycoprotein receptors on hepatocytes (Kim et al., 2007).

Epidermal growth factor receptors (EGFRs) are overexpressed in cancers of the head and neck, breast, and colon, among others (Thomas et al., 2008). Further, the EGFR and its mutant isoform EGFRvIII are also overexpressed in malignant gliomas (Mendelsohn, 2001). EGFs and anti-EGF antibodies such as cetuximab have been used as targeting ligands to target drug-bearing dendrimers to a variety of human carcinomas. Thomas et al. (2008) synthesized a PAMAM dendrimer coupled with multiple EGF molecules as a multivalent targeting agent. This new agent stimulated EGFRs and acted as a superagonist by stimulating cell growth to a greater degree than free EGF, which implied that the dendrimer–EGF conjugate could serve as a topical agent to enhance wound healing (Thomas et al., 2008). The superagonistic mechanism of the dendrimer–EGF conjugate remains to be elucidated. The authors proposed that the conjugate favors cross-linking of the receptor, resulting in a sustained EGFR-mediated signal transduction and increased cell growth. Wu et al. (2006)

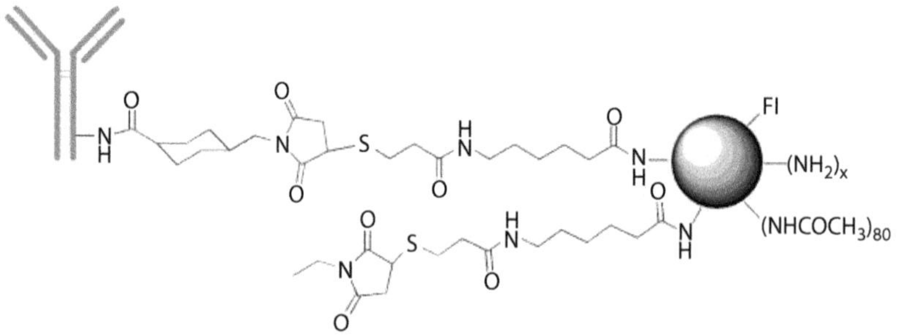

FIGURE 19.11 The structure of J591 antibody-dendrimer conjugates. (Adapted from Patri, A. K., et al. *Bioconjugate Chemistry* 2004, 15, 1174–1181. With permission.)

constructed a targeted delivery system by covalently coupling anti-EGF antibody cetuximab and cytotoxic MTX to a G5.0 PAMAM dendrimer. They attempted to deliver MTX specifically to EGFR-positive brain tumors. Although cetuximab retained its ability to bind EGFRs, the therapeutic efficacy of this new delivery system was not demonstrated, possibly because of the lack of controlled release of the conjugated MTX. Efforts are required to optimize this delivery system.

Galactose recognizes the asialo-glycoprotein receptors that are expressed in a large number by hepatocytes. Galactose was utilized as a targeting ligand to target hepatocyte primaquine phosphate, a liver schizonticide, that was encapsulated by using PPI dendrimers coated with galactose that could be directly delivered to liver cells (Bhadra et al., 2005).

19.4 SUMMARY

Dendrimers provide a unique nanoscale platform to develop effective drug delivery systems. A wide range of drugs can be delivered by dendrimers through either covalent linkage or noncovalent interactions. The presence of a number of end groups on the dendrimer surface enables high drug payloads and assembly of multiple functional entities including targeting ligands for targeted drug delivery. Highly adaptable dendritic structures can be engineered to treat various diseases and develop personalized medicine. In addition to the extensive use of off-the-shelf dendrimers for drug delivery, new dendritic structures are also being developed to obtain additional structural properties to meet specific needs in drug delivery. Dendrimers are capable of evolving with the science of medicine to incorporate state-of-the-art functionalities as they are discovered in order to develop potent nanomedicines. Thus, dendrimer-based drug delivery systems are able to meet the strict regulatory requirements for clinical use.

ACKNOWLEDGMENT

This chapter is dedicated to the memory of Professor Stephanie T. Lopina, who guided and inspired the author in the area of drug delivery.

ABBREVIATIONS

ANS	1-Anilinonaphthalene-8-sulfonic acid
CPT	Camptothecin
DCC	Dicyclohexyl-carbodiimide
EDA	Ethylenediamine
EGFR	Epidermal growth factor receptor
G	Generation
MTX	Methotrexate
NHS	N-Hydroxysuccinimide
PAMAM	Polyamidoamine
PEG	Poly(ethylene glycol)
pK_a	Dissociation constant
PPI	Poly(propylene imine)
PSMA	Prostate-specific membrane antigen
SN-38	7-Ethyl-10-hydroxy-camptothecin

GLOSSARY

Dendrimer: A highly branched, three-dimensional, nanoscale architecture with very low polydispersity and high functionality, comprising a central core, internal branches, and a number of reactive surface groups.

Dendritic polymer: Commonly referred to as a dendrimer.

Electron paramagnetic resonance: A technique for studying chemical species that have unpaired electrons.

Endosome: A vesicular compartment involved in the sorting and transport to lysosomes of material taken up by endocytosis.

Liposome: Usually made of phospholipids for drug delivery.

Lysosomes: A cytoplasmic organelle containing enzymes that break down biological polymers.

Nuclear magnetic resonance: Used to study molecular structures.

PAMAM dendrimer: Commercially known as Starburst™ dendrimer, one of the most investigated dendrimers.

PEG: A polymer of ethylene oxide with a molecular mass below 20,000 g/mol that serves as an important polymer for biomedical applications.

PEGylation: Covalent attachment of PEG polymer chains to another molecule Poly(ethylene oxide): A polymer of ethylene oxide with a molecular mass above 20,000 g/mol that serves as an important polymer for biomedical applications.

Polymeric micelle: Formed by amphiphilic block copolymers.

Unimolecular micelle: A single molecule that functions similarly to a polymeric micelle but is much less dependent on concentration than a polymeric micelle.

REFERENCES

Balogh, L., Swanson, D. R., Tomalia, D. A., Hagnauer, G. L., and McManus, A. T. Dendrimer–silver complexes and nanocomposites as antimicrobial agents. *Nano Letters* 2001, 1, 18–21.

Bhadra, D., Bhadra, S., Jain, S., and Jain, N. K. A PEGylated dendritic nanoparticulate carrier of fluorouracil. *International Journal of Pharmaceutics* 2003, 257, 111–124.

Bhadra, D., Yadav, A. K., Bhadra, S., and Jain, N. K. Glycodendrimeric nanoparticulate carriers of primaquine phosphate for liver targeting. *International Journal of Pharmaceutics* 2005, 295, 221–233.

Bielinska, A. U., Yen, A., Wu, H. L., et al. Application of membrane-based dendrimer/DNA complexes for solid phase transfection *in vitro* and *in vivo*. *Biomaterials* 2000, 21, 877–887.

Braun, C. S., Vetro, J. A., Tomalia, D. A., et al. Structure/function relationships of polyamidoamine/DNA dendrimers as gene delivery vehicles. *Journal of Pharmaceutical Sciences* 2005, 94, 423–436.

Brazeau, G. A., Attia, S., Poxon, S., and Hughes, J. A. *In vitro* myotoxicity of selected cationic macromolecules used in non-viral gene delivery. *Pharmaceutical Research* 1998, 15, 680–684.

Chandrasekar, D., Sistla, R., Ahmad, F. J., Khar, R. K., and Diwan, P. V. Folate coupled poly(ethyleneglycol) conjugates of anionic poly(amidoamine) dendrimer for inflammatory tissue specific drug delivery. *Journal of Biomedical Materials Research A* 2007, 82, 92–103.

Cheng, Y., Wu, Q., Li, Y., and Xu, T. External electrostatic interaction versus internal encapsulation between cationic dendrimers and negatively charged drugs: Which contributes more to solubility enhancement of the drugs? *Journal of Physical Chemistry B* 2008, 112, 8884–8890.

Devarakonda, B., Otto, D. P., Judefeind, A., Hill, R. A., and de Villiers, M. M. Effect of pH on the solubility and release of furosemide from polyamidoamine (PAMAM) dendrimer complexes. *International Journal of Pharmaceutics* 2007, 345, 142–153.

Dhanikula, R. S. and Hildgen, P. Synthesis and evaluation of novel dendrimers with a hydrophilic interior as nanocarriers for drug delivery. *Bioconjugate Chemistry* 2006, 17, 29–41.

Dufes, C., Uchegbu, I. F., and Schaetzlein, A. G. Dendrimers in gene delivery. *Advanced Drug Delivery Reviews* 2005, 57, 2177–2202.

Eichman, J. D., Bielinska, A. U., Kukowska-Latallo, J. F., and Baker, Jr., J. R. The use of PAMAM dendrimers in the efficient transfer of genetic material into cells. *Pharmaceutical Science & Technology Today* 2000, 3, 232–245.

Ghosh, S. and Banthia, A. K. Biocompatibility and antibacterial activity studies of polyamidoamine (PAMAM) dendron, side chain dendritic oligourethane (SCDOU). *Journal of Biomedical Materials Research A* 2004, 71, 1–5.

Guillaudeu, S. J., Fox, M. E., Haidar, Y. M., et al. PEGylated dendrimers with core functionality for biological applications. *Bioconjugate Chemistry* 2008, 19, 461–469.

Haensler, J. and Szoka, Jr., F. C. Polyamidoamine cascade polymers mediate efficient transfection of cells in culture. *Bioconjugate Chemistry* 1993, 4, 372–379.

Hawker, C. J., Wooley, K. L., and Frechet, J. M. J. Unimolecular micelles and globular amphiphiles: Dendritic macromolecules as novel recyclable solubilization agents. *Journal of the Chemical Society, Perkin Transactions 1* 1993, 12, 1287–1297.

Huang, R.-Q., Qu, Y.-H., Ke, W.-L., et al. Efficient gene delivery targeted to the brain using a transferrin-conjugated polyethyleneglycol-modified polyamidoamine dendrimer. *FASEB Journal* 2007, 21, 1117–1125.

Jansen, J. F. G. A., de Brabander van den Berg, E. M. M., and Meijer, E. W. Encapsulation of guest molecules into a dendritic box. *Science* 1994, 266(5188), 1226–1229.

Jansen, J. F. G. A., Meijer, E. W., and de Brabander van den Berg, E. M. M. The dendritic box: Shape-selective liberation of encapsulated guests. *Journal of the American Chemical Society* 1995, 117, 4417–4418.

Kim, K. S., Lei, Y., Stolz, D. B., and Liu, D. Bifunctional compounds for targeted hepatic gene delivery. *Gene Therapy* 2007, 14, 704–708.

Kim, T. I., Seo, H. J., Choi, J. S., et al. PAMAM-PEG-PAMAM: Novel triblock copolymer as a biocompatible and efficient gene delivery carrier. *Biomacromolecules* 2004, 5, 2487–2492.

Klajnert, B., Appelhans, D., Komber, H., et al. The influence of densely organized maltose shells on the biological properties of poly(propylene imine) dendrimers: New effects dependent on hydrogen bonding. *Chemistry—A European Journal* 2008, 14, 7030–7041.

Kojima, C., Kono, K., Maruyama, K., and Takagishi, T. Synthesis of polyamidoamine dendrimers having poly(ethylene glycol) grafts and their ability to encapsulate anticancer drugs. *Bioconjugate Chemistry* 2000, 11, 910–917.

Kolhatkar, R. B., Swaan, P., and Ghandehari, H. Potential oral delivery of 7-ethyl-10-hydroxy-camptothecin (Sn-38) using poly(amidoamine) dendrimers. *Pharmaceutical Research* 2008, 25, 1723–1729.

Kolhe, P., Khandare, J., Pillai, O., et al. Preparation, cellular transport, and activity of polyamidoamine-based dendritic nanodevices with a high drug payload. *Biomaterials* 2006, 27, 660–669.

Kono, K., Kojima, C., Hayashi, N., et al. Preparation and cytotoxic activity of poly(ethylene glycol)-modified poly(amidoamine) dendrimers bearing adriamycin. *Biomaterials* 2008, 29, 1664–1675.

Liu, M. and Frechet, J. M. J. Designing dendrimers for drug delivery. *Pharmaceutical Science & Technology Today* 1999, 2, 393–401.

Luo, D., Haverstick, K., Belcheva, N., Han, E., and Saltzman, W. M. Polyethylene glycol-conjugated PAMAM dendrimer for biocompatible, high-efficiency DNA delivery. *Macromolecules* 2002, 35, 3456–3462.

Majoros, I. J., Myc, A., Thomas, T., Mehta, C. B., and Baker, Jr., J. R. Pamam dendrimer-based multifunctional conjugate for cancer therapy: Synthesis, characterization, and functionality. *Biomacromolecules* 2006, 7, 572–579.

Majoros, I. J., Thomas, T. P., Mehta, C. B., and Baker, Jr., J. R. Poly(amidoamine) dendrimer-based multifunctional engineered nanodevice for cancer therapy. *Journal of Medicinal Chemistry* 2005, 48, 5892–5899.

Mendelsohn, J. The epidermal growth factor receptor as a target for cancer therapy. *Endocrine-Related Cancer* 2001, 8, 3–9.

Morgan, M. T., Nakanishi, Y., Kroll, D. J., et al. Dendrimer-encapsulated camptothecins: Increased solubility, cellular uptake, and cellular retention affords enhanced anticancer activity *in vitro*. *Cancer Research* 2006, 66, 11913–11921.

Naylor, A. M., Goddard III, W. A., Kiefer, G. E., and Tomalia, D. A. Starburst dendrimers. 5. Molecular shape control. *Journal of the American Chemical Society* 1989, 111, 2339–2341.

Newkome, G. R., Yao, Z., Baker, G. R., and Gupta, V. K. Micelles. Part 1. Cascade molecules: A new approach to micelles. A [27]-arborol. *Journal of Organic Chemistry* 1985, 50, 2003–2004.

Paleos, C. M., Tsiourvas, D., and Sideratou, Z. Molecular engineering of dendritic polymers and their application as drug and gene delivery systems. *Molecular Pharmaceutics* 2007, 4, 169–188.

Papagiannaros, A., Dimas, K., Papaioannou, G. T., and Demetzos, C. Doxorubicin—PAMAM dendrimer complex attached to liposomes: Cytotoxic studies against human cancer cell lines. *International Journal of Pharmaceutics* 2005, 302, 29–38.

Patri, A. K., Myc, A., Beals, J., et al. Synthesis and *in vitro* testing of J591 antibody–dendrimer conjugates for targeted prostate cancer therapy. *Bioconjugate Chemistry* 2004, 15, 1174–1181.

Quintana, A., Raczka, E., Piehler, L., et al. Design and function of a dendrimer-based therapeutic nanodevice targeted to tumor cells through the folate receptor. *Pharmaceutical Research* 2002, 19, 1310–1316.

Sarkar, K. and Yang, H. Encapsulation and extended release of anti-cancer anastrozole by stealth nanoparticles. *Drug Delivery* 2008, 15, 343–346.

Schatzlein, A. G., Zinselmeyer, B. H., Elouzi, A., et al. Preferential liver gene expression with polypropylen-imine dendrimers. *Journal of Controlled Release* 2005, 101, 247–258.

Singh, P., Gupta, U., Asthana, A., and Jain, N. K. Folate and folate-PEG-PAMAM dendrimers: Synthesis, characterization, and targeted anticancer drug delivery potential in tumor bearing mice. *Bioconjugate Chemistry* 2008, 19, 2239–2252.

Sonawane, N. D., Szoka, Jr., F. C., and Verkman, A. S. Chloride accumulation and swelling in endosomes enhances DNA transfer by polyamine–DNA polyplexes. *Journal of Biological Chemistry* 2003, 278, 44826–44831.

Svenson, S. and Chauhan, A. S. Dendrimers for enhanced drug solubilization. *Nanomedicine (London)* 2008, 3, 679–702.

Takahashi, T., Harada, A., Emi, N., and Kono, K. Preparation of efficient gene carriers using a polyamidoamine dendron-bearing lipid: Improvement of serum resistance. *Bioconjugate Chemistry* 2005, 16, 1160–1165.

Thomas, T. P., Shukla, R., Kotlyar, A., et al. Dendrimer—epidermal growth factor conjugate displays superagonist activity. *Biomacromolecules* 2008, 9, 603–609.

Tomalia, D. A., Baker, H., Dewald, J., et al. A new class of polymers: Starburst-dendritic macromolecules. *Polymer Journal (Tokyo)* 1985, 17, 117–132.

Turk, M. J., Breur, G. J., Widmer, W. R., et al. Folate-targeted imaging of activated macrophages in rats with adjuvant-induced arthritis. *Arthritis and Rheumatism* 2002, 46, 1947–1955.

Wu, G., Barth, R. F., Yang, W., et al. Targeted delivery of methotrexate to epidermal growth factor receptor-positive brain tumors by means of cetuximab (Imc-C225) dendrimer bioconjugates. *Molecular Cancer Therapeutics* 2006, 5, 52–59.

Yang, H. and Kao, W. J. Synthesis and characterization of nanoscale dendritic RGD clusters for potential applications in tissue engineering and drug delivery. *International Journal of Nanomedicine* 2007, 2, 89–99.

Yang, H. and Kao, W. J. Dendrimers for pharmaceutical and biomedical applications. *Journal of Biomaterials Science: Polymer Edition* 2006, 17, 3–19.

Yang, H. and Lopina, S. T. Penicillin V-conjugated PEG-PAMAM star polymers. *Journal of Biomaterials Science: Polymer Edition* 2003, 14, 1043–1056.

Yang, H. and Lopina, S. T. Extended release of a novel antidepressant, venlafaxine, based on anionic polyamidoamine dendrimers and poly(ethylene glycol)-containing semi-interpenetrating networks. *Journal of Biomedical Materials Research* 2005, 72A, 107–114.

Yang, H. and Lopina, S. T. *In vitro* enzymatic stability of dendritic peptides. *Journal of Biomedical Materials Research* 2006, 76A, 398–407.

Yang, H. and Lopina, S. T. Stealth dendrimers for antiarrhythmic quinidine delivery. *Journal of Materials Science: Materials in Medicine* 2007, 18, 2061–2065.

Yang, H., Lopina, S. T., DiPersio, L. P., and Schmidt, S. P. Stealth dendrimers for drug delivery: Correlation between PEGylation, cytocompatibility, and drug payload. *Journal of Materials Science: Materials in Medicine* 2008, 19, 1991–1997.

Yang, H., Morris, J. J., and Lopina, S. T. Polyethylene glycol-polyamidoamine dendritic micelle as solubility enhancer and the effect of the length of polyethylene glycol arms on the solubility of pyrene in water. *Journal of Colloid and Interface Science* 2004, 273, 148–154.

20 Microsphere
A Novel Drug Delivery System

Abdus Samad, Mohammad Tariq,
Mohammed Intakhab Alam, and M. S. Akhter

CONTENTS

20.1 INTRODUCTION

Microparticles refer to particles with a diameter of 1–1000 µm. Within the broad category of microparticles, *microspheres* specifically refers to spherical microparticles and the subcategory of *microcapsules* applies to microparticles that have a core surrounded by a material that is distinctly different from that of the core. The core may be solid, liquid, or even gas.

20.1.1 BIODEGRADABLE MICROSPHERES AS ORAL SUSTAINED RELEASE DRUG DELIVERY SYSTEM

Biodegradable microparticulate carriers are of interest for oral delivery of drugs to

- Improve bioavailability
- Enhance drug absorption
- Target a particular organ and reduce toxicity
- Improve stomach gastric tolerance of a gastric irritant substance
- Act as carriers for antigens

Conceptually, the phospholipid bilayer of the plasma membrane of the epithelial cells that normally line the intestine (enterocytes) is considered to be the major factor restricting the free movement of substances from the lumen to the bloodstream. The term persorption was suggested by Volkheimer to allow the passage of particles up to 100 µm in diameter (Volkheimer and Schultz, 1968), which subsequently reached the portal venous blood and thoracic lymph (dog). Although this subject is of considerable controversy, new strategies have been developed similar to the "Trojan horse." The active molecules to be delivered are hidden inside hydrophobic, biodegradable microspheres that can be taken up by endocytosis in intestinal cells. The extent and pathway of particle uptake is different in different parts of the intestine. The microfold (M) cells of Peyer's patches (PP) represent a sort of lymphatic island within the intestinal mucosa and are possibly the major gateway through which particles can be adsorbed. Table 20.1 gives the various sites of particle uptake based on the particle size.

20.1.2 PREREQUISITES FOR IDEAL MICROPARTICULATE CARRIERS

The materials utilized for the preparation of a microparticulate system should fulfill the following prerequisites:

- Longer duration of action
- Control of content release

TABLE 20.1
Proposed Mechanisms for Uptake of Particles

Site	Size Range
Persorption	5–150 μm
PP	20 nm–10 μm
Follicle associated epithelium	<750 nm
Enterocyte or endocyte (RES uptake)	<220 nm
Paracellular uptake	100–200 nm
Intestinal macrophage	1 μm

Note: RES, reticuloendothelial system.

- Increase in therapeutic efficacy
- Protection of drug
- Reduction of toxicity
- Biocompatibility
- Sterlizability
- Relative stability
- Water solubility or dispersibility
- Targetability

20.1.3 Materials Used for the Preparation of Microparticulate Systems

A number of different biodegradable and nonbiodegradable substances have been investigated for the preparation of microspheres. These materials include polymers of natural and synthetic origin and modified natural substances:

1. Natural polymers
   a. Carbohydrates: starch, agarose, chitosan (CS), gellan gum, and alginate
   b. Chemically modified carbohydrates: hydroxypropyl methyl cellulose (HPMC), hydroxypropyl ethyl cellulose, ethyl cellulose, polyacryl starch, and polyalkyl dextran
   c. Proteins: albumin, gelatin, and collagen
2. Synthetic polymers
   a. Biodegradable: poly(lactic acid) (PLA), polylactide G, poly(lactic-*co*-glycolic acid) (PLGA), polycaprolactone, and polyanhydrides
   b. Nonbiodegradable: Eudragit L, Eudragit RS, Eudragit RL, poly(methyl methacrylate), and epoxy polymers

20.2 GENERAL METHOD OF PREPARATION

Several techniques are employed for the formulation of microspheres, which are discussed in the following sections. The technique is selected based on parameters such as the polymer, drug (protein, peptide, or nonprotein), duration of therapy, and intended use. Moreover, the method of preparation and its choice are equivocally determined by some formulation- and technology-related factors:

1. Desired particle size range
2. Process should not affect the active pharmaceutical ingredient
3. Release profile should be reproducible
4. No generation of toxic material in the final product

There are several techniques for the preparation of microspheres:

1. Emulsion techniques
   a. Single emulsion
   b. Double emulsion
2. Polymerization techniques
   a. Bulk polymerization
   b. Suspension polymerization
   c. Emulsion polymerization
   d. Interfacial polymerization
3. Phase separation coacervation techniques
4. Spray drying and spray congealing
5. Ionic gelation
6. Solvent extraction or solvent evaporation

20.2.1 EMULSION TECHNIQUES

20.2.1.1 Single Emulsion Technique

The single emulsion technique is a two-step process that employs a natural polymer (i.e., protein and carbohydrate) as a carrier material. The first step is dissolution or dispersion of natural polymers in aqueous media followed by dispersion in nonaqueous media. The second step is cross-linking of the dispersed globule, which is achieved by physical (heating) or chemical means. Cross-linking by physical means is achieved by dispersing the polymer solution or suspension in previously heated nonaqueous media, but it is not suitable for thermolabile drugs (i.e., protein, peptide, and others). The chemical method employs chemical cross-linking agents like formaldehyde, glutaraldehyde (GA), and diacid chloride. However, this technique has the disadvantage of excessive exposure of active ingredients to chemicals if added at the time of preparation (Figure 20.1).

Novel semiinterpenetrating polymer network hydrogel microspheres of CS and hydroxypropyl cellulose were prepared by the emulsion cross-linking method using GA as a cross-linker and chlorothiazide as a drug, which is a diuretic and antihypertensive drug with limited water solubility. The *in vitro* release studies indicated the dependence of the release rate on the extent of cross-linking, drug loading, and amount of hydroxypropyl cellulose used to produce the microspheres; slow release was extended up to 12 h. The release data followed the non-Fickian trend (Rokhade et al., 2008).

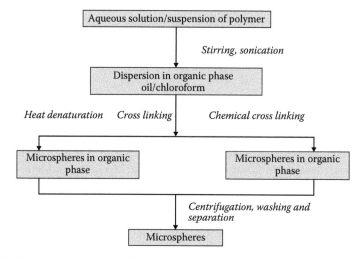

FIGURE 20.1 Schematic representation of microsphere preparation by the single emulsification method.

Drug-loaded magnetic microspheres were prepared by the emulsification/cross-linking method by Saravanan et al. (2008) and characterized by drug loading, magnetite content, and size distribution using optical microscopy, scanning electron microscopy (SEM), Fourier transform infrared analysis, differential scanning calorimetry, and x-ray diffraction methods. Upon variation of the drug/polymer ratio, the microspheres showed different drug loading [9.1, 18.7, and 24.9% (w/w)] and magnetite contents of 27.8–28.9% (w/w) with an average size range of 25–30.6 µm. The microspheres were able to prolong the drug release over 24–30 days. Researchers attempted to deliver diclofenac sodium to a target site by intra-arterial injection of gelatin magnetic microspheres and subsequent localization using an external magnet. The targeting efficiency and therapeutic efficacy of the microspheres were studied *in vivo* in rabbits, and the targeting efficiency was good with 77.7% of the drug recovered at the target site. The microspheres effectively reduced joint swelling in antigen-induced arthritic rabbits without producing gastric ulceration (Saravanan et al., 2008). Ketorolac tromethamine loaded microspheres were prepared by an oil-in-water (O/W) emulsion solvent evaporation technique using different polymers: polycaprolactone, poly-D,L-lactide (Resomer), and PLA. To achieve the release profile of the drug for several days, blends of Resomer and PLA were prepared with polycaprolactone in different ratios. Higher encapsulation efficiency (EE) was obtained with microspheres made with pure Resomer. Differential scanning calorimetric studies revealed drug–polymer interactions. Sinha and Trehan (2008) concluded that a careful selection of different polymers and their ratio can be used to achieve release of ketorolac trimethamine for long periods.

An O/W emulsion solvent evaporation method was used to encapsulate bupivacaine using 50:50 PLGA. The particle size could be controlled by changing the stirring rate and polymer concentration. The EE was affected by the polymer concentration, and the burst effect of bupivacaine released from the particles was affected by the drug/polymer mass ratio. Bupivacaine microspheres (bupi-MS) were evaluated for a dissolution profile and release model with microspheres of low loading (6.41%) and high loading (18.92%). It was observed that the drug release was affected by drug loading, especially the amount of drug crystals attached on the surface of bupi-MS. The drug release profile of low drug loaded bupi-MS agreed with the Higuchi equation and that of high drug loaded bupi-MS agreed with the first-order equation (Zhang et al., 2008).

The drawbacks of this method include the following:

- Tedious procedure
- Use of harsh cross-linking agents, which might possibly induce chemical reactions with the active agent
- Complete removal of unreacted cross-linking agent may be difficult in this process

20.2.1.2 Double Emulsification Technique

This technique involves double or multiple emulsion formation [water-in-oil-in-water (W/O/W)]. It is the method of choice for water-soluble drugs, peptides, and vaccines. Both natural and synthetic polymers can be employed (Figure 20.2).

The aqueous drug solution is emulsified under intense agitation in a lipophilic continuous phase that consists of a polymer solution in an organic solvent. This primary (1°) emulsion was added into the external aqueous phase containing a suitable emulsifying agent. Poly(vinyl alcohol) is the most acceptable emulsifier. Poly(vinyl pyrrolidone) (PVP), gelatin, alginate, methylcellulose, polysorbate, and sodium lauryl sulfate can also be used.

The stability of 1° emulsions depends upon the droplet size. The finer the droplet size is, the more stable is the 1° emulsion. Primary emulsion stability is detrimental to drug loading. The more stable the 1° emulsions are, the more loading capacity, depending upon the internal water volume.

A suitable emulsifying agent is used because during evaporation of the solvent the droplet initially shrinks in size, which leads to coalescence and agglomeration from a secondary (2°) emulsion.

Solvent removal is carried out by solvent evaporation or solvent extraction. Solvent evaporation employs either a reduction in pressure or agitation of the emulsion. Solvent extraction involves the

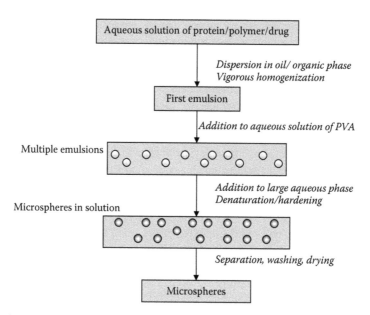

FIGURE 20.2 Schematic representation of microsphere preparation by the double emulsification method.

addition of a large quantity of water into which organic solvents diffuse out; then filtration, washing, and drying are carried out to achieve microspheres.

The double emulsion solvent evaporation technique has been employed for a number of hydrophilic drugs or peptides.

Insulin microspheres were prepared by the W/O/W double emulsion solvent evaporation technique and characterized. The integrity of the encapsulated insulin and the stability of the released insulin were assessed using a wide range of comprehensive analytical techniques. The purpose of this study was to investigate the effect of three zinc salts (i.e., zinc oxide, zinc carbonate, and zinc acetate) on the insulin EE, stability, and *in vitro* release kinetics from PLGA microspheres. The EE of the formulation prepared without the addition of a zinc salt was 69%, and the secondary structure of the encapsulated insulin in this formulation was found to be altered. Further, desamido-insulin and aggregates were observed during *in vitro* release. When insulin was encapsulated with a zinc salt, the EE increased significantly, the secondary structure was unaltered, and no degradation or aggregation products were found. Zinc salts can be useful to increase the EE and stability of insulin in PLGA microspheres prepared by the W/O/W technique (Manoharan and Singh, 2008). Polymer-based microparticles are increasingly becoming of interest for a variety of applications including drug delivery. The copolyester poly(glycol adipate-*co*-ω-pentadecalactone) was evaluated as a colloidal delivery system for encapsulated therapeutic proteins. α-Chymotrypsin, which is a proteolytic enzyme containing microparticles, was prepared via the double W/O/W emulsion solvent evaporation method. It was evaluated for EE and *in vitro* release. This research showed the potential of this polyester as a delivery system for enzymes via microparticles (Gaskell et al., 2008).

20.2.2 POLYMERIZATION TECHNIQUE

20.2.2.1 Bulk Polymerization

Bulk polymerization involves polymerization of monomers or mixture of monomers with or without a drug and with the aid of an initiator or catalyzing agents, which facilitate or accelerate the rate of the reaction. It involves the heating of a mixture of monomers or monomers and initiator with or without a drug. If the polymer is soluble in the monomer, a viscous solution is obtained. If insoluble, the polymer is precipitated out and is molded or fragmented into a microsphere. If the drug is thermostable, it may be added during polymerization; if it is a thermolabile drug, the absorptive method is employed. Bulk polymerization is an exothermic process, so it is not the method of choice for thermolabile drugs; but it has the advantage of producing pure polymer microspheres.

20.2.2.2 Suspension Polymerization and Emulsion Polymerization

Suspension polymerization is also called bead or pearl polymerization. This method involves the dispersion of monomers along with the drug and initiator in aqueous suspension media, which is the most commonly used dispersion media, but nonaqueous media may also be used. Polymers obtained by this method are spherical and typically between 0.01 and 0.5 cm in diameter.

Emulsion polymerization differs from suspension polymerization in that the initiator is insoluble in the monomer and soluble in water. It involves the dispersion of a water-insoluble monomer into water that contains a water-soluble initiator and emulsifying agent. The bulk of the monomer is present as dispersed monomer droplets stabilized by the emulsifier and initiator. Emulsion polymerization is affected by the micelles that serve as the meeting place. It also does not suffer from the heat dissipation problem, and the particles produced by this method are 0.1 μm in size.

Both suspension and emulsion polymerization suffer from the disadvantage of the interaction of the polymer with the monomer and active drug.

20.2.2.3 Interfacial Polymerization

In interfacial polymerization, polymer film formation proceeds via a reaction between two monomers at the interface of two immiscible liquid phases. The formed polymer film covers the dispersed phase that affects the microsphere formation. One monomer is dissolved in the continuous phase while the other is dispersed. The second monomer is emulsified into the continuous phase, which is commonly water. The monomer diffuses out at the interface and polymerizes. Interfacial polymerization is essentially applied to a very rapid reaction such as the reaction between acid chloride and amide. A monolithic matrix (microsphere) or reservoir (microsphere) type carrier formation occurs, which depends on the solubility of the formed polymer in the monomer droplet.

If the polymer is soluble in the droplet, it will lead to formation of a matrix-type carrier and vice versa. The degree of polymerization can be controlled by various factors such as the monomer reactivity, monomer concentration, vesicle composition, and temperature of the system. By controlling the size of the dispersed phase, microspheres of the desired size can be achieved.

This method suffers from many disadvantages that are due to the process:

1. Low drug entrapment due to higher film permeability
2. Higher drug degradation
3. High unreacted monomer residue that is associated with toxicity

20.2.3 Phase Separation Coacervation Technique

Basically this technique was developed for a microreservoir-type carrier system (i.e., microcapsules instead of microspheres) if the drug is hydrophilic. However, the same technique is used for a matrix-type carrier system if the drug is hydrophobic in nature. Microsphere or microcapsule formation involves coacervate formation. Coacervation is affected by decreasing the solubility of the polymer in an organic solvent. Various methods are used to decrease the solubility of polymers. The selection of the method is based on the polymers and processing conditions. The methods are salt addition, nonsolvent addition, incompatible polymer addition, and change in pH (Figure 20.3).

For the preparation of microspheres, a hydrophobic polymer is dissolved in a polymer solution and the final mixture is subjected to coacervate formation. Microparticles in the desired range can be achieved by continuous stirring.

Aspirin-impregnated microspheres of CS/poly(acrylic acid) copolymer were prepared by the coacervation phase separation method, which was induced by the addition of a nonsolvent (NaOH

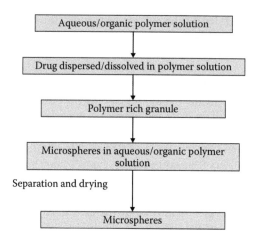

FIGURE 20.3 Schematic representation of microsphere preparation by the phase separation coacervation method.

2.0 M solution) in order to evaluate the release characteristics as a function of the pH, simulating the fluids in the gastrointestinal tract (GIT). The microspheres were cross-linked with GA, reduced with sodium cyanoborohydride, and grafted with poly(acrylic acid). The impregnation of aspirin into CS/poly(acrylic acid) copolymer microspheres was achieved by the dissolution of the drug in water/ethanol (2:1), which was adsorbed by the microspheres for 24 h at 25°C. The efficiency of aspirin impregnation was high (~94%). The approach employed in the production of aspirin-impregnated microspheres using CS/poly(acrylic acid) can be a suitable drug-release control system (Nascimento et al., 2001). A coacervation phase separation technique was used to encapsulate nitrofurantoin with Eudragit RS 100 polymer using variable proportions of polyisobutylene (0–3%) as a protective colloid. The *in vitro* release experiments were carried out over the entire pH range of the GIT, the data obtained from the dissolution profiles were compared in the light of different kinetic models, and the regression coefficients were compared. The *in vivo* studies were performed on human female volunteers. A linear correlation was obtained from *in vitro* and *in vivo* studies (Bedi et al., 1999).

20.2.4 Spray Drying and Spray Congealing

Spray drying and spray congealing are similar processes but differ in the solidification of the polymeric material. Spray drying involves the dispersion of a drug into the polymer solution. A solvent is chosen in which the polymer is soluble but the drug is insoluble. The drug is dispersed with the help of a high-speed homogenizer, and then the dispersion is atomized in hot air. The solvent evaporates immediately from the atomized droplet, which leads to microparticle formation in a size range of 1–100 μm. Cyclone separators are used to separate the microparticle from the hot air.

The spray congealing method employs the dispersion of a drug into a melt instead of a polymer solution. Microsphere formation is achieved by spraying the hot dispersion into a cool air stream. Some materials are solid at room temperature but meltable at reasonable temperatures, such as waxes, alcohol, and polymers, which are used in the spray congealing technique. The spray congealing method was utilized to encapsulate thiamine mononitrate using a mixture of mono- and diglycerol of stearic acid.

Ketoprofen microspheres based on a Eudragit (R) blend were prepared by the spray drying method and evaluated for the influence of the spray drying parameters on the morphology, dimensions, and physical stability of the microspheres. The parameters did not influence the particle properties significantly (Rassu et al., 2008). Rifampin microspheres were prepared by spray drying using either PLA or PLGA polymers in different drug/polymer ratios [90:10–5:95 (w/w)]. The *in vitro* release characteristics, particle size distribution, and cytotoxicity (in an alveolar macrophage cell line) and pharmacokinetics in rats after pulmonary instillation were evaluated (Coowanitwong et al., 2008). PLGA particles showed a steeper change with increasing polymer content (100–20% drug release over 6 h) than PLA particles (88–42% drug release over 6 h). Microspheres with high polymer content showed a lower relative bioavailability (30%).

The potential of waxes for controlling the *in vitro* release of Verapamil HCl was investigated in the preparation of microparticles with the ultrasonic spray congealing technique. Microcrystalline wax, stearyl alcohol, and mixtures of the two were used. A surfactant (soya lecithin) was also added to the formulations. The drug release mechanism from these devices was evaluated using statistical moment analysis. Microparticles with a spherical shape showed good EE and zero-order release for 8 h, without modifying the solid-state properties of the drug. Therefore, waxy microparticles prepared by the ultrasonic spray congealing technique are promising solvent-free devices for controlling the release of Verapamil HCl (Passerini et al., 2003).

20.2.5 Ionic Gelation Method

Microspheres made of gel-type polymers, such as alginates, are produced by dissolving the polymer in an aqueous solution, suspending the active ingredient in the mixture, and extruding through a precision device, producing microdroplets that fall into a hardening bath that is slowly stirred. The hardening bath usually contains a calcium chloride solution, whereby the divalent calcium ions cross-link the polymer and form gelled microspheres. The surface of the microspheres can be further modified by coating them with polycationic polymers, like polylysine or CS after fabrication. The advantages of this process are that it is simple and mild and reversible physical cross-linking by an electrostatic interaction instead of chemical cross-linking is applied to avoid the possible toxicity of the reagents and other undesirable effects.

20.2.6 Solvent Extraction or Solvent Evaporation Method

Solvent extraction methods involve removal of organic solvents by extraction of water-miscible organic solvents such as isopropanol, resulting in hardening of the microspheres. Figure 20.4 summarizes the solvent extraction methods carried out in external aqueous media. This technique is also termed as "in water drying." This technique is basically used for the encapsulation of water-insoluble drugs, and it was reviewed in detail with a polylactide polymer (Jalil and Nixon, 1989,

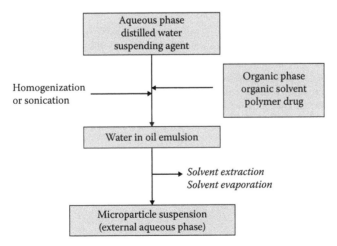

FIGURE 20.4 Schematic representation of microsphere preparation by the solvent extraction or solvent evaporation method.

1990). A solution of a wall-forming polymer was prepared in a water-immiscible organic solvent into which the drug was directly dissolved or with the help of a cosolvent or dispersed.

The drug–polymer solution was poured in a controlled fashion into an aqueous solution containing an emulsifying agent under strong agitation. The emulsifying agent prevents the coalescence of globules. This method has been employed for encapsulation of a few water-insoluble peptides, such as salmon calcitonin (Mehta et al., 1994) and cyclosporine. Water-soluble peptides are not suitable candidates for this technique because, during the emulsification of the drug–polymer solution into the external aqueous phase, the water-soluble drug comes out into the aqueous phase, resulting in no drug entrapment in the microsphere.

20.3 DRUG LOADING IN MICROPARTICULATE SYSTEM

The active components are loaded over the microspheres principally using two methods.

- *In situ* loading (i.e., during preparation of the microspheres)
- Loading in the preformed microspheres

The active component can be loaded by means of physical entrapment, chemical linkage, and surface adsorption; but maximum loading is observed during preparation of microspheres. It may be affected by process variables such as the method of preparation, presence of additives (e.g., cross-linking agent and surfactant stabilizers), and agitation intensity. The percentage of incorporation is relatively less in preformed microspheres, but the major advantage is the absence of the effects of the process variables.

20.4 CHARACTERIZATION OF MICROPARTICULATE SYSTEM

Microspheres are characterized by the particle size and shape, electron spectroscopy for chemical analysis, density determination, isoelectric point, surface carboxylic acid residues, surface amino acid residues, EE, release studies, and angle of contact (Table 20.2).

Microspheres have different microstructures, depending on their method of preparation and the conditions during preparation. These microstructures determine the release property and stability of the carrier (Malmsten, 2002).

20.4.1 Particle Size and Shape

Microspheres obtained from the emulsion technique were studied microscopically for their size and size distribution using a calibrated ocular eyepiece. The effects of the drug concentration, citric acid concentration, and permeation enhancer on the average particle size were studied (Budhian et al., 2007).

The shape and surface morphology of the microspheres is determined by light microscopy, SEM, confocal laser scanning microscopy, and transmission electron microscopy (TEM). The particle size distribution is measured by laser light scattering and a multisize Coulter counter.

20.4.2 Electron Spectroscopy for Chemical Analysis

The chemical analysis of the microspheres is determined with the help of electron spectroscopy to determine the degradation of the surface.

20.4.3 Density Determination

The density of the microspheres is determined by a pycnometer or hydrometer.

TABLE 20.2
Characterization of Microparticulate Systems

Characterization Parameters	Analytical Methods/Instrumentation
Chemical Characterization	
Drug concentration	Different for different drugs
Surface degradation of biodegradable microsphere	Electron spectroscopy for chemical analysis, attenuated total reflectance Fourier transform infrared spectroscopy
Surface carboxylic acid and amino acid residue	Liquid scintillation counter
PH	pH meter
Osmolarity	Osmometer
Physical Characterization	
Microsphere shape and surface morphology	TEM, freeze–fracture electron microscopy
Microsphere size and size distribution	TEM, freeze–fracture electron microscopy, laser light scattering, gel permeation chromatography, gel exclusion
Surface charge	Free-flow electrophoresis
Electrical surface potential and surface pH	Zetasizer and pH sensitive probe
Density determination	Multivolume pycnometer
Isoelectric point	Microelectrophoresis
Capture efficiency	Minicolumn centrifugation, gel exclusion, protamine aggregation, radiolabeling
Drug release	Diffusion cell or dialysis bag
Biological Characterization	
Sterility	Aerobic or anaerobic cultures
Pyrogenicity	Limulus amebocyte lysate test
Animal toxicity	Monitoring survival test, histology, and pathology

20.4.4 ISOELECTRIC POINT

The isoelectric point of the different categories of microspheres is determined by microelectrophoresis.

20.4.5 SURFACE CARBOXYLIC ACID RESIDUES

The surface carboxylic acid residues are measured by a scintillation counter using radioactive glycine.

20.4.6 SURFACE AMINO ACID RESIDUES

The surface amino acid residues are measured with a scintillation counter using a radioactive ^{14}C-acetic acid conjugate.

20.4.7 ENTRAPMENT EFFICIENCY

The total drug present in the microspheres was determined by a method reported by Thanoo et al. (1992). The effects of the drug concentration, citric acid concentration, and permeation enhancer on drug loading were studied.

The entrapment efficiency can be determined following the lysation of washed microspheres. The lysate is then subjected to a determination of the constituents as a monograph requirement. Drug-loaded microspheres are sonicated for 1 h at different pH values to lyse the particles. The extent of drug loading is determined by measuring the absorbance of the solution after filtration and calculation of the concentration of the drug from the calibration curve:

$$\text{Entrapment} = \frac{\text{Actual content}}{\text{Theoretical content}}. \tag{20.1}$$

20.4.8 RELEASE STUDIES

Release studies of microspheres were carried out in phosphate buffer at different pH values, using a rotating paddle apparatus or a dialysis method. The sample is agitated at 50 rpm in the paddle apparatus. Sample are taken at specific time intervals and replaced by the same amount of saline. In the dialysis methods the microspheres are kept in bags. Samples of dialysate are taken at regular time intervals and the same amount is replaced with buffer (Seo et al., 2003).

20.5 DRUG RELEASE KINETICS

The release profile from the microspheres is a cumulative effect of the nature of the polymer used in the preparation as well as the nature of the drug. The kind of drug, its polymorphic form, crystallinity, particle size, solubility, and amount in the pharmaceutical dosage form can influence the release kinetics. Thus, a water-soluble drug incorporated in the matrix is mainly released by diffusion, whereas the self-erosion of the matrix for a low water-soluble drug is the principle release mechanism. The release of a drug from both biodegradable and nonbiodegradable microspheres is influenced by the structure or micromorphology of the carrier. Numerous theoretically possible mechanisms may be considered for the release of drugs from microparticulates:

- Liberation due to polymer erosion or degradation
- Self-diffusion through the pores
- Release from the surface of the polymer
- Pulse delivery initiated by the application of an oscillating or sonic field

Several theories and kinetic models have been proposed to describe the drug release kinetics from modified release dosage forms (Costa and Lobo, 2001). These represent the drug release profiles as a function of time that is related to the amount of drug dissolved from the pharmaceutical dosage system.

20.5.1 MATHEMATICAL MODELS

20.5.1.1 Zero-Order Kinetics

Drug dissolution from a pharmaceutical dosage form that does not disaggregate and release the drug slowly (assuming that the area does not change and no equilibrium conditions are obtained) can be represented by the following equation:

$$f_t = K_0 t = 1 - \frac{W_t}{W_0}, \tag{20.2}$$

where f_t represents the fraction of the drug dissolved in time t, W_0 is the initial amount of drug in the dosage form, W_t is the amount of drug in the dosage form at time t, and K_0 is the apparent dissolution rate constant or the zero-order release rate constant. In this way, a graph of the drug dissolved fraction versus time will be linear if the previously established conditions were fulfilled. This model is an ideal method of drug release in order to achieve a pharmacologically prolonged action.

20.5.1.2 First-Order Kinetics

The equation of the model can be given as follows:

$$\log Q_t = \log Q_0 + \frac{K_1 t}{2.303}, \tag{20.3}$$

where Q_t is the amount of drug released in time t, Q_0 is the initial amount of drug in solution, and K_1 is the first-order release rate constant. In this way, the graph of the decimal logarithm of the released amount of the drug versus time will be linear. The dosage forms following this release kinetics pattern, such as those containing a water-soluble drug in porous matrices, release the drug in a way that is proportional to the amount of drug remaining in the interior, so that the amount of drug released by a unit of time is diminished.

20.5.1.3 Higuchi Model

This model is applicable to the system where water-soluble drugs are dispersing matrices, for example, transdermal systems and matrix tablets, with water-soluble drugs. Initially, the model was proposed to describe the desolation of the drug in suspension from ointment bases. Drug release is described as a diffusion process based on Fick's law of the square root of the time dependence:

$$\text{Drug released (\%)} = K_H t^{1/2}, \tag{20.4}$$

where K_H is Higuchi's dissolution constant. Thus, a plot of the percentage of drug release versus the square root of the time will follow a linear relationship.

20.6 APPLICATION OF MICROSPHERES

Microspheres have a number of applications that are detailed later in this chapter (Table 20.3). Various microspheres employed for the delivery of drugs include semisolids, viscous liquids, and solids and inserts. Acyclovir-loaded CS microparticles showed increased drug bioavailability in the

TABLE 20.3
Applications of Microspheres for the Different Fates

Drug	Polymer	Application	References
Methyl prednisolone	Hyaluronic acid	Ocular	Kyyronen et al., 1992
Acyclovir	CS	Ocular	Genta et al., 1997
Insulin	Starch	Nasal	Farraj et al., 1990
Gentamicin	Starch	Nasal	Farraj et al., 1990
Human growth hormone	Starch	Nasal	Illum et al., 1999
Riboflavin	Adhesive micromatrix system	GI	Akiyma et al., 1998
Furosemide	Adhesive micromatrix system	GI	Akiyama and Nagahara, 1999
Amoxicillin	Adhesive micromatrix system	GI	Akiyama et al., 1994
Vancomycin	Eudragit S 100	Colon	Geary and Schlameus, 1993
Insulin	Eudragit S 100	Colon	Illum, 1999
Nerve growth hormone	HYAFF	Vaginal	Ghezzo et al., 1992
Insulin	HYAFF	Vaginal	Illum, 1999
Salmon calcitonin	HYAFF	Vaginal	Richardson and Armstrong, 1999
Indomethacin	Alginate/sodium CMC/HPMC	Oral	Mathiowitz et al., 1997
Glipizide	Alginate/sodium CMC/HPMC	Oral	Mathiowitz et al., 1997

Note: HYAFF, a new family of biomaterials obtained by esterification of hyaluronic acid with different alcohols.

eye compared to the drug administered alone. Biocompatibles have been developed in a range of products based on microspheres, targeting areas such as the treatment of cancer. One such product, Bead Block™ (Terumo Medical Corporation, Somerset, NJ), is already on the market as an embolization therapy for solid tumors. This is injected into selected vessels to block the blood flow feeding the cancer, causing it to shrink over time.

The company is moving ahead to develop microspheres that not only block blood vessels supplying tumors but also deliver a payload of chemotherapeutic drugs. A pivotal trial of biocompatible drug-eluting microspheres to deliver the drug doxorubicin was started in 2003.

20.6.1 Nasal Drug Delivery

The nasal cavity is a highly vascularized, large subepithelial layer for efficient absorption of drugs. The blood is drained directly from the nose into the systemic circulation, so it avoids the first pass effect. The nasal route is used both for local therapies and, more recently, for the systemic administration of drugs, as well as for the delivery of peptides and vaccines. In lieu of the above advantages, nasal delivery of drugs has certain limitations such as the mucociliary clearance of therapeutic agents from the site of deposition, resulting in a short residence time for absorption. Thus, the microspheres increase the residence time of formulations in the nasal cavity, thereby improving the absorption of drugs. The excellent absorption-enhancing properties of microspheres are now being used extensively for both low molecular weight drugs as well as macromolecular drugs like proteins. The nasal cavity as a site for systemic drug delivery has been investigated extensively; many nasal formulations have already reached commercial status including leutinizing hormone releasing hormone and calcitonin (Illum, 1999; Illum et al., 1987). CS and starch are the two most widely employed polymers for nasal drug delivery. Other microspheres used for nasal administration of peptides and proteins include cross-linked dextran microspheres, which are water insoluble and water absorbable.

Isoniazid–alginate CS was prepared by the complex coacervation method in an emulsion system. Because the encapsulation of isoniazid tends to be limited by its hydrophilic characteristics, this study proposes its microencapsulation by adsorption. The particles were prepared in three steps: (1) preparation of a W/O emulsion, (2) phase separation, and (3) adsorption of the drug. The adsorption observed is probably of a chemical nature; that is, there is an ionic interaction between the drug and the surface of the particles (Lucinda-Silva and Evangelista, 2003).

A microparticulate delivery system based on a thiolated CS conjugate for the nasal application of peptides was developed. All of the microparticulate systems were prepared via the emulsification solvent evaporation technique (Krauland et al., 2006).

CS hydrochloride or CS glutamate microspheres with the model drug carbamazepine for nasal administration were produced using a spray drying technique and characterized in terms of morphology (SEM), drug content, particle size (laser diffraction method), and thermal behavior (differential scanning calorimetry). *In vitro* drug release studies were performed in phosphate buffer (pH 7.0). *In vivo* tests were carried out in sheep using microparticles containing CS glutamate, which were chosen on the basis of the results of *in vitro* studies. The results were compared to those obtained after the nasal administration of carbamazepine (raw material) alone (Gavini et al., 2006).

20.6.2 Ocular Drug Delivery

There are stringent requirements that the drug delivery for the ocular route should be sterile, isotonic, and nonirritant. There are no available marketable sterile ophthalmic products based on these systems. Biodegradable polymers are the polymers of choice for retinal drug delivery. Lactic acid and glycolic acids are biodegradable, and they are produced and eliminated by the body. These polymers decompose into carbon dioxide and water. They are commercially available from either a synthetic or natural origin. PLA and poly(glycolic acid) (PLG) can both be polymerized with good control of the molecular weight and molecular weight distribution. In addition, a composite collagen hydrogel containing protein-encapsulated alginate microspheres was developed for ocular applications using bovine serum albumin. Sustained release of bovine serum albumin was achieved during an 11-day period in neutral phosphate buffer. The composite hydrogel supported human corneal epithelial cell growth and had adequate mechanical strength and excellent optical clarity for possible use as a therapeutic lens for drug delivery or use as a corneal substitute for transplantation into patients who have corneal diseases (Liu et al., 2008). The *in vitro* mucoadhesion of microparticles was tested on a mucous layer under shear stress mimicking the human blink. The resulting microparticles were also formulated in two dosage forms (an aqueous suspension and dry tablet) to test the effect of the formulation on the retention capacity of microparticles on the preocular space of rabbits *in vivo*. The results suggested that the mucoadhesive microdiscs adhered better to the simulated ocular surface than the other types of microparticles. When a dry tablet embedded with mucoadhesive microdiscs was administered in the cul-de-sac of the rabbit eye *in vivo*, these microdiscs exhibited longer retention than the other formulations tested in this work. More than 40% and 17% of mucoadhesive microdiscs remained on the preocular surface at 10 and 30 min after administration, respectively. Fluorescence images from the eye surface showed that mucoadhesive microdiscs remained for at least 1 h in the lower fornix (Choy et al., 2008).

20.6.3 VAGINAL DRUG DELIVERY

The vaginal route is extensively used for the delivery of therapeutic and contraceptive agents to exert a local effect (e.g., antifungal, spermicidal) and for the systemic delivery of drugs (Richardson and Armstrong, 1999). This method of drug delivery is important because it does not have the shortcomings of the oral drug delivery system (DDS) in a few respects. This route has also been explored for the delivery of therapeutic peptides, for example, calcitonin, and microbicidal agents to help prevent the transmission of human immunodeficiency virus and other sexually transmitted diseases.

Absorption of peptides from the vagina can be increased by using absorption enhancers, for example, surfactants and bile salts. The adverse effects of absorption enhancers on the mucosal integrity can be bypassed by employing microspheres within the vaginal cavity.

Vaginal tablets were designed for the local controlled release of acriflavine. The tablets were prepared using drug-loaded CS microspheres and additional excipients such as methylcellulose, sodium alginate, sodium carboxymethyl cellulose (CMC), or Carbopol 974. The microspheres were prepared by a spray drying method, using 1:1 and 1:2 drug/polymer weight ratios. They were characterized in terms of the morphology, EE, and *in vitro* release behavior as well as the minimum inhibitory concentration, minimum bacterial concentration, and killing time (Gavini et al., 2002).

20.6.4 ORAL DRUG DELIVERY

20.6.4.1 Buccal

The oral cavity, in addition to being highly accessible, has a highly permeable mucosa with a rich blood supply that shows short recovery times after stress or damage. Furthermore, oral transmucosal drug delivery bypasses the first pass effect and avoids presystemic elimination in the GIT. These factors make the oromucosal cavity a very attractive and feasible site for systemic drug delivery (Harris and Robinson, 1992).

The composition of the oral epithelium varies according to the site in the oral cavity. The areas exposed to mechanical stress (the gingivae and hard palate) are keratinized similar to the epidermis. The mucosae of the soft palate and the sublingual and buccal regions, however, are not keratinized. The keratinized epithelia contain neutral lipids like ceramides and acylceramides, which have been associated with the barrier function. It is estimated that the permeability of the buccal mucosa is 4–4000 times greater than that of the skin. In general, the permeabilities of the oral mucosa decrease in the order of sublingual is greater than buccal and buccal is greater than palatal. The daily salivary volume secreted in humans is between 0.5 and 2 L, which is sufficient to hydrate oral mucosal dosage forms. This water-rich environment of the oral cavity is the main reason behind the selection of hydrophilic polymeric matrices as vehicles for oral transmucosal DDSs.

CS microparticles for the buccal cavity were prepared by a spray drying technique using the drug chlorhexidine. Their morphological characteristics were studied by SEM, and the *in vitro* release behavior was investigated in pH 7.0 United States Pharmacopoea buffer. The antimicrobial activity of the microparticles was investigated as the minimum inhibitory concentration, minimum bacterial concentration, and killing time (Giunchedi et al., 2002).

Lipid microspheres (LMs) have been recently been used as intravenous carriers for drugs, which are sufficiently soluble in oil. LMs were prepared by transferring the drug to the interfacial surface of the oil and aqueous phases to produce a less irritating intravenous formulation. A probe-type sonicator was used to disperse the drug into the oil phase together with lecithin and Tween 80. A high-pressure homogenization process was used to prepare the LMs and localize the drug at the surfactant layer. The LM loaded with drug consisted of 0.02% drug (Lixin et al., 2006).

20.6.4.2 Gastrointestinal

The development of peroral controlled release DDSs has been hindered by the inability to restrain and localize the DDSs in selected regions of the GIT. Microspheres form an important approach to decrease the gastrointestinal (GI) transit of drugs. Drug properties especially amenable to formulations include a relatively short biological half-life of about 2–8 h, a specific window for the absorption of the drug by an active, saturable absorption process, and small absorption rate constants (Longer et al., 1985). The GI epithelium consists of a single layer of simple, columnar epithelium lying above a collection of cells called the lamina propria and supported by a layer of smooth muscle known as the muscularis mucosae. The cells are held together by tight junctions or the zona occludens. A special type of GI epithelium (PP) of the gut-associated lymphoid tissue is also present. The PP is lined by a specialized epithelium, the follicle-associated epithelium, containing M cells, which have the ability to phagocytize antigens in the intestine. Polymeric microspheres can also be phagocytized by these M cells and hence can be used for vaccination purposes (Carino et al., 1999).

20.6.4.3 Colon

Colon drug delivery has been used for molecules aimed at local treatment of colonic diseases and for delivery of molecules susceptible to enzymatic degradation such as peptides. The mucosal surface of the colon resembles that of the small intestine at birth, but it changes with age, causing the loss of villi and leaving a flat mucosa with deep crypt cells. Therefore, the absorptive capacity of the colon is much less as compared to the small intestine. The mucus layer not only provides a stable pH environment but also acts as a diffusion barrier for the absorption of drugs. There is more mucus production in the elderly because the number of mucous secreting goblet cells increases with age. The colonic mucosal environment is also affected by the colonic microflora as they degrade the mucins. Microspheres can be used during the early stages of colonic cancer (when systemic prevention of possible metastasis in the blood is still not necessary) for enhancing the absorption of peptide drugs and vaccines; for the localized action of steroids and drugs with a high hepatic clearance, for example, budesonide; and for immunosuppressive agents such as cyclosporine. Colon-specific microspheres can be used for the protection of peptide drugs from the enzyme-rich part of the GIT and to release the biologically active drug at the desired site for its maximum absorption.

Calcium pectinate microspheres were prepared by a modified emulsification method using calcium chloride as a cross-linker and the drug methotrexate. It was concluded that calcium pectinate microspheres can be used to effectively localize the release of the drug in the physiological environment of the colon (Chaurasia et al., 2008).

Colon-specific microspheres of 5-fluorouracil for the treatment of colon cancer were prepared and evaluated. Core microspheres of alginate were prepared by the modified emulsification method in liquid paraffin and by cross-linking with calcium chloride (Rahman et al., 2006).

Aminated gelatin microspheres were prepared. The *in vitro* release of the microspheres was evaluated using fluorescein-labeled insulin (rhodamine isothiocyanate–insulin) and fluorescein isothiocyanate–dextran with a molecular weight of 4.4 kDa as a model drug (Wang et al., 2006).

20.6.5 MISCELLANEOUS APPLICATIONS

20.6.5.1 Vesicular Delivery

The mucosal layers in the urinary bladder are different from both the small and large intestine with regards to their structure and thickness. The vesical mucus contains oligosaccharides–glycosaminoglycans that carry a large number of sulfate groups and thus a high negative charge density. Despite these differences, there are certain similarities between the mucus layers in the urinary bladder and intestine because they both contain sugar chains that are completely or partly attached to proteins (Bogataj et al., 1999). Therefore, it is expected that polymers, which show good mucoadhesive strength on the intestinal mucosa, will exhibit some mucoadhesiveness on the vesical mucosa. Various polymers were evaluated for mucoadhesion strength, swelling, and drug release from microspheres applied to the urinary bladder. It was reported that the microspheres containing CMC as a mucoadhesive agent and Eudragit RL as a matrix polymer provided the longest release time from microspheres and showed high mucoadhesion strength.

Rifampicin–PLA microspheres were prepared for lung targeting by a modified emulsion solvent diffusion method. The drug content, particle size distribution, and *in vitro* release properties of the prepared microspheres were evaluated. *In vivo* experiments on rabbits showed remarkable accumulation of microspheres in the lung (Zhang et al., 2000).

PLG microparticles were prepared with the drugs isoniazid and rifampicin in order to improve the compliance of tuberculous chemotherapy. Antitubercular drugs encapsulated in PLG polymers and injected subcutaneously resulted in a sustained release (up to 6 weeks) of drugs in various organs of mice. Further, *Mycobacterium tuberculosis* H37Rv-infected animals given a single dose of PLG microparticles exhibited a better or equivalent clearance of colony forming units in various organs compared to those given a daily administration of free drugs (Dutt and Khuller, 2000).

20.6.5.2 Taste Masking

There are number of drugs with an unpleasant and bitter taste, for example, cefuroxim, quinine, chloramphenicol, roxithromycin, and famotidine. The bitter and unpleasant taste of many effective drugs interrupts the development of oral formulations and their clinical application. Along with the continuous betterment of standards of living, medicines with a bitter or unpleasant taste are no longer acceptable. It is challenging to mask the unpleasantness of drugs that enhances the acceptability of the product, especially in pediatric and geriatric patients.

Various methods have been described for taste masking (Shou and Chen, 2002; Zhang and Hou, 2003). The simplest one is the addition of flavors and sweeteners to mask the distaste sensation or avoid the direct contact of the bitter drug with the taste buds. This can be achieved by coating or granulation (Gao et al., 2006; Hiroyuki et al., 2003; Ishikawa et al., 1999; Kayumba et al., 2007; Lieberman et al., 1989). However, in many cases flavors and sweeteners are overcome by the bitter taste, so this method is not effective enough to mask the taste. Other approaches are capsule dosage forms, complex formations with cyclodextrin (Duchene et al., 1999; Loftsson and Masson 2001), coating with a water-insoluble, pH-dependent water-soluble system (Yajima et al., 1996), or chemical modification. However, these methods require attention because the bioavailability must not change.

During the direct compression of chewing, the coating or granulation of the drug may be destroyed. A new microsphere and microencapsulation approach has been developed to mask the taste. It avoids the direct contact of the drugs with the taste buds (Gouin, 2004; Hashimoto et al., 2002; Nii and Ishil, 2005; Robson et al., 1999, 2000a, 2000b; Sjoqvist et al., 1993). Numerous studies have reported that microparticles remain as such during compression (Raghavendra et al., 2008; Soppimath et al., 2001; Sveinsson et al., 1993).

In the last few decades the popularity of the microsphere and microencapsulation method for taste masking has been increased because of its dual advantages of taste masking and enhanced stability of particles. Gao et al. (2006) developed roxithromycin polymeric microspheres for taste masking. Stearic-acid-coated cefuroxime axetil microspheres were developed for taste masking using the spray drying method. Stearic-acid-coated cefuroxime axetil was formulated as a suspension (Zinnat™; Gooch et al., 1993; Powell et al., 1991; Shalit et al., 1994).

Pediatric, geriatric, and bedridden patients may face difficulty in swallowing tablets or capsules. Orally disintegrating tablets (ODTs) are a good candidate for such patients. Taste masking may not be achieved by ODTs. A microspheric ODT can be used to overcome the problem. Famotidine, an H_2 receptor antagonist, is used to treat peptic ulcers. ODTs of Famotidine are already available on the market (Pepcid RPD®). A Famotidine microspheric ODT for taste masking was prepared by the spray drying method. Eudragid® polymer was investigated in such a preparation because it dissolves at pH 1–3 in the stomach and remains intact in the buccal cavity at pH 5.8–7.4 with good taste masking (Jianchen et al., 2008).

20.6.5.3 Mucosal Immunization

The majority of pathogens initially infect their hosts through mucosal surfaces. Induction of mucosal immunity is therefore likely to make an important contribution to protective immunity. Moreover, mucosal administration of vaccines avoids the use of needles and is thus an attractive approach for development of new generation vaccines. Current research in vaccine development has focused on treatments requiring a single administration, because the major disadvantage of many currently available vaccines is that repeated administrations are required. The ability to provide controlled release of antigens through microspheres has provided the impetus for research in the area of mucosal immunization. Several studies have indicated that active monocytes such as macrophages and T cells play an important role in the pathogenesis of chronic human inflammatory bowel disease, although the etiology remains unclear. Manipulation of these cells appears essential for the treatment of patients with inflammatory bowel disease. Considerable attention has been paid recently to the use of polymer microspheres for the sustained release of various drugs and the targeting of therapeutic agents to their site of action. It was reported that biodegradable poly-D,L-lactic acid microspheres can be efficiently taken up by macrophages and M cells. The effect of a new DDS targeting M cells and macrophages with poly-D,L-lactic acid microspheres and gelatin microspheres was evaluated on colitis models. In the first experiment colitis was induced in mice by 5% dextran sodium sulfate, and microspheres containing dexamethasone (Decadrone, DX; DX microspheres) were orally administered to these mice. Serum levels of DX did not reach a detectable level after administration of DX microspheres.

20.6.5.4 Protein and Peptide Drug Delivery

Protein and peptide drugs offer formidable challenges for peroral delivery because of their relatively large size, enzymatic degradation, and very low permeability across the absorptive epithelial cells. Microspheres provide an interesting noninvasive patient compliant approach to improve the absorption of these drugs. The luminal enzymatic degradation of proteins and peptides can be effectively minimized by direct contact with the absorptive mucosa and avoiding exposition to body fluids and enzymes. Specific enzyme inhibitors can be attached to the surface of microspheres (Bernkop-Schnurch and Dundalek, 1996).

20.6.5.5 Microspheres for Parenteral Delivery

Oral administration of drugs is a widely accepted route of drug delivery. However, there are certain disadvantages and the bioavailability of the drug often varies as a result of GI absorption, the first pass effect, and the hostile environment of the GIT. Long-term parenteral drug delivery can be achieved with the use of biodegradable microspheres, which is a better means of controlling the release of drugs over a long time. Parenteral controlled release formulations using biodegradable microspheres can overcome the problems that arise from the highly unstable liposomes and nanoparticles and can control the release of drugs over a predetermined time span, usually on the order of days to weeks to months. Lupron Depot®, Nutropin Depot®, and Zoladex® are the U.S. Food and Drug Administration approved parenteral products, which are biodegradable microspheres (Sinha et al., 2003).

20.6.5.6 Microspheres for the Bone Marrow Targeting

Certain microparticles have been examined for site-specific imaging of bone marrow such as radio-labeled albumin microspheres and microaggregates (10–30 μm) and fine (1–13 μm) PVP particles. The body distribution of particles after intravenous injection mainly depends on the surface properties of the particles. After injection of these particles into the bloodstream, certain blood components (opsonins) are rapidly absorbed. This absorption of opsonins leads to rapid phagocytic or endocytic uptake into cells of the reticuloendothelial cells, especially into the liver (60–90%), spleen (2–20%), bone marrow (0.1–1%), and varying amounts into the lungs. Thus, with these systems a high proportion of the dose reaches the liver and spleen within a few minutes after intravenous administration and only a small fraction of these particles reaches the bone marrow. Coating the particulates with certain polymers has made it possible to effectively bypass the liver and spleen uptake and obtain a deposition of the microspheres primarily in the bone marrow. Such carrier systems are used in radiodiagnosis and in the treatment of various diseases of the bone marrow. Certain synthetic substances have been shown to exhibit "bone marrow homing" activity. One such substance is polaxamer 407, a nonionic block copolymer containing a central block of hydrophobic polyoxypropylene flanked by blocks of hydrophilic polyoxyethylene (POE).

Coating the model microspheres with polaxamer 407 effectively hindered liver and spleen uptake, and the microspheres were primarily deposited in the bone marrow. Such a carrier system will have applications in radiodiagnosis and for treatment of various bone marrow diseases (Illum and Davis, 1987).

20.6.5.7 Microspheres for Tumor Targeting

Chemotherapy in cancer treatment is associated with serious side effects. As a result, there is great interest in research aimed at bringing down the level of systemic cytotoxicity. With advances in material science, magnetic drug targeting has emerged as one of the viable ways of attaining this. The important applications of microparticles are their possible use as a carrier for antitumor drugs. These microspheres have endocytic activity that favors accumulation of the administered microparticles. Nowadays, stealth microparticles are commonly used for better extravasations. Stealth microspheres are prepared by coating them with polymers such as POE and dialkyl POE and other ingredients like phospholipids. Some of the antitumor drugs that are reportedly either entrapped or adsorbed onto microparticle surfaces are doxorubicin, methoxantrone, aclacinomycin A, and acyclovir.

The targeting and retention of antitumor agents in the tumor area can be improved by forming a DDS that involves the hemodynamics of the tumor. These include balloon occluded arterial infusion therapy, which is administered with gasoconstrictive agents like noradrenalin or angiotensin II.

Magnetic microspheres are an alternative pathway to deliver anticancer drugs. They are a site-specific DDS that is prepared by using albumin and magnetite. Significant improvement in response can be incorporated and obtained with the magnetic albumin microsphere delivery system compared with conventional DDSs. In the presence of a suitable magnetic field, the microspheres are internalized by the endothelial cells of the target tissue in healthy as well as tumor-bearing animals.

Paclitaxel-loaded polymeric microspheres were prepared, which were surface conjugated with antibodies to vascular endothelial growth factor receptor 2, for systemic targeting to angiogenic sites in prostate tumors. It was concluded that anti-vascular endothelial growth factor receptor 2 microspheres containing paclitaxel may offer an effective way of administering a controlled release formulation of the drug to target prostate tumors (Lu et al., 2008).

PVP–poly(vinyl alcohol) magnetic hydrogel microspheres were designed to deliver pingyang-mycin (Bleomycin A5) to rabbit auricular VX2 tumors in the presence of a 0.5-T permanent magnet both during and 24 h after perfusion. A total of 22 New Zealand white rabbits ranging in age from 13 to 16 weeks and weighing 2.5–3.0 kg (2.46 ± 0.2) were successfully implanted with 200–300 mm^2 tumors in one study. The microspheres in conjunction with the magnet delivered pingyangmycin to the tumor and hence may be of use in the future for magnetic drug targeting (Adriane et al., 2006).

In the fight against cancer, new DDSs are attractive to improve the drug targeting of tumors, maximize drug potency, and minimize systemic toxicity. Hong et al. (2006) studied a new DDS comprising microspheres, with unique properties allowing delivery of large amounts of drugs to tumors for a prolonged time, thereby decreasing plasma levels. Liver tumors, unlike the nontumorous liver, draw most of their blood supply from the hepatic artery. Exploiting this property, drug-eluting microspheres and beads loaded with doxorubicin were delivered intra-arterially in an animal model of liver cancer (Vx-2).

ABBREVIATIONS

bupi-MS	Bupivacaine microspheres
CMC	Carboxymethyl cellulose
CS	Chitosan
DDS	Drug delivery system
DX	Dexamethasone
EE	Encapsulation efficiency
GA	Glutaraldehyde
GI	Gastrointestinal
GIT	Gastrointestinal tract
HPMC	Hydroxypropyl methyl cellulose
LM	Lipid microsphere
M	Microfold
ODT	Orally disintegrating tablet
O/W	Oil-in-water
PLA	Poly(lactic acid)
PLG	Poly(glycolic acid)
PLGA	Poly(lactic-co-glycolic acid)
POE	Polyoxyethylene
PP	Peyer's patches
PVP	Poly(vinyl pyrrolidone)
SEM	Scanning electron microscopy
TEM	Transmission electron microscopy
W/O/W	Water-in-oil-in-water

SYMBOLS

$1°$	Primary
$2°$	Secondary
f_t	Fraction of the drug dissolved in time t
K_1	First-order release rate constant
K_H	Higuchi's dissolution constant
K_0	Apparent dissolution rate constant
Q_0	Initial amount of drug in solution
Q_t	Amount of drug released in time t

W_0 Initial amount of the drug in the dosage form
W_t Amount of the drug in the dosage form at time t

REFERENCES

Adriane, K., Huang, J., Ding, G., Chen, J., and Liu, Y. 2006. Self assembled magnetic PVP/PVA hydrogel microspheres: Magnetic drug targeting of VX2 auricular tumours using pingyangmycin. *J. Drug Target.*, 14: 243–253.

Akiyama, Y., Nagahara, N., Nara, E., Kitano, M., Iwasa, S., Yamamoto, I., Azuma, J., and Ogawa, Y. 1998. Evaluation of oral mucoadhesive microspheres in man on the basis of the pharmacokinetics of furosemide and riboflavin, compounds with limited gastrointestinal absorption sites. *J. Pharm. Pharmacol.*, 50: 159–166.

Akiyama, Y., Yoshioka, M., Horibe, H., Inada, Y., Hirai, S., Kitamori, N., and Toguchi, H. 1994. Antihypertensive effect of oral controlled release microspheres containing an ACE inhibitor (Delapril hydrochloride) in rats. *J. Pharm. Pharmacol.*, 46: 661–665.

Akiyama, Y. and Nagahara, N. 1999. Novel formulation approaches to oral mucoadhesive drug delivery systems. In E. Mathiowitz, D.E. Chickering, and C.M. Lehr (Eds.), *Bioadhesive Drug Delivery Systems—Fundamentals, Novel Approaches and Development* (pp. 477–505). New York, Marcel Dekker.

Bedi, S., Baidya, S., Ghosh, L.K., and Gupta, B.K. 1999. Design and biopharmaceutical evaluation of nitrofurantoin-loaded Eudragit RS100 micropellets. *Drug Dev. Ind. Pharm.*, 25: 937–944.

Bernkop-Schnurch, A.S. and Dundalek, K. 1996. Novel bioadesive drug delivery system protecting (poly) peptides from gastric enzymatic degradation. *Int. J. Pharm.*, 138: 75–83.

Bogataj, M., Mrhar, A., and Korosec, L. 1999. Influence of physicochemical and biological parameters on drug release from microspheres adhered on vesical and intestinal mucosa. *Int. J. Pharm.*, 177: 211–220.

Budhian, A., Siegel, S.J., and Winey, K.I. 2007. Haloperidol-loaded PLGA nanoparticles: Systematic study of particle size and drug content. *Int. J. Pharm.*, 36: 367–375.

Carino, P.G., Jacob, J.S., Chen, C.J., Santos, C.A., Hertzog, B.A., and Mathiowitz, E. 1999. Bioadhesive, bioerodible polymers for increased intestinal uptake. In E. Mathiowitz, D.E. Chickering, and C.M. Lehr (Eds.), *Bioadhesive Drug Delivery Systems—Fundamentals, Novel Approaches and Development* (pp. 459–475). New York, Marcel Dekker.

Chaurasia, M., Chourasia, M.K., Jain, N.K., Jain, A., Soni, V., Gupta, Y., and Jain, S.K. 2008. Methotrexate bearing calcium pectinate microspheres: A platform to achieve colon-specific drug release. *Curr. Drug Deliv.*, 5: 215–219.

Choy, Y.B., Park, J.H., McCarey, B.E., Edelhauser, H.F., and Prausnitz, M.R. 2008. Mucoadhesive microdiscs engineered for ophthalmic drug delivery: Effect of particle geometry and formulation on preocular residence time. *Invest. Ophthalmol. Vis. Sci.*, 29: 34–78.

Coowanitwong, I., Arya, V., Kulvanich, P., and Hochhaus, G. 2008. Slow release formulations of inhaled rifampin. *AAPS J.*, 10: 342–348.

Costa, P. and Lobo, J.M.S. 2001. Modeling and comparison of dissolution. *Eur. J. Pharm.*, 13: 123–133.

Duchene, D., Wouessidjewe, D., and Ponchel, G. 1999. Cyclodextrins and carrier systems. *J. Control. Release*, 62: 263–268.

Dutt, M. and Khuller, G.K. 2000. Poly (DL-lactide-*co*-glycolide) microparticles as carriers for antimycobacterial drug rifampicin. *Indian J. Exp. Biol.*, 38: 887–894.

Farraj, N.F., Johansen, B.R., Davis, S.S., and Illum, L. 1990. Nasal administration of insulin using bioadhesive microspheres as a delivery system. *J. Control. Release*, 13: 253–261.

Gao, Y., Cui, F.D., Guan, Y., Yang, L., Wang, Y. S., and Zhang, L.N. 2006. Preparation of roxithromycin-polymeric microspheres by the emulsion solvent diffusion method for taste masking. *Int. J. Pharm.*, 318: 62–69.

Gaskell, E.E., Hobbs, G., Rostron, C., and Hutcheon, G.A. 2008. Encapsulation and release of alpha-chymotrypsin from poly (glycerol adipate-*co*-omega-pentadecalactone) microparticles. *J. Microencapsul.*, 25: 187–195.

Gavini, E., Hegge, A.B., Rassu, G., Sanna, V., Testa, C., Pirisino, G., Karlsen, J., and Giunchedi, P. 2006. Nasal administration of carbamazepine using chitosan microspheres: *In vitro/in vivo* studies. *Int. J. Pharm.*, 307: 9–15.

Gavini, E., Sanna, V., Juliano, C., Bonferoni, M.C., and Giunchedi, P. 2002. Mucoadhesive vaginal tablets as veterinary delivery system for the controlled release of an antimicrobial drug, acriflavine. *AAPS PharmSciTech*, 3(3): 20–25.

Geary, S. and Schlameus, H.W. 1993. Vancomycin and insulin used as models for oral delivery of peptides. *J. Control. Release*, 23: 65–74.

Genta, I., Conti, B., Perugini, P., Pavanetto, F., Spadaro, A., and Puglisi, G. 1997. Bioadhesive microspheres for ophthalmic administration of Acyclovir. *J. Pharm. Pharmacol.*, 49: 737–742.

Ghezzo, E., Benedetti, L., Rochira, N., Biviano, F., and Callegaro, L. 1992. Hyaluronane derivative microsphere as NGF delivery device: Preparation methods and *in vitro* release characterisation. *Int. J. Pharm.*, 87: 21–29.

Giunchedi, P., Juliano, C., Gavini, E., Cossu, M., and Sorrenti, M. 2002. Formulation and *in vivo* evaluation of chlorhexidine buccal tablets prepared using drug-loaded chitosan microspheres. *Eur. J. Pharm. Biopharm.*, 53: 233–239.

Gooch, W.M., McLinn, S.E., Aronovitz, G.H., Pichichero, M.E., Kumar, A., Kaplan, E.L., and Ossi, M.J. 1993. Efficacy of cefuroxime axetil suspension compared with that of penicillin V suspension in children with group A streptococcal pharyngitis. *Antimicrob. Agents Chemother.*, 37: 159–163.

Gouin, S. 2004. Microencapsulation: Industrial appraisal of existing technologies and trends. *Trends Food Sci. Technol.*, 15: 330–347.

Harris, D. and Robinson, J.R. 1992. Drug delivery via the mucous membranes of the oral cavity. *J. Pharm. Sci.*, 81: 1–10.

Hashimoto, Y., Tanaka, M., and Kishimoto, H. 2002. Preparation, characterization and taste-masking properties of polyvinylacetal diethylaminoacetate microspheres containing trimebutine. *J. Pharm. Pharmacol.*, 54: 1323–1328.

Hiroyuki, S., Hiraku, O., Yuri, T., Masanori, I., and Yoshiharu, M. 2003. Development of oral acetaminophen chewable tablets with inhibited bitter taste. *Int. J. Pharm.*, 251: 123–132.

Hong, K., Khwaja, A., Liapi, E., Torbenson, M.S., Georgiades, C.S., and Geschwind, J.F. 2006. New intra-arterial drug delivery system for the treatment of liver cancer: Preclinical assessment in a rabbit model of liver cancer. *Clin. Cancer Res.*, 12: 2563–2567.

Illum, L. 1999. Bioadhesive formulations for nasal delivery. In E. Mathiowitz, D.E. Chickering, and C.M. Lehr (Eds.), *Bioadhesive Drug Delivery Systems—Fundamentals, Novel Approaches and Development* (pp. 519–539). New York, Marcel Dekker.

Illum, L. and Davis, S.S. 1987. Targeting of colloidal particles to the bone marrow. *Life Sci.*, 40: 1553–1560.

Illum, L., Jorgensen, H., Bisgaard, H., Krogsgaard, O., and Rossing N. 1987. Bioadhesive microspheres as a potential nasal drug delivery system. *Int. J. Pharm.*, 39: 189–199.

Ishikawa, T., Watanabe, Y., Utoguchi, N., and Matsumoto, M. 1999. Preparation and evaluation of tablets rapidly disintegrating in saliva containing bitter-tastemasked granules by the compression method. *Chem. Pharm. Bull.*, 47: 1451–1454.

Jalil, R. and Nixon, J.R. 1989. Microencapsulation using poly(L-lactic acid). I: Microcapsule properties affected by the preparative technique. *J. Microencapsul.*, 6: 473–484.

Jalil, R. and Nixon, J.R. 1990. Microencapsulation using poly (L-lactic acid) II: Preparative variables affecting microcapsule properties. *J. Microencapsul.*, 7: 25–39.

Jianchen, X., Bovet, L.L., and Zhao, K. 2008. Taste masking microspheres for orally disintegrating tablets. *Int. J. Pharm.*, 359: 63–69.

Kayumba, P.C., Huyghebaert, N., Cordella, C., Ntawukuliryayo, J.D., Vervaet, C., and Remon, J.P. 2007. Quinine sulphate pellets for flexible pediatric drug dosing: Formulation, development and evaluation of taste masking efficiency using the electronic tongue. *Eur. J. Pharm. Biopharm.*, 66: 460–465.

Krauland, A.H., Guggi, D., and Bernkop-Schnürch, A. 2006. Thiolated chitosan microparticles: A vehicle for nasal peptide drug delivery. *Int. J. Pharm.*, 307: 270–277.

Kyyronen, K., Hume, L., Benedetti, L., Urtti, A., Topp, E., and Stella, V. 1992. Methylprednisolone esters of hyaluronic acid in ophthalmic drug delivery: *In vitro* and *in vivo* release studies. *Int. J. Pharm.*, 80: 161–169.

Lieberman, A., Lachman, L., and Schwartz, J.B. 1989. *Pharmaceutical Dosage Forms: Tablets* (2nd ed., pp. 543–602). New York, Marcel Dekker.

Liu, W., Griffith, M., and Li, F. 2008. Alginate microsphere–collagen composite hydrogel for ocular drug delivery and implantation. *J. Mater. Sci. Mater. Med.*, 19: 3365–3371.

Lixin, W., Haibing, H., Xing, T., Ruiying, S., and Dawei, C. 2006. A less irritant norcantharidin lipid microsphere: Formulation and drug distribution. *Int. J. Pharm.*, 323: 161–167.

Loftsson, T. and Masson, M. 2001. Cyclodextrins in topical drug formulations: Theory and practice. *Int. J. Pharm.*, 225: 15–30.

Longer, M.A., Ch'ng, H.S., and Robinson, J.R. 1985. Bioadhesive polymers as platforms for oral controlled drug delivery. III. Oral delivery of chlorthiazide using a bioadhesive polymer. *J. Pharm. Sci.*, 74: 406–411.

Lu, J., Jackson, J.K., Gleave, M.E., and Burt, H.M. 2008. The preparation and characterization of anti-VEGFR2 conjugated, paclitaxel-loaded PLLA or PLGA microspheres for the systemic targeting of human prostate tumors. *Cancer Chemother. Pharmacol.*, 61: 997–1005.

Lucinda–Silva, R.M. and Evangelista, R.C. 2003. Microspheres of alginate–chitosan containing isoniazid. *J. Microencapsul.*, 20: 145–152.

Malmsten, M. (Ed.). 2002. Polymer particles. In *Surfactants and Polymers in Drug Delivery* (pp. 271–290). New York, Marcel Dekker.

Manoharan, C. and Singh, J. 2008. Insulin loaded PLGA microspheres: Effect of zinc salts on encapsulation, release, and stability. *J. Pharm. Sci.*, 11: 675–681.

Mathiowitz, E., Jacob, J.S., Jong, Y.S., Carino, G.P., Chickering, D.E., Chaturvedi, P., Santos, C.A., Morrell, C., Bassett, M., and Vijayaraghavan, K. 1997. Biologically erodable microspheres as potential oral delivery systems. *Nature*, 386: 410–414.

Mehta, S.K., Dewan, R.K., and Bala, K. 1994. Percolation phenomenon and the study of conductivity, viscosity, and ultrasonic velocity in microemulsions. *Phys. Rev. E. Stat. Phys. Plasmas Fluids Relat. Interdiscip. Top.*, 50: 4759–4762.

Nascimento, A., Laranjeira, M.C., Fávere, V.T., and Josué, A. 2001. Impregnation and release of aspirin from chitosan/poly(acrylic acid) graft copolymer microspheres. *J. Microencapsul.*, 18: 679–684.

Nii, T. and Ishii, F. 2005. Encapsulation efficiency of water-soluble and insoluble drugs in liposomes prepared by the microencapsulation vesicle method. *Int. J. Pharm.*, 298: 198–205.

Passerini, N., Perissutti, B., Albertini, B., Voinovich, D., Moneghini, M., and Rodriguez, L. 2003. Controlled release of verapamil hydrochloride from waxy microparticles prepared by spray congealing. *J. Control. Release*, 88: 263–275.

Powell, D.A., Nahata, M.C., Powell, N.E., and Ossi, M.J. 1991. The safety, efficacy, and tolerability of cefuroxime axetil suspension in infants and children receiving previous intravenous antibiotic therapy. *Ann. Pharmacother.*, 25: 1236–1238.

Raghavendra, C.M., Namdev, B.Sh., Ajit, P.R., Sangamesh, A.P., and Tejraj, M.A. 2008. Formulation and *in-vitro* evaluation of novel starch-based tableted microspheres for controlled release of ampicillin. *Carbohydr. Polym.*, 71: 42–53.

Rahman, Z., Kohli, K., Khar, R.K., Ali, M., Charoo, N.A., and Shamsher, A.A.A. 2006. Characterization of 5-fluorouracil microspheres for colonic delivery. *AAPS PharmSciTech*, 7(2): 1–7.

Rassu, G., Gavini, E., Spada, G., Giunchedi, P., and Marceddu, S. 2008. Ketoprofen spray-dried microspheres based on Eudragit (R) RS and RL: Study of the manufacturing parameters drug. *Dev. Ind. Pharm.*, 5: 1–10.

Richardson, J.L. and Armstrong, T.I. 1999. Vaginal delivery of calcitonin by hyaluronic acid formulations. In E. Mathiowitz, D.E. Chickering, and C.M. Lehr (Eds.), *Bioadhesive Drug Delivery Systems—Fundamentals, Novel Approaches and Development* (pp. 563–599). New York, Marcel Dekker.

Robson, H., Craig, D.Q.M., and Deutsch, D. 1999. An investigation into the release of cefuroxime axetil from taste-masked stearic acid microspheres. Part 1. The influence of the choice of dissolution media on drug release. *Int. J. Pharm.*, 190: 183–192.

Robson, H., Craig, D.Q.M., and Deutsch, D. 2000a. An investigation into the release of cefuroxime axetil from taste-masked stearic acid microspheres. II. The effects of buffer composition on drug release. *Int. J. Pharm.*, 195: 137–145.

Robson, H., Craig, D.Q.M., and Deutsch, D. 2000b. An investigation into the release of cefuroxime axetil from taste-masked stearic acid microspheres. III. The use of DSC and HSDSC as means of characterising the interaction of the microspheres with buffered media. *Int. J. Pharm.*, 201: 211–219.

Rokhade, A.P., Kulkarni, P.V., Mallikarjuna, N.N., and Aminabhavi, T.M. 2008. Preparation and characterization of novel semi interpenetrating polymer network hydrogel microspheres of chitosan and hydroxypropyl cellulose for controlled release of chlorothiazide. *J. Microencapsul.*, 8: 1–10.

Saravanan, M., Anbu, J., Maharajan, G., and Pillai, K.S. 2008. Targeted delivery of diclofenac sodium via gelatin magnetic microspheres formulated for intra-arterial administration. *J. Drug Target.*, 16: 366–378.

Seo, S.A., Khang, G., Rhee, J.M., Kim, J., and Lee, H.B. 2003. Study on *in vitro* release patterns of fentanyl-loaded PLGA microspheres. *J. Microencapsul.*, 20: 569–579.

Shalit, I., Dagan, R., Engelhard, D., Ephros, M., and Cunningham, K. 1994. Cefuroxime efficacy in pneumonia: Sequential short-course iv: Oral suspension therapy. *Israeli J. Med. Sci.*, 30: 684–689.

Shou, H. and Chen, L. 2002. Study on the methods of masking bitter taste of drugs. *Shangdong Tradit. Med. J.*, 21: 302–304.

Sinha, V.R., Goyal, V., Bhinge, J.R., Mittal, B.R., and Trehan A. 2003. Diagnostic microspheres: An overview. *Crit. Rev. Ther. Drug Carrier Syst.*, 20: 433–460.

Sinha, V.R. and Trehan, A. 2008. Development, characterization, and evaluation of ketorolac tromethamine-loaded biodegradable microspheres as a depot system for parenteral delivery. *Drug Deliv.*, 15: 365–372.

Sjoqvist, R., Graffeer, C., and Ekman, J. 1993. *In vitro* validation of the release rate and palatability of remoxipride-modified release suspension. *Pharm. Res.*, 10: 1020–1030.

Soppimath, K.S., Kulkarni, A.R., and Aminabhavi, T.M. 2001. Encapsulation of antihypertensive drugs in cellulose-based matrix microspheres: Characterization and release kinetics of microspheres and tableted microspheres. *J. Microencapsul.*, 18: 397–409.

Sveinsson, S.J., Kristmundsdóttir, T., and Ingvarsdóttir, K. 1993. The effect of tableting on the release characteristics of naproxen and ibuprofen microcapsules. *Int. J. Pharm.*, 92: 29–34.

Thanoo, B.C., Sunny, M.C., and Jayakrishnan, A. 1992. Cross-linked chitosan microspheres: Preparation and evaluation as a matrix for the controlled release of pharmaceuticals. *J. Pharm. Pharmacol.*, 44: 283–286.

Volkheimer, G. and Schultz, F.H. 1968. The phenomenon of persorption. *Digestion*, 1: 213–218.

Wang, J., Tabata, Y., and Moromoto, K. 2006. Aminated gelatin microspheres as a nasal delivery system for peptides drugs: Evaluation of *in vitro* release and *in vivo* insulin absorption in rats. *J. Control. Release*, 113: 31–37.

Yajima, T., Nogata, A., Demachi, M., Umeke, N., Itai, S., Yunoki, N., and Nemoto, M. 1996. Particle design for taste-masking using a spray congealing technique. *Chem. Pharm. Bull.*, 44: 187–191.

Zhang, H., Lu, Y., Zhang, G., Gao, S., Sun, D., and Zhong, Y. 2008. Bupivacaine-loaded biodegradable poly(lactic-*co*-glycolic) acid microspheres I. Optimization of the drug incorporation into the polymer matrix and modelling of drug release. *Int. J. Pharm.*, 351: 244–249.

Zhang, W., Jiang, X., Hu, J., and Fu, C. 2000. Rifampicin polylactic acid microspheres for lung targeting. *J. Microencapsul.*, 17: 785–788.

Zhang, X.B. and Hou, X.P. 2003. Development of methods for masking unpleasant taste of drugs. *Chin. Hosp. Pharm. J.*, 23: 43–45.

21 Colloids in Aerosol Drug Delivery Systems

Nazrul Islam

CONTENTS

21.1 INTRODUCTION

An aerosol is a mixed phase colloidal system that arises when solid particles or liquid droplets are dispersed in a gas. The colloidal particle size range is between 1.0 nm and 1.0 μm; however, some particles, for example, fibers, are larger than this range. Colloid technology is rapidly becoming more rational in terms of applications in different areas such as modern food technology, pharmaceutical and cosmetic preparations, and paints. Colloidal medicinal aerosols are used to improve human health, and aerosol therapy by inhalation has been used for many years. A wide variety of medicinal colloids are administered to the lungs by oral inhalation for the treatment of diverse disease states, especially for the management of asthma and obstructive airway diseases. Direct delivery of drugs into the pulmonary regions of the lung enables lower doses with equivalent therapeutic action compared to oral or parenteral routes because of the large surface area (~100 m^2) of the lungs. The advantages of aerosol formulations over other dosage forms (i.e., parenteral and other liquid dosage forms) are that they are noninvasive, easy to use, less expensive, painless, and user friendly. The inhaled route allows the delivery of small doses of drug directly to the alveoli, attaining a high concentration of drug in the local area and minimizing systemic side effects. This results in a high therapeutic ratio of drugs compared with that of systemic delivery administered by either oral or parenteral routes. This chapter demonstrates aerosol delivery of various medicinal colloids for the optimal management of various disease states.

21.2 HISTORICAL PERSPECTIVES

The concept of colloids was introduced in the late 19th century. Michael Faraday first prepared colloidal gold in 1856 and Thomas Graham invented the term colloid (derived from the Greek *kola*, meaning "glue") in 1861 to demonstrate a colloidal system that exhibited a slow rate of diffusion through a porous membrane. Today colloid science, which are systems involving small particles of one substance suspended in another, is the basis of a variety of scientific and technological advances including food preparation, cosmetics, paints, ceramics, detergents, biological cells, and pharmaceutical and medicinal products. Numerous colloidal preparations are currently used in drug delivery systems, including hydrogels, microspheres, microemulsions, liposomes, micelles, and nanoparticles for sustained release, and target delivery of medical products. However, this chapter will focus on the pulmonary delivery of colloidal medicines as aerosols for the management of various diseases.

Aerosol delivery of medicinal agents can be traced to Ayurvedic medicine over 4000 years ago (Gandevia, 1975; Grossman, 1994). Early medicinal aerosols were basically vapors from volatile aromatic substances such as menthol, thymol, and eucalyptus and smoke produced from burning the leaves of the medicinal plants *Atropa belladona* and *Datura stramonium* to treat diseases of the throat and chest (Muthu, 1922). Initially, *D. stramonium* was substituted with *D. ferox* to treat asthma. Later a mixture of Datura with tobacco was introduced and smoked in a pipe, which was known as Potter's asthma cigarettes; this cigarette smoke may be considered as a colloidal dry powder aerosol system (O'Callaghan et al., 2002). After a long time aerosol delivery was used in the treatment of tuberculosis (Muthu, 1922). In the 1820s delivery of liquid droplets as aerosols was developed and the use of nebulizers in aerosol delivery was established (Gandevia, 1975). In 1900, nebulizers were applied to generate a mist from *D. stramonium* and the improved version of this device (known as a squeeze-bulb nebulizer) was used to deliver adrenaline (Solis-Cohen, 1990). However, nebulizers are not portable, are limited to the treatment of hospitalized patients, and are not very convenient because of the requirement of a compressed gas. Nebulizers were advertised for many ailments like the delivery of aerosolized drugs for the treatment of pharyngitis, bronchitis, pain, asthma, tuberculosis, and sleeplessness. The inhalation of medicinal products by humans has been used professionally since the late 1950s. In 1955, the most important development of delivering colloids, where drugs were formulated as a solid dispersion or solution of an antiasthma drug in a liquefied propellant, was the portable pressurized metered dose inhaler (pMDI; Theil, 1996). Since

1956, pMDIs have become the most generally used devices to deliver asthma drugs (Freedman, 1956). MDIs have increased the popularity of medicinal aerosols because of their improved portability and usability, although these devices have some drawbacks including the need for coordination between actuation and inhalation and the use of chlorofluorocarbons (CFCs), which are environmentally unfriendly. Aerosol delivery of colloidal powder became the choice for drug delivery after the introduction of the dry powder inhaler (DPI) in 1967 (Altounyan, 1967). During the 1980s and 1990s aerosol drug formulations became more complicated as research demonstrated the value of prompt delivery of drugs deep into lungs for diseases like asthma. The first marketed DPI product was sodium chromoglycate that was delivered in micronized powder form from a device known as a Spinhaler. This dry powder aerosol is free from CFCs and independent of breathing coordination, and application of DPIs for treatment of various diseases is now progressing.

21.3 LUNG AS A ROUTE FOR AEROSOL DELIVERY

The lung is the main organ for the filtration of air and the site of gas exchange. Delivery of drugs to the respiratory tract by DPIs has become an effective therapeutic method for treating a wide range of pulmonary disorders, including asthma, bronchitis, and cystic fibrosis (CF). The large surface area (100 m^2) of the respiratory tract offers rapid absorption of drugs (Byron and Patton, 1994; Timsina et al., 1994); the extremely thin (0.1–0.2 µm) absorptive mucosal membrane and good blood supply have demonstrated the potential of the pulmonary route for noninvasive administration of medicinal agents. Respiratory delivery also offers effective therapy with minimum adverse effects by using small doses of drugs through inhalation and allows substantially greater bioavailability of polypeptides (Byron and Patton, 1994). The periphery of the lung gives rapid access to the circulation for systemic delivery. Local action of drugs is achieved by identifying the target receptor of cells. A large quantity of β-adrenoceptors are available in the periphery of the lungs, and cholinergic receptors are situated in the central airway of the lungs. Thus, the lung is the most important route for delivering colloidal solutions of antiasthmatic drugs. The common bacteria *Pseudomonas aeruginosa* live on the sticky mucus in the lungs of patients with CF, and therefore ciliated airways are a target for dry powder antibiotics. Moreover, infective microorganisms in the alveolar region and the periphery of the lungs is the target for appropriate drug-aerosol therapy.

21.4 MECHANISM OF DRUG DEPOSITION FROM AEROSOLS

To achieve a desired therapeutic effect from aerosols, an adequate amount of drug must reach the alveolar sacs of the respiratory airways. The dynamic behavior of aerosol particles is governed by the laws of aerosol kinetics (Newman and Clarke, 1983). The dominant mechanisms of depositing aerosol particles into the respiratory tract include inertial impaction, sedimentation (gravitational deposition), Brownian diffusion, interception, and electrostatic precipitation (Gonda, 1990). Inertial impaction and sedimentation are the most important mechanisms for large particle deposition. The relative contribution of the mechanisms of drug deposition is likely to depend on the characteristics of the inhaled particles, the breathing pattern, and the complex physiology of the respiratory tract. Diffusion is the most important mechanism for depositing smaller particles in the peripheral region of the lung (Yu and Chien, 1997). A brief description of each mechanism of deposition is given below.

21.4.1 INERTIAL IMPACTION

This is considered to be the main deposition mechanism at the tracheal bifurcation or successive branching points of the airways. The airflow changes its direction at the branching of the airways of the respiratory tract. The aerosol particles continue to move in their original direction and impact any obstacle in the way. The deposition of aerosol particles by impaction increases with increasing air velocity, frequency of breathing, and particle size (Carpenter and Kimmel, 1999). Large particles (>5 μm) with high velocity are mainly deposited by impaction (Gonda, 1992).

21.4.2 SEDIMENTATION

Deposition of aerosols by sedimentation occurs when the gravitational force exerted on a particle overcomes the force of the air resistance. Particles of smaller size (0.5–3.0 μm), which have a tendency to escape from deposition by inertial impaction, may be deposited by sedimentation. Deposition of small particles by sedimentation mainly occurs in the smaller airways and alveolar regions, and increased sedimentation is observed during breath holding or slow steady breathing (Gonda, 1992).

21.4.3 DIFFUSION

Deposition of aerosolized particles smaller than 0.5 μm occurs by diffusion that is attributable to Brownian movement. Deposition of aerosols by diffusion is independent of the density of particles but increases with decreasing size. Generally, the deposition of particles larger than 1.0 μm is dominated by inertial impaction and particles smaller than 0.1 μm are deposited by diffusion. Sedimentation and diffusion are both important for particle sizes ranging between 0.1 and 1.0 μm (Brain and Blancard, 1993).

21.4.4 INTERCEPTION

Although particle deposition by interception is not common, the deposition of elongated particles (particles large in one dimension but with small aerodynamic diameters) is believed to occur by this mechanism. Deposition of particles in the respiratory airways by interception is important when the dimensions of the anatomic spaces of airways become comparable to the dimensions of the particles (Gonda, 1992).

21.5 PULMONARY DELIVERY TECHNOLOGY

21.5.1 DRUG FORMULATION AND DELIVERY DEVICES

Aerosol methods for the delivery of drugs into the lower airway of the lungs are formulated as colloidal liquid solutions, suspensions, emulsions, or micronized dry powders, which are aerosolized via some commonly used delivery devices that are now described.

21.5.1.1 Nebulizers

Nebulizers deliver colloidal drug solutions or suspension formulations, and they are frequently used for those drugs that cannot be formulated into other devices such as pMDIs or DPIs. Nebulizers are one of the oldest forms of pulmonary drug delivery devices that deliver a large volume of drugs that are inhaled during normal breathing through a mouthpiece or facemask. Therefore, these devices are useful for patients (young, geriatric, and arthritic patients) who experience difficulties with pMDIs. At present, two categories of nebulizers include air jet and ultrasonic nebulizers that are available on the market. Air jet nebulizers can generate both smaller particles (2–5 μm) and coarse aerosols and deliver medication quickly. Most jet nebulizers operate by forcing pressurized gas (air or oxygen) through a nozzle or jet at high velocity so that the liquid formulation is atomized. In contrast, ultrasonic nebulizers do not require compressed gas. The solution formulation is atomized by an energy source, a piezoelectric crystal transducer, that vibrates at high frequency; these devices generate slightly larger aerosols. However, the overall efficiency of the piezoelectric-driven ultrasonic nebulizers is more or less similar to that of air jet nebulizers. Patients with severe obstructive lung conditions prefer to use nebulizer therapy. Nebulizers are suitable for drugs with high doses and little patient coordination or skill; however, treatment using a nebulizer is time consuming and less efficient. Nebulizers include AeroDose® (Aerogen), AeroEclipse® (Trudel Medical International), Halolite® (Medic-Aid Limited), and Respimat® (Boeringer Ingheim), which are currently available to deliver various types of drugs as aerosols.

21.5.1.2 Pressurized Metered Dose Inhalers

The most commonly used delivery devices are pMDIs. In pMDIs (Figure 21.1), the drug is either dissolved (with or without the aid of a cosolvent) or suspended as micronized particles in liquefied propellants (or a mixture of propellants) with other excipients and presented in a pressurized canister fitted with a metering valve; on actuation, a predetermined amount of drug is released as a spray. The propellants originally used in pMDI formulations were liquefied gases of CFCs, which are not environmentally friendly. To overcome the problem with CFCs, hydrofluoroalkanes that have no effects on the ozone layers are now used in the formulation of pMDIs. Both CFCs and hydrofluoroalkanes are gases at room temperature and pressure; however, they are liquefied by applying high pressure or by lowering the temperature for use in pMDIs.

Presently there are a number of pMDIs available on the market for the management of asthma such as Ventolin® (albuterol, GlaxoSmithKline), Symbicort® (budesonide and formoterol, AstraZeneca), Flovent® (fluticasone, GlaxoSmithKline), and Azmacort® (triamcenolone acetate, Aventis Pharma). There are some breath actuated and microprocessor controlled devices (Autohaler®, Respimat®) that insure the patient receives the drug at the correct dose during inspiration, and in slow inhalation there is an indicator light to inform the patient whether the dose has been inhaled. pMDIs have some disadvantages such as oropharyngeal deposition of drugs that is due to the high velocity of the propellants. The particles aerosolized from the MDIs have a high velocity, which exceeds the patients' inspiratory force; therefore, a large number of particles deposit onto the oropharyngeal areas or the back of the throat. Thus, coordination between valve actuation and inhalation is required and essential to deposit aerosols in the lower airways of the lungs. The inability of many patients to coordinate aerosol actuation and inspiration is a common problem; even after extensive training, many patients cannot operate MDIs appropriately. Several inhalation aids like spacers have been developed to overcome this complexity (Bisgaard, 1998; Ikeda et al., 1999). A spacer fitted into the mouthpiece of the MDIs causes a delay between actuation and inhalation, thereby reducing the need for coordination between actuation and inspiration. However, a large portion of the drug, which would deposit into the

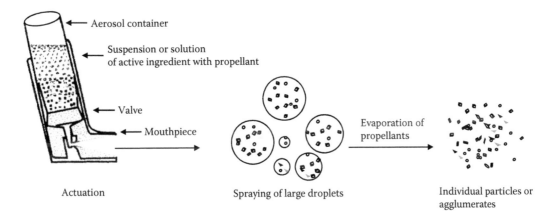

FIGURE 21.1 A schematic of a pMDI representing the formulation and particle delivery mechanism. (Modified from Dalby, R. N., S. L. Tiano, and A. J. Hickey, 2007. In A. J. Hickey (Ed.), *Inhalation Aerosols (Physical and Biological Basis for Therapy)*, 2nd edition, pp. 417–444. New York: Marcel Dekker. With permission.)

upper airway of the lungs from the conventional devices, is retained in the spacer device, resulting in reducing systemic drug absorption and bioavailability.

21.5.1.3 DPI System

DPIs contain a powder formulation of micronized drugs with improved flow properties and dose uniformity (French et al., 1996; Timsina et al., 1994). There are two types of DPI formulations: loose agglomerates of micronized drug particles having controlled flow properties, and carrier-based interactive mixtures (Figure 21.2) that consist of micronized (<5 μm) drug particles mixed with larger carrier particles (i.e., lactose; Hersey, 1975). Applications of other carriers such as mannitol and trehalose (Mao and Blair, 2004; Stahl et al., 2002) have also been reported. It is important to note that both of these formulations need to have good flow properties to insure delivery of a uniform dosage and accurate dose metering of the drug. The redispersion of drug particles depends on the interparticulate forces within the powder formulation and the patient's inspiration. Powder deagglomeration and aerosolization from these formulations are achieved by the patient's inspiratory airflow, which must be sufficient to create an aerosol containing drug particles for lung deposition. Advantages of DPIs over other inhaler systems (i.e., pMDIs) are the independence of breathing coordination with dose actuation, the absence of propellants, and the low innate initial velocity of particles (reducing inertial impaction at the back of the throat). In DPIs, drugs are kept in the solid state and thus exhibit high physicochemical stability. Aerosol delivery of drugs like proteins and peptides, which are highly unstable in liquid formulation, would be superior to other delivery techniques. Delivery of colloidal dry powder drugs to the deep lung by DPIs has become an effective therapeutic method for treating a wide range of pulmonary disorders, including asthma, chronic obstructive pulmonary disease (COPD), bronchitis, and CF. Presently, there are more than 20 different types of DPI devices available on the market and more than 25 are in development (Islam and Gladki, 2008).

There are a wide range of DPIs, which are single or multiple dose, breath activated, and power driven; however, the development of novel devices with new designs continues because the design of the device affects the DPI performance (Coates et al., 2004). Based on the design, DPI devices are classified into three broad categories: first-, second-, and third-generation DPIs. The first-generation DPIs were breath activated single unit doses (capsules) such as the Spinhaler® and Rotahaler®, and the drug delivery issues were related to the particle size and deagglomeration of drug carrier agglomerates or drug carrier mixtures delivered by the patient's inspiratory flow. The second generation of DPIs used better technology, for example, multidose DPIs (Pulmicort®, which measures the dose from a powder reservoir) or multiunit doses (Diskus®, which disperses individual doses that are

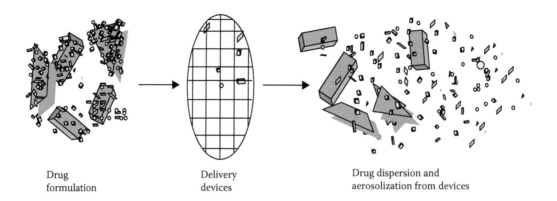

Drug   Delivery   Drug dispersion and
formulation   devices   aerosolization from devices

FIGURE 21.2 A schematic of the pulmonary delivery of a drug in a powder formulation. The formulation consists of micronized drugs adhered on the surface of large carrier particles. Drug particles detach from the surface of large carriers and deposit into the patient airways by inhalation. (Modified from Telko, M. J. and A. J. Hickey, 2005. *Respiratory Care* 50 (9): 1209–1227.)

premetered into blisters, disks, dimples, tubes, and strips). The multiunit dose devices are likely to insure the reproducibility of the powder formulation compared to that of the multidose reservoir. Third-generation DPIs, known as active devices, employ compressed gas or motor-driven impellers (Crowder et al., 2001) or use electronic vibration (Brown et al., 2004) to disperse the drug from the formulation. These devices are more sophisticated and user friendly. Because of the presence of an energy source, active devices enable respiratory force independent dosing precision and reproducible aerosol production. The first approved active device for aerosol delivery of insulin as a DPI formulation was Exubera® (Pfizer); however, the production was discontinued in early 2008 because this large and clumsy device failed to achieve acceptance by physicians and patients.

21.5.2 Factors Affecting Drug Deposition

Particle deposition in the pulmonary regions is dependent on the particle properties including the diameter, shape, and surface properties; the anatomical properties of the airways (healthy vs. unhealthy patients); and patient characteristics, including the breathing patterns and inspiratory force of the patient, which differ among geriatric, young, and normal adult patients. However, the major factors affecting drug deposition are as follows.

21.5.2.1 Particle Size and Shape

The deposition of the inhaled drug in the lung determines the therapeutic outcome of the formulation (Dolovich, 1992) and depends on the characteristics of the inhaled particles, such as the drug particle diameter, shape, electrical charge, density, and hygroscopicity; the physiology of the respiratory tract; and the breathing patterns that influence the frequency, tidal volume, and flow (Byron, 1986; Hickey and Martonen, 1993; Suarez and Hickey, 2000). The drug particle for pulmonary delivery should be below 5.0 μm for successful deposition of the drug deep into the lungs. Particles of 1.0 μm or less are exhaled with very little being deposited. According to Byron (1986), the maximum pulmonary deposition of particles with 1.5–2.5 and 2.5–4 μm diameters occurred with and without breath holding, respectively. However, rapid breathing showed maximum deposition of particles between 1.5 and 2 μm in the tracheobronchial region with breath holding and particles between 2 and 3 μm deposited in the pulmonary region without breath holding (Byron, 1986). Therefore, slow inhalation is desirable to obtain maximum deposition of aerosol particles in the lower airways of the lung.

The specific shape of carriers in controlling the drug dispersion from the DPI formulation is not fully understood. The interactive forces between the drugs and carrier surfaces, and therefore the flow properties of the DPI formulation, may be affected by the shapes of the carriers and drugs (Crowder et al., 2001; Mullins et al., 1992). There was increased dispersion of salbutamol sulfate when elongated particles of lactose carriers were used (Zeng et al., 2000a, 2000b). The authors suggested that the elongated particles may exhibit a much smaller aerodynamic diameter than spherical particles of similar mass or volume. These would detach and deagglomerate more drug particles from the carrier surfaces and thereby increase the fine particle fraction.

21.5.2.2 Patient's Inspiratory Flow and Device Resistance

To insure effective drug delivery into the lower airways of the lungs, the inspiratory flow rate must be sufficient to produce adequate turbulent airflow in any device so that an adequate aerosol cloud of the aerosolized fine particles is produced. Therefore, a balance among the design of an inhaler device, the drug formulation, and the inspiratory flow rate of the patient is required (Steckel and Mueller, 1997). The main limitation with existing devices, particularly DPIs, is that delivery of the drug is often dependent upon the inspiratory flow rates for deagglomeration of drug carrier agglomerates (Islam et al., 2004; Louey and Stewart, 2002; Lucas et al., 1998; Zeng et al., 1998) and effective delivery of the drug powder (Ganderton and Kassem, 1992; Hickey and Concessio, 1997). For example, some DPIs require inspiratory flow of ≥30 L/min to effectively deagglomerate the powder (de Boer et al., 1996). Low-resistance passive DPIs are generally less dependent on the flow rate than high-resistance devices. Inspiratory flow rate played the most important role in determining the dispersion of salbutamol sulfate aerosolized from a Rotahaler® (Srichana et al., 1998). A flow rate of 60 L/min was reported to be advantageous for effective delivery of drugs from the Turbuhaler® (de Boer et al., 1996), and patients could achieve sufficient inspiratory effort to deagglomerate and aerosolize the dose (Li and Edwards, 1997). An increased inspiratory flow rate may help increase the deposition of particles in the upper airways. A slow inhalation rate increases the number of particles that reach the peripheral region of the respiratory tract by impaction. Newman et al. (1981) found that a slow inhalation rate (25 L/min) with breath holding showed maximal deposition of terbutaline sulfate compared to a faster rate (80 L/min) of inhalation.

21.6 AEROSOL DELIVERY OF VARIOUS THERAPEUTICS

21.6.1 CURRENT INHALED THERAPEUTICS

Currently, aerosol drug delivery of colloidal solutions, suspensions, or powders is used for a limited number of therapeutic compounds such as β-adrenoceptor agonists, muscarinic agonists, corticosteroids, and mast cell stabilizers. Certain combinations of drugs have been formulated recently to have synergistic therapeutic benefit. Corticosteroids and long acting β-adrenoceptor agonist colloidal formulations are available as both pMDIs and DPIs (Dhillon and Keating, 2006). Zanamavir, an antiviral agent, has been introduced to the market as an aerosol product for the treatment of influenza (Cass et al., 1999). Aerosol delivery of recombinant human deoxyribonuclease and tobramycin is available as nebulizers for the management of CF (Cass et al., 1999; Kuhn, 2002).

21.6.2 DRUGS ADMINISTERED AS AEROSOLS

Aerosol delivery offers the greatest potential to deliver drugs into the lower airway of the lungs for a wide range of molecules for the treatment of systemic diseases. A list of various drugs administered via the pulmonary route is provided in Table 21.1.

Aerosol delivery of macromolecules is a potential noninvasive way of administering drugs to avoid frequent injections. Lung delivery of insulin has already been established; however, insulin-loaded chitosan nanoparticles (Grenha et al., 2005), nanoparticles of calcitonin (Kawashima et al., 2000), and nanospheres of elcatonin coated with chitosan (Yamamoto et al., 2005) have been demonstrated for successful deep lung delivery. Aerosol delivery of leuprolide has been investigated as both MDI and DPI formulations for the management of prostate cancer (Anon, 2008; Shahiwala and Misra, 2005). Dry powders of other proteins such as parathyroid hormone for osteoporosis (Codrons et al., 2003, 2004), glucagon (Endo et al., 2005), human growth hormone for dwarfism (Bosquillon et al., 2004), and vasoactive intestinal peptide for pulmonary diseases like asthma (Ohmori et al., 2006) have been successfully investigated.

Aerosol delivery of genes directly targets the regions of interest by avoiding problems associated with intravenous delivery. Successful gene delivery into the lungs for CF has been demonstrated recently (Brown et al., 2004; Lentz et al., 2006). Lung delivery of liposomally encapsulated adenoviral vectors containing genes (Laube, 2005; Li et al., 2005), genes complexed with cationic lipids (lipoplex), and polymers (polyplex; Kichler et al., 2001; Wightman et al., 2001); a cationic lipid coupled with plasmid deoxyribonucleic acid (DNA; Deshpande et al., 2002); and protein 53 and cytokine (interleukin 12) have been reported to give therapeutic responses (Densmore, 2006; Gautam et al., 2000; Lungwitz et al., 2005). Based on the mentioned research, there is a potential future of pulmonary gene therapy for various types of clinical applications.

Pulmonary delivery of vaccines is an interesting area of drug delivery, and inhalation of measles vaccine is both safe and effective (Wong-Chew et al., 2006). Nebulized measles vaccine in a human model produced better immunity with reduced side effects compared to that of a subcutaneous injection (Dilraj et al., 2000). DPI formulations of measles vaccine (LiCalsi et al., 2001), mucosal

TABLE 21.1
Drugs Administered as Aerosols for the Treatment of Various Diseases

Indication	Drug Substances	Colloidal Delivery Method	References
Asthma	β-Adrenergic agonists, anticholinergics	Powder	Atabai et al., 2002; Windom et al., 1990
	Corticosteroids	Powder, suspension	Cochrane et al., 2000; Newman, 2003
Bronchospasm	Troventol	Suspension	Lulla and Malhotra, 2007
COPD	Tiotropium	Powder	Bechtold-Peters et al., 2002; Keam and Keating, 2004
	Fluticasone	Powder, suspension	Shapiro et al. 2000; Vanacker Nele et al. 2002
	Salmeterol	Powder	Bennett et al., 2006; Vanacker Nele et al., 2002
CF	Amiloride	Suspension	Chougule et al., 2006; Thomas et al., 1991
	Tobramycin	Powder	Le Conte et al., 1993; Newhouse et al., 2003
	Deoxyribonuclease	Powder	Hodson, 1995; Milla, 1998
	Colistin sulfomethate	Powder	Westerman et al., 2007
Cancer	Doxorubicin	Powder	Haynes et al., 2003; Tian et al., 2004
Diabetes	Insulin	Powder, suspension	Clark and Foulds, 1999; Katz et al., 2001; Laube, 2001
Osteoporosis	Calcitonin	Powder	Kawashima et al., 2000; Yamamoto et al., 2005
Sexual dysfunction	Apomorphine, phosphodiesterase type 5 inhibitors	Powder	Cheatham et al., 2006; Staniforth et al., 2006
Vaccines	Malarial vaccine	Powder	Edwards et al., 2005
	Measles vaccine	Powder	LiCalsi et al., 2001
	Influenza vaccine	Powder	Brito et al., 2007; Smith et al., 2003
	Zanamivir	Powder	Calfee and Hayden, 1998; Cass et al., 2000
	Diptheria toxoid	Powder	Amidi et al., 2007
Endometriosis, pubertus praecox, prostate carcinoma	Leuprolide	Powder, suspension	Anon, 2008; Shahiwala and Misra, 2005
Hormone replacement therapy	Testosterone	Suspension	Davison et al., 2005
Immunosuppressor	CsA	Powder, suspension	Matilainen et al., 2006; Waldrep et al., 1998
Thrombosis and emphysema	Heparin	Powder	Qi et al., 2004; Yang et al., 2004

vaccination for influenza virus (Edwards et al., 2005), malarial vaccine (Smith et al., 2003), and small interference ribonucleic acid (Brito et al., 2007) were recently investigated with significant success. Aerosol delivery of human immunodeficiency virus therapeutic to the infected patients had reduced toxicity and improved patient compliance in a recent study (Cipolla, 2007). Therefore, it seems that pulmonary delivery of various genes will continue to progress and in future the world will see suitable vaccines against many pulmonary pathogens like *Mycobacterium tuberculosis*, respiratory syncytial virus, and severe acute respiratory syndrome.

The inhaled antibiotic rifampicin, rifampicin-loaded poly(lactide-*co*-glycolide) microparticles (Sethuraman and Hickey, 2002), colistin sulfate (Le Brun et al., 2002; Westerman et al., 2004), mucoactive agent nacystelyn (Vanderbist et al., 2001), colistin (Westerman et al., 2007), gentamicin (Crowther Labiris et al., 1999), azithromycin (Hickey et al., 2006), and tobramycin (Geller et al., 2007) were found to be effective against CF. Deep lung delivery of amphotericin B desoxycholate, liposomal amphotericin B, amphotericin B lipid complex, and amphotericin B colloidal dispersion via nebulizers was demonstrated as being valuable in the prophylactic treatment of pulmonary aspergillosis (Ruijgrok et al., 2005). The DPI delivery of antibiotic ofloxacin (Hwang et al., 2008) and levofloxacin (Valle et al., 2008) was effective against tuberculosis, and pulmonary administered ofloxacin improved the treatment efficacy compared to intravenous or oral routes. Inhaled itraconazole, an antifungal drug, showed improved bioavailability in a mouse model (Valle et al., 2008).

Aerosolized chemotherapeutic agents for the direct local treatment of lung tumors are advantageous over other drug delivery systems. Pulmonary delivery of aerosolized 9-nitrocamptothecin and cisplatin in patients with lung cancer was safe and showed a promising antitumor effect (Gagnadoux et al., 2008; Smyth et al., 2008). Other anticancer agents including doxorubicin (Azarmi et al., 2006; Johnston et al., 1995), paclitaxel (Hershey et al., 1999), celecoxib and docetaxel (Fulzele et al., 2006; Haynes et al., 2005), gemcitabine (Koshkina and Kleinerman, 2005), and liposomal camptothecin (Knight et al., 2006) demonstrated reduced toxicity with pulmonary delivery. These researchers indicated the potential of inhalation delivery of anticancer drugs in the treatment of lung cancer; however, further detailed investigations are warranted.

Aerosol delivery of other drugs such as apomorphine for sexual dysfunction (Staniforth et al., 2006), morphine and fentanyl for pain management (Mather et al., 1998), cannabinoids for psychoactive effect (Hung et al., 2001), and ergotamine for migraine headaches (Armer et al., 2007; Pavkov et al., 2005) has been investigated. Furthermore, aerosol delivery of the radiopharmaceutical technetium-99 m (99mTc) with phosphate buffer for lung ventilation imaging purposes is widely used (Ballinger et al., 1993). Inhalation of radiolabeled sulfur colloid aerosol (99mTc sulfur colloid, 0.2 μm) for studying particle uptake by airway surface macrophages has been reported (Alexis et al., 2006).

These findings show the wider application of aerosol delivery of colloidal drugs, with the potential to deliver drugs into the deep lungs for a wide range of molecules in the treatment of systemic diseases.

21.7 ADVANCED TECHNOLOGIES IN AEROSOL DELIVERY

21.7.1 POWDER TECHNOLOGY

A lack of the desired efficiency in traditional methods of powder production led to the development of alternative techniques that produce powders with the desired physicochemical properties, including specific size, density, and morphology, and adhesional properties (Hickey and Concessio, 1997). To produce various particles with improved aerosol delivery, a number of alternative techniques were demonstrated, such as specialized spray drying, ultrasound-assisted crystallization, supercritical fluid technology, and the *in situ* method (Shekunov et al., 2002; York et al., 1999). Particles with low density but large size increased the dispersion of spray-dried porous insulin (Edwards et al., 1997, 1998) and supercritically produced salmeterol xinafoate (Beach et al., 1999; Shekunov et al., 2002, 2003), and tetracycline microparticles produced by the supercritical fluid expansion depressurization method

(Li et al., 2008) showed promising drug delivery. The authors suggested that supercritically produced particles showed less tendency to form agglomerates or formed very loose agglomerates that were easy to aerosolize efficiently with less energy, resulting in increasing drug dispersion. The underlying principle was described as enhanced performance through particle engineering (Shekunov et al., 2003). Moreover, recent particle engineering has developed highly porous particles with large geometric diameters but small aerodynamic diameters, which improves powder dispersion and can improve the efficacy of DPIs (Edwards et al., 1997, 1998). A recent report on the dispersion of amorphous itraconazole nanoparticles produced by the ultrarapid freezing technique showed improved bioavailability in mice (Yang et al., 2008). A brief description on currently available particle technology for the development of suitable particles for deep lung delivery is reviewed by Chow et al. (2007). A number of novel powder production technologies, such as porous particles (Edwards et al., 1997), pulmospheres (Bot et al., 2000; Duddu et al., 2002), nanoparticles (Kawashima et al., 1998), and solid–lipid nanoparticles (SLNs; Pandey and Khuller, 2005) of antitubercular drugs and surface modified particles (Morton, 2006) were demonstrated. However, the efficiency of drug delivery still did not reach the target level. Ely et al. (2007) introduced the application of effervescent carrier particles containing ciprofloxacin nanoparticles and demonstrated active release of the drug to the pulmonary region. Using the spray-drying technique, they produced effervescent carriers with the required size for lung delivery. They showed that the manufactured carrier particles released 56% ciprofloxacin into solution compared with 32% when conventional carrier (lactose) particles were used. This outcome has opened a new window for future research on the pulmonary delivery of a wide range of therapeutic agents to the deep lungs with improved release.

21.7.2 Liposomes and Prolonged Drug Action

To achieve prolonged drug action, novel liposome-loaded pharmaceuticals for aerosol delivery of drugs were developed (Jurima-Romet et al., 1990; Platz et al., 1995). Liposomally encapsulated proteins and peptides, for example, cyclosporin A (CsA; Matilainen et al., 2006), ricin toxin (Griffiths et al., 1997), polymyxin B (Omri et al., 2002), interferon β (IFN-β; Platz et al., 1995), IFN-γ (Moshen and Armer, 2003), interleukin 2 (Heinzer et al., 1999), and glutathione (Jurima-Romet et al., 1990) were successfully delivered as aerosols in order to improve treatment efficacy with reduced toxicity. Aerosol delivery of liposomal encapsulated cannabinoid produced a prolonged psychoactive effect (Hung et al., 2001), and dapsone dry powder prolonged drug release (~16 h) at the site of action in the lungs for the treatment of *Pneumocystis carinii* pneumonia (Chougule et al., 2008). Dapsone was prepared by encapsulating liposomes using the thin film evaporation technique, passing them through a high-pressure homogenizer with subsequent separation by ultracentrifugation, and final spray drying of the powders. Prolonged retention of liposomally encapsulated amiloride hydrochloride for CF (Chougule et al., 2006) and tacrolimus for refractory rejection of lungs after transplantation were observed (Chougule et al., 2007). In another study, the inhaled anti-infective drug ciprofloxacin was encapsulated into liposomes and produced immediate and sustained release of the drug for treatment of respiratory tract infections (Cipolla and Blanchard, 2008). A suspension of tocopherol nanoparticles coupled with biodegradable polymers for delayed release (Bonnet-Gonnet, 2008) and liposomal encapsulated cannabinoid for a prolonged psychoactive effect (Hung et al., 2001) were recently demonstrated. However, extensive research is needed in directing drug deposition to specific sites of the lung that contain the drug's target receptors. Aerosol delivery of liposomal encapsulated retinoic acid to lung tissue via a nebulizer prevented local irritation of lung tissue, prolonged therapeutic levels of drugs, and produced high drug concentrations at the tumor sites with reduced pulmonary toxicity (Gilbert et al., 2002). Furthermore, paraffin wax coated disodium fluorescein and pentamidine aerosols for lung delivery were reported (Pillai et al., 1998). The wax-coated aerosols generated from fluorescein mixed with 99mTc-labeled iron oxide colloid delivered to the canine lung produced a 3.4-fold increase in the absorption half-time of disodium fluorescein compared with uncoated fluorescein. This research proved the wax-coated aerosols may provide a useful delayed release technique with a high drug load for future therapeutic purposes (Pillai et al., 1998). Therefore, we emphasize that inhalation is an effective route of drug delivery to obtain prolonged therapeutic benefits. The aforementioned discussion proves that inhalation of colloidal aerosols is useful in obtaining prolonged therapeutic actions for drugs. Delayed release therapeutics with polymer nanoparticles are discussed in the following section.

21.7.3 POLYMER AND NANOTECHNOLOGY

The introduction of polymers in the development of colloidal nanoparticles for aerosol delivery of drugs has become very popular, and colloidal particles of biodegradable polymers have been demonstrated as a carrier to deliver various drugs into the lungs (Seeger et al., 2004). Fiegel et al. (2004) used biodegradable ether-anhydride polymers to develop microparticles for controlled release of drugs as a dry powder for inhalation. They investigated various ratios of sebacic acid and poly(ethylene glycol) (PEG, 0–30%) to produce polymer carriers with a controlled aerodynamic diameter within the respirable range, and 10% PEG containing particles showed significantly enhanced *in vitro* deposition in the lower stages of a lung model following aerosolization from a Spinhaler® DPI. The authors emphasized that the addition of 10% PEG into the polymers appeared to reduce interparticle adhesion forces because of decreased surface roughness, resulting in increasing deposition (Fiegel et al., 2004). Nebulization to mice in the form of aqueous colloidal dispersions of amorphous CsA nanoparticles produced systemic concentrations of drugs below toxic limits (Tam et al., 2008). The use of a microemulsion containing colloidal nanoparticles of a drug (enhanced green fluorescent protein N-terminal reporter plasmid DNA) was suitable for deep lung delivery by an MDI (Dickinson et al., 2001). Using a poly(ester amine) polymer carrier, aerosol delivery of murine thymoma viral oncogene homolog 1 (Akt1) small interference ribonucleic acid, which was delivered into K-ras through stochastic activation of a latent allele and urethane-induced lung cancer models, was found to be an effective and promising option for the treatment of lung cancer (Xu et al., 2008). Aerosolized liposomal cisplatin was also feasible and safe in patients with lung carcinoma (Wittgen et al., 2007). Elcatonin-loaded surface-modified poly(lactic-*co*-glycolic acid) nanospheres coated with chitosan improved the pulmonary delivery of the drug (Tozuka and Takeuchi, 2008). SLNs incorporated with insulin, which were recently developed by the reverse micelle-double emulsion method, were used as a carrier to deliver insulin via a nebulizer into the rat lung and produced a prolonged hypoglycemic effect with no cytotoxicity; this suggested the potential safety of SLNs for lung delivery at the cellular level (Liu et al., 2008). They also demonstrated that inhaled insulin-SLN produced pharmacological bioavailability of 24.33% and a relative bioavailability of 22.33%, and the results were almost fourfold higher than the control (subcutaneous injection).

An aerosolized colloidal suspension of tocopherol nanoparticles coupled with biodegradable polymers for delayed release of drugs was demonstrated by Bonnet-Gonnet (2008). Pandey and Khuller (2005) developed solid–lipid particles of chemotherapeutics (i.e., rifampicin, isoniazid, and pyrazinamide) nebulized into guinea pigs infected with *Mycobacterium tuberculosis*. They compared the same drug against tuberculosis following oral administration. No tubercule bacilli were detected in the lungs or spleen of infected animals after 7 doses of nebulized drugs for 7 days, whereas similar therapeutic benefits were obtained after oral administration of 46 daily doses. This finding justifies that nebulized solid–lipid based antitubercular drugs improve drug bioavailability with reduced dosing frequency.

Thus, pulmonary delivery of drugs with nanomaterials provides a noninvasive means to provide not only local lung effects but also systemic bioavailability. According to McCallion et al. (1996), the drug-loaded nanoparticles in a colloidal dispersion tend to reach the respirable percentage of aerosolized drugs more easily when compared to the conventional micron-sized drug particles in nebulizers. It has been suggested that nanoparticle colloidal dispersions may increase the drug absorption rate by encouraging more consistent drug distribution throughout the alveoli. More experiments need to be performed in regard to the biological safety of the use of nanomaterials in pulmonary delivery. Therefore, it is anticipated that nanotechnology is an essential technique that is applicable to drug delivery systems. In the near future, various diseases, in addition to pulmonary diseases, may be treated efficiently using aerosolized inhalable pharmaceutical agents loaded with nanoparticles. However, polymeric microparticles must be studied in much more depth before they will be suitable candidates for pulmonary formulations. The selection of polymers for designing nanoparticles and the control of the particle size and surface properties is important for developing efficient lung delivery of various drugs for systemic circulation.

21.7.4 TARGETED DRUG DELIVERY

Inhalation of therapeutics for targeted delivery has not been established yet. Dames and coworkers (2007) developed targeted delivery of colloidal iron oxide nanoparticles (supermagnetic iron oxide) coupled with a target directed magnetic gradient field to deliver drugs into target areas of the lungs for treating lung diseases. The authors initially carried out a computer-aided simulation prior to use with a mice model to investigate targeted aerosol delivery of nanoparticles to treat a local lung infection or tumor. Delivery of some drugs targeting specific receptors was demonstrated in Section 21.7.3. However, further investigation needs to be carried out in this area of research.

21.7.5 DENDRIMERS IN AEROSOL DELIVERY

Colloidal dendrimers are now being used in advanced drug delivery systems and delivery of cationic polyethylenimine and polyamidoamine dendrimers complexed with DNA into the lungs for the treatment of chronic lung diseases. Polyethylenimine–DNA complexes showed higher levels of luciferase gene expression in the lung compared to that of the DNA–polyamidoamine dendrimer complex (Rudolph et al., 2000). Positively charged dendrimers complexed with enoxaparin (a low molecular weight heparin) delivered into the lungs of anesthetized rats produced effective results in preventing deep vein thrombosis (Bai et al., 2007). The authors demonstrated that this complex enhanced the pulmonary absorption of the drug by reducing the negative surface charge density of the drug molecule. This research has opened a new avenue in pulmonary drug delivery.

21.7.6 DEVELOPMENT OF NEW DEVICES FOR PULMONARY DRUG DELIVERY

In recent developments in device technology, research on new inhaler devices to mimic aspects of traditional devices while improving drug delivery, ease of use, and drug formulation is continuing. The use of spacers (Bisgaard, 1998) and other add-on devices (Matida et al., 2004) to improve the performance of existing devices has been reported. Research has focused on new methods of producing pharmaceutical aerosols such as passing a colloidal solution formulation through a series of nozzles to generate a soft mist aerosol as a bolus dose (Iacono et al., 2000; Schuster et al., 1997). Numerous new DPI devices (breath actuated or applied energy system) with extended technology (accurate dose reproducibility with product stability) have been patented recently or are in the application process (Islam and Gladki, 2008).

Based on these investigations performed in the area of drug delivery, it is obvious that deep lung delivery of colloidal drugs in either nanoparticles or polymer conjugates delivered via nebulizers, pMDIs, or DPIs for both systemic and local effects is a promising route of drug aerosol delivery for the treatment of lung diseases as well as systemic disorders.

21.8 FUTURE DIRECTIONS

Pulmonary delivery of various drugs by aerosolization has been used for centuries to treat respiratory tract diseases. Current aerosol therapy is expanding with the advancement of science and technology, especially in the development of nanoparticles to target systemic disorders for the delivery of proteins and peptides, gene therapy, pain management, nanotherapeutics, cancer therapy, and vaccines. In addition to current therapeutics for asthma or COPD, other drugs are in clinical development, such as mucolytics, antituberculars, and antibiotics, as well as drugs for sexual dysfunction and otitis media, fentanyl for cancer pain, tobramycin, opioids for pain, IFNs, α-1 antitrypsin, and human growth hormone for lung delivery. For the treatment of specific diseases (lung and systemic) with costly medicines, it is desirable to provide medicinal colloidal drugs for aerosol delivery to the targeted area in the human respiratory system. Therefore, pulmonary drug delivery of colloidal drugs would extend the new era of drug delivery research with increased patient compliance and reduced total cost of chronic human diseases. With advanced research, it is anticipated that the world will know more about colloidal drug delivery technology as well as the potential applications of deep lung delivery of those drugs. Pulmonary delivery of large molecules for chronic diseases is advancing rapidly and may become successful in the near future. Therefore, deep lung delivery of drugs needs to be focused not only on lung diseases but also on conditions in which rapid onset is desirable such as cancer pain, allergic reactions, brain disorders, and cardiovascular disorders. In the near future researchers will develop more novel therapeutics, efficient delivery devices, and better formulations to deliver drugs into the deep lungs for various types of diseases. The future of drug administration by inhalation will expand immensely and colloid science can play a significant role in advancing the aerosol delivery of drugs for a wide range of diseases. However, it is imperative to establish a strong collaboration among funding agencies, pharmaceutical manufacturers, and academia to reach the goal in deep lung delivery of drug aerosols.

21.9 CONCLUSIONS

Pulmonary administration of medicaments is expanding with the increasing rate of different diseases. There is a promising future for aerosol delivery of colloidal drugs for the treatment of systemic disorders. In the future, minute amounts of potent drugs such as products from biotechnology will require smart devices that efficiently deliver drugs into the deep lungs. The current trend in pulmonary drug delivery and the potential benefits of this route will enable the continued development of reliable aerosol technology with enhanced deposition of drugs into the deep lungs and better patient compliance. Therefore, research in colloid science may contribute to the effective aerosol delivery of various drugs to combat various life threatening diseases in the future.

ACKNOWLEDGMENTS

The author graciously acknowledges the contribution made by Dr. Therese Kairuz and Dr. Tim Dargaville in proofreading this manuscript.

ABBREVIATIONS

CFC   Chlorofluorocarbon
CF   Cystic fibrosis
COPD   Chronic obstructive pulmonary disease
CsA   Cyclosporin A
DNA   Deoxyribonucleic acid
DPI   Dry powder inhaler
IFN   Interferon
MDI   Metered dose inhaler
PEG   Poly(ethylene glycol)
pMDI   Pressurized metered dose inhaler
SLN   Solid–lipid nanoparticle
99mTc   Technetium-99m

REFERENCES

Alexis, N. E., C. J. Lay, K. L. Zeman, M. Geiser, N. Kapp, and W. D. Bennett, 2006. *In vivo* particle uptake by airway macrophages in healthy volunteers. *American Journal of Respiratory Cell and Molecular Biology* 34 (3): 305–313.

Altounyan, R. E. C., 1967. Inhibition of experimental asthma by a new compound—Sodium cromoglycate, Intal. *Acta Allergologica* 22: 487–489.

Amidi, M., H. C. Pellikaan, H. Hirschberg, A. H. de Boer, D. J. A. Crommelin, W. E. Hennink, G. Kersten, and W. Jiskoot, 2007. Diphtheria toxoid-containing microparticulate powder formulations for pulmonary vaccination: Preparation, characterization and evaluation in guinea pigs. *Vaccine* 25 (37–38): 6818–6829.

Anon. 2008. Method of and apparatus for effecting delivery of fine powders. Accessed at http://www.priorartdatabase.com/IPCOM/000166530/

Armer, T. A., S. B. Shrewsbury, S. P. Newman, G. Pitcairn, and N. Ramadan, 2007. Aerosol delivery of ergotamine tartrate via a breath-synchronized plume-control inhaler in humans. *Current Medical Research and Opinion* 23 (12): 3177–3187.

Atabai, K., L. B. Ware, M. E. Snider, P. Koch, B. Daniel, T. J. Nuckton, and M. A. Matthay, 2002. Aerosolized beta(2)-adrenergic agonists achieve therapeutic levels in the pulmonary edema fluid of ventilated patients with acute respiratory failure. *Intensive Care Medicine* 28 (6): 705–711.

Azarmi, S., X. Tao, H. Chen, Z. Wang, W. H. Finlay, R. Loebenberg, and W. H. Roa, 2006. Formulation and cytotoxicity of doxorubicin nanoparticles carried by dry powder aerosol particles. *International Journal of Pharmaceutics* 319 (1–2): 155–161.

Bai, S., C. Thomas, and F. Ahsan, 2007. Dendrimers as a carrier for pulmonary delivery of enoxaparin, a low-molecular weight heparin. *Journal of Pharmaceutical Sciences* 96 (8): 2090–2106.

Ballinger, J. R., T. W. Andrey, I. Boxen, and Z. M. Zhang, 1993. Formulation of technetium-99m-aerosol colloid with improved delivery efficiency for lung ventilation imaging. *Journal of Nuclear Medicine* 34 (2): 268–271.

Beach, S., D. Latham, C. Sidgwick, M. Hanna, and P. York, 1999. Control of the physical form of salmeterol xinafoate. *Organic Process Research and Development* 3 (5): 370–376.

Bechtold-Peters, K., M. Walz, G. Boeck, and R. Doerr, 2002. Tiotropium-containing inhalation powder for the treatment of chronic obstructive pulmonary disease. WO Patent Application, Boehringer Ingelheim Pharma K.-G., Germany.

Bennett, W. D., M. A. Almond, K. L. Zeman, J. G. Johnson, and J. F. Donohue, 2006. Effect of salmeterol on mucociliary and cough clearance in chronic bronchitis. *Pulmonary Pharmacology and Therapeutics* 19 (2): 96–100.

Bisgaard, H., 1998. Automatic actuation of a dry powder inhaler into a nonelectrostatic spacer. *American Journal of Respiratory and Critical Care Medicine* 157 (2): 518–521.

Bonnet-Gonnet, C., 2008. Pharmaceutical formulations comprising colloidal suspensions, for the prolonged release of active principle(s), and their applications, especially therapeutic applications. U.S. Patent Application, Flamel Technologies, Inc., France.

Bosquillon, C., V. Preat, and R. Vanbever, 2004. Pulmonary delivery of growth hormone using dry powders and visualization of its local fate in rats. *Journal of Controlled Release* 96 (2): 233–244.

Bot, A. I., T. E. Tarara, D. J. Smith, S. R. Bot, C. M. Woods, and J. G. Weers, 2000. Novel lipid-based hollow-porous microparticles as a platform for immunoglobulin delivery to the respiratory tract. *Pharmaceutical Research* 17 (3): 275–283.

Brain, J. D. and J. D. Blancard, 1993. Mechanisms of particle deposition and clearance. In F. Moren, M. B. Dolovich, M. T. Newhouse, and S. P. Newman (Eds.), *Aerosols in Medicine. Principles, Diagnosis and Therapy*, 2nd edition, pp. 117–155. Amsterdam: Elsevier Science.

Brito, L., D. Chen, Q. Ge, and D. Treco, 2007. Dry powder compositions for RNA influenza therapeutics. U.S. Patent Application, Nastech Pharmaceutical Company Inc., United States.

Brown, B. A. S., J. A. Rasmussen, D. P. Becker, and D. R. Friend, 2004. A piezo-electronic inhaler for local and systemic applications. *Drug Delivery Technology* 4 (8): 90–93.

Byron, P. R., 1986. Prediction of drug residence times in regions of the human respiratory tract following aerosol inhalation. *Journal of Pharmaceutical Sciences* 75 (5): 433–438.

Byron, P. R. and J. S. Patton, 1994. Drug delivery via the respiratory tract. *Journal of Aerosol Medicine* 7 (1): 49–75.

Calfee, D. P. and F. G. Hayden, 1998. New approaches to influenza chemotherapy. Neuraminidase inhibitors. *Drugs* 56 (4): 537–553.

Carpenter, R. L. and E. C. Kimmel, 1999. Aerosol deposition modeling using ACSL. *Drug and Chemical Toxicology* 22 (1): 73–90.

Cass, L. M., K. A. Gunawardena, M. M. Macmahon, and A. Bye, 2000. Pulmonary function and airway responsiveness in mild to moderate asthmatics given repeated inhaled doses of zanamivir. *Respiratory Medicine* 94 (2): 166–173.

Cass, L. M. R., J. Brown, M. Pickford, S. Fayinka, S. P. Newman, C. J. Johansson, and A. Bye, 1999. Pharmacoscintigraphic evaluation of lung deposition of inhaled zanamivir in healthy volunteers. *Clinical Pharmacokinetics* 36 (Suppl. 1): 21–31.

Cheatham, W. W., A. Leone-Bay, M. Grant, P. B. Fog, and D. C. Diamond, 2006. Pulmonary delivery of inhibitors of phosphodiesterase type 5. WO Patent Application, Mannkind Corporation, United States.

Chougule, M., B. Padhi, and A. Misra, 2007. Nano-liposomal dry powder inhaler of tacrolimus: Preparation, characterization, and pulmonary pharmacokinetics. *International Journal of Nanomedicine* 2 (4): 675–688.

Chougule, M., B. Padhi, and A. Misra, 2008. Development of spray dried liposomal dry powder inhaler of dapsone. *AAPS PharmSciTech* 9 (1): 47–53.

Chougule, M. B., B. K. Padhi, and A. Misra, 2006. Nano-liposomal dry powder inhaler of amiloride hydrochloride. *Journal of Nanoscience and Nanotechnology* 6 (9/10): 3001–3009.

Chow, A. H. L., H. H. Y. Tong, P. Chattopadhyay, and B. Y. Shekunov, 2007. Particle engineering for pulmonary drug delivery. *Pharmaceutical Research* 24 (3): 411–437.

Cipolla, D., 2007. Needle-free delivery of HIV therapeutics through aerosols and inhalants. WO Patent Application, Zogenix, Inc., United States.

Cipolla, D. C. and J. Blanchard, 2008. Dual action, anti-infectives inhaled formulations providing both an immediate and sustained release profile for treatment of respiratory tract infections such as cystic fibrosis. WO Patent Application, Aradigm Corporation, United States.

Clark, A. and G. H. Foulds, 1999. Aerosolized active agent delivery. WO Patent Application, Inhale Therapeutic Systems, Inc., United States.

Coates, M. S., F. F. David, H.-K. Chan, and A. R. Judy, 2004. Effect of design on the performance of a dry powder inhaler using computational fluid dynamics. Part 1: Grid structure and mouthpiece length. *Journal of Pharmaceutical Sciences* 93 (11): 2863–2876.

Cochrane, M. G., M. V. Bala, K. E. Downs, J. Mauskopf, and R. H. Ben-Joseph, 2000. Inhaled corticosteroids for asthma therapy: Patient compliance, devices, and inhalation technique. *Chest* 117 (2): 542–550.

Codrons, V., F. Vanderbist, B. Ucakar, V. Preat, and R. Vanbever, 2004. Impact of formulation and methods of pulmonary delivery on absorption of parathyroid hormone (1–34) from rat lungs. *Journal of Pharmaceutical Sciences* 93 (5): 1241–1252.

Codrons, V., F. Vanderbist, R. K. Verbeeck, M. Arras, D. Lison, V. Preat, and R. Vanbever, 2003. Systemic delivery of parathyroid hormone (1–34) using inhalation dry powders in rats. *Journal of Pharmaceutical Sciences* 92 (5): 938–950.

Crowder, T. M., M. D. Louey, V. V. Sethuraman, H. D. C. Smyth, and A. J. Hickey, 2001. An odyssey in inhaler formulation and design. *Pharmaceutical Technology North America* 25 (7): 99–113.

Crowther Labiris, N. R., A. M. Holbrook, H. Chrystyn, S. M. Macleod, and M. T. Newhouse, 1999. Dry powder versus intravenous and nebulized gentamicin in cystic fibrosis and bronchiectasis. A pilot study. *American Journal of Respiratory and Critical Care Medicine* 160 (5): 1711–1716.

Dalby, R. N., S. L. Tiano, and A. J. Hickey, 2007. Medical devices for the delivery of therapeutic aerosols to the lungs. In A. J. Hickey (Ed.), *Inhalation Aerosols (Physical and Biological Basis for Therapy)*, 2nd edition, pp. 417–444. New York: Marcel Dekker.

Dames, P., B. Gleich, A. Flemmer, K. Hajek, N. Seidl, F. Wiekhorst, D. Eberbeck, et al., 2007. Targeted delivery of magnetic aerosol droplets to the lung. *Nature Nanotechnology* 2 (8): 495–499.

Davison, S., J. Thipphawong, J. Blanchard, K. Liu, R. Morishige, I. Gonda, J. Okikawa, et al., 2005. Pharmacokinetics and acute safety of inhaled testosterone in postmenopausal women. *Journal of Clinical Pharmacology* 45 (2): 177–184.

de Boer, A. H., D. Gjaltema, and P. Hagedoorn, 1996. Inhalation characteristics and their effects on *in vitro* drug delivery from dry powder inhalers Part 2: Effect of peak flow rate (PIFR) and inspiration time on the *in vitro* drug release from three different types of commercial dry powder inhalers. *International Journal of Pharmaceutics* 138 (1): 45–56.

Densmore, C. L., 2006. Advances in noninvasive pulmonary gene therapy. *Current Drug Delivery* 3 (1): 55–63.

Deshpande, D., J. Blanchard, S. Srinivasan, D. Fairbanks, J. Fujimoto, T. Sawa, J. Wiener-Kronish, H. Schreier, and I. Gonda, 2002. Aerosolization of lipoplexes using AERx® pulmonary delivery system. *AAPS PharmSci* 4 (3): article 13. Accessed at http://www.aapsj.org/view.asp?art = ps040313

Dhillon, S. and G. M. Keating, 2006. Beclometasone dipropionate/formoterol: In an HFA-propelled pressurized metered-dose inhaler. *Drugs* 66 (11): 1475–1483.

Dickinson, P. A., I. W. Kellaway, and S. W. Howells, 2001. Particulate compositions based on crosslinked polymers. WO Patent Application, University College Cardiff Consultants Limited, United Kingdom.

Dilraj, A., F. T. Cutts, J. F. de Castro, J. G. Wheeler, D. Brown, C. Roth, H. M. Coovadia, and J. V. Bennett, 2000. Response to different measles vaccine strains given by aerosol and subcutaneous routes to schoolchildren: A randomised trial. *Lancet* 355 (9206): 798–803.

Dolovich, M., 1992. The relevance of aerosol particle size to clinical response. *Journal of Biopharmaceutical Sciences* 3 (1/2): 139–145.

Duddu, S. P., A. S. Steven, H. W. Yulia, E. T. Thomas, R. T. Kevin, R. C. Andrew, A. E. Michael, et al., 2002. Improved lung delivery from a passive dry powder inhaler using an engineered Pulmosphere powder. *Pharmaceutical Research* 19 (5): 689–695.

Edwards, D. A., A. Ben-Jebria, and R. Langer, 1998. Recent advances in pulmonary drug delivery using large, porous inhaled particles. *Journal of Applied Physiology* 85 (2): 379–385.

Edwards, D. A., J. Hanes, G. Caponetti, J. Hrkach, A. Ben-Jebria, M. L. Eskew, J. Mintzes, J. Deaver, N. Lotan, and R. Langer, 1997. Large porous particles for pulmonary drug delivery. *Science* 276 (5320): 1868–1871.

Edwards, D. A., J. Sung, B. Pulliam, E. Wehrenberg-Klee, E. Schwartz, P. Dreyfuss, S. Kulkarni, and E. Lieberman, 2005. Pulmonary delivery of malarial vaccine in the form of particulates. WO Patent Application, President and Fellows of Harvard College, Cambridge, MA.

Ely, L., W. Roa, W. H. Finlay, and R. Loebenberg, 2007. Effervescent dry powder for respiratory drug delivery. *European Journal of Pharmaceutics and Biopharmaceutics* 65 (3): 346–353.

Endo, K., S. Amikawa, A. Matsumoto, N. Sahashi, and S. Onoue, 2005. Erythritol-based dry powder of glucagon for pulmonary administration. *International Journal of Pharmaceutics* 290 (1–2): 63–71.

Fiegel, J., J. Fu, and J. Hanes, 2004. Poly(ether-anhydride) dry powder aerosols for sustained drug delivery in the lungs. *Journal of Controlled Release* 96 (3): 411–423.

Freedman, T. 1956. Medihaler therapy for bronchial asthma: A new type of aerosol therapy. *Postgraduate Medicine* 20 (6): 667–673.

French, D. L., D. A. Edwards, and R. W. Niven, 1996. The influence of formulation on emission, deaggregation and deposition of dry powders for inhalation. *Journal of Aerosol Science* 27 (5): 769–783.

Fulzele S. V., M. S. Shaik, A. Chatterjee, and M. Singh, 2006. Anti-cancer effect of celecoxib and aerosolized docetaxel against human non-small cell lung cancer cell line, A549. *Journal of Pharmacy and Pharmacology* 58 (3): 327–336.

Gagnadoux, F., J. Hureaux, L. Vecellio, T. Urban, A. L. Pape, I. Valo, J. Montharu, et al., 2008. Aerosolized chemotherapy. *Journal of Aerosol Medicine and Pulmonary Drug Delivery* 21 (1): 61–70.

Ganderton, D. and N. M. Kassem, 1992. Dry powder inhalers. In E. D. Ganderton and T. Jones (Eds.), *Advances in Pharmaceutical Sciences*. London: Academic Press.

Gandevia, B., 1975. Historical review of the use of parasympatholytic agents in the treatment of respiratory disorders. *Postgraduate Medical Journal* 51 (7): 13–20.

Gautam, A., C. L. Densmore, and J. C. Waldrep, 2000. Inhibition of experimental lung metastasis by aerosol delivery of PEI–p53 complexes. *Molecular Therapy* 2 (4): 318–323.

Geller, D. E., W. K. Michael, J. Smith, B. N. Sarah, and C. Conrad, 2007. Novel tobramycin inhalation powder in cystic fibrosis subjects: Pharmacokinetics and safety. *Pediatric Pulmonology* 42 (4): 307–313.

Gilbert, B. E., R. Parthasarathy, and K. Mehta, 2002. Liposomal aerosols for delivery of chemotherapeutic retinoids to the lungs. U.S. Patent Application, Research Development Foundation, United States.

Gonda, I., 1990. Aerosols for delivery of therapeutic and diagnostic agents to the respiratory tract. *Critical Reviews in Therapeutic Drug Carrier Systems* 6 (4): 273–313.

Gonda, I., 1992. Targeting by deposition. In A. J. Hickey (Ed.), *Pharmaceutical Inhalation Aerosol Therapy*, pp. 61–82. New York: Marcel Dekker.

Grenha, A., B. Seijo, and C. Remunan-Lopez, 2005. Microencapsulated chitosan nanoparticles for lung protein delivery. *European Journal of Pharmaceutical Sciences* 25 (4–5): 427–437.

Griffiths, G. D., S. C. Bailey, J. L. Hambrook, M. Keyte, P. Jayasekera, J. Miles, and E. Williamson, 1997. Liposomally-encapsulated ricin toxoid vaccine delivered intratracheally elicits a good immune response and protects against a lethal pulmonary dose of ricin toxin. *Vaccine* 15 (17/18): 1933–1939.

Grossman, J., 1994. The evolution of inhaler technology. *Journal of Asthma* 31 (1): 55–64.

Haynes, A., M. S. Shaik, A. Chatterjee, and M. Singh, 2003. Evaluation of an aerosolized selective COX-2 inhibitor as a potentiator of doxorubicin in a non-small-cell lung cancer cell line. *Pharmaceutical Research* 20 (9): 1485–1495.

Haynes, A., M. S. Shaik, A. Chatterjee, and M. Singh, 2005. Formulation and evaluation of aerosolized cele-coxib for the treatment of lung cancer. *Pharmaceutical Research* 22 (3): 427–439.

Heinzer, H., E. Huland, M. Aalamian, and H. Huland, 1999. Treatment of pulmonary metastases from kidney cell carcinoma with inhalational interleukin-2. 10-Year experience Hamburger Unicenter. *Der Urologe A* 38 (5): 466–473.

Hersey, J. A., 1975. Ordered mixing: A new concept in powder mixing practice. *Powder Technology* 11 (1): 41–44.

Hershey, A. E., I. D. Kurzman, L. J. Forrest, C. A. Bohling, M. Stonerook, M. E. Placke, A. R. Imondi, and D. M. Vail, 1999. Inhalation chemotherapy for macroscopic primary or metastatic lung tumors: Proof of principle using dogs with spontaneously occurring tumors as a model. *Clinical Cancer Research* 5 (9): 2653–2659.

Hickey, A. J. and N. M. Concessio, 1997. Descriptors of irregular particle morphology and powder properties. *Advanced Drug Delivery Reviews* 26 (1): 29–39.

Hickey A. J., D. Lu, D. A. Elizabeth, and J. Stout, 2006. Inhaled azithromycin therapy. *Journal of Aerosol Medicine* 19 (1): 54–60.

Hickey, A. J. and T. B. Martonen, 1993. Behavior of hygroscopic pharmaceutical aerosols and the influence of hydrophobic additives. *Pharmaceutical Research* 10 (1): 1–7.

Hodson, M. E., 1995. Aerosolized dornase alfa (rhDNase) for therapy of cystic fibrosis. *American Journal of Respiratory and Critical Care Medicine* 151 (3): S70–S74.

Hung, O., J. Zamecnik, P. N. Shek, and P. Tikuisis, 2001. Pulmonary delivery of liposome-encapsulated can-nabinoids. WO Patent Application, Her Majesty the Queen as Represented by the Minister of National Defence of Canada.

Hwang, S. M., D. D. Kim, S. J. Chung, and C. K. Shim, 2008. Delivery of ofloxacin to the lung and alveolar macrophages via hyaluronan microspheres for the treatment of tuberculosis. *Journal of Controlled Release* 129 (2): 100–106.

Iacono, P., P. Velicitat, E. Guemas, V. Leclerc, and J. J. Thebault, 2000. Improved delivery of ipratropium bro-mide using respimat (a new soft mist inhaler) compared with a conventional metered dose inhaler: Cumulative dose response study in patients with COPD. *Respiratory Medicine* 94 (5): 490–495.

Ikeda, A., K. Nishimura, H. Koyama, M. Tsukino, T. Hajiro, M. Mishima, and T. Izumi, 1999. Comparison of the bronchodilator effects of salbutamol delivered via a metered-dose inhaler with spacer, a dry-powder inhaler, and a jet nebulizer in patients with chronic obstructive pulmonary disease. *Respiration: International Review of Thoracic Diseases* 66 (2): 119–123.

Islam, N. and E. Gladki, 2008. Dry powder inhalers (DPIs)—A review of device reliability and innovation. *International Journal of Pharmaceutics* 360 (1–2): 1–11.

Islam, N., P. Stewart, I. Larson, and P. Hartley, 2004. Lactose surface modification by decantation: Are drug-fine lactose ratios the key to better dispersion of salmeterol xinafoate from lactose-interactive mixtures? *Pharmaceutical Research* 21 (3): 492–499.

Johnston, M. R., R. F. Minchen, and C. A. Dawson, 1995. Lung perfusion with chemotherapy in patients with unresectable metastatic sarcoma to the lung or diffuse bronchioloalveolar carcinoma. *Journal of Thoracic and Cardiovascular Surgery* 110 (2): 368–373.

Jurima-Romet, M., R. F. Barber, J. Demeester, and P. N. Shek, 1990. Distribution studies of liposome-encapsu-lated glutathione administered to the lung. *International Journal of Pharmaceutics* 63 (3): 227–235.

Katz, I. M., J. D. Schroeter, and T. B. Martonen, 2001. Factors affecting the deposition of aerosolized insulin. *Diabetes Technology and Therapeutics* 3 (3): 387–397.

Kawashima, Y., T. Serigano, T. Hino, H. Yamamoto, and H. Takeuchi, 1998. Design of inhalation dry powder of pranlukast hydrate to improve dispersibility by the surface modification with light anhydrous silicic acid (AEROSIL 200). *International Journal of Pharmaceutics* 173 (1/2): 243–251.

Kawashima, Y., H. Takeuchi, H. Yamamoto, and K. Mimura, 2000. Powdered polymeric nanoparticulate system to improve pulmonary delivery of calcitonin with dry powder inhalation. *Proceedings of the International Symposium on Controlled Release of Bioactive Materials* 27: 229–230.

Keam, S. J. and G. M. Keating, 2004. Tiotropium bromide: A review of its use as maintenance therapy in patients with COPD. *Treatments in Respiratory Medicine* 3 (4): 247–268.

Kichler, A., C. Leborgne, E. Coeytaux, and O. Danos, 2001. Polyethylenimine-mediated gene delivery: A mechanistic study. *Journal of Gene Medicine* 3 (2): 135–144.

Knight, J. V., N. Koshkina, B. Gilbert, and C. F. Verschraegen, 2006. Small particle liposome aerosols for deliv-ery of anticancer drugs. U.S. Patent Application.

Koshkina, N. V. and E. S. Kleinerman, 2005. Aerosol gemcitabine inhibits the growth of primary osteosarcoma and osteosarcoma lung metastases. *International Journal of Cancer* 116 (3): 458–463.

Kuhn, R. J., 2002. Pharmaceutical considerations in aerosol drug delivery. *Pharmacotherapy* 22 (3): 80S–85S.

Laube, B. L., 2001. Treating diabetes with aerosolized insulin. *Chest* 120 (Suppl. 3): 99S–106S.

Laube, B. L., 2005. The expanding role of aerosols in systemic drug delivery, gene therapy, and vaccination. *Respiratory Care* 50 (9): 1161–1176.

Le Brun, P. P. H., A. H. de Boer, G. P. M. Mannes, D. M. I. de Fraiture, R. W. Brimicombe, D. J. Touw, A. A. Vinks, H. W. Frijlink, and H. G. M. Heijerman, 2002. Dry powder inhalation of antibiotics in cystic fibrosis therapy: Part 2. Inhalation of a novel colistin dry powder formulation: A feasibility study in healthy volunteers and patients. *European Journal of Pharmaceutics and Biopharmaceutics* 54 (1): 25–32.

Le Conte, P., G. Potel, P. Peltier, D. Horeau, J. Caillon, M. E. Juvin, M. F. Kergueris, D. Bugnon, and D. Baron, 1993. Lung distribution and pharmacokinetics of aerosolized tobramycin. *American Review of Respiratory Disease* 147 (5): 1279–1282.

Lentz, Y. K., T. J. Anchordoquy, and C. S. Lengsfeld, 2006. Rationale for the selection of an aerosol delivery system for gene delivery. *Journal of Aerosol Medicine* 19 (3): 372–384.

Li, H. Y., P. C. Seville, I. J. Williamson, and J. C. Birchall, 2005. The use of amino acids to enhance the aerosolisation of spray-dried powders for pulmonary gene therapy. *Journal of Gene Medicine* 7 (3): 343–353.

Li, W.-I. and D. A. Edwards, 1997. Aerosol particle transport and deaggregation phenomena in the mouth and throat. *Advanced Drug Delivery Reviews* 26 (1): 41–49.

Li, Z., J. Jiang, X. Liu, Y. Xia, S. Zhao, and J. Wang, 2008. Preparation of tetracycline microparticles suitable for inhalation administration by supercritical fluid expansion depressurization. *Chemical Engineering and Processing* 47 (8): 1317–1322.

LiCalsi, C., M. J. Maniaci, T. Christensen, E. Phillips, G. H. Ward, and C. Witham, 2001. A powder formulation of measles vaccine for aerosol delivery. *Vaccine* 19 (17–19): 2629–2636.

Liu, J., T. Gong, H. Fu, C. Wang, X. Wang, Q. Chen, Q. Zhang, Q. He, and Z. Zhang, 2008. Solid lipid nanoparticles for pulmonary delivery of insulin. *International Journal of Pharmaceutics* 356 (1–2): 333–344.

Louey, M. D. and P. J. Stewart, 2002. Particle interactions involved in aerosol dispersion of ternary interactive mixtures. *Pharmaceutical Research* 19 (10): 1524–1531.

Lucas, P., K. Anderson, and J. N. Staniforth, 1998. Protein deposition from dry powder inhalers: Fine particle multiplets as performance modifiers. *Pharmaceutical Research* 15 (4): 562–569.

Lulla, A. and G. Malhotra, 2007. Troventol formulation. WO Patent Application, Cipla Limited, India, and C. R. Turner.

Lungwitz, U., M. Breunig, T. Blunk, and A. Goepferich, 2005. Polyethylenimine-based non-viral gene delivery systems. *European Journal of Pharmaceutics and Biopharmaceutics* 60 (2): 247–266.

Mao, L. and J. Blair, 2004. Effect of additives on the aerosolization properties of spray dried trehalose powders. *Respiratory Delivery of Drugs* 9 (3): 653–656.

Mather, L. E., A. Woodhouse, M. E. Ward, S. J. Farr, R. A. Rubsamen, and L. G. Eltherington, 1998. Pulmonary administration of aerosolized fentanyl: Pharmacokinetic analysis of systemic delivery. *British Journal of Clinical Pharmacology* 46 (1): 37–43.

Matida, E. A., W. H. Finlay, M. Rimkus, B. Grgic, and C. F. Lange, 2004. A new add-on spacer design concept for dry-powder inhalers. *Journal of Aerosol Science* 35 (7): 823–833.

Matilainen, L., K. Jaervinen, T. Toropainen, E. Naesi, S. Auriola, T. Jaervinen, and P. Jarho, 2006. *In vitro* evaluation of the effect of cyclodextrin complexation on pulmonary deposition of a peptide, cyclosporin A. *International Journal of Pharmaceutics* 318 (1–2): 41–48.

McCallion, O. N. M., K. M. G. Taylor, P. A. Bridges, M. Thomas, and A. J. Taylor, 1996. Jet nebulizers for pulmonary drug delivery. *International Journal of Pharmaceutics* 130 (1): 1–11.

Milla, C. E., 1998. Long-term effects of aerosolised rhDNase on pulmonary disease progression in patients with cystic fibrosis. *Thorax* 53 (12): 1014–1017.

Morton, D., 2006. Dry powder inhaler formulations comprising surface-modified particles with anti-adherent additives. WO Patent Application, Vectura Limited, United Kingdom.

Moshen, N. M. and T. A. Armer, 2003. Method to aerosolize interferon-gamma for lung delivery for local and systemic treatments. U.S. Patent Application.

Mullins, M. E., L. P. Michaels, V. Menon, B. Locke, and M. B. Ranade, 1992. Effect of geometry on particle adhesion. *Aerosol Science and Technology* 17 (2): 105–118.

Muthu, D. C., 1922. *Pulmonary tuberculosis: Its Etiology and Treatment—Record of Twenty Two Years Observation and Work in Open Air Sanitoria*. London: Bailliere, Tyndal and Cox.

Newhouse, M. T., P. H. Hirst, S. P. Duddu, Y. H. Walter, T. E. Tarara, A. R. Clark, and J. G. Weers, 2003. Inhalation of a dry powder tobramycin pulmosphere formulation in healthy volunteers. *Chest* 124 (1): 360–366.

Newman, S. P. and S. W. Clarke, 1983. Therapeutic aerosols 1—Physical and practical considerations. *Thorax* 38 (12): 881–886.

Newman, S. P., 2003. Deposition and effects of inhaled corticosteroids. *Clinical Pharmacokinetics* 42 (6): 529–544.

Newman, S. P., D. Pavia, and S. W. Clarke, 1981. How should a pressurized beta-adrenergic bronchodilator be inhaled? *European Journal of Respiratory Diseases* 62 (1): 3–21.

O'Callaghan, C., O. Nerbrink, and M. T. Vidgren, 2002. The history of inhaled drug therapy. *Lung Biology in Health and Disease* 162 (1): 1–20.

Ohmori, Y., S. Onoue, K. Endo, A. Matsumoto, S. Uchida, and S. Yamada, 2006. Development of dry powder inhalation system of novel vasoactive intestinal peptide (VIP) analogue for pulmonary administration. *Life Sciences* 79 (2): 138–143.

Omri, A., E. S. Zacharias, and N. S. Pang, 2002. Enhanced activity of liposomal polymyxin B against *Pseudomonas aeruginosa* in a rat model of lung infection. *Biochemical Pharmacology* 64 (9): 1407–1413.

Pandey, R. and G. K. Khuller, 2005. Solid lipid particle-based inhalable sustained drug delivery system against experimental tuberculosis. *Tuberculosis* 85 (4): 227–234.

Pavkov, R. M., T. A. Armer, and N. M. Mohsen, 2005. Aerosol formulations for delivery of dihydroergotamine to the systemic circulation via pulmonary inhalation. WO Patent Application, Map Pharmaceuticals, Inc., United States.

Pillai, R. S., D. B. Yeates, I. F. Miller, and A. J. Hickey, 1998. Controlled dissolution from wax-coated aerosol particles in canine lungs. *Journal of Applied Physiology* 84 (2): 717–725.

Platz, R. M., N. Kimura, O. Satoh, and L. C. Foster, 1995. Dry powder formulation of interferons. WO Patent Application, Inhale Therapeutic Systems, Inc., United States, and R. M. Platz.

Qi, Y., G. Zhao, D. Liu, Z. Shriver, M. Sundaram, S. Sengupta, G. Venkataraman, R. Langer, and R. Sasisekharan, 2004. Delivery of therapeutic levels of heparin and low-molecular-weight heparin through a pulmonary route. *Proceedings of the National Academy of Sciences of the United States of America* 101 (26): 9867–9872.

Rudolph, C., J. Lausier, S. Naundorf, R. H. Muller, and J. Rosenecker, 2000. *In vivo* gene delivery to the lung using polyethylenimine and fractured polyamidoamine dendrimers. *Journal of Gene Medicine* 2 (4): 269–278.

Ruijgrok, E. J., H. A. Fens Marcel, A. J. M. Bakker-Woudenberg Irma, W. M. van Etten Els, and G. Vulto Arnold, 2005. Nebulization of four commercially available amphotericin B formulations in persistently granulocytopenic rats with invasive pulmonary aspergillosis: Evidence for long-term biological activity. *Journal of Pharmacy and Pharmacology* 57 (10): 1289–1295.

Schuster, J., R. Rubsamen, P. Lloyd, and J. Lloyd, 1997. The AERX aerosol delivery system. *Pharmaceutical Research* 14 (3): 354–357.

Seeger, W., T. Schmehl, T. Gessler, L. A. Dailey, and M. Wittmar, 2004. Biodegradable colloidal particles based on comb polymers, in particular for aerosol therapy. Ger. Patent Application, Transmit Gesellschaft Fuer Technologietransfer MbH, Justus-Liebig-Universitaet Giessen, and Philipps-Universitaet Marburg, Germany.

Sethuraman, V. V. and A. J. Hickey, 2002. Powder properties and their influence on dry powder inhaler delivery of an antitubercular drug. *AAPS PharmSciTech* 3 (4): 1–10.

Shahiwala, A. and A. Misra, 2005. A preliminary pharmacokinetic study of liposomal leuprolide dry powder inhaler: A technical note. *AAPS PharmSciTech* 6 (3): E482–E486.

Shapiro, G., W. Lumry, J. Wolfe, J. Given, M. V. White, A. Woodring, L. Baitinger, K. House, B. Prillaman, and T. Shah, 2000. Combined salmeterol 50 microg and fluticasone propionate 250 microg in the diskus device for the treatment of asthma. *American Journal of Respiratory and Critical Care Medicine* 161 (2, Pt. 1): 527–534.

Shekunov, B. Y., J. C. Feeley, A. H. L. Chow, H. H. Y. Tong, and P. York, 2002. Physical properties of supercritically-processed and micronised powders for respiratory drug delivery. *KONA Powder and Particle Journal* 20: 178–187.

Shekunov, B. Y., J. C. Feeley, A. H. L. Chow, H. H. Y. Tong, and P. York. 2003. Aerosolisation behaviour of micronised and supercritically-processed powders. *Journal of Aerosol Science* 34 (5): 553–568.

Smith, D. J., S. Bot, L. Dellamary, and A. Bot, 2003. Evaluation of novel aerosol formulations designed for mucosal vaccination against influenza virus. *Vaccine* 21 (21–22): 2805–2812.

Smyth, H. D. C., I. Saleem, M. Donovan, and C. F. Verschraegen, 2008. Pulmonary delivery of anti-cancer agents. In R. O. Williams III, D. R. Taft, and J. T. McConville (Eds.), *Advanced Drug Formulation Design to Optimize Therapeutic Outcomes. Drugs and the Pharmaceutical Sciences Series*, Volume 172, pp. 81–111. New York: Informa Healthcare USA, Inc.

Solis-Cohen, S., 1990. The use of adrenal substance in the treatment of asthma. *Journal of Asthma* 27 (6): 401–406.

Srichana, T., A. Brain, G. P. Martin, and C. Marriott, 1998. The determination of drug-carrier interactions in dry powder inhaler formulations. *Journal of Aerosol Science* 29 (Suppl. 1, Pt. 2): S757–S758.

Stahl, K., K. Backstorm, K. Thalberg, A. Axelsson, T. Schaefer, and H. G. Kristensen, 2002. Spray drying and characterization of particles for inhalation. *Respiratory Delivery of Drugs* 8 (2): 565–568.

Staniforth, J. N., D. Morton, M. Tobyn, S. Eason, Q. Harmer, and D. Ganderton, 2006. Pharmaceutical compositions comprising apomorphine for pulmonary inhalation. U.S. Patent Application.

Steckel, H. and B. W. Mueller, 1997. *In vitro* evaluation of dry powder inhalers I: Drug deposition of commonly used devices. *International Journal of Pharmaceutics* 154 (1): 19–29.

Suarez, S. and A. J. Hickey, 2000. Drug properties affecting aerosol behavior. *Respiratory Care* 45 (6): 652–666.

Tam, J. M., J. T. McConville, R. O. Williams III, and K. P. Johnston, 2008. Amorphous cyclosporin nanodispersions for enhanced pulmonary deposition and dissolution. *Journal of Pharmaceutical Sciences* 97 (11): 4915–4933.

Telko, M. J. and A. J. Hickey, 2005. Dry powder inhaler formulation. *Respiratory Care* 50 (9): 1209–1227.

Theil, C. G. 1996. From Susie's question to CFC free: An inventor's perspective on forty years of metered dose inhaler development and regulation. *Respiratory Drug Delivery* 5 (1): 115–123.

Thomas, S. H., M. J. O'Doherty, A. Graham, C. J. Page, P. Blower, D. M. Geddes, and T. O. Nunan, 1991. Pulmonary deposition of nebulised amiloride in cystic fibrosis: Comparison of two nebulisers. *Thorax* 46 (10): 717–721.

Tian, Y., M. E. Klegerman, and A. J. Hickey, 2004. Evaluation of microparticles containing doxorubicin suitable for aerosol delivery to the lungs. *PDA Journal of Pharmaceutical Science and Technology* 58 (5): 266–275.

Timsina, M. P., G. P. Martin, C. Marriott, D. Ganderton, and M. Yianneskis, 1994. Drug delivery to the respiratory tract using dry powder inhalers. *International Journal of Pharmaceutics* 101 (1–2): 1–13.

Tozuka, Y. and H. Takeuchi, 2008. Fine particle design and preparations for pulmonary drug delivery. *Drug Delivery System* 23 (4): 467–473.

Valle, M. J. de Jesus, F. G. Lopez, and A. S. Navarro, 2008. Pulmonary versus systemic delivery of levofloxacin. *Pulmonary Pharmacology and Therapeutics* 21 (2): 298–303.

Vanacker Nele, J., E. Palmans, A. P. Romain, and C. K. Johan, 2002. Effect of combining salmeterol and fluticasone on the progression of airway remodeling. *American Journal of Respiratory and Critical Care Medicine* 166 (8): 1128–1134.

Vanderbist, F., B. Wery, D. Baran, B. Van Gansbeke, A. Schoutens, and A. J. Moes, 2001. Deposition of nacystelyn from a dry powder inhaler in healthy volunteers and cystic fibrosis patients. *Drug Development and Industrial Pharmacy* 27 (3): 205–212.

Waldrep, J. C., J. Arppe, K. A. Jansa, and M. Vidgren, 1998. Experimental pulmonary delivery of cyclosporin A by liposome aerosol. *International Journal of Pharmaceutics* 160 (2): 239–250.

Westerman, E. M., A. H. de Boer, P. P. H. Le Brun, D. J. Touw, H. W. Frijlink, and H. G. M. Heijerman, 2007. Dry powder inhalation of colistin sulphomethate in healthy volunteers: A pilot study. *International Journal of Pharmaceutics* 335 (1–2): 41–45.

Westerman, E. M., P. P. H. Le Brun, D. J. Touw, H. W. Frijlink, and H. G. M. Heijerman, 2004. Effect of nebulized colistin sulphate and colistin sulphomethate on lung function in patients with cystic fibrosis: A pilot study. *Journal of Cystic Fibrosis* 3 (1): 23–28.

Wightman, L., R. Kircheis, V. Rossler, S. Carotta, R. Ruzicka, M. Kursa, and E. Wagner, 2001. Different behavior of branched and linear polyethylenimine for gene delivery *in vitro* and *in vivo*. *Journal of Gene Medicine* 3 (4): 362–372.

Windom, H. H., C. D. Burgess, R. W. Siebers, G. Purdie, N. Pearce, J. Crane, and R. Beasley, 1990. The pulmonary and extrapulmonary effects of inhaled beta-agonists in patients with asthma. *Clinical Pharmacology and Therapeutics* 48 (3): 296–301.

Wittgen, B. P. H., P. W. A. Kunst, K. der Born, A. W. van Wijk, W. Perkins, F. G. Pilkiewicz, R. Perez-Soler, S. Nicholson, G. J. Peters, and P. E. Postmus, 2007. Phase I study of aerosolized SLIT cisplatin in the treatment of patients with carcinoma of the lung. *Clinical Cancer Research* 13 (8): 2414–2421.

Wong-Chew, R. M., R. Islas-Romero, M. de L. Garcia-Garcia, J. A. Beeler, S. Audet, J. I. Santos-Preciado, H. Gans, et al., 2006. Immunogenicity of aerosol measles vaccine given as the primary measles immunization to nine-month-old Mexican children. *Vaccine* 24 (5): 683–690.

Xu, C.-X., D. Jere, H. Jin, S.-H. Chang, Y.-S. Chung, J.-Y. Shin, J.-E. Kim, et al., 2008. Poly(ester amine)-mediated, aerosol-delivered Akt1 small interfering RNA suppresses lung tumorigenesis. *American Journal of Respiratory and Critical Care Medicine* 178 (1): 60–73.

Yamamoto, H., Y. Kuno, S. Sugimoto, H. Takeuchi, and Y. Kawashima, 2005. Surface-modified PLGA nano-sphere with chitosan improved pulmonary delivery of calcitonin by mucoadhesion and opening of the intercellular tight junctions. *Journal of Controlled Release* 102 (2): 373–381.

Yang, T., F. Mustafa, S. Bai, and F. Ahsan, 2004. Pulmonary delivery of low molecular weight heparins. *Pharmaceutical Research* 21 (11): 2009–2016.

Yang, W., J. Tam, D. A. Miller, J. Zhou, J. T. McConville, K. P. Johnston, and R. O. Williams, 2008. High bioavailability from nebulized itraconazole nanoparticle dispersions with biocompatible stabilizers. *International Journal of Pharmaceutics* 361 (1–2): 177–188.

York, P., M. Hanna, and G. O. Humphreys, 1999. Strategies for particle design using supercritical fluid tech-nologies. In Institution of Chemical Engineers (Ed.), *Proceedings of the 14th International Symposium on Industrial Crystallization*, pp. 264–272. Cambridge, U.K.: Institution of Chemical Engineers.

Yu, J. and Y. W. Chien, 1997. Pulmonary drug delivery: Physiologic and mechanistic aspects. *Critical Reviews in Therapeutic Drug Carrier Systems* 14 (4): 395–453.

Zeng, X. M., G. P. Martin, and C. Marriott, 1998. The effect of molecular weight of polyvinylpyrrolidone on the crystallization of co-lyophilized sucrose. *Journal of Pharmacy and Pharmacology* 50 (Suppl.): 65.

Zeng, X. M., G. P. Martin, C. Marriott, and J. Pritchard, 2000a. The influence of carrier morphology on drug delivery by dry powder inhalers. *International Journal of Pharmaceutics* 200 (1): 93–106.

Zeng, X. M., G. P. Martin, C. Marriott, and J. Pritchard, 2000b. The effects of carrier size and morphology on the dispersion of salbutamol sulfate after aerosolization at different flow rates. *Journal of Pharmacy and Pharmacology* 52 (10): 1211–1221.

22 Respiratory Aerosol Dynamics with Applications to Pharmaceutical Drug Delivery

Jinxiang Xi, P. Worth Longest, and Paula J. Anderson

CONTENTS

22.1 INTRODUCTION

The inhalation of medical aerosols is becoming increasingly popular for the treatment of different respiratory airway diseases. In addition, there is a great interest to utilize the inhalation pathway for systemic therapy because of its large surface area that is available for absorption. Therefore, determining the required regional and local distribution of inhaled therapeutic aerosols within the respiratory system is a key issue for effective aerosol drug designs. Regional estimates of aerosol deposition allow for area-averaged predictions of tissue-level doses within prescribed areas of the respiratory tract. However, these regional estimates ignore true particle localization within the considered areas. Local deposition patterns provide a much more accurate view of the potential for component absorption, response, and tissue damage, which are cellular-level processes.

The deposition of inhaled pharmaceutical aerosols throughout the respiratory tract has been considered in a number of experimental and numerical studies. Many *in vivo* studies have been conducted in both healthy subjects (Cheng et al., 1996; Heyder et al., 1986; Hofmann et al., 2001; Jaques and Kim, 2000; Kim and Jaques, 2004; Morawska et al., 1999, 2005; Stahlhofen et al., 1989) and patients with certain respiratory diseases such as cystic fibrosis and asthma (Laube et al., 1998, 2000; Newman, 1988). Stahlhofen et al. (1989) summarized earlier studies regarding the *in vivo* deposition in extrathoracic airways, including the reviews of Chan and Lippmann (1980) and Yu et al. (1981). For most of the studies, particles were administered to the subjects via a mouthpiece. Stahlhofen et al. (1989) proposed a best-fit correlation for filtering efficiency based on available data using an impaction parameter that is similar to the Stokes number. Because of intersubject variability and differences in experimental procedures, considerable scatter was observed among the *in vivo* experiments and the suggested empirical correlation of Stahlhofen et al. (1989). Cheng et al. (1996) measured the nasal and oral deposition of ultrafine aerosols ranging from 4 to 150 nm in 10 normal adult male subjects using magnetic resonance imaging (MRI) scans, and they reported significant intersubject variability in deposition rates. Deposition in the nasal cavity was shown to be a function of the surface area, minimum cross-sectional area, and shape complexity of the nose. Correlations for both nasal and oral deposition were developed as a function of geometric parameters and particle diffusion.

Studies that have considered flow field and particle deposition in human airway replicas include Heenan et al. (2003), Johnstone et al. (2004), Zhang et al. (2006, 2004), Cheng et al. (1997a, 1993), Hanna and Scherer (1986), Kelly et al. (2000), and Zhao and Lieber (1994). Heenan et al. (2003) studied the flow field in the central sagittal plane of an idealized mouth–throat (MT) model using endoscopic particle image velocimetry. Johnstone et al. (2004) measured the flow field using hot-wire anemometry and flow visualization techniques within an extrathoracic model identical to that adopted by Heenan et al. (2003). Both the hot-wire measurements and complementary flow visualization displayed the complex nature inside the extrathoracic airway with regions of highly localized turbulence intensity distal to the oropharynx. Zhang et al. (2004) measured micron aerosol depositions in a highly idealized MT geometry and two standard United States Pharmacopeia (USP; 2005) throats for particle sizes ranging from 2.5 to 6.0 μm. A comparison of measurements with the *in vivo* average curve indicated that current USP throats are not adequate in representing realistic extrathoracic airway geometries, whereas the proposed idealized MT model could largely reproduce the *in vitro* average curve and therefore is a promising alternative to the USP MT. A design optimization of the proposed idealized MT was later performed by Zhang et al. (2006). The results showed that different key dimensions are required for different flow rates in order to adequately reproduce the *in vivo* extrathoracic depositions. Cheng et al. (1993) reported deposition measurements for 1.73- and 1.21-nm radon progeny particles in airway models that were based on MRI and cadavers. A detailed geometric description was provided by Cheng et al. (1997b) for a realistic oral airway cast with a half-mouth opening. Using deposition results for silver aerosols measured in the realistic model, Cheng et al. (1997b) proposed an empirical correlation of the mass transfer Sherwood number as a function of the Reynolds and Schmidt numbers. However, this correlation was based on deposition measurements for a narrow particle range (3.7–5 nm). Application of this correlation to particles outside of this range has not been verified.

In addition to *in vivo* and *in vitro* experiments, a number of numerical studies have investigated aerosol depositions employing either a Eulerian–Lagrangian (Balashazy et al., 2003; Hofmann et al., 2003; Moskal and Gradon, 2002; Robinson et al., 2006) or a Eulerian–Eulerian (Ingham, 1975, 1991; Lee and Lee, 2002; Martonen et al., 1996; Shi et al., 2004; Zhang et al., 2005) model for particle transport and deposition. The Lagrangian transport model tracks individual particles within the Eulerian flow field and can account for a variety of forces on the particle including inertia, diffusion, gravity effects, and near-wall interactions (Longest et al., 2004). The Eulerian transport model typically assumes that particle inertia can be ignored and particles are treated as a dilute chemical species in a multicomponent mixture model (Crowe et al., 1998). Hofmann et al. (2003) employed a Lagrangian particle tracking model for dilute aerosols ranging from 1 to 500 nm in a model of the third to fourth airway junction. The primary deposition mechanism was reportedly dominated by molecular or Brownian diffusion for particles less than 20 nm and by convective diffusion for particles greater than 20 nm. Shi et al. (2004) considered the deposition of ultrafine particles ranging from 1 to 150 nm in the upper respiratory tract. This study showed reasonable agreement with the regional *in vitro* deposition data of Kim and Fisher (1994) and reported deposition enhancement factors (DEFs) on the order of 10 near carinal ridges. Zhang and Kleinstreuer (2003) developed a circular model of the oral airway geometry specified by Cheng et al. (1997b) for local species heat and mass transfer. Their computational geometry is characterized by a 180° bend and circular cross sections, which are comparable to the hydraulic diameters specified by Cheng et al. (1997b). Zhang et al. (2005) used the same circular oral airway model and an ideal model of respiratory generations 0–3 to compare micro- and nanosized particle deposition on a local basis. They found DEFs, defined as the ratio of the local to average deposition concentrations, that ranged from 40 to 2400 for microparticles and 2 to 11 for nanoparticles. Scherer et al. (1994) investigated human olfaction by testing hypotheses about the relative roles of airflow, odorant uptake, and mucus solubility in nasal function.

Several studies have considered the effects of realistic and simplified geometries on airflow fields and particle deposition in respiratory dynamics systems. Nowak et al. (2003) used computational modeling to evaluate particle deposition in a symmetric Weibel A model of the tracheobronchial (TB) region compared to a realistic model from medical computed tomography (CT) data. Their results showed significant differences in flow fields and total particle deposition characteristics between the two models. Recent studies by van Ertbruggen et al. (2005) and Li et al. (2007) suggested that idealized symmetrical bifurcations are not sufficient to simulate deposition in human lungs and that more anatomically accurate models may be required. In addition, Lin et al. (2007) implemented a CT-based intrathoracic airway model and highlighted enhanced local turbulence entering the TB region as a result of the laryngeal jet. Considering the oral airway, Li et al. (1996) compared particle deposition in a 90° bend model and a more realistic geometry reconstructed from a human oral cast extending from the mouth to the larynx. The 90° bend was used to represent the USP throat geometry that is often implemented to test pharmaceutical aerosols. They found that quantitative differences between the flows in these two models are significant with respect to the ability of the air stream to break apart particle agglomerates. Furthermore, much lower micron aerosol deposition was measured by Zhang et al. (2004) with the 90° bend than reported for *in vivo* experimental data. They suggested that the USP 90° bend might be inadequate to mimic realistic extrathoracic geometries. A series of studies by our group have systemically considered the effect of the airway geometry on regional and local depositions for both nano- and microaerosols in the upper respiratory airway (Xi and Longest, 2007, 2008a, 2008b; Xi et al., 2008). Geometric simplifications were shown to have a significant impact on particle deposition as a function of the particle diameter and on local deposition in terms of a DEF.

This chapter will provide an overview of recent developments in respiratory aerosol dynamics with applications to inhaled pharmaceutical aerosols. We first describe a computational technique that translates radiological medical images into three-dimensional (3-D) airway geometries with high-quality computational meshes. Fluid-particle transport models and related efforts to enhance the prediction accuracy are addressed in Section 22.3. Section 22.4 presents several innovative applications that take advantage of specific aerosol properties to improve delivery outcomes, which include large porous particles, soft mist and smart inhalers, and magnetic targeting. The effects of respiratory physiology, such as geometric realism and breathing maneuvers, on drug delivery efficiency will be examined in Section 22.5, with a brief summary following in Section 22.6.

22.2 MODELING OF RESPIRATORY PHYSIOLOGY

Radiological imaging technologies, which include x-rays, CT scans, MRI scans, and ultrasound, have been extensively used as a noninvasive diagnostic method. A 3-D volumetric reconstruction based on these medical images provides an excellent way for displaying anatomy and represents an innovative tool in many biomedical and biomechanical research fields including vascular cardiology, orthopedics, neurology, and respiratory therapy.

For inhalation toxicology and pharmaceutical research, imaged-based modeling represents a significant improvement compared to conventional cadaver casting, which is subject to distortion that is attributable to the shrinkage of mucous membranes or insertion of casting materials. Pritchard and McRobbie (2004) compared a cadaver cast with their MRI oropharyngeal airway model averaged over 20 subjects (10 males, 10 females) and showed significant geometric disparities. The most prominent difference was the greater volume of the cadaver casts, which was 92.7 cm³ for the cast versus 46.7 cm³ (SD = 12.7) for imaging. Generally, a physical model produced from anatomical images results in about a 5% difference in the airway volume and some loss of fine morphological details. Therefore, this approach may adequately approximate internal geometries (McRobbie et al., 2003).

We use one example to illustrate the computational method for developing a surface model of the respiratory tract based on medical images such as CT or MRI data. Figure 22.1a–e illustrates the procedure of translating from CT scans into a high-quality computational mesh of the nasal–laryngeal airway. To construct the 3-D airway model, CT scans of a healthy nonsmoking 53-year-old male were used, which were obtained with a multirow-detector helical CT scanner (GE medical systems, Discovery LS) and the following acquisition parameters: 0.7-mm effective slice spacing, 0.65-mm overlap, 1.2-mm pitch, and 512 × 512 pixel resolution. The multislice CT images were then imported into MIMICS (Materialise, Ann Arbor, MI) and segmented according to the contrast between osseous structures and intra-airway air to convert the raw image data into a set of cross-sectional contours that define the solid geometry (Figure 22.1a–c). Based on these contours, a surface geometry was constructed in Gambit 2.3 (Ansys, Inc.; Figure 22.1d). The surface geometry was then imported into Ansys ICEM 10 (Ansys, Inc.) as an IGES file for meshing. Because of the

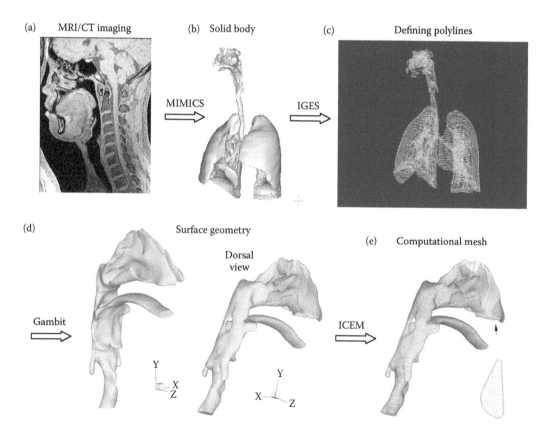

FIGURE 22.1 Procedure for 3-D rendering of CT or MRI images. (a) From the CT scans of an adult healthy subject, (b) a solid model of the airway was isolated using MIMICS. (c) Using the defining polylines, a surface geometry of interest, (d) which is the nasal–oral–laryngeal airway, was constructed in Gambit. (e) Using Ansys ICEM 10, the computational mesh was generated and consisted of unstructured tetrahedral control volumes with a very fine near-wall grid of prism elements.

high complexity of the model geometry, an unstructured tetrahedral mesh was created with high-resolution pentahedral elements in the near-wall region (Figure 22.1e). The main geometric features retained in this example include the nasal passages, nasopharynx (NP), pharynx, and throat. In particular, anatomical details such as the uvula, epiglottal fold, and laryngeal sinus are retained. The resulting model is intended to faithfully represent the anatomy of the extrathoracic airway with only minor surface smoothing.

Existing physical casts can also be digitally reproduced using this procedure (Xi et al., 2008). This is desirable if a direct comparison between model prediction and experimental deposition data in cast are needed for a model validation purposes. Figure 22.2 displays a hollow TB cast implemented by Cohen et al. (1990) and the reconstructed surface geometry of the TB cast. The cast was first scanned by a multirow-detector helical CT scanner. The multislice CT images were then processed as described above and were translated into an almost identical digital model. The airway surface was divided into various sections consistent with the TB cast so that a direct comparison between simulations and measurements could be made in each subdivided segment. Figure 22.3 exemplifies one physical model including the nose, mouth, pharynx, larynx, and TB airways through bifurcation generation G6. This model can be either manufactured into a solid cast by prototyping techniques for *in vitro* studies or meshed with high-quality computational elements for numerical analysis (Longest and Vinchurkar, 2007a).

The limitations of using radiological data include difficulties in reproducing physiological conditions and image artifacts from motion. The radiation dose and cost also limit the use of this approach. Another effect that needs to be taken into account is the typical supine position of the subjects during data acquisition, which is different from normal breathing while awake. Images are also taken at the end of a full inspiration that may not reflect the variation in airway geometry occurring during a breathing cycle. Physiologically accurate airway models are a necessary first step for reliable analysis of environmental toxicant health effects or therapeutic aerosol dosing. The ultimate goal is patient-specific modeling with a user-friendly interface, which one day might be utilized by medical practitioners without extensive computational training.

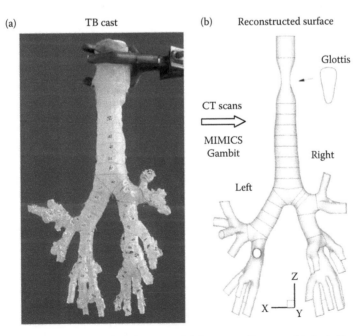

FIGURE 22.2 The development of a TB airway model from a replica cast: (a) the hollow cast utilized by Cohen et al. (1990) showing the divisions of subbranch segments, and (b) a surface geometry was constructed based on CT scans of the cast.

(a)   (b)

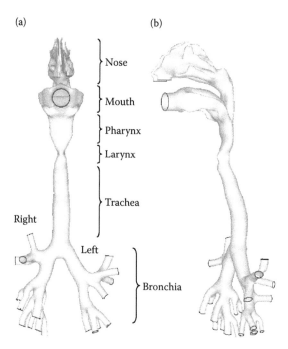

Nose

Mouth

Pharynx

Larynx

Trachea

Right

Left

Bronchia

FIGURE 22.3 A medical image based surface model of the upper respiratory tract including the nose, mouth, pharynx, larynx, and upper TB region as viewed from the (a) front and (b) back. The surface model extends up to respiratory generation G6 along some paths.

22.3 AEROSOL TRANSPORT MODELS AND ENHANCEMENTS

Airflow in the human respiratory tract can be laminar, transitional, or fully turbulent, depending on the breathing activity. There are different scales of turbulence models for simulating turbulent flows: direct numerical simulation, large eddy simulation, Reynolds stress model, and Reynolds-averaged Navier–Stokes models. Selection of an appropriate model for the flow of interest mainly depends on the desired accuracy and the available computational resources. Direct numerical simulation is the most accurate approach and resolves turbulent eddies at all scales, but it also requires the most computational power. Large eddy simulation resolves large-scale energy-containing eddies. Reynolds stress modeling captures the anisotropic turbulence, which is significant near the wall. Even though the lower-order Reynolds-averaged Navier–Stokes models cannot account for turbulence anisotropy, these models are still expected to adequately capture the main features of the flow, provided a very fine mesh is applied in the near-wall boundary region (Xi et al., 2008). The low Reynolds number k-ω model has been implemented in many respiratory simulations based on its ability to accurately predict pressure drop, velocity profiles, and shear stress for transitional and turbulent flows (Ghalichi et al., 1998; Wilcox, 1998). This model was demonstrated to accurately predict particle deposition profiles for transitional and turbulent flows in models of the oral airway (Xi and Longest, 2007, 2008a), nasal passages (Xi and Longest, 2008c), and multipath bifurcations (Longest and Vinchurkar, 2007b; Xi and Longest, 2008b). Moreover, the low Reynolds number k-ω model provided an accurate solution for laminar flow as the turbulent viscosity approaches zero (Wilcox, 1998). Transport equations governing the turbulent kinetic energy (k) and the specific dissipation rate (ω) are provided by Wilcox (1998) and were previously reported by Longest and Xi (2007).

Numerical studies of fine and ultrafine aerosol deposition typically employ either a Eulerian–Eulerian model or a Eulerian–Lagrangian model. The Eulerian–Eulerian particle transport model treats both the continuous and discrete phases as interpenetrating fields. The primary advantage of this model is that it can easily simulate large particle counts and low deposition rates. However, significant modifications are necessary to address inertial (Fernandez de la Mora and Rosner, 1981; Friedlander, 2000) and electrostatic effects (Wang et al., 2002). In contrast, the Lagrangian transport model tracks individual particles within the Eulerian flow field based on Newton's second law of motion. Accordingly, a variety of forces such as inertia, diffusivity, electrostatic effects, and near-wall terms can be accounted for directly (Longest et al., 2004). A major drawback is that the Lagrangian model requires a large number of particles to resolve deposition profiles and therefore can be very computationally expensive.

In this review, a standard chemical species model and a well-tested discrete Lagrangian tracking model were implemented as tools to simulate the dynamics of pharmaceutical colloids. For the chemical species model, the mass transport relation governing the convective–diffusive motion of ultrafine aerosols can be written as

$$\frac{\partial c}{\partial t} + \frac{\partial (u_j c)}{\partial x_j} = \frac{\partial}{\partial x_j}\left[\left(\tilde{D} + \frac{v_T}{Sc_T}\right)\frac{\partial c}{\partial x_j}\right]. \tag{22.1}$$

In the above equation, c represents the mass fraction of nanoparticles, $\tilde{D}$ is the molecular or Brownian diffusion coefficient, and Sc_T is the turbulent Schmidt number taken as 0.9. Assuming dilute concentrations of spherical particles, the Stokes–Einstein equation is used to determine the diffusion coefficients

$$\tilde{D} = \frac{k_B T C_c}{3\pi\mu d_p}, \tag{22.2}$$

where $k_B = 1.38 \times 10^{-16}$ cm^{2}g/s is the Boltzmann constant, d_p is the particle diameter, and C_c is the Cunningham correction factor. The Cunningham correction factor is computed using the expression of Allen and Raabe (1985):

$$C_c = 1 + \frac{\lambda}{d_p}\left(2.34 + 1.05\exp\left(-0.39\frac{d_p}{\lambda}\right)\right), \tag{22.3}$$

where λ is the mean free path of air, assumed to be 65 nm. To approximate the particle deposition on the wall, the boundary condition for the Eulerian transport model is assumed to be $c_{wall} = 0$.

One-way coupled trajectories of monodisperse pharmaceutical particles ranging in diameter (d_p) from 1 nm to approximately 10 μm or more can be calculated on a Lagrangian basis by integrating an appropriate form of the particle transport equation. The appropriate equations for spherical particle motion under these conditions can be expressed as

$$\frac{dv_i}{dt} = \alpha\frac{Du_i}{Dt} + \frac{f}{\tau_p C_c}(u_i - v_i) + g_i(1-\alpha) + f_{i,\text{Brownian}} \tag{22.4a}$$

and

$$\frac{dx_i}{dt} = v_i(t), \tag{22.4b}$$

where v_i and u_i are the components of the particle and local fluid velocity, respectively, and τ_p (i.e., $\rho_p d_p^2/18\mu$) is the characteristic time required for a particle to respond to changes in the flow field.

The drag factor f, which represents the ratio of the drag coefficient C_D to Stokes drag, is based on the expression of Morsi and Alexander (1972). From past simulation experience, we have documented that Fluent's Brownian motion (BM) model does not sufficiently account for nanoparticle depositions (Longest and Xi, 2007). Therefore, the effect of BM on particle trajectories is included as a separate force per unit mass term at each time step based on the approach of Li and Ahmadi (1992) and has been implemented as a user-defined module. The influence of nonuniform fluctuations in the near-wall region is also considered by implementing an anisotropic turbulence model proposed by Matida et al. (2004), which is described as

$$u'_n = f_v x \sqrt{2k/3} \quad \text{and} \quad f_v = 1 - \exp(-0.002 y^+). \tag{22.5}$$

In this equation, ξ is a random number generated from a Gaussian probability density function and f_v is a damping function component normal to the wall for y^+ values of less than approximately 40.

Another enhancement to the Lagrangian tracking is the near-wall treatment of fluid velocity. Evaluation of the particle trajectory equation requires that the fluid velocity u_i be determined at the particle location for each time step. On a computational grid, determining the fluid velocity at the particle location requires spatial interpolation from control-volume centers or nodal values. Based on studies of deposition in the oral airway model using Fluent 6, interpolated fluid velocities in the wall adjacent to the control volumes maintained the value of the control-volume center and did not approach zero at the wall. Longest and Xi (2007) showed that a near-wall interpolation (NWI) algorithm provided an effective approach to accurately predict the deposition of nanoparticles in the respiratory tract. A user-defined routine was developed that linearly interpolates the fluid velocity in near-wall control volumes from nodal values. Velocity values at the wall were taken to be zero. Therefore, implementation of this user-defined NWI routine provided a linear estimate of the fluid velocity at the particle location that goes to zero at the wall boundaries. Similarly, the turbulent kinetic energy values used in Equation 22.5 were also linearly interpolated to approach zero at the wall.

The benefits of applying these two user-defined modules (BM and NWI) are exhibited via the improved agreement with the available *in vitro* data of Cheng et al. (1997a, 1993), as shown in Figure 22.4. Results for the Fluent BM model underpredict the deposition of 1-nm aerosols by a factor of approximately 3 and overpredict the deposition of 100-nm particles by 1 order of

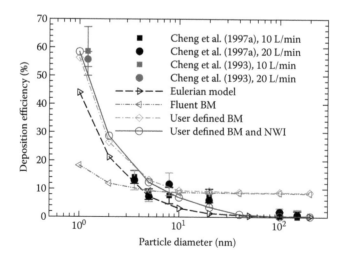

FIGURE 22.4 A comparison of different models in predicting the deposition rates in an oral airway geometry for particles ranging from 1 to 120 nm and a tracheal flow rate of 15 L/min. The Lagrangian model with user-defined BM and NWI routines provides the best match to the available *in vitro* deposition data of Cheng et al. (1993, 1997a).

magnitude. Implementing the user-defined BM routine significantly improves the agreement between the numerical and experimental results for 1 nm, providing a deposition trend similar to that of Eulerian results. The best agreement with the Eulerian results is observed for the user-defined BM and NWI Lagrangian model, which predicted the expected exponential decrease in deposition as particle size is increased. Very similar patterns of local deposition were also obtained between the Eulerian and user-enhanced Lagrangian (BM and NWI) simulations. Therefore, the Lagrangian transport model with appropriate user-defined routines appears to provide an effective approach in simulating the transport and deposition of nanoscale aerosols in the human respiratory tract.

22.4   AEROSOL CHARACTERISTICS AND APPLICATIONS IN DRUG DELIVERY

Factors that determine respiratory dynamics include aerosol characteristics, respiratory geometry and physiology, and breathing conditions. Pharmaceutical aerosols can be categorized according to different aspects such as size, hygroscopicity, and magnetoelectric characteristics. In light of aerosol size, there are ultrafine (<100 nm), fine (100 nm to 1 μm), and coarse (>1 μm) regimes, which have distinct deposition mechanisms. In addition, aerosols with different hygroscopic characteristics exhibit different affinities for moisture, leading to varying degrees of size growth when passing through the humid respiratory tract. If charged or magnetized, the behavior of inhaled drugs may be guided in a desired manner by superimposing an appropriative electrical or magnetic field. This section briefly reviews some innovative applications that take advantage of these aerosol characteristics to help the noninvasive direction of drugs to specific locations within the lung.

22.4.1   Pulmonary Drug Delivery with Large Porous Particles

Large porous particles have recently attracted attention for pulmonary deposition. Characterized by geometric diameters larger than 5 μm and mass densities around 0.1 g/cm^3 or less, porous particles have two major advantages over conventional aerosols (Edwards et al., 1997, 1998; Edwards and Dunbar, 2002). First, increased particle size results in decreased tendency to aggregate, leading to more efficient aerosolization and a higher respirable fraction of inhaled therapeutics. Second, large particles can escape alveolar macrophage clearance, resulting in prolonged bioavailability. Edwards et al. (1997) showed experimentally that the systemic bioavailability of testosterone in porous form with a diameter of 20 μm is approximately 10 times that of conventional therapeutic particles. This attribute can be particularly useful for controlled release inhalation therapies. This ingenious approach to pulmonary delivery arises from a basic aerosol concept, the aerodynamic diameter, which is defined as

$$d_{a} = \sqrt{\frac{\rho}{\rho_{a}}}d, \tag{22.6}$$

where ρ is the particle mass density, d is the geometric diameter, and $\rho_{a} = 1$ g/cm^3. To better understand this concept, consider a spherical particle falling freely under gravity through air, with a settling velocity expressed as

$$v = \left(\frac{g}{18\mu}\right)\rho d^2 = \left(\frac{g}{18\mu}\right)\rho_{a}d_{a}^2 = \left(\frac{g}{18\mu}\right)d_{a}^2. \tag{22.7}$$

The above equation shows that a spherical particle of any mass density will settle with a velocity that depends only on its aerodynamic diameter d_{a}. In other words, as long as the combination $(\sqrt{\rho}d)$ remains the same, particles settle in an identical manner regardless of shape, mass density, or

(a) $d_g = 3$ μm, $Q_{in} = 30$ L/min

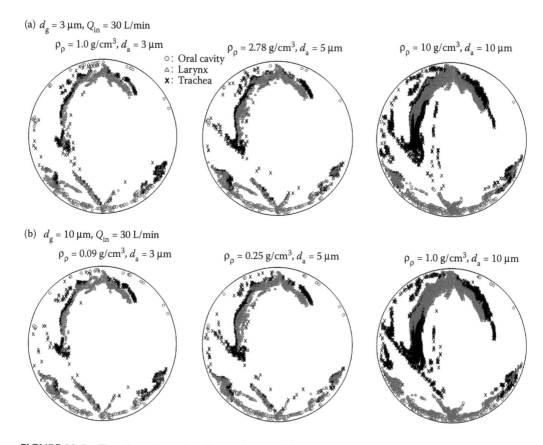

FIGURE 22.5 The released positions (i.e., at the mouth inlet) of particles that deposit in designated regions of the upper respiratory airway for identical geometric diameter (d_g) but varying mass densities: (a) $d_g = 3$ μm and (b) $d_g = 10$ μm. Particles of the same aerodynamic diameter (d_a) exhibit similar behavior.

geometric size. Moreover, the mechanism of inertial impaction also depends uniquely on d_a. Inhaled particles must be within a very narrow range of the aerodynamic diameter (i.e., 1–3 μm) to pass through the mouth and throat region and deposit in the lungs. To formulate particles with a large geometric size while keeping d_a at <3 μm, the mass density should be reduced, which is accomplished by making the particles porous.

To determine whether d_a can adequately quantify a particle's propensity for deposition, we conducted a parametric study by systemically varying the particle mass densities and geometric sizes. Particles were released into a human extrathoracic airway model as shown in Figure 22.5. Local deposition and release positions of particles that deposited in specific lung regions were recorded and compared between particles with identical d_a but different mass densities. For all cases considered, as long as the aerodynamic diameter remains the same, the deposition patterns appear identical for particles with different sizes and densities. Furthermore, particles depositing at a designated site can be traced back to very similar initial released positions for particles with the same aerodynamic diameter but different densities.

22.4.2 Smart Inhaler Devices

Based on observations that regional deposition can usually be traced back to a specific release position at the inlet, Kleinstreuer et al. (2008) put forward an innovative delivery methodology referred to as a smart inhaler system. This system adjusts the air stream, aerosol properties, and aerosol release positions within the mouthpiece, which guide the released aerosols to the targeted airway region. A performance comparison between a conventional pressurized metered dose inhaler (pMDI) and this technique in an idealized extrathoracic airway shows significant improvement for the new approach. Specifically, with controlled inlets, the delivery efficiency to the specified bron-chioles beyond the third generation can increase from approximately 10% to 60–80% (Kleinstreuer et al. 2008). This smart delivery concept provides a potential way for targeting therapeutic agents to diseased lung tissue, which may improve therapeutic efficiency and minimize unwanted side effects. However, the target-specific release positions are expected to vary between different individuals or even one individual within one breathing cycle. Furthermore, particle dispersions due to turbulent vortices and secondary flows may also make these release positions vary irregularly (Xi et al., 2008). Kleinstreuer et al. (2008) suggested that this dispersion effect can be avoided by using aerosols with mildly attractive properties via surfactant coating.

22.4.3 Magnetic Targeting

It is straightforward that the behavior of inhaled ferromagnetic agents could be changed by impos-ing an appropriate magnetic force. Dames et al. (2007) recently demonstrated that it is practical for aerosols to be magnetically targeted *in vivo*. By adding superparamagnetic iron oxide nanoparticles (SPIONs) to colloids, deposition in one lung of a mouse can be greatly enhanced over the other in the presence of a strong magnetic field. The SPIONs are approximately 80 nm in size and biologi-cally inert and have been used for years as contrast agents in MRI. It is interesting to note that it is not a single SPION, which has a low magnetic moment, but rather the SPION assembly that alters the drug droplet trajectories. Dames et al. (2007) showed theoretically that, even with an optimized magnetic design, the magnetic force acting on a single SPION would not be sufficient to effectively guide its trajectory. In contrast, when multiple SPIONs are assembled into an aerosol droplet, which increases the magnetic moment, the assembly becomes guidable within a medically compatible magnetic field. In Dames et al.'s (2007) experiment an aqueous suspension containing SPIONs was nebulized by means of tracheal intubation with the magnetic tip directed toward the right mouse lung and 1 mm away from the tissue. An eightfold higher dose was measured in the right lung while the dose in the left lung was unaltered. The total lung deposition was tripled under the action of the magnetic field compared with the control experiment without a magnetic field.

A potential problem with this method is that the strength of the magnetic field decays quickly with the distance away from the magnetic source, such that a distance of 5 mm resulted in 90% reduction in the magnetic flux. This may reduce the clinical feasibility in humans, because of their relatively large thoracic cavity. It is currently not known if a magnetic field strong enough to guide the SPION-packaged aerosols in human lungs can be achieved in a safe environment.

22.5 EFFECTS OF RESPIRATORY PHYSIOLOGY ON DRUG DELIVERY

22.5.1 Standard Induction Port versus Realistic Mouth–Throat Geometry

Aerosol deposition in the MT region is a critical factor that determines the penetration of inhaled particles into the lungs. For pharmaceutical aerosols, large particles may be filtered out at the MT region and may not efficiently enter deep lung regions. In addition, a significant fraction of smaller particles may also be lost in the MT region because of the aerosol generation technique that is employed (DeHaan and Finlay, 2004; Longest et al., 2007; Stein and Gabrio, 2000). Hence, a better understanding of aerosol deposition in the MT region is needed to improve dosimetry predictions of pulmonary drug delivery.

In addition to the particle size and breathing conditions, the aerosol generation environment or aerosol source may also significantly influence the physical transport mechanisms that contribute to respiratory deposition. *Ambient respiratory aerosols* can be considered to exist in a still environment prior to inhalation and lack initial momentum. These aerosols may include environmental pollutants, some nebulized droplets, and particles from very low inertia dry powder inhaler systems. In contrast, *spray aerosols* are introduced into the MT region through a jet that creates significant momentum and enhanced turbulent effects. Pharmaceutical spray aerosols often arise from liquid-based generation systems such as pMDIs, capillary aerosol generation (CAG), and other microjet systems.

For pMDI spray systems, several researcher groups compared the deposition in USP induction ports (IPs) with deposition in more realistic cast models of the MT geometry (Cheng et al., 2001; Lewis et al., 2006; Stein and Gabrio, 2000; Zhang et al., 2007). These studies typically indicated that pMDI spray deposition in the MT model is greater than the deposition in the USP IP by a factor of approximately 1.5–2.0 (Cheng et al., 2001; Stein and Gabrio 2000; Zhang et al., 2007). Stein and Gabrio (2000) reported a shift in primary deposition from just the first several centimeters of the IP to both the mouth and back of the throat with the MT geometry.

This section evaluates the effects of spray momentum on the deposition of respiratory aerosols in an IP and a realistic MT model. The IP implemented in this study was specified by the 2005 USP as the current best-practice standard for sampling pharmaceutical aerosols into a particle sizing apparatus. CAG was selected as a representative spray aerosol system and the transport and deposition of capillary-generated aerosols were compared with ambient particles of the same size. Two hypotheses will be assessed in this study: (1) the spray system will induce significantly more deposition than an equivalent size distribution of ambient particles, and (2) deposition will increase in the more realistic MT model. By evaluating these hypotheses quantitatively, this study will summarize the findings of Longest et al. (2008) and provide a more accurate prediction of lung delivery for capillary-generated aerosols.

The CAG delivered a solution of 0.6% (w/v) albuterol sulfate (Nephron Pharmaceuticals Corp., Orlando, FL) in water at a mass flow rate of 25 mg/s for 2 s. The tip of the capillary was 57 μm in diameter, which induced a sonic exit velocity, high pressure, and significant turbulence. The aqueous mixture exiting the capillary tip was composed of supersaturated water vapor and liquid droplets. Figure 22.6 shows the velocity profile for the transient CAG flow after 2 s of capillary activation in the USP IP and MT geometries. Considering the IP, there was a significant jet velocity near the

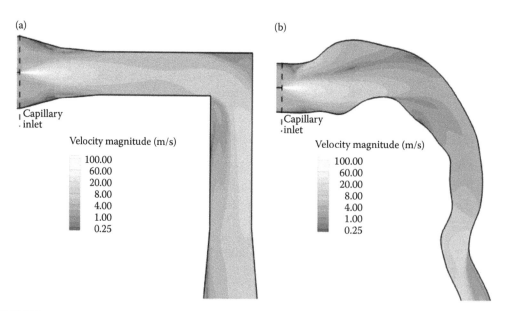

FIGURE 22.6 Contours of the velocity magnitude and velocity vectors at the middle plane and selected cross sections for transient CAG spray conditions after 2 s in the (a) IP and (b) MT geometries.

capillary tip (Figure 22.6a). Within the horizontal portion of the IP model, the turbulent jet velocity decays rapidly because of enhanced turbulent viscosity and interaction with the wall surface. As with the IP model, the jet rapidly dissipated in the MT geometry. However, a significant interaction appeared to be likely between the high momentum jet and the surface of the tongue, which was not present in the IP model.

The *in vitro* deposition of drug mass with transient CAG flow conditions after 2 s is presented in Figure 22.7 for the IP and realistic MT geometries. For all experiments, a minimum of five replicates are performed. The measured fraction of albuterol deposited in the IP model is 15.3% with a standard deviation of 0.9%. Considering computational fluid dynamics (CFD) predictions, the deposition rate of nonevaporating particles and evaporating droplets are 14.7% and 13.8%, respectively, resulting in respective errors of 3.9% and 9.8%. For the MT geometry, the *in vitro* deposition fraction of albuterol is 19.4% with a standard deviation of 1.7%. The comparable CFD model results for the particle and droplet approximations are 20.8% and 20.7%, respectively, resulting in respective errors of 6.7% and 6.3%. As a result, a higher portion of the spray aerosols deposit in the realistic MT geometry than the IP model. In addition, the CFD model with either the particle or droplet approximation matches the experimental result to within an error of 10%. Specifically, little difference is observed between the particle and droplet models in both the IP and MT geometries.

The numerically predicted deposition pattern of the drug mass for a specific CAG aerosol distribution (mass median diameter = 3.32 μm with 12.8% at ≤1 μm and 10.2% at ≥10 μm) is shown in Figure 22.8 in comparison with experimental total deposition values. Considering the total deposition of ambient particles, the deposited mass fraction increases from a value of 4.24% for the IP geometry to 12.2% for the MT, resulting in a difference ratio of 2.9 (Figure 22.8a and b). For the spray aerosol case, the difference ratio between the IP and MT models is 1.4 (Figure 22.8c and d). As a result, the difference in total deposited drug mass between the IP and MT geometries is significantly reduced for the spray system (~40%) in comparison with ambient aerosols (~200%). The spray aerosol increases the deposited mass of drug by a factor of 3.5 in the IP model (Figure 22.8a and c). In comparison, the spray aerosol only increases deposition in the MT by a factor of 1.7

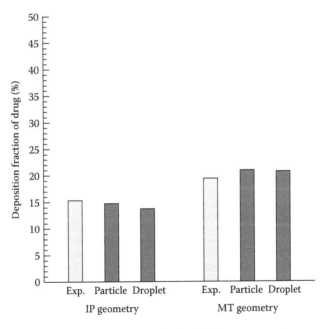

FIGURE 22.7 A comparison of deposited drug mass for the CAG spray system between *in vitro* experiments and CFD model predictions in the IP and MT geometries.

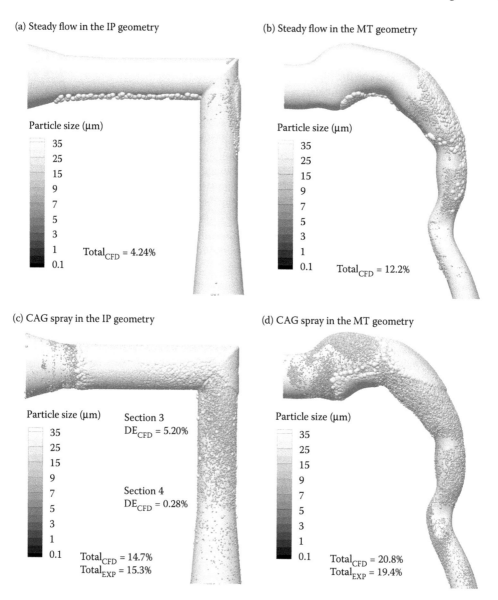

(a) Steady flow in the IP geometry

Particle size (μm)
35
25
15
9
7
5
3
1   Total$_{CFD}$ = 4.24%
0.1

(b) Steady flow in the MT geometry

Particle size (μm)
35
25
15
9
7
5
3
1
0.1   Total$_{CFD}$ = 12.2%

(c) CAG spray in the IP geometry

Particle size (μm)   Section 3
35   DE$_{CFD}$ = 5.20%
25
15
9
7   Section 4
5   DE$_{CFD}$ = 0.28%
3
1
0.1   Total$_{CFD}$ = 14.7%
             Total$_{EXP}$ = 15.3%

(d) CAG spray in the MT geometry

Particle size (μm)
35
25
15
9
7
5
3
1
0.1   Total$_{CFD}$ = 20.8%
             Total$_{EXP}$ = 19.4%

FIGURE 22.8 A comparison of sectional drug mass deposition in the IP and MT geometries for (a, b) steady flow and (c, d) CAG spray aerosols.

(Figure 22.8b and c). As a result, the spray increases the overall deposition more significantly in the IP model (~250%) than in the MT geometry (~70%).

Two observations are noteworthy: (1) deposition was expected to increase for respiratory spray aerosols as a result of spray momentum effects, and (2) enhanced deposition was expected in the MT geometry because of the elevated impaction on the tongue and in the pharyngeal and laryngeal regions. When interpreting previously reported deposition data in an IP model, a higher value is expected in a more realistic MT model. Furthermore, the finding that the spray inertia has a larger effect on the CAG aerosol deposition in the IP compared with the MT geometry has implications for pharmaceutical aerosol testing and the optimization of spray devices. For the IP model, a reduction in spray momentum can reduce deposition from 14.7% to 4.2%, which is a factor of 3.5 (Figure 22.8). However, this same decrease in deposition is not possible in the more realistic MT geometry.

For the MT model, a reduction in spray momentum can only reduce total deposition from 20.8% to 12.2%, which is a factor of 1.7. Therefore, efforts to reduce oral airway deposition that appear highly effective in the IP model may not be as significant in a more realistic MT geometry. This observation highlights the need to evaluate more realistic MT geometries when considered deposition in the oral airway region as a criterion for device performance.

22.5.2 GEOMETRIC EFFECT OF THE ORAL AIRWAY

Respiratory aerosol dynamics in the upper airway has been considered in a number of *in vitro* and numerical studies with casts and models of varying physical realism. However, the effect of geometric simplifications on aerosol regional and local deposition remains unknown for the oral airway and throughout the respiratory tract. In an attempt to quantify this effect, four oral airway models with varying geometric complexities were developed as test samples. The most realistic model considered was reconstructed from CT data in conjunction with measurement of a dental impression with a half-mouth opening (Cheng et al., 1997b). This model was intended to be physiologically accurate with only minor surface smoothing simplifications (Figure 22.9a). The other three models were generated from successive simplifications of the realistic geometry (Figure 22.9b–d). However, geometric parameters such as the airway curvature and equivalent hydraulic diameter are consistent for all models. For further details of these models, refer to Xi and Longest (2008a).

The main geometric features retained in the realistic model include the tooth–cheek chambers, a triangular-shaped glottis, and a dorsal-angled upper trachea. The presence of the NP was also considered. The soft palate (uvula) closes off the NP from the oral cavity during swallowing. For oral inhalation, the soft palate cannot be completely closed off, thereby keeping the oral and nasal cavities connected. A comparison of the realistic model results with and without the NP as shown in Figure 22.9a and b will allow an evaluation of this feature with respect to nanoparticle deposition.

As a simplification, cross-sectional segments of the realistic model without the NP were reconstructed with ellipses of the consistent hydraulic diameter and flow area (Figure 22.9c). The resulting elliptical model is similar to the realistic model in many respects (Figure 22.9c vs. 22.9b). The axial curvatures of both the realistic and elliptical models are identical. The cross-sectional planes in the elliptic model were constructed to approximate the effective flow area at the same locations in the realistic model. The elliptical model consists of a much simpler oral cavity, eliminating the two small tooth–cheek chambers evident in the realistic geometry. This model assumes an oval cross-sectional glottic aperture with a large length/width ratio in contrast to the triangular shape employed for the realistic geometry (Figure 22.9b).

Further geometric simplifications were made to generate the circular model (Figure 22.9d). In this model, all cross-sectional shapes are circular with the equivalent hydraulic diameters that were implemented in the realistic and elliptical models. In contrast to the slanted upper trachea, a vertical smooth tube was adopted in the circular model, as illustrated in Figure 22.9d. Detailed dimensions for the models that were considered, including hydraulic diameters, perimeters, and cross-sectional areas, can be found in Xi and Longest (2007).

The computational fluid-particle dynamics model used in this study was tested for both submicron and micron aerosols by comparing them with the available *in vitro* deposition data of Cheng et al. (1993, 1997a, 1999). Because the oral cavity in this study was based on an oral cast in which the *in vitro* deposition data were obtained, a direct comparison between simulations and measurements was possible. Particles ranging from 1 to 200 nm and 1 to 31 μm were tested under sedentary and light activity conditions. A good match was achieved between the numerically determined deposition results and experimental data in terms of both the magnitude and trend for nanometer and micron aerosols (Xi and Longest, 2007, 2008a). In general, the deposition rate decreases with increasing size for nanoparticles (i.e., diffusion regime) and increases with increasing size for micron particles (i.e., inertia regime). In light of breathing activities, less ultrafine particles deposit

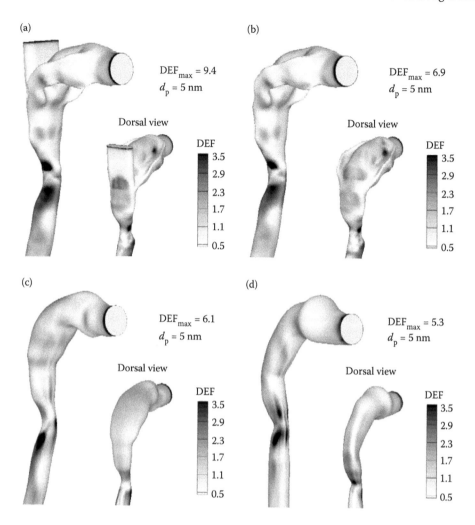

FIGURE 22.9 DEFs for 5-nm aerosols under light activity conditions (Q_{in} = 30 L/min) for (a) realistic with NP, (b) realistic without NP, (c) elliptical, and (d) circular models. (Adapted from Xi, J. and Longest, P. W. 2008a. *ASME Journal of Biomechanical Engineering* 130: article 011008. With permission.)

for higher flow rates because of decreased particle residence times. In contrast, the higher inhalation flow rate (Q_{in}) enhances micron particle deposition because of higher inertia. As expected, we also found that the degree of agreement with the experiments is correlated with airway model realism. That is, the realistic model gives the best match and the circular approximation provides the worst.

Further insight regarding the effects of geometric realism can be gained by examining the local deposition pattern and deposition density of the four models under consideration. Figure 22.9 illustrated the DEF values for light activity conditions (Q_{in} = 30 L/min) and 5-nm aerosols. The realistic geometry with the NP included a 10% nasal–oral flow partition. The DEF parameter denotes local deposition relative to the regionally averaged deposition efficiency. The maximum DEF values (DEF_{max}) for nanoparticles are the same order of magnitude for all models at 5.3–9.4, with the most complicated geometry having the highest DEF value. This is in contrast with the highly concentrated local deposition of micron aerosols greater than 1 μm where DEF_{max} values vary significantly as a function of the geometric complexity (Xi and Longest, 2007). The overall pattern of deposition enhancement appears to be very similar for the four models. Each model displays hot spots in deposition on the lateral walls of the glottis where convective diffusion is high because of the laryngeal jet effect. Another hot spot with a less elevated DEF value is also observed for each model on the dorsal side of the throat. In addition to the laryngeal region, there are enhanced local depositions at the inlet of each model that are attributable to the specification of a constant particle concentration at these locations. However, these elevated contours do not result in the reported maximum DEF values and do not significantly contribute to total deposition.

A further examination of the local deposition in the four models reveals some discernable differences related to geometric simplifications (Figure 22.9). The primary difference is the intensity of the hot spots observed on the lateral walls of the larynx. These hot spots are more pronounced for the two realistic models, especially at the sidewalls of the larynx immediately upstream of the glottis. This discrepancy may result from the more dramatic contraction from the upper larynx toward the triangular-shaped glottis (Figure 22.9a and b) and the resultant higher near-wall gradient in particle concentrations in the realistic model geometries. Another difference among the models is the location of elevated DEF contours in the circular model in comparison to deposition patterns in the other three geometries with sloped tracheas. For the circular model, there are significant hot spots on the dorsal and ventral walls of the larynx, in addition to the lateral walls. In contrast, the other three models exhibit reduced deposition on the dorsal and ventral surfaces. A possible reason for this difference is the orientation of the upper trachea. Xi and Longest (2007) showed significant differences in the secondary flows associated with the trachea angle, resulting in only one pair of vortices in the realistic and elliptical models versus two pairs of vortices in the circular model. This observation indicates the necessity of retaining the angled trachea, which is physiologically realistic, when a local deposition is of interest.

Local depositions of nanoparticles for sizes other than 5 nm and for different breathing conditions were also simulated but not shown. Generally, we found that, for larger nanoparticles, the hot spot region distal to the larynx shrinks and becomes more concentrated whereas deposition becomes more uniform in other areas because of decreased particle diffusivities. For 100-nm aerosols, hot spots are reduced to regions only surrounding the glottis. Further, particles are more evenly distributed for heavier activity conditions.

Based on the deposition results for silver nanoparticles ranging from 3.7 to 5 nm in an oral airway cast, Cheng et al. (1997b) proposed an empirical correlation of the mass transfer Sherwood number (Sh) as a function of the Reynolds and Schmidt numbers. When compared with the numerical predictions in this study (Figure 22.10), good agreement with the correlation is obtained for 5-nm particles, but increasing deviations occur for particles with diameters larger or smaller than 5 nm. Furthermore, the deviations are linear, shifting from the right to left as the particle size increases. Considering that the correlation of Cheng et al. (1997b) was based on measurements for a narrow range of particles (3.7–5 nm), this regular shifting indicates that certain diameter-related influences were not fully included, such as the shift from molecular to convective diffusion and Schmidt number effects. To better account for particle diameters, a new correlation with the form $Sh = a\,Re^{b}\,Sc^{c}$ was developed using the Levenberg–Marquardt nonlinear least-square fitting algorithm (Gill and Murray, 1978). The best-fit correlation was found to be

$$Sh = 0.1654Re^{0.661}Sc^{0.339} \quad 300 < Re < 7500, \quad 2.8 < Sc < 1050, \tag{22.8}$$

where $R^2 = 0.927$ (Figure 22.10b), indicating a greater dependence of the Sherwood number on convective diffusion (Re exponent of 0.661) than on Brownian diffusion (Sc exponent of 0.339). It is interesting that a value of 1 is obtained from the summation of the exponents of these two independent nondimensional parameters. In regard to the geometry effect, no significant difference in the Sherwood number is discernable among the four models based on the small scatter in Figure 22.10a and b.

Based on the analogy between heat and mass transfer, Zhang and Kleinstreuer (2003) developed a similar Sherwood number correlation that implements the parameter $ReSc$ for a circular oral

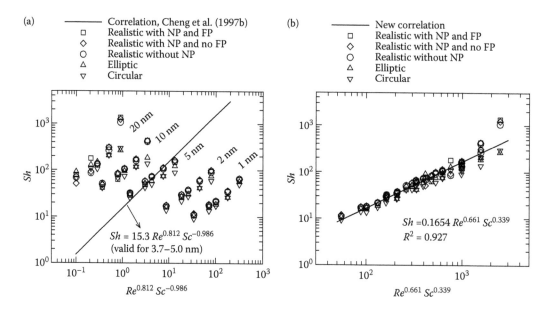

FIGURE 22.10 The variation of the Sherwood number (Sh) as a function of the Reynolds (Re) and Schmidt (Sc) numbers in the four airway models considered (a) in comparison to the empirical correlation of Cheng et al. (1997b). (b) An improved correlation is proposed that captures an expanded dependence of Sh on Re and Sc.

airway model. Applying this parameter to the multiple airway data of trelation $Sh = 0.871(ReSc)^{0.391}$ with $R^2 = 0.86$. This correlation is very similar to the result suggested by Zhang and Kleinstreuer (2003) for their circular airway model. However, the group $ReSc$ may not adequately account for the relative impact of convective diffusion due to secondary motion as suggested by the larger scatter ($R^2 = 0.86$) in comparison to the correlation based on $Re^{0.661}Sc^{0.339}$. Therefore, the new proposed correlation in terms of $Re^{0.661}Sc^{0.339}$ is recommended for the mass transfer calculation of ultrafine particles. This correlation, however, should only be applicable for adult subjects, considering that both the replica casts of Cheng et al. (1993, 1997b) and our computer models are based on adult subjects.

Considering inertial regime particles, localized DEF values have high sensitivity to geometric complexity. As shown in Figure 22.11, the maximum DEF factor observed for the realistic model ($DEF_{max} = 3864$) is 1 order of magnitude greater than the maximum values observed for the remaining models (e.g., $DEF_{max} = 123$ for the circular model shown in Figure 22.11b). This is in contrast with the similarity in DEF values among the four models for submicron (i.e., diffusion regime) particles, where the DFE_{max} ranges from 5.3 to 9.4. For micron particle localizations, enhanced deposition sites or hot spots could be observed in the throat, the sidewalls of the trachea immediately downstream of the throat, as well as the sidewalls of the oropharynx. These features are most pronounced for the realistic model (Figure 22.11a).

In summary, geometric realism significantly influenced the local deposition patterns for both nanometer and micron particles. Geometric factors that significantly contributed to the enhanced particle localization in the realistic model include a triangular-shaped glottis and a dorsal-sloped trachea, and these should be retained in future *in vitro* and numerical studies.

22.5.3 Breathing Maneuvers in Direct Nose–Brain Drug Delivery

The nasal route has recently been considered as a means of delivering topical and systemically acting drugs. The targeted site usually depends on the desired therapeutic outcomes, as listed in

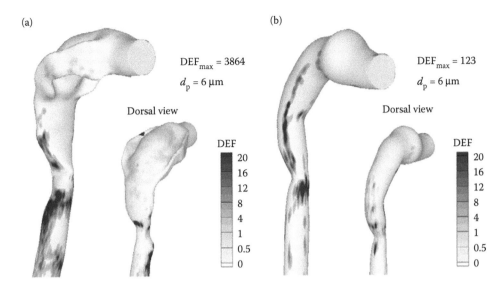

FIGURE 22.11 DEFs for 6-μm aerosols under light activity conditions with both side and dorsal views in (a) realistic and (b) circular geometries. A very high particle localization characterized by a value of $DEF_{max} = 3864$ occurs on the posterior wall of the larynx for the realistic model. (Adapted from Xi, J. and Longest, P. W. 2007. *Annals of Biomedical Engineering* 35: 560–581. With permission.)

Table 22.1. Direct nasal–brain drug delivery has the advantages of bypassing the blood–brain barrier, rapid therapeutic onset, and appeal for the treatment of neurological disorders such as Parkinson or Alzheimer disease. The unique relationship between the nasal and cranial cavity makes this intranasal delivery to the brain feasible. However, because of the convoluted nasal passages, only a minor portion of medications can reach the olfactory region (OR) if administered at the nasal inlets (Figure 22.12). Intranasal intubation may be viable in view of the high cost and potential side effects of neurological medications. This section discusses one such delivery system and the influence of different breathing maneuvers on drug dosimetry.

The delivery system that we considered was a nebulization catheter with a diameter of 1 mm and an exiting velocity of 15 m/s. The catheter was positioned right beneath the targeted OR. Different breathing maneuvers were exercised during the drug administration: inhalation, breath hold, and exhalation. The left panel of Figure 22.13 shows the fluid trajectories of the nebulizer jet air and respiratory airflow under each scenario. Both 150-nm and 1-μm monodisperse droplets were considered. It is interesting that the inhalation maneuver gives rise to the optimal dosage as evidenced by the defined hot spots in the OR (Figure 22.13). Breath hold or exhalation, which were initially considered advantageous, resulted in unfocused depositions throughout the nasal turbinate region for both droplet sizes that we considered.

TABLE 22.1
Targeted Nasal Sites and Desired Therapeutic Outcomes

Desired Outcome	Targeted Site	Example
Locally acting drugs	Middle meatus	Chronic sinusitis
Systemic delivery (rapid onset)	Turbinate and lateral/septal walls	Migraine Heart attack
Direct nose–brain (avoids blood–brain barrier)	OR	Neurological disorders: Parkinson and Alzheimer diseases

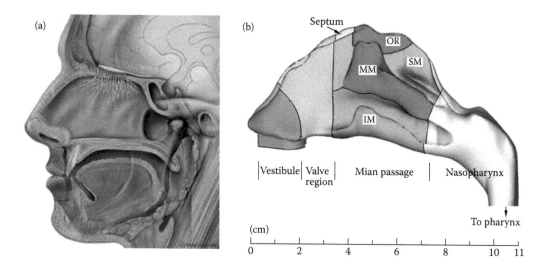

FIGURE 22.12 (a) The olfactory region (OR) is the only site where the nervous system is in direct contact with the surrounding environment. (b) The nasal cavity includes the vestibule, nasal valve region, OR, septum wall (SW), middle meatus (MM), as well as the upper (UP) and lower (LP) passages that include the superior (SM) and inferior (IM) meatus, respectively.

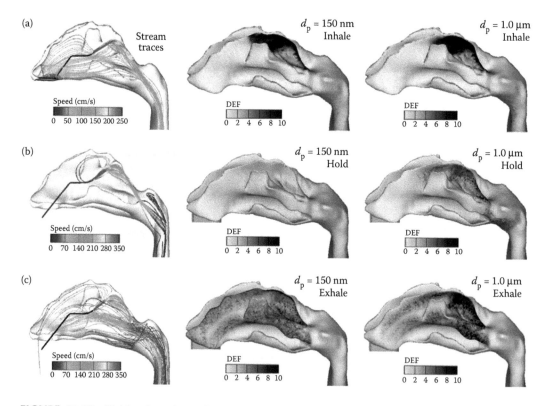

FIGURE 22.13 Fluid trajectories and aerosol deposition patterns under different breathing maneuvers: (a) inhalation, (b) breath hold, and (c) exhalation. Drug administration via an intranasal tube during inhalation gives an optimal dosage to the OR.

22.6 SUMMARY

Potential applications of numerical modeling in pharmaceutical drug delivery, including evaluating the effects of aerosol properties and respiratory physiology, were discussed in this chapter. Efforts in simulating human breathing significantly improved our understanding of the complex factors influencing aerosol delivery in the treatment of respiratory diseases. Specifically, compared with *in vitro* and *in vivo* studies, CFD predictions have the advantage of providing detailed information on airflow and aerosol deposition, like hot spots, which influence uptake and response.

The results of this study demonstrated that hot spot formations are highly sensitive to airway anatomical details that control flow characteristics, such as reversals, secondary motion, and turbulence intensity. Particle localization in terms of the DEF indicated that the realistic oral airway model has a highly elevated deposition hot spot near the throat with a value that is 1–2 orders of magnitude larger than observed for simplified models. The occurrence of particle accumulations and hot spot formations in which localized deposition concentrations greatly exceeded average levels has been widely reported throughout the respiratory tract based on *in vitro* experiments (Martonen, 1986; Oldham et al., 2000; Schlesinger and Lippmann, 1978) and CFD studies (Balashazy and Hofmann, 1993a, 1993b; Balashazy et al., 2003; Zhang and Kleinstreuer, 2002; Zhang et al., 2005; Zhang and Martonen, 1997). Ultimately, a cellular-level dose is a function of a cellular-level particle deposition and accumulation during the clearance process. Cellular-level deposition results are necessary to accurately model subsequent mass transport and accumulation. Moreover, cellular-level deposition values are critical in cases of rapidly absorbed components and for fast-acting relief medications. The results of this study delineate a significant difference in the occurrence of particle localization, hot spot formation, and cellular-level deposition in airway models of varying physical realism. As a result, models that are intended to predict cellular-level dose and response for rapidly absorbed components should preserve a high degree of geometric detail.

Even though CFD is a powerful tool, numerical respiratory analysis in the current stage is still limited by many factors. For example, most studies in the current literature focus on healthy adults for a limited number of individuals. Difficulties associated with applying moving boundaries require investigators to adopt a rigid airway surface. Furthermore, investigators often spend a majority of time trying to preserve realism and neglect intersubject and intrasubject variability. Efficient delivery of therapeutic agents needs to take into account multiple aspects such as the patient's age, the diseases treated, wall surface compliance, transient vocal fold movement, and mucus clearance. Therefore, future studies on respiratory dynamics should be oriented toward (1) improving realism and (2) including a broader population group. Our knowledge of aerosol behavior is currently lacking in subpopulations such as pediatrics, geriatrics, and patients with respiratory diseases. Due to physiological development, aging, or disease states, the airway geometries in these subpopulations are significantly different from a healthy adult. Concentrating on these specific subpopulations will help to clarify intergroup and intersubject variability and will allow for the design of more efficient pharmaceutical formulations and drug delivery protocols.

ABBREVIATIONS

BM	Brownian motion
CAG	Capillary aerosol generation
CFD	Computational fluid dynamics
CT	Computed tomography
DEF	Deposition enhancement factor
IP	Induction port
MRI	Magnetic resonance imaging
MT	Mouth–throat
NP	Nasopharynx

NWI   Near-wall interpolation
OR   Olfactory region
pMDI   Pressurized metered dose inhaler
SPION   Superparamagnetic iron oxide nanoparticle
3-D   Three-dimensional
TB   Tracheobronchial
USP   United States Pharmacopeia

SYMBOLS

c Mass fraction concentration
C_c Cunningham correction factor defined by Equation 22.3
d_p Particle diameter
$\breve{D}$ Brownian diffusion coefficient
k Turbulence kinetic energy
k_B Boltzmann constant (1.38×10^{-16} cm^{2}g/s)
K_c Mass transfer coefficient
λ Mean free path of air
$\dot{m}$ Mass flow rate
μ Dynamic viscosity
v Kinetic molecular viscosity (μ/ρ)
v_T Kinetic eddy viscosity ($\alpha \times k/\omega$)
ω Pseudovorticity, dissipation per unit turbulence kinetic energy
p Time-averaged pressure
P Perimeter of airway cross section normal to the axial direction
Q Flow rate
Q_{in} Inhalation flow rate
ρ Fluid density
r Radial coordinate
Re Reynolds number (ud_h/v)
Sc Schmidt number (v/D)
Sh Sherwood number ($K_c d_h/D$)
St Stokes number ($\rho_p d_p^2 C_c U/18\mu D$)
T Temperature
$\tau_{i,j}$ Shear stress tensor
x_i Position vector in tensor notation
y^+ Dimensionless distance from the wall ($u_\tau y/v$)

REFERENCES

Allen, M. D. and Raabe, O. G. 1985. Slip correction measurements of spherical solid aerosol particles in an improved Millikan apparatus. *Aerosol Science and Technology* 4: 269–286.

Balashazy, I. and Hofmann, W. 1993a. Particle deposition in airway bifurcations—I. Inspiratory flow. *Journal of Aerosol Science* 24: 745–772.

Balashazy, I. and Hofmann, W. 1993b. Particle deposition in airway bifurcations—II. Expiratory flow. *Journal of Aerosol Science* 24: 773–786.

Balashazy, I., Hofmann, W., and Heistracher, T. 2003. Local particle deposition patterns may play a key role in the development of lung cancer. *Journal of Applied Physiology* 94: 1719–1725.

Chan, T. L. and Lippmann, M. 1980. Experimental measurements and empirical modeling of the regional deposition of inhaled particles in humans. *American Industrial Hygiene Association Journal* 41: 399–409.

Cheng, K. H., Cheng, Y. S., Yeh, H. C., Guilmette, R. A., Simpson, S. Q., Yang, S. Q., and Swift, D. L. 1996. *In vivo* measurements of nasal airway dimensions and ultrafine aerosol depositing in human nasal and oral airways. *Journal of Aerosol Science.* 27: 785–801.

Cheng, K. H., Cheng, Y. S., Yeh, H. C., and Swift, D. L. 1997a. An experimental method for measuring aerosol deposition efficiency in the human oral airway. *American Industrial Hygiene Association Journal* 58: 207–213.

Cheng, K. H., Cheng, Y. S., Yeh, H. C., and Swift, D. L. 1997b. Measurements of airway dimensions and calculation of mass transfer characteristics of the human oral passage. *Journal of Biomechanical Engineering* 119: 476–482.

Cheng, Y. S., Fu, C. S., Yazzie, D., and Zhou, Y. 2001. Respiratory deposition patterns of salbutamol pMDI with CFC and HFA-134a formulations in a human airway replica. *Journal of Aerosol Medicine* 14: 255–266.

Cheng, Y. S., Su, Y. F., Yeh, H. C., and Swift, D. L. 1993. Deposition of thoron progeny in human head airways. *Aerosol Science and Technology* 18: 359–375.

Cheng, Y. S., Zhou, Y., and Chen, B. T. 1999. Particle deposition in a cast of human oral airways. *Aerosol Science and Technology* 31: 286–300.

Cohen, B. S., Sussman, R. G., and Lippmann, M. 1990. Ultrafine particle deposition in a human tracheobronchial cast. *Aerosol Science and Technology* 12: 1082–1093.

Crowe, C., Sommerfeld, M., and Tsuji, Y. 1998. *Multiphase Flows with Drops and Bubbles*. CRC Press, Boca Raton, FL.

Dames, P., Gleich, B., Flemmer, A., Hajek, K., Seidl, N., Wiekhorst, F., Eberbeck, D., et al. 2007. Targeted delivery of magnetic aerosol droplets to the lung. *Nature Nanotechnology* 2: 495–499.

DeHaan, W. H. and Finlay, W. H. 2004. Predicting extrathoracic deposition from dry powder inhalers. *Journal of Aerosol Science* 35: 309–331.

Edwards, D. A., Ben-Jebria, A., and Langer, R. 1998. Recent advances in pulmonary drug delivery using large, porous inhaled particles. *Journal of Applied Physiology* 85: 379–385.

Edwards, D. A. and Dunbar, C. 2002. Bioengineering of therapeutic aerosols. *Annual Review of Biomedical Engineering* 4: 93–107.

Edwards, D. A., Hanes, J., Caponetti, G., Hrkach, J., BenJebria, A., Eskew, M. L., Mintzes, J., Deaver, D., Lotan, N., and Langer, R. 1997. Large porous particles for pulmonary drug delivery. *Science* 276(5320): 1868–1871.

Fernandez de la Mora, J. and Rosner, D. E. 1981. Inertial deposition of particles revisited and extended: Eulerian approach to a traditionally Lagrangian problem. *Physicochemical Hydrodynamics* 2: 1–21.

Friedlander, S. K. 2000 *Smoke, Dust and Haze: Fundamentals of Aerosol Dynamics*. Oxford University Press, New York.

Ghalichi, F., Deng, X., Champlain, A. D., Douville, Y., King, M., and Guidoin, R. 1998. Low Reynolds number turbulence modeling of blood flow in arterial stenoses. *Biorheology* 35: 281–294.

Gill, P. E. and Murray, W. 1978. Algorithms for the solution of the nonlinear least-squares problem. *SIAM Journal on Numerical Analysis* 15: 977–992.

Hanna, L. M. and Scherer, P. W. 1986. Measurement of local mass transfer coefficients in a cast model of the human upper respiratory tract. *ASME Journal of Biomechanical Engineering* 108: 12–18.

Heenan, A. F., Matida, E., Pollard, A., and Finlay, W. H. 2003. Experimental measurements and computational modeling of the flow field in an idealized human oropharynx. *Experiments in Fluids* 35: 70–84.

Heyder, J., Gebhart, J., Rudolf, G., Schiller, C. F., and Stahlhofen, W. 1986. Deposition of particles in the human respiratory tract in the size range of 0.005–15 microns. *Journal of Aerosol Science* 17: 811–825.

Hofmann, W., Golser, R., and Balashazy, I. 2003. Inspiratory deposition efficiency of ultrafine particles in a human airway bifurcation model. *Aerosol Science and Technology* 37: 988–994.

Hofmann, W., Morawska, L., and Bergmann, R. 2001. Environmental tobacco smoke deposition in the human respiratory tract: Differences between experimental and theoretical approaches. *Journal of Aerosol Medicine* 14: 317–326.

Ingham, D. B. 1975. Diffusion of aerosols from a stream flowing through a cylindrical tube. *Journal of Aerosol Science* 6: 125–132.

Ingham, D. B. 1991. Diffusion of aerosols in the entrance region of a smooth cylindrical pipe. *Journal of Aerosol Science* 22: 253–257.

Jaques, P. A. and Kim, C. S. 2000. Measurement of total lung deposition of inhaled ultrafine particles in healthy men and women. *Inhalation Toxicology* 12: 715–731.

Johnstone, A., Uddin, M., Pollard, A., Heenan, A., and Finlay, W. H. 2004. The flow inside an idealised form of the human extra-thoracic airway. *Experiments in Fluids* 37: 673–689.

Kelly, J. T., Prasad, A. K., and Wexler, A. S. 2000. Detailed flow patterns in the nasal cavity. *Journal of Applied Physiology* 89(1): 323–337.

Kim, C. S. and Fisher, D. 1994. Deposition of ultrafine particles in the bifurcation airway models. In R. C. Flagan (Ed.), *Abstracts of the 4th International Aerosol Conference*, pp. 888–889. Elsevier, New York.

Kim, C. S. and Jaques, P. A. 2004. Analysis of total respiratory deposition of inhaled ultrafine particles in adult subjects at various breathing patterns. *Aerosol Science and Technology* 38: 525–540.

Kleinstreuer, C., Zhang, Z., and Donohue, J. F. 2008. Targeted drug-aerosol delivery in human respiratory system. *Annual Reviews in Biomedical Engineering* 10: 195–220.

Laube, B. L., Edwards, A. M., Dalby, R. N., Creticos, P. S., and Norman, P. S. 1998. Respiratory pathophysiologic responses: The efficacy of slow versus faster inhalation of cromolyn sodium in protecting against allergen challenge in patients with asthma. *Journal of Allergy and Clinical Immunology* 101: 475–483.

Laube, B. L., Jashnani, R., Dalby, R. N., and Zeitlin, P. L. 2000. Targeting aerosol deposition in patients with cystic fibrosis. *Chest* 118: 1069–1076.

Lee, D. and Lee, J. 2002. Dispersion of aerosol bolus during one respiratory cycle in a model lung airway. *Journal of Aerosol Science* 33: 1219.

Lewis, D., Brambilla, G., Church, T., and Meakin, B. 2006. Comparative *in vitro* performance of a BDP HFA solution MDI using USP and anatomical induction ports. *Respiratory Drug Delivery* 3: 943–946.

Li, A. and Ahmadi, G. 1992. Dispersion and deposition of spherical particles from point sources in a turbulent channel flow. *Aerosol Science and Technology* 16: 209–226.

Li, W. I., Perzl, M., Heyder, J., Langer, R., Brain, J. D., Englmeier, K. H., Niven, R. W., and Edwards, D. A. 1996. Aerodynamics and aerosol particle deaggregation phenomena in model oral–pharyngeal cavities. *Journal of Aerosol Science* 27: 1269–1286.

Li, Z., Kleinstreuer, C., and Zhang, Z. 2007. Simulation of airflow fields and microparticle deposition in realistic human lung airway models. Part II: Particle transport and deposition. *European Journal of Mechanics B Fluids* 26: 632–649.

Lin, C. L., Tawhai, M. H., McLennan, G., and Hoffman, E. A. 2007. Characteristics of the turbulent laryngeal jet and its effect on airflow in the human intra-thoracic airways. *Respiratory Physiology and Neurobiology* 157: 295–309.

Longest, P. W., Hindle, M., Das Choudhuri, S., and Byron, P. R. 2007. Numerical simulations of capillary aerosol generation: CFD model development and comparisons with experimental data. *Aerosol Science and Technology* 41: 952–973.

Longest, P. W., Hindle, M., Das Choudhuri, S., and Xi, J. 2008. Comparison of ambient and spray aerosol deposition in a standard induction port and more realistic mouth–throat geometry. *Journal of Aerosol Science* 39: 572–591.

Longest, P. W., Kleinstreuer, C., and Buchanan, J. R. 2004. Efficient computation of micro-particle dynamics including wall effects. *Computers & Fluids* 33: 577–601.

Longest, P. W. and Vinchurkar, S. 2007a. Effects of mesh style and grid convergence on particle deposition in bifurcating airway models with comparisons to experimental data. *Medical Engineering and Physics* 29: 350–366.

Longest, P. W. and Vinchurkar, S. 2007b. Validating CFD predictions of respiratory aerosol deposition: Effects of upstream transition and turbulence. *Journal of Biomechanics* 40: 305–316.

Longest, P. W. and Xi, J. 2007. Effectiveness of direct Lagrangian tracking models for simulating nanoparticle deposition in the upper airways. *Aerosol Science and Technology* 41: 380–397.

Martonen, T. B. 1986. Surrogate experimental models for studying particle deposition in the human respiratory tract: An overview. In S. D. Lee (Ed.), *Aerosols*. Lewis Publishers, Chesea, MI.

Martonen, T. B., Zhang, Z., and Yang, Y. 1996. Particle diffusion with entrance effects in a smooth-walled cylinder. *Journal of Aerosol Science* 27: 139–150.

Matida, E. A., Finlay, W. H., and Grgic, L. B. 2004. Improved numerical simulation of aerosol deposition in an idealized mouth–throat. *Journal of Aerosol Science* 35: 1–19.

McRobbie, D. W., Pritchard, S., Quest, R. A. 2003. Studies of the human oropharyngeal airspaces using magnetic resonance imaging. I. Validation of a threedimensional MRI method for producing ex vivo virtual and physical casts of the oropharyngeal airways during inspiration. *J Aerosol Med.* 16: 401–415.

Morawska, L., Barron, W., and Hitchins, J. 1999. Experimental deposition of environmental tobacco smoke submicron particulate matter in the human respiratory tract. *American Industrial Hygiene Association Journal* 60: 334–339.

Morawska, L., Hofmann, W., Hitchins-Loveday, J., Swanson, C., and Mengersen, K. 2005. Experimental study of the deposition of combustion aerosols in the human respiratory tract. *Journal of Aerosol Science* 36: 939–957.

Morsi, S. A. and Alexander, A. J. 1972. An investigation of particle trajectories in two-phase flow systems. *Journal of Fluid Mechanics* 55: 193–208.

Moskal, A. and Gradon, L. 2002. Temporal and spatial deposition of aerosol particles in the upper human airways during breathing cycles. *Journal of Aerosol Science* 33: 1525.

Newman, S. P. 1988. Deposition of carbenicillin aerosol in cystic fibrosis: Effects of nebulizer system and breathing pattern. *Thorax* 43: 318–322.

Nowak, N., Kakade, P. P., and Annapragada, A. V. 2003. Computational fluid dynamics simulation of airflow and aerosol deposition in human lungs. *Annals of Biomedical Engineering* 31: 374–390.

Oldham, M. J., Phalen, R. F., and Heistracher, T. 2000. Computational fluid dynamic predictions and experimental results for particle deposition in an airway model. *Aerosol Science and Technology* 32: 61–71.

Pritchard, S. E. and McRobbie, D. W. 2004. Studies of the human oropharyngeal airspaces using magnetic resonance imaging. II. The use of three-dimensional gated MRI to determine the influence of mouthpiece diameter and resistance of inhalation devices on the oropharyngeal airspace geometry. *J Aerosol Med.* 17: 310–324.

Robinson, R. J., Oldham, M. J., Clinkenbeard, R. E., and Rai, P. 2006. Experimental and numerical smoke carcinogen deposition in a multi-generation human replica tracheobronchial model. *Annals of Biomedical Engineering* 34: 373–383.

Scherer, P. W., Keyhani, K., and Mozell, M. M. 1994. Nasal dosimetry modeling for humans. *Inhalation Toxicology* 6: 85–97.

Schlesinger, R. B. and Lippmann, M. 1978. Selective particle deposition and bronchogenic carcinoma. *Environmental Research* 15: 424–431.

Shi, H., Kleinstreuer, C., Zhang, Z., and Kim, C. S. 2004. Nanoparticle transport and deposition in bifurcating tubes with different inlet conditions. *Physics of Fluids* 16: 2199–2213.

Stahlhofen, W., Rudolf, G., and James, A. C. 1989. Intercomparison of experimental regional aerosol deposition data. *Journal of Aerosol Medicine* 2: 285–308.

Stein, S. W. and Gabrio, B. J. 2000. Understanding throat deposition during cascade impactor testing. In R. N. Dalby, P. R. Byron, S. J. Farr, and J. Peart (Eds.), *Respiratory Drug Delivery VII*, pp. 287–290. Serentec Press, Raleigh, NC.

United States Pharmacopeia. 2005. Physical tests and determinations: Aerosols, nasal sprays, metered-dose inhalers, and dry powder inhalers. United States Pharmacopeia First Supplement, 28-NF, General Chapter (601). United States Pharmacopeial Convention, Rockville, MD, pp. 3298–3316.

van Ertbruggen, C., Hirsch, C., and Paiva, M. 2005. Anatomically based three-dimensional model of airways to simulate flow and particle transport using computational fluid dynamics. *Journal of Applied Physiology* 98: 970–980.

Wang, J., Flagan, R. C., and Seinfeld, J. H. 2002. Diffusional losses in particle sampling systems containing bends and elbows. *Journal of Aerosol Science* 33: 843–851.

Wilcox, D. C. 1998. *Turbulence Modeling for CFD*, 2nd edition. DCW Industries, Inc., La Cañada, CA.

Xi, J. and Longest, P. W. 2007. Transport and deposition of micro-aerosols in realistic and simplified models of the oral airway. *Annals of Biomedical Engineering* 35: 560–581.

Xi, J. and Longest, P. W. 2008a. Effects of oral airway geometry characteristics on the diffusional deposition of inhaled nanoparticles. *ASME Journal of Biomechanical Engineering* 130: article 011008.

Xi, J. and Longest, P. W. 2008b. Evaluation of a drift–flux model for simulating submicrometer aerosol dynamics in human upper tracheobronchial airways. *Annals of Biomedical Engineering* 36: 1714–1734.

Xi, J. and Longest, P. W. 2008c. Numerical predictions of submicrometer aerosol deposition in the nasal cavity using a novel drift–flux approach. *International Journal of Heat and Mass Transfer* 51: 5562–5577.

Xi, J., Longest, P. W., and Martonen, T. B. 2008. Effects of the laryngeal jet on nano- and microparticle transport and deposition in an approximate model of the upper tracheobronchial airways. *Journal of Applied Physiology* 104: 1761–1777.

Yu, C. P., Diu, C. K., and Soong, T. T. 1981. Statistical analysis of aerosol deposition in nose and mouth. *American Industrial Hygiene Association Journal* 42: 726–733.

Zhang, Y., Chia, T. L., and Finlay, W. H. 2006. Experimental measurement and numerical study of particle deposition in highly idealized mouth–throat models. *Aerosol Science and Technology* 40: 361–372.

Zhang, Y., Finlay, W. H., and Matida, E. A. 2004. Particle deposition measurements and numerical simulations in a highly idealized mouth–throat. *Journal of Aerosol Science* 35: 789–803.

Zhang, Y., Gilbertson, K., and Finlay, W. H. 2007. *In vivo–in vitro* comparison of deposition in three mouth–throat models with Qvar and Turbuhaler inhalers. *Journal of Aerosol Medicine* 20: 227–235.

Zhang, Z. and Kleinstreuer, C. 2002. Transient airflow structures and particle transport in a sequentially branching lung airway model. *Physics of Fluids* 14: 862–880.

Zhang, Z. and Kleinstreuer, C. 2003. Species heat and mass transfer in a human upper airway model. *International Journal of Heat and Mass Transfer* 46: 4755–4768.

Zhang, Z., Kleinstreuer, C., Donohue, J. F., and Kim, C. S. 2005. Comparison of micro- and nano-size particle depositions in a human upper airway model. *Journal of Aerosol Science* 36: 211–233.

Zhang, Z. and Martonen, T. B. 1997. Deposition of ultrafine aerosols in human tracheobronchial airways. *Inhalation Toxicology* 9: 99–110.

Zhao, Y. and Lieber, B. B. 1994. Steady expiratory flow in a model symmetrical bifurcation. *Journal of Biomechanical Engineering* 116: 318–323.

23 Colloidal Carriers for Drug Delivery in Dental Tissue Engineering

Nader Kalaji, Nida Sheibat-Othman, and Hatem Fessi

CONTENTS

23.1 INTRODUCTION

The functional life of the tooth is determined by the vitality of the dentin–pulp complex. The importance of maintaining the vitality of the dental pulp after pathological changes in the hard tissue of a tooth caused by mechanical trauma or chronic irritation (carious lesions) is thus well accepted. The pulp maintains tissue homeostasis after tooth development and underpins the defense reactions taking place in response to injury and the reparative events leading to tissue regeneration (Smith, 2003).

Typically, pulp capping is applied to protect the exposed pulp (to avoid bacterial invasion of the pulp from the external environment) to allow it to recover and preserve its vitality and its functional and biologic activity (Olsson et al., 2006). The better understanding of the molecules involved in pathological processes such as growth factors and extracellular matrix molecules offers significant benefits in tissue engineering (TE). Growth factors are a group of molecules responsible for signaling a variety of cellular processes and various aspects of tissue regeneration. Controlled tissue repair and regeneration is stimulated by the application of recombinant growth factors in pulp capping (Lovschall et al., 2001; Tziafas et al., 2000) by forming a mineralized barrier (reparative dental bridge). This new therapy in TE is called vital pulp therapy (Iohara et al., 2004; Nakashima, 2005; Tziafas et al., 2000).

The application of growth factors in TE necessitates the use of a delivery system. Most growth factors have a relatively low half-life. For instance, the half-life of transforming growth factor β1 (TGF-β1) is of the order of few minutes in its free form during which it is capable of signaling regenerative events. Therefore, it is important to protect growth factors against denaturation during the treatment. The direct application of growth factors would give unsatisfactory results because the applied dose would be very high at the beginning of the treatment and might have some side effects. The drug might also diffuse to nontarget sites. The subsequent enzymatic digestion or deactivation of the growth factor leads to a rapid decrease of the available dose that becomes insufficient for the treatment and fails to induce the intended effects on target cells and tissues. Controlled release provides a sustained effect where a protected bioactive factor is released to act for a longer time. The use of a delivery system also allows the incorporation of a reduced amount of the growth factors that is crucial because of their high price. A supplementary advantage of drug delivery is the possibility of transporting the bioactive factors to the desired medium.

These benefits are conditioned by the use of biocompatible carriers for drug delivery and an adequate encapsulation technique that preserves the protein bioactivity during the preparation and release. Notable efforts are made to provide longer-term release of growth factors in TE and to ensure appropriate concentrations of therapeutic molecules. A wide range of drug delivery carriers has been applied to accommodate different molecules to regenerate dental tissues by this new pulp therapy.

This chapter reviews the biocompatible carriers used for drug delivery in dental TE with a specific emphasis on colloidal carriers. A brief presentation of the teeth structure is proposed followed by a description of the different types of growth factors involved in teeth development and healing. Then, the different materials used in dental TE are summarized as well as a promising drug delivery system: microspheres produced by the water-in-oil-in-water (W/O/W) method. Finally, the main applications of scaffolds or capsules loaded with growth factors in dental TE are reviewed.

23.2 TEETH STRUCTURE

Enamel (cf. Figure 23.1) is the hardest part of the tooth and the highest mineralized tissue of the body (approximately 96% mineral with the remainder water plus organic material). It protects the exposed part of the tooth, called the crown. It is as hard as steel with a Knoop hardness number of 343 (Summitt et al., 2001). Therefore, materials used to replace enamel are strong glass-, ceramic-, or metal-based substitutes (e.g., amalgam and composites).

Dentin is the tissue underneath the enamel that forms the main mass of the tooth (composed of 70% mineral, 20% organic material, and 10% water). It supports the tooth enamel and absorbs the pressure of eating because it is less mineralized and less brittle than the enamel. It consists of a number of microfibers, called dentinal tubules, imbedded in a dense homogeneous matrix of collagenous proteins (unlike enamel that does not contain collagen). The formation of dentin, known as dentinogenesis, is initiated by odontoblasts.

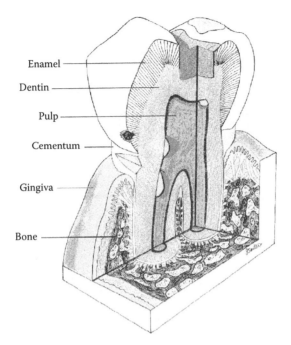

FIGURE 23.1 Structure of the human tooth. (From Summitt, J.B., Robbins, W., and Schwartz, R.S. 2001. *Fundamentals of Operative Dentistry, A Contemporary Approach*, 2nd ed. Quintessence Publishing, Chicago. With permission.)

Cementum is an anatomic part that covers the dentin outside the root (under the gum line) and is attached to the jawbone with a few elastic collagen fibers. It is as hard as bone but not as hard as the tooth enamel and dentin (composed of approximately 45% inorganic material, 33% organic material, and 22% water). Cementum is typically thinner and more permeable than enamel. Therefore, it is subject to abrasion upon root exposure to the oral environment. It is formed by cementoblasts.

Dental pulp is the most internal structure of a tooth, surrounded by the dentin. It is found in the soft center of the tooth inside the pulp chamber and the root canal. It is a living soft tissue containing nerves and blood vessels that nourish the tooth.

The tooth is therefore composed of distinct yet connected tissues. For tooth treatment, multidirectional engineering perspectives are to be achieved using composite materials that satisfy the numerous functionality requirements. The anatomy of the tooth also reveals the importance of preserving the dental pulp during the treatment because of its crucial functions such as dentin formation by the odontoblasts (as well as formation of reparative dentin), nutrition of the surrounding tissues, and sensing (trauma, pressure, and temperature).

23.3 GROWTH FACTORS

Growth factors are peptide molecules that have a central role in the regulation of a variety of cellular processes. They act as signaling molecules between cells to stimulate and inhibit growth and proliferation and modulate the differentiation state.

The interaction and binding between growth factors and specific receptors on the cell surface lead to a chain of intracellular signals as a result of signal transduction to the cell nucleus. The effect of growth factors on gene expression in the cell nucleus influences the cell behavior and activity, thus affecting intra- and extracellular events. Therefore, growth factors may regulate the genes controlling cell proliferation, differentiation, or products of the cell secretion. Because of their high potential, they are employed in very low concentrations (Smith, 2003).

23.3.1 Families of Growth Factors

Different families of growth factors exist. A noncomprehensive list of growth factors is provided in Table 23.1. There are at least 100 members in the TGF-β family and 22 members in the fibroblast growth factor (FGF) family. Growth factors can be found in a variety of cells. For example, the principal sources for TGF-β are activated T-helper cells and natural killer cells and sources for platelet-derived growth factors (PDGFs) are platelets, endothelial cells, and placenta. Some growth factors act upon specific types of cells while others are versatile and act upon a variety of cells. Most of the signaling molecules involved in tooth development belong to the TGF-β and FGF families.

23.3.2 Growth Factors Involved in Tooth Development

The early development of teeth resembles (morphologically and molecularly) other organs such as hair, nails, feathers, and glands. Teeth develop from the ectoderm via sequential and reciprocal interactions between epithelial and mesenchymal tissues during which the ectoderm undergoes distinct bud, cap, bell, and differentiation stages.

At the molecular level, these interactions are mediated by signaling molecules mainly belonging to the TGF-β, FGF, Hedgehog, and Wnt families (Miletich and Sharpe, 2003). In addition to FGF and members of the bone morphogenetic protein (BMP)/TGF-β family, nerve growth factor (NGF) was found, by immunohistochemistry or *in situ* hybridization, to be expressed in the developing tooth germ (Lesot et al., 2001; Miletich and Sharpe, 2003; Unda et al., 2001). In addition, insulin-like growth factor 1 might promote proliferation of many cell types particularly in bone growth. It stimulated cytological (but not functional) odontoblast differentiation and might have a primary role in the growth and differentiation of pulp cells (Lesot et al., 2001; Lovschall et al., 2001).

In vitro experiments have shown that members of the TGF-β family can induce polarization and stimulate matrix secretion by preodontoblasts (Lesot et al., 2001; Smith et al., 2001a; Unda et al., 2001). The combined use of FGF1 and TGF-β1 promoted cytological and functional differentiation of odontoblasts, whereas FGF2 combined with TGF-β1 only stimulated cytological differentiation (Unda et al., 2001). None of these factors alone induced the terminal differentiation of odontoblasts.

BMP constitutes a subfamily of TGF-β consisting of about 20 members (note that BMP1 does not belong to the TGF-β family of proteins). Members of the BMP family are implicated throughout embryonic tooth development, initiation, morphogenesis, cytodifferentiation, and matrix secretion. They were used to stimulate differentiation of the bovine and human adult pulp cells into odontoblasts in monolayer cultures and in three-dimensional pellet cultures (Iohara et al., 2004). In addition, they were utilized to stimulate differentiation of odontoblasts in organ cultures of mouse dental papillae cells and induction of reparative and regenerative dentin formation (Lesot et al., 2001; Nakashima, 1994a).

23.3.3 Role of Growth Factors during Dental Repair

Many growth factors and extracellular matrix proteins normally expressed during primary dentinogenesis are assumed to play a role in dental repair and dentin regeneration (Iohara et al., 2004; Tziafas et al., 2000; Unda et al., 2001). Injured endothelial cells are also known to release signaling molecules that participate in initiating the inflammatory reaction and the healing process. These molecules also appear to be involved in the recruitment of odontoblast-like cells at the injury site.

Among these molecules, growth factors are assumed to play a role in signaling the biological events responsible for the processes of reactionary and reparative dentinogenesis (cf. Figure 23.2; Smith, 2003). In reparative dentinogenesis, recruitment of progenitor or stem cells leads to differentiation of a new generation of odontoblast-like cells to replace the dead cells due to the injury.

TABLE 23.1
Families of Growth Factors

Superfamily	Family	Some Known Functions
TGF-β	TGF-β exists in three isoforms: TGF-β1, TGF-β2, and TGF-β3	Proliferation, differentiation of most cells: tissue regeneration, embryonic development, regulation of the immune system
	Inhibins (inhibin/activin)	Regulation of the menstrual cycle
	BMP	Osteogenic activity; induces cartilage and bone formation
PDGF	PDGF	Regulation of cell growth and division; angiogenesis (growth of blood vessels); migration and proliferation of endothelial and mesenchymal stem cells
	VEGFs	Vasculogenesis (formation of the embryonic circulatory system) and angiogenesis
	CTGFs	Skeletogenesis; wound healing; mediates profibrotic action of TGF-β
EGF	EGF	Growth, proliferation, and differentiation of epithelial cells, cancer cells
	TGFα	Epithelial development, neural cell proliferation
Other large peptide growth factor families	FGF1–FGF23: aFGF (FGF-1) and bFGF (FGF-2)	Mesoderm induction and patterning; cell growth, migration, and differentiation; angiogenesis, wound healing, and embryonic development
	IGFs	Regulation of most cells: bone and cartilage metabolism; muscle; nerve cells
	NGF	Differentiation and survival of particular nerve cells
	TNF-α and TNF-β, classified either as proinflammatory cytokines or sometimes as growth factors	Regulation of immune cells

Source: Smith, A.J. 2003. Vitality of the dentin–pulp complex in health and disease: Growth factors as key mediators. *J Dent Educ* 6: 678–689; Issa, M.J.P., Tiossi, R., Pitol, D.L., and Mello, S.A.S. 2006. TGF-β and new bone formation. *Int J Morphol* 2: 399–405.

Note: VEGFs—Vascular EGFs; CTGFs—connective tissue growth factors; aFGF or FGF-1—acidic FGF; IGFs—insulin-like growth factors; TNF-α and TNF-β—tumor necrosis factors α and β.

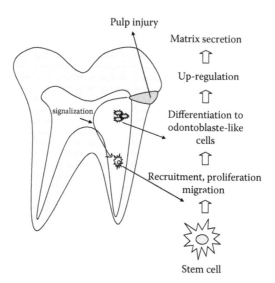

FIGURE 23.2 Schematic representation of reparative dentinogenesis.

Induction of cytodifferentiation of odontoblast-like cells from these progenitor cells is determined by the signaling role of growth factors.

Signaling can proceed with different growth factors because many growth factors are versatile and can act on different cells. However, in tooth healing, only a few growth factors are efficient. Even though a number of growth factors stimulate reparative responses in pulp capping, the reparative dentin matrix secreted can be of variable structure ranging from a tubular matrix to osteodentin-like matrices (About and Mitsiadis, 2001; Tran-Hung et al., 2007).

The growth factors TGF-β, FGF, and BMP were applied on exposed pulps *in vitro* (Six et al., 2002; Sloan and Smith, 1999) or *in vivo* (Lesot et al., 2001; Nakashima, 1994b; Unda et al., 2001); they demonstrated the potential of signaling reparative dentinogenic events. Transdentinal or direct application of TGF-1, FGF, and BMP-7 to the odontoblasts of unexposed pulps in cultured tooth slices demonstrated the ability of these growth factors to signal reactionary dentinogenesis (Sloan and Smith, 1999; Sloan et al., 2000). Application of soluble isolated dentin matrix extracts, containing endogenous growth factors, in unexposed or exposed cavity preparations confirmed these findings (Smith, 2001a, 2001b). Smith (2001b) and Lucchini et al. (2002) found that TGFs were able to reproduce the normal gradients of odontoblast differentiation and morphology of the tissues in dental papilla isolated from tooth germs. Unda et al. (2001) showed that the addition of FGF-β1 to preodontoblasts induces their terminal differentiation by synergistically acting with TGF-β.

23.4 MATERIALS USED IN DENTIN REGENERATION

Efforts are being made to create TE scaffolds that mimic the chemical composition, physical structure, and biological function of the native extracellular matrix for cell proliferation and differentiation. Porous structures are being researched for cellular adaptation and sufficient nutrient permeation for which control of the pore size, porosity, and interconnectivity is fundamental. Solid scaffolds, with a pore size of 50–300 μm and a porosity of about 90%, are used in applications where mechanical strength is important and necessary, especially in bone regeneration. Hydrogel scaffolds can also be used if the pore size allows cell and tissue penetration. Gradual degradation of the scaffold to nontoxic products within a predefined time could be researched in some applications.

The activity of these biomaterials or scaffolds can be improved by the synchronous incorporation of bioactive molecules. The scaffold might contain a drug that can be incorporated by different methods:

1. Imbibing or coating the scaffold with the drug (followed by drying or lyophilization)
2. Incorporating the drug in collagen or agarose beads
3. Encapsulating the drug in colloidal microspheres that are then incorporated directly into a hydrogel scaffold or first in a gel that is then introduced in a solid scaffold

Distinct release profiles are obtained in each case. Controlled delivery of growth factors is researched since it has been demonstrated to promote pulp cell activation for tissue regeneration.

The scaffolds can also contain stem cells to undergo differentiation and regenerate the tissue. The triad notion is therefore used in TE that implies the use of extracellular matrix scaffolds, responding progenitor or stem cells, and indicative morphogenetic signals (using growth factors) in order to converge to the formation of regenerative tissues suitable for rapidly established functionality after intervention (Kuo et al., 2007).

Consequently, in addition to the requirements of biocompatibility and appropriate shape, esthetics, porosity, mechanical properties, and potential biodegradability, the materials used to prepare the scaffold should ensure the ability to incorporate cells and growth factors. Note that the incorporation of cells is not indispensable because the dental pulp contains progenitor cells, which can proliferate and differentiate into dentin-forming odontoblasts, a process that can be stimulated by the use of growth factors (Nakashima, 1994a).

This chapter focuses on the colloidal materials used as delivery systems in dental TE (dentin regeneration or pulp healing), so the materials used as scaffolds and the incorporation of cells will not be discussed. The different natural and synthetic (organic or inorganic) materials used (with different shapes and patterns) to deliver growth factors *in vivo* or *in vitro* are summarized below. A good review of controlled drug delivery systems in TE is given by Biondi et al. (2008).

23.4.1 Cements

Before the emergence of the idea of TE, including the application of growth factors and cells in pulp therapy, capping by cements was practiced. These cements might be beneficial when applied to an exposed pulp and allow the pulp to recover and maintain its vitality and function. They can be bioactive themselves.

The most widely used capping agent to keep teeth alive since the early 1940s is calcium hydroxide [$Ca(OH)_2$] because of its low-grade irritation of the traumatized pulp tissue. It was used for the first time by Hermann in 1930 (Ingle and Bakland, 2002). However, the use of $Ca(OH)_2$ is not without disadvantages including its gradual degradation, tunnel defects in dentinal bridges, and inappropriate sealing properties (Hasheminia et al., 2008). Its high pH might create superficial obliteration zones. It has also been reported that $Ca(OH)_2$ has a high toxic effect on cells in tissue culture and provokes some pulp tissue destruction (Asgary et al., 2008).

Mineral trioxide aggregate (MTA) was successfully introduced in 1995 for pulp capping (Hasheminia et al., 2008), and it is approved by the U.S. Food and Drug Administration (FDA) for use in pulpal therapy. MTA has superior biocompatibility and sealing ability and is less cytotoxic compared to the conventional root-end filling materials. It prevents microleakage, resists bacterial leakage, and promotes regeneration of the original tissues allowing continued pulp vitality in teeth. When compared to $Ca(OH)_2$, it has shown less inflammation and more predicable results after pulp capping (Asgary et al., 2008) and provided a superior seal when compared to amalgam. MTA's disadvantages are its delayed setting time, poor handling characteristics, and off-white color, as well as its higher price.

Numerous efforts continue to be made to develop new materials that can induce reparative dentinal bridge formation with less undesirable effects. The application of adhesive systems as pulp capping materials is suggested for their suitable physical and mechanical properties (de Souza Costa et al., 2000).

23.4.2 AGAROSE BEADS

Agarose is a gelatinous porous medium that is obtained from agar (a natural polymer of the sugar galactose), and is used in many applications such as gel electrophoresis. It consists of linear polysaccharide chains cross-linked to provide a porous structure and mechanical stability. The degree of cross-linking defines the amount of water adsorbable by the agarose beads (they may contain up to 80% water).

The pores between crosslinked chains are typically large enough to allow proteins with molecular weights up to 1 MDa to penetrate, whereas small molecules move freely through the network. Therefore, agarose beads can be used as solid elastic protein carriers.

23.4.3 COLLAGEN SPONGE

Collagen is a protein-based natural polymer. It is the main protein of connective tissue in animals and possesses great tensile strength (present in skin, cartilage, tendons, and ligaments). A collagen sponge is a collagen-based porous biodegradable device used for cell culture. In TE, cells can penetrate into the sponge and proliferate three-dimensionally. The properties of the collagen sponge are determined by its cross-linking, degradation, porosity, and sterilization. The collagen sponge is the most intensively tested drug delivery carrier in animal and clinical studies (as wound dressing materials, matrices for bone and cartilage repair, surgical tampons, and implants for drug delivery; Chvapil, 1976). It is the only FDA-approved carrier for many proteins (such as BMP; Chen et al., 2005). However, because of its xenogeneic origin, it might have some disadvantages.

23.4.4 HYDROGELS

Hydrogels are networks of hydrophilic polymer chains that are cross-linked. They are water insoluble but can absorb an important amount of water (up to 99%). Natural (such as agarose, methyl cellulose, and hylaronan) and synthetic polymers [poly(vinyl alcohol) (PVA), sodium polyacrylate, and acrylate polymers] can be used to form a hydrogel via chemical or physical cross-linking. Cross-linking avoids dissolution of the hydrophilic polymer chains and determines the mechanical properties of the hydrogel and its flexibility. A hydrogel can be used as a scaffold to repair tissues if the macropores are large enough to contain human cells. However, the main use of biodegradable hydrogels is as peptide and protein delivery matrices because of their biocompatibility and the possibility of manipulating the permeability of the solutes. It is important to note that the release of drugs from hydrogels is relatively faster and less controlled than microspheres.

Drug-loaded hydrogels are prepared by incorporating the drug either before or after cross-linking. Encapsulating the protein before cross-linking usually gives higher encapsulation efficiencies and lower release rates. Chemical cross-linking ensures high mechanical strength but, because of the chemical reaction, the protein might denature. For this reason, physical cross-linking is proposed in some works. In physical cross-linking, there is no need for a cross-linking agent and only physical interactions take place. For instance, the dextran-*co*-gelatin hydrogel is based on physical interactions (Chen et al., 2005).

The most widely investigated system to form hydrogels is poly(ethylene glycol) (PEG) because of the ability to confer significant functionalities of the native extracellular matrix and the possibility of controlling their chemical and mechanical properties. Gelatin (a protein produced by partial hydrolysis of collagen extracted from skin and bones of animals) is also commonly applied to form hydrogels because of its bioadhesive properties, which allow the maintaining of the drug delivery system at the target and to its biodegradability. Gelatin can also adsorb different growth factors and has considerable desorption when biodegraded. For instance, acidic proteins can be adsorbed to basic gelatin with isoelectric points (Chen et al., 2005). Dextran and alginate hydrogels are also commonly used.

23.4.5 BIODEGRADABLE POLYMERS

Natural and synthetic polymers can be used to produce scaffolds, hydrogels, and microspheres. Natural polymers such as bovine serum albumin (BSA), human serum albumin, collagen, gelatin, hemoglobin, carboxymethyl cellulose, and alginic acid are expensive and their purity and reproducibility are not assured (Jain, 2000). Biodegradable synthetic polymers, such as poly(*p*-dioxanone), polylactones, polyanhydrides, polyurethanes, polyphosphazenes, and polyesters, and thermoplastic aliphatic polyesters, such as polylactide [also called poly(lactic acid) or PLA], polyglycolide [also called poly(glycolic acid) or PGA], and their copolymer poly(lactide-*co*-glycolide) (PLGA), have a lower cost and more reproducible properties with the possibility of controlling their mechanical properties, degradation kinetics, and physical structures to suit various applications. Some common biodegradable polymers used in scaffold or microsphere preparation are discussed below and classified as natural or synthetic polymers.

23.4.5.1 Biodegradable Natural Polymers

Dextrans are the most abundant and naturally occurring (by certain bacteria) biodegradable glucose polymers. They are hydrophilic, relatively inert, and nontoxic. Dextran-derived materials are considered to be compatible matrices for protein and bioactive drugs because of their hydrophilic properties and ability to control drug dissolution and permeability (Chen et al., 2007). They have also been used as plasma volume expanders over the last 50 years.

Chitosan {poly[-(1,4)-2-amino-2-deoxy-D-glucopiranose]} is a natural hydrophilic cationic polysaccharide biopolymer obtained by the hydrolysis of chitin, and it can be found in arthropod exoskeletons or shellfish (the second most abundant polysaccharide after cellulose). The solubility of the polymer and its rheological properties are determined by the degree of hydrolysis (deacetylation). It is biodegradable, nontoxic, and biocompatible but is not bioactive itself. In the pharmaceutical industry, chitosan microspheres are promising in oral and mucosal administrations (Niu et al., 2009). They are also introduced in scaffolds to make the material stronger and antiwash-out.

Hyaluronic acid, or hyaluronan, is a natural polysaccharide that is abundant in cartilage, vitreous, and skin. It is a versatile, biocompatible, and bioresorbable material, which is enzymatically degraded. It is known to contribute to cell proliferation and migration especially in skin tissue repair.

23.4.5.2 Biodegradable Synthetic Polymers

PLA and PGA are biodegradable, thermoplastic, aliphatic polyesters. They can be used to form fibers and films. The glass transition temperature of PLA is between 60°C and 65°C. Because of the methyl group in PLA, it is hydrophobic and has slow degradation (>24 months; Gunatillake and Adhikari, 2003). PGA has a glass transition temperature between 35°C and 40°C. It is crystalline and insoluble in most organic solvents. It is insoluble in water which increases its degradation time (6–12 months). By integrating different ratios of PLA and PGA in the PLGA copolymer and by controlling the copolymer molecular weight, distinct release and degradation profiles can be obtained to satisfy various applications in TE and to accommodate different growth factors (Moioli et al., 2006). The use of PLGA is approved by the FDA. These copolymers have been used to prepare various drug-loaded devices (vaccines, peptides, proteins, and micromolecules) because of their excellent biocompatibility and biodegradability. PLA, PGA, and their copolymers have been used to form both scaffolds and microspheres. A good review of PLGA-based drug delivery systems is given by Jain (2000).

One of the main disadvantages of PLGA is the need for an organic solvent to dissolve the polymer to prepare the scaffold or the microspheres. In addition, the degradation of PLGA leads to a reduction in the pH, which might affect protein integrity during the release. Even if degradation of PLGA-based scaffolds might be of benefit in some applications, it is annoying in others because this affects the mechanical strength of the scaffold. Note however that PLGA-based scaffolds support cellular viability, attachment, and proliferation during *in vitro* culture.

Polycaprolactone (PCL) is a biodegradable polyester with a low glass transition temperature of about −60°C. It is approved by the FDA as a drug delivery device in the human body. It degrades slower than PLA (approximate degradation time is >24 months; Gunatillake and Adhikari, 2003) because of its higher resistance to hydrolysis. PCL-based scaffolds were found to support the viability, proliferation, and differentiation of cells. Because of their lower degradation rate, PCL matrices maintain a mechanically stable three-dimensional architecture during *in vitro* culture (Weert et al., 2000). Li et al. (2005) used PCL to form a nanofibrous scaffold that was seeded with bone marrow derived human mesenchymal stem cells to mimic the architecture of natural tissue. The ultrafine nanofibers provided improved mechanical strength and more extensive substrate for cellular attachment as compared to larger fibers.

23.5 PREPARATION OF MICROSPHERES

The main objectives of encapsulating the drug in microspheres consist of the protection against degradation and controlling the release rate. Microspheres employed as drug delivery devices are known to improve the bioavailability of the drugs and to give controlled release, which allows a reduction of the side effects of the drugs because of the controlled dose. They can be designated to ensure drug targeting to specific tissues or for self-regulated therapeutic action. Therefore, many pharmaceutical delivery systems are based on the use of microspheres.

In many applications microspheres are first prepared and sterilized and then loaded with the growth factor and lyophilized (e.g., for BMP, Woo et al., 2001; for TGF, Zhang et al., 2007). This avoids the degradation of the growth factor during the synthesis of microspheres or during sterilization. The growth factor in this case is incorporated by adsorption on the polymer matrix.

The formulation and sterilization conditions can be adapted to minimize the denaturation of the growth factor during the preparation. *In situ* encapsulation of the growth factor leads to better protection of the drug from environmental invasion and ensuring a controlled release profile. The use of organic solvents in the formulation must be chosen wisely. Weert et al. (2000) presented the main parameters that might lead to the denaturation of proteins during encapsulation that happen to be principally the interfaces between the drug and the organic solvent, which lead to aggregation of the protein, added to the effects of heat and loss of hydration that affect the protein stability. They outline that even when the recovery of native protein is high, the presence of low amounts of degraded protein may be harmful and cause side effects.

The sterilization should also be optimized to maintain the bioactivity of the drug and the microsphere architecture to preserve the predefined release kinetics. Different sterilization techniques can be used: ultraviolet (UV) light, ethylene oxide gas, radiofrequency glow discharge, and γ irradiation. UV light sterilization induced surface damage of basic FGF (bFGF)-encapsulated PLGA microspheres and reduced the rate of growth factor release (Moioli et al., 2006).

Drug release from such matrices takes place by diffusion through the polymer matrix (through the pores of the microspheres and through the polymer) and by erosion of the polymer material in case of simultaneous degradation of the microsphere. The release medium is supposed to penetrate inside the matrix during the release. Weakly adsorbed drugs might be desorbed rapidly from the surface of the matrix which causes a burst effect.

Different objectives have to be kept in mind during the preparation of microspheres or capsules:

- The stability of the drug during the preparation and the release (avoid the protein denaturation) needs to be ensured.
- The colloidal stability of the microspheres or capsules (avoid aggregation of particles) must be ensured.

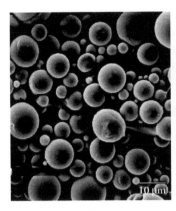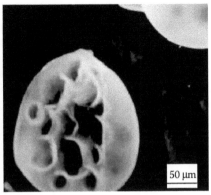

FIGURE 23.3 Scanning electron microscope micrographs of microparticles produced with the W/O/W double emulsion: (left) the exterior structure and (right) the interior structure.

- The encapsulation efficiency has to be maximized and the loading should be appropriate to the application.
- The choice of the method of encapsulation is determined by the solubility of the drug and the duration of the treatment (determined by the release profile).
- The polymers selected for pharmaceutical applications should be biocompatible and drug compatible and should possess the desired mechanical properties and biodegradation kinetics (related to the molecular weight of the polymer).

Various encapsulation methods are practiced to produce pharmaceutical carriers, such as the single-emulsion process, double-emulsion process, phase separation (coacervation), and spray drying. The oil-in-water (O/W) and W/O/W techniques are usually used for the production of nano- and microspheres. The O/W technique is best adapted to encapsulate water-insoluble drugs such as steroids and the W/O/W technique is adapted for water-soluble drugs such as peptides, proteins, and vaccines. Solvent evaporation and extraction methods are applied for solvent evacuation from the microspheres.

Because of their hydrophilic nature, growth factors are usually encapsulated by the W/O/W method. This fabrication procedure for microspheres allows better protection of the growth factor against denaturation because it reduces the contact of the organic solvent.

In the W/O/W technique, the polymer is dissolved in an organic solvent (immiscible in water, such as dichloromethane) that consists of the organic phase. The growth factor is dissolved in distilled water (internal aqueous phase). Both phases are emulsified to produce a stable W/O emulsion. This emulsion is then introduced under stirring in an aqueous phase (external aqueous phase) containing a colloidal stabilizing agent (e.g., PVA). This forms the W/O/W emulsion. Afterward, the solvent is extracted from the particles and is evaporated. This leads to the solidification of the polymer, thus protecting the internal aqueous phase and the growth factor forming the solid nano- or microspheres. These microspheres are usually collected on a filter, rinsed and dried, or lyophilized. Figure 23.3 shows micrographs of microspheres produced by a W/O/W double emulsion obtained by a scanning electron microscope. Spherical microspheres obtained by this method have a smooth surface. The porosity of the internal structure is directly related to the ratio of the internal aqueous phase to the oil phase (Deloge et al., 2009).

23.6 *IN VITRO* AND *IN VIVO* DELIVERY OF GROWTH FACTORS IN DENTAL TISSUE ENGINEERING

This section reviews the main applications of growth factor delivery systems in dental TE. These carriers can be classified into three categories. First, the scaffolds can be impregnated with growth factors and applied on the tissue. Second, agarose beads or collagen can be soaked into a drug solution and then placed on the tissue. Third, colloidal microspheres are used to encapsulate the growth factor (either *in situ* or just impregnated after microsphere preparation). These microspheres are then applied directly for tissue regeneration or incorporated into a scaffold that usually contains a gel to retain the microspheres. Note that, the use of colloidal microspheres presents several advantages related to the good characterization of their physicochemical properties and controlled release rate.

A recent review of drug delivery scaffolds used in dental, oral, and craniofacial TE is provided by Moioli et al. (2007).

23.6.1 DRUG- OR CELL-LOADED SCAFFOLDS

The scaffold can be impregnated with a drug solution and then dried or lyophilized and used as a drug carrier. For instance, derivatized hyaluronic acid has been tested as a delivery scaffold for BMP-2 (Kim and Valentini, 2001). The release profile of BMP-2 from derived hyaluronic acid scaffolds was slower (32% of total encapsulated BMP-2 released in 28 days) than in both collagen gels (14 days) and PLA scaffolds. The released BMP-2 conserved its bioactive form in all the delivery systems that were studied. Tziafas et al. (1998) used Millipore filters as implants impregnated with bFGF and TGF-β1 to evaluate the effects of the growth factors on dog dental pulp cells *in vivo*. A deposition of a tubular matrix surrounding the implants and highly elongated odontoblast-like cells was observed after 3 weeks of implantation of the Millipore filters.

The incorporation of growth factors is not systematic in all scaffolds because they might be bioactive themselves or contain stem cells, which allows cell regeneration. Kuo et al. (2007) developed a gelatin–chrondroitin–hyaluronan tricopolymer to form scaffolds containing swine dental bud cells for tooth regeneration. The dental bud cells survived and regenerated tooth structures in the swine jaw, including dental pulp, odontoblasts, dentin, cementum, and periodontal ligaments, but the regenerated teeth were smaller than normal. The authors suggest adapting the scaffold size for a better control of the morphology of the regenerated tooth. Duailibi et al. (2008) found that using PGA/PLLA and PLGA scaffolds seeded with cultured 4-day postnatal rat tooth bud cells can form organized and bioengineered dental tissues containing dentin, enamel, pulp, and periodontal tissues. Xu et al. (2008) developed solid and porous calcium phosphate cement composite scaffolds by incorporating chitosan and mannitol in the scaffold. Because of its rapid dissolution, mannitol provides initial macropores and the subsequent degradation of chitosan offers the possibility of increasing the scaffold porosity to integrate cell ingrowth. Chitosan also adds mechanical strength to the scaffold. These scaffolds can act as carriers for osteogenic cell and growth factor delivery for bone TE.

23.6.2 DRUG-LOADED AGAROSE BEADS OR COLLAGEN

One of the early delivery systems of growth factors consisted of using agarose beads that were soaked into a drug solution and then placed on the tissue. Sloan and Smith (1999) used agarose beads for the delivery of TGF-β. They noted that the presence of empty agarose beads did not induce tissue changes. Local stimulation of the extracellular matrix by TGF-β isoforms was demonstrated. TGF-β1 and TGF-β3 stimulated dentinogenic effects in odontoblasts *in vitro* when applied in agarose beads. Sloan and Smith used agarose beads for the delivery of BMP-7. They showed that the use of agarose beads enhances predentine growth. The use of BMP-7 stimulated secretion of extracellular matrix *in vitro* but with a lower rate than TGF-β1 at similar concentrations.

Similarly to agarose beads, growth factors can be applied after impregnation in collagen that does not operate as scaffolds during the release study. For instance, Six et al. (2002) studied the ability of BMP-7 to induce dentinogenesis in the dental pulp of rat upper first molars and studied the effect of the dose on the treatment. A collagen carrier was applied with and without BMP-7. The results were compared to those obtained with direct pulp capping by $Ca(OH)_2$. They concluded that BMP-7 has the potential to provide an alternative to conventional endodontic treatments. Bergenholtz et al. (2006) used an absorbable collagen sponge as a carrier for recombinant human BMP-2 to evaluate its benefits in bone healing. Evidence for the enhancing effect of bone formation by recombinant human BMP/absorbable collagen sponge carrier could not be found.

23.6.3 DRUG-LOADED MICROSPHERES

Only micron-sized spheres are used in dental TE, as revealed by the literature investigated in this work. In the early applications using microspheres for drug delivery in TE, blank microspheres were first produced and sterilized and then loaded with the drug and lyophilized. By enhanced optimization of the preparation conditions (choice of solvent, temperature, and sterilization method), the encapsulation of drugs during capsule formation (*in situ* encapsulation) can now be done while ensuring the protection of the protein against denaturation. The W/O/W encapsulation technique has mainly been used to prepare such microspheres because of the hydrophilic nature of most growth factors.

23.6.3.1 Drug-Loaded Microspheres Alone

The release rate from the microspheres reviewed in this section was evaluated by applying pure microspheres in the release medium (either *in vitro* or *in vivo*). Note that the same microspheres could be integrated in hydrogels or scaffolds, but the release rate would be altered and has to be preevaluated.

The first generation of drug-loaded microspheres formed by *in situ* encapsulation was bigger in size. Capsules with a diameter of few millimeters were developed by Isobe et al. (1996) to encapsulate BMP by W/O/W using dichloromethane to dissolve PLGA and an aqueous gelatin solution for particle stabilization. They concluded that the use of biodegradable polymers was a promising method to induce cartilage and bone formation (70% of BMP was released from the capsules in 5 days). Similar capsules were produced by Weber et al. (2002) who gave them the name foamspheres (because of their big size). They compared the release results to a gelatin-based hydrogel and found that the slow release of BMP from foamspheres led to the formation of larger amounts of bone than the hydrogels with their faster release of BMP.

Péan et al. (1998) studied the formation of PLGA microspheres by a W/O/W emulsion solvent evaporation/extraction method to encapsulate NGF and tried to optimize the formulation parameters in order to control the release rate and reduce the intensity of the burst effect. A mixture of dichloromethane and acetone was used as an organic solution to dissolve the polymer, and PVA was used as a colloidal particle stabilizer. The use of salt and the type of PLGA affected the morphology of the microspheres, which provides a means to control the release rate. However, employing a salt had a negative effect on the protein stability.

Lu et al. (2000) encapsulated TGF-B1 (and BSA) in PLGA/PEG by a W/O/W double-emulsion solvent extraction technique (using dichloromethane as a PLGA solvent and PVA to stabilize the produced microspheres). They found that increasing the PEG ratio in the polymer and decreasing the pH resulted in decreased protein release rate.

Cleland et al. (2001) used PLGA microspheres to encapsulate epidermal growth factor (EGF) for the treatment of ischemia. The microspheres were produced by a spray freeze-drying technique to yield a powder, using ethyl acetate as a polymer solvent. Then disks were produced using these microspheres. A lower release rate was obtained using the disks (40 days) than the microspheres (21 days) that was attributable to the lower surface area to volume of the disks compared to the microspheres. The release profile from the disks was characterized by two subsequent phases of release, indicating an important requirement of PLGA hydrolysis to generate sufficient pores for diffusion of EGF.

Moioli et al. (2006) encapsulated TGF-β3 in PLGA microspheres by a W/O/W double-emulsion solvent extraction technique using dichloromethane to dissolve PLGA and PVA as a colloidal stabilizer of the microspheres. The authors outlined that UV light sterilization reduces the release of TGF-β3, which was explained either by direct degradation of TGF-β3 because of the light or degradation of the polymer surface that causes a decrease in the amount of encapsulated TGF-β3. TGF-β3 was released during about 42 days, and a burst effect was observed.

Niu et al. (2009) recently created chitosan microspheres to encapsulate BSA and oligopeptide. The microspheres were produced according to an emulsion-ion cross-linking method using acetic acid as a solvent and Span 80 and triphosphate for stabilization. The microspheres were collected after precipitation and washed with petroleum ether and isopropyl alcohol. The encapsulation efficiency was higher than 80% for both proteins. They found that the released oligopeptide conserved its biological activity. BSA had a faster release rate than oligopeptide (6–8 days for both proteins).

23.6.3.2   Drug-Loaded Microspheres within Implants or Cements

The incorporation of growth factors in implants or cements gives the possibility for these implants to become bioactive in order to stimulate tissue healing. The use of microspheres ensures a controlled release of the growth factors and its protection against denaturation whereas the implant assures the maintenance of the desired mechanical properties and impermeability.

Woo et al. (2001) compared immediate and sustained release of BMP from implants of sodium carboxymethyl cellulose (an anionic water-soluble polymer derived from cellulose) containing PLGA microspheres. Blank microspheres were formed by a W/O/W double-emulsion method followed by solvent evaporation and extraction. The microspheres were then collected, rinsed, dried, sterilized, and loaded with BMP to evaluate the induced local bone growth. Release of BMP decreased quickly from the immediate release system after 7 days and the released quantity was insufficient to maintain the osteoinductive effects of BMP *in vivo*. However, both systems resulted in faster and more complete bone healing in the animal model.

For the encapsulation of TGF-β1, Zhang et al. (2007) prefabricated blank microspheres by a W/O/W double-emulsion solvent evaporation technique and then impregnated the microspheres with TGF-β1. The produced microspheres were incorporated in calcium phosphate cements and used in pulp capping. The PLGA–calcium phosphate composite enhanced tertiary dentin formation that was further enhanced by the addition of TGF-β1. The introduction of biodegradable microspheres in the cements creates new pores after their degradation that gives the possibility for new tissue ingrowth and replacement, given that the degradation rate of calcium phosphate cements is very low.

23.6.3.3   Drug-Loaded Microspheres within Scaffolds

The combination of microspheres and scaffolds comes from the necessity to ensure distinct functionalities during the same treatment (mechanical properties, permeability, release rate, and protection of the drug). Employing scaffolds is necessary to ensure the desired mechanical properties and porosity for cell development. Drug-loaded microspheres are incorporated into scaffolds to enhance tissue regeneration as a result of the controlled release.

Kim et al. (2003) developed a porous chitosan scaffold containing microspheres loaded with TGF-β1 for cartilage TE. Acetic acid was used to dissolve the chitosan, to which TGF-β1 was added. Microspheres were formed by adding this solution to a hydrogenated caster oil solution. Tripolyphosphate was then used to stabilize the produced microspheres. The microspheres were subsequently washed with isopropyl alcohol and water prior to lyophilization. Then they were dispersed in 90% aqueous ethanol and incorporated into the previously formed chitosan scaffolds (by freeze-drying), after which the scaffolds were lyophilized. The chondrocytes rapidly propagated on the chitosan scaffolds, which presents a potential to be used as implants to treat cartilage defects.

DeFail et al. (2006) encapsulated TGF in PLGA microspheres by a double-emulsion technique for applications in TE of articular cartilage (using dichloromethane to dissolve PLGA and PVA as a colloidal stabilizer). The produced microspheres were then incorporated into a PEG–hydrogel scaffold. TGF-β1-loaded PLGA microspheres showed a high burst release whereas hydrogels containing microspheres showed a delayed burst release. This shows that the combination of microspheres and scaffolds affects the release rate and has to be optimized before *in vivo* applications.

Chen et al. (2005, 2007) proposed the use of hybrid hydrogel scaffolds containing microspheres as new BMP carriers. They developed dextran-*co*-gelatin hydrogel microspheres by radical cross-linking and low-dose γ irradiation. BMP was either incorporated in the microspheres before their preparation (in a dextran–glycidyl methacrylate solution before the addition of the gelatin) or after their formation. They found that this matrix could prolong the release of BMP and promote the proliferation and osteoblastic differentiation of periodontal ligament cells better than using BMP aqueous solution.

Liu et al. (2007) fabricated biodegradable chitosan–gelatin microspheres containing bFGF that were then incorporated into a porous chitosan–gelatin scaffold to improve skin regeneration efficacy and to promote vascularization. The chitosan–gelatin microspheres were prepared by the W/O emulsion method using olive oil and acetone. The microspheres were washed with acetone and isopropyl alcohol, collected by centrifugation, and air dried. They were then impregnated with a bFGF solution and incorporated into a chitosan–gelatin scaffold. The scaffold pore sizes varied from 95 to 160 μm, which is appropriate for TE. The scaffolds incorporating the bFGF-loaded microspheres significantly increased the cell proliferation when compared to empty chitosan–gelatin scaffolds.

Kikuchi et al. (2007) used collagen sponges as a scaffold and gelatin hydrogel microspheres as a carrier for FGF2. Microspheres were prepared from a gelatin solution through cross-linking of gelatin using glutaraldehyde. After dropwise addition in olive oil, acetone was added and the microspheres were collected by centrifugation and added into a Tween 80 solution. They were then impregnated with FGF2 (not *in situ* encapsulation). These FGF2-loaded microspheres were incorporated into a collagen sponge scaffold and used for the formation of dentin in exposed rat molar pulps. These results were compared to FGF2-loaded collagen scaffolds that did not contain gelatin hydrogel microspheres. They noted that, during the early phase of pulp healing, pulp cell proliferation and invasion of vessels into dentin defects above the exposed pulp were induced in both groups; but in the later phase, the induction of dentin formation was distinctly different between the two types of FGF2 release. A controlled release profile was obtained when the FGF2 was incorporated in the gelatin hydrogel microspheres and induced the formation of dentin-like particles with dentin defects above exposed pulp. A noncontrolled release was obtained with the collagen sponge scaffold that induced excessive reparative dentin formation in the residual dental pulp (although dentin defects were not noted).

23.7 CONCLUSIONS

This chapter presented a review of the biocompatible carriers used in dental TE with emphasis on colloidal carriers (biodegradable polymer-based microspheres, gels such as hydrogels and alginate gels, gelatinous carriers). Dental TE is practiced in conservative dentistry to restore the original tooth (structure, function, and esthetics) while preserving pulp health.

The use of microsphere colloids as delivery systems for growth factors was found to be promising in pulp healing and dentin regeneration. These colloids improve the stability of growth factors and provide sustained release that ensures continuous dosing. They proved reproducible delivery of growth factors because of the possibility to produce microspheres with specific physicochemical and colloidal properties in large volumes (Kalaji et al., 2009).

The ultimate objective of dental TE is to completely regenerate in the jaw a previously lost or missing tooth with similar physical properties and functions that would provide a remedy for all problems encountered with false teeth and implants. The term regenerative TE has recently been used to designate this branch of dental TE involving the triad (three key elements): responding progenitor or stem cells, indicative morphogenetic signals (growth factors), and extracellular matrix scaffolds (Kuo et al., 2007).

A variety of dental tissues have to be regenerated to form the whole tooth. This necessitates a rigorous study of the scaffold structure, the incorporation of the stem cells, and the growth factors

(Duailibi et al., 2008). The scaffolds have to contain hybrid functional microspheres with distinct release rates of a variety of growth factors, which are necessary to maintain the different tissue ingrowths, and the molecules necessary to overcome changes in the release medium (such as buffers).

ABBREVIATIONS

bFGF	Basic FGF
BMP	Bone morphogenetic protein
BSA	Bovine serum albumin
$Ca(OH)_2$	Calcium hydroxide
EGF	Epidermal growth factor
FDA	Food and Drug Administration
FGF	Fibroblast growth factor
MTA	Mineral trioxide aggregate
NGF	Nerve growth factor
O/W	Oil-in-water
PCL	Polycaprolactone
PDGF	Platelet-derived growth factor
PEG	Poly(ethylene glycol)
PGA	Poly(glycolic acid)
PLA	Poly(lactic acid)
PLGA	Poly(lactide-*co*-glycolide)
PVA	Poly(vinyl alcohol)
TE	Tissue engineering
TGF-α	Transforming growth factor α
TGF-β	Transforming growth factor β
UV	Ultraviolet
W/O/W	Water-in-oil-in-water

REFERENCES

About, I. and Mitsiadis, T.A. 2001. Molecular aspects of tooth pathogenesis and repair: *In vivo* and *in vitro* models. *Adv Dent Res* 15: 59–62.

Asgary, S., Eghbal, M.J., Parirokh, M., Ghanavati, F., and Rahimi, H. 2008. A comparative study of histologic response to different pulp capping materials and a novel endodontic cement. *Oral Surg Oral Med Oral Pathol Oral Radiol Endodont* 106: 609–614.

Bergenholtz, G., Wikesjo, M.E., Sorensen, R.G., Xiropaidis, A.V., and Wozney, J.M. 2006. Observations on healing following endodontic surgery in nonhuman primates (*Macaca fascicularis*): Effects of rhBMP-2. *Oral Surg Oral Med Oral Pathol Oral Radiol Endodont* 101: 116–125.

Biondi, M., Ungaro, F., Quaglia, F., and Netti, P.A. 2008. Controlled drug delivery in tissue engineering. *Adv Drug Deliv Rev* 60: 229–242.

Chen, F.M., Wu, Z.F., Wu, H., Zhang, Y.J., Xin, S.N., and Jin, Y. 2005. Preparation and biological characteristics of recombinant human bone morphogenetic protein-2-loaded dextran-*co*-gelatin hydrogel microspheres, *in vitro* and *in vivo* studies. *Pharmacology* 75: 133–144.

Chen, F.M., Zhao, Y.M., Sun, H.H., Jin, T., Wang, Q.T., Zhou, W., Wu, Z.F., and Jin, Y. 2007. Novel glycidyl methacrylated dextran (Dex-GMA)/gelatin hydrogel scaffolds containing microspheres loaded with bone morphogenetic proteins: Formulation and characteristics. *J Control Release* 118: 65–77.

Chvapil, M. 1976. Collagen sponge: Theory and practice of medical applications. *J Biomed Mater Res* 11: 721–741.

Cleland, J.L., Duenasa, E.T., Park, A., Daugherty, A., Kahn, J., Kowalski, J., and Cuthbertson, A. 2001. Development of poly-(D,L-lactide-*co*-glycolide) microsphere formulations containing recombinant human vascular endothelial growth factor to promote local angiogenesis. *J Control Release* 72: 13–24.

De Souza Costa, C.A., Hebling, J., and Hanks, C.T. 2000. Current status of pulp capping with dentin adhesive systems: A review. *Dent Mater* 16: 188–197.

DeFail, A.J., Chu, C.R., Izzo, N., and Marra, K.G. 2006. Controlled release of bioactive TGF-β1 from microspheres embedded within biodegradable hydrogels. *Biomaterials* 27: 1579–1585.

Deloge, A., Kalaji, N., Sheibat-Othman, N., Lin, V.S., Farge, P., and Fessi, H. 2009. Investigation of the preparation conditions on the morphology and release kinetics of biodegradable particles: A mathematical approach. *J Nanosci Nanotechnol* 8: 1–8.

Duailibi, S.E., Duailibi, M.T., Zhang, W., Asrican, R., Vacanti, J.P., and Yelick, P.C. 2008. Bioengineered dental tissues grown in the rat jaw. *J Dent Res* 87: 745–750.

Gunatillake, P. and Adhikari, R. 2003. Biodegradable synthetic polymers for tissue engineering. *Eur Cells Mater* 5: 1–16.

Hasheminia, S.M., Feizi, G., Razavi, S.M., Feiziandfard, M., Gutknecht, N., and Mir, M. 2008. A comparative study of three treatment methods of direct pulp capping in canine teeth of cats: A histologic evaluation. *Lasers Med Sci* [online only] DOI: 10.1007/s10103-008-0584-9.

Ingle, J.I. and Bakland, L.K. 2002. *Endodontics*, 5th ed. Marcel Decker, London.

Iohara, K., Nakashima, M., Ito, M., Ishikawa, M., Nakasima, A., and Akamine, A. 2004. Dentin regeneration by dental pulp stem cell therapy with recombinant human bone morphogenetic protein 2. *J Dent Res* 83: 590–595.

Isobe, M., Yamazaki, Y., Oida, S., Ishihara, K., Nakabayashi, N., and Amagasa, T. 1996. Bone morphogenetic protein encapsulated with a biodegradable and biocompatible polymer. *J Biomed Mater Res* 32: 433–438.

Issa, M.J.P., Tiossi, R., Pitol, D.L., and Mello, S.A.S. 2006. TGF-β and new bone formation. *Int J Morphol* 24: 399–405.

Jain, R.A. 2000. The manufacturing techniques of various drug loaded biodegradable poly(lactide-*co*-glycolide) (PLGA) devices. *Biomaterials* 21: 2475–2490.

Kalaji, N., Sheibat-Othman, N., Saadaoui, H., Elaissari, H., and Fessi, H. 2009. Colloidal and physicochemical characterization of protein-containing poly(lactide-*co*-glycolide) (PLGA) microspheres before and after drying. *e-Polymers* 2009: article 010.

Kikuchi, N., Kitamura, C., Morotomi, T., Inuyama, Y., Ishimatsu, H., Tabata, Y., Nishihara, T., and Terashita, M. 2007. Formation of dentin-like particles in dentin defects above exposed pulp by controlled release of fibroblast growth factor 2 from gelatin hydrogels. *J Endodont* 33: 1198–1202.

Kim, H.D. and Valentini, R.F. 2001. Retention and activity of BMP-2 in hyaluronic acid-based scaffolds *in vitro*. *J Biomed Mater Res Part B: Appl Biomater* 59: 573–584.

Kim, S.E., Park, J.H., Cho, Y.W., Chung, H., Jeong, S.Y., Lee, E.B., and Kwon, I.C. 2003. Porous chitosan scaffold containing microspheres loaded with transforming growth factor-β: Implications for cartilage tissue engineering. *J Control Release* 91: 365–374.

Kuo, T.F., Huang, A.T., Chang, H., Lin, F.H., Chen, S.T., Chen, R.S., Chou, C.H., Lin, H.C., Chiang, H., and Chen, M.H. 2007. Regeneration of dentin–pulp complex with cementum and periodontal ligament formation using dental bud cells in gelatin–chondroitin–hyaluronan tri-copolymer scaffold in swine. *J Biomed Mater Res* 86A: 1062–1068.

Lesot, H., Lisi, S., Peterkova, R., Peterka, M., Mitolo, V., and Ruch, J.V. 2001. Epigenetic signals during odontoblast differentiation. *Adv Dent Res* 15: 8–13.

Li, W.J., Tuli, R., Huang, X., Laquerriere, P., and Tuan, R.S. 2005. Multilineage differentiation of human mesenchymal stem cells in a three-dimensional nanofibrous scaffold. *Biomaterials* 26: 5158–5166.

Liu, H., Fan, H., Cui, Y., Chen, Y., Yao, K., and Goh, J.C.H. 2007. Effects of the controlled-released basic fibroblast growth factor from chitosan–gelatin microspheres on human fibroblasts cultured on a chitosan–gelatin scaffold. *Biomacromolecules* 8: 1446–1455.

Lovschall, H., Fejerskov, O., and Flyvbjerg, A. 2001. Pulp-capping with recombinant human insulin-like growth factor I (rhIGF-1) in rat molars. *Adv Dent Res* 15: 108–112.

Lu, L., Stamatas, G.N., and Mikos, A.G. 2000. Controlled release of transforming growth factor β1 from biodegradable polymer microparticles. *J Biomed Mater Res* 50: 440–451.

Lucchini, M., Romeas, A., Couble, M.L., Bleicher, F., Magloire, H., and Farges, J.C. 2002. TGFβ1 signaling and stimulation of osteoadherin in human odontoblasts *in vitro*. *Connect Tissue Res* 43: 345–353.

Miletich, I. and Sharpe, P.T. 2003. Normal and abnormal dental development. *Hum Mol Genet* 12: R69–R73.

Moioli, E.K., Clark, P.A., Xin, X., Lal, S., and Mao, J.J. 2007. Matrices and scaffolds for drug delivery in dental, oral and craniofacial tissue engineering. *Adv Drug Deliv Rev* 59: 308–324.

Moioli, E.K., Hong, L., Guardado, J., Clark, P.A., and Mao, J.J. 2006. Sustained release of TGFβ3 from PLGA microspheres and its effect on early osteogenic differentiation of human mesenchymal stem cells. *Tissue Eng* 12: 537–546.

Nakashima, M. 1994a. Induction of dentin formation on canine amputated pulp by recombinant human bone morphogenetic proteins (BMP)-2 and -4. *J Dent Res* 73: 1515–1522.

Nakashima, M. 1994b. Induction of dentine in amputated pulp of dogs by recombinant human bone morphogenetic proteins-2 and -4 with collagen matrix. *Arch Oral Biol* 39: 1085–1089.

Nakashima, M. 2005. Bone morphogenetic proteins in dentin regeneration for potential use in endodontic therapy. *Cytokine Growth Factor Rev* 16: 369–376.

Niu, X., Feng, Q., Wang, M., Guo, X., and Zheng, Q. 2009. Preparation and characterization of chitosan microspheres for controlled release of synthetic oligopeptide derived from BMP-2. *J Microencapsul* 26: 297–305.

Olsson, H., Petersson, K., and Rohlin, M. 2006. Formation of a hard tissue barrier after pulp cappings in humans. A systematic review. *Int Endodont J* 39: 429–442.

Péan, J.M., Venier-Julienne, M.C., Boury, F., Menei, P., Denizot, B., and Benoit, J.P. 1998. NGF release from poly(D,L-lactide-*co*-glycolide) microspheres. Effect of some formulation parameters on encapsulated NGF stability. *J Control Release* 56: 175–187.

Six, N., Lasfargues, J.J., and Goldberg, M. 2002. Differential repair responses in the coronal and radicular areas of the exposed rat molar pulp induced by recombinant human bone morphogenetic protein 7 (osteogenic protein 1). *Arch Oral Biol* 47: 177–187.

Sloan, A.J., Rutherford, R.B., and Smith, A.J. 2000. Stimulation of the rat dentine–pulp complex by bone morphogenetic protein-7 *in vitro*. *Arch Oral Biol* 45: 173–177.

Sloan, A.J. and Smith, A.J. 1999. Stimulation of the dentine–pulp complex of rat incisor teeth by transforming growth factor-β isoforms 1–3 *in vitro*. *Arch Oral Biol* 44: 149–156.

Smith, A.J. 2003. Vitality of the dentin–pulp complex in health and disease: Growth factors as key mediators. *J Dent Ed* 67: 678–689.

Smith, A.J., Murray, P.E., Sloan, A.J., Matthews, J.B., and Zhao, S. 2001a. Trans-dentinal stimulation of tertiary dentinogenesis. *Adv Dent Res* 15: 51–54.

Smith, A.J., Tobias, R.S., and Murray, P.E. 2001b. Transdentinal stimulation of reactionary dentinogenesis in ferrets by dentine matrix components. *J Dent* 29: 341–346.

Summitt, J.B., Robbins, W., and Schwartz, R.S. 2001. *Fundamentals of Operative Dentistry, A Contemporary Approach*, 2nd ed. Quintessence Publishing, Chicago.

Tran-Hung, L., Laurent, P., Camps, J., and About, I. 2007. Quantification of angiogenic growth factors released by human dental cells after injury. *Arch Oral Biol* 52: 9–13.

Tziafas, D., Alvanoub, A., Papadimitriou, S., Gasic, J., and Komnenou, A. 1998. Effects of recombinant basic fibroblast growth factor, insulin-like growth factor-II and transforming growth factor-β1 on dog dental pulp cells *in vivo*. *Arch Oral Biol* 43: 431–444.

Tziafas, D., Smith, A.J., and Lesot, H. 2000. Designing new treatment strategies in vital pulp therapy. *J Dent* 28: 77–92.

Unda, F.J., Martin, A., Hernandez, C., Perez-Nanclares, G., Hilario, E., and Arechaga, J. 2001. FGFs-1 and -2, and TGFβ1 as inductive signals modulating *in vitro* odontoblast differentiation. *Adv Dent Res* 15: 34–38.

Weber, F.E., Eyrich, G., Gratz, K.W., Maly, F.E., and Sailer, H.F. 2002. Slow and continuous application of human recombinant bone morphogenetic protein via biodegradable poly(lactide-*co*-glycolide) foam-spheres. *Int J Oral Maxillofac Surg* 31: 60–65.

Weert, M., Hennink, W.E., and Jiskoot, W. 2000. Protein instability in poly(lactic-*co*-glycolic acid) microparticles. *Pharm Res* 17: 1159–1167.

Woo, B.H., Fink, B.F., Page, R., Schrier, J.A., Jo, Y.W., Jiang, G., DeLuca, M., Vasconez, H.C., and DeLuca1, P. 2001. Enhancement of bone growth by sustained delivery of recombinant human bone morphogenetic protein-2 in a polymeric matrix. *Pharm Res* 18: 1747–1753.

Xu, H.K., Weir, M.D., and Simon, C.G. 2008. Injectable and strong nano-apatite scaffolds for cell/growth factor delivery and bone regeneration. *Dent Mater* 24: 1212–1222.

Zhang, W., Walboomers, X.F., and Jansen, J.A. 2007. The formation of tertiary dentin after pulp capping with a calcium phosphate cement, loaded with PLGA microparticles containing TGF-β 1. *J Biomed Mater Res* 85A: 439–444.

24 Classification and Application of Colloidal Drug Delivery Systems
Passive or Active Tumor Targeting

H. Yesim Karasulu, Burcak Karaca, and Ercüment Karasulu

CONTENTS

24.1 INTRODUCTION

Cancer represents one of the most important health problems of the developing world. There are 25,000,000 cancer patients all over the world, and thousands of new cases were diagnosed each year in the last century. The development of cytotoxic treatments was revolutionary for the treatment of cancer. Cytotoxic treatments have let us achieve cures for certain type of neoplasms (germ cell tumors, lymphomas, etc.). In addition, application of cytotoxic agents in the adjuvant setting has provided an obvious overall survival advantage compared to that obtained with surgical treatment alone (Alison, 2002; Jemal et al., 2008).

Cancer is actually a group of diseases characterized by the ability of cells to grow in an autonomous manner and the invasion and spread of the cells from the primary site to other sites in the body (metastasis). Carcinogens are well known agents that cause general deoxyribonucleic acid (DNA) damage in the cell. Thus, cancer may be accepted as a genetic disease at the cellular level and alterations in DNA are the cornerstone of the disease mechanism. The accumulation of damage to DNA over time explains the multistep process of cancer development. Thus, as we live longer, there will be enough time for accumulation of changes in our DNA that may lead to cancer. That certainly is the reason why we are faced with many more cancer patients today compared to the past (Hanahan and Weinberg, 2000). Because DNA is the cornerstone of disease in cancer, most of the conventional cytotoxic drugs are designed to work on DNA either by causing DNA damage that will result in apoptosis or by blocking DNA synthesis.

24.1.1 Principles of Conventional Cytotoxic Agents in Cancer Treatment

The earliest therapeutic strategy to treat cancer was to remove as much cancerous tissue as possible out of the body (Chabner and Longo, 2001). However, it is obvious that surgery is limited to certain types of cancers at certain stages. As our understanding of cancer increased, treatment modalities focused at the cellular level of the disease gained importance. Thus, the development of cytotoxic treatment for cancer had significant importance in providing longer survival for cancer patients in recent years (Helfand, 1990; Unger, 1996).

The main goals of cytotoxic agents are to prevent proliferation (cytostatic effect) and to cause apoptosis (cytotoxic effect) in cancer cells. The aim for all agents is to achieve an effective treatment result with minimum side effects. This is indicated as the "therapeutic index" (Figure 24.1). It represents the value of the difference between the minimum effective dose and the maximum tolerated dose (MTD). A larger value of this difference means the agent is better. Many conventional cytotoxic treatments are applied at MTDs.

Conventional cytotoxic agents are chemicals that target DNA, ribonucleic acid, and some proteins in the cancer cell to inhibit proliferation and to cause apoptosis in rapidly dividing cancer cells; thus, they have broad specificity. More limitations were noticed as these cytotoxic agents were used more frequently for cancer treatment in the last years, mainly in patients with the advanced stage of the disease, where the adverse effects of this broad treatment may preclude its potential side effects (Bonadonna, 1990; Brigger et al., 2002).

The development of cytotoxic agents faces many challenges, such as the narrow therapeutic index of some drugs. Another major problem encountered in cancer treatment is that the cytotoxic agents are nonselective between normal and cancer cells. This nonselectivity results in a wide

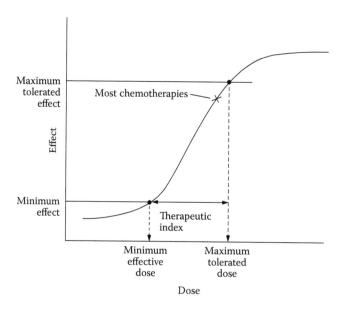

FIGURE 24.1 The therapeutic index profile for cytotoxic agents.

spectrum of side effects and limits the dose of the chemotherapeutic agents (Bonadonna, 1990; Helfand, 1990). The occurrence or development of drug resistance may also limit the efficacy of this treatment. Therefore, there is an urgent need for developing new cytotoxic agents that combine safety, efficacy, and convenience. We now have a new aspect for cancer treatment because of the plethora of potential targets in human cancer cells that may be utilized for the development of novel targeted treatment with lesser side effects. The new trend in daily cancer practice is to provide efficacious treatment with diminished side effects that can be achieved by targeted therapies (Azarmi et al., 2006; Couvreur et al., 2006; Jones and Leroux, 1999).

24.1.2   CONVENTIONAL CHEMOTHERAPY

Despite the emerging and explosive increase in the understanding of the molecular basis of cancer, the current treatment of malignancy is still based on broad specific anticancer agents. Although the ultimate goal of cytotoxic treatment is to kill rapidly proliferating cancer cells, it causes a number of side effects, such as alopecia, ulcers, and low blood counts. Often applied MTDs of conventional cytotoxic agents induce toxicity in sensitive tissues and require a time pause for drug application in order to give normal cells a chance to recover. Conventional cytotoxic treatments have had some very important results in cancer treatment and caused extended life expectancy for many patients. However, there is an unquestionable demand for safer and more effective anticancer drugs that specifically target cancer cells and overcome resistance pathways (Bonadonna, 1990; Chabner and Longo, 2001).

Conventional cytotoxic agents are mainly divided into two subgroups according to their action in the cell cycle. The first group is cell-cycle independent and alkylating agents; antitumor antibiotics are examples of this group. The second group is cell-cycle dependent drugs that are generally younger than the other group and are mostly effective on dividing cells. This group includes antimetabolites, antitubulin agents, and topoisomerase inhibitors (Pecorino, 2005; Reddy and Kaelin, 2002).

24.1.3   DRUG RESISTANCE: AN IMPORTANT LIMITATION FOR CYTOTOXIC TREATMENT

Cancer is a genetic disease at the molecular level. Although it is believed that cancer arises from one stem cell, because the cells in cancerous tissue go on to rapidly divide and form a mass, there are a number of molecular changes within each cell cycle that result in a heterogeneous population. Moreover, when conventional cytotoxic agents are applied to these heterogeneous cell groups, they develop different kinds of genetic mutations and some of them become resistant to the treatment (Gottesman, 2002; Jemal et al., 2008; Pecorino, 2005; Ueda et al., 1987). This is the most important challenge of conventional cytotoxic treatment.

After the process of carcinogenesis begin within a cell, overexpression of some pivotal molecules that may play a role in cell proliferation or antiapoptotic systems occurs. These molecules are

Cancer cells may become resistant by an efflux system for the drug, by decreasing the intake of the drug, by increasing the number of molecules that play a pivotal role in cell proliferation or antiapoptotic systems within the cell, or by alternating drug metabolism and DNA repair processes (Gottesman, 2002). One of the most investigated reasons for drug resistance mechanisms is the efflux system within the cancer cell that causes the drug to be pumped out of the cell. This is due to the overexpression of P-glycoprotein that is encoded by a multidrug resistance 1 gene (Ueda et al., 1987). The P-glycoprotein function through binding cytotoxic drugs; then adenosine triphosphate is hydrolyzed, a conformational change occurs within the protein, and finally the drug is released from the cancer cell. This resistance mechanism can widely apply to cancer cells exposed to many of the conventional cytotoxic agents. It is one of the most challenging problems of cytotoxic treatment, and new pharmaceutical drug forms are urgently needed to overcome this problem.

After the process of carcinogenesis begin within a cell, overexpression of some pivotal molecules that may play a role in cell proliferation or antiapoptotic systems occurs. These molecules are accepted as ideal targets for new anticancer drugs focused on blocking their function. Drug resistance is generally classified in two forms:

1. Biochemical or absolute resistance: This represents a situation in which the cancer cell cannot be killed at any dose of the drug.
2. Relative resistance: The cancer cell may be killed with higher doses of the drug.

Absolute resistance is a very important issue regarding why most cancer types are not cured by cytotoxic treatment. However, relative resistance may be solved by some new pharmaceutical forms of drugs. Thus, modern oncology is looking for newly developed targeted drugs that may overcome the resistance mechanisms and have diminished side effects while maintaining the therapeutic index (Pecorino, 2005; Reddy and Kaelin, 2002; Unger, 1996).

24.2 AN INTRODUCTION TO COLLOIDAL DRUG DELIVERY SYSTEMS

The use of colloidal drug delivery systems (CDDSs) for cancer treatment has been increased in recent years. These carrier systems are able to increase the therapeutic efficacy of some conventional cytotoxic drugs by changing the pharmacokinetics and biodistribution of the drugs. This leads to decreased drug toxicity and increased efficacy of the drug (Allen and Cullis, 2004; Gao et al., 2007; Vasir and Labhasetwar, 2005).

The word *colloid* comes from the Greek language etymologically, which means "glue." This term was first used in 1861. However, it is not easy to give a general definition for colloids. Within pharmaceuticals, they attract attention as "drug carriers" because of their potential in delivering any type of drug. Hydrophilic, lipophilic, small, or large molecules may be loaded easily to colloidal systems and administered through parenteral or nonparenteral methods. These systems have been used to solve many biomedical and pharmaceutical problems. They also provide some stability issues for drug molecules and are very effective in drug targeting (Bonacucina et al., 2008; Danielsson and Lindman, 1981; Peppas-Brannon and Blanchette, 2004; Shinoda and Friberg, 1975).

All of these systems carry an important feature related to their small size: they have a high surface area compared with their volume, resulting in a rise to adsorption. A chemical alteration in the surface properties of one component may effect the interactions between the constituents of the system and change the performance of the resulting system. A physical modification of the interphase can enhance the performance. Another important property of the system is the particle shape. The shape is related to the surface and determines the presence of attractive forces within dispersive and dispersant phases, which also affect different physical features such as flow and osmotic pressure (Bonacucina et al., 2008; Gao et al., 2007; Peppas-Brannon and Blanchette, 2004; Shinoda and Friberg, 1975).

One way of classifying colloids is to group them according to thermodynamic classification:

- Lyophilic or hydrophilic when water is the dispersion medium used
- Lyophobic or hydrophobic

This distinction is based on whether their nature may be subdivided as solvent-loving or solvent-fearing substances. Lyophilic colloids show an affinity for the solvent that is higher than the mutual affinity between particles and little or no particle aggregation. These dispersions form spontaneously from macroscopic phases and are thermodynamically stable with respect to both enlargement of particles through their aggregation and disintegration to individual molecules. Some examples of lyophilic colloids are block copolymer micelles and microemulsions. However, lyophobic colloids are thermodynamically unstable and tend to aggregate because of the lack of solvation and high interfacial free energy such as emulsions, suspensions, and aerosols (Danielsson and Lindman, 1981; Pecorino, 2005).

TABLE 24.1
Some Benefits and Drawbacks of CDDSs

	Formulation Aspect	Biopharmaceutical Aspect
Benefits	Injectability	Controlled or sustained release
	Increased dissolution rate	Improved bioavailability
	Solubilization of poorly soluble drugs	Protection for active agent
	High surface area	Targeting
	Increased drug stability	Extravascular or transdermal transport
Drawbacks	Low drug loading	Complicates pharmacokinetic analysis
	Low colloidal stability	Excipients may influence the drug pharmacokinetics
	Needs excipients	Complex manufacture and quality control

Although significant progress has been made in drug targeting, there are still some problems to overcome. Systemic toxicity and damage to healthy cells as well as drug resistance are the challenges of cytotoxic treatment (Martino and Kaler, 1990; Mehta et al., 1990; Vasir et al., 2005). Therefore, pharmaceutical science is working for solutions either by inventing new targeted therapies or by finding new solutions for conventional drugs. Further efforts are needed to enlarge our understanding of molecular targets and to discover new ones (Kayser et al., 2005; Krauel et al., 2007; Wang et al., 2004).

The benefits and drawbacks of using a CDDS may be classified into two broad categories: formulation (physicochemical) and biopharmaceutical aspects (Table 24.1).

CDDSs are used as intravenously injected carriers for drug delivery to specific organs or targeted sites within the human body (Kurihara et al., 1996; Yamaguchi et al., 1984). In colloidal systems the size of the particle is very important in the distribution of the drug in the human body (Rabinow, 2004). Generally, a large particle is easily removed by the liver and spleen. The stability of a small particle is higher than the a large particle in drug delivery devices. Reducing the size of colloidal particle carriers in the range of 100 to 400 nm enhances the stability of these systems and creates the chance of escaping from the vascular system via cavities in the lining of blood vessels (Claudia and Rainer, 2002; Hultin et al., 1995; Rahimnejad et al., 2006). Therefore, the larger colloidal particles given by an intravenous (i.v.) route are rapidly taken up by the mononuclear phagocyte system (MPS; Yamaguchi et al., 1984) whereas small particles (<400 nm) can significantly extravasate into tumors because of the enhanced permeability of the tumor vasculature (Hultin et al., 1995). A smaller particle size (<200 nm) CDDS could allow for more favorable biodisposition. Moreover, the poly(ethylene glycol) (PEG) modification (PEGylation) of these systems provides longer circulation times and enhanced accumulation at the tumor site (Butsele et al., 2007). Figure 24.2 was obtained from an examination in various tumor models in mice, clearly indicating that the size of the long-circulating liposome was an important factor for extravasation (Maruyama et al., 1999). Because of the increased circulation time of the liposomes containing distearoyl phosphatidylcholine and PEG and the leaky structure of the microvasculature in the solid tumor tissue, those liposomes have been shown to accumulate preferentially in tumor tissue. Thus, under physiological tumor conditions, only small liposomes ranging from 100 to 200 nm in diameter with a prolonged circulation half-life encounter more opportunities to extravasate through discontinuous capillaries, as well as to escape the gaps between adjacent endothelial cells and openings at the vessel termini during tumor angiogenesis. It is conceivable that long-circulating liposomes could also take advantage of the enhanced permeability and retention (EPR) effect for efficient targeting binding in the tumor.

CDDSs can be also divided into three convenient classes: "hard," "soft," and "macromolecular" colloidal carriers (Table 24.2; Boyd, 2005).

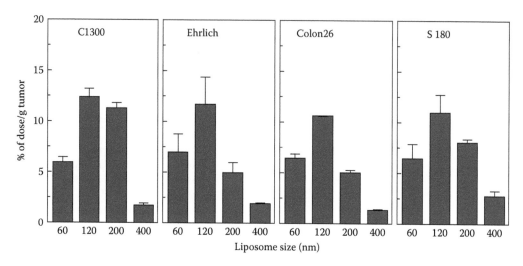

FIGURE 24.2 The effect of liposome size on the extravasation of PEG liposome into tumor tissue. Liposomes were injected into tumor-bearing mice via the tail vein and biodistribution was estimated at 6 h after injection. Tumor-bearing mice were prepared by subcutaneously inoculating tumor cells (~1 × 10⁷) and tested when the tumor mass volume reached about 1000–1500 mm³. (From Maruyama, K., et al. 1999. *Advanced Drug Delivery Review*, 40, 89–102. With permission.)

TABLE 24.2
Classification of CDDSs

Colloidal Carriers		
Hard	**Soft**	**Macromolecular**
Solid drug nanoparticles	Micelles	Dendrimers
Solid lipid nanoparticles	Submicron emulsions and microemulsions	Protein conjugates
Polymeric nanoparticles	Liposomes	Polymeric micelles
	Physical Form	
Crystalline materials formed by precipitation	Self-assemble into structural form; core is essentially liquid	Molecular materials approaching colloidal dimensions
	Drug Release	
Drug primarily released by erosion or degradation mechanism	Drug released by diffusion or partition control	Often a prodrug requiring cleavage of drug from the carrier

Source: From Boyd, B. B. 2005. *Drug Delivery Report*, Autumn/Winter: 63–69. With permission.

Even though CDDSs are quite complex, their use in the pharmaceutical field is growing day by day. Emulsions, microemulsions, and micelles are used to modify drug solubility; and nanoparticles, nanosuspensions, liposomes, microspheres, and micellar systems have provided new options for delivering drugs either perorally or parenterally. Recent developments in CDDSs are addressing many biomedical and pharmaceutical problems. It is assumed that nanotechnology will serve to develop these systems even further in order to achieve an ideal "cytotoxic drug" (Bromberg and Ron, 1998; Cho et al., 2007; Primo et al., 2008).

24.3 DRUG TARGETING

One of the major problems facing cancer therapy is to achieve good specificity of antineoplastic agents for their intended site of action in the body. As a result of their toxicity toward healthy tissues, many anticancer drugs are often administered at doses that are subtherapeutic. Therefore, the effective cancer therapy is to selectively destroy cancer cells while sparing normal tissues. Drug targeting has evolved as the most desirable but elusive goal for drug delivery. By altering the pharmacokinetics and biodistribution of drugs and restricting their action to the targeted tissue, increased drug efficacy with concomitant reduction of their toxic effects can be achieved. The encapsulation of the chemotherapy agents within colloidal systems usually improves drug efficiency and leads to a decrease of the toxicity because the carrier exits the blood circulation in tissues where capillary junctions have been disrupted and are not tightly bound, for example, tumor growth areas. A number of drugs have now been successfully encapsulated in liposomes, microparticles, nanoparticles, and particularly lipidic–colloidal systems such as emulsions and microemulsions; in most cases, it appears to improve therapeutic efficacy and largely decreases toxicity (Au et al., 2001, 2002; Azevedo et al., 2005; Formariz et al., 2006; Hwang et al., 2004; Junping et al., 2003; Kawakami et al., 2005; Shiokawa et al., 2005; Thorpe, 2004).

24.3.1 ANATOMICAL BARRIERS OF TUMOR TISSUE

CDDSs are generally eliminated by the MPS. The MPS is composed of cells (macrophages and monocytes) that remove senescent cells from the blood circulation and provide phagocyte cells to inflammatory sites following their recruitment by cytokines or complement proteins. The principal phagocyte cells are found in the liver (Kuppfer cells), spleen, and bone marrow. CDDSs can be removed from blood circulation by various routes such as transcytosis, which constitutes a transport process across the endothelial cells as observed, for example, for the blood–brain barrier or via intracellular transfer. When the CDDSs are small (<100 nm), a possible transfer passage through fenestrates in the endothelium of the liver and through the permeable vascular endothelium in lymph nodes has been described (Lu et al., 2005; Vonarbourg et al., 2006).

The size of the CDDSs as well as their surface characteristics are the key for the biological fate of these systems, because these parameters can prevent their uptake by MPS macrophages. A high curvature and hydrophilic surface are needed to reduce opsonization reactions and subsequent clearance by macrophages (Azevedo et al., 2005; Brigger et al., 2002; Butsele et al., 2007; Maruyama et al., 1999).

In vivo target-specific drug delivery is not simple. Colloid-based delivery systems are one of the most promising approaches that overcome the limitation. The unique vascular structural changes associated with the pathophysiology of a tumor may provide opportunities for the use of CDDSs. The abnormal tumor vasculature breaks all of the rules of normal blood vessel construction. In general, tumor vessels are inherently leaky, because of the wide interendothelial junctions, large number of fenestrates, and transendothelial channels formed by vesicles and discontinuous or absent basement membranes. However, the blood vessels in most normal tissues are nonfenestrated capillaries. These blood vessels are composed of a single layer of endothelial cells with tight junctions. The extravasation of intravenously administered macromolecules or CDDSs such as liposomes is restricted to sites where the endothelial barrier has open fenestration because normal tissues have tight endothelial junctions. The endothelial barrier may prevent liposomes from traversing intact vessels (Au et al., 2001, 2002; Cho et al., 2007; Thorpe, 2004). An illustration of the anatomical barriers of tumor tissue and extravasations of liposomes from the blood are is provided in Figure 24.3. Generally, the capillary permeability of the endothelial barrier in newly vascularized tumors is significantly greater than that of normal tissues. The extravasation of circulating molecules from blood vessels to the tumor region is a function of both local blood flow and microvascular permeability. Thus, the amount of tissue accumulation of a drug is directly proportional to the area under

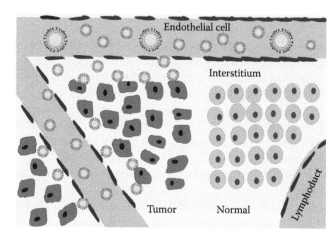

FIGURE 24.3 A pictorial presentation of the anatomical barriers of solid tumor tissue and extravasation of liposomes through capillaries from blood circulation. (From Maruyama, K., et al. 1999. *Advanced Drug Delivery Review*, 40, 89–102. With permission.)

the curve (AUC) of the plasma clearance. In this regard, a low AUC value attributable to rapid uptake of immunoliposomes (ILPs) by the MPS leads to a low level of tumor accumulation, even if the tumor vasculature is leakier than the normal tissue. In addition, because of little or no lymphatic drainage in tumor tissues, accumulated macromolecules are retained in the tumor interstitium for a prolonged period of time. Such a phenomenon, termed the EPR effect, occurs universally among tumors (Maruyama et al., 1999).

24.3.2 APPROACHES TO DRUG TARGETING

Chemotherapeutic and radiotherapeutic options are designed to kill cells in cancer treatment, so the specificity of drug action gains paramount importance. These strategies are based on the basic principle of preferentially killing cancer cells without having any significant toxic effect on normal cells (Azarmi et al., 2006; Boyd, 2005; Peppas-Brannon and Blanchette, 2004). Recently, ILPs, in which brain-specific targetors were covalently conjugated to PEG-modified liposomes (PEGylated liposomes) as a drug carrier via the tips of its functional PEG strands, proved to be successful in brain drug delivery. Its advantages over the drug–targetor direct combination technique were the larger drug loading capacity, disguise of the limiting characteristics of drugs with the physical nature of the liposome, and reduction of drug degradation *in vivo*. The surface modification of PEG enabled the liposomes to escape the arrest of the MPS so as to prolong its half-life in plasma and increase the AUC (Lu et al., 2005).

A vast array of methods, which can be further classified into three key approaches (passive, active, and physical), have been explored for targeting drugs by means of designing innovative colloidal systems (Table 24.3). Some of the methods like catheterization, direct injection, prodrugs, or chemical delivery systems have the inherent capability to deliver drugs or their appropriate modifications to the specific site of action. However, others require the design of a suitable carrier system to be able to deliver the drug to its target site. Following is a description of such approaches to drug targeting (Azarmi et al., 2006, 2008; Byrne et al., 2008; Fonsatti et al., 2003; Peppas-Brannon and Blanchette, 2004; Sasatsu et al., 2008; Vasir et al., 2005).

24.3.2.1 Passive Targeting Approaches

Passive targeting refers to the accumulation of a drug or drug-carrier system at a particular site because of physicochemical or pharmacological factors. Drug or drug-carrier systems can be

TABLE 24.3
Drug Targeting Approaches

Passive Targeting	Active Targeting	Physical Targeting
1. Pathophysiological factors	1. Biochemical targets	1. Ultrasound targeting
• Inflammation or infection	• Organs	• Low ultrasound energies
• EPR effect	• Cellular	2. Magnetic targeting
2. Physicochemical factors	• Organelles	• Magnetic field
• Size	*Intracellular*	
• Molecular weight	2. Physical or external stimuli	
3. Anatomical opportunities	• Ultrasound	
• Catheterization	• Magnetic field	
• Direct injection	3. Pretargeting or sandwich targeting	
4. Chemical approaches	4. Promoter or transcriptional targeting	
• Prodrugs		
• Chemical delivery systems		

passively targeted, making use of the pathophysiological and anatomical opportunities (Au et al., 2001, 2002; Butsele et al., 2007).

The EPR effect has been predominantly used for passive targeting of drugs with molecular weights of more than 40 kDa and for low molecular weight drugs presented in drug carriers such as polymeric drug conjugates, liposomes, polymeric nanoparticles, and micellar systems to solid tumors. To take advantage of this pathophysiological opportunity, the targeted drug or drug carriers should have long circulation times and should not lose therapeutic activity while in circulation. Other factors that influence EPR include the size of tumors, degree of tumor vascularization, and angiogenesis. Thus, the stage of the disease is critical for drug targeting using the EPR effect (Vasir and Labhasetwar, 2005).

Sasatsu et al. (2008) performed an interesting study for passive tumor targeting. A novel probe, poly(D,L-lactic acid) (PLA)-pyrene, was prepared by reductive amination of PLA-aldehyde and aminopyrene. Methoxy-PEG amine-*block*-PLA copolymer (PLA-*b*-MeO-PEG) nanoparticles loaded with PLA-pyrene were formulated and examined on retention of PLA-pyrene in the nanoparticles and biodisposition in normal and sarcoma-180 solid tumor bearing mice. The biodisposition profiles of the probe in the plasma and major organs are provided in Figure 24.4. After i.v. injection in normal rats, the plasma level of PLA-pyrene was very high for the initial 8 h and accumulated gradually into organs, especially the spleen and liver. After i.v. injection in tumor-bearing mice, PLA-pyrene had similar biodistribution profiles and was accumulated well in the tumor, suggesting that PLA-*b*-MeO-PEG nanoparticles should be delivered efficiently to solid tumors. PLA-pyrene might be a useful probe of the nanoparticles themselves. In addition, Sasatsu et al. demonstrated that PLA-*b*-MeO-PEG nanoparticles should be a useful drug carrier for passive tumor targeting.

24.3.2.2 Active Targeting Approaches

Passive targeting approaches are limited in their scope, so tremendous effort has been directed toward the development of active approaches for drug targeting. Active targeting employs a specific modification of a drug or drug-carrier nanosystem with "active" agents having selective affinity for recognizing and interacting with a specific cell, tissue, or organ in the body. Direct coupling of drugs to a targeting ligand restricts the coupling capacity to a few drug molecules. In contrast, coupling of drug-carrier nanosystems to ligands allows the import of thousands of drug molecules by means of one receptor-targeted ligand (Byrne et al., 2008; Fonsatti et al., 2003; Thorpe, 2004).

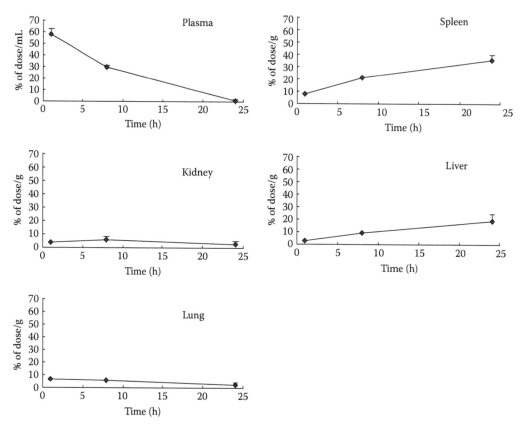

FIGURE 24.4 Biodistribution profiles of the probe in plasma and organs after i.v. injection of PLA-pyrene-loaded nanoparticles to normal mice at a dose of 20 mg/kg. The results are expressed as mean ± SE (*n* = 4). (From Sasatsu, M., Onishi, H., and Machida, Y. 2008. *International Journal of Pharmaceutics*, 358, 271–277. With permission.)

Active targeting involves the use of peripherally conjugated targeting moieties for enhanced delivery of nanoparticle systems. The targeting moieties are important to the mechanism of cellular uptake. Long circulation times will allow for effective transport of the nanoparticles to the tumor site through the EPR effect, and the targeting molecule can increase endocytosis of the nanoparticles. If the nanoparticle attaches to vascular endothelial cells via a noninternalizing epitope, high local concentrations of the drug will be available on the outer surface of the target cell. Although this has a higher efficiency than the free drug released into circulation, only a fraction of the released drug will be delivered to the target cell (Byrne et al., 2008; Vasir and Labhasetwar, 2005).

In active targeting by ILPs, two anatomical compartments are considerable for targeting sites. One is located in a readily accessible site in the intravascular compartment, and another is a much less accessible target site located in the extravascular compartment. However, the active targeting with ILPs was determined by two kinetically competing processes, such as binding to the target site and uptake by the MPS. To overcome these contradictions, Maruyama et al. (1999) designed a new type of long-circulating ILP, which was PEG-ILP attached antibodies at the distal end of the PEG chain, the so-called pendant-type ILP [transferrin (TF)-PEG-ILP]. TF was used for targeting to the solid tumor tissue. TF-PEG-ILP was internalized into tumor cells with receptor-mediated endocytosis, after extravasation into tumor tissue (Figure 24.5). Active targeting to tumor tissue with the pendant-type ILP is particularly important for many highly toxic anticancer drugs utilized in cancer chemotherapy.

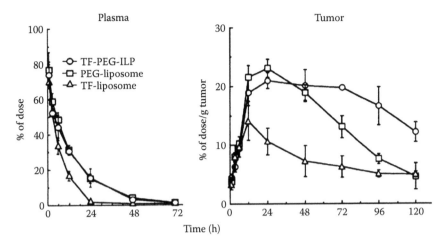

FIGURE 24.5 The time course of blood residence and liver uptake of TF conjugating pendant-type immunoliposomes in B16 tumor-bearing mice. (From Maruyama, K., et al. 1999. *Advanced Drug Delivery Review*, 40, 89–102. With permission.)

24.3.2.3 Physical Targeting Approaches

This is a new targeting strategy that makes use of an external stimulus to target the release of the drug at a specific site in the body. The magnetic targeting approach involves i.v. injection of a therapeutic agent that is bound or encapsulated in a magnetic drug carrier, which can then be directed and preferentially localized in the tumor tissue upon application of an external localized magnetic field. Such drug carriers include magnetic liposomes, microspheres, nanospheres, and a colloidal iron oxide solution (magnetic ferrofluids). When using the ultrasound technique, the low ultrasound energies required for this kind of targeting increase the intracellular uptake of the drug, whereas energies greater than the cavitation threshold can severely damage the cell membranes. Ultrasound waves focused at the tumor tissue can be used to trigger the release of anticancer agents from polymeric micelles and thus allow effective intracellular uptake of the encapsulated drug (Vasir and Labhasetwar, 2005).

24.3.3 DELIVERY OF SPECIFIC ANTICANCER AGENTS AS COLLOIDAL DRUG DELIVERY SYSTEM

Anticancer drugs can be associated with CDDSs such as polymeric micelles, microemulsions, nanoparticles, and liposomes. Numerous studies were performed on body distribution after using the systemic route of colloidal systems. Over the last few decades, CDDSs and especially nanoparticles have received much attention. Nanoparticles can be administered via different routes such as parenteral, oral, intraocular, transdermal, or pulmonary inhalation (Azarmi et al., 2006, 2008; Brigger et al., 2002; Cho et al., 2007; Primo et al., 2008; Vonarbourg et al., 2006). An interesting nanoparticle formulation was prepared by Azarmi et al. (2006). In their study doxorubicin (DOX)-loaded nanoparticles were incorporated into inhalable carrier particles. The cytotoxic effects of free DOX, carrier particles containing blank nanoparticles, or DOX-loaded nanoparticles on H460 and A549 lung cancer cells were assessed using a colorimetric 2,3-bis(2-methoxy-4-nitro-5-sulfophenyl)-2H-tetrazolium-5-carbox-anilide cell viability assay. The DOX nanoparticles showed higher cytotoxicity at the highest tested concentration compared with the blank nanoparticles and the free DOX in both cell lines. This study supports the nanoparticle approach of treatment in lung cancer in dry powder aerosol form.

A prime example of targeting membrane type-1 matrix metalloproteinase (MT1-MMP) involves Fab$_{222\text{-}1D8}'$ fragments of antihuman MT1-MMP monoclonal antibody conjugated to DOX ILPs,

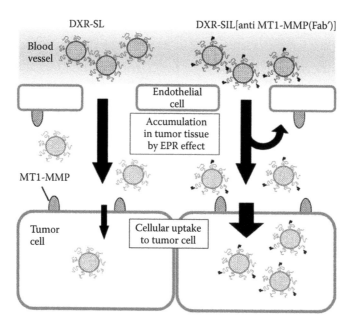

FIGURE 24.6 A schematic diagram demonstrating a possible mechanism for the accumulation of targeted {DXR-SIL [anti-MT1-MMP(Fab′)]} and nontargeted (DXR-SIL) nanoparticles. Tumor accumulation of the targeted nanoparticle is comparable to the nontargeted nanoparticle and can be explained by the EPR effect. However, after the nanoparticles accumulate in the tissue, the targeted nanoparticle is more efficiently internalized via receptor-induced endocytosis. (From Hatakeyama, H., et al. 2007. *International Journal of Pharmaceutics*, 342, 194–200. With permission.)

which are targeted to tumor-bearing mice inoculated with HT1080 cells (Hatakeyama et al., 2007). The Fab′ fragments of an antibody against MT1-MMP were modified at the distal end of PEG of DOX-encapsulating liposomes (DXR), producing DXR-sterically stabilized ILPs {DXR-SIL [anti-MT1-MMP (Fab′)]}. These ILPs were administered into the tumor-bearing mice. After 12 days, the nontargeted liposome (DXR-SIL) treatment showed a decreased tumor volume in only one out of six mice and three of six mice died, presumably from the side effect of DOX-filled nontargeted liposomes. However, in the targeted liposomes, tumor volume decreased by a factor of 2 by day 12, only one mouse experienced notable body weight changes, and no deaths occurred. To measure DOX distribution in the tumor, [^{14}C]-DOX was administered and the radioactivities in the blood and tumor were tested after 48 h. The blood concentrations of [^{14}C]-DOX for targeted and nontargeted liposomes were comparable, indicating that the targeting moiety did not alter tumor accumulation. However, there was more efficient internalization of the targeted nanoparticles compared to the nontargeted nanoparticles, as demonstrated in Figure 24.6.

Another successful example of a CDDS is microemulsions, and several anticancer drugs can be successfully encapsulated into microemulsions (Hwang et al., 2004; Junping et al., 2003; Karasulu, 2008; Karasulu et al., 2004, 2007; Shiokawa et al., 2005; Terek et al., 2006). An interesting study was performed by Hwang et al. (2004). The primary aim of this study was to develop a parenteral formulation of all-*trans*-retinoic acid (ATRA) by overcoming its solubility limitation by utilizing a phospholipid-based microemulsion system as a carrier. The pharmacokinetic profile of ATRA on human cancer HL-60 and MCF-7 cell lines were also similar (Table 24.4) between free ATRA and a microemulsion formulation of ATRA, suggesting that the anticancer activity was not impaired by loading in a microemulsion. This study demonstrated that phospholipid-based microemulsions may provide an alternative parenteral formulation of ATRA, which has significant antitumoral activity for many types of cancer. Other important research was carried out with injectable microemulsions

TABLE 24.4
Noncompartmental Pharmacokinetic Parameters of ATRA
after i.v. Administration of Sodium ATRA or ATRA
Microemulsion, Equivalent to 4 mg/kg as ATRA, to rats ($n = 5$)

Parameters	Formulation	
	Sodium ATRA	Microemulsion
$T_{1/2}$ (h)	1.31 ± 0.50	1.20 ± 0.39
AUC (μg h/mL)	8.27 ± 2.27	9.86 ± 1.39
MRT (h)	1.89 ± 0.13	1.73 ± 0.13
Vss (mL/kg)	850.26 ± 145.91	651.18 ± 106.71[a]
CL (mL/h)	450.73 ± 178.47	376.84 ± 63.56

Source: From Hwang, S.R. et al. 2004. *International Journal of Pharmaceutics,*
276, 175–183. With permission.

Note: $T_{1/2}$ (h); MRT—mean reaction time, Vss (mL/kg)—Apparent volume of distribution at steady state, CL (mL/h)—Total clearance.

[a] $p < 0.05$, when compared with sodium ATRA.

of vincristine (M-VCR); its pharmacokinetics, acute toxicity, and antitumor effects were evaluated in C57BL/6 mice bearing mouse murine histocytoma M5076 tumors (Junping et al., 2003). The plasma AUC of M-VCR was significantly greater than that of free VCR (F-VCR; Table 24.5). M-VCR had lower acute toxicity and greater potential antitumor effects than F-VCR in C57BL/6 mice with M5076 tumors. M-VCR is as a useful tumor-targeting microemulsion drug delivery system.

In addition to these data, Shiokawa and coworkers (2005) reported a novel microemulsion formulation for a tumor-targeted drug carrier of the lipophilic antitumor antibiotic aclacinomycin A. Their findings suggested that a folate-linked microemulsion is feasible for tumor-targeted aclacinomycin A delivery. The study showed that folate modification with a sufficiently long PEG chain on emulsions is an effective way of targeting the emulsion to tumor cells.

Much effort has been made to achieve lymphatic targeting of drugs using colloidal carriers. Nishioka and Yoshino's (2001) study described the recent advance of the targeted delivery of drugs to the lymph node. To evaluate the lymphatic targeting ability, the performance of polyisobutylcyanoacrylate nanocapsules following intramuscular administration were compared with three conventional colloidal formulations, including an egg phosphatidylcholine (EPC) emulsion, phospholipid (PL) emulsion, and EPC liposome, and an oily solution of 12-(9-anthroxy) stearic acid (ASA). The ASA concentration–time profiles in the injection site, lymph nodes, and blood are shown in Figure 24.7. Clearly, polyisobutylcyanoacrylate nanocapsules are retained in the regional lymph nodes far longer than other colloidal carriers following intramuscular administration.

Thus, without question, targeting strategies involving CDDSs, which can alter a drug's biodistribution to avoid toxicity and maximize its effectiveness, can enhance the prospects that new anticancer drugs will reach the patients.

24.4 CONCLUSION

One of the major problems encountered in cancer treatment is that the cytotoxic agents are nonselective between normal and cancer cells. This nonselectivity results in a wide spectrum of side effects and limits the dose of the chemotherapeutic agents. Modern oncology seeks solutions for this challenge to daily oncologic practice. Thus, effective drug delivery systems are appearing as a result of more accurate targeting of pathological tissues. Therefore, CDDSs offer opportunities to achieve drug targeting with newly discovered disease-specific targets. Advances in identification

TABLE 24.5
Comparison of Pharmacokinetic Parameters of M-VCR and F-VCR Administered in C57BL76 Mice

	F-VCR[a]	M-VCR[a]
C_0 (μg/mL)	1.11 ± 0.16	0.85 ± 0.12
K_{12} (L/h)	31.74 ± 4.24	0.30 ± 0.01
K_{21} (L/h)	6.78 ± 1.52	0.21 ± 0.08
K_e (L/h)	1.37 ± 0.28	0.07 ± 0.01
$T_{1/2}×$ (h)	0.02 ± 0.01	1.25 ± 0.31*
$T_{1/2β}$ (h)	2.96 ± 0.43	25.76 ± 3.88*
V_1 (L/kg)	1.82 ± 0.37	2.35 ± 0.41
V_2 (L/kg)	8.51 ± 2.32	3.52 ± 0.59
Vss (L/kg)	10.33 ± 3.45	5.88 ± 1.21
CL (L/h/kg)	3.28 ± 0.25	0.17 ± 0.09*

Source: From Junping, W., et al. 2003. *International Journal of Pharmaceutics*, 251, 13–21. With permission.

Note: The VCR i.v. dose was 2 mg/kg. C_0—initial concentration; Vss (mL/kg)—Apparent volume of distribution at steady state, CL (mL/h)—Total clearance.

*The results are given as mean ± SD (n = 3).

[a] $p < 0.01$, compared with those of F-VCR.

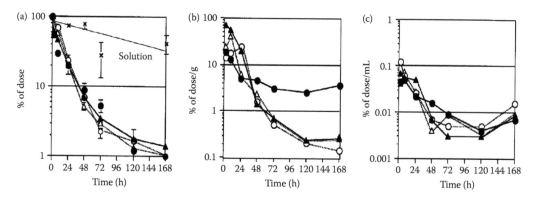

FIGURE 24.7 Comparisons of ASA concentration–time profiles of various colloidal formulations in the injection site, lymph node, and blood after intramuscular administration to rats. (a) Injection site (right thigh muscle), (b) lymph node (right iliac lymph node), and (c) blood. (●) Nanocapsules (○) EPC liposome, (▲) PL emulsion, (△) EPC emulsion, and (X) oily solution. (From Nishioka, Y. and Yoshino, H. 2001. *Advanced Drug Delivery Reviews*, 47, 55–64. With permission.)

of tumor-specific targets and development of different drug delivery approaches for tumor targeting have opened an important avenue for the development of a successful targeted drug delivery modality for cancer therapy. Incorporation of CDDSs to conventional cytotoxic treatment might allow a reduction in doses and may thus diminish adverse effects while maintaining the therapeutic effect in cancer patients.

ABBREVIATIONS

ASA	12-(9-Anthroxy) stearic acid
ATRA	All-trans-retinoic acid
AUC	Area under the curve
CDDS	Colloidal drug delivery system
DOX	Doxorubicin
DXR-SIL [anti-MT1-MMP(Fab′)]	Doxorubicin–sterically stabilized immunoliposomes
EPC	egg phosphatidylcholine
EPR	Enhanced permeability and retention
F-VCR	Free vincristine
ILP	Immunoliposome
i.v.	Intravenous
MPS	Mononuclear phagocyte system
MT1-MMP	Membrane type-1 matrix metalloproteinase
MTD	Maximum tolerated dose
M-VCR	Microemulsions of vincristine
PEG	Poly(ethylene glycol)
PL	phospholipid
PLA	Poly(D,L-lactic acid)
TF	Transferrin
TF-PEG-ILP	Pendant-type immunoliposome

REFERENCES

Alison, M. R. 2002. *The Cancer Handbook*. London: Nature Publishing Group.

Allen, T. M. and Cullis, P. R. 2004. Drug delivery systems: Entering the mainstream. *Science* 303, 1818–1822.

Au, J. L., Jang, S. H., and Wientjes, M. G. 2002. Clinical aspects of drug delivery to tumors. *Journal of Controlled Release*, 78, 81–95.

Au, J. L., Jang, S. H., Zheng, J., Chen, C. T., Song, S., Hu, L., Wientjes, M. G. 2001. Determinants of drug delivery and transport to solid tumors. *Journal of Controlled Release*, 74, 31–46.

Azarmi, S., Löbenberg, R., and Roa, W. 2008. Targeted delivery of nanoparticles for the treatment of lung diseases. *Advanced Drug Delivery Reviews*, 60, 863–875.

Azarmi, S., Tao, X., Chen, H., Wang, Z., Finlay, W. H., Löbenberg, R., Roa W. 2006. Formulation and cytotoxicity of doxorubicin nanoparticles carried by dry powder aerosol particles. *International Journal of Pharmaceutics*, 319, 155–161.

Azevedo, C. H. M., Carvalho, J. P., Valduga, C. J., and Maranhao, R. C. 2005. Plasma kinetics and uptake by the tumor of a cholesterol-rich microemulsion (LDE) associated to etoposide oleate in patients with ovarian carcinoma. *Gynecologic Oncology*, 97, 178–182.

Bonacucina, G., Misici-Falzi, M., Cepsi, M., and Palmieri, G. F. 2008. Characterization of micellar systems by the use of acoustic spectroscopy. *Journal of Pharmaceutical Sciences*, 97, 2217–2227.

Bonadonna, G. 1990. Does chemotherapy fulfill its expectations in cancer treatment? *Annals of Oncology*, 1, 11–21.

Boyd, B. B. 2005. Colloidal drug delivery. *Drug Delivery Report*, Autumn/Winter: 63–69.

Brigger, I., Dubernet, C., and Couvreur, P. 2002. Nanoparticles in cancer therapy and diagnosis. *Advanced Drug Delivery Reviews*, 5, 631–651.

Bromberg, L. V. and Ron, E. S. 1998. Temperature-responsive gel and hermogelling polymer matrices for protein and peptide delivery. *Advanced Drug Delivery Reviews*, 31, 197–221.

Butsele, K. V., Jerome, R., and Jerome, C. 2007. Functional amphiphilic and biodegradable copolymers for intravenous vectorisation. *Polymer*, 48, 7431–7443.

Byrne, J. D., Betancourt, T., and Brannon-Peppas L. 2008. Active targeting schemes for nanoparticle systems in cancer therapeutics. *Advanced Drug Delivery Reviews*, 60, 1615–1626.

Chabner, B. A. and Longo, D. L. 2001. *Cancer Chemotherapy and Biotherapy—Principles and Practice*, 3rd edition. Philadelphia, PA: Lippincott Williams & Wilkins.

Cho, Y. W., Park, S. A., Han, T. H., Son, D. H., Park, J. S., Oh, S. J., Moon, D. H., et al. 2007. *In vivo* tumor targeting and radionuclide imaging with self-assembled nanoparticles: Mechanisms, key factors, and their implications. *Biomaterials*, 28, 1236–1246.

Claudia, J. and Rainer, H. M. 2002. Production and characterization of a budesonide nanosuspension for pulmonary administration. *Pharmaceutical Research*, 19, 189–194.

Couvreur, P., Gref, R., Andrieux, K., and Malvy, C. 2006. Nanotechnologies for drug delivery. Application to cancer and autoimmune diseases. *Progress in Solid State Chemistry*, 34, 231–235.

Danielsson, I. and Lindman, B. 1981. The definition of microemulsion. *Colloids and Surfaces*, 3, 391–392.

Fonsatti, E., Altomonte, M., Arslan, P., and Maio, M. 2003. Endoglin (CD105): A target for anti-angiogenetic cancer therapy. *Current Drug Targets*, 4, 291–296.

Formariz, T. P., Sarmento, V. H. V., Silva-Junior, A. A., Scarpa, M. V., Santilli, C. V., and Oliveira, A. G. 2006. Doxorubicin biocompatible o/w microemulsion stabilized by mixed surfactant containing soya phosphatidylcholine. *Colloids and Surfaces B: Biointerfaces*, 51, 54–61.

Gao, L., Zhang D., Chen, M., Duan, C., Dai, W., and Jia, L. 2007. Pharmacokinetics and tissue distribution of oridonin. *International Journal of Pharmaceutics*, 355, 321–327.

Gottesman, M. M. 2002. Mechanisms of cancer drug resistance. *Annual Review of Medicine*, 53, 615–627.

Hanahan, D. and Weinberg, R. A. 2000. The hallmarks of cancer. *Cell*, 100, 57–70.

Hatakeyama, H., Akita, H., Ishida, E., Hashimoto, K., Kobayashi, H., Aoki, T., Yasuda, J., et al. 2007. Tumor targeting of doxorubicin by anti-MT1-MMP antibody-modified PEG liposomes. *International Journal of Pharmaceutics*, 342, 194–200.

Helfand, S. C. 1990. Clinical management of the cancer patient. Principles and applications of chemotherapy. *Veterinary Clinics of North America. Small Animal Practice*, 20, 987–1013.

Hultin, M., Carneheim, C., Rosenqvist, K., and Olivecrona, T. 1995. Intravenous lipid emulsions: Removal mechanisms as compared to chylomicrons. *Journal of Lipid Research*, 36, 2174–2184.

Hwang, S. R., Lim, S.-J., Park, J.-S., and Kim, C.-K. 2004. Phospholipid-based microemulsion formulation of all-trans-retinoic acid for parenteral administration. *International Journal of Pharmaceutics*, 276, 175–183.

Jemal, A., Siegel, R., Ward, E., Hao Y., Xu J., Murray T., Thun M. J. 2008. Cancer statistics. *CA Cancer Journal for Clinicians*, 58, 71–96.

Jones, M. C. and Leroux, J. C. 1999. Poymeric micelles—A new generation of colloidal drug carriers. *European Journal of Pharmaceutics and Biopharmaceutics*, 48, 101–111.

Junping, W., Takayama, K., Nagai, T., and Maitan, Y. 2003. Pharmacokinetics and antitumor effects of vincristine carried by microemulsions composed of PEG-lipid, oleic acid, vitamin E and cholesterol. *International Journal of Pharmaceutics*, 251, 13–21.

Karasulu, H. Y. 2008. Microemulsions as novel drug carriers: The formation, stability, applications and toxicity. *Expert Opinion on Drug Delivery*, 5, 119–135.

Karasulu, H. Y., Karabulut, B., Goker, E., Guneri, T., and Gabor, F. 2007. Controlled release of methotrexate from w/o microemulsion and its *in vitro* anti-tumor activity. *Drug Delivery*, 14, 225–233.

Karasulu, H. Y., Karabulut, B., Kantarci, G., Ozgüney, I., Sezgin, C., Sanli, U. A., Göker, E. 2004. Preparation of arsenic trioxide loaded microemulsion and its enhanced cytotoxicity on MCF-7 breast carcinoma cell line. *Drug Delivery*, 11, 345–350.

Kawakami, S., Opanasopit, P., Yokoyama, M., Chansri, N., Yamamoto, T., Okano, T., Yamashita, F., Hashida, M. 2005. Biodistribution characteristics of all trans retinoic acid incorporated in liposomes and polymeric micelles following intravenous administration. *Journal of Pharmaceutical Science*, 94, 2606–2615.

Kayser, O., Lemke, A., and Trejo, H. 2005. The impact of nanobiotechnology on the development of new drug delivery systems. *Current Pharmaceutical Biotechnology*, 6, 3–5.

Krauel, K., Girvan, L., Hook, S., and Rades, T. 2007. Characterisation of colloidal drug delivery systems from the naked eye to cryo-FESEM. *Micron*, 38, 796–803.

Kurihara, A., Shibayama, Y., Mizota, A., Yasuno, A., Ikeda, M., Sasagawa, K., Kobayash, T., Hisaoka M. 1996. Enhanced tumor delivery and antitumor activity of palmitoyl rhizoxin using stable lipid emulsions in mice. *Pharmaceutical Research*, 13, 305–310.

Lu, W., Tan, Y. Z., Hu, K. L., and Jiang, X. G. 2005. Cationic albumin conjugated PEGylated nanoparticle with its transcytosis ability and little toxicity against blood–brain barrier. *International Journal of Pharmaceutics*, 295, 247–260.

Martino, A. and Kaler, E. W. 1990. Phase behavior and microstructure of non-aqueous microemulsions. *Journal of Physical Chemistry*, 94, 1627–1631.

Maruyama, K., Ishida O., Takizawa, T., and Moribe, K. 1999. Possibility of active targeting to tumor tissues with liposomes. *Advanced Drug Delivery Review*, 40, 89–102.

Mehta, S. K., Kavaljit, X. X., and Bala, K. 1999. Phase behavior structural effects, volumetric and transport properties in non-aqueous microemulsions. *Physical Reviews*, E-59, 4317–4325.

Nishioka, Y. and Yoshino, H. 2001. Lymphatic targeting with nanoparticulate system. *Advanced Drug Delivery Reviews*, 47, 55–64.

Pecorino, L. 2005. *Molecular Biology of Cancer: Mechanisms, Targets and Therapeutics*. New York: Oxford University Press.

Peppas-Brannon, L. and Blanchette, J. O. 2004. Nanoparticle and targeted systems for cancer therapy. *Advanced Drug Delivery Reviews*, 56, 1649–1659.

Primo, F. L., Rodrigues, M. M. A., Simioni, A. R., Bentley, V. L. B., Morais, P. C., and Tedesca, A. C. 2008. *In vitro* studies of cutaneous retention of magnetic nanoemulsion loaded with zinc phthalocyanine for synergic use in skin cancer treatment. *Journal of Magnetism and Magnetic Materials*, 320, e211–214.

Rabinow, B. E. 2004. Nanosuspensions in drug delivery. *Nature Reviews Drug Discovery*, 3, 785–796.

Rahimnejad, M., Jahanshahi, M., and Najafpour, G. D. 2006. Production of biological nanoparticles from bovine serum albumin for drug delivery. *African Journal of Biotechnology*, 5, 1918–1923.

Reddy, A. and Kaelin, W. G. 2002. Using cancer genetics to guide the selection of anticancer drug targets. *Current Opinion in Pharmacology*, 2, 366–373.

Sasatsu, M., Onishi, H., and Machida, Y. 2008. Preparation and biodisposition of methoxypolyethylene glycol amine–poly(DL-lactic acid) copolymer nanoparticles loaded with pyrene-ended poly(DL-lactic acid). *International Journal of Pharmaceutics*, 358, 271–277.

Shinoda, K. and Friberg, S. 1975. Microemulsions colloidal aspects. *Advances in Colloid and Interface Science*, 4, 281–300.

Shiokawa, T., Hattori, Y., Kawano, K., Ohguchi, Y., Kawakami, H., Toma, K., Maitani, Y. 2005. Effect of polyethylene glycol linker chain length of folate-linked microemulsions loading aclacinomycin A on targeting ability and antitumour effect *in vitro* and *in vivo*. *Clinical Cancer Research*, 11, 2018–2025.

Terek, M. C., Karabulut, B., Selvi, N., Akman, L., Karasulu, Y., Ozguney, I., Sanli, A. U., Uslu, R., Ozsaran, A. 2006. Arsenic trioxide-loaded, microemulsion-enhanced cytotoxicity on MDAH 2774 ovarian carcinoma cell line. *International Journal of Gynecological Cancer*, 16, 532–537.

Thorpe, P. E. 2004. Vascular targeting agents as cancer therapeutics. *Clinical Cancer Research*, 10, 415–427.

Ueda, K., Clark, D. P., Chen, C. J., Robinson, I. B., Gottesman, M. M., and Pastan I. 1987. The human multidrug resistance (*mdr 1*) gene: c DNA cloning and transcription initiation. *Journal of Biology and Chemistry*, 262, 505–508.

Unger, C. 1996. Current concepts of treatment in medical oncology: New anticancer drugs. *Journal of Cancer Research and Clinical Oncology*, 122, 189–198.

Vasir, J. K. and Labhasetwar, V. 2005. Targeted drug delivery in cancer therapy. *Technology in Cancer Research and Treatment*, 4, 363–374.

Vasir, J. K., Reddy, M. K., and Labhasetwar, V. D. 2005. Nanosystems in drug targeting. *Opportunities and Challenges Current Nanoscience*, 47, 47–64.

Vonarbourg, A., Passirani, C., Saulnier, P., and Benoit, J. P. 2006. Parameters influencing the stealthiness of colloidal drug delivery systems. *Biomaterials*, 27, 4356–4373.

Wang, X., Dai, J., Chen, Z., Zhang, T., Xia, G., Nagai, T., Zhang, Q. 2004. Bioavailability and pharmacokinetics of cyclosporine A-loaded pH-sensitive nanoparticles for oral administration. *Journal of Controlled Release*, 97, 421–429.

Yamaguchi, H., Watanabe, K., Hayashi, M., and Awazu, S. 1984. Effect of egg yolk phospholipids plasma elimination and tissue distribution of coenzyme Q_{10} administered in an emulsion to rats. *Journal of Pharmacy and Pharmacology*, 36, 768–769.

25 Nanocarriers for Imaging Applications

Vandana Patravale and Medha Joshi

CONTENTS

25.1 INTRODUCTION

A new era dawned in the history of medical diagnosis after the discovery of the x-ray by Roentgen. Since the inception of x-ray technology for medical imaging, the invention of noninvasive methodologies and the exploration of their applications in various fields such as clinical diagnosis, cellular biology, and drug discovery have been continuous. The noninvasive imaging of various organs, tissues, or cells or various pathologies has been conducted with contrast agents. These contrast agents have become a mainstay in modern medicinal and biological research, and many reports now discuss the future scope and commercialization prospects. The sales of medical imaging contrast agents reached $1.41 billion in 2003 and are expected to rise to $2.58 billion by 2009 in the United States alone.

A contrast agent helps to establish the appropriate signal intensity from an area of interest, thus differentiating certain structures from surrounding tissues and highlighting the abnormalities, regardless of the imaging modality. It is difficult to detect and locate pathologies without a proper imaging agent. With the advent of contrast agents such as fluorescent probes, it is now possible to selectively view specific biological events and processes in both living and nonviable systems with improved detection limits, imaging modalities, and engineered biomarker functionality.

The real success of medical imaging depends upon the selective and targeted delivery of these imaging agents or contrast agents to the organ or tissue of interest. The imaging agents themselves do not possess the ability to selectively accumulate in the region under investigation and often require a suitable carrier to achieve this task.

Various carriers ranging from simple emulsions to innovative nanoparticulate carriers have been explored for the delivery of imaging agents in the last decade. This chapter focuses on the potential and applications of various pharmaceutical nanocarriers and their advantages in the delivery of imaging agents used for various imaging modalities.

The appropriate signal intensity from an area of interest is aided by the use of a contrast agent, which differentiates certain structures from the surrounding tissues and highlights the abnormalities, whatever imaging modality is utilized. Nonenhanced imaging is seldom used except when relatively large tissue areas are involved in pathological process. It is difficult to detect and locate pathologies without a proper imaging agent. For a target to be diagnostically identified, there must be a sufficient amount of activity above the background activity of normal tissue; generally, this target to background ratio is 1.5. A contrast agent encapsulated in a nanocarrier is absorbed by tissues (i.e., macrophage-rich tissues). This adsorption process is dependent on the size of the nanocarriers. The small particles enter blood or lymphatic capillaries whereas the large ones are retained in the tissues. The two main routes of administration of contrast agents are systemic and via local circulation. The performance of the contrast agent can be optimized by varying its physicochemical properties or that of the contrast carrier so that the rate of its disappearance from the injection site upon local administration can be determined. The major drawback of systemic administration is the increase of the exposure of nontarget organs to a potentially toxic contrast agent. A specific contrast agent concentration must be reached for successful imaging, and it varies with different diagnostic moieties. For example, the tissue concentration is relatively high in magnetic resonance imaging (MRI, 10e4 M). Nanocarriers have been used to overcome this for efficient delivery of contrast agents to areas of interest (Torchilin, 1997b). The term nanocarriers is derived from the Greek word nano, meaning dwarf; their dimensions and tolerances are in the 0.1- to 100-nm range. Various colloidal nanocarriers such as liposomes; nanoemulsions; polymeric, lipid, and magnetic nanoparticles; dendrimers; quantum dots (QDs); and ferrofluids have been suggested for the accumulation of contrast agent in the required zone. We will discuss all of them in detail along with a brief review of their applications as suggested in the literature. The desirable attributes of nanocarriers that are required for enhanced imaging are small size, stability, biocompatibility and biodegradability, large surface to volume ratio, site-specific delivery, ease of production and clinical use, sensitivity, early detection, and possibility of long-circulating "stealth." These attributes lower the cost of therapy and enhance the applicability with standard commercially available imaging modalities.

The various imaging modalities used currently include γ scintigraphy, MR, computed tomography (CT), and ultrasonography (US; Torchilin, 2005). Examples of the contrast agents used with different imaging modalities and their required sensitivities are summarized in Table 25.1.

TABLE 25.1
Contrast Agents Used with Different Imaging Modalities

Imaging Modality	Contrast Agent	Sensitivity
γ Scintigraphy	^{111}In, ^{99m}Tc, ^{67}Ga	1 ng mL^{-1} to 100 mcg mL^{-1}
MR	Gd, Mn, iron oxide	1–10 mcg mL^{-1}
CT	Iodine, bromine, barium	1 mg mL^{-1}
US	Gas (air, argon, nitrogen)	1 mcg mL^{-1}
PET	^{18}F, ^{11}C, ^{15}O	<1 pg mL^{-1}
SPECT	^{111}In chelates, ^{99m}Tc	<1 pg mL^{-1}

25.1.1 Scintigraphic Imaging

The radiopharmaceutical in scintigraphic imaging is labeled with a γ-irradiation (photon) emitting radionuclide. Photons with the appropriate energy (100 ± 500 keV) can penetrate tissues and can be detected outside the body (Boerman et al., 2000). The *in vivo* distribution of a γ-ray emitting radiopharmaceutical can be visualized using a γ camera. The advantages offered by scintigraphic imaging over other higher resolution cross-sectional imaging modalities such as CT and MRI are rapid imaging and screening of the whole body for sites of abnormal uptake. It is a very sensitive method, and a relatively low dose of radionuclide is required because the γ label is highly penetrating. The amount of injected activity that reaches a particular location in the body can be quantified noninvasively and requires the following:

1. Correction of isotope decay
2. Correction for the attenuation that occurs when the parts of the body absorb the emitted radiation before it can exit the body
3. Subtraction of the amount of activity in the blood contained by a particular part or organ of the body

25.1.2 Magnetic Resonance Imaging (MR/MRI)

MRI provides detailed images of the body in any plane. It is based on the principle of nuclear magnetic resonance, where measurements are made of the changes in magnetization of hydrogen protons in water molecules sitting in a magnetic field after a pulse of radiofrequencies hits them. The protons in various tissues react differently, resulting in a picture of the anatomical structures. After turning off the radiofrequency, the protons relax to the ground state. The relaxation is achieved in two ways: longitudinal relaxation (T_1) and spin–spin relaxation (T_2). The values of T_1 and T_2 depend on the chemical and magnetic environment in which a particular nucleus is situated, and the structures and tissues within the body have different T_1 and T_2 values (Tilcock, 1999a). In MR images derived from changes in T_1, regions that are associated with a contrast agent (nearby water molecules) have increased signal intensity compared with regions not associated with a contrast agent. The inverse is true for T_2-weighted images. Regions associated with a superparamagnetic center (such as an iron oxide particle) have reduced signal intensity in an MR image compared with areas without contrast agents. MRI provides much greater contrast between the different soft tissues of the body than CT, making it especially useful in neurological, musculoskeletal, cardiovascular, and oncological imaging. These images can be enhanced by adding contrast agents that sharpen the contrast by affecting the behavior of protons in their proximity.

The contrast agents used in scintigraphic imaging and their salient features are summarized in Table 25.2. Other radionucleotides are also used, such as redium 186, krypton 81 m, iodine 123,

TABLE 25.2
Salient Features of Contrast Agents Used in Scintigraphic Imaging

Contrast Agent	Features
Gallium 67 (^{67}Ga)	Excessive high-energy emission (397 and 300 keV)
	78-h half-life
Indium 111 (^{111}In)	Energy emission (247 and 172 keV)
	67-h half-life
	Expensive as produced by cyclotron
Technetium 99m (^{99m}Tc)	Most extensively (70%) used isotope
	6-h half-life
	Ideal radiation energy 140 keV
	Inexpensive
	Wide availability
	Safe decay products
Gadolinium(III)	High magnetic moment ($m^2 = 63$ BM^2)
	Has the most unpaired electrons of any stable ion
	Stability constant of Gd(III)-DTPA complex is high (log $K_{Gd} = 22.4$),
	water soluble, and nontoxic
Manganese(II)	High magnetic moments
	Five unpaired electrons
Iron(III)	High magnetic moments
	Five unpaired electrons

xenon 133, iodine 131, and selenium 75 (Hardy and Wilson, 1981). MRI free transition metals, for example, copper(II) chloride, manganese(II) chloride, and ferric ammonium citrate, have also been reported (Pautler and Koretsky, 2002).

25.1.3 COMPUTED TOMOGRAPHY (CT)

CT is also known as computed axial tomography or body section roentgenography. It uses digital geometry processing to generate a three-dimensional (3-D) image of the inside of an object from a large series of two-dimensional (2-D) x-ray images taken around a single axis of rotation (Krause, 1999). The contrast in CT depends on the differential attenuation of the x-ray beam by different tissues or structures. The attenuation varies with the energy of the beam and the tissue mass attenuation coefficient that depends on the average proton number Z of the tissue. With higher Z, the probability of interaction between the incident photon and the electron orbital of a particular atom within the tissue increases and therefore higher attenuation is achieved (Tilcock, 1999a). Synchrotron x-ray tomographic microscopy is a 3-D scanning technique that allows noninvasive high definition scans of objects with details as fine as 1000th of a millimeter, meaning it has 2000–3000 times the resolution of a traditional medical CT scan.

25.1.4 ULTRASONOGRAPHY (US)

US uses the principle of the ultrasound-based diagnostic imaging technique. It can visualize muscles and internal organs and their size, structures, and possible pathologies or lesions, as well as blood flow. All acoustic energies with a frequency above human hearing (20,000 Hz or 20 Kz) are used. The typical operating range of a US diagnostic is 2–18 MHz. The frequency is chosen so that it is a trade-off between the spatial resolution of the image and the imaging depth: lower frequencies produce less resolution but image deeper into the body.

TABLE 25.3
Nanocarriers Used with Different Techniques

Imaging Technique	Liposomes	Dendrimers	Polymeric Nanoparticles	Lipid Nanoparticles	Emulsions	Polymeric Micelles
γ Scintigraphy	√	—	√	√	—	√
MR	√	√	√	√	√	√
CT	√	√	√	—	√	√
US	√	—	—	—	√	—
PET	√	—	√	—	—	—
SPECT	√	—	—	—	—	—

25.1.5 POSITRON EMISSION TOMOGRAPHY (PET)

Positron emission tomography (PET) is a nuclear medicine imaging technique. It produces a 3-D image or map of functional processes in the body. Pairs of γ-rays are emitted indirectly by a positron-emitting radionuclide (tracer), which is introduced into the body on a biologically active molecule. Computer analysis then reconstructs the images of the tracer concentration in 3-D space within the body. Modern scanners are capable of reconstruction with the help of a CT x-ray scan that is performed on the patient during the same session and in the same machine.

25.1.6 SINGLE PHOTON EMISSION COMPUTED TOMOGRAPHY (SPECT)

Single photon emission computed tomography (SPECT or SPET) is another nuclear medicine tomographic imaging technique that uses γ-rays. The principle is similar to conventional nuclear medicine planar imaging using a γ camera. However, it can provide true 3-D information. The information is presented as cross-sectional slices, which can be easily reformatted or manipulated as required. SPECT imaging is performed initially by using a γ camera to acquire multiple 2-D images (also called projections) from multiple angles. A computer is then used to reconstruct the 3-D data set by applying a tomographic reconstruction algorithm to the multiple projections. This data set can then be manipulated to show thin slices along any chosen axis of the body, similar to those obtained from other tomographic techniques, such as MRI, CT, and PET. Because SPECT acquisition is very similar to planar γ-camera imaging, the same radiopharmaceuticals may be used.

A summary of different nanocarriers used as carriers for different contrast agents for various imaging modalities is provided in Table 25.3, and the requirements of nanocarriers to carry out efficient imaging in various imaging modalities are listed in Table 25.4.

TABLE 25.4
Requirements from Nanocarriers

Imaging Technique	Requirement from a Contrast Agent
γ Scintigraphy	Optimum energy and half-life
MR	Increase in relaxivity
CT	Enhanced attenuation
US	Enhanced echogenicity
PET	Optimum half-life
SPECT	Optimum energy and half-life

25.2 LIPOSOMES

Liposomes are spherical, self-closed structures formed by one or several concentric liquid bilayers with an aqueous phase inside and between the lipid bilayers. The different types of liposomal vesicles are as follows (Phillips, 1999):

1. *Multilamellar vesicles (MLVs):* These range in size from 500 to 5000 nm and consist of several concentric bilayers.
2. *Small unilamellar vesicles:* These are around 100 nm in size and formed by a single bilayer.
3. *Large unilamellar vesicles (LUVs):* LUVs range in size from 200 to 800 nm.
4. *Long-circulating liposomes:* These liposomes are modified (usually surface-grafted with certain polymers) so that they can stay in the blood much longer (for hours) than nonmodified liposomes.
5. *Immunoliposomes:* Immunoliposomes carry antibodies attached to their surfaces and are able to accumulate in the area within the body where an attached antibody recognizes and binds its antigen.

Liposomes have several attractive features that make them an ideal carrier for diagnostic agents: excellent biocompatibility, ability to entrap hydrophilic as well as hydrophobic molecules, ability to protect the encapsulated agent from chemical and enzymatic degradation, and amenability for surface modification that enables targeting to various tissues (Torchilin, 2005). There are numerous published reviews describing the encapsulation of different contrast agents into liposomes (Al-Jamal and Kostarelos, 2007; Boerman et al., 2000; Caride, 1985, 1990; Dagar et al., 2003; Goins, 2001; Hamoudeh et al., 2008; Krause, 1999; Laverman et al., 2003; Mitra et al., 2006; Mulder et al., 2006b; Oku et al., 1993; Osborne et al., 1979; Phillips, 1999; Storm and Crommelin, 1998; Torchilin, 1994a, 1996, 1997a, 1997b, 2005, 2006, 2007).

Diagnostic agents are usually incorporated in the liposomes by the lipid film hydration method or the reverse-phase evaporation method that yields MLVs. Upon sonication or extrusion under high pressure the MLVs are converted into unilamellar vesicles (ULVs). The ULVs with sizes ranging from 0.1 to 0.4 μm are known as LUVs, and they have an important role in targeting (Phillips, 1999).

25.2.1 LIPOSOMES AS CARRIERS FOR CONTRAST AGENTS USED IN γ SCINTIGRAPHY AND MAGNETIC RESONANCE

For efficient contrast in γ scintigraphy the contrast agent needs to be associated with the liposome (Boerman et al., 1998; Phillips, 1999). The ideal requirements of the γ-imaging liposome label for clinical diagnostic applications (Phillips and Goins, 1995) include the following:

1. Shelf stability of the preformed liposome
2. High labeling efficiency
3. Convenient labeling at room temperature
4. Use of readily available isotopes that have good physical imaging characteristics and half-life
5. *In vitro* and *in vivo* stability
6. Retention of the label at the site of initial uptake
7. Universal applicability to all types of liposomes

There are different ways by which liposomes can be labeled by radioisotopes (Phillips, 1999):

1. Labeling by encapsulation during manufacture
2. Liposome surface labeling after manufacture
3. Incorporation into the lipid bilayer after manufacture

4. Surface labeling of the preformed liposome incorporating a lipid chelator conjugate during manufacture
5. Aqueous phase loading of the preformed liposomes or "after loading"

Technetium was encapsulated in MLVs and LUVs with labeling efficiencies of 30% and 5%, respectively (Caride, 1981; Espinola et al., 1979). The quantity of liposome-associated reporter metal can be enhanced by using membrane-anchored chelating polymers that are capable of carrying metal atoms, or the liposome surface may be modified to enhance the signal intensity from the liposomal reporter metal by using different polymers (Torchilin, 1994a, 1994b). These polymers may be amphiphilic or water soluble. In MRI the metal atoms chelated into these groups are directly exposed to the water environment, resulting in the enhancement of the signal intensity of the paramagnetic ions that further leads to enhancement of the vesicle contrast properties. Association of amphiphilic poly(ethylene glycol) (PEG) with the liposomal surface causes an increased concentration of PEG-associated water protons in the vicinity of chelated gadolinium (Gd) ions, leading to enhanced relaxivity. Coating with PEG makes the liposomes circulate a long time, which can be utilized for blood pool imaging (Torchilin, 2005). A liposomal drug delivery system that is responsive to pH stimuli can be used for the detection of pathological areas with decreased pH (Lokling et al., 2004a, 2004b, 2004c).

Various methods have been suggested in the literature to improve the labeling efficiencies such as surface labeling (Barratt et al., 1984; Love et al., 1989b; Richardson et al., 1977, 1978), incorporation of lipid diethylenetriaminepentaacetic acid (DTPA; Ahkong and Tilcock, 1992; Goto et al., 1989; Hnatowich et al., 1981; Tilcock et al., 1994a), and hexamethylpropyleneamine oxime (HMPAO) glutathione (Awasthi et al., 1998b; Boerman et al., 1997a, 1997b; Neirinckx et al., 1988; Ogihara-Umeda et al., 1996; Oyen et al., 1996a; Phillips et al., 1992; Rudolph et al., 1991; Tilcock et al., 1994b). An amphiphilic chelating polymer such as N,α-(DTPA-polylysyl)glutaryl phosphatidyl ethanolamine has been incorporated to increase the loading of the contrast agent inside liposomes (Torchilin, 2000). These polymers markedly increase the number of chelated Gd or In atoms attached to a single lipid anchor (Tilcock et al., 1989; Unger et al., 1990). A hydrophobic group, which can anchor the chelating moiety on the liposome surface during or after liposome preparation, can be incorporated by chemical derivatization of DTPA or other chelators (Kabalka et al., 1991; Schwendener et al., 1989). Various chelators and anchors are listed in Table 25.5.

Administration of free metal ions can lead to toxicity in biological systems, so suitable ligands or chelates are used to bind the metal ions to form nontoxic complexes. These chelators inhibit the uptake of free metal ions in biological systems, thus preventing toxicity.

MRI utilizes liposomes encapsulated with ferrite particles called ferrosomes (mainly to decrease the toxicity but sacrificing the relaxivity) and manganese complexes called memsomes (Unger et al., 1991, 1993, 1994a). Liposomes are also used as vesicles to carry membrane-bound paramagnetic complexes. PEGylated liposomes loaded with Gd-containing chelates have been used as long-circulating blood pool agents. MRI contrast agents have also been used to image the lymph nodes using liposomes containing Gd-DTPA. Liposomes containing Gd have been conjugated to antibodies and targeted to a specific organ system. Liposomes containing Gd-HPDO3A [1,4,7-tris(carboxymethyl)-10-(2′-hydroxypropyl)-1,4,7,10-tetraazacycl ododecane] or MnDPDP [N,N'-dipyridoxylethylenediamine-N,N'-diacetate-5,5′-bis(phosphate)] have been proposed to image the liver or the hepatobiliary system.

Bertini et al. (2004) evaluated PEG-stabilized paramagnetic liposomes in murine B16-F10 melanomas. A threefold higher relaxivity compared to the conventional paramagnetic complexes gadolinium-tetraazacyclododecanetetraacetic acid (Gd-DOTA) and Gd-DTPA was observed for liposomes. They also showed prolonged and persistent signal enhancement without the application of molecular targeting. The tumor area was enhanced to an average of 33% after 2 h, 43% after 20 h, and 25% after 54 h (Figure 25.1) in striking contrast to the results with Gd-DTPA (maximum enhancement of 29% was achieved after 20 min, which rapidly disappeared; Figure 25.1).

An excellent review describing the use of metal ions in MRI applications was recently published (Delli Castelli et al., 2008). There are also several reviews describing the application of liposomes

TABLE 25.5
Different Modifications, Chelators, and Hydrophobic Anchors Used in the Preparation of Liposomes Loaded with Contrast Agent Used in Scintigraphic Imaging

Contrast Agent	Modification, Chelator, or Anchor	Reference
^{111}In	PEGylated PLL-based PAP mAb 2C5	Erdogan et al. (2006b)
	Avidin and [^{99m}Tc]-biotin liposomes	Medina et al. (2004)
	IDLPL	Syrigos et al. (2003), Harrington et al. (2000a, 2000b)
	Acetylacetone chelate	Beaumier and Hwang (1982)
	PEG	Boerman et al. (1998)
	Bound inulin	Essien and Hwang (1988)
	Oxine	Kassis and Taube (1987)
	5% PEG-distearoyl PE with encapsulated glutathione and deferoxamine	Awasthi et al. (1998a)
	Nitrilotriacetic acid	Hwang et al. (1982)
	Desferoxime	Gabizon et al. (1990)
	Ionophore (A23187)	Mauk and Gamble (1979)
^{99m}Tc	HMPAO	Gaal et al. (2002)
	PEG	Boerman et al. (1998, 1995)
	DTPA	Espinola et al. (1979)
	Stealth® liposomes HMPAO	Oyen et al. (1996b)
	DPPE-DTTA	Tilcock et al. (1993)
	HMPAO	Goins et al. (1993)
	Contains 5% PEG-DSPE with encapsulated glutathione and deferoxamine	Awasthi et al. (1998a)
	Stearylamine-DTPA	Hnatowich et al. (1981)
	HYNIC derivative of DSPE	Laverman et al. (1999)
	HYNIC derivative of DSPE labeled with IgG	Dams et al. (1999)
Gallium	PEG	Boerman et al. (1998)
	Long circulation liposomes	Boerman et al. (2000)
	Deferoxamine (67 Ga-DF)	Gabizon et al. (1988)
	Stearylamine-DTPA	Hnatowich et al. (1981)
	Nitrilotriacetic acid	Ogihara-Umeda and Kojima (1988)
	PEG	Woodle (1993)
	DTPA-SA	Schwendener et al. (1989)
	Membrane-bound complexes of manganese (memsomes)	Unger et al. (1993)
Manganese(II)	Gd-HPDO3A complex	Terreno et al. (2008)
	DTPA-PLL-NGPE	Erdogan et al. (2008)
Gadolinium(III)	DPPGOG	Wang et al. (2008)
	PE-DTPA conjugate	De Cuyper et al. (2007)
	DTPA bisoctadecylamide and GdAcAc	Zielhuis et al. (2006)
	Membrane-incorporated PAPs	Erdogan et al. (2008, 2006a), Weissig et al. (2000)
	PPLs	Storrs et al. (1995)
	Gd-DTPA-PE as an amphiphilic paramagnetic label	
	Immunoliposomes and PEGylated immunoliposomes	Torchilin (1994a)
	Gd-BOPTA	Schwendener et al. (1994)

continued

TABLE 25.5 (continued)
Different Modifications, Chelators, and Hydrophobic Anchors Used in the Preparation of Liposomes Loaded with Contrast Agent Used in Scintigraphic Imaging

Contrast Agent	Modification, Chelator, or Anchor	Reference
	Mannan (cholesterol-aminoethylcarbamylmethyl mannan)-coated liposomes	Kunimasa et al. (1992), Suga et al. (2001)
	PEGlyated liposomes	Bulte et al. (1999), Meincke et al. (2008), Plassat et al. (2007)
	PE-DTPA conjugate	De Cuyper et al. (2007), Soenen et al. (2007)
Iron oxide	Distearoyl-*sn*-glycero-3-phosphoethanolamine-*N*-[methoxy(PEG)-2000] coated with liposome	Martina et al. (2005)

Note: PPL—Paramagnetic polymerized liposomes; PAP—polychelating amphiphilic polymer; IDLPL—DTPA-labeled PEGylated liposomes; DSPE: distearoylphosphatidyl-ethanolamine; HYNIC: 6-Hydrazinopyridine-3-carboxylic acid; HMPAO: hexamethylpropylene amine oxime; HPDO3A—10-(2-hydroxypropyl)-1,4,7,10-tetraazacyclodo-decane-1,4,7-triacetic acid; DTPA-PLL-NGPE: Diethylenetriaminepentaacetic acid-polylysyl-N-glutaryl-phosphatidyl ethanolamine; BOPTA—benzoyloxypropionictetraacetate; DPPE-DTTA—dipalmitoylphosphatidylethanolamine-diethylenetriaminetetraacetic acid; DPPGOG—phosphatidylglyceroglycerol; GdAcAc—gadoliniumacetylacetonate; SA: stearylamine.

in MR. Perez-Mayoral et al. (2008) explored the utility of pH-sensitive liposomes that use the pH dependence of their relaxivity or magnetization transfer rate constant (chemical exchange saturation transfer) for imaging. Specific reviews by Tilcock (1999a), Torchilin (1997b), and Unger et al. (1993) are also worth reading.

A summary of different types of liposomes and their modifications used in conjugation with MRI is provided in Table 25.6.

25.2.1.1 Factors Affecting Delivery of γ-Scintigraphic Agent Containing Liposomes

There are a number of points that must be considered while delivering the γ-scintigraphic agent to various targets in the body.

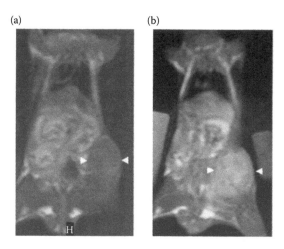

(a)   (b)

FIGURE 25.1 The T_1-weighted images of a mouse (a) before and (b) 20 h after the injection of PEG-stabilized paramagnetic liposomes. (Adapted from Bertini, I., Bianchini, F., Calorini, L. et al. 2004. *Magn Reson Med* 52: 669–672. With permission.)

TABLE 25.6

Type of Liposome or Modification Used in Imaging of Different Diseases Used in Conjugation with MRI

Type of Liposome or Modification	Imaging Application	Reference
Gd-DTPA liposomes targeted to CD105	Tumor angiogenesis	Zhang et al. (2009)
Gd-DTPA liposomes	Liver and biliary tract imaging	Najafi et al. (1987)
Gd-DTPA liposomes	Intravascular contrast agents	Unger et al. (1988)
Mn-, Gd-, and Fe-DTPA-SA complex liposomes	Liver and upper abdomen	Schwendener et al. (1989)
Gd-DTPA liposomes	Hepatoma	Unger et al. (1989)
Mn^{2+} liposomes	Liver	Bacic et al. (1990)
Gadolinium-, manganese-, and iron-DTPA-SA liposomes	Upper abdomen (Gd-DTPA-SA liposomes)	Schwendener et al. (1990)
Gd-DTPA liposomes	Liver	Kabalka et al. (1991), Unger et al. (1990)
Mannan-coated Gd-DTPA liposomes	Liver	Kunimasa et al. (1992)
Gd-DTPA liposomes	Lymph nodes	Trubetskoy et al. (1995)
Gd-DTPA-SA liposomes and Gd-BOPTA liposomes	Liver, spleen, vascular system	Schwendener et al. (1994)
Gd-DTPA-PE liposomes	Lymph nodes	Trubetskoy et al. (1995)
Gd(BME-DTTA) liposomes	Myocardium	Chu et al. (1997)
NGPE-PE Gd-containing PEG-coated liposomes	Blood pool	Torchilin et al. (2000)
NGPE-PE polylysine Gd-DTPA-containing PEG-coated liposomes	Blood pool	Weissig et al. (2000)
Gd-DOTA liposomes	Intragastric	Faas et al. (2001)
Gd-DTPA-mannan-cholesterol-coated liposomes	Pulmonary perfusion	Suga et al. (2001)
pH-sensitive DPPE/DSPG liposomal Gd-DTPA-BMA PEG-coated liposomes	Blood pool imaging	Lokling et al. (2004a, 2004c)
Liposome loading with Gd via the membrane-incorporated PAPs + PEG + anticancer mAb 2C5	Cancer	Erdogan et al. (2006a)
Annexin V was linked to Gd-DTPA-coated liposomes	Myocardium	Hiller et al. (2006)
Cationic liposomes containing the paramagnetic contrast agent Gd-DTPA	Brain	Masotti et al. (2006)
Paramagnetic liposomes	Atherosclerotic plaques	Mulder et al. (2006a)
MAGfect, a novel liposome formulation containing a lipidic Gd contrast agent for MRI, Gd-DOTA-Chol 1	Labeling and visualization of cells	Oliver et al. (2006)
Gd-DTPA-BSA lipids in PEGylated liposomes	Apoptosis	Van Tilborg et al. (2006)
2C5-modified Gd-PAP-containing PEGylated liposomes	Tumor	Erdogan et al. (2008)
Gd-DOTA-DSA liposomes	Cellular and tumor imaging	Kamaly et al. (2008)
Liposomes containing SPIO particles	Tumor	Pauser et al. (1997)
Maghemite liposomes	Angiography	Martina et al. (2005)
Iron oxide liposomes	Cell labeling	Meincke et al. (2008)

Note: DSPG—1,2-Distearoyl-sn-Glycero-3[Phospho-rac-(1-glycerol); PAP—polychelating amphiphilic polymer; BOPTA—benzoyloxypropionictetraacetate; BSA—bis(stearylamide); DPPE—dipalmitoylphosphatidylethanolamine; DTTA—diethylenetriaminetetraacetic acid; Gd-DOTA-DSA—gadolinium 2-{4,7-bis-carboxymethyl-10-[(N,N-distearylamidomethyl-N'-amidomethyl]-1,4,7,10 tetraazacyclododec-1-yl}acetic acid.

25.2.1.1.1 Size and Surface of the Liposomes

Liposomes greater than 250 nm are removed rapidly from the circulation by the reticuloendothelial system (RES) and have poor accumulation at the site of action. Therefore, liposomes with a smaller size (around 180 nm) are preferred. The surface charge also has an effect on the accumulation of liposomes. Goins et al. (1994) observed that neutral liposomes accumulate in tumors more effectively than negatively charged liposomes.

25.2.1.1.2 Composition of Liposomes

ULVs smaller than 100 nm and having a high percentage of cholesterol have a prolonged residence in the circulation. Modification of the surface of liposomes with the addition of hydrophilic groups such as PEG also prolongs the circulation (Mori et al., 1991; Senior et al., 1991).

25.2.1.1.3 Circulation Half-Life

A shorter circulation half-life of around 4 h may be ideal for liposomes to allow them to accumulate in the inflamed region and to simultaneously provide enough background clearance to target inflammation. Longer-circulating liposomes may be required for targeting chronic inflammation (Awasthi et al., 1998a).

25.2.1.1.4 Active Background Removal

Avidin biotin affinity mechanisms have been used to remove liposomes from blood circulation and to localize the liposomes at the site of diagnosis (Goins et al., 1998; Laverman et al., 2000b; Ogihara-Umeda et al., 1993; Zavaleta et al., 2007).

25.2.1.1.5 Immunological Effects of Liposomes

In two studies, decreased circulation was persistent upon repeated injection of PEG surface-modified technetium 99m (^{99m}Tc) liposomes (Dams et al., 1998, 2000a; Laverman et al., 2000a), an effect mediated by a serum factor. The liposomal formulation must be designed while taking this into consideration.

Imaging of tumors, infections, inflammation, blood, lymph, and myocardial infarction (MI) are possible with liposomes containing the scintigraphic agent (Boerman et al., 1998, 2000; Bogdanov et al., 1995; Briele et al., 1990; Phillips, 1999; Thakur, 1977; Torchilin, 2007).

25.2.1.2 Scintigraphic or Metallic Contrast Agent Containing Liposomes for Liver and Spleen Imaging

The liver and spleen are the first organs where unmodified liposomes are accumulated for passive targeting. The tumors and metastases of these organs can be easily detected using liposomes loaded with appropriate contrast agents (Boerman et al., 2000; Phillips, 1999).

25.2.1.3 Scintigraphic or Metallic Contrast Agent Containing Liposomes for Tumor Detection

The mechanism by which the accumulation of liposomes occurs in tumors is the enhanced-permeation and retention (EPR) effect or the extravasation through leaky tumor capillaries. Various types of cancers have been detected using scintigraphic contrast agents that contain liposomes, for example, melanoma, sarcoma, lymphoma, human tumor xenograft in mice, lung cancer, Kaposi sarcoma, and lymphoma in acquired immunodeficiency syndrome patients. VesCan (Vestar Inc.) indium 111 (^{111}In)-labeled liposomes are already in clinical trials (Torchilin, 1997b). Richardson et al.'s (1978) study was the first report of liposomes used for imaging studies, and Presant et al.'s (1990) study was the first clinical report. In this study, tumors were detected in 22 of 24 participants using ^{111}In-labeled liposomes. The utility of gallium-loaded small ULVs in the detection of tumors was also established (Ogihara et al., 1986a, 1986b). ^{99m}Tc-PEG liposomes were successful in determining the extent and severity of Crohn's colitis (Brouwers et al., 2000a). The ability of vasoactive intestinal peptide (VIP) containing ^{99}Tc liposomes to detect breast cancer was also established by Dagar et al. (2001, 2003).

25.2.1.4 Scintigraphic or Metallic Contrast Agent Containing Liposomes for Detection of Infection and Inflammation

The accumulation of liposomes at an infection and inflammation site occurs via nonspecific leakage out of the vasculature that is attributable to the increased permeability of the capillaries within the inflamed tissue (Love et al., 1989a) and their uptake by phagocytic cells at the site. An LTB4 receptor antagonist, [99]Tcm-RP517, which contains the hydrazine nicotinamide (HYNIC) moiety for scintigraphic detection of infection and inflammation in rabbits, was reported (Brouwers et al., 2000b). Improved detection of infections by avidin-induced clearance of ^{99m}Tc-biotin-PEG liposomes was also described (Laverman et al., 2000b). The utility of ^{99m}Tc-PEG liposomes and ^{99m}Tc-HYNIC immunoglobulin G (IgG) was demonstrated in an experimental model of chronic osteomyelitis (Dams et al., 2000b). The review by Boerman et al. (2001) is excellent to acquire more knowledge on this subject.

25.2.1.5 Scintigraphic or Metallic Contrast Agent Containing Liposomes for Blood Pool and Lymph Imaging

The most commonly used contrast agent for this application is [99m]Tc. ^{99m}Tc-labeled red cells are commonly used for studying the motion of the heart and for the determination of the percentage of blood ejected from the heart. Blood pool imaging is particularly important for diagnosis of artherosclerotic lesions, thrombi, and tumors. It is also used to detect sites of gastrointestinal bleeding. Because the association between ^{99m}Tc and red blood cells is not very strong, ^{99m}Tc tries to dissociate and the free ^{99m}Tc interferes with the detection of gastrointestinal bleeding. Therefore, ^{99m}Tc liposomes labeled with HMPAO glutathione are used because of the greater stability of HMPAO glutathione. Detection of lymph nodes is also possible using liposomes. The first lymph node draining a tumor bed is known as the sentinel lymph node, and it has important diagnostic and therapeutic implications it is sometimes the first and only metastasis from the cancer.

In a study by Philips et al. (2001), biotin-coated liposomes encapsulating a blue dye labeled with ^{99m}Tc of 130-nm average size were administered subcutaneously followed by injection of avidin. As avidin progressed toward the lymphatic vessels, it caused aggregation of the biotin-coated liposomes that were in the process of migrating through the lymphatic vessels (Figures 25.2 and 25.3).

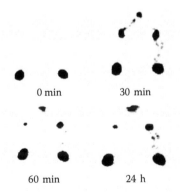

0 min   30 min

60 min   24 h

FIGURE 25.2 Scintigrams showing anterior views of the legs and lower abdomen of a rabbit at different time points after s.c. injection of ^{99m}Tc-labeled blue-biotin liposomes in both hind feet; this was followed by s.c. injection of avidin in only the left hind foot. Increased retention of ^{99m}Tc activity in the left popliteal node compared with the right popliteal node is clearly visualized at 60 min and 24 h. (Adapted from Phillips, W. T., Klipper, R., and Goins, B. 2001. *J Nucl Med* 42: 446–451. With permission.)

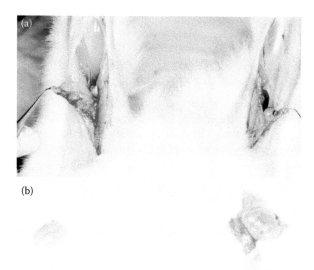

(b)

FIGURE 25.3 (a) The posterior aspect of the hind legs of a rabbit showing exposed popliteal nodes at necropsy at 24 h. The feet of the rabbit are located superior to the picture. (b) Excised experimental (right side of image) and control (left side of image) popliteal nodes. The blue coloration of the popliteal node in the left foot (on right side of image) that received avidin can be readily observed. (Adapted from Phillips, W. T., Klipper, R., and Goins, B. 2001. *J Nucl Med* 42: 446–451. With permission.)

25.2.1.6 Scintigraphic Contrast Agent Containing Liposomes for Detection of Myocardial Infarction

^{99m}Tc-encapsulated liposomes have been shown to detect MI (Caride and Zaret, 1977). The uptake of liposomes bearing positive, negative, or no net charge on their membrane and containing a radioactive tracer, [^{99m}Tc] DTPA, was studied in 12 intact dogs 24 h after the induction of MI and compared to the relative regional myocardial blood flow determined from radioactive microspheres. Positively charged and neutral liposomes are concentrated in infarcted regions against a flow gradient, whereas negative liposomes are passively distributed according to regional blood flow. Because positively charged and neutral liposomes concentrate in infarct areas and have the ability to incorporate pharmacologic agents in their aqueous or lipid phase, they may serve as vehicles for delivery of imaging agents or therapeutic agents to infarcted zones of low flow. Various other liposomal constructs, such as PEGylated liposomes and liposomes containing AM monoclonal antibody, dextran stearate (SA), and chelating agent DTPA-phosphatidylethanolamine (PE), as well as an *N*-glutaryl-phosphatidyl ethanolamine (NGPE)-modified chelating polymer DTPA-PLL-NGPE (Diethylenetriaminepentaacetic acid-polylysyl-*N*-glutaryl-phosphatidyl ethanolamine) have also been shown to target MI (Torchilin, 1994b).

25.2.2 LIPOSOMES AS CARRIERS FOR CONTRAST AGENTS USED IN COMPUTED TOMOGRAPHY

Because the attenuation values of different tissues and organs are almost similar, a contrast agent is required for CT imaging. The contrast agents used in CT have a high atomic number, for example, barium, bismuth, strontium, lead, and iodine. Iodine was recently substituted with metal ions such as gadolinium, ytterbium, dysprosium, or xenon (Zwicker et al., 1993). These contrast agents were incorporated into liposomes to improve their pharmacokinetics. Table 25.7 summarizes the liposomes and modified liposomes used as carriers for contrast agents in CT.

25.2.3 LIPOSOMES AS CARRIERS FOR CONTRAST AGENTS USED IN ULTRASONOGRAPHY

The application of liposomes for ultrasonic imaging has been reviewed by many authors (Bloch et al., 2004; Dagar et al., 2003; Dayton and Ferrara, 2002; Maruyama et al., 2007; Tilcock, 1999b). The

TABLE 25.7
Types of Liposomes as Carriers for Contrast Agents Used in CT

Type of Liposome	Organ Detected	Reference
Positively charged liposomes diatrizoic acid salts (renografin)	Over 100 HU, spleen imaging	Havron et al. (1981)
Brominated radioopaque liposomes	Liver and spleen	Caride et al. (1982)
Diatrizoate (Hypaque®, Renografin®)	Liver and spleen	Ryan et al. (1983)
Diatrizoate liposomes	Blood pool and reticuloendothelial structures	Seltzer et al. (1984a)
Iosefamate liposomes	Liver and spleen	Seltzer et al. (1984b)
Ethiodol liposomes	Liver	Jendrasiak et al. (1985)
Proliposomes with water-soluble contrast media	Spleen	Payne and Whitehouse (1987)
Liposomes containing iodine-125-labeled diatrizoate, 125-I-labeled iotrol	Liver and spleen	Zalutsky et al. (1987)
Iopromide	Hepatic	Capellier et al. (1988), Musu et al. (1988)
Iopamidol-carrying liposomes	Liver and spleen	Krause et al. (1993), Musu et al. (1988), Sachse et al. (1993)
Iodine-containing liposomes	Liver tumors	Henze et al. (1989)
Iotrolan liposomes	Liver and spleen	Kofi Adzamli et al. (1990)
Ioxaglate	Liver and spleen	Revel et al. (1990)
Iodine-125-iotrolan	Liver and spleen	White et al. (1990)
Iomeprol-containing liposomes	Liver tumors	Fouillet et al. (1995)
Iopromide-containing liposomes	Liver	Erdogan et al. (1998), Krause et al. (1996), Leike et al. (1996), Schmiedl et al. (1995)
Iotrolan	Hepatosplenic	Seltzer et al. (1995)
Iodixanol liposomes	Liver	Desser et al. (1999), Dick et al. (1996), Leander, (1996), Leander et al. (2001)
PEG-coated iopromide-carrying liposomes	Blood pool imaging	Sachse et al. (1997)
Liposomes containing free and encapsulated iomeprol	Liver	Petersein et al. (1999)
Iopromide-carrying liposomes	Blood pool	Schmiedl et al. (1999), Leike et al. (2001)
Iohexol	Blood pool	Burke et al. (2007), Kao et al. (2003)
Iodine	Blood pool	Mukundan et al. (2006), Zheng et al. (2006)
Iodine	Small liver metastases	Montet et al. (2007)

Note: HU—Hounsfield units.

contrast agents used for ultrasound imaging using liposomes are broadly classified into two categories: gas-filled liposomes and inherently acoustically reflective (echogenic) liposomes (Dagar et al., 2003).

25.2.3.1 Gas-Filled Liposomes

The commonly used gases in these types of liposomes are nitrogen and carbon dioxide (Unger et al., 1992, 1994b). Carbon dioxide is generally used to prepare these kinds of liposomes by entrapping gas bubbles inside the liposomes or to form gas directly inside the liposomes as a consequence of a chemical reaction (Torchilin, 1995). The aqueous core in these gas-filled liposomes is entirely replaced by either a water-soluble or insoluble gas (Unger et al., 1994b). Examples of water-soluble gas are carbon dioxide (Kimura et al., 1998) and nitrogen. There are several ways to produce these water-soluble gas-filled liposomes that have a precursor inside the liposome that is activated by various means such as pH and temperature (Unger et al., 1999; Unger and Wu, 2001). They can also be prepared by a basic pressurization and depressurization technique. The disadvantages of gas-filled liposomes are that they are unstable and have a short half-life. Water-insoluble gases such as perfluorates are entrapped in liposomes. They have a more stable and long-lasting effect because of their low solubility in water (Maresca et al., 1998). Examples of different perfluorate gas-filled liposomes are MRX-113® perfluorobutane-filled liposomes (Unger et al., 1994b), MRX-408® perfluorobutane-filled liposomes targeted to thrombi, and MRX-115® perfluoropropane-filled liposomes (Fritz et al., 1997). These kinds of liposomes are prepared by shaking an aqueous solution composed of lipid in the presence of a gas at a temperature below the gel state of the liquid crystalline state phase transition temperature of the lipid. The control of the size and stability of these gas-filled liposomes is difficult. When injected intravenously, these 4–5 µm particles become trapped in pulmonary capillaries and cannot pass from the right to the left side of the heart. Because of the large particle size, the circulation half-life is limited to a few minutes. These problems have been overcome by preparing liposomes that do not require the gas to be acoustically reflective, which are inherently acoustically reflective (echogenic) liposomes.

25.2.3.2 Echogenic Liposomes

These liposomes are inherently acoustically reflective because of their multiple rigid bilayer membranes that are formed by controlling the lipid composition and production method. These liposomes can also be made into immunoliposomes by conjugating a relevant antibody to them (Demos et al., 1997), for example, antifibrinogen antibody and anti-intracellular adhesion molecule-1 antibody for targeting artherosclerotic plaques (Demos et al., 1998, 1999) and human hepatocarcinoma-specific monoclonal antibody HAb18 for the detection of human hepatocellular carcinoma (Bian et al., 2004). A second generation of these liposomes has been prepared, which are acoustically reflective VIP sterically stabilized liposomes (Onyuksel et al., 2000; Rubinstein et al., 1999). They have a long-circulation half-life and because of the VIP they can be targeted to cancer and arthritis, where the vasculature is leaky and VIP receptors are overexpressed.

25.2.4　Liposomes as Carriers for Contrast Agents Used in Positron Emission Tomography

There are only few examples in the literature where liposomes were applied for imaging with PET (Hamoudeh et al., 2008). Oku (1999) reviewed most of the work in this field. Long-circulating liposomes encapsulating [2-^{18}F]-2-fluoro-2-deoxyglucose were administered to tumor-bearing mice, and a PET scan was performed. Small long-circulating liposomes (100 nm) tend to accumulate in tumor tissues of tumor-bearing mice compared with conventional liposomes. The effect of size on trafficking of long-circulating liposomes was investigated. Large liposomes (>300 nm) accumulated in the liver and spleen in a time-dependent manner. In contrast, small ones (<200 nm) were transiently accumulated in the liver right after injection. The accumulation of liposomes decreased in a time-dependent manner, which suggested that the majority of small long-circulating liposomes remain in the bloodstream; however, a few extravasate once into the interstitial spaces in the liver and reenter the bloodstream. The trafficking of so-called long-circulating liposomes, which are liposomes modified with ganglioside GM1, palmityl glucuronide (PGlcUA), and PEG, in tumor-bearing mice was examined. The accumulation of all three kinds of long-circulating liposomes in the liver decreased in a time-dependent manner, and PGlcUA liposomes could efficiently avoid liver trapping. Tumor accumulation of liposomes was obvious for PGlcUA liposomes and PEG liposomes immediately after injection, but not for GM1 liposomes. Finally, the trafficking of differently charged liposomes was investigated in normal mice. The accumulation of positively charged liposomes containing 1,2-dimyristyloxypropyl-3-dimethyl-hydroxyethyl bromide was different from that of neutral and negatively charged dicetyl phosphate liposomes. The agglutinability and serum protein binding to positively charged liposomes were marked, suggesting that these factors affect the high accumulation of 1,2-dimyristoyloxypropyl-3-dimethyl-hydroxyethyl ammonium bromide liposomes in the liver. Noninvasive PET analysis of liposomal trafficking is beneficial for obtaining information about liposomal drug delivery, and long-circulating liposomes might be useful for diagnostic tumor imaging by PET. Other studies include the imaging of the myocardium using coenzyme Q10 entrapped in liposomes (Ishiwata et al., 1985), tumor imaging (Cattel et al., 2003; Oku et al., 1995; Zijlstra et al., 2003), liver imaging (Oku, 1996; Oku et al., 1996a), blood pool imaging (Oku et al., 1996b), and infection imaging (Spyridonidis and Markou, 2005).

25.2.5 Liposomes as Carriers for Contrast Agents Used in Single Photon Emission Computed Tomography

Liposomes are infrequently used for SPECT imaging (Goins and Phillips, 2001; Hamoudeh et al., 2008). In a study by Khalifa et al. (1997), phospholipid liposomes were labeled with 74 MBq ^{111}In and injected intravenously. The distribution of radioactive material in the brain was imaged using a dedicated neuro-SPECT imager. Images were taken 1 h postinjection and repeated at 24, 48, and 72 h. In addition, whole-body images were obtained using a γ camera and blood was taken for radio-activity determination. At 72 h postinjection, excellent tumor demarcation was seen in seven of eight patients. The obtained images correlated well with the corresponding CT images. Blood radio-activity levels gradually declined over 72 h. Tumor uptake continued to rise throughout this time and, together with the steady fall in normal brain tissue, the tumor–brain contrast gradually increased (maximum 7:5). The images of the whole-body indicated that up to 50% of the injected dose was taken up by the liver. No toxicity was observed in injected liposomes. Although the total percentage of uptake was low (1.1%), the tumor/brain contrast ratios, together with the SPET images, suggest the potential for tumor-specific targeting. In a study by Chang et al. (2007), the biodistribution and pharmacokinetics of [188]Re-N,N-bis(2-mercaptoethyl)-N',N'-diethylethylenediamine ([188]Re-BMEDA)-labeled PEGylated liposomes (RBLPLs) and unencapsulated [188]Re-BMEDA were investigated in murine C26 colon tumor bearing mice after intravenous (i.v.) injection. Micro-SPECT and CT scans evaluated the distribution and tumor targeting potential of RBLPLs in mice. The results indicated the highest uptake of liposomes in tumors as well as a 7.1-fold higher tumor/muscle ratio of RBLPLs than that of [188]Re-BMEDA. These results suggested the potential benefit and advantage of [188]Re-labeled nanoliposomes for imaging and treatment of malignant diseases. Similar results were obtained by Chen et al. (2007) in their studies on RBLPLs as imaging agents in a C26 colon carcinoma ascites mouse model.

25.3 DENDRIMERS

Dendrimers are highly branched synthetic macromolecules with nanoscopic dimensions (2–10 nm) and tunable sizes consisting of a vast array of types, chemical structures, and functional groups (Bosman et al., 1999). Two types of dendrimers are commercially available: polypropyleneimine (PPI) polyamidoamine (PAMAM; Roberts et al., 1990) and PPI diaminobutane (DAB) dendrimers (Malik et al., 2000). Schering (Berlin, Germany) has also synthesized Gadomers that are a lysine-based class of dendritic agents. Gadomer-17® consists of 24 N-monosubstituted Gd(III)–DO3A moieties. There are two approaches for the synthesis of dendrimers: convergent and divergent. In the convergent approach, the dendritic wedges are synthesized first and subsequently attached to a multifunctional core; in the divergent approach, synthesis is started from a multifunctional core. The yield obtained by the convergent process is lower but the purity is higher than the divergent one (Langereis et al., 2007).

25.3.1 DENDRIMERS AS CARRIERS FOR CONTRAST AGENTS USED IN MAGNETIC RESONANCE

These dendrimers are highly water soluble and possess a unique surface topology of primary amine groups. Because of the defined structure and large number of surface amino groups available with these dendrimers, they can be used as substrates to attach large numbers of chelating agents for creating a macromolecular MR contrast agent or attachment of an antibody molecule (Barth et al., 1994; Kobayashi et al., 1999, 2000; Singh et al., 1994; Wiener et al., 1994). The possibility of attaching a discrete number of MRI labels and targeting units at well-defined positions within the same macromolecular structure is a unique feature of dendrimers that provides an opportunity to improve both the sensitivity and specificity of MRI (Langereis et al., 2007). There is abundant literature available for the applications of dendrimers as carriers for contrast agents in MRI. The topic has also been reviewed by the experts in the field (Fischer and Vogtle, 1999; Gajbhiye et al., 2007; Helms and Meijer, 2006; Jang and Kataoka, 2005; Kobayashi and Brechbiel, 2003, 2004, 2005; Langereis et al., 2007; Svenson and Tomalia, 2005; Venditto et al., 2005). Most of the literature is focused on the applications of dendrimers in the delivery of the low molecular weight (MW) Gd(III) complexes Gd(III)-DTPA and Gd(III)-DOTA. DOTA is a marginally larger molecule and is noteworthy for forming highly kinetically and thermodynamically stable metal complexes with lanthanides such as Gd(III) somewhat more than DTPA. However, DTPA generally has far more rapid complex formation rates that may or may not be an important factor in this application. When these chelates were formed with Gd(III), the differences in R1 relaxivity of both reagents were negligible (4.7 mM^{-1} s at 20 MHz and 25°C). By attaching these Gd complexes to dendrimers, the contrast is increased with a consequent reduction in the required dose (Langereis et al., 2007). The size of the dendrimer is dependent on its generation, the spacer used in the synthesis of the dendrimer, and the free groups on the dendrimer and charge. These properties affect the pharmacokinetic properties of the dendrimer thus deciding the quality of the organ to be imaged by MRI; for example, a sixth-generation dendritic MRI contrast agent (MW = 139 kg mol^{-1}) displayed an R1 of 34 mM^{-1} s^{-1} (0.6 T, 20°C), which was 6 times higher than the R1 of Gd(III)-DTPA (MW = 0.55 kg mol^{-1}, R1 = 5.4 mM^{-1} s^{-1}; Wiener et al., 1994). A series of Gd(III)-DTPA-functionalized PPI dendrimers was reported by Kobayashi et al. (2003). They demonstrated that the R1 increased almost linearly with the MW of the dendrimer without reaching a plateau value, eventually resulting in an R1 value of 29 mM^{-1} s^{-1} (1.5 T, 20°C) for the fifth generation of the dendritic contrast agent. However, Langereis et al. (2004) proved that the linker between the Gd(III)-DTPA complex and the dendrimer has a large affect on the overall relaxivity. Recent *in vitro* studies have shown that amine-terminated PPI and PAMAM dendrimers are cytotoxic, in particular the higher generations of protonated (cationic) dendrimers [half-maximal inhibitory concentration for DAB-dendr $(NH_2)_{64}$ <5 µg mL^{-1}; Zinselmeyer et al., 2002]. However, PPI and PAMAM dendrimers functionalized with carboxylate end groups at the periphery are neither cytotoxic nor hemolytic up to a concentration of 2 mg mL^{-1}. This suggests that the overall toxicity of dendritic structures is strongly determined by the functionalities along the periphery. PAMAM dendrimers (up to the fifth generation) unmodified or modified with chemically inert surface moieties do not appear to be toxic in mice (Roberts et al., 1996). Furthermore, peptide-functionalized polylysine dendrimers were also found to be biocompatible (Sadler and Tam, 2002).

Several *in vivo* MRI studies have shown that the higher generations of dendritic MRI contrast agents, in contrast to low-MW Gd(III) chelates, remain in high concentrations in the bloodstream for longer periods of time. This results in improved visualization of vascular structures. Because high-MW contrast agents show little extravasation and intravascular retention, they are commonly referred to as blood pool agents whereas low-MW contrast agents are referred to as extravascular agents.

The pharmacokinetic parameters of various dendrimers prepared by Kobayashi and Brechbiel (2005) are summarized in Table 25.8.

TABLE 25.8
Pharmacokinetic Characteristics of Contrast Agent-Loaded Dendrimers

Property of Dendrimer	Important Finding
Gd-DTPA-terminated higher generation	Increased relaxivities (both R1 and R2) compared to Gd-DTPA
PAMAM, 7–12 nm, G5 versus G1	Prolonged blood residence times, blood contrast agent
PAMAM, 3–6 nm	Quickly excreted through the kidney, kidney contrast agent
PAMAM, 7–9 nm	Slowly excreted via the kidney
PAMAM, 10–14 nm	Minimal renal excretion
Hydrophobic variants, PPI DAB dendrimer	Accumulated in the liver, liver contrast agents
Larger hydrophilic agents	Lymphatic imaging
PAMAM, 3–5 nm	Quickly distributed in soft tissues
PAMAM, 8–10 nm	Minimal leakage from tumor vessels
PAMAM, 14–15 nm	Quickly taken up by RES

Note: G; generation.

25.3.1.1 Contrast Agent Loaded Dendrimers in Detection of Kidney Function

Kobayashi and Brechbiel (2005) changed the dendrimer core from PAMAM to a less hydrophilic one and the results indicated greater liver accumulation of the contrast agent. Therefore, these dendrimer-based macromolecules can also be used as liver contrast agents. After a thorough comparison of different generations of dendrimers that are suitable for imaging renal function, it was concluded that nanosized agents were nearly exclusively retained in the blood vessels or urinary tract with minimal perfusion into extravascular tissue. Thus, after filtering through the glomerulus, these agents were concentrated and the formation of a high-intensity band at the layer of the proximal tubules was observed. The absence or delayed formation of this high-intensity band correlated well to the renal tubular function in an acute tubular necrosis mice model generated by the injection of cisplatin or mechanical ischemia. In addition, the enhancement pattern was predictive of the pathogenesis of acute or chronic renal failure such as heavy metals, ischemia, obstruction, and sepsis.

25.3.1.2 Dendrimers in the Detection of Sentinel Lymph Nodes and Lymphatics

In a study by Kobayashi et al. (2006), dendrimers carrying Gd-labeled contrast agents with diameters of less than 1–12 nm were tested to determine the size that provides fast and efficient delivery of the contrast agent to the lymph nodes in a mouse model bearing lymphatic metastases. A comparison was made of PAMAM-G2, PAMAM-G4, PAMAM-G6, PAMAM-G8, and DAB-G5 Gd-dendrimer agents in particular and Gadomer-17 and Gd-DTPA. Of these, the G6 Gd-dendrimer showed the highest concentration in lymphatics and lymph nodes. The peak concentration was achieved at 24–36 min postinjection.

25.3.1.3 Contrast Agent Loaded Dendrimers in Blood Pool Imaging

The best dendrimers for blood pool imaging proved to be 11–13 nm in size. PEG conjugation along with this dendrimer size range further enhances the efficacy of imaging. The coinjection of lysine with a G4-based dendrimer accelerates renal excretion of the contrast agent, thus reducing its toxicity. The avidin chase system is similar (Kobayashi and Brechbiel, 2005).

25.3.1.4 Contrast Agent Loaded Dendrimers in Liver Imaging

The dendrimer core has to be hydrophobic for the dendrimer-loaded contrast agent to be used for liver imaging purposes. Hence, a DAB core is preferred over PAMAM (Kobayashi and Brechbiel, 2005).

25.3.1.5 Contrast Agent Loaded Dendrimers in Tumor Imaging

Efficient tumor imaging is possible with dendrimers by attaching a large number of GD(III) atoms or tumor-specific ligands and antibodies (Sipkins et al., 1998) or ligands (Wiener et al., 1997) or by using an avidin biotin system (Kobayashi et al., 2001; Xu et al., 2007; Zhu et al., 2008).

25.3.2 DENDRIMERS AS CARRIERS FOR CONTRAST AGENTS USED IN COMPUTED TOMOGRAPHY

The use of dendrimers carrying contrast agents for use in CT is a relatively new concept, and there are only a few recent literature reports. In the first study by Yordanov et al. (2002), water-soluble iodinated dendritic nanoparticles of G-4-3-[(N',N'-dimethylaminoacetyl)amino]-α-ethyl-2,4,6-tri-iodobenzenepropanoic acid (DMAA-IPA)$_{37}$ were prepared and characterized. Another study by the same group (Yordanov et al., 2005) reported the synthesis and characterization of a new water-soluble triiodo amino acid, DMAA-IPA, and its Starburst PAMAM generation 4.0 dendrimer conjugate, G-4-(DMAA-IPA)$_{37}$. In an interesting study by Fu et al. (2006), dendritic-iodinated contrast agents with PEG cores were synthesized for CT imaging. This study showed the synthetic feasibility, desired basic characteristics, and potential utility for CT contrast enhancement that is achieved with a new type of iodinated, large molecular PEG-core dendritic construct.

25.4 POLYMERIC NANOPARTICLES

These are solid colloidal particles ranging in size from 1 to 1000 nm and consisting of various biocompatible polymeric matrices to which a therapeutic moiety can be adsorbed, entrapped, or covalently attached (Lockman et al., 2002). Biodegradable and biocompatible synthetic polymers such as poly(D,L-lactide-co-glycolide) and polyalkylcyanoacrylates are generally preferred for obtaining nanoparticles. However, polysaccharides such as curdlan and macromolecules such as chitosan, albumin, and gelatin have been very well described in the literature for fabrication of nanoparticles (Covreur and Vauthier, 2006; Lockman et al., 2002). Because of their particulate nature, polymeric nanoparticles are rapidly cleared by cells of the mononuclear phagocyte system after i.v. injection. Moreover, similar to liposomes, the size, surface properties, composition, concentration, and hydrophilicity or hydrophobicity of nanoparticles play a major role in their *in vivo* performance.

25.4.1 POLYMERIC NANOPARTICLES AS CARRIERS FOR CONTRAST AGENTS USED IN γ SCINTIGRAPHY

The application of polymeric nanoparticles for delivery of γ-scintigraphic agents has not been exploited to its full potential thus far. In a study by Douglas and Davis (1986), poly(butyl 2-cyano-acrylate) nanoparticles were radiolabeled with a ^{99m}Tc–dextran complex with a labeling efficiency of 18%. The *in vitro* studies suggested that the radiolabeled nanoparticles degraded slowly and the release was unaffected by the presence of plasma protein, which indicates their utility in *in vivo* research. When injected intravenously in rabbits, more than 60% of the particles accumulated in the liver and spleen region whereas about 30% stayed in the blood (Douglas et al., 1987). Polycyanoacrylate nanoparticles radiolabeled with iodine (^{125}I or ^{131}I), indium, or technetium showed high labeling yields of >80%. The *in vivo* studies revealed that 60–75% of them accumulated in the RES (Ghanem et al., 1993). The influence of the labeling procedure on the biodistribution of ^{99m}Tc-radiolabeled chitosan nanoparticles was determined by Banerjee et al. (2005). This was observed when Tc(IV) was reduced by sodium borohydride and not by stannous chloride during the radiolabeling process, and the particles circulated a long time. Self-assembled nanoparticles of amphiphilic chitosan derivatives radiolabeled with ^{131}I accumulated in tumor tissues through the EPR effect (Cho et al., 2008).

25.4.2 POLYMERIC NANOPARTICLES AS CARRIERS FOR CONTRAST AGENTS USED IN MAGNETIC RESONANCE

Polymer-coated magnetite nanoparticles were prepared by Zaitsev et al. (1999). They found that inclusion of the magnetite particle into a hydrophilic polymeric shell increased the stability of the dispersion. A core–shell morphology was the basis for the preparation of 120-nm diameter particles capable of imaging the heart and gastrointestinal tract (GIT) in a rat animal model. The shell consisted of a hydrophobic polymer that modulates the access to the core (Reynolds et al., 2000). Diblock copolymers of bicyclo[2.2.1]hept-5-ene 2-carboxylic acid 2-cyanoethyl ester and bicyclo[2.2.1]hept-2-ene, consisting of both anchoring and steric stabilizing blocks, were prepared by Belfield and Li (2006) by ring-opening metathesis polymerization. Cyanoester groups were incorporated into the norbornene polymers to chelate and stabilize the iron oxide magnetic nanoparticles. These nanostructures were stable. Multifunctional polymeric nanoparticles consisting of a surface-localized tumor vasculature targeting the F3 peptide and encapsulated photodynamic therapy and imaging agents were prepared by Reddy et al. (2006). Significant contrast enhancement was achieved in rat 9L gliomas. An increase in the survival rate was also observed when the rats were treated with photodynamic therapy.

25.4.3 POLYMERIC NANOPARTICLES AS CARRIERS FOR CONTRAST AGENTS USED IN COMPUTED TOMOGRAPHY

Radiopaque-iodinated 30- to 350-nm polymeric nanoparticles were prepared by emulsion polymerization of the monomer 2-methacryloyloxyethyl (2,3,5-triiodobenzoate) in the presence of sodium dodecyl sulfate as a surfactant and potassium persulfate as an initiator. In dog models, the CT scans revealed enhanced lymph node, liver, kidney, and spleen imaging (Galperin et al., 2007). Polymer-coated bismuth sulfide nanoparticles had a higher biological half-life (Rabin et al., 2006), and lower volumes were required for contrast imaging compared to the conventional agents.

25.4.4 POLYMERIC NANOPARTICLES AS CARRIERS FOR CONTRAST AGENTS USED IN POSITRON EMISSION TOMOGRAPHY

A core–shell structure containing nanoparticles was prepared with PEG in the outer shell and an inner hydrophilic shell bearing reactive functional groups and a central hydrophobic core (Fukukawa et al., 2008). PET imaging showed that increasing the PEG shell thickness of the nanoparticles increased blood circulation and decreased accumulation in excretory organs. Similar results were obtained by using poly(methyl methacrylate-*co*-methacryloxysuccinimide-*graft*-PEG) copolymers that form poly(methyl methacrylate)-core/PEG-shell nanoparticles upon hydrophobic collapse in water (Pressly et al., 2007).

25.5 LIPID NANOPARTICLES

Lipid nanoparticles, namely, solid lipid nanoparticles (SLNs) and nanostructured lipid carriers (NLCs), evolved to overcome the disadvantages of polymeric nanoparticles and to maintain the advantages of liposomes such as the solid physical state of the polymeric nanoparticles and biodegradability, biocompatibility, and ease of manufacture of lipid emulsions (Wissing et al., 2004). The lipid nanoparticles described here are SLNs. SLNs are colloidal particles of a lipid matrix that is solid at body temperature. They were first introduced by Müller and Lucks (1993, published as a German patent) and produced by high-pressure homogenization and by Gasco (1993) by diluting a warm microemulsion. NLCs are composed of a binary mixture of a solid lipid and a spatially different liquid lipid as the carrier. The major advantage of NLCs is the increased drug load.

25.5.1 Lipid Nanoparticles as Carriers for Contrast Agents Used in Imaging

Magnetite was the first imaging agent that was incorporated in an SLN (Müller et al., 1996). The cytotoxicity of this magnetite-loaded SLN was compared with magnetite-loaded polylactide/glycolide particles to determine the toxicological acceptance as an i.v. formulation for MRI and as a potential carrier for drug targeting. The magnetite-loaded SLNs were the least cytotoxic with an effective concentration (50% effective dose) above 10% whereas the polymeric nanoparticles were in the range of 0.15–0.38%. However, no *in vivo* studies were carried out to complement this observation.

In vivo studies incorporating iron oxide in SLNs (Peira et al., 2003) were carried out in rats. These iron oxide nanoparticles had relaxometric properties similar to Endorem®. *In vivo* MRI of the central nervous system with both SLN and Endorem showed that supermagnetic SLNs have slower blood clearance than Endorem. The retention in the central nervous system continued till the end of the experimental observation period of 135 min. These observations also boosted other findings like blood–brain targeting by SLNs. Another *in vivo* study (Ballot et al., 2006) using a radiopharmaceutical, ^{99m}Tc/[188]Re-labeled lipid nanocapsules, in rats revealed its long circulation time. These lipid nanocapsules showed a biodistribution close to those of classical PEG-coated particles and good stability of the [188]Re/^{99m}Tc-SSS (name of the chelating complex) labeling. Radiolabeled SLNs intravenously injected in rats indicated that SLN formulations remain in the blood in contrast to many other colloidal drug carriers (Weyhers et al., 2006). The body distribution of circulating SLNs that rapidly release the marker is generally similar to the administered reference at later time points after the injection. Luminescent lipophilic CdSe/ZnS core–shell QDs were encapsulated into SLNs to prepare fluorescent nanocomposite particles (Liu et al., 2008). The fluorescence measurements showed that the encapsulated QDs maintain their high fluorescence and narrow or symmetric emission spectra. Assembling many QDs in a single nanocomposite particle increases the fluorescence signal and the signal/background ratio compared to individual QDs. In addition, the QD-loaded SLN was stable and slow to photobleach. These QDs were also biocompatible with fluorescence stability and had good potential in biological imaging applications.

25.6 MAGNETIC NANOPARTICLES

Superparamagnetic nanoparticles consisting of an inorganic core of iron oxide (magnetite, Fe_3O_4), maghemite, or other insoluble ferrites coated or uncoated with polymers such as dextran, PEG, poly(ethylene oxide) (PEO), poloxamer, and polaxamines are used as contrast agents. These are classified according to their size, which can affect their plasma half-life, and biodistribution into the following categories:

- Superparamagnetic iron oxides (SPIOs) with sizes greater than 50 nm (coating included)
- Ultrasmall SPIOs (USPIOs) or long-circulating dextran-coated iron oxide particles with sizes less than 50 nm

The study by Weissleder et al. reported on highly lymphotropic superparamagnetic nanoparticles (Combidex®). They gain access to lymph nodes by means of interstitial–lymphatic fluid transport. These were used in conjunction with high-resolution MRI.

Examples of marketed contrast agents are summarized in Table 25.9.

25.6.1 Ultrasmall Superparamagnetic Iron Oxide in Magnetic Resonance Imaging for Lymph Node Metastases Detection

Highly lymphotropic superparamagnetic nanoparticles (Combidex size = 30 nm) can reveal small nodal metastases in humans that are otherwise undetectable with conventional MRI and only detectable via biopsy.

TABLE 25.9
Examples of Marketed Contrast Agents

Brand Name	Generic Name	Type	Admin. Route	Composition	Coating	Size	Application	Developer	Remarks
Lumirem®, GastroMark®	Ferumoxsil	SPIO	Oral	Iron oxide particles	Silicon	300 nm	GIT imaging	Advanced Magnetics, Cambridge, USA	In Europe *Lumirem®* is approved for rectal administration to delineate the lower intestinal system.
Endorem®, Feridex®	Ferumoxide	SPIO	i.v.	Magnetite nanoparticles	Dextran	150 nm	Liver–spleen diseases	Advanced Magnetics, Cambridge, MA	—
Resovist®	Ferrixan (ferucarbotran)	SPIO	i.v. bolus	Iron oxide particles	Carboxydextran	—	Liver lesions	Schering, Berlin, Germany	Phase III in U.S., for sale in Europe and Japan
Sinerem®, Combidex®	Ferumoxtran	USPIO	i.v.	Magnetite nanoparticles	Dextran	30 nm	Brain tumor imaging, angiography, liver diseases, lymph nodes	Advanced Magnetics, Cambridge, MA	Sinerem® is in phase III of clinical trials.
Gadomer-17	(Gd-DTPA)-17, 24 cascade polymer	SPIO	i.v.	Gadomer-17, 24	—	—	MR angiography and tumor differentiation	Schering AG	Dendrimer
Abdoscan®	Ferristene	SPIO	Oral	Fe^{2+}/Fe^{3+}	—	300 nm	Bowel marking	Amersham	—
MION-46	Monocrystalline iron oxide nanoparticles	USPIO	s.c., i.v., intra-arterial	Iron oxide particles	Dextran	20 nm	Lymphography		—

25.6.1.1 Mechanism of Action of Lymphotropic Superparamagnetic Iron Oxide

After systemic injection, the model lymphotropic superparamagnetic nanoparticles gain access to the interstitium through fenestrations and are drained through lymphatic vessels. Disturbances in lymph flow or nodal architecture caused by metastases lead to abnormal patterns of accumulation of lymphotropic superparamagnetic nanoparticles, which are detectable by MRI (Figure 25.4). Harisinghani et al. (2003) found that lymph node metastases can be accurately diagnosed by high-resolution MRI with lymphotropic superparamagnetic nanoparticles but not by conventional MRI alone. Even very small metastases that are less than 2 mm in diameter can be identified within normal-sized lymph nodes. Such microscopic tumor deposits are below the threshold of detection of any other imaging technique. (For comparison, the limit of PET detection of tumor deposits in the pelvis is often 6–10 mm. See Figure 25.5; Harisinghani et al., 2003.)

25.7 FERROFLUIDS

Ferrofluids are colloidal solutions of iron oxide magnetic nanoparticles surrounded by a polymeric layer coated with affinity molecules, such as antibodies, for capturing cells and other biological targets from blood or other fluid and tissue samples. Ferrofluid particles are so small (25–100 nm in radius) that their behavior in liquids is as a solution rather than a suspension. When the coated ferrofluid particles are mixed with a sample containing cells or other analytes, they interact intimately and completely. These properties enabled the development of specialized reagents and systems with extremely high sensitivity and efficiency (Sahoo and Labhasetwar, 2003).

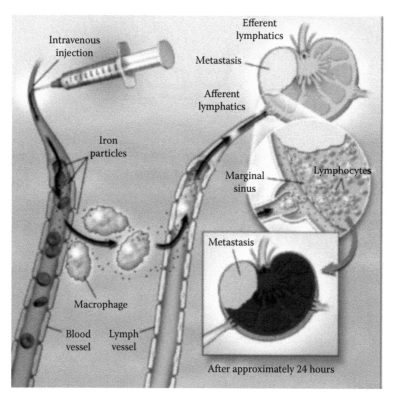

FIGURE 25.4 The mechanism of action of a lymphotropic SPIO. (Adapted from Harisinghani, M. G., Barentsz, J., Hahn, P. F., et al. 2003. *N Engl J Med* 348: 2491–2499. With permission.)

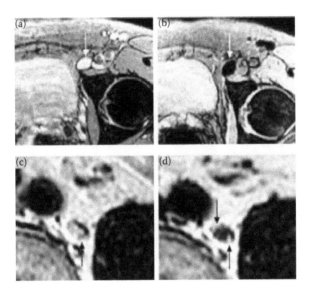

FIGURE 25.5 (a) A conventional MRI obtained 24 h after the administration of lymphotropic superparamagnetic nanoparticles. (b) The homogeneous decrease in signal intensity due to the accumulation of lymphotropic superparamagnetic nanoparticles in a normal lymph node in the left iliac region (arrow). (c) A conventional MRI shows high signal intensity in a retroperitoneal node with micrometastases (arrow). (d) MRI with lymphotropic superparamagnetic nanoparticles demonstrates two hyperintense foci (arrows) within the node, corresponding to 2-mm metastases. (Adapted from Harisinghani, M. G., Barentsz, J., Hahn, P. F. et al. 2003. *N Engl J Med* 348: 2491–2499. With permission.)

25.8 EMULSIONS

Different types of emulsions such as macroemulsions and microemulsions have been used for diagnostic imaging. Microemulsions are optically isotropic and thermodynamically stable mixtures of water, oil, and an amphiphile. The research groups of Lanza and Wickline are the most active groups in this field (Mulder et al., 2006b). Most of the applications of macro- or microemulsions have been in the field of perfluorocarbon (PFC) delivery (Mattrey, 1994) used for MRI and gas bubbles for ultrasound. The area is very well protected by patents.

25.8.1 EMULSIONS AS CARRIERS FOR CONTRAST AGENTS USED IN MAGNETIC RESONANCE IMAGING

The emulsions of PFC are often targeted and investigated for molecular imaging. Important potential target pathologies include inflammation, atherosclerosis, tumor-related angiogenesis, and thrombi. Detection and differentiation from normal tissue involves binding the ligands specific for the epitopes of interest onto the emulsion particles. The incorporation of a paramagnetic material, for example, a Gd complex, into the lipid monolayer of such targeted emulsion droplets provided contrast agents that are useful for both ultrasound and MRI modalities (Krafft et al., 2003). We discuss some specific examples below.

25.8.1.1 Emulsion-Encapsulated Perfluorocarbon Nanoparticles in Magnetic Resonance Imaging for Fibrin Detection

Antifibrin antibody-tagged biotinylated emulsion-encapsulated Gd-DTPA PFC nanoparticles were checked for clotting efficacy *in vitro* (Yu et al., 2000). There were numerous Gd-DTPA complexes on the surface of these emulsion-coated PFC nanoparticles. After binding to fibrin clots, scanning electron microscopy of the treated clots revealed dense accumulation of nanoparticles on the clot surfaces, as revealed in Figure 25.6.

0.5 μm

FIGURE 25.6 Scanning electron micrographs of clots. (a) An untreated clot showing a dense network of fibrin strands. (b) A treated clot. The clot was treated following the standard three-step incubation protocol described in the text. Fibrin strands were covered with numerous paramagnetic nanoparticles. (c) A partially blocked clot. (Adapted from Yu, X., Song, S. K., Chen, J. et al. 2000. *Magn Reson Med* 44: 867–872. With permission.)

A clot was pretreated with nonbiotinylated antifibrin antibody before the standard three-step treatment. As a result, a significant fraction of the fibrin-binding sites was unavailable for binding to the targeted contrast agent. Therefore, the amount of paramagnetic nanoparticles was significantly reduced compared to the treated clot.

A decrease in T_1 and T_2 (20–40%) relaxation times was also observed because of the abundance of Gd-DTPA complexes carried by each nanoparticle (Figure 25.7).

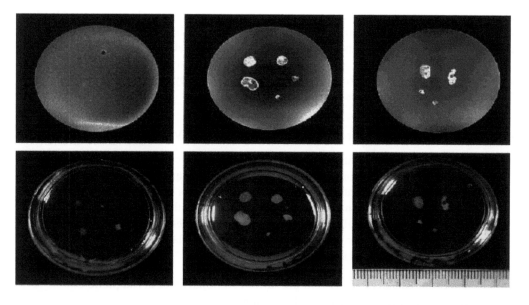

FIGURE 25.7 Scanning electron micrographs of clots showing T_1-weighted spin-echo images (upper panels) of clots embedded in agarose gel on a petri dish. Each small grid in the scalar bar represents 1 mm. Untreated clots (left) showed no contrast with agarose gel, but treated (center) and partially blocked (right) clots showed excellent contrast. A direct comparison with the corresponding optical images (lower panels) indicates that MR images can also accurately delineate the morphology of the clots. (Adapted from Yu, X., Song, S. K., Chen, J. et al. 2000. *Magn Reson Med* 44: 867–872. With permission.)

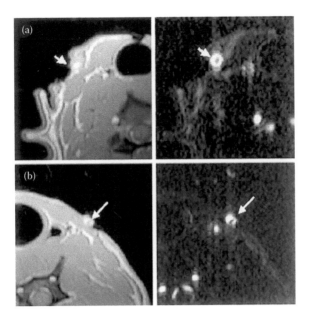

FIGURE 25.8 (a) Thrombi in an external jugular vein targeted with fibrin-specific paramagnetic nanoparticles demonstrating dramatic T_1-weighted contrast enhancement in the gradient-echo image (arrow) on the left with the flow deficit (arrow) of the thrombus in a corresponding phase-contrast image on the right (3-D phase-contrast angiogram). (b) A control thrombus in the contralateral external jugular vein imaged as in (a). (a, left) Thrombi in the external jugular vein targeted with fibrin-specific paramagnetic nanoparticles demonstrating dramatic T_1-weighted contrast enhancement in the gradient-echo image (arrow). (a, right) A corresponding 3-D phase-contrast angiogram with the flow deficit (arrow) of the thrombus. (b, left) The control thrombus in the contralateral external jugular vein and (b, right) a corresponding 3-D phase-contrast angiogram. (Adapted from Flacke, S., Fischer, S., Scott, M. J., Fuhrhop, R. J., Allen, J. S., Mclean, M., Winter, P., et al. 2001. *Circulation* 104: 1280–1285.)

Clots as small as 0.5 mm were detected with enhanced resolution and reduced T_1 and T_2 with or without treatment with a targeted contrast agent, which proves their utility in imaging artheroscle-rotic plaques (Figure 25.8; Yu et al., 2000).

25.8.1.2 Emulsion-Encapsulated Perfluorocarbon Nanoparticles in Magnetic Resonance Imaging for Fibrin Detection *In Vivo* Studies

A similar emulsion was prepared that contained nanoparticles of Gd-DTPA-bis-oleate (Flacke et al., 2001). Nanoparticles were present as a thin layer over the clot surface; this was confirmed by higher-resolution scans and scanning electron microscopy. *In vivo* contrast enhancement under open-circulation conditions was assessed in dogs. There was a significant enhancement in the contrast/noise ratio between the targeted clot (20 mol% Gd-DTPA nanoparticles) and blood in comparison to the targeted clot and the control clot. These results suggest that molecular imaging of fibrin-targeted paramagnetic nanoparticles provides sensitive detection and localization of fibrin. This in turn allows early direct identification of vulnerable plaques, leading to early thera-peutic decisions.

Other examples of emulsions utilized for MRI are provided in Table 25.10.

25.8.2 Emulsions as Carriers for Contrast Agents Used in Computed Tomography

Emulsification of the imaging agent considerably reduces its dose required for imaging compared to the conventional agent, for example, AG 60.99, an emulsion of poppy seed oil. This emulsion was

TABLE 25.10
Emulsions as Carriers for Contrast Agents Used in MRI

Emulsion + Agent	Application	Reference
PFC	Whole-body imaging of oxygen	Clark et al. (1984)
FTBA	Vascular system	Joseph et al. (1985a)
PFC	Vascular system	Joseph et al. (1985b)
Fluorine	Liver, tumor, and abscess	Longmaid Iii et al. (1985)
Fluorine/perfluorooctylbromide	RES	Ratner et al. (1987), Noth et al. (1995)
CEA antibody conjugated with perfluorochemical FTBA	Tumor	Shimizu et al. (1987)
PFC	Liver	Busse et al. (1988)
PFC	Blood oxygenation in the brain	Eidelberg et al. (1988)
Perfluorooctylbromide	Damage of spleen	Ratner et al. (1988)
PFC emulsions (oxypherol and fluosol-DA)	Tumors	Mason et al. (1989)
Oxypherol, a perfluorinated blood substitute comprising an emulsion of 25% (w/v) FTBA	Blood flow to tumor	Ceckler et al. (1990)
Combinations of ferric ammonium citrate, ferrous sulfate, Gd-DPTA	Retroperitoneal structure	Li et al. (1990)
FMIQ emulsion coupled with anti-CEA antibody and FTPA	Tumor	Mishima et al. (1991)
FTPA	Liver and spleen	Meyer et al. (1992)
FTBA	Vascular volume and changes in vascular volume with growth	Meyer et al. (1993)
PFC	Tumor	Itoh et al. (1995)
Water-in-PFC emulsions	Lungs	Fan et al. (2006), Huang et al. (2002, 2004)

Note: PFC—perflurocarbon; FTBA—Perfluorotributylamine; FTPA—perfluorotripropylamine; CEA—carcinoembryonic antigen; FMIQ—perfluoromethyldecahydroisoquinoline.

found to be specific for CT imaging of the liver and spleen. The emulsion opacified the hepatic parenchyma well enough to enhance visualization of the bile ducts (Alfidi and Laval, 1976). Similar results were obtained when this emulsion was tested in monkeys (Vermess et al., 1977) as well as clinically (Vermess et al., 1981). An ethiodized oil emulsion (EOE-13) even detected tumors in the liver, spleen, colon, and rectum (Reed et al., 1986) and small lesions, which were undetectable on the preliminary CT scan (Vermess et al., 1979). The emulsion was able to selectively accumulate in the liver and spleen (Vermess et al., 1982b). The agents were also found to be promising in clinical trials (Vermess et al., 1980). They detected lesions of less than 1-cm diameter that were present in the liver and spleen (Vermess et al., 1982a). An emulsion of ethyl monoiodostearate (compound 208E) had an effect similar to that of EOE-13 as a liver- and spleen-specific imaging agent (Miller et al., 1983a). The EOE-based CT examination was 2 times more sensitive than scintigraphy for detection of small hepatic metastases (Miller et al., 1983b).

Fat emulsions have also been used with CT for examination of bladder (Ahlberg et al., 1981) and liver (Kawata et al., 1984) tumors. An oral emulsion of corn oil was found to be good for evaluation of the stomach, duodenum, and pancreas; in patients suspected of having solid tumors; and in thin people (Raptopoulos et al., 1987).

Emulsions of perfluoroctylbromide were found to be useful in detecting V2 carcinoma in rabbits and were also found to be reticuloendothelial specific (Mattrey et al., 1982). However, their toxic dose must be elucidated (Young et al., 1981). These emulsions were very effective clinically (Thomas et al., 1982). In a study by Lewis et al. (1982), 23 oncologic patients were demonstrated to have hepatic metastases. These emulsions were also effective in imaging heart and vascular structures in periods from minutes to hours (Mattrey et al., 1984).

Emulsions of Ethiodol® and Pluronic® were utilized to image the aorta, vena cava, pulmonary, and femoral vessels and distorted vessels around tumors for 20 min after infusion (Cassel et al., 1982).

When used in combination with CT, perflubron was demonstrated for imaging lymph nodes and thus can be helpful in staging cancer. The preclinical and clinical data proved that the perflubron emulsion is safe and can enhance axillary lymph nodes on CT images following injections in the hand (Hanna et al., 1994). Rapid (10 min) and long-term (5 days) contrast was obtained with an emulsion of 1-bromoperfluorooctane and perfluoro[1-(4-methylcyclohexyl) piperidine] stabilized with egg yolk phospholipids. Different organs such as the liver, spleen, adrenals, heart, and abdominal aorta were imaged. The higher concentration of the emulsion in the circulation gave rise to the rapid detection, whereas the long-term contrast was due to the uptake of the emulsion in the RES (Sklifas et al., 2007).

25.8.3 EMULSIONS AS CARRIERS FOR CONTRAST AGENTS USED IN ULTRASONOGRAPHY

A fluosol emulsion consisting of perfluorodecalin and perfluorotripropylamine (FTPA) was tested for its liver-specific ultrasound contrast. The echogenicity of the rabbit liver was increased as compared to the controls receiving Ringer's solution (Mattrey et al., 1983). The utility of fluosol as an ultrasound contrast agent for contrast enhancement of the liver, spleen, and tumors was evaluated clinically (Mattrey et al., 1987).

Lipidol emulsions were also used as a contrast agent for visualization of the spleen and liver (Yoshida et al., 1984). However, contrasting results were also reported denying the use of lipid emulsions as a hepatic contrast agent (Fink et al., 1985). Detection of swelling and normal lymph nodes surrounding the upper GIT was also demonstrated using a 10% oil-in-water type emulsion that was fed orally (Aibe et al., 1986). Acute venous thrombi were detectable when using PFC emulsions administered intravenously (Coley et al., 1994). PFC emulsions that are biotinylated, lipid-coated, and targeted were used for the diagnosis of thrombi. Avidin-induced aggregation was utilized to produce marked enhancement of backscatter, and the method was highly specific and sensitive (Lanza et al., 1996, 1997). EchoGen® (perflenapent) emulsion enhanced the signal up to 20 min, thus making the detection of renal artery stenosis and intrarenal vascular disorders, characterization of some indeterminate renal masses, imaging of the liver and vasculature (Robbin and Eisenfeld, 1998), and renal transplant assessment possible (Correas et al., 1997a). Similar results were obtained with a perflenapent emulsion (Correas et al., 1997b). These emulsions were well tolerated in patients undergoing echocardiography and ultrasound of other target organs (Quay and Eisenfeld, 1997).

25.9 POLYMERIC MICELLES

Polymeric micelles emerged as an alternative to liposomes because of the size restriction on them; that is, they can be prepared in smaller sizes than liposomes. Polymeric micelles are self-assembled aggregates of block polymers with 20- to 100-nm sizes (Trubetskoy, 1999). Contrast agents are attached to the polymeric micelles via covalent or noncovalent bonds. Polymeric micelles offer numerous advantages as a carrier for contrast agents such as increased water solubility, enhanced permeability, altered biodistribution, increased bioavailability, and reduced side effects. These polymeric micelles have a hydrophobic core and hydrophilic corona. The physicochemical properties such as size, charge, and surface properties can be easily changed by adding new ingredients to the amphiphilic mixture at the time of micelle preparation or by changing the method of production of micelles (Torchilin, 1999). Chelated radioactive metals such as ^{111}In and ^{99m}Tc are used for scintigraphy, whereas chelated paramagnetic metals such as Gd for MRI and iodine for CT are the most commonly studied contrast agents with polymeric micelles. Many reviews have been published dealing with polymeric micelles for the delivery of contrast agents (Torchilin, 1999, 2000, 2002; Trubetskoy,

1999). The three most common applications of such polymeric micelles are blood pool imaging because of their long-circulating nature, tumor accumulation because of circulation longevity, and lymph node detection after subcutaneous (s.c.) injection. These can be applied in different types and stages of cancer detection and the diagnosis of certain heart conditions (Trubetskoy, 1999).

25.9.1 POLYMERIC MICELLES AS CARRIERS FOR CONTRAST AGENTS USED IN γ SCINTIGRAPHY

Chelated radioactive metals such as ^{111}In and ^{99m}Tc for scintigraphy have been associated with polymeric micelles. PEO-based polymers have been demonstrated to form 10- to 50-nm polymeric micelles. Percutaneous lymphography has been performed using these polymeric micelles (Trubetskoy et al., 1996). Polymer–lipid amphiphilic compounds such as diacyl lipid-PEG have been used as polymeric micelles to incorporate ^{111}In. These ^{111}In-labeled PEG-PE micelles were found to have a longer circulation half-life. Because of their small size, imaging of lymph nodes was also possible (Torchilin, 2001).

25.9.2 POLYMERIC MICELLES AS CARRIERS FOR CONTRAST AGENTS USED IN MAGNETIC RESONANCE IMAGING

Chelated contrast agents such as Gd, Mn, or Dy aqueous ions are of major interest for delivery with polymeric micelles. Different methods and polymers are used to prepare the polymeric micelles. PE-PEO was used to prepare 10- to 30-nm polymeric micelles. These micelles targeted the lymphatic system upon percutaneous administration as determined by MRI (Trubetskoy and Torchilin, 1995). Magnetite particles encapsulated inside the hydrophobic core of a polymeric micelle of a poly(ε-caprolactone)-b-PEG (PCL-b-PEG) copolymer, whose surface was stabilized by a PEG shell, were reported (Ai et al., 2005). These micelles were able to achieve an ultrasensitive MRI detection limit of 5.2 g L^{-1}. The polymeric micelles containing iron accumulated in the tumor by the EPR effect, making it possible to visualize the tumors. The tumor images were negatively enhanced as early as 1 h after the injection, which was not the case with a 10 times higher dose of Feridex®. Similar results were obtained when polymeric micelles encapsulating ferric hydroxide particles were used. There was also a report of pH-responsive polymeric micelles of PEG-b-poly(methacrylic acid), with calcium phosphate crystals entrapping Gd-DTPA. Polymeric micelles formed from cationic polymers (polyallylamine or protamine) and anionic block copolymers [PEG-b-poly(aspartic acid) derivative] that bound Gd ions provided high contrasts in MRI by shortening the longitudinal relaxation time of protons of water. They were found to be highly selective to image tumor tissue (Nakamura et al., 2006). Polymeric micelles made targetable to αV3 integrins using cyclic Arg-Gly-Asp peptide on the surface and loaded with SPIO nanoparticles were able to image cancerous cells specifically (Nasongkla et al., 2006). Folate-coated SPIO Fe$_3$O$_4$ contained in an amphiphilic block copolymer of PEG and PCL attached to the distal ends of PEG (folate-PEG-PCL) was prepared by Hong et al. (2008a, 2008b). The system was demonstrated as targeting hepatic tumors and therefore can be used for their imaging.

25.9.3 POLYMERIC MICELLES AS CARRIERS FOR CONTRAST AGENTS USED IN COMPUTED TOMOGRAPHY

Iodine encapsulated into polymeric micelles has been used as a contrast agent for CT to attain high iodine loads. Trubetskoy et al. (1997) used hydrophobization of poly-L-lysine (PLL) for the synthesis of a core-forming block. They used unpolar tribenzoic acid N-hydroxysuccinimide ester for modification of the PLL block in the PEO-b-PLL polymer. The 100-nm polymeric micelles that were created were able to successfully micellize in aqueous solution, and stable micelles were produced incorporating 45% iodine. These micelles were demonstrated for imaging lymph on percutaneous injection in rabbits and as blood pool imaging agents in rats and rabbits. Subcutaneous injection of these iodine-containing micelles in the rabbit hind paw allowed the detection of popliteal lymph nodes within 2 h after injection. Upon i.v. injection the blood pool, liver, and spleen were opacified.

The circulation time was enhanced with half of the dose staying in the circulation 24 h after administration. Liver and blood vessels were also detectable during the first 0.5 h after administration.

25.10 QUANTUM DOTS (NANOCRYSTALS, QUANTUM DOTS, AND NANODOTS)

QDs have been used as an alternative to conventional fluorescent markers, which have the following disadvantages:

- Requirement of color matched lasers
- Fluorescence bleaching
- Lack of discriminatory capacity of multiple dyes

QDs are crystalline clumps of a few hundred atoms, which are coated with an insulating outer shell of a different material (Figure 25.9). They can be attached to biological materials, such as cells, proteins, and nucleic acids. QDs are so bright that it is possible to detect a cell carrying a single crystal. They are inorganic and thus very stable. Their inert surface coating makes them less toxic than organic dyes. They can be used *in vitro* as well as *in vivo* (Gao et al., 2005).

QDs are semiconductor nanoparticles with a selenide core and a zinc sulfide shell. They are extremely small in size, and the electrons are very compact. They emit light when in contact with

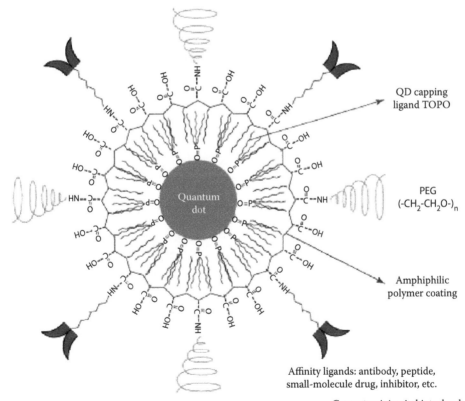

QD capping ligand TOPO

PEG (-CH$_2$-CH$_2$-O-)$_n$

Amphiphilic polymer coating

Affinity ligands: antibody, peptide, small-molecule drug, inhibitor, etc.

Current opinion in biotechnology

FIGURE 25.9 A schematic representation of QDs. (Adapted from Gao, X., Yang, L., Petros, J., Marshall, F. J. S., and Nie, S. 2005. *Curr Opin Biotechnol* 16: 63–72. With permission.)

infrared light. QDs may emit light in 10 different colors, and their color changes with size. The polymer structure can be used to finely tune color.

25.10.1   QUANTUM DOTS FOR CANCER DETECTION

QDs have great potential in the detection of cancer and are more effective than conventional fluorescent dyes and fluorescent proteins. In an interesting study, Gao et al. (2004) successfully demonstrated the ability of QDs entrapped in block polymers for imaging of prostate cancers growing in nude mice. QDs were accumulated in tumor cells by the EPR effect and because of the presence of antibodies attached to the QDs (Figure 25.10). Thus, it is possible to achieve ultrasensitive and multicolor fluorescence imaging of cancer cells under *in vivo* conditions.

25.10.2   QUANTUM DOTS IN SENTINEL LYMPH NODE IMAGING

The potential of QDs in imaging of sentinel lymph nodes was successfully established by Kim et al. (2004). They designed near-infrared (NIR) type II QDs with an oligomeric phosphine coating. Interestingly, mapping of sentinel lymph nodes as deep as 1 cm was possible after injection of a very small amount of NIR QDs (400 pmol; Figure 25.11). Furthermore, the imaging was possible using a very small fluorescence excitation dose of intensity 5 mW cm^{-2}.

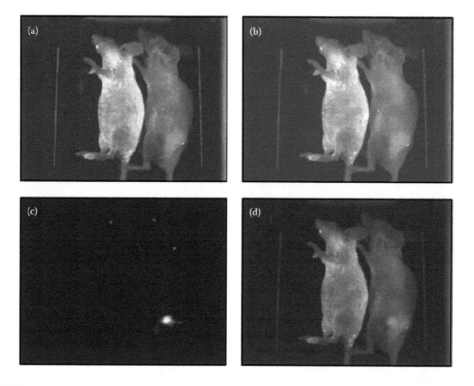

FIGURE 25.10   Spectral imaging of QD–prostate-specific membrane antigen antibody conjugates in live animals harboring C4-2 tumor xenografts. (Right) Orange-red fluorescence signals indicate a prostate tumor growing in a live mouse. (Left) Control studies using a healthy mouse (no tumor) and the same amount of QD injection showing no localized fluorescence signals. (a) Original image, (b) unmixed autofluorescence image, (c) unmixed QD image, and (d) superimposed image. After *in vivo* imaging, histological and immunocytochemical examinations confirmed that the QD signals came from an underlying tumor. Note that QDs in deep organs such as the liver and spleen were not detected because of the limited penetration depth of visible light. (Adapted from Gao, X., Cui, Y., Levenson, R. M. et al. *Nat Biotechnol* 22: 969–976. With permission.)

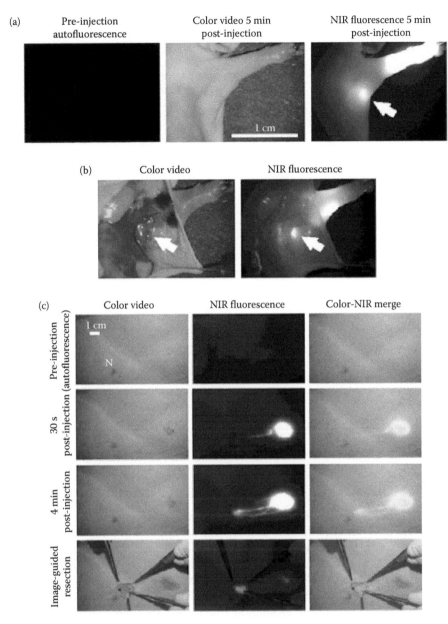

FIGURE 25.11 The NIR QD sentinel lymph node mapping in the mouse and pig. (a) Images of mouse injected intradermally with 10 pmol of NIR QDs in the left paw. (Left) Preinjection NIR autofluorescence image, (middle) 5-min postinjection white light color video image, and (right) 5-min postinjection NIR fluorescence image. An arrow indicates the putative axillary sentinel lymph node. Fluorescence images have identical exposure times and normalization. (b) Images of the mouse shown 5 min after reinjection with 1% isosulfan blue and exposure of the actual sentinel lymph node. (Left) A color video and (right) NIR fluorescence images. Isosulfan blue and NIR QDs were localized in the same lymph node (arrows). (c) Images of the surgical field in a pig injected intradermally with 400 pmol of NIR QDs in the right groin. Four time points are shown from top to bottom: before injection (autofluorescence), 30 s after injection, 4 min after injection, and during image-guided resection. For each time point, (left) color video, (middle) NIR fluorescence, and (right) color NIR merge images are shown. Fluorescence images have identical exposure times and normalization. To create the merged image, the NIR fluorescence image was pseudocolored lime green and superimposed on the color video image. The position of the nipple (N) is indicated. (Adapted from Kim, S., Taik, Y. L., Soltesz, E. G.,et al. 2004. *Nat Biotechnol* 22: 93–97. With permission.)

TABLE 25.11

Nanocarriers in Clinical Development

Carrier	Stage of Development	Technique	Contrast Agent, Imaging Agent, or Radiolabeled	Limitations of Use
Liposomes	Preclinical	SPECT	99 mTc	Preparation steps have to be carefully controlled to achieve reproducible properties such as size and entrapment efficiency.
		MRI	Gadolinium	
		PET	([2–18 F)FDG	
QDs or nanocrystals	Preclinical	Optical or fluorescence	QDs	Further safety studies are required because QDs are very stable.
			QDs micelles	
			QDs conjugates	
Magnetic nanoparticles	Clinical	MRI	Iron oxide-dextran	Toxicity may occur due to cellular internalization and membrane disruption.
	Preclinical		Iron oxide-polyacrylamide	
	Cellular		Iron oxide-SLN	
			Iron oxide-insulin	
Dendrimers	Preclinical	MRI	Gadolinium	Positive charge on dendrimer surface may lead to toxicity and immunogenicity

25.11 CONCLUSION

Nanocarriers hold the promise of delivering a technological breakthrough. They are moving quite quickly from concept to reality in the field of diagnostic applications as evidenced from a number of nanocarriers in clinical development (Table 25.11). The advantage of nanocarriers is their flexibility of modification or adaptation to meet the needs of pathological conditions for exact diagnostic applications. Targeted imaging with MRI and ultrasound has great potential as a tool for elucidating the molecular mechanisms responsible for disease. The noninvasiveness of this approach, in conjunction with the ability to visualize anatomical information with these modalities, creates an opportunity for serial characterization of disease progress and for monitoring drug delivery and treatment efficacy in one step.

ABBREVIATIONS

CT	Computed tomography
DAB	Diaminobutane
DMAA-IPA	3-[(N′,N′-Dimethylaminoacetyl)amino]-α-ethyl-2,4,6-triiodobenzenepropanoic acid
DOTA	Tetraazacyclododecanetetraacetic acid
DPDP	N,N'-dipyridoxylethylenediamine-N,N'-diacetate-5,5'-bis(phosphate)
DTPA	Diethylenetriaminepentaacetic acid
EOE	Ethiodized oil emulsion

EPR	Enhanced permeation and retention
FTPA	Perfluorotripropylamine
^{67}Ga	Gallium 67
GIT	Gastrointestinal tract
HAb18	Hepatocarcinoma-specific monoclonal antibody
HMPAO	Hexamethylpropyleneamine oxime
HP-DO3A	1,4,7-tris(carboxymethyl)-10-(2'-hydroxypropyl)-1,4,7,10-tetraazacycl ododecane
HYNIC	Hydrazine nicotinamide
IgG	Immunoglobulin G
^{111}In	Indium 111
i.v.	Intravenous
LUV	Large unilamellar vesicle
MI	Myocardial infarction
MLV	Multilamellar vesicle
MRI	Magnetic resonance imaging
MW	Molecular weight
NGPE	*N*-Glutaryl-phosphatidyl ethanolamine
NIR	Near-infrared
NLC	Nanostructured lipid carrier
PAMAM	Polyamidoamine
PCL	Poly(ε-caprolactone)
PE	Phosphatidylethanolamine
PFC	Perfluorocarbon
PEG	Poly(ethylene glycol)
PEO	Poly(ethylene oxide)
PET	Positron emission tomography
PGlcUA	Palmityl glucuronide
PLL	Poly-L-lysine
PPI	Polypropyleneimine
QD	Quantum dot
RBLPL	^{188}Re-*N,N*-bis(2- mercaptoethyl)-*N',N'*-diethylethylenediamine-labeled PEGylated liposome
^{188}Re-BMEDA	^{188}Re-*N,N*-bis(2- mercaptoethyl)-*N',N'*-diethylethylenediamine
RES	Reticuloendothelial system
SPECT or SPET	Single photon emission computed tomography
s.c.	subcutaneous
SLNs	Solid lipid nanoparticle
SPIO	Superparamagnetic iron oxide
^{99m}Tc	Technetium 99 m
2-D	Two-dimensional
3-D	Three-dimensional
ULV	Unilamellar vesicle
US	Ultrasonography
USPIO	Ultrasmall superparamagnetic iron oxide
VIP	Vasoactive intestinal peptide

SYMBOLS

T_1 Longitudinal relaxation
T_2 Spin–spin relaxation

REFERENCES

Ahkong, Q. F. and Tilcock, C. 1992. Attachment of ^{99m}Tc to lipid vesicles containing the lipophilic chelate dipalmitoylphosphatidylethanolamine-DTTA. *Int J Rad Appl Instrum B* 19: 831–840.

Ahlberg, N. E., Berlin, T., and Calissendorff, B. 1981. Intravesical fat emulsion at computed tomography of bladder tumors. *Acta Radiol* 22: 645–647.

Ai, H., Flask, C., Weinberg, B., Shuai, X., Pagel, M. D., Farrell, D., Duerk, J., and Gao, J. 2005. Magnetite-loaded polymeric micelles as ultrasensitive magnetic-resonance probes. *Adv Mater* 17: 1949–1952.

Aibe, T., Ito, T., and Yoshida, T. 1986. Endoscopic ultrasonography of lymph nodes surrounding the upper GI tract. *Scand J Gastroenterol* S21: 164–169.

Alfidi, R. J. and Laval, J. M. 1976. AG 60.99: A promising contrast agent for computed tomography of the liver and spleen. *Radiology* 121: 491.

Al-Jamal, W. T. and Kostarelos, K. 2007. Liposome–nanoparticle hybrids for multimodal diagnostic and therapeutic applications. *Nanomedicine* 2: 85–98.

Awasthi, V., Goins, B., Klipper, R., Loredo, R., Korvick, D., and Phillips, W. T. 1998a. Imaging experimental osteomyelitis using radiolabeled liposomes. *J Nucl Med* 39: 1089–1094.

Awasthi, V. D., Goins, B., Klipper, R., and Phillips, W. T. 1998b. Dual radiolabeled liposomes: Biodistribution studies and localization of focal sites of infection in rats. *Nucl Med Biol* 25: 155–160.

Bacic, G., Niesman, M. R., Magin, R. L., and Swartz, H. M. 1990. NMR and ESR study of liposome delivery of Mn^{2+} to murine liver. *Magn Reson Med* 13: 44–61.

Ballot, S., Noiret, N., Hindre, F., Denizot, B., Garin, E., Rajerison, H., and Benoit, J. P. 2006. ^{99m}Tc/^{188}Re-labelled lipid nanocapsules as promising radiotracers for imaging and therapy: Formulation and biodistribution. *Eur J Nucl Med Mol Imaging* 33: 602–607.

Banerjee, T., Singh, A. K., Sharma, R. K., and Maitra, A. N. 2005. Labeling efficiency and biodistribution of technetium-99 m labeled nanoparticles: Interference by colloidal tin oxide particles. *Int J Pharm* 289: 189–195.

Barratt, G. M., Tuzel, N. S., and Ryman, B. E. 1984. The labelling of ^{99m}Tc liposomal membranes with radioactive technetium. In G. Gregoriadis (Ed.), *Liposome Technology*. Boca Raton, FL: CRC Press.

Barth, R. F., Adams, D. M., Soloway, A. H., Alam, F., and Darby, M. V. 1994. Boronated starburst dendrimer–monoclonal antibody immunoconjugates: Evaluation as a potential delivery system for neutron capture therapy. *Bioconj Chem* 5: 58–66.

Beaumier, P. L. and Hwang, K. J. 1982. An efficient method for loading indium-111 into liposomes using acetylacetone. *J Nucl Med* 23: 810–815.

Belfield, K. D. and Li, Z. 2006. Norbornene-functionalized diblock copolymers via ring-opening metathesis polymerization for magnetic nanoparticle stabilization. *Chem Mater* 18: 5929–5936.

Bertini, I., Bianchini, F., Calorini, L., Colagrande, S., Fragai, M., Franchi, A., Gallo, O., Gavazzi, C., and Luchinat, C. 2004. Persistent contrast enhancement by sterically stabilized paramagnetic liposomes in murine melanoma. *Magn Reson Med* 52: 669–672.

Bian, A. N., Gao, Y. H., Tan, K. B., Liu, P., Zeng, G. J., Zhang, X., and Liu, Z. 2004. Preparation of human hepatocellular carcinoma-targeted liposome microbubbles and their immunological properties. *World J Gastroenterol* 10: 3424–3427.

Bloch, S. H., Dayton, P. A., and Ferrara, K. W. 2004. Targeted imaging using ultrasound contrast agents. *IEEE Eng Med Biol* 23: 18–27.

Boerman, O. C., Dams, E. T. M., Oyen, W. J. G., Corstens, F. H. M., and Storm, G. 2001. Radiopharmaceuticals for scintigraphic imaging of infection and inflammation. *Inflamm Res* 50: 55–64.

Boerman, O. C., Laverman, P., Oyen, W. J. G., Corstens, F. H. M., and Storm, G. 2000. Radiolabeled liposomes for scintigraphic imaging. *Prog Lipid Res* 39: 461–475.

Boerman, O. C., Oyen, W. J. G., Corstens, F. H. M., and Storm, G. 1998. Liposomes for scintigraphic imaging: Optimization of *in vivo* behavior. *Q J Nucl Med* 42: 271–279.

Boerman, O. C., Oyen, W. J., Storm, G., Corvo, M. L., Van Bloois, L., Van Der Meer, J. W., and Corstens, F. H. 1997a. Technetium-99 m labelled liposomes to image experimental arthritis. *Ann Rheum Dis* 56: 369–373.

Boerman, O. C., Oyen, W. J., Van Bloois, L., Koenders, E. B., Van Der Meer, J. W., Corstens, F. H., and Storm, G. 1997b. Optimization of technetium-99 m-labeled PEG liposomes to image focal infection: Effects of particle size and circulation time. *J Nucl Med* 38: 489–493.

Boerman, O. C., Storm, G., Oyen, W. J. G., Van Bloois, L., Van Der Meer, J. W. M., Claessens, R. A. M. J., Crommelin, D. J. A., and Corstens, F. H. M. 1995. Sterically stabilized liposomes labelled with indium-111 to image focal infection. *J Nucl Med* 36: 1639–1644.

Bogdanov, Jr., A. A., Weissleder, R., and Brady, T. J. 1995. Long-circulating blood pool imaging agent. *Adv Drug Deliv Rev* 16: 335–348.

Bosman, A. W., Janssen, H. M., and Meijer, E. W. 1999. About dendrimers: Structure, physical properties, and applications. *Chem Rev* 99: 1665–1688.

Briele, B., Hotze, A., Oehr, P., Biersack, H. J., Rosanowski, F., Gorgulla, W., Hertberhold, C., and Harlapp, J. P. 1990. Tumour imaging with labelled liposomes. *Lancet* 336: 875–876.

Brouwers, A. H., De Jong, D. J., Dams, E. T., Oyen, W. J., Boerman, O. C., Laverman, P., Naber, T. H., Storm, G., and Corstens, F. H. 2000a. Tc-99 m-PEG-liposomes for the evaluation of colitis in Crohn's disease. *J Drug Target* 8: 225–233.

Brouwers, A. H., Laverman, P., Boerman, O. C., Oyen, W. J., Barrett, J. A., Harris, T. D., Edwards, D. S., and Corstens, F. H. 2000b. A 99Tcm-labelled leukotriene B4 receptor antagonist for scintigraphic detection of infection in rabbits. *Nucl Med Commun* 21: 1043–1050.

Bulte, J. W. M., De Cuyper, M., Despres, D., and Frank, J. A. 1999. Short- vs. long-circulating magnetoliposomes as bone marrow-seeking MR contrast agents. *J Magn Reson Imaging* 9: 329–335.

Burke, S. J., Annapragada, A., Hoffman, E. A., Chen, E., Ghaghada, K. B., Sieren, J., and Van Beek, E. J. R. 2007. Imaging of pulmonary embolism and T-PA therapy effects using MDCT and liposomal iohexol blood pool agent. Preliminary results in a rabbit model. *Acad Radiol* 14: 355–362.

Busse, L. J., Pratt, R. G., and Thomas, S. R. 1988. Deconvolution of chemical shift spectra in two- or three-dimensional [19F] MR imaging. *J Comput Assisted Tomogr* 12: 824–835.

Capellier, M., Besson, P., and Coustaut, D. 1988. Radiopaque liposomes for computed tomography opacification. Paper presented at the IEEE Engineering in Medicine and Biology Society Annual Conference, New York.

Caride, V. J. 1981. Liposome for diagnostic imaging. In R. P. Spencer (Ed.), *Radiopharmaceuticals Structure—Activity Relationship*, pp. 321–330. New York: Grune & Stratton.

Caride, V. J. 1985. Liposomes as carriers of imaging agents. *Crit Rev Ther Drug Carrier Syst* 1: 121–153.

Caride, V. J. 1990. Technical and biological considerations on the use of radiolabeled liposomes for diagnostic imaging. *Int J Radiat Appl Inst Part B Nucl Med Biol* 17: 35–39.

Caride, V. J., Sostman, H. D., and Twickler, J. 1982. Brominated radiopaque liposomes: Contrast agent for computed tomography of liver and spleen. A preliminary report. *Invest Radiol* 17: 381–385.

Caride, V. J. and Zaret, B. L. 1977. Liposome accumulation in regions of experimental myocardial infarction. *Science* 198: 735–738.

Cassel, D. M., Young, S. W., and Brody, W. R. 1982. Radiographic blood pool contrast agents for vascular and tumor imaging with projection radiography and computed tomography. *J Comput Assisted Tomogr* 6: 141–146.

Cattel, L., Ceruti, M., and Dosio, F. 2003. From conventional to stealth liposomes: A new frontier in cancer chemotherapy. *Tumor* 89: 237–249.

Ceckler, T. L., Gibson, S. L., Hilf, R., and Bryant, R. G. 1990. *In situ* assessment of tumor vascularity using fluorine NMR imaging. *Magn Reson Med* 13: 416–433.

Chang, Y. J., Chang, C. H., Chang, T. J., Yu, C. Y., Chen, L. C., Jan, M. L., Luo, T. Y., Lee, T. W., and Ting, G. 2007. Biodistribution, pharmacokinetics and microSPECT/CT imaging of [188]Re-BMEDA liposome in a C26 murine colon carcinoma solid tumor animal model. *Anticancer Res* 27: 2217–2225.

Chen, L. C., Chang, C. H., Yu, C. Y., Chang, Y. J., Hsu, W. C., Ho, C. L., Yeh, C. H., Luo, T. Y., Lee, T. W., and Ting, G. 2007. Biodistribution, pharmacokinetics and imaging of [188]Re-BMEDA-labeled PEGylated liposomes after intraperitoneal injection in a C26 colon carcinoma ascites mouse model. *Nucl Med Biol* 34: 415–423.

Cho, Y. W., Kim, Y. S., Kim, I. S., Park, R. W., Oh, S. J., Moon, D. H., Kim, S. Y., and Kwon, I. C. 2008. Tumoral accumulation of long-circulating, self-assembled nanoparticles and its visualization by gamma scintigraphy. *Macromol Res* 16: 15–20.

Chu, W. J., Simor, T., and Elgavish, G. A. 1997. *In vivo* characterization of Gd(BME-DTTA), a myocardial MRI contrast agent: Tissue distribution of its MRI intensity enhancement, and its effect on heart function. *NMR Biomed* 10: 87–92.

Clark, Jr., L. C., Ackerman, J. L., Thomas, S. R., Millard, R. W., Hoffman, R. E., Pratt, R. G., Ragle-Cole, H., Kinsey, R. A., and Janakiraman, R. 1984. Perfluorinated organic liquids and emulsions as biocompatible NMR imaging agents for 19F and dissolved oxygen. *Adv Exp Med Biol* 180: 835–845.

Coley, B. D., Trambert, M. A., and Mattrey, R. F. 1994. Perfluorocarbon-enhanced sonography: Value in detecting acute venous thrombosis in rabbits. *Am J Roentgenol* 163: 961–964.

Correas, J. M., Helenon, O., Menassa, L., Mejean, A., Boyer, J. C., and Moreau, J. F. 1997b. Potential of perflenapent emulsion, an ultrasound contrast agent, in the management of urinary tract disease. *Ultrasound Med Biol* 23: S138.

Correas, J. M., Helenon, O., and Moreau, J. F. 1997a. Current assessment of renal ultrasound and the future role of contrast enhancement. *Appl Radiol* 26: 20–24.

Covreur, P. and Vauthier, C. 2006. Nanotechnology: Intelligent design to treat complex disease. *Pharm Res* 23: 1417–1450.

Dagar, S., Rubinstein, I., Önyüksel, H., and Nejat, D. 2003. Liposomes in ultrasound and gamma scintigraphic imaging. *Methods Enzymol* 373: 198–214.

Dagar, S., Sekosan, M., Lee, B. S., Rubinstein, I., and Onyuksel, H. 2001. VIP receptors as molecular targets of breast cancer: Implications for targeted imaging and drug delivery. *J Control Release* 74: 129–134.

Dams, E. T., Laverman, P., Oyen, W. J., Storm, G., Scherphof, G. L., Van Der Meer, J. W., Corstens, F. H., and Boerman, O. C. 2000a. Accelerated blood clearance and altered biodistribution of repeated injections of sterically stabilized liposomes. *J Pharmacol Exp Ther* 292: 1071–1079.

Dams, E. T., Nijhof, M. W., Boerman, O. C., Laverman, P., Storm, G., Buma, P., Lemmens, J. A., Van Der Meer, J. W., Corstens, F. H., and Oyen, W. J. 2000b. Scintigraphic evaluation of experimental chronic osteomyelitis. *J Nucl Med* 41: 896–902.

Dams, E. T., Reijnen, M. M., Oyen, W. J., Boerman, O. C., Laverman, P., Storm, G., Van Der Meer, J. W., Corstens, F. H., and Van Goor, H. 1999. Imaging experimental intraabdominal abscesses with ^{99m}Tc-PEG liposomes and ^{99m}Tc-HYNIC IgG. *Ann Surg* 229: 551–557.

Dams, E. T. M., Oyen, W. J. G., Boerman, O. C., Laverman, P., Storm, G., and Meer, J. W. V. D. 1998. Effect of multiple injections on pharmacokinetics and biodistribution of sterically stabilized liposomes. *J Nucl Med* 5(Suppl.): 242.

Dayton, P. A. and Ferrara, K. W. 2002. Targeted imaging using ultrasound. *J Magn Reson Imaging* 16: 362–377.

De Cuyper, M., Soenen, S. J. H., Coenegrachts, K., and Beek, L. T. 2007. Surface functionalization of magnetoliposomes in view of improving iron oxide-based magnetic resonance imaging contrast agents: Anchoring of gadolinium ions to a lipophilic chelate. *Anal Biochem* 367: 266–273.

Delli Castelli, D., Gianolio, E., Geninatti Crich, S., Terreno, E., and Aime, S. 2008. Metal containing nanosized systems for MR-molecular imaging applications. *Coord Chem Rev* 252: 2424–2443.

Demos, S. M., Alkan-Onyuksel, H., Kane, B. J., Ramani, K., Nagaraj, A., Greene, R., Klegerman, M., and Mcpherson, D. D. 1999. *In vivo* targeting of acoustically reflective liposomes for intravascular and transvascular ultrasonic enhancement. *J Am Coll Cardiol* 33: 867–875.

Demos, S. M., Dagar, S., Klegerman, M., Nagaraj, A., Mcpherson, D. D., and Onyuksel, H. 1998. *In vitro* targeting of acoustically reflective immunoliposomes to fibrin under various flow conditions. *J Drug Target* 5: 507–518.

Demos, S. M., Onyuksel, H., Gilbert, J., Roth, S. I., Kane, B., Jungblut, P., Pinto, J. V., Mcpherson, D. D., and Klegerman, M. E. 1997. *In vitro* targeting of antibody-conjugated echogenic liposomes for site-specific ultrasonic image enhancement. *J Pharm Sci* 86: 167–171.

Desser, T. S., Rubin, D. L., Muller, H., Mcintire, G. L., Bacon, E. R., and Toner, J. L. 1999. Blood pool and liver enhancement in CT with liposomal iodixanol: Comparison with iohexol. *Acad Radiol* 6: 176–183.

Dick, A., Adam, G., Tacke, J., Prescher, A., Southon, T. E., and Gunther, R. W. 1996. Computed tomography of experimental liver abscesses using a new liposomal contrast agent. *Invest Radiol* 31: 194–203.

Douglas, S. J. and Davis, S. S. 1986. Radiolabelling of poly(butyl 2-cyanoacrylate) nanoparticles with a technetium-99 m-dextran complex. *J Label Compd Radiopharm* 23: 495–504.

Douglas, S. J., Davis, S. S., and Illum, L. 1987. Biodistribution of poly(butyl 2-cyanoacrylate) nanoparticles in rabbits. *Int J Pharm* 34: 145–152.

Eidelberg, D., Johnson, G., Barnes, D., Tofts, P. S., Delpy, D., Plummer, D., and Mcdonald, W. I. 1988. 19F NMR imaging of blood oxygenation in the brain. *Magn Reson Med* 6: 344–352.

Erdogan, S., Medarova, Z. O., Roby, A., Moore, A., and Torchilin, V. P. 2008. Enhanced tumor MR imaging with gadolinium-loaded polychelating polymer-containing tumor-targeted liposomes. *J Magn Reson Imaging* 27: 574–580.

Erdogan, S., Ozer, A. Y., Ercan, M. T., Aydin, K., and Hincal, A. A. 1998. Biodistribution and computed tomography studies on iopromide liposomes. *STP Pharma Sci* 8: 133–137.

Erdogan, S., Roby, A., Sawant, R., Hurley, J., and Torchilin, V. P. 2006a. Gadolinium-loaded polychelating polymer-containing cancer cell-specific immunoliposomes. *J Liposome Res* 16: 45–55.

Erdogan, S., Roby, A., and Torchilin, V. P. 2006b. Enhanced tumor visualization by γ-scintigraphy with ^{111}In-labeled polychelating-polymer-containing immunoliposomes. *Mol Pharm* 3: 525–530.

Espinola, L. G., Beaucaire, J., Gottschalk, A., and Caride, V. J. 1979. Radiolabelled liposomes as metabolic and scanning tracers in mice. II. In-111 oxine compared with Tc-99 m DTPA, entrapped in multilamellar lipid vesicles. *J Nucl Med* 20: 434–440.

Essien, H. and Hwang, K. J. 1988. Preparation of liposomes entrapping a high specific activity of $^{111}In^{3+}$-bound inulin. *Biochim Biophys Acta Biomembr* 944: 329–336.

Faas, H., Schwizer, W., Feinle, C., Lengsfeld H., De Smidt, C., Boesiger, P., Fried, M., and Rades, T. 2001. Monitoring the intragastric distribution of a colloidal drug carrier model by magnetic resonance imaging 460. *Pharm Res* 18: 460–466.

Fan, X., River, J. N., Muresan, A. S., Popescu, C., Zamora, M., Culp, R. M., and Karczmar, G. S. 2006. MRI of perfluorocarbon emulsion kinetics in rodent mammary tumours. *Phys Med Biol* 51: 211–220.

Fink, I. J., Miller, D. L., and Shawker, T. H. 1985. Lipid emulsions as contrast agents for hepatic sonography: An experimental study in rabbits. *Ultrason Imaging* 7: 191–197.

Fischer, M. and Vogtle, F. 1999. Dendrimers: From design to application—A progress report. *Angew Chem* 38: 884–905.

Flacke, S., Fischer, S., Scott, M. J., Fuhrhop, R. J., Allen, J. S., Mclean, M., Winter, P., et al. 2001. Novel MRI contrast agent for molecular imaging of fibrin implications for detecting vulnerable plaques. *Circulation* 104: 1280–1285.

Fouillet, X., Tournier, H., Khan, H., Sabitha, S., Burkhardt, S., Terrier, F., and Schneider, M. 1995. Enhancement of computed tomography liver contrast using iomeprol-containing liposomes and detection of small liver tumors in rats. *Acad Radiol* 2: 576–583.

Fritz, T. A., Unger, E. C., Sutherland, G., and Sahn, D. 1997. Phase I clinical trials of MRX-115. A new ultrasound contrast agent. *Invest Radiol* 32: 735–740.

Fu, Y., Nitecki, D. E., Maltby, D., Simon, G. H., Berejnoi, K., Raatschen, H. J., Yeh, B. M., Shames, D. M., and Brasch, R. C. 2006. Dendritic iodinated contrast agents with PEG-cores for CT imaging: Synthesis and preliminary characterization. *Bioconj Chem* 17: 1043–1056.

Fukukawa, K. I., Rossin, R., Hagooly, A., Pressly, E. D., Hunt, J. N., Messmore, B. W., Wooley, K. L., Welch, M. J., and Hawker, C. J. 2008. Synthesis and characterization of core–shell star copolymers for *in vivo* PET imaging applications. *Biomacromolecules* 9: 1329–1339.

Gaal, J., Mezes, A., Siro, B., Varga, J., Galuska, L., Janoky, G., Garai, I., Bajnok, L., and Suranyi, P. 2002. ^{99m}Tc-HMPAO labelled leukocyte scintigraphy in patients with rheumatoid arthritis: A comparison with disease activity. *Nucl Med Commun* 23: 39–46.

Gabizon, A., Huberty, J., Straubinger, R. M., Price, D. C., and Papahadjopoulos, D. 1988. An improved method for *in vivo* tracing and imaging of liposomes using a gallium 67–deferoxamine complex. *J Liposome Res* 1: 123–135.

Gabizon, A., Price, D. C., Huberty, J., Bresalier, R. S., and Papahadjopoulos, D. 1990. Effect of liposome composition and other factors on the targeting of liposomes to experimental tumors: Biodistribution and imaging studies. *Cancer Res* 50: 6371–6378.

Gajbhiye, V., Kumar, P. V., Tekade, R. K., and Jain, N. K. 2007. Pharmaceutical and biomedical potential of PEGylated dendrimers. *Curr Pharm Des* 13: 415–429.

Galperin, A., Margel, D., Baniel, J., Dank, G., Biton, H., and Margel, S. 2007. Radiopaque iodinated polymeric nanoparticles for X-ray imaging applications. *Biomaterials* 28: 4461–4468.

Gao, X., Cui, Y., Levenson, R. M., Chung, L. W. K., and Nie, S. 2004. *In vivo* cancer targeting and imaging with semiconductor quantum dots. *Nat Biotechnol* 22: 969–976.

Gao, X., Yang, L., Petros, J., Marshall, F. J. S., and Nie, S. 2005. *In vivo* molecular and cellular imaging with quantum dots. *Curr Opin Biotechnol* 16: 63–72.

Gasco, M. R. 1993. Method for producing solid lipid microspheres having a narrow distribution. U.S. Patent 188,837.

Ghanem, G. E., Joubran, C., Arnould, R., Lejeune, F., and Fruhling, J. 1993. Labelled polycyanoacrylate nanoparticles for human *in vivo* use. *Appl Radiat Isotopes* 44: 1219–1224.

Goins, B., Awasthi, V. D., Klipper, R., and Phillips, W. T. 1998. Use of a technetium-99m (Tc-99m)-labeled biotin liposome/avidin system in a rabbit colitis model to improve early image detection. *J Nucl Med* 39: 125–126.

Goins, B., Klipper, R., Rudolph, A. S., Cliff, R. O., Blumhardt, R., and Phillips, W. T. 1993. Biodistribution and imaging studies of technetium-99m-labeled liposomes in rats with focal infection. *J Nucl Med* 34: 2160–2168.

Goins, B., Klipper, R., Rudolph, A. S., and Phillips, W. T. 1994. Use of technetium-99m-liposomes in tumor imaging. *J Nucl Med* 35: 1491–1498.

Goins, B. A. and Phillips, W. T. 2001. The use of scintigraphic imaging during liposome drug development. *Journal of Pharmacy Practice* 14: 397—406.

Goto, R., Kubo, H., and Okada, S. 1989. Liposomes prepared from synthetic amphiphiles. I. Their technetium labeling and stability. *Chem Pharm Bull* 37: 1351–1354.

Hamoudeh, M., Kamleh, M. A., Diab, R., and Fessi, H. 2008. Radionuclides delivery systems for nuclear imaging and radiotherapy of cancer. *Adv Drug Deliv Rev* 60: 1329–1346.

Hanna, G., Saewert, D., Shorr, J., Flaim, K., Leese, P., Kopperman, M., and Wolf, G. 1994. Preclinical and clinical studies on lymph node imaging using perflubron emulsion. *Artif Cells Blood Substit Immobil Biotechnol* 22: 1429–1439.

Hardy, J. G. and Wilson, C. G. 1981. Radionuclide imaging in pharmaceutical, physiological and pharmacological research. *Clin Phys Physiol Meas* 2: 71–121.

Harisinghani, M. G., Barentsz, J., Hahn, P. F., Deserno, W. M., Tabatabaei, S., Van De Kaa, C. H., De La Rosette, J., and Weissleder, R. 2003. Noninvasive detection of clinically occult lymph-node metastases in prostate cancer. *N Engl J Med* 348: 2491–2499.

Harrington, K. J., Rowlinson-Busza, G., Syrigos, K. N., Abra, R. M., Uster, P. S., Peters, A. M., and Stewart, J. S. W. 2000a. Influence of tumour size on uptake of ^{111}In-DTPA-labelled PEGylated liposomes in a human tumour xenograft model. *Br J Cancer* 83: 684–688.

Harrington, K. J., Rowlinson-Busza, G., Syrigos, K. N., Uster, P. S., Abra, R. M., and Stewart, J. S. W. 2000b. Biodistribution and pharmacokinetics of ^{111}In-DTPA-labelled PEGylated liposomes in a human tumour xenograft model: Implications for novel targeting strategies. *Br J Cancer* 83: 232–238.

Havron, A., Seltzer, S. E., Davis, M. A., and Shulkin, P. 1981. Radiopaque liposomes: A promising new contrast material for computed tomography of the spleen. *Radiology* 140: 507–511.

Helms, B. and Meijer, E. W. 2006. Dendrimers at work. *Science* 313: 929–930.

Henze, A., Freise, J., Magerstedt, P., and Majewski, A. 1989. Radio-opaque liposomes for the improved visualisation of focal liver disease by computerized tomography. *Comput Med Imaging Graph* 13: 455–462.

Hiller, K. H., Waller, C., Nabrendorf, M., Bauer, W. R., and Jakob, P. M. 2006. Assessment of cardiovascular apoptosis in the isolated rat heart by magnetic resonance molecular imaging. *Mol Imaging* 5: 115–121.

Hnatowich, D. J., Friedman, B., Clancy, B., and Novak, M. 1981. Labeling of preformed liposomes with Ga-67 and Tc-99m by chelation. *J Nucl Med* 22: 810–814.

Hong, G., Yuan, R., Liang, B., Shen, J., Yang, X., and Shuai, X. 2008a. Folate-functionalized polymeric micelle as hepatic carcinoma-targeted, MRI-ultrasensitive delivery system of antitumor drugs. *Biomed Microdevices* 10: 693–700.

Hong, G. B., Zhou, J. X., Shen, J., Yuan, R. X., Shuai, X. T., and Liang, B. L. 2008b. Preparation of a folate-mediated tumor targeting ultraparamagnetic polymeric micelles and its *in vitro* experimental study. *Chin J Radiol* 42: 19–23.

Huang, M. Q., Basse, P. H., Yang, Q., Horner, J. A., Hichens, T. K., and Ho, C. 2004. MRI detection of tumor in mouse lung using partial liquid ventilation with a perfluorocarbon-in-water emulsion. *J Magn Reson Imaging* 22: 645–652.

Huang, M. Q., Ye, Q., Williams, D. S., and Ho, C. 2002. MRI of lungs using partial liquid ventilation with water-in-perfluorocarbon emulsions. *Magn Reson Med* 48: 487–492.

Hwang, K. J., Merriam, J. E., Beaumier, P. L., and Luk, K. S. 1982. Encapsulation, with high efficiency, of radioactive metal ions in liposomes. *Biochim Biophys Acta* 716: 101–109.

Ishiwata, K., Miura, Y., and Takahashi, T. 1985. 11C-Coenzyme Q10: A new myocardial imaging tracer for positron emission tomography. *Eur J Nucl Med* 11: 162–165.

Itoh, Y. H., Kandori, Y., Takahashi, M., Fritz-Zieroth, B., and Miyata, N. 1995. 19F-MR imaging of transplanted tumor: Perfluorochemicals as a fluorine tumor imaging agent. *Nippon Acta Radiol* 55: 345–347.

Jang, W. D. and Kataoka, K. 2005. Bioinspired applications of functional dendrimers. *J Drug Deliv Sci Technol* 15: 19–30.

Jendrasiak, G. L., Frey, G. D., and Heim, Jr., R. C. 1985. Liposomes as carriers of iodolipid radiocontrast agents for CT scanning of the liver. *Invest Radiol* 20: 995–1002.

Joseph, P. M., Fishman, J. E., Mukherji, B., and Sloviter, H. A. 1985a. *In vivo* 19F NMR imaging of the cardiovascular system. *J Comput Assisted Tomogr* 9: 1012–1019.

Joseph, P. M., Yuasa, J., and Kundel, H. L. 1985b. Magnetic resonance imaging of fluorine in rats infused with artificial blood. *Invest Radiol* 20: 504–509.

Kabalka, G. W., Davis, M. A., Holmberg, E., Maruyama, K., and Huang, L. 1991. Gadolinium-labeled liposomes containing amphiphilic Gd-DTPA derivatives of varying chain length: Targeted MRI contrast enhancement agents for the liver. *J Magn Reson Imaging* 9: 373–377.

Kamaly, N., Kalber, T., Ahmad, A., Oliver, M. H., So, P. W., Herlihy, A. H., Bell, J. D., Jorgensen, M. R., and Miller, A. D. 2008. Bimodal paramagnetic and fluorescent liposomes for cellular and tumor magnetic resonance imaging. *Bioconj Chem* 19: 118–129.

Kao, C. Y., Hoffman, E. A., Beck, K. C., Bellamkonda, R. V., and Annapragada, A. V. 2003. Long-residence-time nano-scale liposomal iohexol for x-ray-based blood pool imaging. *Acad Radiol* 10: 475–483.

Kassis, A. I. and Taube, R. A. 1987. Efficient radiolabeling of mammalian cells using ^{111}In-tagged liposomes. *Nucl Med Biol* 14: 33–35.

Kawata, R., Sakata, K., and Kunieda, T. 1984. Quantitative evaluation of fatty liver by computed tomography in rabbits. *Am J Roentgenol* 142: 741–746.

Khalifa, A., Dodds, D., Rampling, R., Paterson, J., and Murray, T. 1997. Liposomal distribution in malignant glioma: Possibilities for therapy. *Nucl Med Commun* 18: 17–23.

Kim, S., Taik, Y. L., Soltesz, E. G., De Grand, A. M., Lee, J., Nakayama, A., Parker, J. A. et al. 2004. Near-infrared fluorescent type II quantum dots for sentinel lymph node mapping. *Nat Biotechnol* 22: 93–97.

Kimura, A., Sakai, A., Tsukishiro, S. I., Beppu, S., and Fujiwara, H. 1998. Preparation and characterization of echogenic liposome as an ultrasound contrast agent: Size-dependency and stabilizing effect of cholesterol on the echogenicity of gas-entrapping liposome. *Chem Pharm Bull* 46: 1493–1496.

Kobayashi, H. and Brechbiel, M. W. 2003. Dendrimer-based macromolecular MRI contrast agents: Characteristics and application. *Mol Imaging* 2: 1–10.

Kobayashi, H. and Brechbiel, M. W. 2004. Dendrimer-based nanosized MRI contrast agents. *Curr Pharm Biotechnol* 5: 539–549.

Kobayashi, H. and Brechbiel, M. W. 2005. Nano-sized MRI contrast agents with dendrimer cores. *Adv Drug Deliv Rev* 57: 2271–2286.

Kobayashi, H., Kawamoto, S., Bernardo, M., Brechbiel, M. W., Knopp, M. V., and Choyke, P. L. 2006. Delivery of gadolinium-labeled nanoparticles to the sentinel lymph node: Comparison of the sentinel node visualization and estimations of intra-nodal gadolinium concentration by magnetic resonance imaging. *J Control Release* 111: 343–351.

Kobayashi, H., Kawamoto, S., Jo, S. K., Bryant, Jr., H. L., Brechbiel, M. W., and Star, R. A. 2003. Macromolecular MRI contrast agents with small dendrimers: Pharmacokinetic differences between sizes and cores. *Bioconj Chem* 14: 388–394.

Kobayashi, H., Kawamoto, S., Saga, T., Sato, N., Ishimori, T., Konishi, J., Ono, K., Togashi, K., and Brechbiel, M. W. 2001. Avidin-dendrimer-(1B4M-Gd)254: A tumor-targeting therapeutic agent for gadolinium neutron capture therapy of intraperitoneal disseminated tumor which can be monitored by MRI. *Bioconj Chem* 12: 587–593.

Kobayashi, H., Sato, N., Saga, T., Nakamoto, Y., Ishimori, T., Toyama, S., Togashi, K., Konishi, J., and Brechbiel, M. W. 2000. Monoclonal antibody–dendrimer conjugates enable radiolabeling of antibody with markedly high specific activity with minimal loss of immunoreactivity. *Eur J Nucl Med* 27: 1334–1339.

Kobayashi, H., Wu, C., Kim, M. K., Paik, C. H., Carrasquillo, J. A., and Brechbiel, M. W. 1999. Evaluation of the *in vivo* biodistribution of indium-111 and yttrium-88 labeled dendrimer-1B4M-DTPA and its conjugation with anti-Tac monoclonal antibody. *Bioconj Chem* 10: 103–111.

Kofi Adzamli, I., Seltzer, S. E., Slifkin, M., Blau, M., and Adams, D. F. 1990. Production and characterization of improved liposomes containing radiographic contrast media. *Invest Radiol* 25: 1217–1223.

Krafft, M. P., Chittofrati, A., and Reiss, J. G. 2003. Emulsions and microemulsions with a fluorocarbon phase. *Cur Opin Colloid Interface Sci* 8: 251–258.

Krause, W. 1999. Delivery of diagnostic agents in computed tomography. *Adv Drug Deliv Rev* 37: 159–173.

Krause, W., Leike, J., Sachse, A., and Schuhmann-Giampieri, G. 1993. Characterization of iopromide liposomes. *Invest Radiol* 28: 1028–1032.

Krause, W., Leike, J., Schuhmann-Giampieri, G., Sachse, A., Schmiedl, U., and Strunk, H. 1996. Iopromide-carrying liposomes as a contrast agent for the liver. *Acad Radiol* 3: S235–S237.

Kunimasa, J., Inui, K., Hori, R., Kawamura, Y., and Endo, K. 1992. Mannan-coated liposome delivery of gadolinium-diethylenetriaminepentaacetic acid, a contrast agent for use in magnetic resonance imaging. *Chem Pharm Bull* 40: 2565–2567.

Langereis, S., Delussanet, Q. G., Vangenderen, M. H. P., Backes, W. H., and Meijer, E. W. 2004. Multivalent contrast agents based on gadolinium-diethylenetriaminepentaacetic acid-terminated poly(propylene imine) dendrimers for magnetic resonance imaging. *Macromolecules* 37: 3084–3091.

Langereis, S., Dirksen, A., Hackeng, T. M., Van Genderen, M. H. P., and Meijer, E. W. 2007. Dendrimers and magnetic resonance imaging. *New J Chem* 31: 1152–1160.

Lanza, G. M., Wallace, K. D., Fischer, S. E., Christy, D. H., Scott, M. J., Trousil, R. L., Cacheris, W. P., Miller, J. G., Gaffney, P. J., and Wickline, S. A. 1997. High-frequency ultrasonic detection of thrombi with a targeted contrast system. *Ultrasound Med Biol* 23: 863–870.

Lanza, G. M., Wallace, K. D., Scott, M. J., Cacheris, W. P., Abendschein, D. R., Christy, D. H., Sharkey, A. M., Miller, J. G., Gaffney, P. J., and Wickline, S. A. 1996. A novel site-targeted ultrasonic contrast agent with broad biomedical application. *Circulation* 94: 3334–3340.

Laverman, P., Boerman, O. C., Storm, G., and Nejat, D. 2003. Radiolabeling of liposomes for scintigraphic imaging. *Methods Enzymol* 373: 234–248.

Laverman, P., Brouwers, A. H., Dams, E. T., Oyen, W. J., Storm, G., Van Rooijen, N., Corstens, F. H., and Boerman, O. C. 2000a. Preclinical and clinical evidence for disappearance of long-circulating characteristics of polyethylene glycol liposomes at low lipid dose. *J Pharmacol Exp Ther* 293: 996–1001.

Laverman, P., Dams, E. T., Oyen, W. J., Storm, G., Koenders, E. B., Prevost, R., Van Der Meer, J. W., Corstens, F. H., and Boerman, O. C. 1999. A novel method to label liposomes with ^{99m}Tc by the hydrazino nicotinyl derivative. *J Nucl Med* 40: 192–197.

Laverman, P., Zalipsky, S., Oyen, W. J., Dams, E. T., Storm, G., Mullah, N., Corstens, F. H., and Boerman, O. C. 2000b. Improved imaging of infections by avidin-induced clearance of ^{99m}Tc-biotin-PEG liposomes. *J Nucl Med* 41: 912–918.

Leander, P. 1996. A new liposomal contrast medium for CT of the liver—An imaging study in a rabbit tumour model. *Acta Radiol* 37: 63–68.

Leander, P., Hoglund, P., Børseth, A., Kloster, Y., and Berg, A. 2001. A new liposomal liver-specific contrast agent for CT: First human phase-I clinical trial assessing efficacy and safety. *Eur Radiol* 11: 698–704.

Leike, J., Sachse, A., Ehritt, C., and Krause, W. 1996. Biodistribution and CT-imaging characteristics of iopromide-carrying liposomes in rats. *J Liposome Res* 6: 665–680.

Leike, J. U., Sachse, A., and Rupp, K. 2001. Characterization of continuously extruded iopromide-carrying liposomes for computed tomography blood-pool imaging. *Invest Radiol* 36: 303–308.

Lewis, E., Aufderheide, J. F., and Bernardino, M. E. 1982. CT detection of hepatic metastases with ethiodized oil emulsion 13. *J Comput Assisted Tomogr* 6: 1108–1114.

Li, K. C. P., Ang, P. G. P., Tart, R. P., Storm, B. L., Rolfes, R., and Ho-Tai, P. C. K. 1990. Paramagnetic oil emulsions as oral magnetic resonance imaging contrast agents. *J Magn Reson Imaging* 8: 589–598.

Liu, W., He, Z., Liang, J., Zhu, Y., Xu, H., and Yang, X. 2008. Preparation and characterization of novel fluorescent nanocomposite particles: CdSe/ZnS core–shell quantum dots loaded solid lipid nanoparticles. *J Biomed Mater Res A* 84: 1018–1025.

Lockman, P., Mumper, R., Kahn, M., and Allen, D. 2002. Nanoparticle technology for drug delivery across the blood–brain barrier. *Drug Dev Ind Pharm* 28: 1–13.

Lokling, K. E., Fossheim, S. L., Klaveness, J., and Skurtveit, R. 2004a. Biodistribution of pH-responsive liposomes for MRI and a novel approach to improve the pH-responsiveness. *J Control Release* 98: 87–95.

Lokling, K. E., Skurtveit, R., Bjornerud, A., and Fossheim, S. L. 2004b. Novel pH-sensitive paramagnetic liposomes with improved MR properties. *Magn Reson Med* 51: 688–696.

Lokling, K. E., Skurtveit, R., Dyrstad, K., Klaveness, J., and Fossheim, S. L. 2004c. Tuning the MR properties of blood-stable pH-responsive paramagnetic liposomes. *Int J Pharm* 274: 75–83.

Longmaid Iii, H. E., Adams, D. F., and Neirinckx, R. D. 1985. *In vivo* 19F NMR imaging of liver, tumor, and abscess in rats. Preliminary results. *Invest Radiol* 20: 141–145.

Love, W. G., Amos, N., Kellaway, I. W., and Williams, B. D. 1989a. Specific accumulation of technetium-99m radiolabelled, negative liposomes in the inflamed paws of rats with adjuvant induced arthritis: Effect of liposome size. *Ann Rheum Dis* 48: 143–148.

Love, W. G., Amos, N., Williams, B. D., and Kellaway, I. W. 1989b. Effect of liposome surface charge on the stability of technetium (^{99m}Tc) radiolabelled liposomes. *J Microencapsulation* 6: 105–113.

Malik, N., Wiwattanapatapee, R., Klopsch, R., Lorenz, K., Frey, H., Weener, J. W., Meijer, E. W., Paulus, W., and Duncan, R. 2000. Dendrimers: Relationship between structure and biocompatibility *in vitro*, and preliminary studies on the biodistribution of 125I-labelled polyamidoamine dendrimers *in vivo*. *J Control Release* 65: 133–148.

Maresca, G., Summaria, V., Colagrande, C., Manfredi, R., and Calliada, F. 1998. New prospects for ultrasound contrast agents. *Eur J Radiol* 27: S171–S178.

Martina, M. S., Fortin, J. P., Menager, C., Clement, O., Barratt, G., Grabielle-Madelmont, C., Gazeau, F., Cabuil, V., and Lesieur, S. 2005. Generation of superparamagnetic liposomes revealed as highly efficient MRI contrast agents for *in vivo* imaging. *J Am Chem Soc* 127: 10676–10685.

Maruyama, K., Suzuki, R., Takizawa, T., Utoguchi, N., and Negishi, Y. 2007. Drug and gene delivery by "bubble liposomes" and ultrasound. *Yakugaku Zasshi* 127: 781–787.

Mason, R. P., Antich, P. P., Babcock, E. E., Gerberich, J. L., and Nunnally, R. L. 1989. Perfluorocarbon imaging *in vivo*: A 19F MRI study in tumor-bearing mice. *J Magn Reson Imaging* 7: 475–485.

Masotti, A., Mangiola, A., Sabatino, G., Maira, G., Denaro, L., Conti, F., Ortaggi, G., and Capuani, G. 2006. Intracerebral diffusion of paramagnetic cationic liposomes containing Gd(DTPA)2- followed by MRI spectroscopy: Assessment of pattern diffusion and time steadiness of a non-viral vector model. *Int J Immunopathol Pharmacol* 19: 379–390.

Mattrey, R. F. 1994. The potential role of perfluorochemicals (PFCS) in diagnostic imaging. *Artif Cells Blood Substit Immobil Biotechnol* 22: 295–313.

Mattrey, R. F., Leopold, G. R., and Van Sonnenberg, E. 1983. Perfluorochemicals as liver- and spleen-seeking ultrasound contrast agents. *J Ultrasound Med* 2: 173–176.

Mattrey, R. F., Long, D. M., Multer, F., Mitten, R., and Higgins, C. B. 1982. Perfluoroctylbromide: A reticuloendothelial-specific and tumor-imaging agent for computed tomography. *Radiology* 145: 755–758.

Mattrey, R. F., Long, D. M., and Peck, W. W. 1984. Perfluoroctylbromide as a blood pool contrast agent for liver, spleen, and vascular imaging in computed tomography. *J Comput Assisted Tomogr* 8: 739–744.

Mattrey, R. F., Strich, G., and Shelton, R. E. 1987. Perfluorochemicals as US contrast agents for tumor imaging and hepatosplenography: Preliminary clinical results. *Radiology* 163: 339–343.

Mauk, M. R. and Gamble, R. C. 1979. Preparation of lipid vesicles containing high levels of entrapped radioactive cations. *Anal Biochem* 94: 302–307.

Medina, L. A., Klipper, R., Phillips, W. T., and Goins, B. 2004. Pharmacokinetics and biodistribution of [^{111}In]-avidin and [^{99m}Tc]-biotin-liposomes injected in the pleural space for the targeting of mediastinal nodes. *Nucl Med Biol* 31: 41–51.

Meincke, M., Schlorf, T., Kossei, E., Jansen, O., Glueer, C. C., and Mentlein, R. 2008. Iron oxide-loaded liposomes for MR imaging. *Frontiers Biosci* 13: 4002–4008.

Meyer, K. L., Carvlin, M. J., Mukherji, B., Sloviter, H. A., and Joseph, P. M. 1992. Fluorinated blood substitute retention in the rat measured by fluorine-19 magnetic resonance imaging. *Invest Radiol* 27: 620–627.

Meyer, K. L., Joseph, P. M., Mukherji, B., Livolsi, V. A., and Lin, R. 1993. Measurement of vascular volume in experimental rat tumors by 19F magnetic resonance imaging. *Invest Radiol* 28: 710–719.

Miller, D. L., O'Leary, T., and Vucich, J. J. 1983a. Experimental evaluation of five liver-spleen specific CT contrast agents. *J Comput Assisted Tomogr* 7: 1022–1028.

Miller, D. L., Rosenbaum, R. C., and Sugarbaker, P. H. 1983b. Detection of hepatic metastases: Comparison of EOE-13 computed tomography and scintigraphy. *Am J Roentgenol* 141: 931–935.

Mishima, H., Kobayashi, T., Shimizu, M., Tamaki, Y., Baba, M., Shimano, T., Itoh, S., Yamazaki, M., Iriguchi, N., and Takahashi, M. 1991. *In vivo* F-19 chemical shift imaging with FTPA and antibody-coupled FMIQ. *J Magn Reson Imaging* 1: 705–709.

Mitra, A., Nan, A., Line, B. R., and Ghandehari, H. 2006. Nanocarriers for nuclear imaging and radiotherapy of cancer. *Curr Pharm Des* 12: 4729–4749.

Montet, X., Pastor, C. M., Vallee, J. P., Becker, C. D., Geissbuhler, A., Morel, D. R., and Meda, P. 2007. Improved visualization of vessels and hepatic tumors by micro-computed tomography (CT) using iodinated liposomes. *Invest Radiol* 42: 652–658.

Mori, A., Klibanov, A. L., Torchilin, V. P., and Huang, L. 1991. Influence of the steric barrier activity of amphipathic poly(ethyleneglycol) and ganglioside GM1 on the circulation time of liposomes and on the target binding of immunoliposomes *in vivo*. *FEBS Lett* 284: 263–266.

Mukundan, Jr., S., Ghaghada, K. B., Badea, C. T., Kao, C. Y., Hedlund, L. W., Provenzale, J. M., Johnson, G. A., Chen, E., Bellamkonda, R. V., and Annapragada, A. 2006. A liposomal nanoscale contrast agent for preclinical CT in mice. *Am J Roentgenol* 186: 300–307.

Mulder, W. J. M., Douma, K., Koning, G. A., Van Zandvoort, M. A., Lutgens, E., Daemen, M. J., Nicolay, K., and Strijkers, G. J. 2006a. Liposome-enhanced MRI of neointimal lesions in the ApoE-KO mouse. *Magn Reson Med* 55: 1170–1174.

Mulder, W. J. M., Strijkers, G. J., Van Tilborg, G. A. F., Griffioen, A. W., and Nicolay, K. 2006b. Lipid-based nanoparticles for contrast-enhanced MRI and molecular imaging. *NMR Biomed* 19: 142–164.

Müller, R. H. and Lucks, J. S. 1993. Arzneistoffträger aus festen Lipidteilchen, Feste Lipidnanosphären (SLN) [Medication vehicles made of solid lipid particles (solid lipid nanospheres–SLN)]. Eur. Patent 0605497.

Müller, R. H., Maaen, S., Weyhers, H., Specht, F., and Lucks, J. S. 1996. Cytotoxicity of magnetite-loaded polylactide, polylactide/glycolide particles and solid lipid nanoparticles. *Int J Pharm* 138: 85–94.

Musu, C., Felder, E., Lamy, B., and Schneider, M. 1988. A liposomal contrast agent: Preliminary communication. *Invest Radiol* 23: S126–S129.

Najafi, A., Amparo, E. G., and Johnson, Jr., R. F. 1987. Gadolinium labeled pharmaceuticals as potential MRI contrast agents for liver and biliary tract. *J Label Compd Radiopharm* 24: 1131–1141.

Nakamura, E., Makino, K., Okano, T., Yamamoto, T., and Yokoyama, M. 2006. A polymeric micelle MRI contrast agent with changeable relaxivity. *J Control Release* 114: 325–333.

Nasongkla, N., Bey, E., Ren, J., Ai, H., Khemtong, C., Guthi, J. S., Chin, S. F., Sherry, A. D., Boothman, D. A., and Gao, J. 2006. Multifunctional polymeric micelles as cancer-targeted, MRI-ultrasensitive drug delivery systems. *Nano Lett* 6: 2427–2430.

Neirinckx, R. D., Burke, J. F., Harrison, R. C., Forster, A. M., Andersen, A. R., and Lassen, N. A. 1988. The retention mechanism of technetium-99m-HM-PAO: Intracellular reaction with glutathione. *J Cereb Blood Flow Metab* 8: S4–S12.

Noth, U., Morrissey, S. P., Deichmann, R., Adolf, H., Schwarzbauer, C., Lutz, J., and Haase, A. 1995. *In vivo* measurement of partial oxygen pressure in large vessels and in the reticuloendothelial system using fast 19F-MRI. *Magn Reson Med* 34: 738–745.

Ogihara, I., Kojima, S., and Jay, M. 1986a. Differential uptake of gallium-67-labeled liposomes between tumors and inflammatory lesions in rats. *J Nucl Med* 27: 1300–1307.

Ogihara, I., Kojima, S., and Jay, M. 1986b. Tumor uptake of 67 Ga-carrying liposomes. *Eur J Nucl Med* 11: 405–411.

Ogihara-Umeda, I. and Kojima, S. 1988. Increased delivery of gallium-67 to tumors using serum-stable liposomes. *J Nucl Med* 29: 516–523.

Ogihara-Umeda, I., Sasaki, T., Kojima, S., and Nishigori, H. 1996. Optimal radiolabeled liposomes for tumor imaging. *J Nucl Med* 37: 326–332.

Ogihara-Umeda, I., Sasaki, T., and Nishigori, H. 1993. Active removal of radioactivity in the blood circulation using biotin-bearing liposomes and avidin for rapid tumour imaging. *Eur J Nucl Med* 20: 170–172.

Oku, N., Namba, Y., Takeda, A. and Okada, S. 1993. Tumor imaging with technetium-99m-DTPA encapsulated in RES-avoiding liposomes. *Nuclear Medicine and Biology* 20: 407--412.

Oku, N. 1996. Basic study on functional liposomes and their application. *Cell Mol Biol Lett* 1: 417–427.

Oku, N. 1999. Delivery of contrast agents for positron emission tomography imaging by liposomes. *Adv Drug Deliv Rev* 37: 53–61.

Oku, N., Tokudome, Y., Namba, Y., Saito, N., Endo, M., Hasegawa, Y., Kawai, M., Tsukada, H., and Okada, S. 1996a. Effect of serum protein binding on real-time trafficking of liposomes with different charges analyzed by positron emission tomography. *Biochim Biophys Acta Biomembr* 1280: 149–154.

Oku, N., Tokudome, Y., Tsukada, H., Kosugi, T., Namba, Y., and Okada, S. 1996b. *In vivo* trafficking of long-circulating liposomes in tumour-bearing mice determined by positron emission tomography. *Biopharm Drug Dis* 17: 435–441.

Oku, N., Tokudome, Y., Tsukada, H., and Okada, S. 1995. Real-time analysis of liposomal trafficking in tumor-bearing mice by use of positron emission tomography. *Biochim Biophys Acta Biomembr* 1238: 86–90.

Oliver, M., Ahmad, A., Kamaly, N., Perouzel, E., Caussin, A., Keller, M., Herlihy, A., Bell, J., Miller, A. D., and Jorgensen, M. R. 2006. MAGfect: A novel liposome formulation for MRI labelling and visualization of cells. *Org Biomol Chem* 4: 3489–3497.

Onyuksel, H., Bodalia, B., Sethi, V., Dagar, S., and Rubinsteina, I. 2000. Surface-active properties of vasoactive intestinal peptide. *Peptides* 21: 419–423.

Osborne, M. P., Richardson, V. J., Jeyasing, K., and Ryman, B. E. 1979. Radionuclide-labelled liposomes—A new lymph node imaging agent. *Int J Nucl Med Biol* 6: 75–83.

Oyen, W. J., Boerman, O. C., Storm, G., Van Bloois, L., Koenders, E. B., Crommelin, D. J., Van Der Meer, J. W., and Corstens, F. H. 1996a. Labelled stealth liposomes in experimental infection: An alternative to leukocyte scintigraphy *Nucl Med Commun* 17: 742–748.

Oyen, W. J. G., Boerman, O. C., Storm, G., Van Bloois, L., Koenders, E. B., Claessens, R. A. M. J., Perenboom, R. M., Crommelin, D. J. A., Van Der Meer, J. W. M., and Corstens, F. H. M. 1996b. Detecting infection and inflammation with technetium-99m-labeled Stealth® liposomes. *J Nucl Med* 37: 1392–1397.

Pauser, S., Reszka, R., Wagner, S., Wolf, K. J., Buhr, H. J., and Berger, G. 1997. Liposome-encapsulated superparamagnetic iron oxide particles as markers in an MRI-guided search for tumor-specific drug carriers. *Anti-Cancer Drug Des* 12: 125–135.

Pautler, R. G. and Koretsky, A. P. 2002. Tracing odor-induced activation in the olfactory bulbs of mice using manganese-enhanced magnetic resonance imaging. *NeuroImage* 16: 441–448.

Payne, N. I. and Whitehouse, G. H. 1987. Delineation of the spleen by a combination of proliposomes with water-soluble contrast media: An experimental study using computed tomography. *Br J Radiol* 60: 535–541.

Peira, E., Marzola, P., Podio, V., Aime, S., Sbarbati, A., and Gasco, M. R. 2003. *In vitro* and *in vivo* study of solid lipid nanoparticles loaded with superparamagnetic iron oxide. *J Drug Target* 11: 19–24.

Perez-Mayoral, E., Negri, V., Soler-Padros, J., Cerdan, S., and Ballesteros, P. 2008. Chemistry of paramagnetic and diamagnetic contrast agents for magnetic resonance imaging and spectroscopy. pH responsive contrast agents. *Eur J Radiol* 67: 453–458.

Petersein, J., Franke, B., Fouillet, X., and Hamm, B. 1999. Evaluation of liposomal contrast agents for liver CT in healthy rabbits. *Invest Radiol* 34: 401–409.

Phillips, W. T. 1999. Delivery of gamma-imaging agents by liposomes. *Adv Drug Deliv Rev* 37: 13–32.

Phillips, W. T. and Goins, B. 1995. Targeted delivery of imaging agents by liposomes. In V. P. Torchilin (Ed.), *Handbook of Targeted Delivery of Imaging Agents*, pp. 331–339. Boca Raton, FL: CRC Press.

Phillips, W. T., Klipper, R., and Goins, B. 2001. Use of ^{99m}Tc-labeled liposomes encapsulating blue dye for identification of the sentinel lymph node. *J Nucl Med* 42: 446–451.

Phillips, W. T., Rudolph, A. S., Goins, B., Timmons, J. H., Klipper, R., and Blumhardt, R. 1992. A simple method for producing a technetium-99m-labeled liposome which is stable *in vivo*. *Int J Radiat Appl Instrum B* 19: 539–547.

Plassat, V., Martina, M. S., Barratt, G., Menager, C., and Lesieur, S. 2007. Sterically stabilized superparamagnetic liposomes for MR imaging and cancer therapy: Pharmacokinetics and biodistribution. *Int J Pharm* 344: 118–127.

Presant, C. A., Ksionski, G., and Crossley, R. 1990. ^{111}In-Labeled liposomes for tumor imaging: Clinical results of the international liposome imaging study. *J Liposome Res* 1: 431–436.

Pressly, E. D., Rossin, R., Hagooly, A., Fukukawa, K. I., Messmore, B. W., Welch, M. J., Wooley, K. L. et al. 2007. Structural effects on the biodistribution and positron emission tomography (PET) imaging of well-defined 64Cu-labeled nanoparticles comprised of amphiphilic block graft copolymers. *Biomacromolecules* 8: 3126–3134.

Quay, S. C. and Eisenfeld, A. J. 1997. Safety assessment of the use of perflenapent emulsion for contrast enhancement of echocardiography and diagnostic radiology ultrasound studies. *Clin Cardiol* 20: 119–126.

Rabin, O., Perez, J. M., Grimm, J., Wojtkiewicz, G., and Weissleder, R. 2006. An X-ray computed tomography imaging agent based on long-circulating bismuth sulphide nanoparticles. *Nat Mater* 5: 118–122.

Raptopoulos, V., Davis, M. A., Davidoff, A., Karellas, A., Hays, D., D'Orsi, C. J., and Smith, E. H. 1987. Fat-density oral contrast agent for abdominal CT. *Radiology* 164: 653–656.

Ratner, A. V., Hurd, R., Muller, H. H., Bradley-Simpson, B., Pitts, W., Shibata, D., Sotak, C., and Young, S. W. 1987. 19F Magnetic resonance imaging of the reticuloendothelial system. *Magn Reson Med* 5: 548–554.

Ratner, A. V., Muller, H. H., Bradley-Simpson, B., Hirst, D., Pitts, W., and Young, S. W. 1988. Detection of acute radiation damage to the spleen in mice by using fluorine-19 MR imaging. *Am J Roentgenol* 151: 477–480.

Reddy, G. R., Bhojani, M. S., Mcconville, P., Moody, J., Moffat, B. A., Hall, D. E., Kim, G. et al. 2006. Vascular targeted nanoparticles for imaging and treatment of brain tumors. *Clin Cancer Res* 12: 6677–6686.

Reed, W. P., Haney, P. J., and Elias, E. G. 1986. Ethiodized oil emulsion enhanced computerized tomography in the preoperative assessment of metastases to the liver from the colon and rectum. *Surg Gyn Obstet* 162: 131–136.

Revel, D., Corot, C., Carrillon, Y., Dandis, G., Eloy, R., and Amiel, M. 1990. Ioxaglate-carrying liposomes: Computed tomographic study as hepatosplenic contrast agent in rabbits. *Invest Radiol* 25: S95–S97.

Reynolds, C. H., Annan, N., Beshah, K., Huber, J. H., Shaber, S. H., Lenkinski, R. E., and Wortman, J. A. 2000. Gadolinium-loaded nanoparticles: New contrast agents for magnetic resonance imaging. *J Am Chem Soc* 122: 8940–8945.

Richardson, V. J., Jeyasingh, K., Jewkes, R. F., Ryman, B. E., and Tattersall, M. H. 1977. Properties of [^{99m}Tc] technetium-labelled liposomes in normal and tumour-bearing rats. *Biochem Soc Trans* 5: 290–291.

Richardson, V. J., Jeyasingh, K., Jewkes, R. F., Ryman, B. E., and Tattersall, M. H. 1978. Possible tumor localization of Tc-99m-labeled liposomes: Effects of lipid composition, charge, and liposome size. *J Nucl Med* 19: 1049–1054.

Robbin, M. L. and Eisenfeld, A. J. 1998. Perflenapent emulsion: A US contrast agent for diagnostic radiology—Multicenter, double-blind comparison with a placebo. *Radiology* 207: 717–722.

Roberts, J. C., Adams, Y. E., Tomalia, D., Mercer-Smith, J. A., and Lavallee, D. K. 1990. Using starburst dendrimers as linker molecules to radiolabel antibodies. *Bioconj Chem* 1: 305–308.

Roberts, J. C., Bhalgat, M. K., and Zera, R. T. 1996. Preliminary biological evaluation of polyamidoamine (PAMAM) starburst dendrimers. *J Biomed Mater Res* 30: 53–65.

Rubinstein, I., Patel, M., Ikezaki, H., Dagar, S., and Onyuksel, H. 1999. Conformation and vasoreactivity of VIP in phospholipids: Effects of calmodulin. *Peptides* 20: 1497–1501.

Rudolph, A. S., Klipper, R. W., Goins, B., and Phillips, W. T. 1991. *In vivo* biodistribution of a radiolabeled blood substitute: ^{99m}Tc-labeled liposome-encapsulated hemoglobin in an anesthetized rabbit. *Proc Natl Acad Sci USA* 88: 10976–10980.

Ryan, P. J., Davis, M. A., and Melchior, D. L. 1983. The preparation and characterization of liposomes containing X-ray contrast agents. *Biochim Biophys Acta* 756: 106–110.

Sachse, A., Leike, J. U., Rossling, G. L., Wagner, S. E., and Krause, W. 1993. Preparation and evaluation of lyophilized iopromide-carrying liposomes for liver tumor detection. *Invest Radiol* 28: 838–844.

Sachse, A., Leike, J. U., Schneider, T., Wagner, S. E., Roßling, G. L., Krause, W., and Brandl, M. 1997. Biodistribution and computed tomography blood-pool imaging properties of polyethylene glycol-coated iopromide-carrying liposomes. *Invest Radiol* 32: 44–50.

Sadler, K. and Tam, J. P. 2002. Peptide dendrimers: Applications and synthesis. *J Biotechnol* 90: 195–229.

Sahoo, S. K. and Labhasetwar, V. 2003. Nanotech approaches to drug delivery and imaging. *Drug Discov Today* 8: 1112–1120.

Schmiedl, U. P., Krause, W., Leike, J., Nelson, J. A., and Schuhmann-Giampieri, G. 1995. Liver contrast enhancement in primates using iopromide liposomes. *Acad Radiol* 2: 967–972.

Schmiedl, U. P., Krause, W., Leike, J., and Sachse, A. 1999. CT blood pool enhancement in primates with iopromide-carrying liposomes containing soy phosphatidyl glycerol. *Acad Radiol* 6: 164–169.

Schwendener, R. A., Tilcock, C., Khar, R. N., Goto, R., Trubetskoy, V. S., and Torchilin, V. P. 1994. Liposomes as carriers for paramagnetic gadolinium chelates as organ specific contrast agents for magnetic resonance imaging (MRI). *J Liposome Res* 4: 837–859.

Schwendener, R. A., Wuthrich, R., Duewell, S., Wehrli, E., and Von Schulthess, G. K. 1990. A pharmacokinetic and MRI study of unilamellar gadolinium-, manganese-, and iron-DTPA-stearate liposomes as organ-specific contrast agents. *Invest Radiol* 25: 922–932.

Schwendener, R. A., Wuthrich, R., Duewell, S., Westera, G., and Von Schulthess, G. K. 1989. Small unilamellar liposomes as magnetic resonance contrast agents loaded with paramagnetic Mn-, Gd- and Fe-DTPA-stearate complexes. *Int J Pharm* 49: 249–259.

Seltzer, S. E., Blau, M., Herman, L. W., Hooshmand, R. L., Herman, L. A., Adams, D. F., Minchey, S. R., and Janoff, A. S. 1995. Contrast material-carrying liposomes: Biodistribution, clearance, and imaging characteristics. *Radiology* 194: 775–781.

Seltzer, S. E., Davis, M. A., Adams, D. F., Shulkin, P. M., Landis, W. J., and Havron, A. 1984a. Liposomes carrying diatrizoate. Characterization of biophysical properties and imaging applications. *Invest Radiol* 19: 142–151.

Seltzer, S. E., Shulkin, P. M., and Adams, D. F. 1984b. Usefulness of liposomes carrying iosefamate for CT opacification of liver and spleen. *Am J Roentgenol* 143: 575–579.

Senior, J., Delgado, C., Fisher, D., Tilcock, C., and Gregoriadis, G. 1991. Influence of surface hydrophilicity of liposomes on their interaction with plasma protein and clearance from the circulation: Studies with poly(ethylene glycol)-coated vesicles. *Biochim Biophys Acta* 1062: 77–82.

Shimizu, M., Kobayashi, T., Morimoto, H., Matsuura, N., Shimano, T., Nomura, N., Itoh, S. et al. 1987. Tumor imaging with anti-CEA antibody labeled 19F emulsion. *Magn Reson Med* 5: 290–295.

Singh, P., Moll, F., Lin, S. H., Ferzli, C., Yu, K. S., Koski, R. K., Saul, R. G., and Cronin, P. 1994. Starburst dendrimers: Enhanced performance and flexibility for immunoassays. *Clin Chem* 40: 1845–1849.

Sipkins, D. A., Cheresh, D. A., Kazemi, M. R., Nevin, L. M., Bednarski, M. D., and Li, K. C. 1998. Detection of tumor angiogenesis *in vivo* by alphaVbeta3-targeted magnetic resonance imaging. *Nat Med* 4: 623–626.

Sklifas, A. N., Chaplygina, S. L., Shekhtman, D. G., and Kukushkin, N. I. 2007. A bromoperfluorooctane-based emulsion: Radiographic contrast properties and mechanisms. *Biophysics* 52: 253–255.

Soenen, S. J. H., Desender, L., and De Cuyper, M. 2007. Complexation of gadolinium(III) ions on top of nanometre-sized magnetoliposomes. *Int J Environ Anal Chem* 87: 783–796.

Spyridonidis, T. and Markou, P. 2005. Current trends in infection imaging. *Hellenic J Nucl Med* 8: 12–18.

Storm, G. and Crommelin, D. J. A. 1998. Liposomes: Quo vadis. *Pharm Sci Technol Today* 1: 19–31.

Storrs, R. W., Tropper, F. D., Li, H. Y., Song, C. K., Sipkins, D. A., Kuniyoshi, J. K., Bednarski, M. D., Strauss, H. W., and Li, K. C. 1995. Paramagnetic polymerized liposomes as new recirculating MR contrast agents. *J Magn Reson Imaging* 5: 719–724.

Suga, K., Mikawa, M., Ogasawara, N., Okazaki, H., and Matsunaga, N. 2001. Potential of Gd-DTPA-mannan liposome particles as a pulmonary perfusion MRI contrast agent: An initial animal study. *Invest Radiol* 36: 136–145.

Svenson, S. and Tomalia, D. A. 2005. Dendrimers in biomedical applications—Reflections on the field. *Adv Drug Deliv Rev* 57: 2106–2129.

Syrigos, K. N., Vile, R. G., Peters, A. M., and Harrington, K. J. 2003. Biodistribution and pharmacokinetics of ^{111}In-DTPA-labelled PEGylated liposomes after intraperitoneal injection. *Acta Oncol* 42: 147–153.

Terreno, E., Sanino, A., Carrera, C., Castelli, D. D., Giovenzana, G. B., Lombardi, A., Mazzon, R., Milone, L., Visigalli, M., and Aime, S. 2008. Determination of water permeability of paramagnetic liposomes of interest in MRI field. *J Inorg Biochem* 102: 1112–1119.

Thakur, M. L. 1977. Gallium 67 and indium 111 radiopharmaceuticals. *Int J Appl Radiat Isotopes* 28: 183–201.

Thomas, J. L., Bernardino, M. E., Vermess, M., Barnes, P. A., Fuller, L. M., Hagemeister, F. B., Doppman, J., Fisher, R. I., and Longo, D. L. 1982. EOE-13 in the detection of hepatosplenic lymphoma. *Radiology* 145: 629–634.

Tilcock, C. 1999. Delivery of contrast agents for magnetic resonance imaging, computed tomography, nuclear medicine and ultrasound. *Adv Drug Deliv Rev* 37: 33–51.

Tilcock, C., Ahkong, Q. F., and Fisher, D. 1993. Polymer-derivatized technetium ^{99m}Tc-labeled liposomal blood pool agents for nuclear medicine applications. *Biochim Biophys Acta Biomembr* 1148: 77–84.

Tilcock, C., Ahkong, Q. F., and Fisher, D. 1994a. ^{99m}Tc-Labeling of lipid vesicles containing the lipophilic chelator PE-DTTA: Effect of tin-to-chelate ratio, chelate content and surface polymer on labeling efficiency and biodistribution behavior. *Nucl Med Biol* 21: 89–96.

Tilcock, C., Unger, E., Cullis, P., and Macdougall, P. 1989. Liposomal Gd-DTPA: Preparation and characterization of relaxivity. *Radiology* 171: 77–80.

Tilcock, C., Yap, M., Szucs, M., and Utkhede, D. 1994b. PEG-coated lipid vesicles with encapsulated technetium-99m as blood pool agents for nuclear medicine. *Nucl Med Biol* 21: 165–170.

Torchilin, V. 1995. *Handbook of Trageted Delivery of Imaging Agents.* Boca Raton, FL: CRC Press.

Torchilin, V. P. 1994a. Immunoliposomes and PEGylated immunoliposomes: Possible use for targeted delivery of imaging agents. *Immunomethods* 4: 244–258.

Torchilin, V. P. 1994b. Targeted delivery of diagnostic agents by surface-modified liposomes. *J Control Release* 28: 45–58.

Torchilin, V. P. 1996. Liposomes as delivery agents for medical imaging. *Mol Med Today* 2: 242–249.

Torchilin, V. P. 1997a. Pharmacokinetic considerations in the development of labeled liposomes and micelles for diagnostic imaging. *Q J Nucl Med* 41: 141–153.

Torchilin, V. P. 1997b. Surface-modified liposomes in gamma- and MR-imaging. *Adv Drug Deliv Rev* 24: 301–313.

Torchilin, V. P. 1999. Polymeric micelles in diagnostic imaging. *Colloid Surf B* 16: 305–319.

Torchilin, V. P. 2000. Polymeric contrast agents for medical imaging. *Curr Pharm Biotechnol* 1: 183–215.

Torchilin, V. P. 2001. Structure and design of polymeric surfactant-based drug delivery systems. *J Control Release* 73: 137–172.

Torchilin, V. P. 2002. PEG-based micelles as carriers of contrast agents for different imaging modalities. *Adv Drug Deliv Rev* 54: 235–252.

Torchilin, V. P. 2005. Recent advances with liposomes as pharmaceutical carriers. *Nat Rev Drug Discov* 4: 145–160.

Torchilin, V. P. 2006. Multifunctional nanocarriers. *Adv Drug Deliv Rev* 58: 1532–1555.

Torchilin, V. P. 2007. Targeted pharmaceutical nanocarriers for cancer therapy and imaging. *AAPS J* 9: article 15.

Torchilin, V., Babich, J., and Weissig, V. 2000. Liposomes and micelles to target the blood pool for imaging purposes. *J Liposome Res* 10: 483–499.

Trubetskoy, V. S. 1999. Polymeric micelles as carriers of diagnostic agents. *Adv Drug Deliv Rev* 37: 81–88.

Trubetskoy, V. S., Cannillo, J. A., Milshtein, A., Wolf, G. L., and Torchilin, V. P. 1995. Controlled delivery of Gd-containing liposomes to lymph nodes: Surface modification may enhance MRI contrast properties. *J Magn Reson Imaging* 13: 31–37.

Trubetskoy, V. S., Frank-Kamenetsky, M. D., Whiteman, K. R., Wolf, G. L., and Torchilin, V. P. 1996. Stable polymeric micelles: Lymphangiographic contrast media for gamma scintigraphy and magnetic resonance imaging. *Acad Radiol* 3: 232–238.

Trubetskoy, V. S., Scott Gazelle, G., Wolf, G. L., and Torcffllin, V. P. 1997. Block-copolymer of polyethylene glycol and polylysine as a carrier of organic iodine: Design of long-circulating particulate contrast medium for X-ray computed tomography. *J Drug Target* 4: 381–388.

Trubetskoy, V. S. and Torchilin, V. P. 1995. Use of polyoxyethylene-lipid conjugates as long-circulating carriers for delivery of therapeutic and diagnostic agents. *Adv Drug Deliv Rev* 16: 311–320.

Unger, E., Cardenas, D., Zerella, A., Fajardo, L. L., and Tilcock, C. 1990. Biodistribution and clearance of liposomal gadolinium-DPTA. *Invest Radiol* 25: 638–644.

Unger, E., Needleman, O., Cullis, P., and Tilcock, C. 1988. Gadolinium-DTPA liposomes as a potential MRI contrast agent. *Invest Radiol* 23: 928–932.

Unger, E., Fritz, T., Wu, G., Shen, D., Kulik, B., New, T., Crowell, M., Wilke, N., Tilcock, C., and Goto, R. 1994a. Liposomal MR contrast agents. *J Liposome Res* 4: 811–836.

Unger, E., Shen, D., Fritz, T., Kulik, B., Lund, P., Wu, G. L., Yellowhair, D., Ramaswami, R., and Matsunaga, T. 1994b. Gas-filled lipid bilayers as ultrasound contrast agents. *Invest Radiol* 29: 111–125.

Unger, E., Shen, D. K., Wu, G., and Fritz, T. 1991. Liposomes as MR contrast agents: Pros and cons. *Magn Reson Med* 22: 304–308.

Unger, E. C., Fritz, T. A., Matsunaga, T., Ramaswami, V., Yellowhair, D., and Wu, G. 1999. Methods of preparing gas-filled liposomes. U.S. Patent 5,935,553.

Unger, E. C., Lund, P. J., Shen, D. K., Fritz, T. A., Yellowhair, D., and New, T. E. 1992. Nitrogen-filled liposomes as a vascular US contrast agent: Preliminary evaluation. *Radiology* 185: 453–456.

Unger, E. C., Macdougall, P., Cullis, P., and Tilcock, C. 1989. Liposomal Gd-DTPA: Effect of encapsulation on enhancement of hepatoma model by MRI. *J Magn Reson Imaging* 7: 417–423.

Unger, E. C., Shen, D. K., and Fritz, T. A. 1993. Status of liposomes as MR contrast agents. *J Magn Reson Imaging* 3: 195–198.

Unger, E. C. and Wu, G. 2001. Gas filled liposomes and their use as ultrasonic contrast agents. U.S. Patent 6,315,981.

Van Tilborg, G. A. F., Mulder, W. J. M., Deckers, N., Storm, G., Reutelingsperger, C. P. M., Strijkers, G. J., and Nicolay, K. 2006. Annexin A5-functionalized bimodal lipid-based contrast agents for the detection of apoptosis. *Bioconj Chem* 17: 741–749.

Venditto, V. J., Regino, C. A. S., and Brechbiel, M. W. 2005. PAMAM dendrimer based macromolecules as improved contrast agents. *Mol Pharm* 2: 302–311.

Vermess, M., Adamson, R. H., Doppman, J. L., and Girton, M. 1977. Computed tomographic demonstration of hepatic tumor with the aid of intravenous iodinated fat emulsion: An experimental study. *Radiology* 125: 711–715.

Vermess, M., Bernardino, M. E., and Doppman, J. L. 1981. Use of intravenous liposoluble contrast material for the examination of the liver and spleen in lymphoma. *J Comput Assisted Tomogr* 5: 709–713.

Vermess, M., Chatterji, D. C., and Doppman, J. L. 1979. Development and experimental evaluation of a contrast medium for computed tomographic examination of the liver and spleen. *J Comput Assisted Tomogr* 3: 25–31.

Vermess, M., Doppman, J. L., and Sugarbaker, P. 1980. Clinical trials with a new intravenous liposoluble contrast material for computed tomography of the liver and spleen. *Radiology* 137: 217–222.

Vermess, M., Doppman, J. L., and Sugarbaker, P. H. 1982a. Computed tomography of the liver and spleen with intravenous lipid contrast material: Review of 60 examinations. *Am J Roentgenol* 138: 1063–1071.

Vermess, M., Lau, D. H. M., and Adams, M. D. 1982b. Biodistribution study of ethiodized oil emulsion 13 for computed tomography of the liver and spleen. *J Comput Assisted Tomogr* 6: 1115–1119.

Wang, T., Hossanna, M., Reinl, H. M., Peller, M., Eibl, H., Reiser, M., Issels, R. D., and Lindner, L. H. 2008. *In vitro* characterization of phosphatidylglyceroglycerol-based thermosensitive liposomes with encapsulated 1H MR T1-shortening gadodiamide. *Contrast Media Mol Imaging* 3: 19–26.

Weissig, V., Babich, J., and Torchilin, V. 2000. Long-circulating gadolinium-loaded liposomes: Potential use for magnetic resonance imaging of the blood pool. *Colloid Surf B* 18: 293–299.

Weyhers, H., Löbenberg, R., Mehnert, W., Souto, E. B., Kreuter, J., and Muller, R. H. 2006. *In vivo* distribution of 125I-radiolabelled solid lipid nanoparticles. *Pharm Ind* 68: 889–894.

White, C., Slifkin, M., Seltzer, S. E., Blau, M., Adzamli, I. K., and Adams, D. F. 1990. Biodistribution and clearance of contrast-carrying MREV lipsomes. *Invest Radiol* 25: 1125–1129.

Wiener, E. C., Brechbiel, M. W., Brothers, H., Magin, R. L., Gansow, O. A., Tomalia, D. A., and Lauterbur, P. C. 1994. Dendrimer-based metal chelates: A new class of magnetic resonance imaging contrast agents. *Magn Reson Med* 31: 1–8.

Wiener, E. C., Konda, S., Shadron, A., Brechbiel, M., and Gansow, O. 1997. Targeting dendrimer-chelates to tumors and tumor cells expressing the high-affinity folate receptor. *Invest Radiol* 32: 748–754.

Wissing, S. A., Kayser, O., and Muller, R. H. 2004. Solid lipid nanoparticles for parenteral drug delivery. *Adv Drug Deliv Rev* 56: 1257–1272.

Woodle, M. C. 1993. 67Gallium-labeled liposomes with prolonged circulation: Preparation and potential as nuclear imaging agents. *Nucl Med Biol* 20: 149–155.

Xu, H., Regino, C. A. S., Koyama, Y., Hama, Y., Gunn, A. J., Bernardo, M., Kobayashi, H., Choyke, P. L., and Brechbiel, M. W. 2007. Preparation and preliminary evaluation of a biotin-targeted, lectin-targeted dendrimer-based probe for dual-modality magnetic resonance and fluorescence imaging. *Bioconj Chem* 18: 1474–1482.

Yordanov, A. T., Lodder, A. L., Woller, E. K., Cloninger, M. J., Patronas, N., Milenic, D., and Brechbiel, M. W. 2002. Novel iodinated dendritic nanoparticles for computed tomography (CT) imaging. *Nano Lett* 2: 595–599.

Yordanov, A. T., Mollov, N., Lodder, A. L., Woller, E., Cloninger, M., Walbridge, S., Milbnic, D., and Brechbiel, M. W. 2005. A water-soluble triiodo amino acid and its dendrimer conjugate for computerized tomography (CT) imaging. *J Serb Chem Soc* 70: 163–170.

Yoshida, Y., Suematsu, T., Sugimura, K., Hirata, Y., Iwasawa, Y., Ushio, K., and Nishiyama, S. 1984. Lipiodol emulsion (Lip. (E)20) as a new contrast agent for selective visualization of the liver and spleen. Basic analysis and experimental studies in dogs by X-ray CT and ultrasound. *Nippon Acta Radiol* 44: 735–737.

Young, S. W., Enzmann, D. R., Long, D. M., and Muller, H. H. 1981. Perfluoroctylbromide contrast enhancement of malignant neoplasms: Preliminary observations. *Am J Roentgenol* 137: 141–146.

Yu, X., Song, S. K., Chen, J., Scott, M. J., Fuhrhop, R. J., Hall, C. S., Gaffney, P. J., Wickline, S. A., and Lanza, G. M. 2000. High-resolution MRI characterization of human thrombus using a novel fibrin-targeted paramagnetic nanoparticle contrast agent. *Magn Reson Med* 44: 867–872.

Zaitsev, V. S., Filimonov, D. S., Presnyakov, I. A., Gambino, R. J., and Chu, B. 1999. Physical and chemical properties of magnetite and magnetite-polymer nanoparticles and their colloidal dispersions. *J Colloid Interface Sci* 212: 49–57.

Zalutsky, M. R., Noska, M. A., and Seltzer, S. E. 1987. Characterization of liposomes containing iodine-125-labeled radiographic contrast agents. *Invest Radiol* 22: 141–148.

Zavaleta, C. L., Phillips, W. T., Soundararajan, A., and Goins, B. A. 2007. Use of avidin/biotin-liposome system for enhanced peritoneal drug delivery in an ovarian cancer model. *Int J Pharm* 337: 316–328.

Zhang, D., Feng, X. Y., Henning, T. D., Wen, L., Lu, W. Y., Pan, H., Wu, X., and Zou, L. G. 2009. MR imaging of tumor angiogenesis using sterically stabilized Gd-DTPA liposomes targeted to CD105. *Eur J Radiol* 70: 180–189.

Zheng, J., Liu, J., Jaffray, D., and Allen, C. 2006. *In vivo* tracking of liposomes using CT and MR imaging. Paper presented at the 2006 NSTI Nanotechnology Conference and Trade Show—NSTI Nanotech 2006 Technical Proceedings.

Zhu, W., Okollie, B., Bhujwalla, Z. M., and Artemov, D. 2008. PAMAM dendrimer-based contrast agents for MR imaging of Her-2/neu receptors by a three-step pretargeting approach. *Magn Reson Med* 59: 679–685.

Zielhuis, S. W., Seppenwoolde, J. H., Mateus, V. A. P., Bakker, C. J. G., Krijger, G. C., Storm, G., Zonnenberg, B. A., Van Het Schip, A. D., Koning, G. A., and Nijsen, J. F. W. 2006. Lanthanide-loaded liposomes for multimodality imaging and therapy. *Cancer Biother Radiopharm* 21: 520–527.

Zijlstra, S., Gunawan, J., and Burchert, W. 2003. Synthesis and evaluation of a 18F-labelled recombinant annexin-V derivative, for identification and quantification of apoptotic cells with PET. *Appl Radiat Isotopes* 58: 201–207.

Zinselmeyer, B. H., Mackay, S. P., Schatzlein, A. G., and Uchegbu, I. F. 2002. The lower-generation polypropylenimine dendrimers are effective gene-transfer agents. *Pharm Res* 19: 960–967.

Zwicker, C., Langer, M., Urich, V., and Felix, R. 1993. CT contrast administration of iodine, gadolinium and ytterbium. *In-vitro* studies and animal experiments. *RoFo* 158: 255–259.

Index